SOURCE BOOK ON MARAGING STEELS

Other Source Books in This Series:

Materials for Elevated-Temperature Applications (in preparation)
Powder Metallurgy
Wear Control Technology
Selection and Fabrication of Aluminum Alloys
Industrial Alloy and Engineering Data
Materials Selection (Volumes I and II)
Nitriding
Ductile Iron
Forming of Steel Sheet
Stainless Steels
Heat Treating (Volumes I and II)
Cold Forming
Failure Analysis

Metals Science Source Book Series:

The Metallurgical Evolution of Stainless Steels (in preparation)
Hydrogen Damage

SOURCE BOOK ON MARAGING STEELS

A comprehensive collection of outstanding articles from the periodical and reference literature

Compiled by
Consulting Editor:

Raymond F. Decker
Vice President
Inco Limited

American Society for Metals
Metals Park, Ohio 44073

Library of Congress Cataloging in Publication Data

Main entry under title:

Source book on maraging steels.

Includes bibliographical references and index.
1. Maraging steel. I. Decker, Raymond Frank.
TN757.N5S66 669'.142 79-13743
ISBN 0-87170-079-4

PRINTED IN THE UNITED STATES OF AMERICA

Contributors to This Source Book*

A. M. ADAIR
Air Force Materials Laboratory
Wright Patterson A.F.B.

JOHN A. BAILEY
North Carolina State University

H. G. BARON
Ministry of Technology (England)

C. J. BARTON
United States Steel Corp.

C. BATHIAS
Aerospatiale (France)

C. M. BERGER
Laussig Associates

C. G. BIEBER
International Nickel Co., Inc.

A. J. BIRKLE
United States Steel Corp.

J. G. BLAUEL

T. BONISZEWSKI
British Welding Research Assn.

F. E. BRINE
Royal Armament Research & Development Establishment (England)

K. BROWN
Australian Atomic Energy Commission

C. S. CARTER
Boeing Co.

J. M. CHILTON
United States Steel Corp.

M. C. COLEMAN
University of Aston (England)

PHILIP P. CRIMMINS
Aerojet-General Corp.

C. R. CUPP
International Nickel Co., Inc.

D. S. DABKOWSKI
United States Steel Corp.

R. F. DECKER
International Nickel Co., Inc.

F. D. DUFFEY
Newport News Shipbuilding & Dry Dock Co.

J. T. EASH
International Nickel Co., Inc.

MORRIS E. FINE
Northwestern University

S. FLOREEN
International Nickel Co., Inc.

R. M. GERMAN
Sandia Laboratories

J. B. GILMOUR
Department of Energy, Mines and Resources (Canada)

ALFRED GOLDBERG
University of California

A. J. GOLDMAN
Yale University

ELWOOD G. HANEY
Mellon Institute

ROBERT E. HERFERT
Northrop Corp.

P. P. HYDREAN
International Nickel Co., Inc.

*Affiliations given were applicable at date of contribution.

S. ISSEROW

S. JEELANI
North Carolina State University

M. F. JORDAN
University of Aston (England)

DAVID KALISH
Lockheed-Georgia Co.

N. KENYON
International Nickel Co., Inc.

R. J. KNOTH
International Nickel Co., Inc.

F. H. LANG
International Nickel Co., Inc.

EDWARD A. LAUCHNER
Northrop Corp.

H. M. LEDBETTER
National Bureau of Standards

HARRIS MARCUS
Northwestern University

J. H. OLSON
International Nickel Co., Inc.

R. N. PARKINS
University of Newcastle upon Tyne (England)

D. L. PASQUINE
International Nickel Co., Inc.

J. P. PAULINA
United States Steel Corp.

H. W. PAXTON
Carnegie-Mellon University

R. M. PELLOUX
Massachusetts Institute of Technology

D. T. PETERS
International Nickel Co., Inc.

W. A. PETERSEN
International Nickel Co., Inc.

L. F. PORTER
United States Steel Corp.

H. J. RACK
Lockheed-Georgia Co.

D. T. READ
National Bureau of Standards

J. A. ROBERSON
Aerospace Research Laboratories
Wright Patterson A.F.B.

C. ROBERTS
Military Engineering Experimental
Establishment (England)

E. P. SADOWSKI
International Nickel Co., Inc.

G. SAUL
Air Force Materials Laboratory
Wright Patterson A.F.B.

G. SCHULZE

LYLE H. SCHWARTZ
Northwestern University

H. R. SMITH

P. D. SMITH
Australian Atomic Energy Commission

J. E. SMUGERESKY
Sandia Laboratories

G. R. SPEICH
United States Steel Corp.

W. A. SPITZIG
United States Steel Corp.

A. J. STAVROS
Bethlehem Steel Corp.

W. SUTAR
Newport News Shipbuilding & Dry Dock Co.

B. C. SYRETT
Department of Energy, Mines and Resources
(Canada)

WALTER S. TENNER
Aerojet-General Corp.

G. W. TUFFNELL
International Nickel Co., Inc.

D. WEBBER
Military Engineering Experimental
Establishment (England)

R. D. WELTZIN
IBM General Systems Division

CONTENTS

Preface

SOURCE BOOK ON MARAGING STEELS is unique in at least three respects. It is, to the best of our knowledge, the first book of record devoted exclusively to the subject of maraging steels. As such, it is also the first comprehensive collection of engineering and processing information on this important new family of steels. Finally, Dr. Raymond F. Decker, the book's consulting editor, is a co-inventor of the steels under consideration — a most appropriate, though rarely achievable, combination of editor and subject, and a distinction to which no previous book in this series can lay claim. Dr. Decker's comments on the development of maraging steels, which follow this preface, intimately reflect his role as an active participant in a highly significant chapter in the history of contemporary metallurgy.

Although certain exceptional properties of maraging steels, notably their ultrahigh yield and tensile strengths and superior fracture toughness, have been widely publicized, equally important, though lesser-known, properties have not received comparable coverage. These properties, by and large, are of principal interest to engineers engaged in processing and manufacturing, although they might also influence design parameters. Among them are simplicity of heat treatment (including elimination of quenching), machinability similar to or better than that of conventional steels of the same hardness, good weldability even in the fully hardened condition (and with no need for preheating), and good hot or cold deformation properties. Mechanical properties are maintained during and after low-temperature brazing, and dimensional stability during and after aging is outstanding, permitting finish machining *before* age-hardening.

SOURCE BOOK ON MARAGING STEELS offers comprehensive coverage of all significant engineering properties of maraging steels, their processing characteristics, and their behavior in corrosive media. The book is arranged in thirteen principal sections, a brief summary of which follows.

Introduction. The opening section of the book presents the first major technical paper reporting the development of the new maraging steels and describes the "remarkable synergistic effect" derived from additions of cobalt and molybdenum.

Heat Treatment and Thermomechanical Treatment. Not the least remarkable characteristic of maraging steels is their response to solution annealing, followed by aging to develop full strength and hardness. Five articles review the details of heat treatment, possible embrittling effects, grain refinement, and the enhancement of fatigue behavior through thermomechanical treatment.

Mechanical Properties. In this section, eight highly informative articles provide general coverage of mechanical properties, with special emphasis on fatigue behavior and fracture toughness. The concluding article relates metallographic features and notch toughness.

Cryogenic and Elevated-Temperature Properties. The first article in this section reports on selected elastic properties at temperatures ranging from room temperature to liquid-helium temperature. A second article examines elevated-temperature strength, creep-rupture behavior, and the effects of exposure at elevated temperatures on subsequent room-temperature properties.

Stress-Corrosion Cracking Behavior. The cracking of high-strength steels subjected to tensile loads in corrosive environments is a major consideration in many critical applications. The subject is thoroughly explored in the four authoritative articles that comprise this section.

Environmental Cracking Behavior. Environmental cracking is often related to stress-corrosion cracking, with special emphasis given to the role of hydrogen in the cracking phenomenon. Two articles review the cracking behavior of 18% nickel maraging steels.

Castings. A high-strength cast maraging steel alloy with excellent ductility, notch tensile and impact properties was developed by The International Nickel Co. and is reported on at length in an article by Sadowski and Decker selected for this section.

Welding. Unlike many of the ultrahigh-strength steels, the maraging steels have good weldability characteristics. The six articles in this section cover the welding processes of principal interest, often providing data that can be directly related to processing. The engineering properties of welds also receive ample coverage.

Machining. The effect of cutting speed and tool wear on the surface integrity of annealed 18% nickel maraging steel machined under dry and lubricated conditions is reported on in an article that helps establish machining guidelines.

Electroplating. This section presents a "how to" article on the cadmium plating of maraging steel by an aircraft manufacturer determined to avoid hydrogen embrittlement of critical components.

Powder Metallurgy. Two grades of maraging steel, the 250 and 350 alloys, have been produced by powder metallurgy and are reported on in this section. Although some refinements can be anticipated, the powder metallurgy products are generally satisfactory.

Physical Metallurgy. For the physical metallurgist and others concerned primarily with structural features and alloy development, the maraging steels are among the most interesting of ferrous materials and reflect a major development of recent years. Five articles in this section review structure and strengthening mechanisms, aging phenomena and austenite reversion, precipitation, and the bases for strength and toughness in maraging steels.

Appendix. The concluding section of the book presents a comprehensive summary of the engineering properties of maraging steels and, by way of historical interest, the first general article on 25% nickel maraging steels ever published, based on information furnished by Clarence Bieber, the metallurgist credited with their development.

The American Society for Metals extends its grateful appreciation to Dr. Raymond F. Decker, who served as consulting editor of this book and who guided the selection and organization of the articles it contains. Dr. Decker also prepared his notes on the development of maraging steels specifically for inclusion in this book and brought to his assignment both a knowledge of his subject and an enthusiasm for its potential that comprise a unique service to the engineering community. We extend a special note of thanks to The Metallurgical Society of AIME, The Metals Society (London), the Society of Manufacturing Engineers, the many other technical publishers whose materials appear in this book, and the authors upon whose outstanding articles the value of this collection depends.

Paul M. Unterweiser
Staff Editor
Manager, Publications Development
American Society for Metals

William H. Cubberly
Director of Reference Publications
American Society for Metals

Cover chart: Metals Handbook, 9th Edition, Volume 1

Notes on the Development of Maraging Steels

R. F. DECKER

Publication of this book marks an appropriate occasion for reflecting on the origin, development and future of maraging steels. Their origin was by no means accidental, but the application we initially envisioned for them at that time never became a reality. Twenty years ago, our primary target was the development of a material suitable for submarine hulls, and the first maraging steels were intended for this application. They never qualified, because they were unable to provide the very high tear energies needed.

At the same time, however, the "space race of the 1960's" was very much alive, and lightweight rocket components were in critical demand. As a consequence, the development of maraging steels moved at a hectic pace, and applications were found in small rocket-motor cases, in launch-pad components, in Lunar Excursion Modules, and in Lunar Rover Vehicle parts. An equally pleasant surprise was the number of industrial applications that emerged, notably in tools and dies, where ease of fabrication and heat treatment was most fully realized. The various industrial applications and some of the major processing and cost advantages of maraging steels are listed in Tables 1 and 2, respectively.

For whatever solace it may offer inventors, the fact that maraging steels missed the submarine-hull target had its compensations. For it proved that if the invention is different enough and represents a significant increment of improvement over the prior art, it will garner unexpected applications afield from the original target. Ironically, maraging steel did eventually make its way in one deep-submergence hull; it was used in Deep Quest! The advantages of maraging steels, as we now understand them, are listed in Table 3.

Origin of the Steels. Mention of "increment of improvement" calls to mind my introduction to maraging steels or, more properly, their precursors. These derived from the work of Clarence Bieber, the father of these steels. It was from him that I first learned of them four months after I started research on superalloys under his tutelage at Inco, having recently completed my doctoral studies at the University of Michigan. Clarence and I were attending a vacuum-melting conference at the University Heights campus of New York University and, for brief diversion, were strolling through their Hall of Fame. The imposing scene — the Hall's open-air, semicircular corridor overlooking the Palisades and studded with statuary commemorating America's greatest figures of the past — turned our conversation to R&D and, more specifically, how we could contribute something quite significant — something bearing that large increment of improvement over the metallurgical state-of-the-art of that year, 1958. Clarence recalled that he had studied several 25% nickel alloys that transformed to martensite during aging, and thereupon attained high hardness. Why, he wondered, should we not initiate some research to pursue this earlier work? Needless to say — we *did*.

It was significant that we were not steel metallurgists, but rather were practitioners of the gamma-prime $Ni_3(Al,Ti)$ age-hardening type of superalloys. Clarence's teachings in composition, melting practice, hot working and heat treatment were derived from afield and were foreign to conventional steel metallurgy. Thus, a store of knowledge was transferred across fields for cross-fertilization, and unconventional ideas were tried.

After this early exploratory project work, Clarence brought the results to the attention of Inco R&D executives, who organized a matrix project team under the skilled management of Dr. John Eash. We had very formalized targets to meet for the aim market — submarine hulls. Talent was added from the steel, welding and corrosion technologies, and progress was fast. This progress on Clarence's work

Table 1. Applications of Maraging Steels

Production Tooling:

Pinion shafts for plastic extrusion presses
Flexible drive shafts
Wear-resistant index plates
Plastic molds
Hot-forging dies
Carbide die holders
Cold-heading dies and cases
Extrusion-press rams, dies and containers
Drawbolt heads for boring and milling machines
Aluminum and zinc die-casting dies
Gears in machine tools
Cams and cam followers
Stub shafts
Pistons
Splined shafts
Clutch disks
Autofrettage equipment
Springs
Forging-hammer piston rods
Diesel fuel-pump pins
Router bits

Hydrospace:

Deep Quest pressure hull

Military:

Cannon recoil springs
Rocket-motor cases

Aerospace and Aircraft:

Large rocket-motor cases
Smaller, special-purpose rocket-motor cases
Load cells
Universal flexures
Gimbal-ring pivots
Helicopter flexible drive shafts
Jet-engine shafts
Anchor rails for mobile service tower of Saturn 1B
Explosive bolts for Lunar Excursion Module (LEM)
Shock absorbers for Lunar Rover
Landing gear
Arresting hooks
Wing hinge for swing-wing planes

Auto-Racing Cars:

Drive shafts
Gearbox and axle-yoke ends
Connecting rods

Other:

Pump impellers and casings
Belleville springs
Tensile-test equipment
Hydraulic hoses
Cable sockets

Table 2. Savings in Processing and Costs Incurred by Using Maraging Steels

Comparison of Production Sequences

Maraging Steels

1 As-rolled or as-forged stock (280-320 HV)
2 Rough machine
3 Stress-relief anneal, if stock has been heavily or nonuniformly machined, by heating at 800-900 °C; air cool
4 Machine to finished dimensions
5 Harden to 500-800 HV by heating at 450-500 °C; air cool
6 Final surface finish

Conventional High-Tensile Steel

1 As-rolled or as-forged stock (260-650 HV)
2 Soften to 230-320 HV, by heating at 600-700 °C; air cool
3 Rough machine
4 Stress relieve if heavily or unevenly machined
5 Machine to oversize dimensions to accommodate decarburization and distortion
6 Harden by heating to 800-1050 °C and cooling at a controlled rate — usually water quench or oil quench
7 Temper to 500-600 HV, by heating at 200-600 °C
8 Finish machine to remove hardened oversize material (see 5 above)
9 Final surface finish

Savings by Use of Maraging Steels

1 At least three machining or heat treating sequences
2 Heat treating costs — fewer treatments and no quenchant
3 Setup time
4 Cost of machining hard material

was noted in the first general article on maraging steel in *Metal Progress* (see page 378).

Why "Maraging"? In the 25% nickel steels, we employed aging of the austenite, which we called the *first* age, to trigger transformation to martensite. Later, we aged the newly formed martensite, calling this the *second* age. Clarence then found that he could get better toughness by lowering nickel content to 20% and by simply applying one aging treatment to the transformed martensite. This created some confusion over use of the terminology, *first* and *second* age. To overcome the problem, I coined

Table 3. Special Advantages of Maraging Steels

1 Ultrahigh yield and tensile strengths.
2 High notched tensile strength — up to 200 tons/in.^{2} (310 kg/mm^2).
3 Exceptionally high resistance to crack propagation. K_{Ic} values up to 160,000 lb/in.$^{3/2}$ (570 kg/mm$^{3/2}$).
4 Maintain strength at temperatures of at least 350 °C.
5 Softened and solution treated at 750-900 °C, followed by air cooling.
6 Age hardened by simple low-temperature heat treatment (maraging) at 450-500 °C, followed by air cooling. Quenching is unnecessary and decarburization no problem.
7 Fewer thermal treatments enable furnace and handling costs to be reduced.
8 Consistently very small dimensional changes during maraging — possible to finish-machine before age hardening.
9 The ductile nickel martensite of maraging steels is much less prone to cracking than the low-ductility as-quenched martensite of conventional steels, which are overhardened before tempering.
10 Especially low carbon content — no reliance on this element to provide tensile strength or fatigue resistance.
11 Machinability in all conditions is similar to or better than that of conventional steels of same hardness.
12 Can be surface hardened by nitriding.
13 Good weldability, even in fully hardened condition. Preheat is unnecessary, and properties are restored by maraging.
14 Properties are maintained during and after low-temperature brazing.
15 Hot and cold deformation are practicable by most techniques.

the words *ausaging* for aging of the austenite and *maraging* for aging of the martensite. The second term was quickly adopted as the generic term for the alloy system.

Maraging With Cobalt and Molybdenum. By spring of 1960, our rate of progress diminished. We were stymied by toughness and stress-corrosion problems due to precipitation of titanium compounds at grain boundaries. Other melting, ingot, and weld-deposit problems were equally serious.

Preparation of a lecture for Polytechnic Institute of Brooklyn on precipitation-hardening stainless steels stimulated my thoughts on age-hardening agents other than titanium and aluminum. Mild strengthening effects (of the order of 20 ksi) in these precipitation-hardening stainless steels had been attributed in the literature to copper, beryllium, cobalt and molybdenum. Fortunately, Inco's R&D management had provided the vehicle to check feasibility, the free-wheeling Creative Research Program outside the structured project and project-team system. It was realized that cobalt and molybdenum were particularly attractive, and so it was recorded on May 11, 1960, that "the alloys with these elements should air melt more readily, be more adaptable for weld filler materials (less duplex M + γ from segregation) and may be more ductile than titanium-aluminum-containing materials. Better castings should be a result of less segregation."

Good fortune came soon thereafter. On June 23, 1960, several 200-g melts were melted. The eighth melt was the tenth heat listed in Table 3 of the paper by Decker, Eash and Goldman (see page 4). It contained 18.9% Ni, 7% Co and 3.8% Mo, and its strong maraging response amazed us. Later investigations revealed that high nickel was essential for this vigorous Co-Mo interaction, and that our discovery was a Ni-Co-Mo synergism. With lower nickel content, such as that in the 4% Ni precipitation-hardening stainless steels, phase balance is poor and coarse embrittling precipitates occur, accounting for lower strength and toughness.

On July 18, 1960, scale-up was commenced with 30 to 45-lb air melts. The very first melt was heat IV-6 of Table 4 of the paper by Decker, Eash and Goldman. Aim composition was 18.5% Ni, 7% Co, 5% Mo, 0.4% Ti, 0.1% Al, 0.003% B, 0.02% Zr and less than 0.03% C, calcium treated to leave less than 0.03% Ca, 0.1% Mn and 0.1% Si. That is virtually the *nominal composition* of the 18% Ni 250 maraging steel, the most common of current commercial grades. With pressure from the missile industry, we further scaled up successfully to a commercial 20,000-lb air melt in December, 1960 (see Table 10 of the paper by Decker, Eash and Goldman).

Commercialization and Teamwork. Although the early creative period was the most exciting of my career, I have grown to appreciate more the developmental years 1960-1964. Literally hundreds of organizations and thousands of people contributed. Included were not only Inco's product-development team, but also alloy producers, fabricators and users; industries, government labs, universities and technical societies; scientists, lawyers, engineers, executives and manufacturing and sales people; from the U. S., the U. K., West Germany, France, Japan, Russia, Canada and Israel. The papers and authors deservedly included in this volume represent many of the critical findings. But there were many others on the team! I regret that they remain unnamed in this book, but I hope they gained much satisfaction from their respective roles.

Market Growth. In hindsight, the commercialization and market growth of maraging steel were atypical of Inco's other new products. The tonnage grew very rapidly and reached a level of about 5000 tons per year by 1964. However, this rapid growth halted, and the tonnage is probably below that level now, still almost all in the 18% Ni 250 and 300 grades. Exact statistics are not available. This rapid initial growth pleased us, but the lack of further growth frustrated us. The limits of growth undoubtedly can be attributed to the high cost of the alloy system and the limits found in stress-corrosion cracking and toughness. Nevertheless, dollar volume has been very significant. Total sales revenue to the alloy producers has amounted to more than $300-million to date. Added value revenue to the fabricators of parts has been of the same order of magnitude.

Advances in Research. The understanding of mechanisms for attainment of toughness with high strength is now very sophisticated when compared to that of 1960. As disclosed by papers in this volume, the optimum structure of a maraging steel includes:

a) an extremely fine array (less than 1000 Å size and interparticle distance) of Ni_3Mo, its volume fraction increased by cobalt;
b) freedom from segregation of tramp elements to prior austenite grain boundaries and interphase boundaries, by controlled purity and titanium and molybdenum additions;
c) yet, a freedom from coarse and/or planar arrays of TiC, or Ti_2S, or sigma.

It was our fond hope in the 1960's to find a thermal or thermomechanical treatment that would improve toughness and stress-corrosion resistance. In fact, the alloy producers were very successful at this, and reduced thermal embrittlement during hot forging and rolling. Several other treatments, such as grain refinement by thermal cycling, partial reversion to austenite and microduplexing, sometimes provided promising improvements in properties, but these were not reproducible. Several authors in this volume have studied these treatments and noted the lack of outstanding effects on properties.

Advances in Alloy Development. Alloy development was intense during the 1960's. The scope of Inco's efforts on Ni-Co-Mo, cobalt-free, and low-alloy aging grades is reflected in the alloy compositions given in Table 4. Certain key findings on the role of molybdenum, thermal embrittlement, alloy segregation-delamination and trace elements led to improved compositions, such as IN 763, for wrought

Table 4. Maraging Steels and Aging Steels

Alloy	Nominal composition, wt % (rem Fe; low C)									Yield strength, ksi
	Ni	Co	Mo	Ti	Cr	Si	Cu	Cb	Other	
Ni-Ti Types										
25Ni	25	—	—	1.5	—	—	—	—	—	250
20Ni	20	—	—	1.5	—	—	—	—	—	250
Ni-Co-Mo Types										
18Ni250	18	8	5	0.4	—	—	—	—	—	250
18Ni200	18	8	3	0.2	—	—	—	—	—	200
18Ni300	18	9	5	0.7	—	—	—	—	—	300
350 alloy	18	12	4	1.6	—	—	—	—	—	350
400 alloy	13	15	10	0.2	—	—	—	—	—	380
IN 763	18	15	3	0.1	—	—	—	—	(0.5V)	250
17Ni cast(a)	17	10	5	0.3	—	—	—	—	—	240
IN 0180(a)	18	10	1.7	0.1	—	0.5	—	—	—	180
230 alloy(a)	18	15	3	0.1	—	—	—	—	—	230
Ni-Cr Types										
12-5-3	12	—	3	0.2	3	—	—	—	(0.3Al)	180
IN 733	10	—	—	0.3	12	—	—	—	(0.7Al)	220
IN 736	10	—	2	0.2	10	—	—	—	(0.3Al)	180
IN 833 (CA-6N)(a)	7	—	—	—	12	1	—	—	—	130
Low-Alloy Aging Steels										
IN 863(a)	3	—	—	—	3	0.5	—	—	—	140
IN 0335(a)	3	—	0.5	—	3	0.5	—	—	—	140
IN 787	0.8	—	0.2	—	0.6	0.3	1.1	0.03	—	90
IN 866(a)	1	—	0.2	—	0.5	0.4	1.3	—	(1.8Mn)	90

(a) Cast alloys; other alloys listed are wrought.

heavy sections, and the 230 and IN 0180 alloys for castings. Cobalt-free Ni-Cr alloys, such as 12-5-3, IN 733 and IN 736, evolved. Silicon hardening proved advantageous in cast IN 833. Several of these post-1961 maraging grades have found modest application. Low-alloy aging steels (sometimes called "feraging"), such as IN 787 and cast alloys IN 863, IN 866 and IN 0335, were inspired by the necessity for lower costs and the maraging impetus. In terms of commercial success, IN 787 is reaching an annual tonnage about equal to that of the 18% Ni maraging steels.

The Outlook. As I review the maraging-steel scene it is apparent that the 1970's were static. This is especially evident when I contrast the lack of advances in properties of the Ni-Co-Mo grades with the many advances in those of my first specialty, the superalloys. The temperature capability of superalloys has been advanced steadily, from 1450 °F during World War II to 2150 °F today.

In straddling these two fields, I might be led to wonder: are maraging steels fated to technological dormancy? My intuition says no. The world is not static, nor are the needs and ideas of technology. We can be certain that new needs and ideas will surface. Within the last few months, a severe cobalt shortage has emerged as a real threat. Perhaps this new emergency will be met with one of the "shelf alloys" listed in Table 4 or will foster the rejuvenation of alloy designs — hopefully, those alloy designs of lower cost.

Today, we have more alloy designers in practice and in reserve than ever before. Many newcomers have entered the field of metallurgy with new understanding of deformation mechanisms. New technologies of electron microscopy and surface analysis provide tremendous powers of insight to the new trailblazers. Small wonder that breakthroughs can be expected. The surprises to come will be in the form and simplicity of these breakthroughs!

SECTION I:
Introduction

18% Nickel Maraging Steel

BY R. F. DECKER, J. T. EASH AND A. J. GOLDMAN*

ABSTRACT. An unusually high combination of strength and toughness can be developed in low-carbon 18% Ni steel containing cobalt and molybdenum. These alloys have a martensitic structure that is very ductile and formable in the annealed or hot rolled condition. Upon reheating to 900 F ("maraging"), they age harden and produce yield strengths of 250 to 310,000 psi, reduction of area of 60%, notch-tensile strength/tensile strength ratio of over 1.4, up to 26 ft-lb Charpy impact resistance and a very low drop-weight, nil-ductility temperature. The new alloy possesses a remarkable resistance to stress-corrosion.

Investigations of the effects of composition, working and heat treatment have indicated the composition limits desirable for plate and sheet with a range of properties of immediate commercial interest. Production of the material has been extended to commercial practice, a 20,000 lb basic-arc air heat having been made.

THE AUSTENITE-MARTENSITE transformation in iron-nickel alloys has been studied by several investigators (1–4). The transformation characteristics illustrated in Fig. 1 are from the diagram by Jones and Pumphrey (1).

The martensite at 20 to 30% Ni forms by diffusionless shear as in conventional steels. However, the nature of the transformation and product contrast sharply with that of conventional carbon steels in that:

a. Section-size effects are small due to an insensitivity of the martensite reaction to cooling rate and the lack of higher temperature-diffusion controlled austenite decomposition to carbide phases. Thus rapid quenching is not required and hardenability is no problem.

b. The transformation can proceed both athermally and isothermally.

c. The martensite structure exhibits no tetragonality; but rather, is body-centered cubic.

d. The martensite is only moderately hard (Rockwell C-25) and very tough contrasted to the high hardness and brittleness of untempered carbon steel martensites.

e. Tempering does not occur upon reheating of the martensite. As in c and d above, this is due to absence of carbon (4, 5). Finally, the hysteresis of the transformation seen in Fig. 1 allows considerable reheating of the martensite for aging (in the presence of alloying additions) before reversion to austenite occurs. This step is designated *maraging*.

It was desired to further strengthen this iron-nickel martensite to a level of engineering interest, but with a minimum loss in toughness. Therefore, a program was undertaken for study of the effect of various elements on solid-solution hardening and maraging of the iron-nickel martensite.

* R. F. Decker and J. T. Eash are associated with the Research Laboratory of The International Nickel Co., Inc., Bayonne, New Jersey; A. J. Goldman is a Graduate Student, Yale University, New Haven, Conn. Manuscript received August 21, 1961.

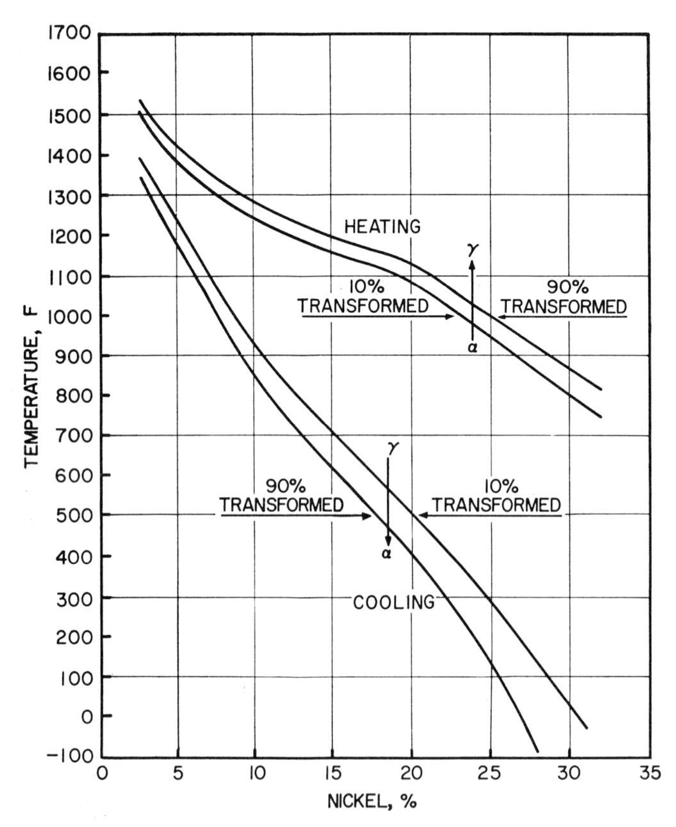

Fig. 1. The Fe-Ni transformation diagram (1).

Pilling and Merica (6) had shown that with aluminum and titanium additions, iron-nickel alloys could be ausaged (aged in austenite) to both harden the austenite and to induce transformation to martensite. Bieber (7) had demonstrated that unusually high strength and hardness could be obtained in iron-nickel martensite when alloyed with aluminum and/or titanium. Many other works, too numerous to list but reviewed in Geisler (8) and Lena (9), have dealt with the subject of age hardening reactions in iron-base alloys.

Experimental Procedures

The early experimentation was on hardness of 200 g melts from which a remarkable synergistic maraging effect from cobalt and molybdenum was discovered. This led to more extensive engineering experiments on 30-lb laboratory melts. Finally, a 20,000-lb basic electric-arc heat was air melted and rolled to plate to assess feasibility of commercial manufacture. Melting and working histories for all materials are given in Table 1. Reference is made to testing methods in Table 2.

Results and Discussion

Hardness

The singular coaction of cobalt and molybdenum was first observed in the compositions listed in Table 3. Maraging studies were made on each alloy to include times of 1 to 10 hr and temperatures of 800 to 1000 F. The maximum hardness obtained on each alloy and the associated heat treatment are given in Table 3. All specimens had been annealed initially at 1600 F and transformed to martensite.

TABLE 1. Processing of Materials

Size of heat	Melting practice				Working history
	Furnace	Base charge	Slag	Deoxidation	
200 g	Induction—argon blanket, alumina crucible	Electro Fe, Ni, Co; FeMo	—	None	Forged from 1800 F, soaked 1 hr at 2300 F, forged at 1800 F. Total reduction—80%.
30 lb	Induction—argon blanket, magnesia crucible	Electro Fe, Ni, Co; FeMo, 0.05 C	—	CO boil, SiMn, CaSi, Al, Ti, B, Zr Residuals—0.1 Si, Mn, Al; 0.01 Ca, 0.003 B, 0.01 Zr, 0.01–0.03 C	Forged from 2300 F from 4 x 4-in. ingot to 2 x 2-in. bar with 1 hr intermediate soak at 2300 F. Rolled from 1800 F to ⅝-in. plate and ¾-in. diam bar, finishing near 1500 F.
30 lb	Vacuum induction, magnesia crucible	Electro Fe, Ni, Co; Mo, 0.05 C (held ½ hr at 2900 F at 2 microns pressure)	—	CO boil, Al, Ti, B, Zr Residuals—0.05 Si, Mn; 0.1 Al, 0.003 B, 0.01 Zr, 0.01 C	Forged from 2300 F from 4 x 4-in. ingot to 2 x 2-in. bar with 1 hr intermediate soak at 2300 F. Rolled from 1800 F to ⅝-in. plate and ¾-in. diam bar, finishing near 1500 F.
20,000 lb	Basic—electric arc	Electro Fe, Ni, Co; FeMo, 0.50 C (Oxygen lance to 0.02 C)	3 lime 1 flourspar + Grain Al	CaSi, Al, Ti, B, Zr	23 x 42 x 60-in. ingot, stripped at 1500 F, transferred to soaking pit and heated to 2230 F, bloomed from 23 to 16-in., reheated to 2280 F, bloomed to 4 x 36-in. plate, finishing at about 1950 F, hot sheared and overhauled, rolled from 2150 F to 1-in. plate, finishing at about 1750 F, and to ½-in. plate, finishing at about 1600 F, plates sandblasted and dipped in light acid bath.

TABLE 2. Test Descriptions

Property	Description of test specimen and procedure	
Yield strength (0.2% offset) Ultimate tensile strength Elongation Reduction in area or thickness	Bar Sheet Strain rate	1-in. gage length of 0.250 in. diameter reduced section, strain rate—0.01 in. per in. per min to yield, then 0.05 in. per in. per min 0.08 x ½ x 2 in. reduced section (Ref. 12) same as above
Notch-tensile strength	Bar	0.5 in. major diameter, 0.350 in. minor diameter, 0.0008 to 0.0010 in. root radius, K_t (notch acuity factor) > 10 0.3 in. major diameter, 0.212 in. minor diameter, 0.0002 to 0.0007 in. root radius, $K_t > 10$ Both machined after full heat treatment
Sheet toughness	Sheet	1 in. and 2 in. edge notch, 0.0006 in. root radius, 60° notch (Ref. 12) $K_t > 10$ Machined after full heat treatment
NDT (Nil-ductility temperature)	Plate	½ x 2 x 5 in.—Deposit Murex weld bead, then full heat treatment, notch bead, weight drop 10 to 15 ft, as prescribed in Ref. 13, 14
Stress-corrosion	3-pt loaded U-bend	$\frac{1}{16}$ x $\frac{3}{16}$ x 3 in., beam loaded to yield stress, in Bayonne atmosphere and aerated artificial sea water Bent 150°, heat treated, closed final 30° to give plastic deformation over yield point, bolt insulated by paraffin, tested in aerated artificial sea water
Charpy V-notch impact energy	0.394 x 0.394 x 2.165 in., 0.315 in. section under notch, 45° notch, 0.010 in. root radius	
Density	Water immersion method	
Linear coefficient of expansion	Leitz dilatometer	

TABLE 3. Compositions and Aging Responses of 200-g Melts*

Ni	Co	Mo	As-annealed hardness, Rockwell C	Aging treatment	Aged hardness, Rockwell C
20	—	—	25.0	No response	25.0
20.1	—	2.0	27.0	10 hr at 800 F	30.5
19.4	—	4.7	27.5	10 hr at 900 F	40.0
19.0	—	8.0†	30.5	10 hr at 900 F	46.5
18.4‡	—	10.2†	31.0	3 hr at 900 F	47.5
20.4	1.7	1.8	27.5	10 hr at 800 F	31.0
19.9	3.4	1.9	28.5	10 hr at 800 F	32.0
20.0	5.0	2.2	28.0	10 hr at 800 F	35.0
18.9	7.0	3.8	29.0	3 hr at 900 F	43.0
18.9	7.0	5.4	30.0	10 hr at 800 F	50.0
18.5‡	7.5‡	8.3‡	30.0	3 hr at 900 F	53.0
18.5	7.0	—	25.5	No response	25.5**

* *Fe bal, all annealed 1 hr at 1600 F, air cooled.*
† *Refrigerated 16 hr at −100 F after anneal to convert all austenite to martensite (high molybdenum content depressed M_s temperature).*
‡ *Aim analysis.*
** *Data from a 30-lb melt.*

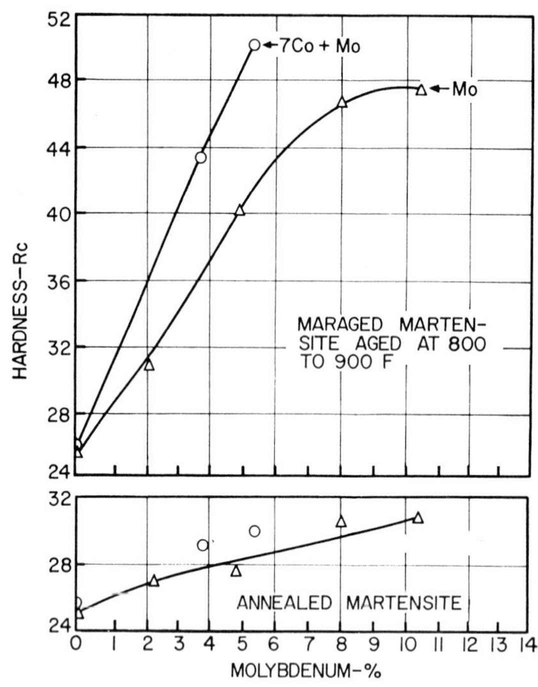

FIG. 2. Effect of molybdenum and molybdenum + 7% Co on maximum hardness of 18.5 to 20.1 Ni-Fe alloys. Solution annealed 1 hr at 1600 F, air cooled; maraged 3 to 10 hr at 800 to 900 F.

It can be noted in Fig. 2 that the basic hardness of the iron-nickel martensite was Rockwell C-25 and that this did not respond to heat treatment at 800 to 900 F. Age hardening resulted with molybdenum additions. Also plotted in Fig. 2 is a curve for maraged hardness for molybdenum with an addition of 7% Co. A strong additional effect from the presence of cobalt can be seen. No response to maraging was found in an alloy with 7% Co and no Mo.

Figure 3 reveals the synergistic effect of cobalt with molybdenum on maraging between 800 and 900 F. Maraged hardness increased linerly as the product, cobalt times molybdenum, increased, between the

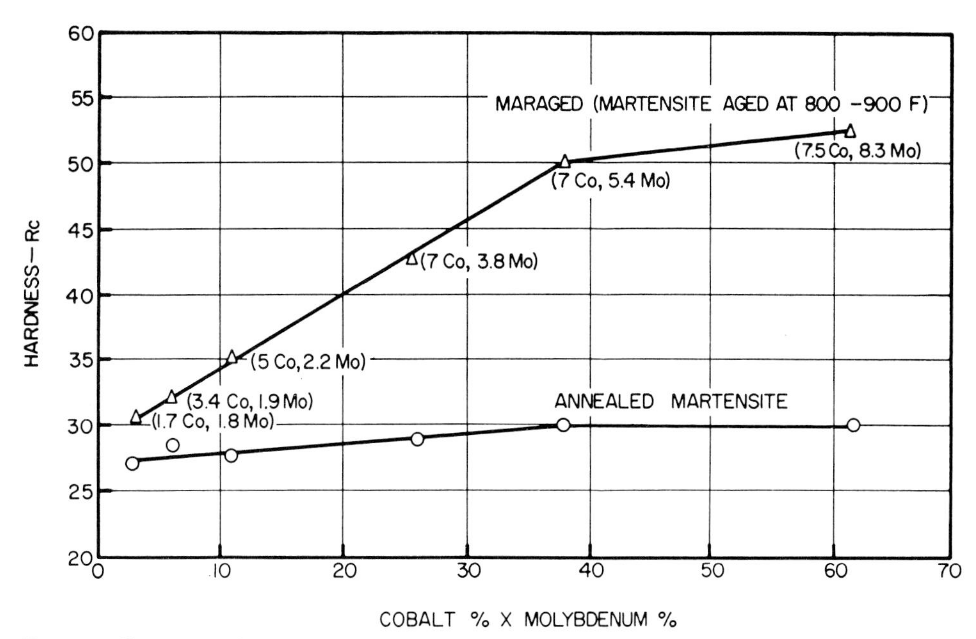

FIG. 3. Effect of product of cobalt X molybdenum on maximum hardness of 200-g melts of 18.5 to 20.1 Ni-Fe alloys. Solution annealed 1 hr at 1600 F, air cooled; maraged 3 to 10 hr at 800 to 900 F.

limits of 1.7 Co, 1.8 Mo to 7 Co, 5.4 Mo. It is possible by increasing this product to obtain alloys that show only a change from Rockwell C-25, in the absence of both cobalt and molybdenum, to Rockwell C-30 in the annealed martensitic condition, but then exhibit very strong maraging to give hardnesses above Rockwell C-50.

With this balance of strong maraging effect and weak solid-solution hardening, a tough martensite matrix is maintained even with high strength. This appears to be the key to the remarkable toughness of the nickel-cobalt-molybdenum steels as covered in the following discussion.

Tensile Properties

The initial tensile tests were on 30-lb air melts with 7% Co and 5% Mo. These had small (0.4%) titanium additions for reasons outlined below under Titanium Effects. These tests indicated unusually high strength coupled with high ductility and low notch sensitivity. Additional features of low nil-ductility temperature (NDT), as determined by the drop weight test, and outstanding stress-corrosion resistance gave a combination of properties unprecedented in ferrous materials. With these early promising results in hand, an extensive program of composition studies was conducted.

Cobalt and Molybdenum Effects. Variation in the molybdenum content between 3 and 4.9% in alloys containing 18.5 Ni, 7.5 Co, 0.4 Ti, balance iron, produced an increase in yield strength of about 20,000 psi for each per cent increase of molybdenum, giving values up to 256,000 psi (Fig. 4 and Table 4). Toughness was exceptional, with the ratio of notch-tensile strength to ultimate strength (*NTS/TS*) decreasing slightly from 1.6 to 1.5 as molybdenum increased. Exceptionally high values of reduction of area, up to 59%, and Charpy impact resistance of 26 ft-lbs were secured even at a 250,000 psi yield strength. The data obtained without annealing revealed the possibility of simple heat treatments to be described later under Direct Maraging.

Extensive mechanical and physical test results on 10 to 30-lb air heats at this promising level of about 5% Mo are

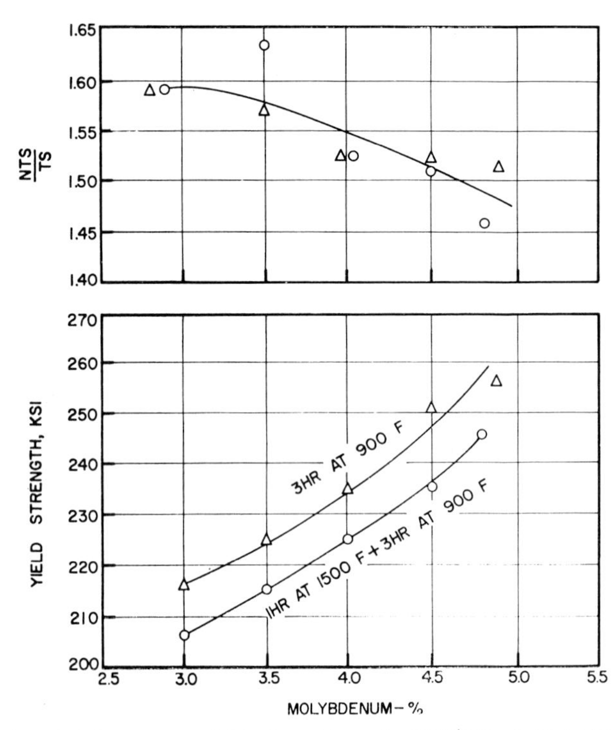

FIG. 4. Effect of molybdenum on yield strength and *NTS/TS* of 18.5 Ni, 7.5 Co, 0.4 Ti-Fe; 30-lb air melts.

TABLE 4. Effect of Molybdenum
(18.5 Ni, 7.5 Co, 0.4 Ti, bal Fe, 30-lb air melt)

Heat	Mo, %	Heat treatment	Tensile strength, ksi		Elong, %	R.A., %	Charpy V-notch impact energy, ft-lb (70 F)	Notch-tensile* strength, ksi	NTS/TS†
			Yield	Ultimate					
IV-1	3.0	1/1500 + 3/900‡	206	214	16	64	51	340	1.59
		3/900**	217	222	14	64	38	353	1.59
IV-2	3.5	1/1500 + 3/900	215	225	14	61	35	367	1.63
		3/900	226	233	14	60	38	365	1.57
IV-3	4.0	1/1500 + 3/900	225	237	13	56	30	363	1.53
		3/900	235	242	13	62	28	371	1.53
IV-4	4.5	1/1500 + 3/900	235	247	14	54	32	374	1.51
		3/900	252	257	12	59	26	391	1.52
IV-5	4.8	1/1500 + 3/900	246	256	11	54	—	372	1.46
IV-6	4.9	3/900	256	260	10	58	—	396	1.52

* *0.3 in. major diameter round bar.*
† *Notched tensile strength to ultimate tensile strength ratio.*
‡ *1 hr. at 1500 F, AC + 3 hr at 900 F, AC.*
** *3 hr at 900 F, AC.*

summarized in Table 5. Outstanding for this yield strength level of 250,000 psi are:

1. Passing of drop weight test at −80 F. For equivalent nil-ductility temperature, this represents a jump in yield strength of about 80,000 psi over low alloy steels.

2. Excellent stress-corrosion resistance. Low alloy steel at the same yield and

TABLE 5. Nickel-Cobalt-Molybdenum Steel

Composition						
Ni	Co	Mo	Ti	C	Mn	Si
17–19	7–8.5	4.6–5.1	0.3–0.5	0.01–0.03	0.1 max	0.1 max

S	P	Al	B	Zr	Ca
0.01 max	0.01 max	0.1 (added)	0.003 (added)	0.02 (added)	0.05 (added)

Physical and Mechanical Properties

Physical Properties
- Avg. coefficient of thermal expansion (70 to 900 F), per °F .... 5.6×10^{-6}
- Modulus of elasticity, psi .... 26.5 to 27.5×10^{6}
- Density, g per cu cm 8.0

Mechanical Properties
- Proportional limit, psi .... 200,000 to 220,000
- Yield strength (0.2% offset), psi .... 240,000 to 268,000
- Ultimate strength, psi .... 250,000 to 275,000
- Elongation, % .... 10 to 12
- Reduction of area, % .... 48 to 58
- Notch-tensile strength (0.3-in. bar, $K_t > 10$), psi .... air melt, 370,000 to 385,000; vac melt, 393,000 to 415,000
- Charpy V-Notch impact energy (at R.T.), ft-lb .... air melt, 18.5 to 26; vac melt, 25 to 30
- Fatigue-endurance limit, psi .... 95 to 100,000

Nil-ductility temperature (based on ½-in. plate) .... below −80 F

Stress Corrosion
- In aerated artificial sea water
  - 3-point loaded specimens, 246,000 psi yield strength, load level—unbroken in 210 days (in test)
  - U-bend specimens, up to 272,000 psi yield strength, load level—some failures in 35 to 45 days; some unbroken after 210 days (in test)
- In Bayonne atmosphere
  - 3-point loaded specimens, 246,000 psi yield strength, stress level—unbroken 150 days

Properties as Annealed
- Hardness .... 28 to 30 Rc
- Yield strength .... 95,000 psi
- Ultimate strength .... 140,000 psi
- Elongation 17%
- Reduction of area 75%

Heat Treatment for Above Properties
- Anneal: 1500 F, 1 hr, air cool (optional)
- Marage: 900 F, 3 hr, air cool

Melting
- Readily air melted in induction and basic electric arc furnaces.

Hot working
- Readily forged, bloomed, and hot rolled
- Soaking temperature, F .... 2200 to 2300
- Blooming temperature range, F .... 2100 to 2300
- Recommended intermediate soaking for homogenization, F .... 2200 to 2300
- Forging temperature range, F .... 1850 to 2200
- Hot rolling temperature range, F .... 1900 to 1500
- Finishing temperature, F 1500

Cold working
- Cold rolled up to 90% reduction
- Hardness increased from Rockwell C-28 to C-35

Welding
- Sound crack-free welds produced in 1-in. fully heat treated plate by coated electrode, submerged-arc, and metal-inert gas welding.
- Hardness of weld deposit and heat-affected-zone after aging at 900 F was Rockwell C-47 to C-50.

Properties of butt welds at present stage of investigation:

Tensile strength, psi		Elong, %	Red. Area, %	Notch-strength, psi
Yield	Ultimate			
222 to 229,000	250 to 260,000	5 to 8	17 to 20	250 to 315,000

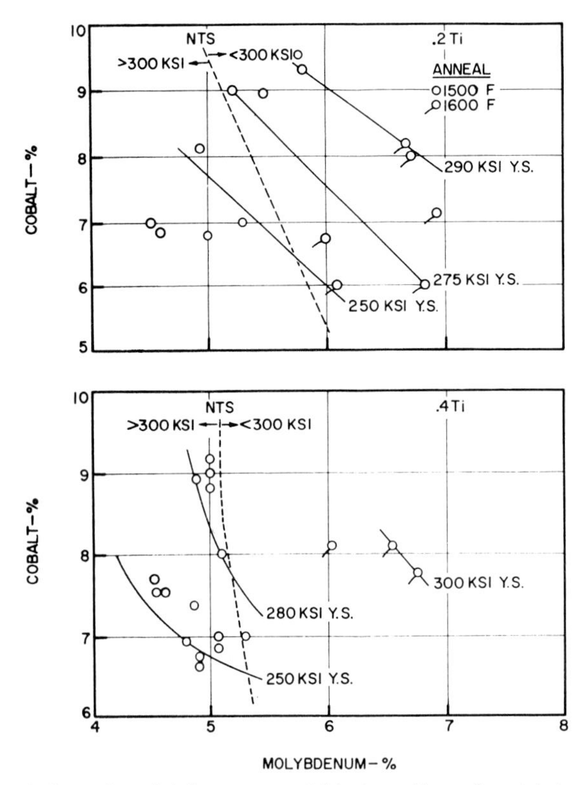

FIG. 5. Effect of cobalt and molybdenum on yield strength and notch-tensile strength (0.5 in. major diameter) of 18.5 Ni-Fe alloys, 30-lb air melts. Annealed, maraged 3 hr at 900 F.

stress level failed in 2 days with 3-point loading in sea water. Furthermore, low alloy steels subjected to atmospheric stress-corrosion at lower yield and stress levels failed (10).

3. High notch-tensile strengths of 370,000 to 385,000 psi. This is a gain of about 80,000 psi over alloy steels at the same yield strength.

4. Excellent weldability. No heat-affected-zone cracking has been found, even in fully hardened, restrained plate of up to 4-in. sections (no preheat). The heat-affected-zone can be restored to full strength by a simple post-weld marage.

5. Low coefficient of expansion.

Increasing cobalt above 7% increased the yield strength about 15,000 to 20,000 psi for each per cent of cobalt (see Fig. 5). The coaction of cobalt and molybdenum on yield strength is shown in Fig. 5 for 18.5 nickel alloys at two titanium levels, 0.2 and 0.4%. Superimposed are curves defining the limits for 300,000-psi notch-tensile strength obtained on 0.5-in. major diameter bars. The annealing temperature needs to be raised to 1600 F when molybdenum content exceeds about 5.5 to 5.8% to dissolve all phases in the austenite. This fact plus, more significantly, the curves on notch-tensile strength define compositional areas for increase of yield strength from 250,000 psi with minimum loss of toughness. This can best be done by increase of cobalt content.

Titanium Effects. Titanium proved to be a desirable supplemental hardener for cobalt and molybdenum-containing alloys. Data in Fig. 6 illustrate this supplementation in 7 Co, 5 Mo alloys. As titanium was increased from 0.1 to 0.7%, the yield strength increased from 220,000 to 280,000 psi. This increase was about 100,000 psi for each per cent of titanium. The *NTS/TS*

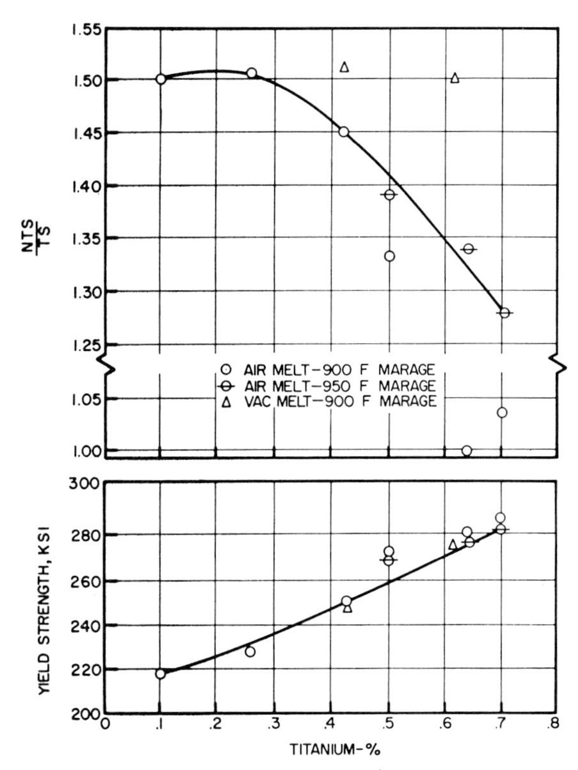

FIG. 6. Effect of titanium on yield strength and *NTS/TS* of 18.5 Ni, 7 to 7.5 Co, 5 Mo-Fe alloys 30 lb melts. Annealed 1 hr at 1500 F, maraged.

ratio decreased in air melts containing more than 0.4% Ti. This drop-off was very sharp for maraging at 900 F, dropping to a *NTS/TS* of 1 at 0.6 to 0.7% Ti. Maraging at 950 F gave higher notch-tensile strength in the air melts at 0.5 Ti and above. A titanium content of 0.7% was found to be the upper limit to preserve the excellent air melting characteristics of these alloys. Above this level, dross and films developed. It should be noted, however, that hardnesses above Rockwell C-60 were obtained with maraging alone when titanium supplementation was boosted above 1%.

Titanium also neutralizes residual carbon and nitrogen by removing them from solution in the martensite.

Silicon and Manganese Effects. Silicon and manganese were detrimental to the alloys. Most of the heats contained less than 0.10 Mn and less than 0.15 Si. Heat VI-2 in Table 6 with 0.23 Mn and 0.27 Si reached the same yield strength level as the low manganese and silicon heats (Heat VI-1); but had much lower notch-tensile strength, 217,000 psi. Heats VI-3, with 1% Mn, and VI-4, with 1.9% Mn, illustrate that replacing nickel with manganese led to lower notch-toughness.

Other Residuals. Carbon was present, usually in the range of 0.01 to 0.03%. Carbon up to 0.03% was not detrimental to notch-tensile strength at the 250,000 psi yield strength level. A heat with 0.038 C had slightly a lower and one with 0.05% C had a significantly lower notch-tensile strength. There was some evidence that increasing carbon from 0.005 to 0.03 raised the yield strength, Table 6.

Aluminum was added (0.1%) throughout as a deoxidizer. Larger amounts supplement hardening but decrease toughness.

Calcium content of the air melts was

TABLE 6. Effect of Silicon, Manganese, Carbon and Aluminum
(Heat Treatment: 1 hr at 1500 F, air cool, 3 hr at 900 F, air cool)

Heat	Ni	Si	Mn	C	Al	Tensile strength, psi		Elong, %	R.A., %	Notch-tensile strength, ksi*	Ratio, NTS/TS
						Yield	Ultimate				
Effect of Silicon and Manganese (On 7 Co, 5 Mo, 0.04 Ti bal Fe)											
VI-1	18.5	0.11	0.07	0.02†	0.1†	255	265	11	52	350	1.32
VI-2	18.3	0.27	0.23	0.02†	0.1†	264	270	7	36	217	0.80
VI-3	17.0	0.18	1.0	0.02†	0.1†	271	279	6	30	227	0.81
VI-4	15.0	0.16	1.9	0.02†	0.1†	279	288	5	26	178	0.62
Effect of Carbon (On 7 Co, 5 Mo, .4 Ti, bal Fe)											
VI-5	18.7	0.10†	0.10†	0.011	0.1†	242	254	11	57	381	1.50
VI-6	18.5	0.10†	0.10†	0.023	0.1†	248	256	12	54	374	1.46
VI-7	18.2	0.10†	0.10†	0.028	0.1†	258	266	11	52	374	1.41
VI-8	18.3	0.10†	0.10†	0.030	0.1†	256	264	10	49	362	1.37
VI-9	18.2	0.10†	0.10†	0.038	0.1†	263	268	10	44	356	1.34
VI-10	18.2	0.10†	0.10†	0.050	0.1†	243	256	11	48	322	1.26
Effect of Aluminum (On 7 Co, 5 Mo, 0.2 Ti, bal Fe)											
VI-11	18.4	0.10†	0.10†	0.02	0.1†	226	240	12	58	325	1.35
VI-12	18.5†	0.10†	0.10†	0.02	0.5†	263	275	12	48	277	1.0

* *0.5-in. major diam round bar.*
† *Aim analysis.*

usually 0.01 to 0.02%. The effect on toughness is unknown.

Boron content was usually about 0.003% and zirconium content 0.01%. These were purposefully added since earlier work on titanium hardened alloys proved that these retarded grain-boundary precipitation, thereby improving toughness and stress-corrosion resistance. This is probably due to grain-boundary segregation of these elements, which have atomic radii incompatible with interstitial or substitutional solution.

Heat Treatment Studies

Dimensional and Structural Studies. Heat treatment was studied using an alloy containing 18.5 Ni, 7 Co, 5 Mo, 0.4 Ti, balance iron. The dimensional and structural changes are illustrated in Fig. 7 by a dilatometer trace of change in length vs time in the heat treating cycle. The as-rolled material at room temperature was martensitic. When this was heated to the solution annealing temperature of 1500 F, it started reversion to austenite at the A_s and finished this reversion at the A_f temperature, still below 1500 F. At 1500 F little change in dimension occurred. Upon air cooling after solution annealing, no transformations took place in the austenite until the M_s temperature was reached at about 310 F. Here the austenite started transformation to martensite. As an aside, cobalt raised M_s by 10 F/per cent and molybdenum lowered M_s by 40 F/per cent when they were co-present. Transformation finished at around 210 F, so that the sample was substantially martensitic (more than 99% martensite by x-ray determinations) at room temperature. The martensite was then heated to 900 F for maraging. None of the martensite reverted to austenite and very little dimensional change took place. On cooling from 900 F in air, no further transformations came about. Final

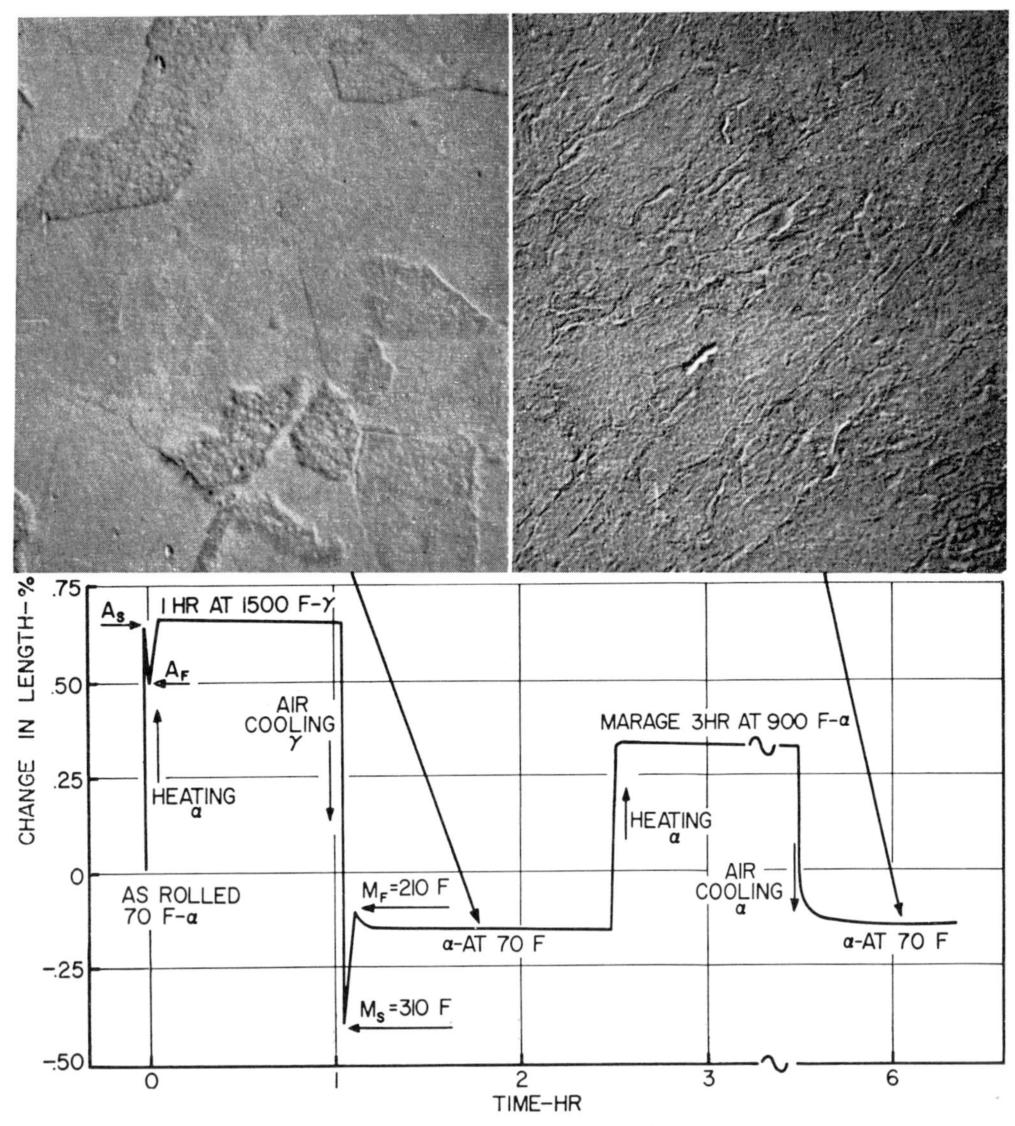

FIG. 7. A Leitz dilatometer trace of change in length with time during a heat treating cycle of 1 hr at 1500 F, air cooled, maraged 3 hr at 900 F, air cooled. Alloy of 18.5 Ni, 7 Co, 5 Mo, 0.4 Ti-Fe. Electron micrographs (×7500) were from parlodion replicas from the metal surface etched with 50 ml HCl, 25 ml HNO_3, 1 g $CuCl_2$, 150 ml H_2O.

length was very close to that before maraging. This, plus the lack of quenching stresses and retained austenite, make these alloys relatively free of dimensional changes during and after heat treatment.

Hardness vs aging time is shown in Fig. 8. An abrupt initial hardening response in the first 2 min was followed by a more gradual increase over a period of about 3 hr. Overaging was slight, even up to 200 hr at 900 F.

It was noted during heat treatment and hot working operations that the alloys were free-scaling; in fact, more so than alloy steels. Of course, another advantage of this alloy system to heat treaters is the lack of decarburization problems.

The electron microstructures in the

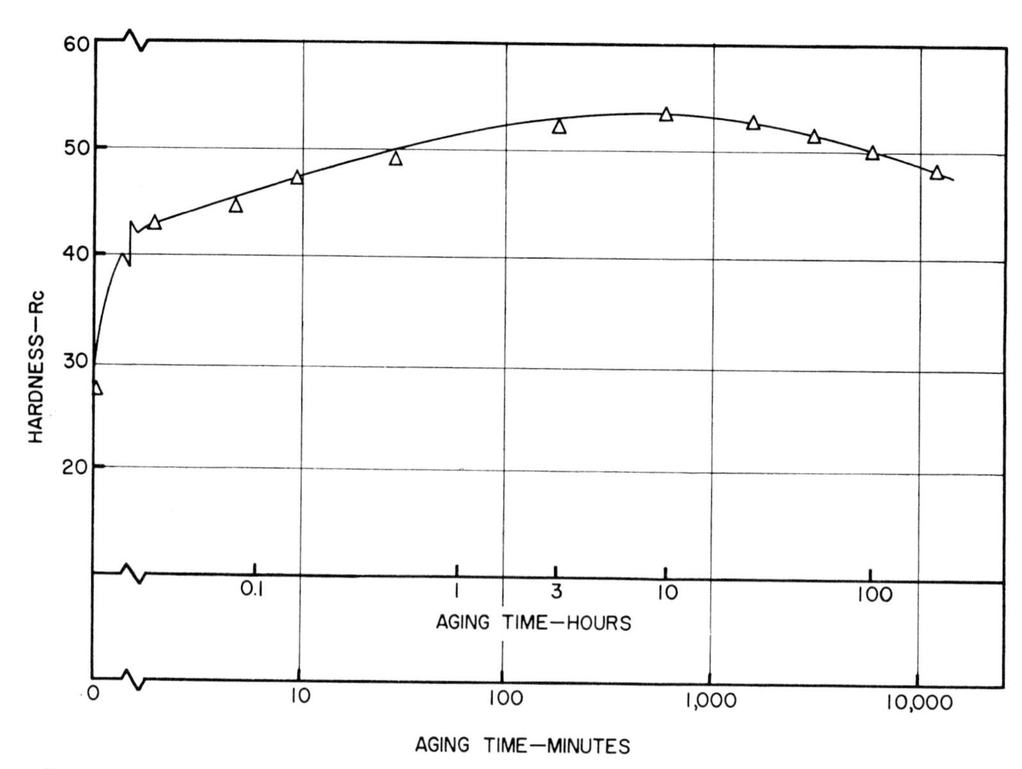

FIG. 8. Hardness vs maraging time at 900 F for a 19 Ni, 7 Co, 4.9 Mo, 0.4 Ti-Fe alloy. Annealed 1 hr at 1500 F, air cooled.

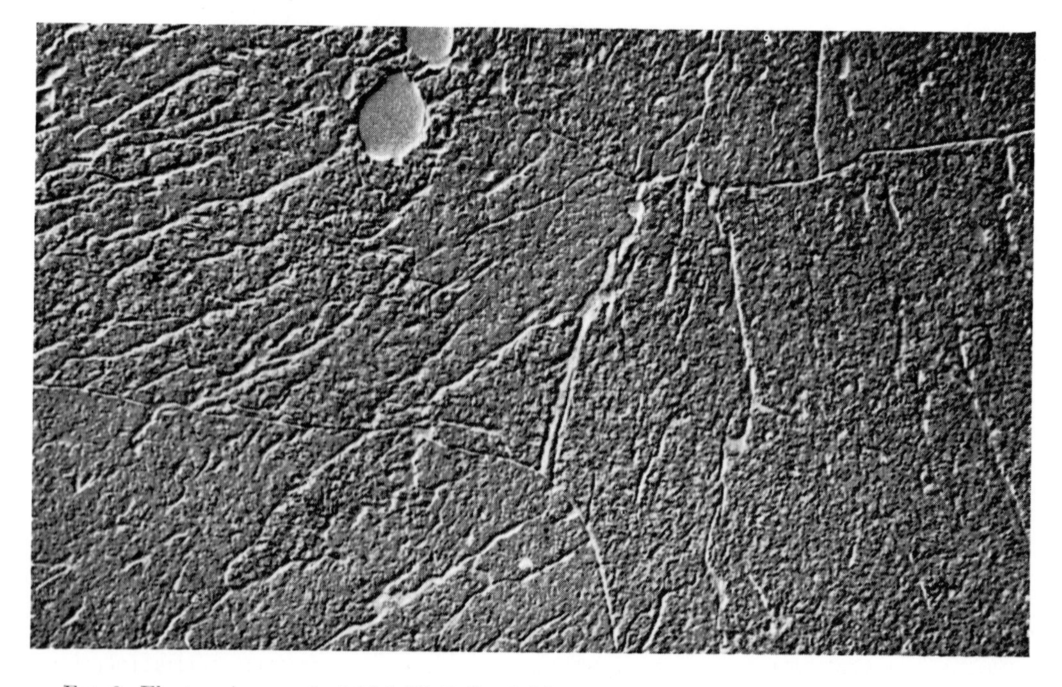

FIG. 9. Electromicrograph of 18.5 Ni, 7 Co, 5 Mo, 0.4 Ti-Fe alloy, treated 1 hr at 1500 F, air cooled 10 hr at 950 F. Etch same as in Fig. 7.

Fig. 10. Demonstrations of formability of annealed martensitic Ni-Co-Mo steel, (a) bend and twist specimens, (b) cold rolling of annealed $\frac{1}{2}$-in. plate to 0.10 in. sheet with no intermediate anneal

annealed and aged conditions are given in Fig. 7. The aged sample appears slightly acicular, while the annealed does not. After a treatment of 3 hr at 900 F, the precipitate is not clearly visible, even in the electron micrograph. An electron micrograph prepared after 10 hr at 950 F (see Fig. 9) reveals precipitate.

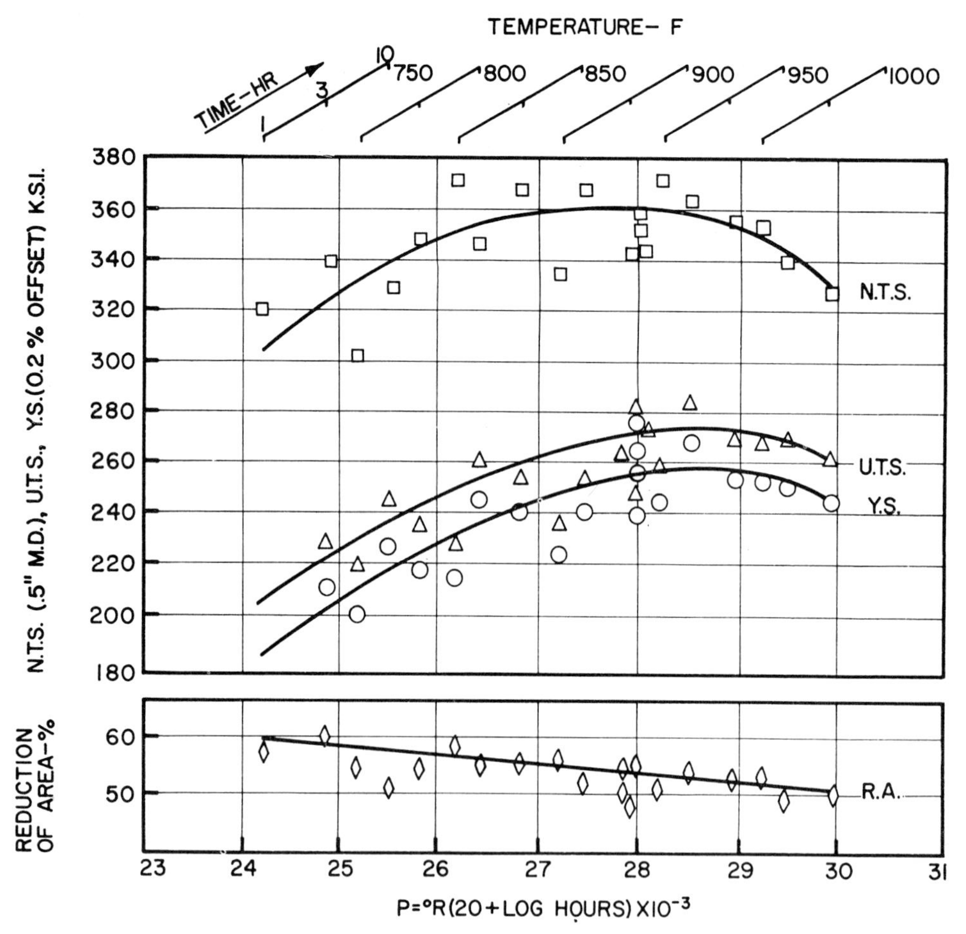

FIG. 11. Effect of maraging time and temperature (as correlated by P) on mechanical properties of 18.5 Ni, 7 Co, 5 Mo, 0.4 Ti-Fe alloy; 30-lb air melts.

Extraction-replica techniques with electron diffraction gave evidence of an ordered phase based on Fe_2CoNi. This ties in with the remarkable rate of initial hardening which is not consistent with normal diffusion reactions. However, it is not clear yet if this ordered phase is the precipitate seen in Fig. 9. The possibility exists that the visible phase is from a diffusion-controlled reaction, which provides the slower increase in hardness from 5 min to 3 hr.

Mechanical Properties

The yield strength of the alloy in the annealed condition was 95,000 psi, and the ductility and formability were high (note the bend specimen and cold worked specimens in Fig. 10).

There is considerable latitude in maraging temperature and time to give high strength and toughness as indicated in Fig. 11. These data correlate well with a Larsen-Miller parameter (11). Time and temperature are also indicated for convenience. The yield strength and tensile strength of the alloy increased as P was raised from 24 (1 hr at 750 F) to 28.5 (10 hr at 900 F). At higher temperatures and longer times there was a drop in strength. A high degree of notch insensitivity appeared under all conditions.

TABLE 7. Effect of Direct Maraging and Vacuum Melting at Yield Strength Levels From 200 to 300 Ksi

Heat*	Melting	Co	Mo	Ti	Heat treatment	Tensile strength, psi		Elong, %	R.A., %	Notch-tensile strength, ksi†	Ratio, NTS/TS
						Yield	Ultimate				
VII-1	Air	7.5	3.0	0.4	SA + MA‡	206	214	16	64	340	1.59
					MA**	217	222	14	64	353	1.59
VII-2	Vac	7.0	3.2	0.4	SA + MA	205	215	15	64	347	1.61
VII-3	Vac	7.1	3.1	0.4	SA + MA	207	217	14	64	344	1.50
VII-4	Air	7.5	4.6	0.4	SA + MA	248	256	12	54	374	1.46
VII-5	Vac	7.5	4.8	0.4	SA + MA	247	259	12	56	395	1.52
					MA	267	273	12	58	415	1.52
VII-6	Air	6.8	5.1	0.6	SA + MA	280	285	10	43	283	0.99
					SA + MA (950 F)	277	282	8	48	378	1.34
VII-7	Vac	7.5	4.9	0.6	SA + MA	275	285	12	58	426	1.50
					MA	295	297	12	60	449	1.51
VII-8	Air	8.9	5.0	0.4	SA + MA	279	286	9	44	324	1.13
VII-9	Vac	9.0	4.9	0.4	SA + MA	278	289	13	56	409	1.42
					MA	303	306	12	60	439	1.44

* *All steels contained 18.5 Nickel.*
† *0.3 in. major diameter round bar.*
‡ *1 hr at 1500 F, air cool, 3 hr at 900 F.*
** *3 hr at 900 F.*

The notch-tensile strength values followed a curve parallel to the yield strength curve. Ductility as measured by reduction of area was very insensitive to heat treatment. Austenite determinations by x-ray diffraction have shown that the decrease in yield strength above $P = 28.5$ cannot be totally rationalized on the basis of reverted austenite and must be due partially to overaging. However, there was some reversion of the martensite to austenite at 1000 F. A sample held 100 hr at 1000 F had 30% reverted austenite.

Direct Maraging. In many cases, as illustrated in Fig. 4 and Table 7, omission of the solution anneal at 1500 F improved both yield and notch-tensile strengths. The high ratio of *NTS-TS* also was maintained. This operated regardless of strength level (Table 7). Thus, the heat treatment can be as simple as a marage of 3 hr at 900 F on hot rolled material.

Effect of Marforming. The martensite can be heavily cold worked (defined as marforming) and this can increase strength after maraging. As noted in the specimens of Fig. 10, the low carbon content of the martensite provides a very low work-hardening rate and excellent formability. Figure 12 shows the results of an experiment in cold rolling an annealed bar over 90%. The hardness increased only from Rockwell C-28 to about C-34. No cracking was experienced; in fact, the small heavily cold reduced rod was then flattened in a forge hammer to $\frac{1}{16}$-in. sheet with ease.

To demonstrate the effect of cold work prior to maraging, a series of bars were annealed 1 hr at 1500 F, cold rolled 50 to 90% and then maraged 3 hr at 900 F. Figure 13 shows that the yield strength increased from 220,000 psi to 300,000 psi. Other comparisons of tensile properties are given in Table 8. There was no decrease in ductility. Notch-tensile strength was also benefited and values up to 447,000 psi were obtained.

Vacuum Melting

Outstanding features of cobalt and molybdenum alloys are the excellent fluidity, castability and lack of dross and

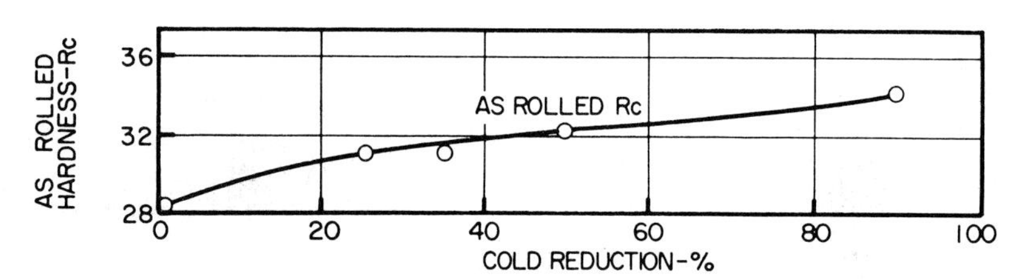

FIG. 12. Effect of marforming on hardness of 18.5 Ni, 7 Co, 5 Mo, 0.4 Ti-Fe alloy. Bar annealed 1 hr at 1500 F and air cooled before cold rolling.

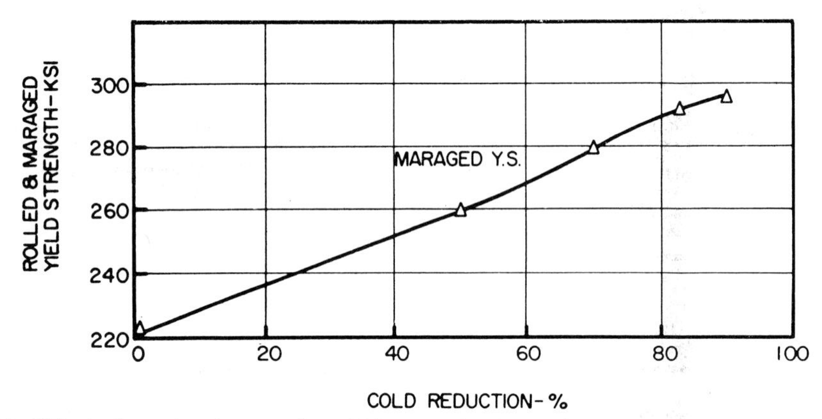

FIG. 13. Effect of marforming on the yield strength of 18.5 Ni, 7 Co, 5 Mo, 0.1 Ti-Fe alloy. Specimens annealed 1 hr at 1500 F, air cooled, cold rolled and maraged 3 hr at 900 F.

TABLE 8. Effect of Marforming

(Heated 1 hr at 1500 F, marformed, 3 hr at 900 F)

Heat*	Melting	Co	Mo	Ti	Mar-form (70 F)	Tensile strength, psi Yield	Tensile strength, psi Ultimate	Elong, %	R.A., %	Notch-tensile strength, ksi†	Ratio, NTS/TS
VIII-1	Air	6.8	4.9	0.4	0%	242	254	11	57	381	1.50
					50%	259	263	12	58	392	1.50
VIII-2	Air (test ingot from 20,000-lb heat)	6.9	4.6	0.2	0%	238	241	12	58	367	1.52
					50%	247	250	14	58	396	1.58
					67%	261	264	10	54	386	1.46
VIII-3	Vac	7.5	4.9	0.6	0	275	285	12	58	426	1.50
					50%	295	301	13	58	427	1.42
VIII-4	Vac	9.0	4.9	0.4	0	278	289	13	56	409	1.42
					50%	288	294	14	58	447	1.52

* *30-lb heats, each steel contained 18.5 Ni.*
† *0.3-in. major diameter round bar.*

films in air melting. The unique combination of properties noted in the above sections was repeatedly obtained with air melts. Still, at high strength levels, pronounced improvements in toughness were obtained with vacuum-induction melting.

Referring to Table 7, the benefit of vacuum melting was negligible in steels having 200,000 psi yield strength and moderate at the 250,000 psi level; however, at high strength levels of 280,000 to 300,000 psi yield there was a gain of some

TABLE 9. Sheet Properties*

(Heated 1 hr at 1500 F, marformed, 3 hr at 900 F)

Heat†	Melting	Co	Mo	Ti	Marform (70 F)	Specimen, in. Width	Thickness	σ *YS*, ksi (a)	σ *a*, ksi (b)	σ *NS*, ksi (c)	K_c, ksi in. (d)	G_c, in.-lb/in.^{2} (e)	β (f)	Critical crack index, in. (g)
IX-1	Air	6.9	4.8	0.4	50%	1	0.079	270	215	192	185	1280	5.9	0.15
IX-2	Vac	7.5	4.8	0.4	50%	1	0.078	286	302	265	>212	>1690	>7.1	>0.18
						2	0.077		272	224	246	2290	9.7	0.24
IX-3	Vac	7.4	4.8	0.6	50%	1	0.062	288	305	268	>213	>1710	>8.8	>0.17
						2	0.062		255	228	225	1905	9.7	0.19
IX-4	Vac	8.8	4.9	0.4	50%	1	0.068	300	271	258	172	1116	4.9	0.11
						2	0.068		213	205	178	1200	5.2	0.11
IX-5	Vac	7.5	4.9	0.6	50%	1	0.061	306	316	276	>244	>1825	>8.7	>0.17
						2	0.078		217	211	181	1195	4.5	0.11
IX-6	Vac	9.0	4.9	0.4	50%	1	0.042	309	275	263	174	1100	7.6	0.10
						2	0.039		202	197	180	1178	8.8	0.11
					70%	1	0.050	326	247	239	149	813	4.2	0.07
						2	0.050		191	183	156	888	4.7	0.08

* *Edge notch, see Table 2 for description, also Ref. 12.*
† *30-lb heats, each steel contained 18.5 Ni.*
(a) σ YS, 0.2% offset yield strength.
(b) σ a, net fracture stress (maximum load divided by fast-crack area).
(c) σNS, notch strength (maximum load divided by original area).
(d) K_c, fracture toughness at point of crack growth instability = $\sigma\sqrt{q_2 W}$, where q_2 is stress distribution factor and W is specimen width.
(e) G_c, critical strain energy release rate for unstable crack extension = $\frac{K_c^2}{E}$, where E is Young's modulus.
(f) β, measure of ratio of plastic zone size to sheet thickness = $\frac{K_c^2}{B\sigma YS^2}$, where B is specimen thickness.
(g) Critical crack index, critical length (just short of rapid crack propagation) of a through-crack in a very wide place sheet carrying a stress equal to the yield strength normal to the crack = $\frac{K_c^2}{\pi\sigma YS^2}$.

50,000 to 140,000 psi in the notch-strength. The retention of high *NTS/TS* ratio at high yield strength also is shown in Fig. 6.

The mechanism of this benefit is not established. It may be one or a combination of 1) low (0.05%) silicon and manganese compared to air melts, 2) volatilization of subversive elements such as bismuth and 3) low gas content.

Sheet Properties

Edge-notch tensile results on sheets (see Table 9) point out the high toughness which can be maintained up to and above a 300,00 psi yield strength. G_c values greater than 1100 in.-lb per sq in. were maintained with yield strengths of 300,000 to 309,000 psi. The result at 326,000 psi yield strength illustrates that there is potential for higher strengths coupled with toughness.

The bar properties in Tables 7 and 8 and sheet properties in Table 9 illustrate that the principles of alloying with cobalt and titanium and the findings on direct maraging or marforming can be used to reach a 300,00 psi yield stiength consistently. Most notable are the direct maraged results at this strength level. Values for Charpy V-notch impact-energy at room temperature were 22 to 26 ft-lb. on three direct maraged vacuum heats at the 300,000 psi yield strength level. Notch-tensile strengths were as high as 449,000 psi.

An aim composition could be:

Ni	Co	Mo	Ti
18.5	9	5	0.6

The recommended treatments are:

(a) Direct maraging treatment of 3 hr at 900 F, or

(b) 1 hr at 1500 F, marform, marage 3 hr at 900 F.

TABLE 10. Test Results on 20,000-lb Basic Electric Arc Air Melt*

Form		Yield strength (0.2% offset), ksi	Ultimate tensile strength ksi	Elong, %	R.A., %	Notch-tensile strength, ksi†	Ratio, NTS/TS
½-in. plate	Long. center	243	248	12	52	361	1.46
	Tran. center	244	251	8	45	354	1.41
	Long. edge	239	245	13	58	365	1.49
	Tran. edge	246	253	11	48	356	1.41
30 lb test ingot processed same as lab. heats	Long. center	238	241	12	58	367	1.52
	Tran. center					363	1.51
						364	1.53

* *Ladle Analysis: 18.6 Ni, 6.9 Co, 4.6 Mo, 0.22 Ti, 0.025 C, 0.07 Al, <0.05 Mn, 0.08 Si, 0.003 B, <0.01 Zr, 0.007 S, 0.003 P, bal Fe. Heat treatment: 1 hr at 1500 F, air cool, 3 hr at 900 F.*

† *0.3-in. major diameter round bar.*

Commercial 20,000 lb-Heat

Laboratory results have been transferred to commercial practice with good success. A 20,000-lb heat of the steel was produced in a basic arc furnace using a basic slag and normal operating procedures (see Table 1). An ingot, 23 by 42 in., was poured and processed to plates of ½ and 1 in. thicknesses. The results on the ½ in. plate are summarized in Table 10. In the annealed and maraged condition, the plate had a yield strength of 238,000 to 246,000 psi, which is higher than predicted from experimental results, Fig. 6, for the composition. Excellent transverse properties were obtained. Charpy V-notch impact-energy was 22 ft-lb at 70 F.

Conclusions

1. The iron-nickel martensite can be hardened to a level of engineering interest by maraging when cobalt and molybdenum are co-present.

2. The strength increases with: (a) cobalt-molybdenum product, (b) supplementation by titanium, (c) marforming prior to maraging, (d) direct maraging.

3. Toughness is maintained by: (a) use of alloying elements which introduce strong maraging, but weak solid-solution hardening of martensite, (b) minimizing content of elements such as silicon which do not comply with (a), (c) vacuum melting for highest strength levels.

4. Maraging nickel-cobalt-molybdenum steels develop a unique combination of properties including: (a) Useful yield strengths to and above 300,00 psi, (b) high toughness and impact energy even at a 300,000 psi yield strength level, (c) low nil-ductility temperature, (d) exceptional stress-corrosion resistance, (e) excellent formability, (f) simple heat treatment, (g) adaptability to conventional air melting and steel hot working practices, (h) excellent weldability, (i) low distortion during heat treatment, (j) Free scaling and lack of decarburization problems.

5. These maraged steels should find engineering applications as plate, sheet, bar, and shapes where high strength coupled with unmatched toughness are desired. Thus, the way should be open for advanced engineering design concepts heretofore unattainable.

Acknowledgments

An alloy development is a complex problem involving the efforts of an entire team of scientists and technicians. Activities include melting, heat treating, mechanical and physical testing, phase studies plus many other evaluations. Accordingly, numerous staff members of the Research Laboratory, The International Nickel Co., Inc. have played their part in this research. Although there is not space here to list these contributors, plus those who have made administrative contributions, it is fitting to recognize some specific contributions.

The research of C. G. Bieber on aluminum-titanium-columbium hardened iron-nickel alloys revealed the great potential of the iron-nickel martensite and pointed the way for this development. Furthermore, the contributions of Dr. R. B. G. Yeo and W. A. Fragetta were important factors. The work of Dr. S. Floreen, Dr. B. H. Heise and J. S. Iwanski guided the discussion on mechanisms. The corrosion studies by Dr. H. R. Copson and Dr. F. S. Lang and the property evaluations by J. E. Chard were also instrumental in this development.

REFERENCES

1. F. W. Jones and W. I. Pumphrey, "Free Energy and Metastable States in the Iron-Nickel and Iron-Manganese Systems," Journal, Iron and Steel Institute, Vol. 163, 1949, p. 121.
2. L. Kaufman and M. Cohen, "The Martensitic Transformation in the Iron-Nickel System," *Transactions* of the American Institute of Mining, Metallurgical and Petroleum Engineers, Vol. 206, 1956, p. 1393.
3. E. S. Machlin and M. Cohen, "Isothermal Mode of the Martensite Transformation," *Transactions* of the American Institute of Mining, Metallurgical and Petroleum Engineers, Vol. 194, 1952, p. 489.
4. G. V. Kurdjumov, "Phenomena Occurring in the Quenching and Tempering of Steels," Journal, Iron and Steel Institute, Vol., 195, 1960, p. 26.
5. W. L. Fink and E. D. Campbell, "Influence of Heat Treatment and Carbon Content on the Structure of Pure Iron-Carbon Alloys," *Transactions* of the American Society for Steel Treating, Vol. 9, 1926, p. 717.
6. N. B. Pilling and P. D. Merica, U. S. Patent 2,048,164, July 21, 1936.
7. C. G. Bieber, Metal Progress, Vol. 78, 1960, p. 99.
8. A. H. Geisler, "Precipitation from Solid Solutions of Metals," Phase Transformations in Solids, John Wiley & Sons, Inc., New York, 1951, p. 387.
9. A. J. Lena, "Precipitation Reactions in Iron-Base Alloys," Precipitation from Solid Solution, American Society for Metals, Cleveland, Ohio, 1959, p. 244.
10. E. H. Phelps and A. W. Loginow, "Stress Corrosion of Steels for Aircraft and Missiles," Corrosion, Vol. 16, July 1960, p. 97.
11. F. R. Larson and J. Miller, "A Time-Temperature Relationship for Rupture and Creep Stresses," *Transactions* of the American Society of Mechanical Engineers, Vol. 74, 1952, p. 765.
12. "Fracture Testing of High-Strength Sheet Materials," ASTM Bulletin, No. 239, Jan 1960, p. 29 and No. 240, Feb 1960, p. 18.
13. P. P. Puzak, M. E. Schuster and W. S. Pellini, "Crack-Starter Tests of Ship Fracture and Project Steels," The Welding Journal Research Supplement, Vol. 33, Oct 1954, p. 481s.
14. P. P. Puzak and A. J. Babecki, "Normalization Procedures for NRL Drop-Weight Test," The Welding Journal Research Supplement, Vol. 38, May 1959, p. 209s.

SECTION II:
Heat Treatment and Thermo-mechanical Treatment

Heat Treatment of 18% Ni Maraging Steel

BY S. FLOREEN AND R. F. DECKER*

ABSTRACT. Studies were made of the effects of annealing, warm or cold working, and aging heat treatments on the properties of an 18% Ni maraging steel. It was found that annealing must be done above approximately 1350 F to achieve satisfactory hardening during aging. Annealing at higher temperatures up to 2300 F tended to decrease the hardening during aging to a slight degree, with no loss in notched tensile strength. Warm or cold rolling prior to aging produced increases in the final strength. Marked hardening occurred within several minutes during aging at 900 to 1000 F, but only after several hundred hours in the 600 to 700 F range. A heat treatment of 3 hr at 900 F ± 25 F was most satisfactory. Aging at 900 F for longer than 10 hr caused softening due to the reversion of the body-centered cubic matrix to austenite.

The aging kinetics suggested that lattice defects were influencing the reaction rate. Electron diffraction evidence indicated the formation of an ordered bcc structure. It appeared that hardening was due to either short-range order, or to coherent precipitates having long range order.

INTRODUCTION

THE RECENT DISCOVERY of 18% Ni maraging steels has provided a class of alloys possessing high strengths and toughness (1). These steels contain 18% Ni, 7 to 9% Co, 3 to 5% Mo, 0.2 to 0.8% Ti, and less than 0.03% C max. They are austenitic at elevated temperatures, and transform to martensite on cooling to room temperature. Within the limits of the measuring techniques employed, this martensite is body-centered cubic. Reheating the steels to 900 F, the "maraging" treatment produces 200,000 to 300,000 psi yield strength, notch tensile strength/tensile strength ratios over 1.4, and 45 to 60% reduction in area. In addition, the alloys possess low drop weight nil ductility temperatures and good resistance to stress-corrosion cracking in industrial and marine atmospheres.

* The authors are associated with the Research Laboratory of The International Nickel Co., Inc., Bayonne, N. J. Manuscript received June 6, 1962.

The primary purpose of the present work was to evaluate the effects of initial plastic deformation, annealing treatments and maraging treatments on the mechanical properties of this steel. In addition, subsidiary studies were made of the hardening mechanism responsible for the high strengths.

EXPERIMENTAL PROCEDURE

The composition of the alloy studied was as follows: 0.028 C, <0.05 Mn, 0.09 Si, 0.22 Ti, 18.4 Ni, 7.0 Co, 4.5 Mo, 0.07 Al, 0.003 B, <0.01 Zr, 0.008 S, and 0.003 P. The alloy was made as a 20,000-lb basic-electric arc heat. The material used in the present study was taken from ½-in. thick plate, and ½-in. and 1-in. diameter bar stock that was hot rolled with a finishing temperature below 1700 F.

The alloy was studied by Vickers hardness measurements augmented by tensile tests and structural studies. Preliminary work had shown that the hardness values correlated quite well with yield strength. Thus the hardness measurments furnished a simple and reasonably reliable method of evaluating the mechanical properties.

The effects of annealing temperature were studied on ½-in. cubes annealed for 1 hr at temperatures ranging from 1200 to 2300 F. Samples for studying warm or cold rolling were prepared by rolling plate various amounts and cutting cubes from the plate.

The maraging behavior of the alloy after these various initial treatments was then evaluated by heat treating the samples in the temperature range of 600 to 1000 F for times ranging from ½ min to 336 hr. A separate specimen was used for each maraging time and temperature to eliminate any effects of reheating or re-cooling a single specimen, and to provide a complete set of specimens that could be re-examined.

In addition to the hardness specimens, a number of smooth and notched tensile specimens were given various heat treatments. The smooth tensile specimens were ¼-in. diameter, 1¼-in. gage length round specimens. The notched specimens had a 0.300-in. major diameter, and 0.212-in. root diameter, with notch root radii of approximately 0.0005 in. and notch accuity factors greater than ten. The specimens were machned after heat treatment.

Most of the heat treatments were done

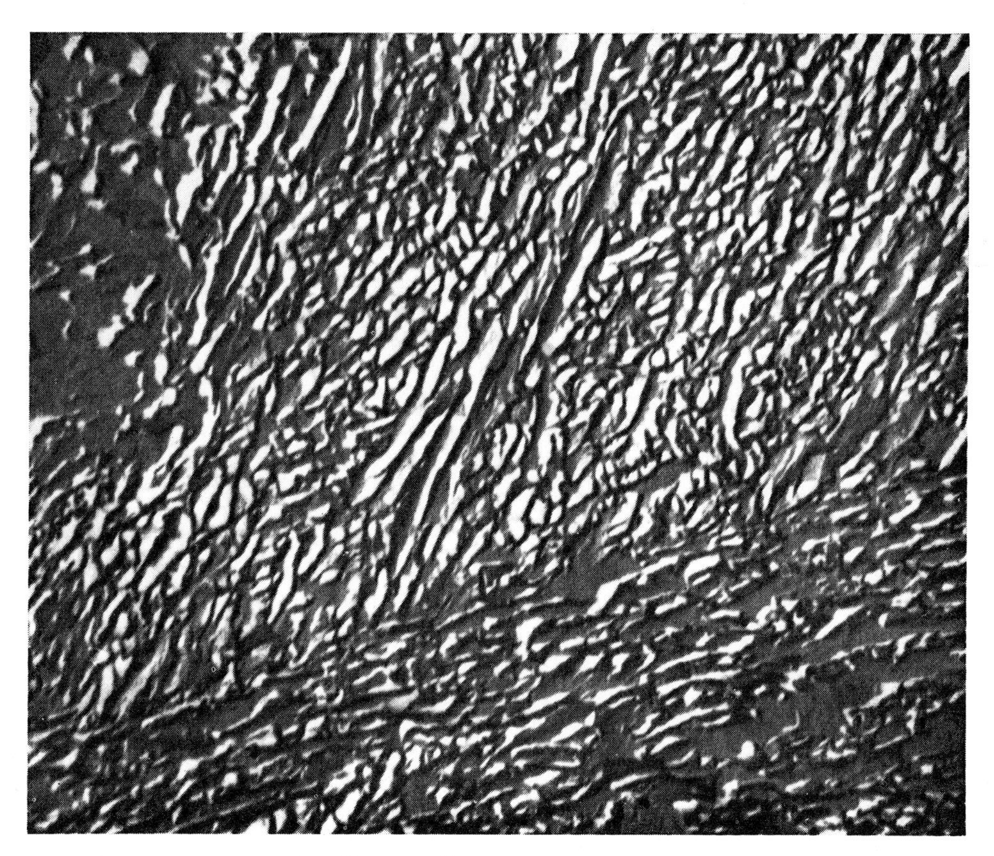

Fig. 1. Microstructure produced by annealing 1 hr at 1200 F. Replica electron micrograph. Austenite plus bcc phase. Etchant: 50 cc HCl, 25 cc HNO_3, 1 g $CuCl_2$, 150 cc H_2O. 15,000×.

FIG. 2. Microstructure produced by annealing 1 hr at 1500 F; 100% bcc phase. Etchant: Same as Fig. 1. 1000×.

in air, except for the very high temperature annealing treatments which were done in argon, and the very short time maraging tests which were done in a lead bath to insure rapid heating. All of the specimens were air cooled, as prior tests had indicated no effects due to cooling rate.

A number of light and electron metallographic examinations, x-ray and electron diffraction analyses also were made using conventional methods.

RESULTS

Annealing Behavior

When the alloy was annealed at 1200 F, the structure contained approximately 50% austenite after cooling to room temperature. This austenite had the fine, lamellar-appearing structure shown in Fig. 1. Apparently at this annealing temperature the austenite that formed during annealing became partially stabilized and did not retransform to the body-centered cubic structure on cooling to room temperature. This retained austenite may have been due to the transformation of the metastable body-centered cubic α_2 matrix formed on cooling into the equilibrium α and γ phases (2). During this transformation there was probably some partitioning of the alloying elements between the α and γ phases so that some of the austenite was enriched and did not transform back to α_2 on cooling to room temperature.

As the annealing temperature was raised, the amount of austenite retained on cooling to room temperature decreased until only the body-centered cubic phase was present. The minimum annealing temperature required to eliminate all of the austenite was approximately 1350 F. As shown by the subsequent maraging

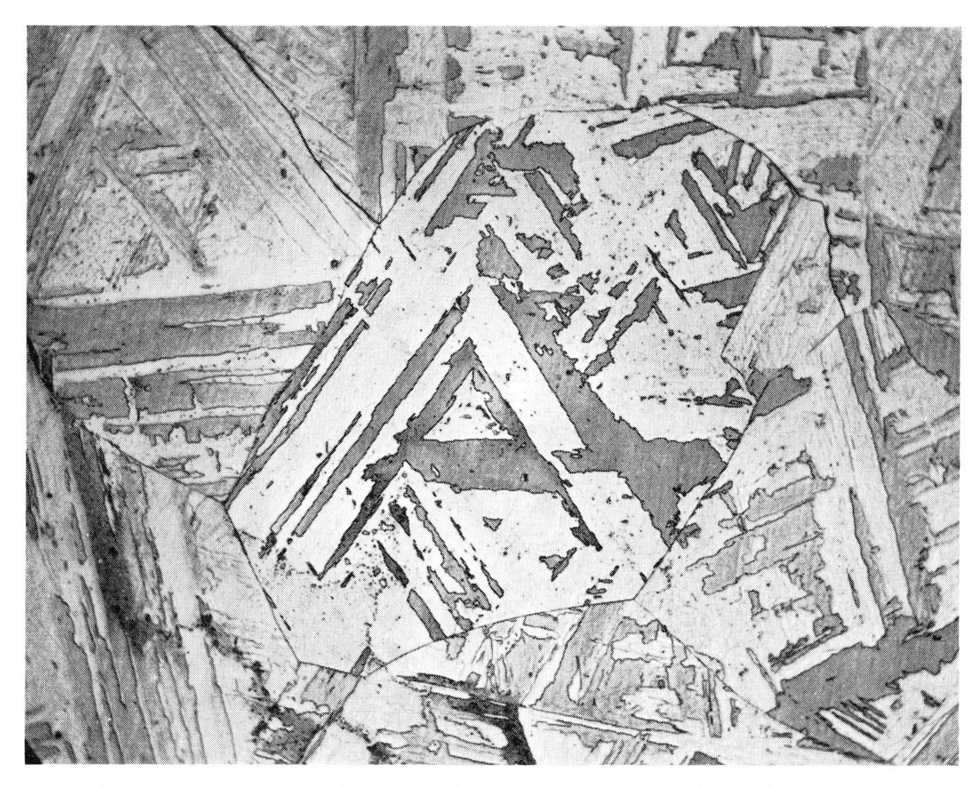

FIG. 3. Microstructure produced by annealing 1 hr at 2300 F; 100% bcc phase. Etchant: Same as Fig. 1. 250X.

results, it was essential to eliminate this austenite to achieve satisfactory hardening during maraging.

The austenite ASTM grain size was 6 to 8 after annealing at temperatures up to 1800 F. Above this temperature, grain growth occurred, with ASTM grain size No. 4 resulting after annealing at 2100 F, and No. 0 after annealing at 2300 F.

With high annealing temperatures a Widmanstatten morphology was evident in the microstructures of the body-centered cubic α_2 phase. Figure 2 shows the microstructure after annealing at 1500 F, and Fig. 3 shows the well developed Widmanstatten structure produced by annealing at 2300 F. Similar structures have also been found in an iron-18% Ni binary alloy (2).

It should be emphasized that the structures in Fig. 2 and 3 were entirely body-centered cubic. Microprobe analyses revealed no differences in chemical composition between the dark platelets and light background in areas such as exemplified by Fig. 3. Microhardness measurements showed no hardness differences between these areas. The M_s temperature after the 2300 F anneal was approximately 400 F. Variations in cooling rate, ranging from furnace cooling to quenching in liquid nitrogen, did not affect the Widmanstatten morphology. It thus appears that the Widmanstatten structure was the result of a martensitic transformation. A detailed study of this transformation is now in progress.

The tensile properties after annealing at 1500 and 2300 F are listed in Table 1. In the annealed condition the alloy had excellent formability, and could be cold-worked as much as 98% reduction in area without intermediate annealing (1). The tensile properties of the annealed material

TABLE 1. Tensile Properties of 18% Ni Steel after Various Heat Treatments

One smooth and one notched specimen tested for each condition. Notched specimens not tested in two cases. All specimens air cooled after heat treatments.

Heat treatment temperature (F), time (hours)	0.2% Y.S. (ksi)	U.T.S. (ksi)	Elong., % in 1 in.	Red. of area, %	Notch tensile strength (ksi)	NTS/UTS
1500 F—1 hr	95	140	17	75	—	—
2300 F—1 hr	103	151	18	66	242	1.60
1500 F—1 hr + 50% cold roll	177	188	13	62	—	—
2300 F—1 hr + 50% cold roll	186	195	12	55	302	1.55
1200 F—1 hr + 900 F—3 hr	150	173	24	63	259	1.50
1400 F—1 hr + 900 F—3 hr	244	248	12	55	359	1.45
1500 F—1 hr + 900 F—3 hr	239	245	13	58	365	1.49
1600 F—1 hr + 900 F—3 hr	225	236	12	53	348	1.48
1700 F—1 hr + 900 F—3 hr	226	236	12	55	356	1.51
1800 F—1 hr + 900 F—3 hr	228	239	12	55	358	1.50
2100 F—1 hr + 900 F—3 hr	222	235	12	55	355	1.51
2300 F—1 hr + 900 F—3 hr	222	232	10	44	352	1.52
1500 F—1 hr + 700 F—3 hr	175	189	16	57	296	1.57
1500 F—1 hr + 50% cold roll + 900 F—3 hr	246	251	11	57	379	1.51
2300 F—1 hr + 50% cold roll + 900 F—3 hr	253	262	10	48	378	1.44
Hot rolled + 900 F—3 hr	245	250	12	57	387	1.55
Hot rolled + 800 F—24 hr	243	253	14	54	364	1.44

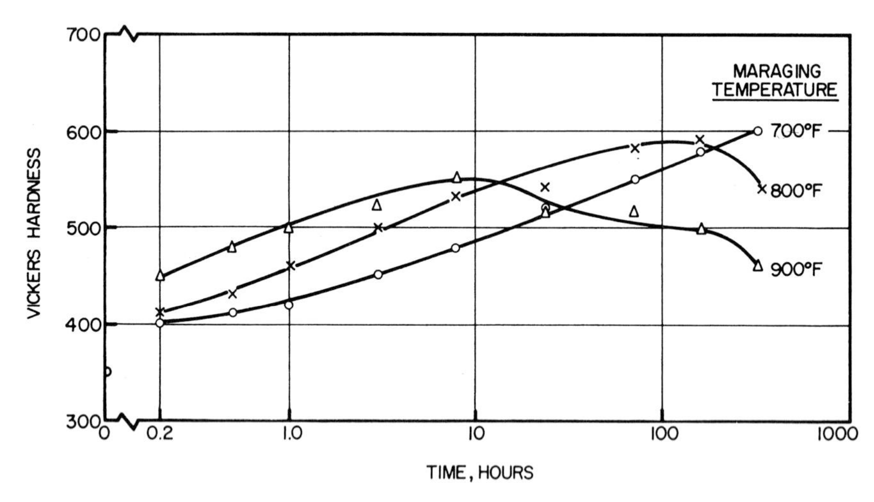

FIG. 4. Effect of maraging on hardness of 18% Ni steel. Initially annealed at 1500 F

after a reduction of 50% are included in Table 1.

Maraging Behavior

The hardness of the specimens initially annealed at 1500 F and then maraged are shown in Fig. 4. More extensive data for specimens initially annealed at 1800 F are shown in Fig. 5. The curves show a rapid initial rate of hardening at 900 to 1000 F, while the response at 600 to 700 F was slower.

The general shapes of these curves are comparable to those of precipitation hardenable alloys. In the present alloy, however, overaging in the conventional sense may not occur. Instead, a drop in hardness with prolonged maraging was always associated with the reformation of austenite. The general appearance of the austenite formed after prolonged maraging was similar to the fine, lamellar type of morphology exemplified by Fig. 1. The amount of austenite that was observed after any given maraging treatment varied somewhat with the initial conditions. Prior plastic deformation tended to increase the amount of austenite, while higher annealing temperatures reduced the austenite content.

The effects of various initial treatments on the hardness after maraging at 900 F are shown in Figs. 6 and 7. Changing the initial condition of the alloy prior to maraging produced some distinct changes in hardening behavior. The specimen annealed at 1200 F contained about 50% austenite and was not hardened by the maraging treatment. Raising the annealing temperature from 1500 to 2300 F resulted in progressively lower hardness after maraging. Also, as shown in Fig. 7, cold rolling 25% prior to maraging did not cause any change in the hardening response.

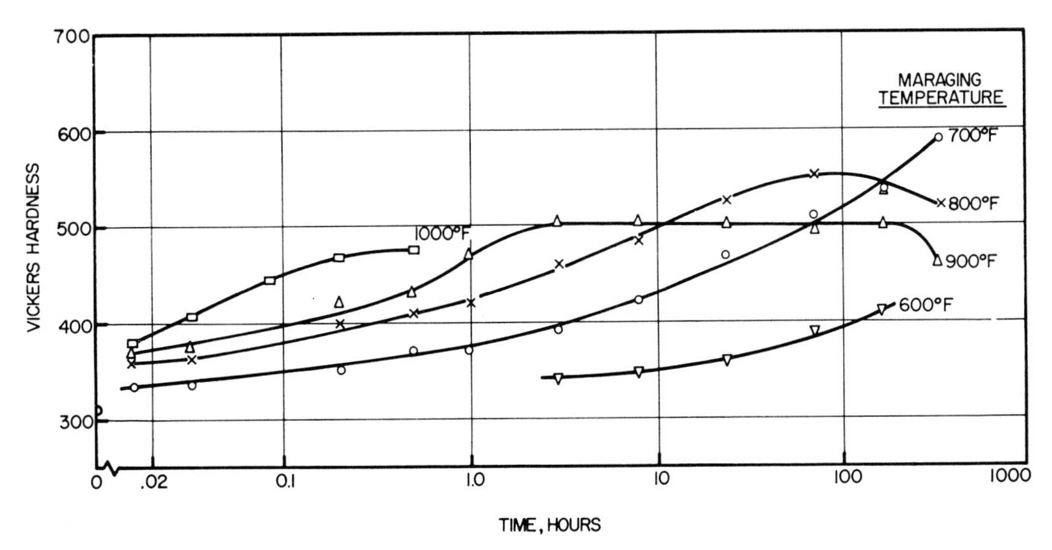

Fig. 5. Effect of maraging on hardness of 18% Ni steel. Initially annealed at 1800 F

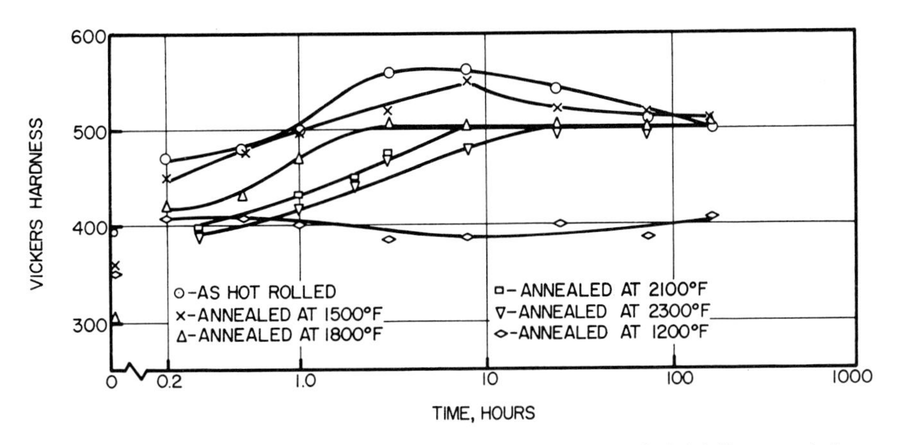

Fig. 6. Effect of maraging at 900 F on hardness of 18% Ni steel. Initially annealed at various temperatures.

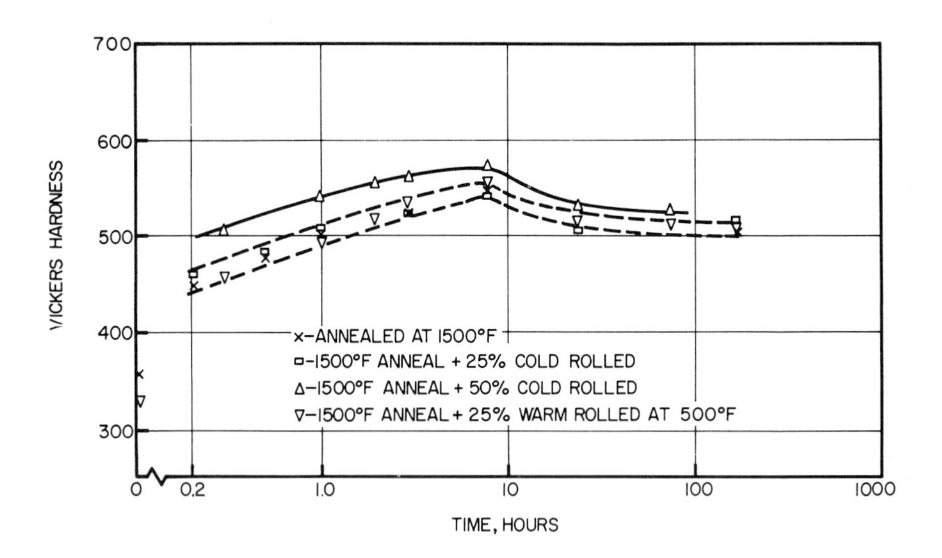

FIG. 7. Effect of maraging at 900 F on hardness of 18% Ni steel. Initially cold or warm rolled

This lack of sensitivity to cold work may have been related to the low rate of work hardening of the alloy. With 50% cold work prior to maraging, however, there was a significant increase in maraged hardness. This has been found to be especially noticeable in sheet materials, where 50% cold work can add about 40,000 psi to the maraged yield strength.

Warm rolling above the M_s temperature (~310 F) altered the maraging behavior to a small degree. In this case, the alloy was cooled directly from 1500 F to 500 F and then rolled at 500 F. In this way the austenitic structure was retained at 500 F, because the alloy was not cooled below the M_s temperature. Some martensite may have formed during rolling, however. The small response to this warm rolling treatment indicates that this type of treatment probably would not yield large increases in strength.

The tensile properties obtained after various heat treatments are included in Table 1. The general findings were in accord with the hardness results. The "standard" treatment, 1 hr at 1500 F, air cool + 3 hr at 900 F, gave a yield strength of 239 ksi and NTS/TS of 1.49. Raising the annealing temperature to 1600 F lowered the yield strength 14 ksi, apparently from more complete removal of warm work. Further increase of the annealing temperature had little effect on strength or notched tensile strength. Higher yield and tensile strengths were achieved by maraging the hot-rolled alloys, or by extensive cold rolling prior to maraging. Some gain in strength was also noted by maraging at 800 F for 24 hr. It might be expected that maraging for very long times at 700 F would furnish some further increase in strength, but the time required would be too long to be of any practical interest. The strength properties after maraging for 3 hr at 700 F were lower than those obtained after maraging for 3 hr at 900 F.

Double Maraging

Several experiments were conducted to study the effects of double maraging heat treatments. Specimens were initially maraged at one temperature, and then maraged for various times at a second temperature. The results showed that when the specimens were maraged at a lower temperature and then at 900 F the

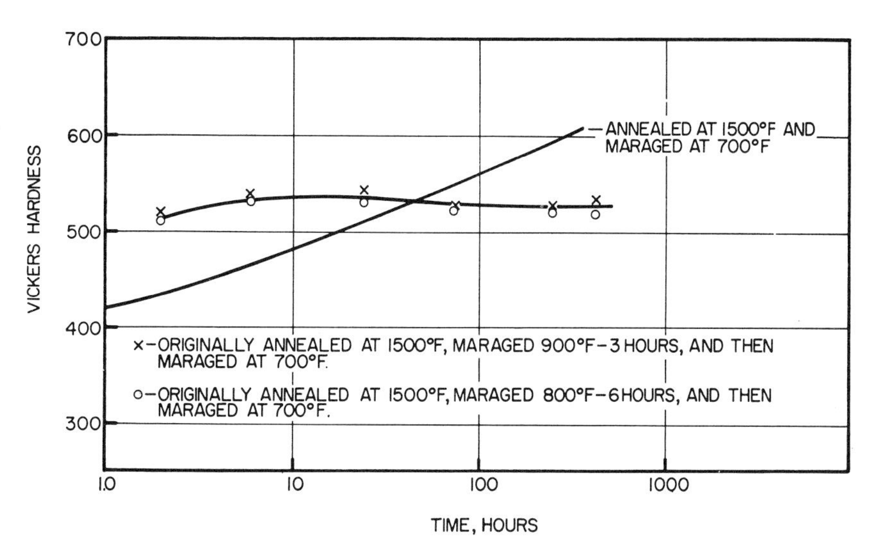

FIG. 8. Double maraging results. Comparison of hardnesses obtained at 700 F by single marage with those obtained after preliminary maraging at high temperatures.

resultant hardness curves at 900 F closely followed the normal 900 F maraging curve.

Of more interest are the results in Fig. 8, showing that when specimens were first maraged at 800 F or 900 F to produce about 525 Vickers hardness and then held at 700 F, there was essentially no change in hardness after 400 hr at 700 F. The hardness curves, in fact, crossed that obtained as a result of maraging only at 700 F. After 400 hr the hardness values were much lower than would have been expected from a single 700 F marage. It is possible, of course, that an increase in hardness would eventually have occurred after longer times at 700 F, but it is surprising that no significant changes occurred up to 400 hr.

From a practical standpoint the results indicate that there is no special advantage to be gained by double maraging treatments. One possible benefit of the apparent stability of the maraged structure at lower maraging temperatures might occur in elevated temperature service applications in the 700 F range. In such cases the alloy, when maraged at 900 F, might not be susceptible to the long-time hardening and embrittlement that occurs in many alloys.

Overheating After Maraging

The maraging results showed that softening could occur after long-time high-temperature exposures. In some applications a maraged section might be partially reverted to austenite by overheating due to incorrect heat treatment, or over-temperature service. To estimate the effects of overheating, a set of specimens was annealed at 1500 F, maraged at 900 F for 3 hr, and then overheated for 5 or 30 minutes at temperatures ranging from 1000 to 1400 F. The hardnesses and amounts of austenite in the specimens after air cooling to room temperature were then measured. The specimens were then re-annealed at 1500 F, re-maraged at 900 F for 3 hr and the hardnesses re-measured.

The results of these tests, listed in Table 2, indicated that the original hardness was essentially restored by re-annealing and re-maraging. With overheating there was a progressive decrease in hardness with increasing time and temperature. The amount of austenite increased and then decreased with increasing temperatures, probably because of changes in the stabilization of the austenite. Thus, at the lower overheating temperatures the austenite that formed was probably due to the $\alpha_2 \rightarrow \alpha + \gamma$ reaction and was sufficiently enriched in alloy content so that it did not transform back to the body-centered cubic structure on cooling. At higher overheating temperatures the austenite probably formed by the $\alpha_2 \rightarrow \gamma$ reaction, and most of the austenite transformed back to α_2 on cooling to room temperature. However, when the specimens were re-annealed and re-maraged the hardnesses returned substantially to the average initial hardness before overheating.

TABLE 2. Effects of Overheating on Specimens Initially Annealed 1 Hr at 1500 F and Maraged 3 Hr at 900 F. (Initial Heat Treatment Produced a Vickers Hardness of 520)

Overheat temp, F (1)	Time, min	Vickers hardness	Per cent γ (2)	Vickers hardness after second anneal and marage (3)
1000	5	501	3	512
	30	497	6	508
1100	5	478	8	520
	30	452	35	518
1200	5	395	35	517
	30	366	25	512
1400	5	355	8	517
	30	346	7	512

Notes:
(1) *One specimen used for each heat treatment.*
(2) *Determined by X-ray diffraction.*
(3) *The second anneal and marage were 1 hr at 1500 F followed by 3 hr at 900 F.*

Maraging Mechanism Study

A number of studies of the maraged specimens were made to try to determine the cause of hardening. X-ray and electron diffraction measurements showed a slight decrease in the lattice parameter of the body-centered cubic structure after maraging. Electron diffraction determinations indicated that ordering was taking place during maraging. Because of the similarity in the scattering factors of iron, nickel and cobalt, however, the resolution of the superlattice structure was poor, and the present findings are only tentative. Neutron diffraction will apparently be required to establish clearly the nature of the ordering reaction. Light microscopy, replica electron microscopy and transmission electron microscopy failed to establish the presence of precipitates.

Kinetic analysis of the maraging data showed that the hardness results could be expressed by the simple relationship:

$$H_t - H_0 = Kt^n$$

where
H_t = hardness at time t
H_0 = initial hardness
K = a constant
t = maraging time
n = time constant

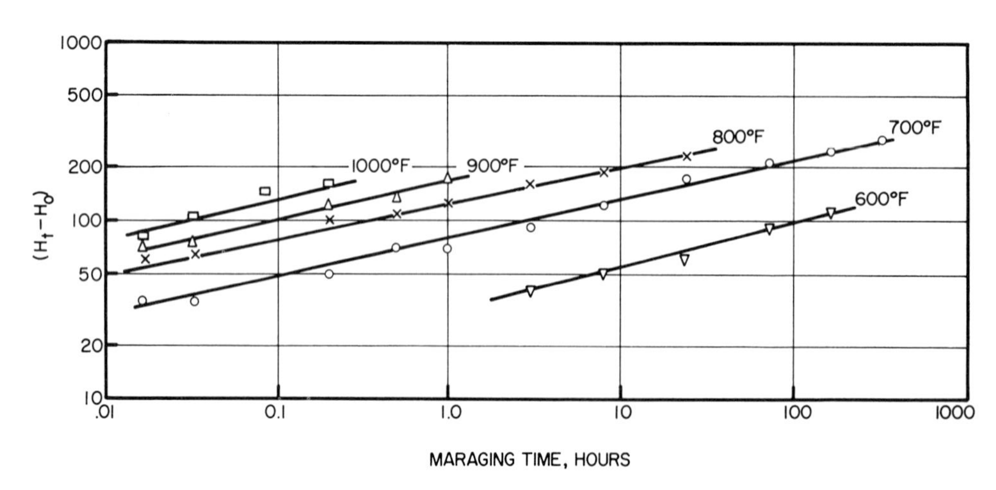

FIG. 9. Maraged Vickers hardness (H_t) minus initial Vickers hardness (H_0) vs maraging time at 600 F to 1000 F for specimens initially annealed at 1800 F.

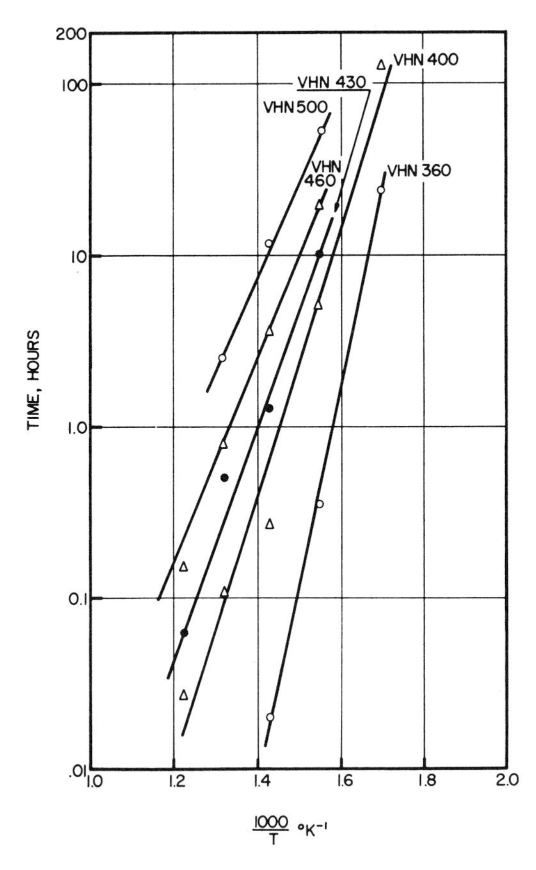

FIG. 10. Time to reach various hardness levels vs reciprocal maraging temperature for specimens initially annealed at 1800 F.

The hardness data for the various initial conditions were plotted according to this equation for time periods before extensive formation of austenite took place. Figure 9 is an example of such a plot for the specimens initially annealed at 1800 F, showing the fit of the data. The value of the time constant n, from the average slope of these curves, was 0.22.

The hardness data for other initial conditions could also be plotted in this fashion. The average values of n varied from 0.19 to 0.30 with changing initial conditions. No systematic variation in n with initial cold work or annealing temperature was detected.

Several attempts were made to determine the activation energy for the hardening reaction. It became apparent that there was no clearly defined activation energy, Fig. 10. In this figure the times to reach various hardness values were plotted versus the reciprocal maraging temperature for the specimens initially annealed at 1800 F. The slope of each curve should have been proportional to the activation energy, and if the hardening was characterized by a single activation energy the curves should have been parallel. Instead, they showed a progressive change in slope with increasing hardness. The resultant activation energy values from these slopes varied from approximately 50 k cal/mole to 25 k cal/mole with increasing hardness.

The data for the other initial conditions did not show this continuous decrease, but instead showed erratic variations in the apparent activation between 20 and 50 k cal/mole with changing hardness. No single activation energy could be observed with these specimens, however.

Discussion

Practical Implications

The toughness (NTS/TS ratio) of 18% Ni maraging steel is very insensitive to wide variations in annealing and maraging conditions. Even with omission of the annealing treatment, wide variations of annealing temperature from 1200 to 2300 F, or extremes from under-maraging to over-maraging, an NTS/TS ratio greater than 1.4 is maintained. Thus severe embrittlement of the alloy by heat treating variations appears to be improbable.

Strength is more sensitive to heat treatment than toughness, yet there is still broad latitude in annealing and maraging conditions to yield reproducible high strengths. First, the stabilization of austenite by annealing below 1350 F or by maraging at too high a temperature should be avoided to obtain optimum strength. Prior residual warm and cold work can be utilized to supplement the maraged strength. The major effect of annealing on strength is in its influence on residual work. Higher strengths can be obtained by complete retention of warm or cold work (no subsequent anneal). Although this study revealed little sensitivity of toughness of bar stock to annealing temperature, this must be qualified by recent practical experience. There is some evidence that toughness can be optimized by more careful selection of time and temperature. The annealing temperature depends upon the mill form of the material; but is generally in the range of 1 hr at 1500 F to 1700 F, air cool.

Further leeway in the heat treatment is offered by the nature of the $\gamma \rightarrow \alpha_2$ reaction. This is insensitive to section size and cooling rate so that heavy sections can be transformed uniformly during air cooling from either hot working or annealing temperatures.

There also appears to be a considerable latitude in the maraging treatment because of the relatively flat nature of the hardness-time curve at 900 F. For most purposes maraging for 3 hr at 900 F would be the normal heat treatment, but varying the maraging time at 900 F from 2 to 6 hr does not significantly alter the hardness. Maraging could also be done at other temperatures, but the increase in the rate of reversion to austenite at higher temperatures, and the sluggishness of the reaction at lower temperatures, indicate that 900 $\pm$ 25 F is probably the optimum maraging temperature.

For certain specific applications the treatment may be varied from the recommended ones. Higher strengths are possible, for example, by extensive cold work prior to maraging. Conversely, the strengths can be lowered by higher annealing temperatures and/or maraging at lower temperatures. For applications requiring elevated temperature strength the use of higher annealing temperatures is advantageous in retarding austenite reversion and thus maintaining the strength properties for longer time periods.

Mechanisms of Maraging

No definite evidence of precipitation during maraging has been found thus far. It cannot be concluded that no precipitate is formed, but it is probable that, if a precipitate is formed, it has approximately the same lattice parameter as the matrix and is coherent with the matrix.

Instances have been found where coherent precipitates produced rather potent strengthening effects. Williams (3), for example, has explained the hardening of iron-chromium binary alloys by the presence of a matrix and a precipitate that isotropically strained to a common lattice parameter as a result of coherency. One point that argues against this mechanism is the excellent ductility and toughness of the present alloy after hardening. The iron-chromium alloys of Williams were brittle, and in general one might expect that the presence of high elastic strains would seriously reduce the toughness. Thus there is some reason to doubt whether coherency straining could be the sole hardening mechanism.

Another possible source of hardening would be a coherent precipitate or zone having long range order in a disordered matrix. Williams (4) has shown that, neglecting coherency hardening, the increase in shear stress would be:

$$\tau = \frac{f\gamma}{b}$$

Where f is the volume fraction of the precipitate, γ is the antiphase boundary energy, and b the Burger's vector.

The present results would require values of γ of several hundred ergs/cm^2, depending upon the values of f. Values of γ of this magnitude seem unreasonably high. It is possible, however, that some hardening due to coherency strains is present, which would lower the required value of γ. The resultant hardening would then be due to the sum of the coherency and ordering effects. This combination of hardening mechanisms has been postulated by Williams to account for the strengthening due to precipitation in nickel-aluminum alloys (5).

In view of the evidence of ordering that was obtained, it is tempting to ascribe the hardening to such a mechanism. The maraging kinetics support this hypothesis to some extent. Dienes (6) has pointed out that there should not be a single activation energy for ordering reactions. Instead, the apparent activation energy should vary with the degree of order. If the hardness of the present alloy is a function of the degree of order, the rather erratic variations in the apparent activation energy may be consistent with this suggestion.

It may not be wise to attach too much significance to the activation energy variations in attempting to establish the hardening mechanism. The reason for this is that excess lattice imperfections probably play a significant role in the maraging kinetics. The principal source of these imperfections is probably the martensite transformation. Transmission electron micrographs of the annealed structures showed a high density of dislocations. The low values of the time constant n would be consistent with enhanced diffusion due to such imperfections.

The double maraging results also suggest that imperfections influence the kinetics. The lack of response of the alloy to long time maraging at 700 F after preliminary maraging at 800 or 900 F can be explained on the basis that the preliminary maraging treatments removed or tied up enough excess imperfections so that the subsequent reaction at 700 F was significantly retarded. In contrast to the double maraging results, the degree of order changes much more rapidly when an alloy is initially ordered than when it is initially disordered (7, 8). Thus if the hardness is a function of the degree of order the double maraging results also indicate that the kinetics are significantly influenced by lattice imperfections.

The rapid rate of hardening at the relatively low temperatures used in maraging thus may be possible for two reasons. First, the presence of excess lattice imperfections enhances the diffusion rates. Secondly, only relatively short diffusion distances may be necessary to create local precipitates or zones having long range order.

Some hardening could also result from short range ordering. Fisher (9) has shown the increase in the critical resolved shear stress with short range ordering to be:

$$\tau = \frac{\gamma}{b}$$

where γ is the energy of the distorted interface produced by slip through the ordered zone.

In the present case a value of γ on the order of 100 ergs/cm^2 would produce the observed increase in strength. This value may not be unreasonable. In view of the evidence of ordering that was obtained, it would appear that short range ordering could also be a hardening mechanism in the present alloy.

There is little basis for choosing between mechanisms of short range order and coherent precipitates of long range order or combinations of these mechanisms at this time. Tentatively, it is felt that an ordering reaction is involved, but further work is necessary to clarify the situation.

Conclusions

1. In annealing of 18% Ni maraging steel, temperatures in the range 1100 to 1300 F should be avoided to prevent retained austenite. Temperatures as high as 1800 F can be tolerated without loss of toughness.

2. Warm working of the austenite or cold working of the martensite (marforming) before maraging can be used to add an increment to the final maraged strength. With the proper finishing temperature off of the hot rolling mill, the anneal can be omitted to gain this benefit of warm work.

3. Marked hardening during maraging occurs in several minutes at 900 or 1000 F. The recommended maraging treatment is

3 to 6 hr at 900 $\pm$ 25 F. Large amounts of retained austenite are present after overheating the steel for 5 or 30 minutes in the 1000 to 1400 F range. The initial hardness can be restored by re-annealing and remaraging.

4. Maraging at a lower temperature followed by a second maraging at 900 F produces hardness values equal to the values produced by a single maraging at 900 F. Conversely, specimens first maraged at 800 or 900 F and then maraged at 700 F show no hardness changes after 500 hr at 700 F.

5. Electron diffraction studies of the maraged alloy gave evidence of ordering.

6. The change in hardness during maraging could be expressed in the form $(H_t - H_0) = Kt^n$, where H_t is the hardness after time t, H_0 is the initial hardness, and n and K are constants. The values of n were in the range 0.2 to 0.35, depending upon the initial condition prior to maraging. The apparent activation energies varied with hardness, with values ranging between 20 and 50 k cal/mole. The results suggest that excess imperfections play an important part in maraging kinetics.

7. Tentatively, short range ordering or coherent zones, or precipitates having long range order, appear responsible for the hardening reaction. Further studies, by neutron diffraction for example, appear necessary to evaluate the kinetic behavior and hardening mechanism.

ACKNOWLEDGMENTS

We are indebted to K. G. Carroll, B. E. Heise, and J. S. Iwanski for conducting the x-ray diffraction and electron microscopy studies.

REFERENCES

1. R. F. Decker, J. T. Eash, and A. J. Goldman, "18% Ni Maraging Steel", *Transactions Quarterly*, ASM, Vol. 55, March 1962, p. 58.
2. N. P. Allen and C. C. Earley, "The Transformations $\alpha \rightarrow \gamma$ and $\gamma \rightarrow \alpha$ in Iron Rich Binary Iron-Nickel Alloys," *J. Iron and Steel Inst.*, Vol. 166, 1950, p. 281.
3. R. O. Williams, "Theory of Precipitation Hardening-Isotropically Strained Systems," *Acta Met*, Vol. 5, 1957, p. 385.
4. R. O. Williams, "Origin of Strengthening in Precipitation Ordered Particles," *Acta Met*, Vol. 5, 1957, p. 241.
5. R. O. Williams, "Aging of Nickel Base Aluminum Alloys" *Trans. AIME*, Vol. 215, 1959, p. 1026.
6. G. J. Dienes, "Kinetics of Order-Disorder Transformations," *Acta Met*, Vol. 3, 1955, p. 549.
7. C. Sykes and H. Evans, "The Transformation in the Copper-Gold Alloy Cu_3Au," *J. Inst. Met*, Vol. 58, 1936, p. 255.
8. J. M. Cowley, "X-Ray Measurements of Order in Single Crystals of Cu_3Au," *J. App. Phys.*, Vol. 21, 1950, p. 24.
9. J. C. Fisher, "On the Strength of Solid Solution Alloys," *Acta Met*, Vol. 2, 1954, p. 9.

Thermal Embrittlement of 18 Ni (350) Maraging Steel

DAVID KALISH AND H. J. RACK

The embrittlement of as-solutionized 18 Ni(350) Maraging steel was monitored as a function of heat treatment variables by means of Charpy impact tests. The processing parameters of interest were annealing temperatures in the range of 1900° to 2400°F, intermediate holding temperatures in the range of 1300° to 1800°F, and the quenching rate. The changes in fracture mode with heat treatment were characterized by replica and scanning electron microscopy. The severity of thermal embrittlement increases with decreasing cooling rate from the annealing treatment upon direct quenching to room temperature. Intermediate isothermal holding, particularly at 1500° to 1600°F, further accentuates the embrittlement. A large grain size is beneficial to the toughness when rapid direct quenches from the annealing range are imposed but is detrimental upon air cooling or intermediate holding. The major loss in toughness may be associated with the diffusion of interstitial impurity atoms (C + N) to the austenite grain boundaries during cooling or intermediate isothermal holding below 2000°F. An advanced stage of the embrittlement is characterized by the discrete precipitation of Ti(C,N) platelets on these boundaries. Thermal embrittlement is accompanied by change in fracture mode from transgranular dimpled rupture to intergranular quasi-cleavage.

THE 18-pct Ni maraging steels are recognized for their high fracture toughness as compared to quench and temper steels at tensile strengths of 200,000 to 300,000 psi.[1] However, maraging steels show a considerable loss in fracture resistance, as measured in terms of the Charpy impact energy or the plane-strain fracture toughness (K_{Ic}), when subjected to certain heat treatment schedules.[2,3] This degradation of toughness, termed "thermal embrittlement", was found to occur in 18 Ni (250) Maraging steel upon heating above 2000°F followed by slow cooling or by interrupted cooling and holding in the temperature range of 1500° to 1800°F.[2] The source of the embrittlement has been attributed to the precipitation of the intermetallic compound Ti (C, N) on austenite grain boundaries at temperatures between 1500° to 1800°F. Some Ti_2S was observed on the fracture faces of embrittled specimens but this compound was not considered to be important in the embrittlement. It has also been suggested that austenite grain growth, during the prior heating at 2000° to 2400°F, contributes to the embrittlement although the individual contributions of the precipitation and of the grain growth have not been established.

Thermal embrittlement has been documented in as-solutionized specimens, at tensile strengths of 140,000 to 150,000 psi, and in aged specimens, at tensile strengths of 230,000 to 250,000 psi. For example, embrittlement led to a decrease in the Charpy V-notch impact energy from about 67 to 7 ft-lb. in one set of unaged samples and from about 19 to 3 ft-lb. in the aged condition. In both heat treatment conditions the embrittled steel was reported to exhibit an intergranular fracture mode.

The potential loss in fracture toughness of 18 Ni maraging steel due to thermal embrittlement is particularly significant in the fabrication of large forgings or extrusions and in the higher strength grades of this type of alloy. The requirement for finish forging temperatures above 1800°F, coupled with relatively slow cooling rates in heavy sections provides a natural framework for developing an embrittled condition. Moreover, as the strength of maraging steel is increased up to 350,000 psi, through modification of the Ni-Co-Mo-Ti content,[4] the fracture toughness requirements become more stringent and one cannot afford the slightest degradation from the normal fracture resistance.

The purpose of this paper is to examine the causes and extent of thermal embrittlement in an 18 Ni (350) Maraging steel. The heat treatment limits are defined for which a loss of notched impact resistance is incurred. Fracture toughness measurements, scanning and replica electron fractography, and a diffusion analysis are used to arrive at the mechanism of thermal embrittlement.

EXPERIMENTAL PROCEDURES

Two heats of the 18 Ni (350) Maraging steel were used, Table I. The alloys were double-consumable electrode vacuum melted and supplied as $\frac{1}{2}$ and $\frac{5}{8}$ in. thick plates, finish rolled at about 1800°F and mill annealed at 1500°F. The minor differences in chemical composition between the two heats did not affect the mechanical properties so that the results are reported without reference to the particular heat from which the specimens were taken.

All annealing and solutionizing heat treatments were performed in a prepurified ultra-high purity argon atmosphere. At least 0.050 in. was removed from each surface in machining the specimens after heat treatment. The impact energy, in units of foot-pounds, was

Table I. Chemical Composition of 18 Ni(350) Maraging Steel

	Element, Wt Pct										
Heat	Ni	Co	Mo	Ti	Al	C	N	S	P	Si	Mn
A	18.3	11.9	4.72	1.42	0.13	0.007	0.0016	0.005	0.002	<0.01	<0.01
B	18.5	11.9	4.69	1.52	0.09	0.008	0.0013	0.006	0.003	0.02	0.02

DAVID KALISH and H. J. RACK are Scientists, Lockheed-Georgia Co., Marietta, Ga. This paper was presented at the 1970 Spring Meeting of The Metallurgical Society of AIME in Las Vegas, Nev.
Manuscript submitted April 12, 1971.

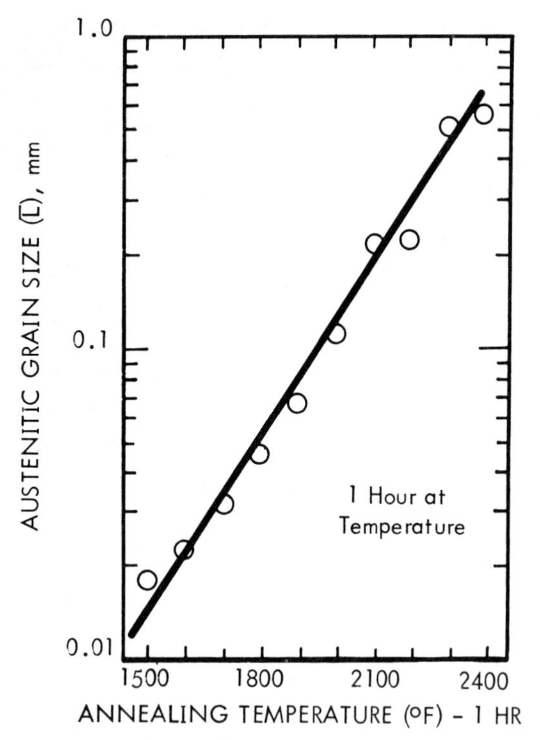

Fig. 1—Influence of annealing temperature on austenitic grain size.

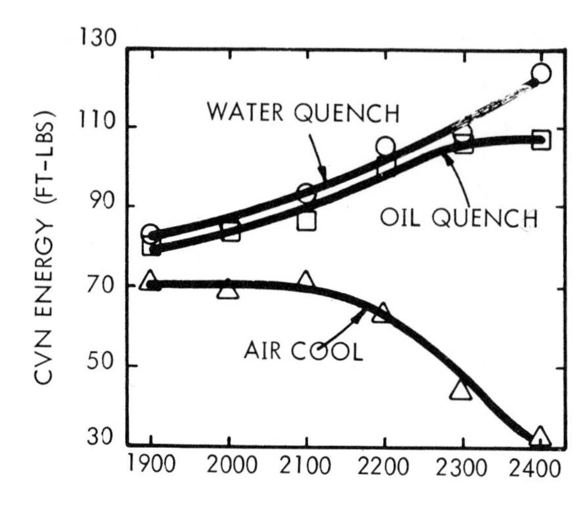

Fig. 2—Influence of annealing temperature on Charpy V-notch impact energy.

measured with standard ASTM Charpy V-notch specimens.

Fracture surfaces were examined by both replica and scanning electron microscopy techniques. Carbon films were first shadowed with platinum and then floated off the fracture surface in a solution of 5HCl-$5HNO_3$-90 methanol. The replication process extracted surface particles which were then identified by electron diffraction. The scanning electron microscopy observation angle was normal to the specimen surface.

The unaged heat treatment condition was selected for studying the degree of thermal embrittlement in order to avoid the possible complicating effects on the fracture toughness of the precipitation reactions during aging.[5] The much higher fracture toughness of the as-solutionized vs the aged martensite in the normal conditions, about 60 ft-lb. vs 8 ft-lb., provides a more sensitive baseline for reflecting embrittlement reactions. The results of Barton *et al.*[2] and Spaeder[3] support this criterion and also indicate that the Charpy V-notch impact test is suitable for studying the phenomenon in question.

EXPERIMENTAL RESULTS

Austenite Grain Size and Quench Rate

The 18 Ni (350) Maraging steel shows an exponential grain growth behavior, Fig. 1, over the range of austenitizing and annealing temperatures of interest in this study. The conventional heat treatment of 1500°F for 1 hr produces an average austenite intercept grain size ($\bar{L}$)* of 18 μ (ASTM grain size number 8). Annealing at 2400°F for 1 hr increases $\bar{L}$ to 680 μ (ASTM No. -3).

In order to separate the grain size contribution to embrittlement from that of precipitation on the austenite grain boundaries, impact specimens were annealed at temperatures from 1900° to 2400°F for 1 hr and then either water quenched, oil quenched, or air cooled. The impact energies for the air cooled series, Fig. 2, shows the expected embrittlement for annealing above 2100°F. For example, air cooling from 2400°F gives an impact energy of 33 ft-lb. as compared to about 60 ft-lb. after solutionizing at 1500°F and air cooling (not shown here). Air cooling from 1900° to 2100°F, with $\bar{L}$ ranging from 80 to 190 μ, gives impact energies of about 70 ft-lb. The increase in grain size for these austenitizing temperatures did not lower the toughness but instead gave slightly higher CVN energy values than for austenitizing at 1500°F. Furthermore, the water quenched and oil quenched series show an increase in impact energy with increasing grain size for the entire range of annealing temperatures.

Annealing Temperature

The influence of combined annealing treatments on the impact energy of the as-solutionized steel is shown in Fig. 3. Charpy test specimens were annealed first for 1 hr at temperatures from 1800° to 2400°F, hot quenched to temperatures of 1300° to 1800°F, held for 4 hr at the intermediate annealing temperature and then air cooled. Thermal embrittlement occurs when the first annealing temperature is above 1800°F. When a subsequent intermediate anneal is used, as in Fig. 3, the degree of embrittlement increases with an increase in the first annealing temperature. The most severe embrittlement is found when the intermediate anneal is 1500° to 1600°F.

Fracture Modes

In order to correlate the fracture mode with the degree of thermal embrittlement, two high temperature annealing treatments were selected, *i.e.*, 2000°

*$\bar{L} = L/N$ = the length of a test line on a micrograph divided by the number of grain boundary intersections.

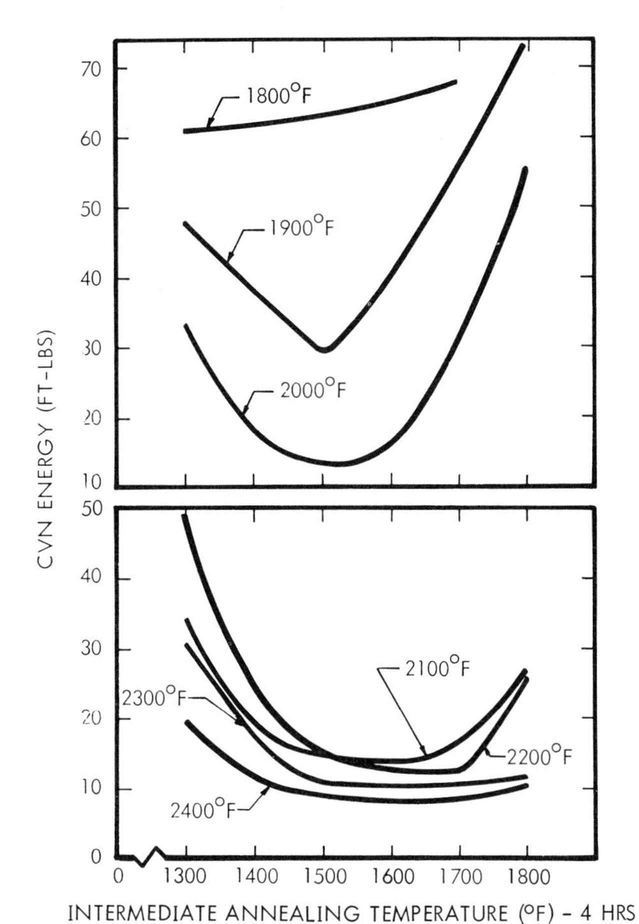

Fig. 3—Influence of combined annealing treatments on the impact energy of unaged 18 Ni (350) Maraging steel.

and 2400°F for 1 hr. With each of these first anneals the fracture characteristics were examined for specimens directly cooled to room temperature and for specimens given intermediate anneals.

a) 2400°F SERIES

The lowest impact energy measured, 9 ft-lb., was produced by the combination anneal of 2400°F-1600°F. A low magnification scanning electron fractograph, Fig. 4, shows this fracture to be intergranular. The size of the fracture facets agrees with the austenite grain size, *i.e.* of the order of 680 μ. Although the fracture tended to follow the prior austenite grain boundaries, higher magnification fractographs show that each intergranular face fractures by quasicleavage, Figs. 4(*b*) and 4(*c*). The ridges in Fig. 4 are formed by individual cleavage facets joining together. The replica fractograph, Fig. 4(*c*), reveals a small quantity of stringers of a dark block-like precipitate which were extracted and identified by electron diffraction as Ti_2S; no other precipitate such as Ti (C, N) could be indexed from the electron diffraction patterns. With the 2400°F first anneal, all intermediate anneals give relatively low impact energy, an intergranular fracture, and very few precipitates on the fracture surface. The few Ti_2S particles that are observed seem to be undissolved particles that have fractured during a prior hot working operation and

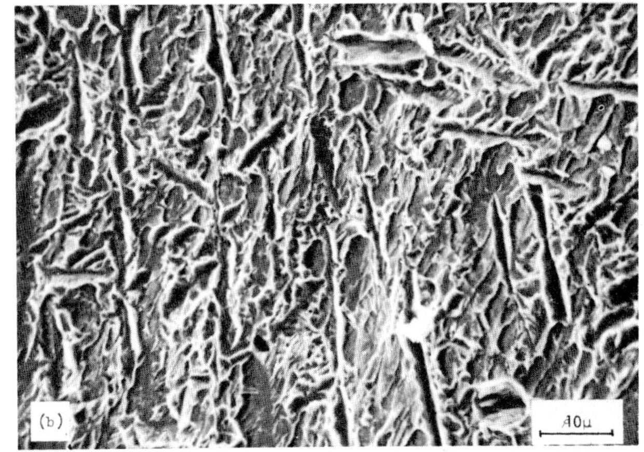

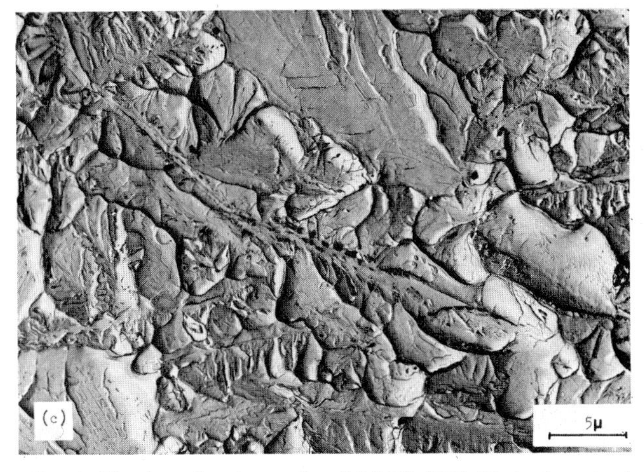

Fig. 4—Electron fractographs of 18 Ni (350) Maraging steel annealed at 2400° F (1 hr)—hot quenched to 1600° F (4 hr)—air cooled.

which serve to eventually pin the austenite grain boundaries during the first anneal.

An increase in the impact resistance, by a factor of 3 to 4 up to 33 ft-lb., occurs if the 1600°F intermediate anneal is eliminated and the steel is directly air cooled from 2400°F. Now the fracture path is a combination of transgranular and intergranular, Fig. 5(*a*), although the austenite grain size is unchanged ($\bar{L}$ = 680 μ). The intergranular facets have a crystallographic nature as indicated by a quasicleavage microfracture mode similar to that shown in Fig. 4. The transgranular facets are formed by microvoid coalescence giving the usual dimpled rupture appearance,

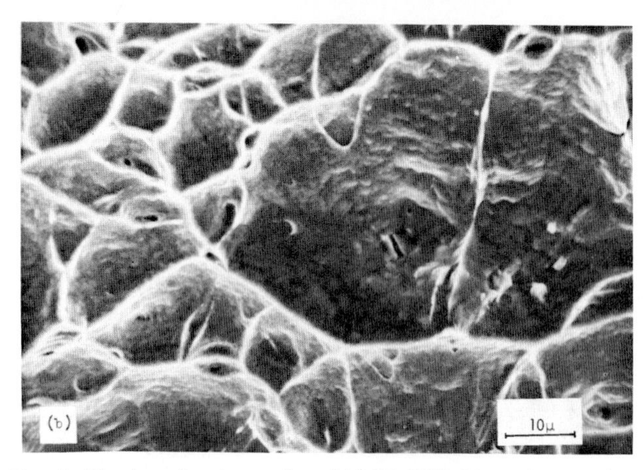

Fig. 5—Electron fractographs of 18 Ni (350) Maraging steel annealed at 2400°F (1 hr)—air cooled.

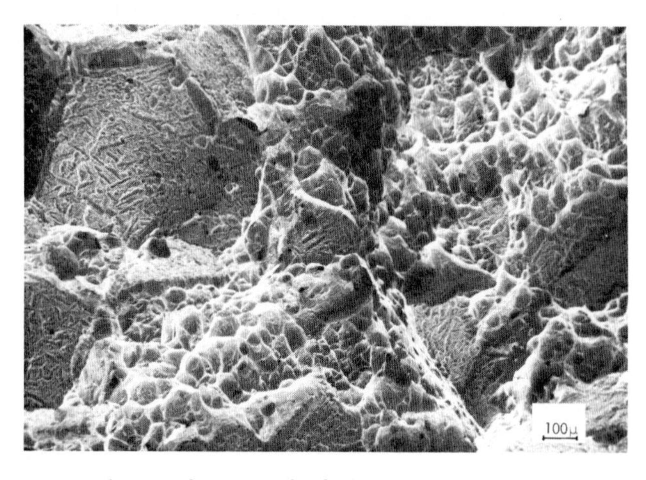

Fig. 6—Electron fractograph of 18 Ni (350) Maraging steel annealed at 2400°F (1 hr)—oil quenched.

Fig. 5(*b*). Both the quasicleavage and dimpled rupture fracture faces contain a small amount of Ti_2S similar to the intermediate anneal condition.

Oil quenching directly from 2400°F increases the impact resistance, again by a factor of more than 3, up to 107 ft-lb. This heat treatment gives a predominantly transgranular fracture path, Fig. 6, formed by microvoid coalescence. The impact energy absorbed after rapid (oil or water) quenching from 2400°F is thus an order of magnitude greater than when an intermediate anneal is employed, and the

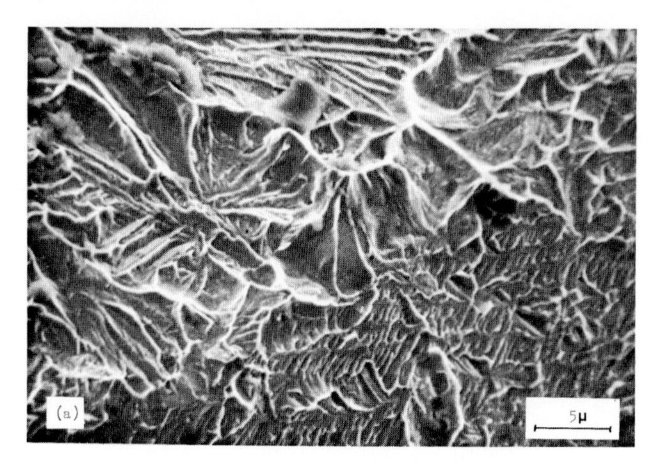

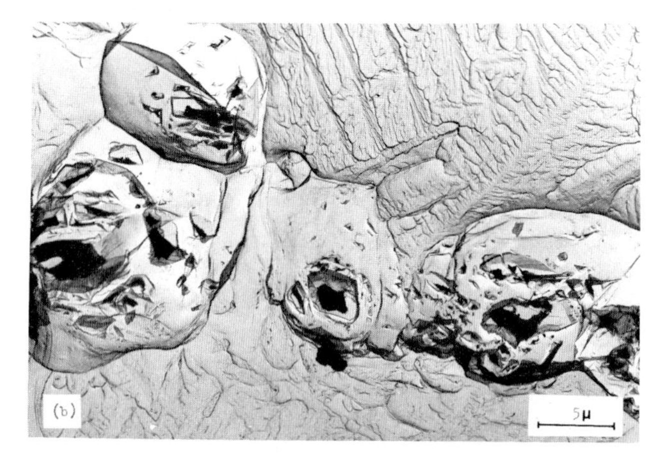

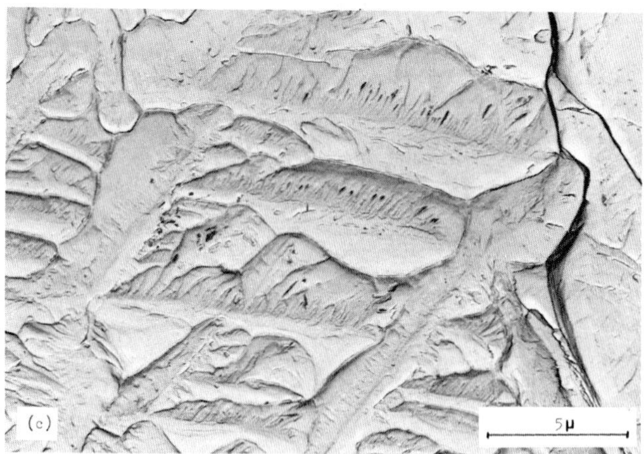

Fig. 7—Electron fractographs of 18 Ni (350) Maraging steel annealed at 2000°F (1 hr)—hot quenched to 1500°F (4 hr)—air cooled.

fracture mode changes from intergranular to transgranular. This all occurs with an extraordinarily large (but constant) austenite grain size and without the precipitation of Ti (C, N) on the austenite grain boundaries.

b) 2000°F SERIES

A first anneal of 2000°F develops an austenite grain size of $\overline{L}$ = 120 μ, or one-fifth that of the 2400°F treatment. Here an intermediate anneal can also be very detrimental to the impact resistance but not all intermediate anneals act in this way, Fig. 3. A 2000°F-

1500°F combination does lower the CVN energy to 13 ft-lb. and produces a typical intergranular fracture (not shown), formed by quasicleavage, Figs. 7(*a*) and 7(*b*). However, the fracture faces now contain both block-like particles (presumably undissolved compounds) and a fine feathery precipitate identified by electron diffraction as Ti (C, N), Fig. 7(*c*).

An increase in the intermediate anneal to 1800°F raises the impact resistance to 56 ft-lb. and changes the fracture mode to predominantly transgranular, Fig. 8. The transgranular dimpled rupture regions in the 2000°F series again contains the block-like precipitate (but no feathery precipitate). However, in the 2000°F treatment, these large particles were identified as both Ti_2S and Ti (C, N). Apparently, some Ti (C, N) exists within the austenite grains, and the 2000°F first anneal is not sufficient to dissolve all of this carbonitride while the 2400°F first anneal is able to do so. Whenever dimpled rupture occurs, in either the 2000° or 2400°F series, the dimples are about 10 to 30 μ in diam, irrespective of the austenite grain size.

Air cooling directly from the 2000°F first anneal gives an essentially transgranular dimpled rupture, Fig. 9, containing only large block-like particles [Ti_2S and Ti (C, N)] and increases the CVN energy to 69 ft-lb. Increasing the cooling rate, by either oil or water quenching, is not as effective in improving the fracture resistance, Fig. 3, with the 2000°F anneal ($\overline{L}$ = 120 μ) as with the 2400°F anneal ($\overline{L}$ = 680 μ).

MECHANISM OF THERMAL EMBRITTLEMENT

General Description

Thermal embrittlement of maraging steel appears to occur as a result of solute segregation to austenite grain boundaries. As in the case of temper embrittlement of low alloy steels,[6] the increase in solute concentration at the grain boundary reduces the interfacial energy and promotes intergranular low energy fracture. The precipitation of Ti (C, N) at the austenite grain boundaries appears to be an advanced stage of the intergranular embrittlement, the major source of embrittlement arising from the concentration of carbon and nitrogen solute atoms at the grain boundaries.

The driving force for segregation to the grain boundaries may be analogous to that for temper embrittlement;[6] due to either a) the misfit strain of impurity atoms in the austenite lattice, or b) the chemical interaction of certain embrittling elements with major alloying elements which have previously concentrated at the grain boundaries. The large misfit of interstitial carbon (or nitrogen) in the octahedral interstices in austenite and the atomic diameter difference of 17 pct between titanium (~2.91Å) and iron (~2.48Å)

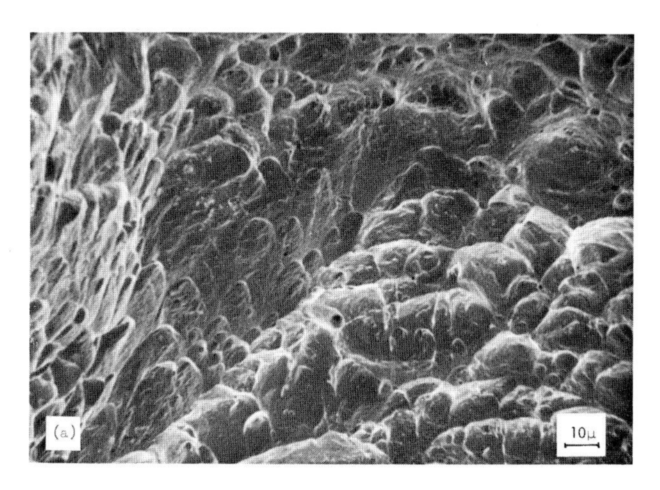

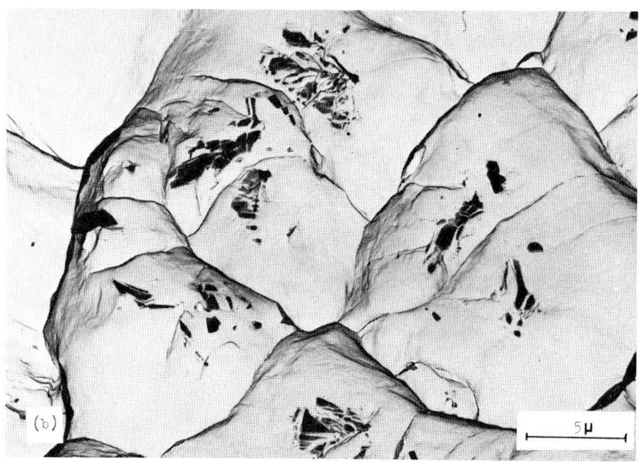

Fig. 8—Electron fractographs of 18 Ni (350) Maraging steel annealed at 2000°F (1 hr)—hot quenched to 1800°F (4 hr)—air cooled.

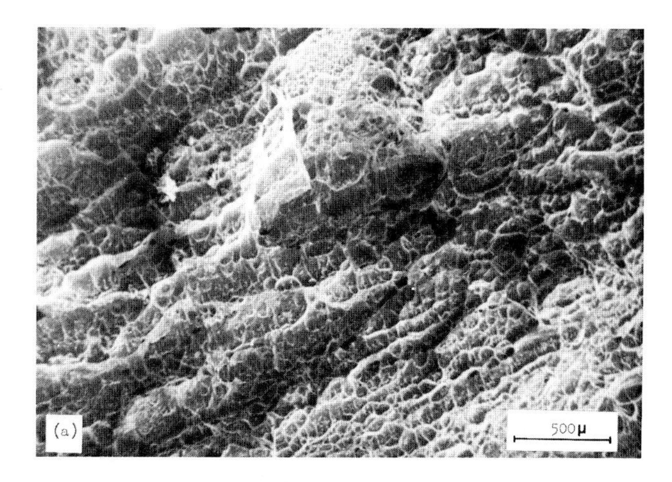

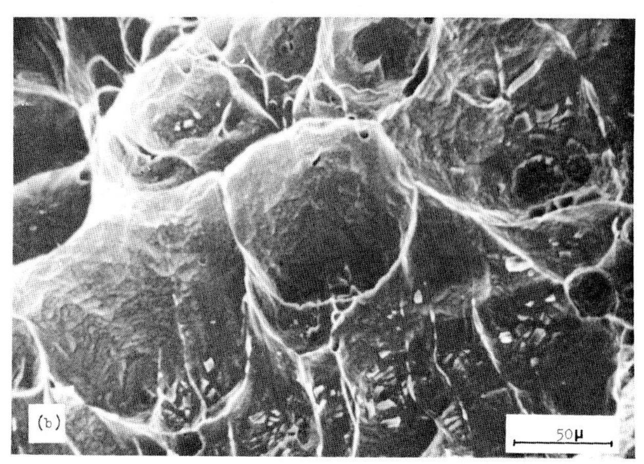

Fig. 9—Electron fractographs of 18 Ni (350) Maraging steel annealed at 2000°F (1 hr)—air cooled.

may provide the strain energy to account for the segregation of these elements to the austenite boundaries.

On the other hand, during solidification of a steel, there is a certain amount of natural segregation of alloying elements; the 18 Ni maraging steels have been particularly noted for titanium segregation. During the ingot breakdown, high grain boundary diffusion rates will spread the alloying elements over the entire austenite grain boundary network. In the case of low alloy steels subject to temper embrittlement, it has been noted that alloying elements tend to segregate at austenite grain boundaries in order of their increasing affinity for carbon.[6] The potency of titanium in this regard suggests that titanium will indeed be concentrated at the austenite grain boundaries in maraging steel. During cooling through the austenite range the chemical interaction between titanium and carbon (or nitrogen) will promote the segregation of the interstitial impurities to the grain boundaries of maraging steel. The embrittlement temperature range is determined by the lowest temperature for adequate impurity solute diffusion or by an interfering martensitic reaction whereby other sites such as dislocation complexes are provided for the clustering and precipitation of carbon. The upper end of the embrittlement range is determined by the temperature at which impurity atoms are "boiled off" the grain boundaries and driven back into solution in the austenite; the results in Fig. 3 suggest that this temperature may be around 2000°F.

The role of the high temperature anneal (1900° to 2400°F) is to dissolve Ti (C, N) particles found within the austenite grains. If this first anneal is below 1900°F, then carbon and nitrogen interstitials will precipitate as Ti (C, N) and natural grain growth will leave most of these particles within austenite grains rather than at the grain boundaries. The unaged martensite will then fracture in a predominantly transgranular mode and absorb about 60 ft-lb. impact energy.

When the first anneal is above 1800°F, the increased solubility of Ti (C, N) provides a sufficient amount of solute carbon and nitrogen for the embrittlement process. At the same time, a high temperature anneal develops a large austenite grain size, which can be either beneficial or detrimental to the fracture resistance, depending upon the subsequent heat treatment. If the steel is rapidly quenched, the highest annealing temperature (2400°F) is most favorable because the "boiling off" of interstitials is most effective and the diffusion distance (proportional to the austenite grain size) is the longest for them to return to the grain boundaries during the quench. This gives the highest toughness (>100 ft-lb.) in unaged martensite despite the very large austenite grain size. On the other hand, a large austenite grain size can be detrimental if the embrittlement process proceeds to an advanced stage as may be characterized either by considerable segregation of the impurities to the grain boundaries (*e.g.* with air cooling) or by the precipitation of Ti (C, N) (*e.g.* with subsequent intermediate annealing). The decrease in the total grain boundary area with increasing grain size now results in a higher concentration of embrittling elements at the boundaries. The reduction in the interfacial energy of the austenite grains will depend upon the concentration of embrittling elements; the steel with the largest grain size will then have the lowest toughness (<10 ft-lb.) and a completely intergranular fracture.

Corroborative Calculations

One can calculate the impurity atom concentration per unit of austenite grain boundary area in order to substantiate the proposed embrittlement mechanism. The air cooled and water quenched series of impact tests, Fig. 2, are considered here because they offer opposite trends in fracture toughness and fracture mode with grain size. Yet in both series there is no evidence for the feathery Ti (C, N) precipitate either on the austenite grain boundaries (intergranular fracture) or within the grains (transgranular fracture).

Initially it is assumed that the affinity of titanium (previously concentrated at the grain boundaries) for carbon and nitrogen dominates over the thermal "boiling off" effect as the temperature is lowered through the range of 2000° to 1900°F. The absence of embrittlement for intermediate anneals above 1800° to 1900°F, Fig. 3, supports this assumption. Although this transition will actually be gradual over some finite temperature range, 2000°F is used as the critical temperature for the purposes of these calculations. This assumption, as others to follow, may cause an error in the absolute value of the calculated concentration of impurities at the grain boundaries, but will not significantly affect the trend in the concentration with which we are primarily concerned.

The maximum distance within the austenite grains from which carbon (nitrogen) diffuses to the grain boundaries is controlled by the cooling rate when the heat transfer process is limited by the surface resistance.[7]

$$\frac{dT}{dt} = \alpha \frac{h}{k} \frac{A}{V} (T - T_q) \qquad [1]$$

where α is the thermal diffusivity, h the heat transfer coefficient of the quenching medium, k the thermal conductivity, A the specimen surface area, V the specimen volume, T_q the quench temperature, and T the annealing temperature.

The austenite grain boundaries may be treated as infinite sinks for the carbon because of the titanium-carbon coupling occurring at the boundaries and the overall low carbon concentration in these alloys. The square of the diffusion distance, d, on quenching is:[8,9]

$$d^2 = (Dt)_q = \int_{T_q}^{T} D(T) \frac{dt}{dT}\, dT \qquad [2]$$

Combining Eqs. [1] and [2]:

$$(Dt)_q = \frac{1}{\alpha} \frac{k}{h} \frac{V}{A} \int_{T_q}^{T} D(T) \frac{1}{T - T_q}\, dT \qquad [3]$$

In the present situation the impurity segregation process is terminated at the M_s temperature (*i.e.* T_q = 323°F) since many other sinks for carbon, *e.g.* dislocations, are introduced as martensite forms. The temperature from which segregation ensues is 2000°F. When the austenitizing temperature is below 2000°F, say 1900°F, the actual austenitizing temperature is used. The quench sensitivity parameter, $H = h/k$, for steel is[7] 8.202×10^{-3} cm^{-1} during air cooling and 0.394 cm^{-1} during water quenching. The ratio of the specimen volume to surface area is 0.286 cm. The combinations of quench sensitivity and specimen dimensions used in this study fulfill the criterion[6] for using Eq. [1] to describe the unsteady heat flow conditions. The thermal diffusivity of austenite in a 32 pct Ni steel quenched from 1380°F is about 0.04 sq cm per sec.[10]

The diffusivity of carbon in austenite is given by:[11]

$$D(T) = 0.47 \exp(-37{,}000/RT) \quad [4]$$

A similar value for the activation energy, Q = 33,700 cal per g-atom has been reported for the diffusion of carbon in an 18 Cr-9 Ni stainless steel.[12] Moreover, diffusion of nitrogen in austenite[13,14] gives values of D_0 = 0.2 sq cm per sec and Q = 36,000 cal per g-atom so that Eq. [4] is thought to be a fair estimate of the diffusion processes in austenitic maraging steel.

The fraction of the wt pct C, f_c, that reaches the austenite grain boundaries is obtained by assuming all diffusion occurs in a direction normal to the nearest grain boundary and that the carbon is uniformly dissolved in the austenite above 2000°F. If the austenite grains are essentially equiaxed (*e.g.* the grains have the shape of tetrakaidecahedrons of edge length $a = 0.5925\bar{L}$)

$$f_c = \frac{\bar{L}^3 - L'^3}{\bar{L}^3} \quad [5]$$

where L' is the mean intercept length of a hypothetical small grain ($a' = 0.5925L'$) within the actual austenite grain, from which no carbon (nitrogen) reaches the austenite grain boundary. This inner grain is related to the true grain size by:

$$a' = 0.5925(\bar{L} - 2d) \quad [6]$$

Combining Eqs. [5] and [6]:

$$f_c = \frac{6d\bar{L}^2 - 12\bar{L}d^2 + 8d^3}{\bar{L}^3} \quad [7]$$

where d is calculated from Eqs. [2] and [3].

The grain boundary area per unit volume, S_v, is:

$$S_v = \frac{2}{\bar{L}} \quad [8]$$

Finally, the concentration of carbon (nitrogen) per unit of grain boundary area is:

$$C_c^{GB} = \frac{\bar{L}}{2} f_c W_c P \frac{N_0}{M} \quad [9]$$

where W_c is the weight fraction of carbon and nitrogen in the alloy, P the density, N_0 Avogadro's number, and M the molecular weight of the interstitial atoms. The results of this calculation for the air cooled and water quenched series are presented in Fig. 10. It is seen that C_c^{GB} for the air cooled series is inversely related to the impact energy, Fig. 2. Where the toughness does not change with austenitizing temperature, there is essentially no change in the concentration of impurities of the grain boundaries, and when the toughness decreases with increasing austenitizing temperature, C_c^{GB} increases. Throughout the air cooled series, there is a substantial fraction of intergranular fracture and so the correlation between impurity segregation and fracture toughness is in excellent agreement with the proposed mechanism of embrittlement. Water quenching gives a much smaller impurity concentration at the grain boundaries than does air cooling, hence the higher toughness. The C_c^{GB} is nearly constant with increasing austenitizing temperature although the toughness does increase. Perhaps in the water quench case, the change in the "boiling off" effect with temperature must be taken into account since the degree of impurity segregation upon quenching is so low. Slight initial concentrations could be as important as the total amount diffusing to the boundaries below 2000°F and could give a negative slope to the C_c^{GB} vs austenitizing temperature curve, Fig. 10.

Certain assumptions embodied in the preceding calculations were verified by determining the impact energy for constant grain size specimens quenched from various annealing temperatures. A grain size of 360 μ was established, at temperatures from 2000° through 2300°F, by varying the annealing time. The CVN impact energy was found to be independent of the anneal-

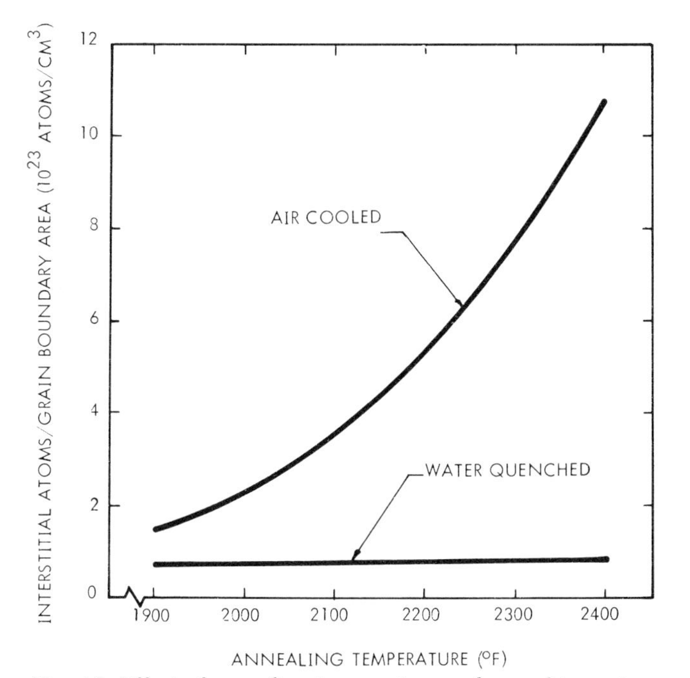

Fig. 10—Effect of annealing temperature and quenching rate on the concentration of interstitial atoms segregated at austenite grain boundaries.

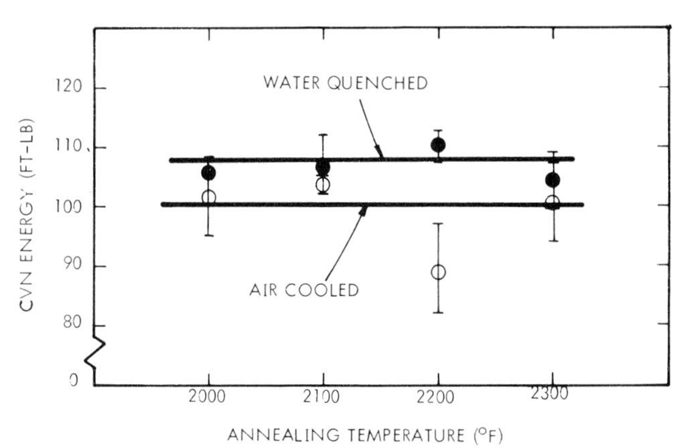

Fig. 11—Influence of annealing temperature at constant grain size on impact energy of unaged 18 Ni (350) Maraging steel.

ing temperature from which the specimens were quenched but dependent upon the quenching rate, Fig. 11. This confirms that the predominant role of annealing temperature, above 2000°F, is to establish the distance over which a given impurity atom must diffuse to arrive at a grain boundary. Negligible segregation to the boundaries occurs during the annealing itself. Finally, Fig. 11 supports the view that the effect of quenching rate, through temperatures below 2000°F, is to determine the volume of steel from which impurity atoms can reach a grain boundary at a given quench rate.

CONCLUSIONS

1) The austenite grain boundaries of 18 Ni (350) Maraging steel become embrittled, to some extent, during all cooling treatments through the austenite phase field.

2) The severity of thermal embrittlement increases with decreasing cooling rate from the austenitizing temperature. The most severe embrittlement occurs upon austenitizing at 1900° to 2400°F followed by intermediate isothermal holding at 1500° to 1600°F.

3) A large austenite grain size is beneficial to the toughness when direct rapid quenches from the annealing range are imposed but is detrimental upon air cooling or intermediate holding.

4) Thermal embrittlement of the unaged martensite is manifested by a decrease in the Charpy V-notch impact energy and by a change in fracture mode from transgranular dimpled rupture to intergranular quasi-cleavage.

5) The CVN energy of the unaged martensite may be varied from 9 to over 100 ft-lb. by altering the annealing conditions.

6) The major loss in toughness is associated with the diffusion of interstitial impurity atoms (carbon and nitrogen) to the austenite grain boundaries during cooling or intermediate isothermal holding below 2000°F. An advanced stage of the embrittlement is characterized by the discrete precipitation of Ti (C, N) platelets on these boundaries.

ACKNOWLEDGMENTS

This work was performed in the Physical Metallurgy Team of the Materials Sciences Laboratory and was funded by the Lockheed-Georgia Company Independent Research Program. The authors are indebted to J. W. Elling and K. D. Fike for assistance in experimental parts of this study.

REFERENCES

1. J. E. Campbell: DMIC Report S-28, June 1969, pp. 8-16.
2. C. J. Barton, B. G. Reisdorf, P. H. Salmon Cox, J. M. Chilton, and C. E. Oskin: AFML-TR-67-34, 1967.
3. G. J. Spaeder: *Met. Trans.,* 1970, vol. 1, pp. 2011-14.
4. G. W. Tuffnell and R. L. Cairns: *ASM Trans. Quart.,* 1968, vol. 61, pp. 798-806.
5. H. J. Rack and David Kalish: *Met. Trans.,* to be published.
6. J. M. Capus: *Am. Soc. Testing Mater., Spec. Tech. Publ.* No. 407, 1968, pp. 3-19.
7. R. Schuhmann, Jr.: *Met. Eng.,* vol. 1, Chap. 8, Addison-Wesley Press, 1952.
8. J. Crank: *Mathematics of Diffusion,* pp. 42-61, Oxford, 1956.
9. G. R. Speich: *Trans. TMS-AIME,* 1969, vol. 245, pp. 2553-64.
10. H. Scott: *Trans. ASM,* 1934, vol. 22, pp. 68-96.
11. C. Wells, W. Batz, and R. F. Mehl: *Trans. AIME,* 1950, vol. 188, pp. 553-60.
12. B. A. Nevzorov and O. V. Starkor: *Fiz.-Chim. Mekh. Mater.,* 1967, vol. 3, pp. 257-60.
13. J. D. Fast and M. B. Verrijp: *J. Iron Steel Inst.,* 1954, vol. 176, pp. 24-27.
14. K. Bohnekamp: *Arch. Eisenhuettenw.,* 1967, vol. 38, pp. 229-32.

Effects of Repeated Thermal Cycling on the Microstructure of 300-Grade Maraging Steel

ALFRED GOLDBERG

ABSTRACT. Investigations were made of the influence of thermal cycling on transformation behavior, microstructure, and hardness in a 300-grade, 18% nickel maraging steel. Heating rates ranged from 0.075 to 500 C/sec. After cycling 10 times between ambient temperature and 815 C with a heating rate of 4.4 C/sec, an increase in M_s (martensite start temperature) from 197 to 377 C and a decrease in A_s (austenite start temperature) from 655 to 545 C were observed. At this relatively slow heating rate the greatest changes in these values and in the shape of the corresponding dilatometric curves were obtained during the first few cycles. However, when specimens were cycled at fast heating rates of about 300 to 500 C/sec, changes did not occur until after six cycles. This difference in behavior is explained in terms of diffusion-controlled and diffusionless reactions for the martensite-to-austenite reversion at the slow and fast heating rates, respectively.

Repeated cycling continuously increased both the amount of retained austenite and the degradation in aged hardness. On the basis of these observations and the changes observed in M_s it is suggested that cycling causes large variations in composition. Additional evidence indicates that these variations occur within regions less than 1 μ. The martensite-to-austenite reversion produced by continuous heating results in a lamellar pearlitic-like structure, which becomes spheroidal with repeated cycling.

ALTHOUGH numerous studies have been reported on the development of retained austenite in maraging steels (1–13), the effect of this retained austenite on subsequent transformation behavior has not been noted. It has been suggested that the formation of austenite on heating a maraging steel is initiated at precipitate sites by local re-solution of the precipitates that normally form in the martensite during aging (2). These precipitates are rich in Mo, Ti, and Ni (2, 4, 14), elements which lower the M_s when dissolved in austenite (15–17). As a consequence, the reverted austenite may not necessarily retransform on cooling and may therefore be retained. On subsequent aging the mechanical properties are thus degraded as compared to properties of this alloy in the annealed martensitic and aged condition.

If the reverted austenite is enriched in certain elements, then the residual martensite must be depleted in these same elements. Also, if any of the reverted austenite tranforms on cooling then the newly formed martensite should be similarly enriched in these elements. It may be expected that the extent of the segregation resulting from such partitioning of alloying elements will be reflected in the transformation behavior obtained on subsequent reheating. Furthermore, assuming that the same phenomena occur during each heating and cooling cycle, the amount of segregation and retained austenite should increase with repeated cycling. Studies involving numerous reheating cycles have have been reported in connection with welding of maraging steels, but the cycling was a consequence of successive welding passes so that a given region was not subjected to the same thermal conditions during each successive cycle. Therefore the effect of repeated cycling under a set of fixed conditions was not represented. Studies of reversion, *per se*, have been concerned only with single thermal cycles, the interest being in the effect of time at temperature (2, 5, 7) or the effect of maximum temperature reached (10, 18). Several studies have indicated that at sufficiently high heating rates, generally above those used in welding, the reversion is diffusionless (10, 19). Under these conditions, at least for a single thermal cycle, changes in composition do not develop and retained austenite is absent. The effect of repeated cycling at these high rates, to our knowledge, has not been reported.

The present investigation was undertaken to determine the effects of repeated thermal cycling, under different heating conditions, on a maraging steel's transformation behavior, microstructure, and hardness. The transformation temperatures were obtained by dilatometric techniques.

The author is Research Metallurgist, Process and Materials Development Div., Chemistry Dept., University of California Lawrence Radiation Laboratory, Livermore, Calif. Manuscript received 29 September 1967.

MATERIAL

A 6-in. diameter ingot of 300-grade maraging steel, made by the consumable electrode vacuum-melting process, was obtained from Vasco, Latrobe, Pa. Its composition in weight per cent was: 18.40 Ni, 8.86 Co, 4.76 Mo, 0.60 Ti, 0.09 Al, 0.02 C, 0.04 Si, 0.05 Cr, 0.06 V, 0.008 S, 0.016 Zr, 0.003 B, 0.05 Ca (added), and the balance Fe. After an initial 3-hr soaking period in a slightly oxidizing atmosphere at 1260 C, the ingot was forged into square bars 1 in. on a side and round bars $\frac{5}{8}$ in. in diameter. Both shapes were annealed for 2 hr at 815 C and surface-machined to remove all scale. The square bars were cold rolled into strips about 0.285 in. thick, then sectioned and machined into cylindrical specimens $\frac{1}{4}$ in. in diameter by 1 in. long. The round bars were swaged and drawn to 0.078 in. in diameter and used for wire specimens. Before testing, the specimens were annealed in vacuum for 2 hr at 815 C. When they were cooled to ambient temperature a martensitic structure developed. Specimens given this treatment are said to be in the annealed martensitic condition, which was the initial condition of all test specimens unless otherwise stated.

EXPERIMENTAL PROCEDURE

Cylindrical specimens were used for the slow-heating dilatometric studies (Fig. 1). They were heated in a conventional resistance-wound furnace. The change in length of the specimen was transmitted to an LVDT (linear variable differential transformer) by the relative displacement of two concentric vertical quartz tubes. The specimen rested on the bottom of the outer tube, while the inner tube rested on top of the specimen. Temperatures were measured by a chromel-alumel thermocouple intrinsically welded to the specimen. The LVDT output was demodulated to provide a dc signal whose polarity depended on the direction of the core displacement from the preset zero. This signal and that of the thermocouple output were fed to a recorder.

For the rapid-heating dilatometric studies at rates of 300 to 500 C/sec, some modification of the above dilatometric technique was required. This involved the use of a wire specimen which could be resistance heated (Fig. 2). The specimen was suspended between two Inconel power-terminal lugs, the lower of which was connected to two flexible leads. The dimensional change of the specimen was obtained by suspending the iron core of an LVDT from the bottom end of the specimen, while the top end of the specimen and the LVDT casing were attached to a common framework. In order to minimize the temperature gradient along the 6-in. length of the wire specimen, the ends of the specimen were reduced in diameter. This increased the resistance heating at the ends and reduced the heat loss from the bulk of the wire to the lugs.

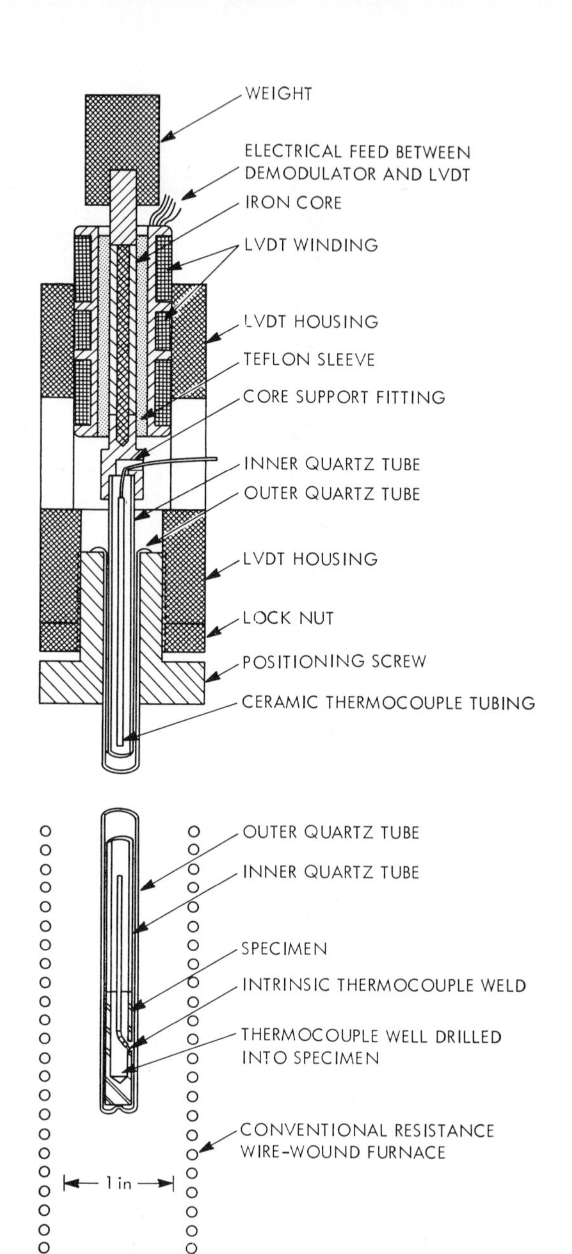

Fig. 1. Vertical cross section of dilatometer used for studies with externally heated cylindrical specimens.

The temperature of the wire specimen was measured with the technique described by Boedtker and Duwez (20), whereby the regions along the specimen between three intrinsically welded thermocouple wires form part of a Wheatstone bridge circuit used to nullify the emf drop resulting from the dc voltage source that heats the specimen. A 3-v battery, a reversing switch, terminal clips, and a resistance panel box form the other part of the bridge circuit. The resistance box contains a selector switch and six variable resistors for six thermocouples used to study temperature gradients along the wire.

All testing was done in air. Although slight oxidation resulted at the highest test temperature, there appeared to be no difference between curves from a vacuum dilatometer and those from air dila-

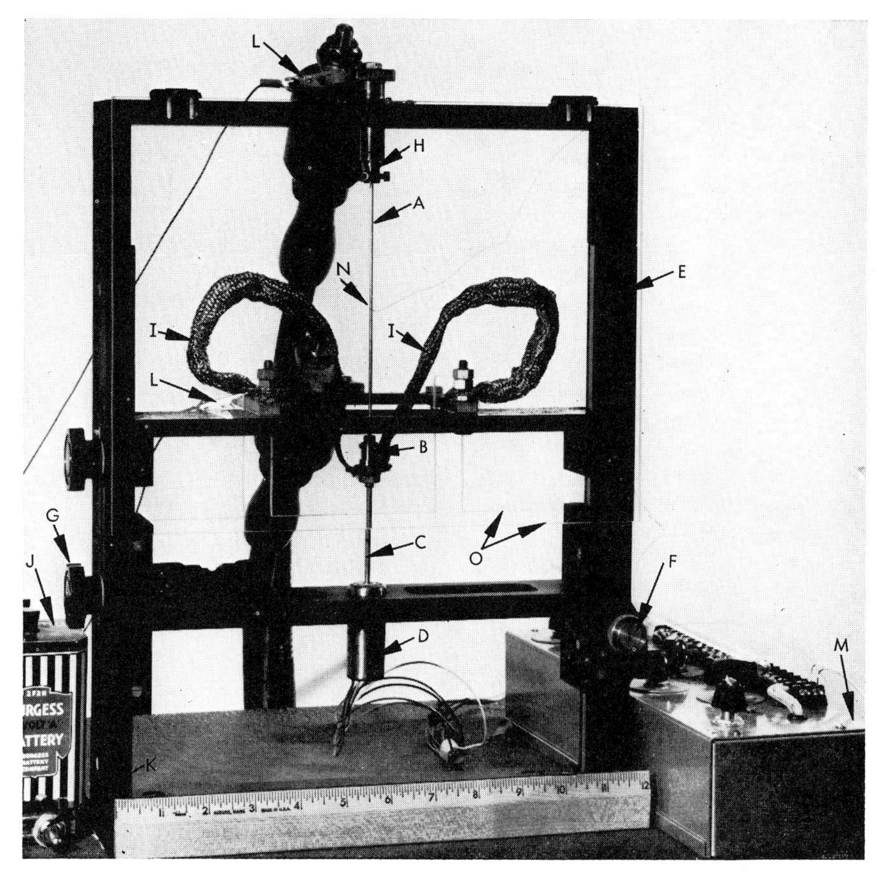

FIG. 2. Dilatometer for resistance-heated wire specimen. **A, specimen; B,** lower terminal power lug; C, stainless steel tubing **connecting** LVDT core to specimen; D, LVDT casing; E, frame; **F,** coarse rack-and-pinion gears; G, fine rack-and-pinion gears; H, top terminal power lug; I, flexible battery-type strap leads; J, 3-v battery for balancing thermocouple bridge circuit; K, reversing switch; L, terminal clips for battery; M, resistance panel box; N, intrinsically welded thermocouple (three 3-mil wires welded about 0.005 in. apart); O, plastic shields front and back to minimize air disturbances during balancing.

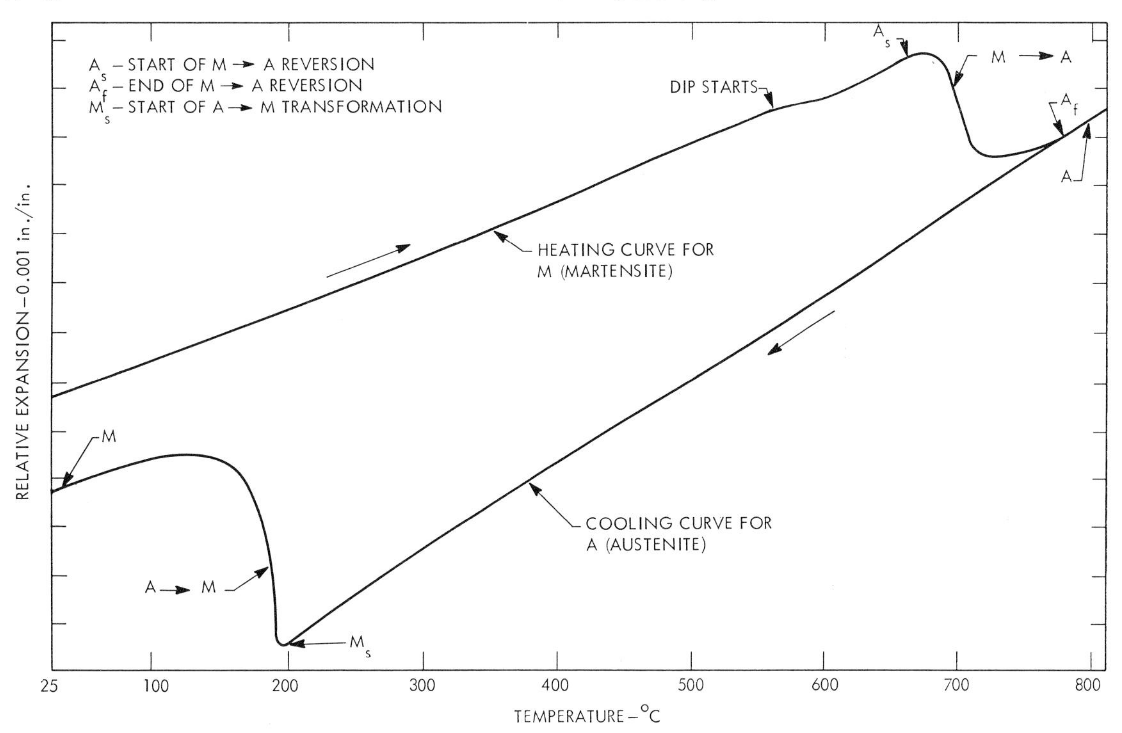

FIG. 3. Dilatometric heating and cooling curve for a 300-grade, 18% Ni maraging steel initially in the annealed martensitic condition. Specimen was heated at 4.4 C/sec from 25 to 815 C, held for 2 hr at 815 C, and then air cooled.

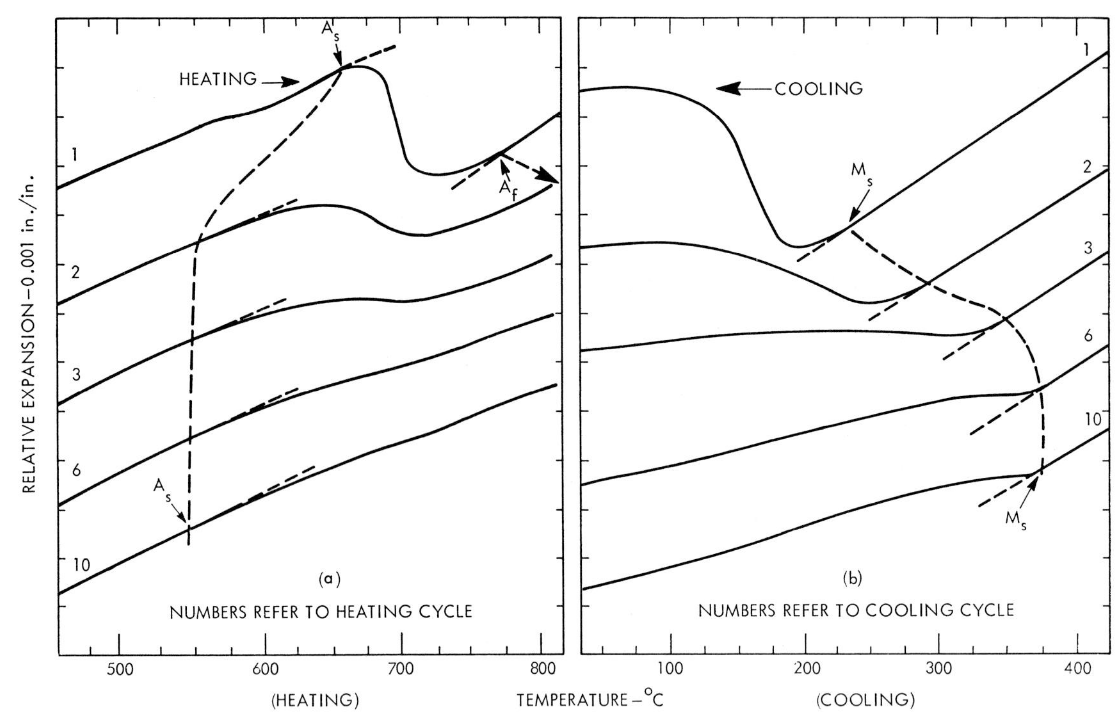

FIG. 4 Effect of the number of thermal cycles between ambient temperature and 815 C on the dilatometric heating and cooling curves. The cylindrical specimen, initially in the annealed martensitic condition, was heated at 4 4 C/sec and air cooled for each cycle.

tometers. Furthermore, no significant differences were observed between the dilatometric results from two specimens annealed for 48 hr at 815 C, one in air and the other in vacuum.

RESULTS AND DISCUSSION

DILATOMETRIC CURVES

Figure 3 shows a typical dilatometric heating and cooling curve for a cylindrical specimen heated at 4.4 C/sec from about 25 to 815 C, held for 2 hr, and cooled in air. The shape of the deviations on the heating portion of the curve and the temperatures over which they exist have been shown to depend greatly on heating rate in tests where the heating rates ranged from 0.0021 to 700 C/sec (19). The slight dip below A_s is attributed to precipitation. The large contraction between A_s (austenite start temperature) and A_f (austenite finish temperature) is a consequence of the $M \rightarrow A$ (martensite-to-austenite) transformation. Large variations in composition may be present in the specimen as a result of this transformation and the previous precipitation. For the heating rate used the A is still heterogeneous on reaching 815 C, about 45 C above A_f. A minimum time of almost 1 hr at 815 C is required to eliminate this segregation. The specimen from which the curve in Fig. 3 was obtained was held for 2 hr at 815 C prior to cooling. The expansion observed on cooling is due to the $A \rightarrow M$ transformation, which is initiated at M_s (martensite start temperature) and is completed somewhat above 25 C.

The shortening of the specimen, which is indicated by the displacement of the end of the curve by nearly 0.002 in./in. with respect to the start, is a real effect. The change in dimensions depends on the orientation of the specimen relative to the rolling direction of the strips from which the specimens were obtained. Any shortening of a specimen is accompanied by a corresponding growth in diameter; the cross section in fact changes from circular to elliptical. There is no change in density with this change in shape.

Figure 4 shows changes in the dilatometric curves as a result of repeated thermal cycling by heating at a rate of 4.4 C/sec to 815 C and immediately air cooling. Portions of the heating and cooling curves are shown for the 1st, 2nd, 3rd, 6th, and 10th cycles. Significant differences between successive curves may be seen in both the heating and cooling results only over the first three cycles. Differences between subsequent cycles become increasingly smaller. The magnitudes of the contraction dip associated with the $M \rightarrow A$ transformation on heating and the expansion hump associated with the $A \rightarrow M$ transformation on cooling diminish with each additional cycle, especially during the first few cycles. The A_s decreases from 655 to 545 C over the 10 cycles. The A_f shifts from 770 C for the first cycle to above 815 C for the second cycle, suggesting that some bcc phase is still present on reaching 815 C for all but the first cycle. The M_s increases from 235 to 377 C from the 1st to the 10th cycle. This is to be compared to an M_s of 197 C obtained for the homogeneous alloy

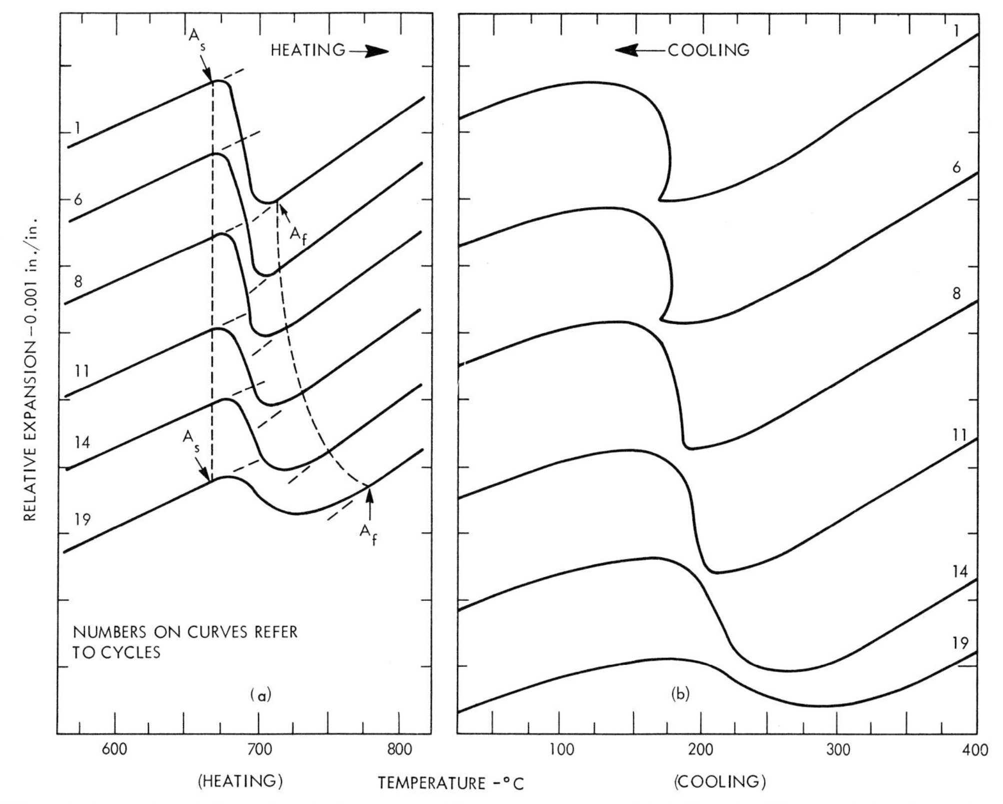

FIG. 5. Effect of the number of thermal cycles between ambient temperature and 815 C on the dilatometric heating and cooling curves. The resistance-heated wire specimen, initially in the annealed martensitic condition, was heated at 300 to 500 C/sec and air cooled for each cycle.

cooled from 815 C. The nature of the cooling curves between M_s and ambient temperature as compared to that shown in Fig. 3 suggests the presence of retained A for all cases. X-ray analyses of cycled specimens show that the amount of retained A increases with the number of cycles. The large changes observed in both the transformation temperatures and the ranges over which the transformations occur as a result of cycling are indicative of the development of extensive chemical segregation.

The decrease in A_s and the presence of retained A may be explained by the nucleation and growth of this A at precipitates which are rich in Ni, Mo, and Ti (2), elements that depress the $A \rightarrow M$ transformation temperature range. The increase in A_f and M_s may similarly be attributed to the formation of A and its subsequent transformation to M in regions that had been depleted in these same alloying elements. Furthermore, to account for the dilatometric and x-ray results, the process of enrichment and depletion must continue during each successive thermal cycle. This can occur by repeated precipitation in the martensite during each heating cycle, so that additional retained A is repeatedly formed at precipitates while the martensite regions, in turn, become continuously depleted in alloying elements that lower the M_s.

If the $M \rightarrow A$ transformation is diffusionless at high heating rates, above 60 C/sec (19), one would expect that cycling at such rates would bring about, at most, only minor changes in composition. Some of the heating and cooling curves obtained by cycling a resistance-heated 6-in. wire specimen 19 times between ambient temperature and 815 C are shown in Fig. 5. The wire was heated at rates from 300 to 500 C/sec, and air cooled. No significant differences were noted in either the heating or cooling curves during the first six cycles. The A_s remains virtually unchanged up to about the 14th cycle, and shows only a very slight drop from the 14th to the 19th cycle. The A_f increases from 710 to 780 C over the 19 cycles; however, this increase occurs largely after the 6th cycle. The changes in the cooling curves as a result of repeated cycling are consistent with the changes in the corresponding heating curves. That is, the range over which the $A \rightarrow M$ transformation occurs is shifted to higher temperatures after the 6th cycle as the number of cycles increases. At low temperatures, from 25 to 125 C, the cooling curves appear to be virtually parallel to each other. This is consistent with the observation that, on heating, the A_s changes very little with cycling.

If, as for the case when cycling with the low heating rate, the increase in M_s and A_f with high heating rates were attributable to the formation of regions depleted in certain alloying elements, then corresponding regions enriched in these elements should also have formed. The presence of such enriched regions should have caused a drop in both the $A \rightarrow M$ transformation temperature range and in the A_s, but these drops were not observed. This apparent contradiction can be rationalized by assuming that any enriched A which forms during a thermal cycle does not undergo any subsequent transformation on further cycling. This implies that the A which had nucleated at precipitates was initially enriched enough in alloying elements so as to be permanently stabilized. Furthermore, it would appear that these elements do not significantly diffuse out of the retained A formed at the high heating rates, after repeated cycling.

X-ray diffraction of a specimen cycled six times at high heating rates did not show any retained A. Thus, at most only a few per cent, if any, of this phase was then present. After 19 cycles, over 15% retained A is estimated to be present. These two results are consistent with the trends observed in the dilatometric curves, namely, that no change is apparent during the first 6 cycles whereas changes (in A_s and $A \rightarrow M$) occur after the 6th cycle and up to the final 19th cycle. We therefore suggest that the phenomena leading to segregation and the development of retained A, which occur most prominently during the initial cycling periods with low heating rates, are delayed during the early cycling periods when high heating rates are used.

The delay in the observed changes that take place on cycling at high heating rates may possibly be explained on two different bases: 1) an incubation period necessary for precipitation, and 2) strain-induced precipitation.

In the first case, assuming that some diffusion occurs during each cycle and that it is additive, enough diffusion may have taken place by about the 7th cycle to permit the formation of stable-size nuclei of the alloy-rich precipitate. Once this occurs, subsequent growth is more rapid with a corresponding depletion of the matrix in precipitate elements consistent with the rise in A_f between the 6th and 19th cycles. Concurrent with precipitate formation, reversion to A at precipitate sites also takes place following the 6th cycle.

In the second case, high internal strains and/or excess vacancies may develop as a result of repeated transformations and thereby accelerate the diffusion process necessary for precipitation. In partial support of this, it was noted that some external bending of the wire occurred as a result of cycling.

MICROSTRUCTURE AND X-RAY DIFFRACTION

Several techniques were used in an attempt to obtain more definitive information on the extent to which local variations in composition may have developed during thermal cycling. An electron microprobe analysis was made with a focused beam of about 1 μ diameter, monitoring Co, Mo, and Ni. No significant variations in any of these elements were detected, thus implying that regions either depleted or enriched in alloying elements are finer than 1 μ.

M_s data on Ni-base alloys suggest that appreciable changes in composition should have occurred as a result of cycling. A difference of about 160 C in M_s was reported for two maraging steels where the main difference in composition was 4.8 wt % Mo (5). On the basis of results reported by Decker for an Fe–20Ni-base alloy (15), the increase in M_s from 197 to 377 C, which is observed here, should correspond to a decrease in Ni or Mo by about 8.5 or 8.1 wt %, respectively, or to an increase in Co of 10.8 wt % when each element is taken separately. Due to the opposite effect of Co on the M_s, as compared to Mo and Ni, composition changes in our case would be expected to be even somewhat greater. The possible presence of Co in the precipitates has been suggested (4, 21).

Banerjee and Hauser (4) reported differences of +3.0, +0.4, +1.0, and −0.3 wt % of Mo, Ti, Ni, and Co, respectively, in reverted austenite relative to the matrix composition for a maraging steel plate; the material, however, was initially segregated, and individual austenite stringers were of the order of 10^2 to 10^3 μ long. Somewhat similar differences in composition were reported for the different microconstituents of weld deposits; here, however, composition variations within the heat-affected zone of the base metal were not generally reported, presumably due to the inability of obtaining this information because of the fineness of the base structure in contrast to that of the coarse welds (9). The size of the microconstituents shown for the base metal appears to be about 10^{-1} μ. It therefore appears to us that one can be misled by the relatively small variations in composition that have been reported in the literature for initially homogeneous maraging alloys that have undergone the $M \rightarrow A$ reversion. These values are considerably smaller than one would infer from changes obtained in the M_s and A_s.

Figure 6 shows electron micrographs obtained from specimens in the annealed martensitic condition and after cycling 1, 3, 10, and 30 times between ambient temperature and 815 C, with a heating rate of 4.4 C/sec followed by air cooling. The features of the martensitic structure depicted in the replica of the annealed specimen are similar to those shown for an Fe–Ni alloy (22) and for annealed 18% Ni maraging steels (1, 21). This etch-pit-like structure may also be seen in the matrix of the replicas of the cycled specimens.

A measure of the relative amounts of M and A was obtained with an XRD counter unit. The ratios of the relative intensities of the $(111)_{fcc}$ to $(110)_{bcc}$ reflections for α and β values of 90° for specimens in the annealed condition and after 1, 3, 10, and 30 cycles are, respectively, 0, 0.08, 0.43, 0.83, and 0.98.

These values represent the ratios of the product of

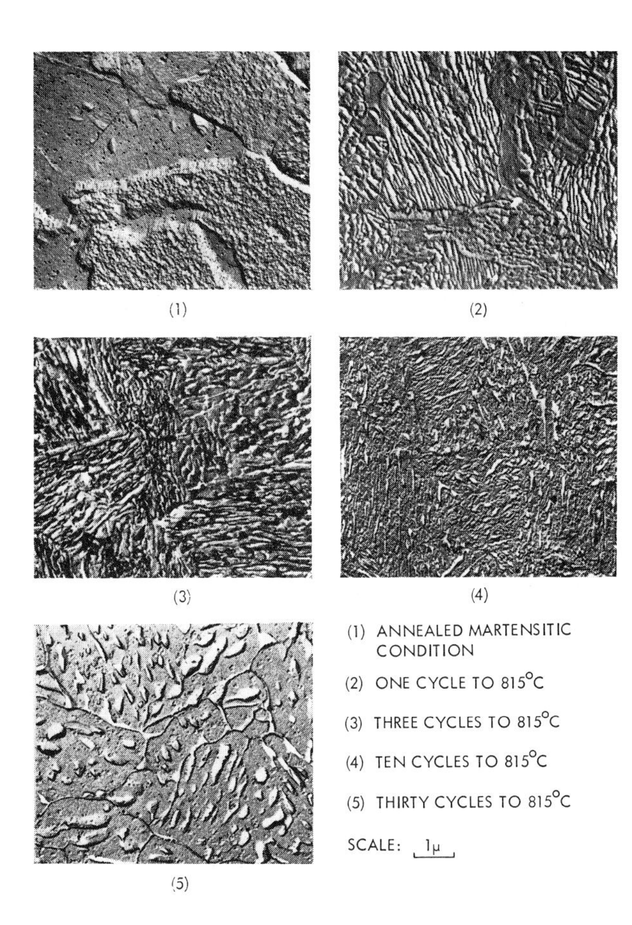

FIG. 6. Effect of the number of thermal cycles on the electron microstructure of maraging steel specimens heated at 4.4 C/sec to 815 C and air cooled during each cycle. Direct carbon replica, nital etch.

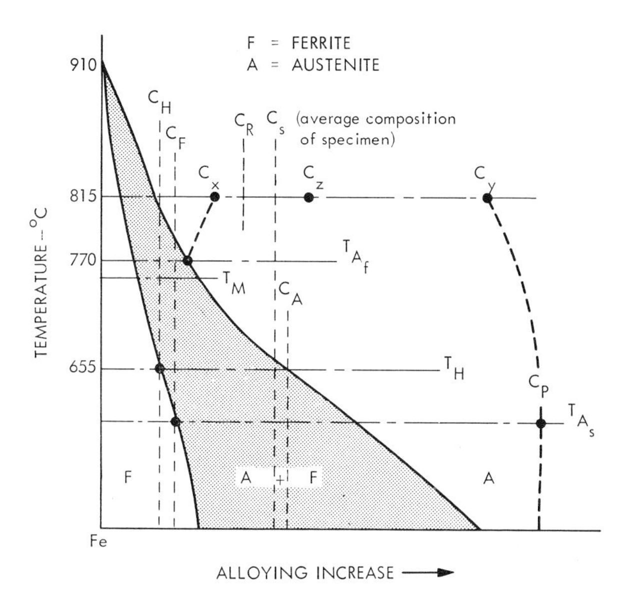

FIG. 7. Hypothetical Fe-alloy phase diagram to illustrate the range in composition developed in austenite which forms on heating a maraging steel specimen to 815 C. C refers to composition, T to temperature, F to ferrite, A to austenite.

the peak height and width at half maximum. No correction factors were used; furthermore, the intensities varied with α and β rotations due to preferred orientation. Thus, these values should be taken only as a quantitative approximation of the increase in the A/M ratio with cycling. It may be noted, however, that the nearly one-to-one ratio for the specimen cycled 30 times is consistent with the relative proportion of the two microconstituents (the globular-like structure and the matrix) seen in the photograph of the corresponding replica (Fig. 6, view 5). In a separate study on reversion in this alloy, evidence was presented which suggested that the reverted A shows up as white or light areas in contrast to the gray areas observed for martensite (23).

Optical examination at ×1000 showed that the specimens of Fig. 6 have microstructures resembling those commonly observed in steels. At this magnification a very fine lamellar pearlitic-like structure could be seen in the specimens subjected to one and three cycles. The specimen cycled 30 times showed a spheroidal structure—not unlike that of tempered martensite—which is consistent with the globular structure of the corresponding replica (Fig. 6, view 5). The specimen cycled 10 times had a structure intermediate between those of the specimens cycled 3 times and 30 times; its mixed rod-like and globular-like structure is clearly seen in the replica (Fig. 6, view 4). It is also interesting to note that during the early stages of cycling the proportion of lamellar regions increases with cycling (compare views 2 and 3 of Fig. 6). The lamellar structure suggests the decomposition of one phase simultaneously into two phases.

On the basis of the dilatometric curves, x-ray diffraction results and microstructural observations, we wish to propose that the following phenomena occur when the maraging steel is repeatedly cycled between ambient temperature and 815 C with a heating rate of 4.4 C/sec and air cooling. The sluggishness of the $M \rightarrow A$ transformation permits the M to enter well within the two-phase equilibrium region on heating. For purposes of our discussion this behavior may be illustrated with a phase diagram based on the Fe – Ni system, as shown in Fig. 7. At A_s, A (austenite) of some alloy-rich composition C_p is nucleated at precipitate sites. This is accompanied by the formation of F (ferrite) of composition C_f. The simultaneous cooperative growth of the two phases, A and F, must occur in preferential crystallographic directions in the M (martensite) to result in a lamellar formation. At some higher temperature, say T_H, this newly formed F, in turn, will start to decompose into F and A of different compositions C_H and C_A. At some temperature T_E, which may or may not be above T_H, the precipitates which act as nuclei for the formation of alloy-rich A are all consumed. The composition of the A which subsequently forms will depend on whether it forms by the growth of existing lamellae or by the development of new nuclei. If it forms by growth of existing lamellae, the composition should be between C_p and the phase boundary between A and $A + F$. If it forms by development of new nuclei, the composition should be represented by the phase boundary between A and $A + F$. At A_f (770 C) the alloy is all austenitic but quite heterogeneous. Of course, it is to be expected that, during heating, diffusion has been occurring in the transformation products in the direction that would lead to equilibrium. At 815 C the composition of the A ranges between C_x and C_y with the A regions of different composition arranged in the original lamellar growth form. On cooling, those lamellae which consist of A relatively poor in alloying elements (i.e., between C_x and C_z) will transform to M, while those represented between C_z and C_y will remain as retained A. On reheating, this process is repeated. More retained A forms, while the remaining M and/or F is further depleted in alloying elements with a corresponding shift of A_f and M_s to higher temperatures. In a subsequent cycle the average composition of the residual M or F, for example, would be represented by C_R.

Areas that are void of any lamellae, such as may be seen in view 2 of Fig. 6, may be explained by the presence of a direct transformation of M to A without any composition change. This may occur between T_M and A_f of Fig. 7. Other possibilities exist. With increased exposure at elevated temperatures, due to subsequent cycling, the tendency to decrease surface energy will result in spheroidization of the lamellae. That this has occurred may be seen by comparing the structure after 3, 10, and 30 cycles (Fig. 6, views 3, 4, and 5).

Although the dilatometric and electron microscopic observations strongly indicate the presence of large variations in composition within a sample, we could not ascertain quantitatively the extent of such variations because of the fineness of the microconstituents. With the use of magnetic techniques (24), we previously noted that exceptionally high but ill-defined Curie temperatures were obtained on cooling specimens after heating to within the temperature range of 550 to 700 C, even though no $A \rightarrow M$ reaction was observed (23). On the basis of the Curie-point data for the Fe – Co – Ni system (25) we had suggested (23) that the localized Ni + Co content may become as high as 60% when the 18% Ni maraging steel was heated to about 680 C (just above where reversion to A on precipitates is initiated on heating). It is interesting to note that by an extraction technique a 44% Ni content was observed in particles of A for a 9% Ni steel (26).

It might be suggested that in part, the stability of the A arises from transformation-induced deformation which might result from either the volume changes obtained on transformation or from differences in the expansion coefficients between the bcc and fcc phases and precipitates. However, on the basis of a study we are now conducting on the effect of plastic deformation on transformation, we believe that such an effect would be relatively minor.

HARDNESS MEASUREMENTS

It might be expected that both the presence of large amounts of retained A and the extensive depletion of alloying elements from the residual M would cause a significant degradation in the mechanical properties. With this in mind, we examined the effect of cycling on hardness. Specimens initially in the annealed martensitic condition were subjected to several different cycling treatments and then aged for 3 hr at the usual aging temperature of 482 C. Table 1 lists the hardness measurements obtained, first after cycling and then after aging. Measurements were made on a Leitz microhardness tester with a 200-g load. All values represent an average of at least four readings. A value of DPH 326 ± 5 was obtained for the annealed martensitic condition; on aging 3 hr at 482 C the hardness increased to 595 ± 10. Specimens that were aged before cycling were also examined. Prior aging had no significant effect on the differences of the final aged hardness; therefore, the results for these tests are omitted.

TABLE 1. Effect of Thermal Cycling on Hardness

Heating rate	No. cycles	DPH microhardness	
		After cycling	After aging
4.4 C/sec to 815 C	1	378	591
	2	385	483
	3	370	446
	4	369	426
	6	348	409
	10	334	395
	30	321	368
500 C/sec to 815 C	1	330	605
	6	327	585
	10	329	585
	19	377	544
0.075 C/sec to 815 C	1	345	591
	2	351	536
	4	332	490

A value of DPH 378 is obtained when the specimen is cooled immediately after heating at a rate of 4.4 C/sec to 815 C. On aging the hardness is increased to DPH 591. Thus, the first cycle did not detract from the aged hardness. With repeated cycling, however, the hardness values for both the cycled and the cycled-and-aged conditions drop substantially. The largest drop in the aged hardness between successive cycles occurs between the first and second cycle; the value drops from DPH 591 to 483. This drop is consistent with the observation that the greatest difference in the dilatometric curves (Fig. 4) occurs between these two cycles. The microstructure and x-ray results also show the greatest differences between the 1st and 3rd cycles (the 2nd cycle was not examined here). The hard-

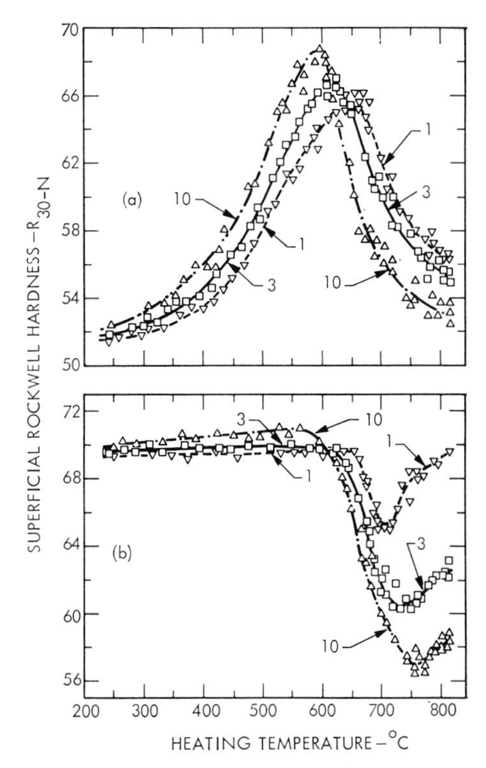

FIG. 8. Effect of the number of thermal cycles and the maximum temperature reached on the hardness values obtained at ambient temperature: (a) after cycling, and (b) after cycling and then aging for 3 hr at 482 C Specimens were 4½ in. long, gradient-heated during each cycle at rates ranging from 1 C/sec (at the cool end) to 4.4 C/sec (at the hot end) and air cooled.

ness value on aging after the 30th cycle drops to DPH 368.

The effects of cycling at the very high heating rates were next examined. Relatively high hardness values were found for specimens that were cycled to 815 C with heating rates of about 500 C/sec and then aged. The data obtained after 1, 6, 10, and 19 cycles are listed in Table 1. No significant change in hardness was experienced during the first 10 cycles when comparing either the cycled hardness with the initial hardness or the cycled-and-aged hardness with the standard aged hardness. This is consistent with the concept of a diffusionless $M \rightarrow A$ transformation at these high heating rates, and with the observation that the heating and cooling curves remain relatively unchanged during the first six cycles (Fig. 5) and change only slightly during the next four cycles. After 19 cycles the cycled hardness increases from about DPH 330 to 377. This increase in hardness suggests that some precipitation takes place at least during the last few cycles. The corresponding cycled-and-aged hardness is somewhat low, having dropped from DPH 585 after 10 cycles to 544 after 19 cycles, consistent with the development of retained A in the later cycles.

The influence of thermal cycling at very low heating rates was also examined. Several specimens were heated for one, two, and four cycles to 815 C at a rate of 0.075 C/sec. Table 1 shows the cycled and cycled-and-aged hardness values. Reduction of the cycling rate from 4.4 to 0.075 C/sec allowed more homogenization to take place as 815 C was approached during each cycle. Consequently, the loss in the as-aged hardness was generally less for the lower of these two cycling rates, at least for heating to 815 C. For example, the cycled-and-aged hardness values after four cycles with heating rates of 4.4 and 0.075 C/sec were DPH 426 and 490, respectively.

An attempt was next made to obtain information which would show how the maximum temperature reached during cycling affected the hardness. Several strips $\frac{3}{32}$ by $\frac{3}{8}$ by $4\frac{1}{2}$ in. were subjected to gradient heating; the heating rate at the hot end, which reached a maximum temperature of 815 C, was about 4.4 C/sec. Hardness values were taken on the Rockwell superficial 30-N scale, before and after aging following the 1st, 3rd, and 10th cycles. These values are plotted as a function of maximum temperature in Fig. 8. The temperatures were obtained by interpolation between readings taken from welded thermocouples spaced at $\frac{3}{4}$-in. intervals, and are thus somewhat approximate. The heating rate at the low temperature end (225 C) is estimated to be about 1 C/sec.

The increase in hardness to the peak values shown in Fig. 8a is attributed to precipitation (aging effects). The drop in hardness beyond the peak is attributed to overaging, reversion, and re-solution of precipitates in the reverted A. The hardness values at temperatures below the peak temperature are generally displaced to higher values with each successive cycle due to additional effects of aging. The reverse trend is true on the high temperature side of the peaks, where increased retained A and the presence of M further depleted of alloying elements are expected as a result of successive cycling.

With reference to the cycled-and-aged hardness, Fig. 8b, the hardness values obtained by cycling almost to 660 C are essentially due to the final 3-hr aging treatment at 482 C, although these are modified slightly by the aging effects from prior cycling. For the single cycle a sharp drop in the aged hardness is obtained above 660 C. This can be attributed to the presence of retained A and M depleted in alloying elements, both of which increase with an increase in temperature above 660 C. A minimum hardness is reached after cycling to a temperature of about 700 C. The subsequent increase in hardness must be due to a decrease in the amount of retained A and to an increase in homogenization as the temperature is raised above about 700 C. This temperature corresponds closely to the temperature where, after cooling individual specimens from different temperatures, a maximum amount of retained A is detected by x-ray diffraction (23). This temperature also corresponds closely to the end of the sharp contraction associated with the $M \rightarrow A$

reversion (curve 1 of Fig. 4A). Additional cycling causes the minimum hardness value to drop and shift to higher temperatures, paralleling the behavior depicted both in the dilatometric curves of Fig. 4 and in our x-ray results.

The results described above indicate that a large degradation in the potential final aged hardness may result from thermal cycling. The extent of such degradation depends on the heating rate, the maximum temperature reached, and the number of heating and cooling cycles used. This degradation may be especially serious in the heat-affected zone of weldments subjected to numerous welding passes. Although we did not examine the elimination of the microsegregation caused by cycling, we feel that because of the fineness of this segregation it would be eliminated by the normal solution annealing treatment at 815 C.

SUMMARY AND CONCLUSIONS

Dilatometry is a simple and useful method for determining the extent to which retained austenite and variations in composition may develop in a maraging steel specimen subjected to thermal cycling. From the combined results obtained by dilatometric, x-ray diffraction, magnetic, microscopic, and hardness studies, it is concluded that repeated thermal cycling may lead to extensive variations in composition within a specimen. An electron microprobe survey of cycled specimens suggests that such variations are contained within regions that are smaller than 1 μ. This is consistent with the spacing observed between austenite regions in the replicas of cycled specimens; these regions (lamellae or globules) are about 0.05 to 0.5 μ apart. The combination of retained austenite and depletion of alloying elements in martensite leads to a large degradation in the aged hardness.

The martensite-to-austenite reversion is believed to be initiated at precipitates rich in Ni, Mo, and Ti, elements that lower the M_s. It is proposed that on continuous heating, at a heating rate of 4.4 C/sec, reversion occurs by the decomposition of martensite simultaneously into austenite and ferrite, which are respectively enriched and depleted in the above elements. The growth of these two phases is crystallographically oriented and results in a lamellar, pearlitic-like product in which the ferrite, on further heating, reverts back to austenite of low alloy content. On subsequent cooling, this austenite transforms to martensite. The lamellar structure, which now consists of alloy-poor martensite and alloy-rich retained austenite, is still maintained. On additional cycling this process is repeated, resulting in further alloy depletion of the martensite and an increase in the amount of retained austenite. Due to the driving force of surface tension the austenite lamellae eventually spheroidize.

The amount of retained austenite, the extent of composition variations, and the degradation in aged hardness all increase with an increase in the number of cycles. At slow and intermediate heating rates, where the reactions are diffusion-controlled, the greatest changes take place during the first few cycles. At high heating rates where the reactions are predominantly diffusionless, very little change, if any, is noted during the first few cycles; significant changes are not observed until after some six cycles have occurred. It is suggested that the six cycles are required for the incubation period necessary to form stable-size precipitates that act as nuclei for the formation of alloy-rich austenite.

On thermally cycling a gradient-heated specimen to the same peak temperature, a minimum in the aged hardness is obtained which decreases in value and shifts to higher heating temperatures with an increase in the number of cycles. This minimum probably occurs at the temperature that results in a maximum amount of retained austenite.

ACKNOWLEDGMENTS

The author wishes to thank David G. O'Connor for his assistance with the dilatometric studies and Earle O. Snell, Jr., for his metallographic work. The instrumentation was made possible through the efforts of William U. Dent. The author is particularly grateful to Professor Oleg D. Sherby of Stanford University for his many stimulating discussions related to the preparation of this paper, and to Lee W. Roberts, Jr., for encouraging this study and reviewing the manuscript. The work was done under the auspices of the U. S. Atomic Energy Commission.

REFERENCES

1. R. F. Decker, J. T. Eash, and A. J. Goldman, 18% Nickel Maraging Steel, ASM Trans Quart, 55 (1962) 58.
2. G. E. Pellisier, The Physical Metallurgy and Properties of Maraging Steels, Defense Materials Information Center, Battelle Memorial Institute, Rept. 210 (October 26, 1964) 173.
3. A. M. Hall, Review of Maraging Steel Development, Cobalt, 24 (Sept. 1964) 138.
4. B. R. Banerjee and J. J. Hauser, Hardening Mechanisms and Delamination Studies of 18 Percent Nickel Marage Steels, Wright-Patterson Air Force Base, Tech. Rept. AFML-TR-66-166 (March 1965).
5. B. G. Reisdorf and A. J. Baker, The Kinetics and Mechanisms of the Strengthening of Maraging Steels, Wright-Patterson Air Force Base, Tech. Rept. AFML-TR-64-390 (Jan. 1965).
6. D. L. Corn, An Evaluation of Maraging Steel Sheet, J Metals, 16 (1964) 814.
7. D. T. Peters and C. R. Cupp, The Kinetics of Aging Reactions in 18 Percent Ni Maraging Steels, Trans AIME, 236 (1966) 1420.
8. S. Floreen and R. F. Decker, Heat Treatment of 18% Ni Maraging Steel, ASM Trans Quart, 55 (1962) 518.
9. B. G. Reisdorf, A. J. Birkle, and P. H. Salmon Cox, An Investigation of the Mechanical Properties and Microstruc-

tures of 18 Ni (250) Maraging Steel Weldments, Wright-Patterson Air Force Base, Tech. Rept. AFML-TR-65-364 (Nov. 1965).

10. W. A. Petersen, Weld Heat-Affected Zone of 18% Nickel Maraging Steel, Welding J, 43 (1964) 428-S.
11. F. D. Duffey and W. Sutar, Submerged-Arc Welding of 18% Nickel Maraging Steel, Welding J, 44 (1965) 251-S.
12. C. M. Adams, Jr., and R. E. Travis, Welding of 18% Ni – Co – Mo Maraging Alloys, Welding J, 43 (1964) 193-S.
13. D. A. Canonico, Gas Metal-Arc Welding of 18% Nickel Maraging Steel, Welding J, 43 (1964) 433-S.
14. A. J. Baker and P. R. Swann, The Hardening Mechanism in Maraging Steel, ASM Trans Quart, 57 (1964) 1008.
15. R. F. Decker, Maraging Steel: Structural Property Relations, Proc. Symp. on Relations Between the Structure and Mechanical Properties of Metals, Natl. Phys. Lab., Teddington, England, 1963, p. 647.
16. R. B. G. Yeo, The Effects of Some Alloying Elements on the Transformation in Iron-Nickel Alloys, J Iron Steel Inst, 202 (1964) 104.
17. A. P. Gulyaev and N. I. Kavehevskaya, Martensitic Transformation Alloys with Martensite Which Ages, translated from Russian in Metal Sci and Heat Treat (Nov.–Dec., 1964) 660.
18. P. Legendre, Some Properties of Maraging-Type Steels, Cobalt, 29 (1965) 171.
19. Alfred Goldberg and David G. O'Connor, Influence of Heating Rate on Transformations in an 18% Nickel Maraging Steel, Nature, 213 (1967) 170.
20. O. A. Boedtker and P. Duwez, A Delay Time in the Alpha to Gamma Transformation in Iron, California Institute of Technology Rept. AEC-AT(04-3)221 and Nonr-220(30) (March 1962).
21. B. G. Reisdorf and R. P. Wei, A Preliminary Investigation of Fracture Toughness and Microstructures of Maraging Steels, United States Steel Corp., Monroeville, Pa., Tech. Rept. Proj. No. 89.25-013(1) (June 14, 1963).
22. S. Shaperio and G. Krauss, Replication of Fine Structure in Martensite, Trans AIME, 236 (1966) 1371.
23. Alfred Goldberg, The Effect of Heating Temperature on Dilatometric Curves, Microstructure and Hardness in a Maraging Steel (in preparation).
24. A. Goldberg, W. U. Dent, and J. H. Miller, A Magnetic Device for Detecting Phase Changes at Elevated Temperatures, J Sci Inst, 44 (1967) 200.
25. R. M. Bozorth, Ferromagnetism, Van Nostrand (1951) 160–180.
26. C. Crussard, M. Krön, A. Constant, and J. Plateau, discussion The Characteristics of 9 Per Cent Nickel Low Carbon Steel, ASM Trans Quart, 55 (1962) 1021.

The Effects of Thermal Treatment on the Austenitic Grain Size and Mechanical Properties of 18 Pct Ni Maraging Steels

G. SAUL, J. A. ROBERSON, AND A. M. ADAIR

A technique has been developed for controlling the austenitic grain size of 18 pct Ni maraging steels by thermal treatment alone. This treatment has been applied to two different grades, 250 and 300, of maraging steel, and a large grain size, ASTM 2, was reduced to ASTM 7 in both cases. The process of grain size refinement requires thermal cycling from a temperature below M_f to a temperature considerably above the austenitizing temperature. The minimum austenitic grain size attainable depends on the prior strain in the material as well as the thermal treatment. While significant grain size refinement can be attained by one cycle to the proper temperature, the attainment of the minimum uniform grain size requires several cycles. The effects of austenitic grain size on tensile properties have been investigated both at room temperature and at elevated temperatures. The prior austenitic grain size has a small but measurable effect on the mechanical behavior of aged material at room temperature. The austenitic grain size has a significant effect on the ultimate tensile strength at 1600°F.

THE effects of thermal treatment on the structure and properties of maraging steel have been studied by many investigators. Goldberg[1] has shown that repeated thermal cycling between room temperature and 1500°F increases the amount of retained austenite and reduces the aged hardness of maraging steel because chemical segregation occurs during cycling. Goldberg and O'Connor[2] have shown that solute segregation may be suppressed by high heating and cooling rates. Pellissier[3] has reported that impurity segregation occurs in maraging steel at temperatures above about 2000°F. Tuffnell, Pasquine, and Olson[4] investigated the effects of various heat treatments on the fatigue behavior of maraging steel and found that a duplex annealing treatment produced significant improvements under some conditions. The martensitic $\alpha \rightarrow \gamma$ transition during heating has been reported by Hall[5] to be a reversible shearing process. This observation seems to be in agreement with that of Goldberg and O'Connor. The α to γ transition has been discussed in greater detail by Hall, Simon, and Moon,[6] and they report that grain size refinement in maraging steel cannot be accomplished without plastic strain to cause recrystallization of the austenite. These and other investigators have shown that the reactions which occur during heating of maraging steel are complex and variable. The reactions which may occur depend on both the heating rate and the prior history of the material in question.

The present work was concerned with the effects of thermal treatment on the prior austenitic grain size and mechanical properties of two grades, 250 and 300, of 18 pct Ni maraging steels. Two different methods were used to vary the prior austenitic grain size. One method was conventional thermomechanical processing, which effectively resulted in a wide range of grain sizes. The second method involved a refinement of relatively large prior austenite grains by thermal treatment only, a method which has previously been reported to be ineffective. This latter method is considered to be particularly desirable, since a high temperature homogenization treatment used in previous experiments[7] had caused excessive grain growth in the 250 grade steel.

G. SAUL and A. M. ADAIR are Metallurgists, Air Force Materials Laboratory, Wright Patterson Air Force Base, Ohio. J. A. ROBERSON is Metallurgist, Aerospace Research Laboratories, Wright Patterson Air Force Base.

Manuscript submitted June 11, 1969.

EXPERIMENTAL PROCEDURES

250 Grade Maraging Steel.* The 250 grade tensile specimens were machined from cross rolled stock, and then they were given thermal treatments prior to testing. All heat treating was done in evacuated quartz capsules. After the specimens had been subjected to a primary treatment selected to promote grain growth, they were cooled to room temperature before further processing. Some of them were then thermally cycled, and all of them were subjected to the standard solution annealing treatment, 1500°F-1 hr. All specimens received the standard aging treatment, 900°F-3 hr. All tensile tests were conducted in air at room temperature at a strain rate of 0.02 per min. The gage section of the specimens was 0.100 in. thick by 0.180 in. wide and 1.00 in. long. Smaller sample coupons were used for determining the temperature range and number of cycles required for optimizing grain size refinement. All heating was done in air for various times and various cooling rates were used.

*Composition in wt percent: 18.1 Ni, 8.6 Co, 5.3 Mo, 0.4 Ti, 0.07 Mn, 0.05 Cr, 0.05 Si, 0.002 B, 0.1 Cu, 0.05 Al, 0.016 C, 0.002 Zr, Bal Fe.

300 Grade Maraging Steel.* This steel was received

*Composition in wt percent: 19.94 Ni, 9.53 Co, 4.93 Mo, 0.69 Ti, 0.11 Al, 0.01 C, 0.05 Si, 0.02 Mn, 0.005 S, 0.005 P.

as $3\frac{1}{4}$-in.-round bar stock. The grain size was about ASTM 5. It was subsequently hot extruded and cold swaged to produce suitable specimen stock. Cylindrical tensile specimens were machined with a gage length of 2.00 in. and a gage section diam of 0.252 in. These specimens were given various thermal treatments to produce a range of grain sizes and they were tested in air at 1600°F and various strain rates. Other specimens were annealed with predetermined thermal gradients to investigate recrystallization behavior. The pertinent details of the methods of thermal grain refinement are discussed in the next section.

RESULTS AND DISCUSSION

250 Grade Maraging Steel. Most of the grain size variation was accomplished by heating in a temperature range above 1700°F. The temperatures used are indicated on Figs. 1 and 2; the samples were held at the indicated temperature for 1 hr, except that samples 2000-12 were heated for 12 hr. The grain sizes of sample TC1, ASTM 6.5-7, and TC4, ASTM 2, were obtained by thermal cycling samples which had been held at 2000°F for 12 hr. The samples were cycled between 1700°F and room temperature five times and one time, respectively. The grain size before thermal cycling was ASTM 1. During subsequent experiments it was found that the grain size refinement temperature of 1700°F was somewhat critical. Some benefit could be obtained by cycling between 1675°F and room temperature, but 1700°F was found to be the optimum temperature. Grain growth begins to become a problem above 1700°F.

A brief examination of the texture of various specimens by X-ray diffraction did not indicate that any significant changes occurred during the 1 hr grain growth annealing treatments. Some changes did occur during the 2000°F-12 hr treatment, and the texture produced by this treatment was apparently unaffected by thermal cycling. A tendency for 110 planes to be aligned parallel to the rolling plane was found in each specimen examined. It seems improbable that the minor variations which were detected would seriously affect the yield strength or ultimate tensile strength of this steel.

The effects of prior austenitic grain size on the yield stress and ultimate tensile strength of the 250 grade steel are shown in Figs. 1 and 2. It may be seen that prior austenitic grain size has a small but measurable effect on these properties, especially at small grain sizes. These results are similar to, but somewhat higher than, those obtained by Floreen and Decker[8] at the same temperatures. A direct comparison may not be appropriate since Floreen and Decker did not report grain size as a function of annealing temperature. The work of Tuffnell *et al.*, also shows that the ultimate tensile strength of 18 pct Ni maraging steel is not strongly sensitive to either annealing temperature or grain size. The reason for this weak sensitivity is probably that the barriers to slip are much more closely spaced than the grain boundaries. On cooling, each austenite grain transforms into a number of martensite plates or blocks. The thickness of the blocks is much less than the grain diameter, although the length of the first blocks to form is roughly comparable to the prior austenite grain diameter. These blocks are bounded by extremely dense dislocation tangles[7] which may serve as barriers to dislocation motion.

While a slight increase in tensile properties accompanied the grain refinement of specimens TC1 and TC4, their values were less than those observed for similar grain sizes produced in a more conventional manner. This difference may be due to changes in texture accompanying the 12 hr-2000°F anneal prior to thermal cycling, although detailed pole figure determinations were not attempted. The lower properties of TC1 and TC4 may also be due to differences in grain boundary segregation between the two methods of grain size control.

The processes which occur during heating of maraging steel are much more numerous and complex than those which occur on cooling. The results of a typical differential thermal analysis trace are shown in Fig. 3. In this experiment, a sample and a Pt standard with similar heat capacity are placed in identical containers in a heating chamber. The chamber is heated contin-

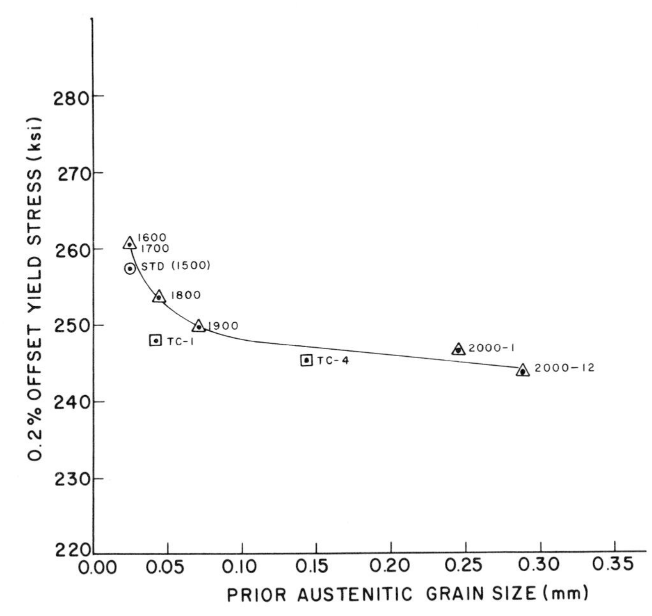

Fig. 1—The effect of prior austenitic grain size on the yield stress of 250 grade maraging steel. The austenitizing times were 1 hr except as indicated.

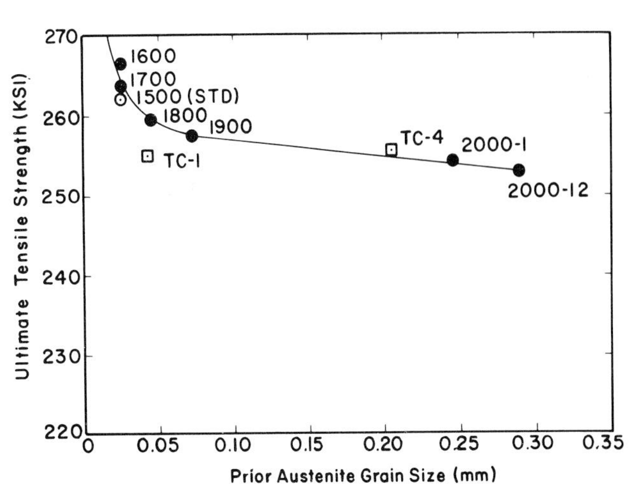

Fig. 2—The effect of prior austenitic grain size on the ultimate tensile strength of 250 grade maraging steel.

ually while the temperatures of both the standard and the sample are monitored by thermocouples. Their temperature difference is plotted as a function of the specimen temperature. Excursions in ΔT ideally represent the occurrence of exothermic or endothermic processes in the specimen. Complexities in the analysis of such experiments arise because of the difficulty of preparing a standard which has the same heat capacity, surface area, and thermal conductivity as the specimen. However, a cursory examination of Fig. 3 will reveal the features listed in Table I.

The precise location of the points of inflection of these curves depends somewhat on the rate of temperature change. The above temperature ranges are, therefore, only approximate. Heating rates used in the present work ranged from about 6° to 12°F per min with a few runs as high as 20°F per min. Peters and Cupp[9] have reported that the aging reaction in 18 pct Ni maraging steel proceeds in three stages: recovery of martensite, two precipitation reactions, and formation of nickel rich austenite. They report all of these processes as occurring at 1050°F or below. Peters[10] has subsequently reported that two different molybdenum-rich precipitates form in 18 pct Ni maraging steel. One of the precipitates nucleates in the matrix of the martensite and the other nucleates on dislocations. He reports that the matrix precipitate is unstable with respect to the dislocation nucleated precipitate. The formation of the nickel-rich austenite detected by Peters and Cupp was time dependent indicating that it was a diffusional process. Kessler and Pitsch[11] have combined metallographic analysis with DTA in a study of the reversion of Fe-Ni alloys, and they have established that the α to γ transformation proceeds by a shear process followed by a diffusional process. Goldberg and O'Connor[2] have shown that some of the reactions can be completely suppressed at high heating rates. Goldberg[1] has shown that the nature of the reactions which do occur can be influenced by the heating rate.

In the present work the heating rates were probably great enough to suppress some of the diffusional processes. One of the DTA specimens was heated to 1265°F and allowed to cool to 400°F. The DTA plot showed no evidence of the martensitic reaction on cooling. Since the complete transformation to austenite requires the occurrence of a diffusional process,[11] it seems reasonable to believe that the M_S temperature for our sample was lowered. When this sample was reheated from 400°F the reactions in zones H_1 and H_2 were absent, but the higher temperature reactions were evident. The reactions occurring in zone H_1 or before are probably those reported by Peters and Cupp. The process indicated in zone H_4 of Fig. 3 is believed to be the very rapid resolution of precipitated compounds. The process indicated by zone H_5 is believed to be recrystallization which occurs as sessile dislocations and dislocation tangles are annealed out of the austenite.

The martensitic α to γ transformation has not been studied in as much detail as the more familiar γ to α transformation. However, several studies of martensite reversion in Fe ~ 33 pct Ni alloys have been reported recently.[11-16] The observations reported may not be directly comparable to the processes which occur in maraging steel because the nickel-rich alloys have a twinned rather than massive bcc martensite and because of the effects of the nickel content on the transformation temperature. Nonetheless, it is reported that martensite reversion in these alloys produces an fcc phase which is heavily distorted and may be chemically inhomogeneous.[11-13] At high heating rates, the reversion reaction is apparently very similar to the reaction which occurs during cooling. Our work demonstrates that maraging steel may be recrystallized to a finer grain size by heating it to some suitable temperature above the transformation temperature. It seems reasonable to believe that this occurs as a result of the shear strain produced by the martensitic α to γ transformation. This process may be repeated until a limiting minimum grain size is obtained.

The further reduction in austenite grain size obtained by thermal cycling vs a single heating is believed to be related to the reduction in martensite platelet size with a reduction in prior austenite grain size. The strain energy available for forming nuclei during recrystallization, resulting from the sessile dislocations and dislocation tangles remaining after transformation to austenite, decreases as the martensite platelet size becomes finer, and apparently becomes ineffective at a prior austenite grain size of ASTM 7. This explanation is also considered to be applicable to the 300 grade steel.

300 Grade Maraging Steel. The optimum temperature for producing austenite grain size refinement in this steel was determined by annealing coarse grained samples in a predetermined temperature gradient. A grain coarsened $\frac{1}{2}$ in round, 7 in. long maraging steel specimen was heated so that a uniform thermal grad-

Table I. Interpretation of DTA Results

Zone	Span, °F	Reaction	Probable Cause
H_1	920° to 1080°	exothermic	formation of precipitate
H_2	1080° to 1180°	endothermic	martensitic α to γ transformation
H_3	1300° to 1370°	endothermic	diffusion controlled transformation
H_4	1370° to 1480°	exothermic	resolution of precipitate
H_5	1720°	exothermic	release of energy during recrystallization
C_1	455° to 380°	exothermic	martensitic γ to α transformation

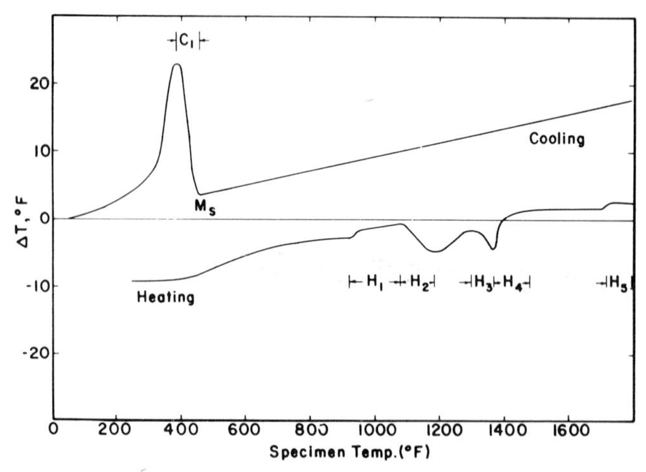

Fig. 3—Composite of differential thermal analysis traces for 250 grade maraging steel. Heating rates from 6° to 12° F per min.

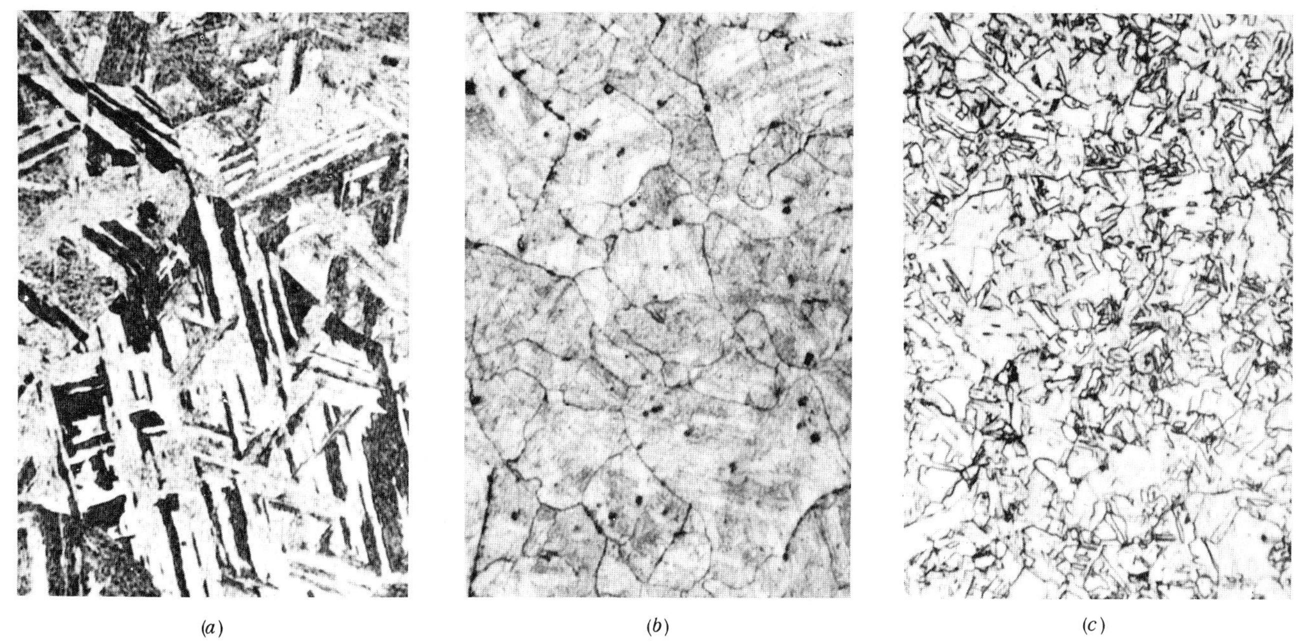
(a) (b) (c)

Fig. 4—Microstructures developed in 300 grade steel during thermal cycling. Magnification 140 times. (a) Specimen held at 2100°F for 1 hr and water quenched, grain size 1.5. (b) Same specimen shown in (a) after heating to 1880°F for 10 min and quenching, grain size 4. (c) Same specimen shown in (b) after three more cycles between room temperature and 1880°F, grain size 7.

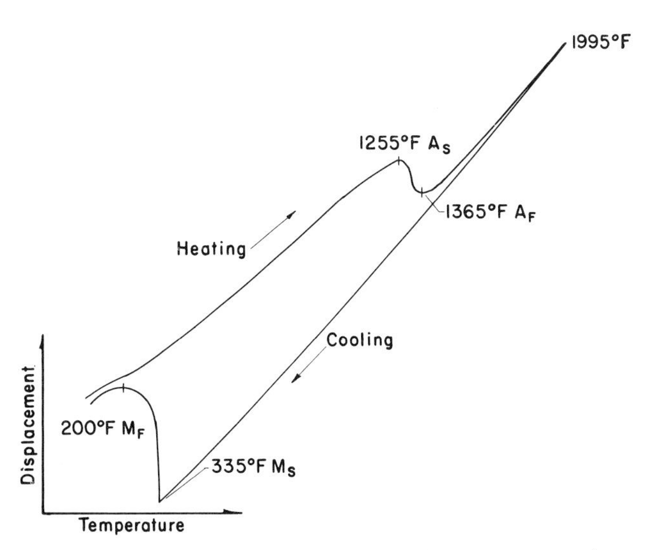

Fig. 5—Dilatometer trace of 300 grade maraging steel.

ient existed along its length from 1400° to 2338°F. The rod was held in the temperature gradient for 1 hr and water quenched. The rod exhibited coarse grains at either end and finer grains in the center. The finest grains were evident at a dimension representing 1880°F and gradually enlarged for about 1 in. on either side of this mark. The temperature range thus indicated was from 1746° to 1980°F. The cooler end still had the original grain size while the hot end had experienced grain growth. The optimum refining temperature of 1880°F was used in further experiments with small coupons involving heating and cooling rates and multiple cycling. Various heating and cooling rates all produced grain refinement. To simulate a large section, a coupon was heated in a large furnace from ambient temperature to 1880°F which took several hours. After holding at temperature for 10 min, the power was turned off and the specimen was furnace cooled which took overnight. Grain refinement was again evident.

Multiple heating cycles to the critical temperature are required to produce the smallest attainable grain size. A 1 in. round of coarse grained maraging steel was heated to 1880°F, held for a short time, and cooled. Between cycles a disk was taken from the round for microexamination. As many as six cycles and examinations were performed on a single bar. Heating rates again had minor influence on the structure. The grain size produced by one thermal refinement cycle, although finer than the original structure, is not as fine as can be achieved. Three cycles are required to produce the minimum grain size obtainable. A coarse grain specimen, grain size ASTM −1 to 1 was cycled, and the grain size was decreased to ASTM 4, ASTM $5\frac{1}{2}$, and finally ASTM 7 after three cycles. Fig. 4 is a typical illustration. Additional cycling did not further decrease the grain size. Between cycles, the specimen must be cooled to below the M_f temperature of about 200°F or refinement will not occur. Experiments were performed in which a rod shaped specimen was cooled in a temperature gradient furnace. The specimen was heated to 2200°F for an hour to produce coarse grains. The specimen was then cycled three times between the gradient furnace, 100°F at the cold and 400°F at the warm end, and a furnace held at 1880°F. The transfers between furnaces were made quickly. The portion of the bar which had been cooled to the M_f temperature or lower exhibited uniform grain refinement. The portion which had been cooled into the temperature range between M_s and M_f exhibited only partial refinement. The portion of the bar which had not been cooled to the M_s temperature still exhibited the coarse structure produced by the 2200°F heating.

The transformation temperatures for this steel

were determined from the dilatometer traces shown in Fig. 5. It may be seen that the M_s and M_f temperatures are appreciably lower for this steel than for the 250 steel. However, the A_s and A_f temperatures are roughly comparable. The dilatometer trace indicates that a net contraction occurred in the longitudinal direction during the heating cycle as has been reported by Goldberg[17] for a similar 300 grade steel.

The effects of austenitic grain size and strain rate on the ultimate tensile strength of this steel at 1600°F are shown in Fig. 6. At each strain rate investigated, an increase in grain size caused an increase in ultimate tensile strength. It was not possible to measure variations in the yield stress because of the lack of a suitable extensometer. It will be noted that the lines in Fig. 6 have a positive rather than a negative slope, which indicates that the grain boundaries do not serve as dislocation barriers. These results indicate that grain boundary motion, presumably by sliding, is an important factor in the deformation of this steel at high temperature and low strain rates. At higher strain rates the effect of grain diameter is less significant. Grain boundary sliding is usually accompanied by some shearing within the grains. The higher stress levels required to sustain high strain rates causes grain distortion to become a more significant deformation process thereby reducing the importance of grain boundary effects. Also, less time is available at the higher deformation rates for thermally activated processes to make a contribution to the deformation process.

The possible contribution of other structural variables, such as a change in annealing texture during grain growth, cannot be determined from the results shown in Fig. 6. However, if any such effects exist, they must be relatively unimportant. A change in grain diameter from about 0.03 to -.7 mm has a very large effect, 24,000 psi or 150 pct, on the ultimate tensile strength at a strain rate of 8.3×10^{-4} per sec whereas the same change in grain diameter causes an increase of only 4000 psi or 8.4 pct at a strain rate of 8.0 per sec. Variations in annealing texture would not be expected to produce such a large change in strain rate sensitivity. However, this increase in grain diameter would cause a decrease of about 95 pct in total grain boundary area and would thus produce an increase in flow stress, providing grain boundary sliding was the mode of deformation.

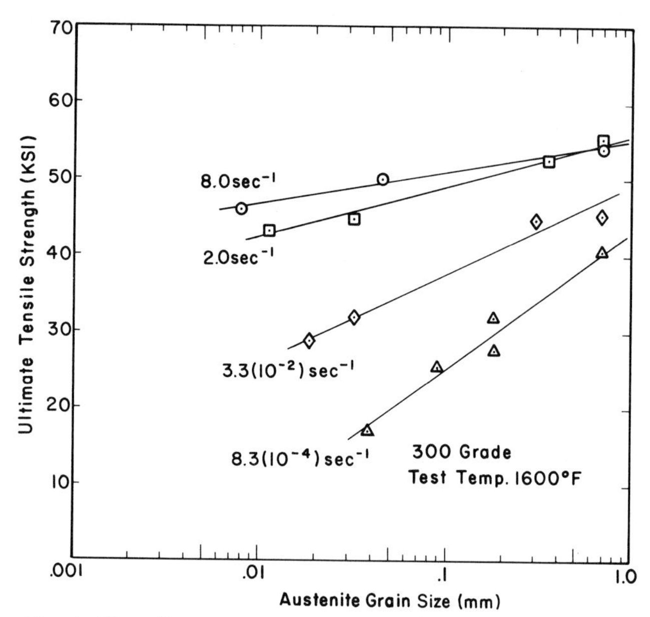

Fig. 6—The effects of strain rate and austenitic grain size on the ultimate tensile strength of 300 grade maraging steel.

SUMMARY

The austenitic grain size of 18 pct Ni maraging steel can be reduced significantly by thermal treatment alone. The optimum temperature for grain refinement is very sensitive to alloy composition and must be determined for each alloy. The room temperature tensile properties of this steel are not strongly affected by prior austenitic grain size; however, small but measurable improvements in tensile properties can be attained when coarse grain material is refined by thermal cycling. The ultimate tensile strength of maraging steel at elevated temperature is strongly affected by both grain size and strain rate. The nature of the effect indicates that grain boundary shear is an important deformation mode.

REFERENCES

1. A. Goldberg: *ASM Trans. Quart*, 1968, vol. 55, p. 26.
2. A. Goldberg and D. G. O'Connor: *Nature*, 1967, vol. 213, p. 170.
3. G. E. Pellissier: *Eng. Fracture Mechanics*, 1968, vol. 1, p. 55.
4. G. W. Tuffnell, D. L. Pasquine, and J. H. Olson: *ASM Trans. Quart.*, 1966, vol. 59, p. 769.
5. A. M. Hall: *Metals Handbook*, vol. 2, 8th ed., p. 255, Am. Soc. Metals, Novelty, Ohio.
6. A. M. Hall, R. C. Simon, and D. P. Moon: *DMIC Processes and Properties Handbook*, p. I-1, Battelle Memorial Institute, Columbus, Ohio, 1968.
7. J. A. Roberson and A. M. Adair: *Trans. TMS-AIME*, 1969, vol. 245, p. 1937.
8. S. Floreen and R. F. Decker: *ASM Trans. Quart.*, 1962, vol. 55, p. 518.
9. D. T. Peters and C. R. Cupp: *Trans. TMS-AIME*, 1966, vol. 236, p. 1420.
10. D. T Peters: *Trans. TMS-AIME*, 1967, vol. 239, p. 1981.
11. H. Kessler and W. Pitsch: *Acta Met.*, 1967, vol. 15, p. 401.
12. S. Shapiro and G. Krauss: *Trans. TMS-AIME*, 1967, vol. 239, p. 1408.
13. S. Jana and C. M. Wayman: *Trans. TMS-AIME*, 1967, vol. 239, p. 1187.
14. R. W. Rohde, J. R. Holland, and R. A. Graham: *Trans. TMS-AIME*, 1968, vol. 242, p. 2017.
15. W. Pitsch: *Trans. TMS-AIME*, 1968, vol. 242, p. 2019.
16. S. Shapiro and G. Krauss: *Trans. TMS-AIME*, 1968, vol. 242, p. 2021.
17. A. Goldberg: *ASM Trans. Quart.*, 1969, vol. 62, p. 219.

Improved Fatigue Resistance of 18Ni (350) Maraging Steel Through Thermomechanical Treatments

H. J. RACK AND DAVID KALISH

The influence of thermomechanical treatments, ausforming and marforming, on the fatigue resistance of 18 Ni (350) maraging steel has been examined. Although the low cycle fatigue resistance of this material is essentially unaffected by these treatments, an increase of 30 pct in the low-stress, high-cycle fatigue resistance can be achieved. This increase can be explained by considering the influence of processing on the resulting precipitate and dislocation substructures. Differences in texture, residual stress level and inclusion morphology have no effect on the improved fatigue resistance.

THE low-stress, high cycle fatigue resistance of 18 Ni maraging steels compares quite favorably with that of quench-and-temper steels at equivalent tensile strengths from 200 to 300 ksi (1379 to 2069 MN/m^2).[1] However, a further increase in the monotonic strength of 18 Ni maraging steel, up to 350 ksi (2410 MN/m^2), does not result in a concomitant improvement in fatigue strength.[2] Landgraf[3] showed that a number of high strength steels, including maraging steels, undergo appreciable cyclic softening. Such metallurgical structures, which are unstable under cyclic straining, exhibit early crack initiation and decreased fatigue resistance, particularly in high-cycle fatigue. The microstructural basis for cyclic softening in maraging steels is probably associated with the relative ease with which the partially coherent precipitate particles are sheared by dislocations, thereby creating soft regions in which strain concentrates and crack initiation occurs.[4]

Thermomechanical treatments have received considerable attention in carbon steels, primarily because of the increases in monotonic strength that can be attained; benefits to fatigue resistance and the effects on other properties unfolding as a consequence of the initial interest in strengthening.[5] Since early studies on maraging steels have shown that comparable thermomechanical treatments can increase the monotonic strength by only 5 to 10 pct, the fatigue behavior of thermomechanically treated maraging steels has received little attention.

The objective of this study was therefore to determine if thermomechanical treatments could improve the fatigue resistance of a steel, not hardened by carbon, where the precipitation processes have some analogy to those in age-hardening aluminum alloys.[6-8]

EXPERIMENTAL PROCEDURE

The maraging steel used was double-consumable vacuum remelted, of composition (wt pct) 18.51 Ni, 11.89 Co, 4.67 Mo, 1.53 Ti, 0.09 Al, 0.008 C, bal. Fe, in the form of $\frac{5}{8}$ in. plate. This material was the basis of previous studies of maraging steel.[8-11]

H. J. RACK is a Member of Technical Staff, Sandia Laboratories, Albuquerque, New Mexico. 87115. DAVID KALISH is a Member of Technical Staff, Bell Telephone Laboratories, Norcross, Georgia. 30371.

Manuscript submitted July 23, 1973.

Two types of thermomechanical treatments were explored. One involved rolling the unaged martensite at room temperature (marforming or strain aging). The other involved hot rolling the metastable austenite prior to transforming to martensite (ausforming). The ausforming temperature was selected by preliminary experiments involving austenitizing, hot rolling (70 pct reduction in thickness) above the M_S temperature[8] and directly up-quenching to 950°F (510°C) for 30 min. Optical microscopy revealed that no stress-induced martensite was formed at 750°F (399°C). An ausforming temperature of 750°F (399°C) was thus selected and agrees with that used by Bush.[12] All rolling blanks were solution treated at 1500°F (816°C) for one hour and either air-cooled to room temperature for marforming or quenched directly to 750°F (399°C) for ausforming. Two degrees of deformation were employed, 25 and 70 pct reduction in thickness.

Fatigue and tensile blanks were rough machined following the deformation and then aged for 3 h at 950°F (510°C). Tension-tension fatigue specimens, Fig. 1, and standard one-inch gauge length flat tensile specimens were final machined from the aged blanks. Tensile tests were performed on a 150,000 lb. Baldwin testing machine at an initial strain rate of 6×10^{-4} s^{-1}. The fatigue tests were carried out on an MTS fatigue machine at 30 cps in laboratory air (75°F (24°C), 60 pct relative humidity). The length dimen-

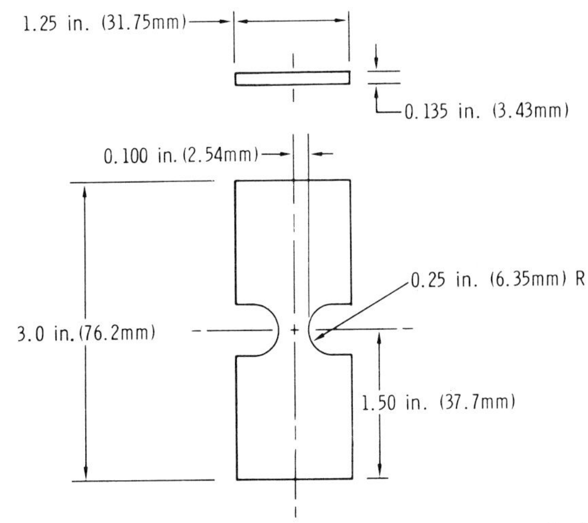

Fig. 1—Tension-tension fatigue specimens, K_t = 1.2. Surface finish 10 rms.

sions of the tensile and fatigue specimens were parallel to the rolling direction.

Samples normal to the rolling direction were prepared for transmission electron microscopy by wafering from the aged blanks, hand grinding to 0.001-in. and thinning in a jet polishing apparatus using a chrome-acetic acid electrolyte.[13] After perforation and washing in dilute acetic acid and ethyl alcohol, the foils were dried and examined in a JEOLCO 200 electron microscope operated at 200 KV. A detailed description of the crystallographic procedures used to identify the precipitate phases present is given elsewhere.[14]

Selected fracture surfaces were also examined with a JEOLCO scanning electron microscope; the observation angle generally being within 15 deg of the fracture surface.

The deformation texture was determined from the {110} and {200} X-ray pole figures obtained by the Schulz technique utilizing a Siemens texture goniometer with Mn-filtered Fe-radiation. At least five mils were removed from the rolled surface by electropolishing prior to the texture measurements. Both defocusing and background corrections were performed before automatic plotting of the pole figures. The former utilized a random aluminum sample while the latter involved background measurements taken on opposite sides of the principal X-ray peak. Finally, the level of residual stress was determined, using V-filtered Cr-radiation, by measuring the change in (211) interplanar spacing parallel and inclined to the surface. The change in d_{211} was related to the variation in residual stress level, σ_ϕ by:

$$\sigma_\phi = P(2\theta - 2\phi_\Psi) \quad [1]$$

where

$$P = \frac{E \cot_\perp \theta}{2 \sin^2 \psi (1 + \nu)} \quad [2]$$

and E and ν were assumed to be 27×10^6 psi (186.2×10^3 MN/m^2) and 0.3, respectively. Variations of Ψ between 45 and 60 pct did not yield noticeable differences in the residual stress levels. Therefore the stress levels are reported without respect to the observation angle. Furthermore the reported residual stress levels should be considered as comparative rather than absolute since the calculated values of the stress constant, P, were not verified experimentally.[15] However the data was corrected for variations in geometrical absorption and Lorentz-polarization factor prior to calculating the residual stress level.[16]

EXPERIMENTAL RESULTS

Mechanical Behavior

The influence of thermomechanical treatments on the tensile properties of 18 Ni (350) maraging steel are summarized in Table I. Some investigators[17,18] have indicated that cold rolling prior to aging can result in a substantial increase in the yield strength of maraging steel. However, the present results confirm the conclusions of Kula and Hickey[19] that the tensile property changes in maraging steels due to ausforming or cold rolling, are small in comparison with those observed in quench and temper steels. The apparent loss of ductility between the undeformed and deformed conditions may be due, in part, to the former being determined on 0.25-in. round tensile specimens, whereas the latter used the flat tensile samples described above.

Table I. Influence of Thermomechanical Treatment of Tensile Properties of 18Ni (350) Maraging Steel*

Condition†	0.2 Pct YS ksi (MN/m²)	UTS ksi (MN/m²)	Elong pct	Red. Area pct
1500°F (816°C)-1 h, AC (from Ref. 7)	331.2 (2283.5)	345.2 (2380.1)	8.9	37.2
1500°F (816°C)-1 h, AC + 25 pct red. thick, 75°F (24°C)†	336.0 (2316.6)	343.9 (2371.1)	5.8	21.5
1500°F (816°C)-1h, AC + 70 pct red. thick, 75°F (24°C)	355.3 (2449.7)	360.3 (2484.2)	5.6	25.0
1500°F (816°C)-1 h, HQ** 25 pct red. thick, 750°F (399°C)-AC	339.2 (2338.7)	347.2 (2393.9)	6.7	28.0
1500°F (816°C)-1 h, HQ + 70 pct red. thick, 750°F (399°C), AC	350.2 (2414.5)	358.2 (2469.7)	6.2	26.7

*Strain rate = 6×10^{-4}/s.
†All conditions subsequently aged for 3 h at 950°F (510°C).
**HQ–hot quench to ausforming temperature: 750°F (399°C).

Results of the fatigue tests are presented in Fig. 2. The thermomechanical treatments do not affect the low-cycle, high-stress fatigue resistance of 18 Ni (350) maraging steel. This is not surprising since this fatigue regime is dominated by Stage II crack propagation and is generally thought to be insensitive to microstructural details.[20,21]

The high-cycle, low-stress fatigue resistance, Fig. 2, is clearly dependent on thermomechanical history. At an expected lifetime of 10^7 cycles, a 30 pct improvement in the alternating stress to cause failure is achieved by appropriate processing, Table II. The increase in fatigue resistance is not proportional to an increase in monotonic strength and is not dependent on the type of thermomechanical treatment. The controlling variable appears to be the degree of deformation. However, the scatter in the fatigue strength may be a function of the deformation temperature, Fig. 2. Ausforming at 750°F (399°C), results in less scatter than marforming at room temperature. Consequently, the more desirable thermomechanical treatment for improving the high-cycle fatigue resistance, from the standpoints of minimizing the scatter and maximizing the alternating stress for failure, involves large ausforming deformations.

Microstructures

The structure associated with solution treating and aging for 3 h at 950°F (510°C) (no deformation) is described in detail elsewhere.[8] It consists of a low dislocation density ferrite matrix and two precipitates: rod-like Ni_3(Mo, Ti) and spherical σ-FeTi, Fig. 3; the Ni_3(Mo, Ti) being the predominant precipitate. Reverted austensite has also begun to form at prior austenite boundaries (Point G).

Two distinct changes in microstructure were produced by the thermomechanical treatments, Figs. 4 thru 7. First, the precipitate structure is coarser with only the rod-like $Ni_3(Mo, Ti)$ being observed. Baker and Swann[22] and Shimizur and Okamoto[23] have shown that $Ni_3(Mo, Ti)$ preferentially nucleates at dislocations, while σ-FeTi homogeneously nucleates in the matrix.[8] Deformation prior to aging is expected to increase the dislocation density and should, thereby, enhance $Ni_3(Mo, Ti)$ precipitation while suppressing σ-FeTi formation. Although the precipitates produced by deforming and aging are larger than those formed during simple aging, the change in interparticle spacing is small. Spitzig[24] showed that a 20Å change in interparticle spacing can account for the 20 ksi (138 MN/m^2) change in yield strength listed in Table I. The former is well within the error in measurements normally associated with particle spacings derived from transmission electron microscopy.

The second microstructural change observed involved the production of a stable dislocation cell structure. The development of a dislocation cell structure was primarily dependent upon the amount of deformation. Both marforming and ausforming 25 pct and aging resulted in an increased dislocation density, compared to the undeformed material, and a dislocation cell structure. Larger deformations by marforming or ausforming yields even finer cell structrues. After ausforming, the cells tend to be slightly larger and more nearly equiaxed than after marforming.

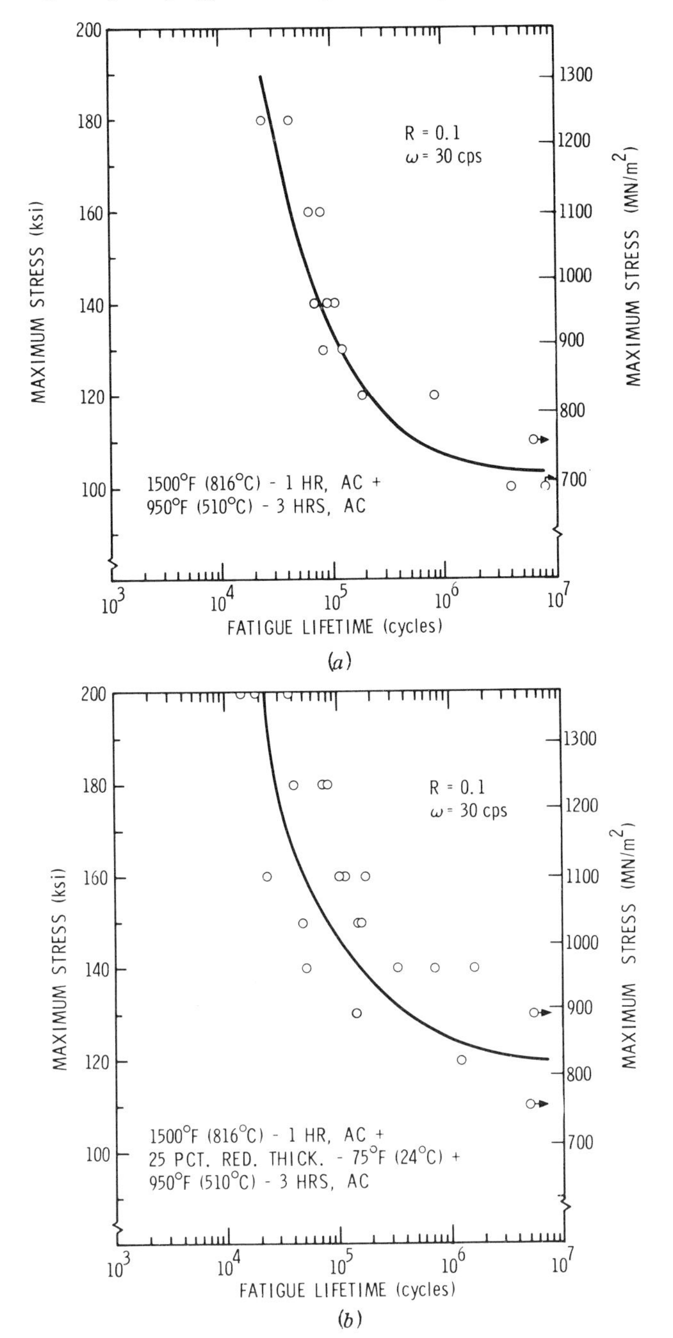

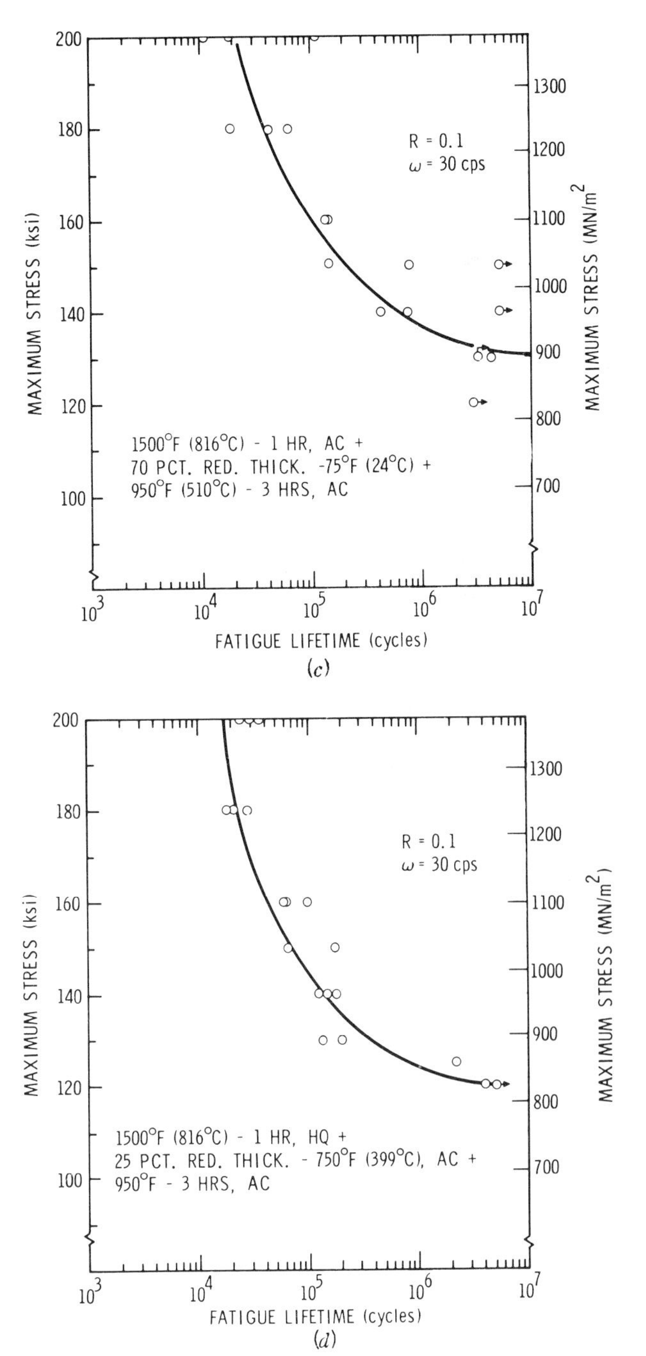

Fig. 2—Fatigue resistance of 18Ni (350) maraging steel following various thermomechanical treatments. (*a*) 1500°F (816°C)—1 h, AC+950°F (510°C)—3 h, AC; (*b*) 1500°F (816°C)—1 h, AC+25 pct Red. Thick. at 75°F (24°C)+950°F (510°C)—3 h, AC; (*c*) 1500°F (816°C)—1 h, AC+70 pct Red. Thick. at 75°F (24°C)+950°F (510°C)—3 h, AC; (*d*) 1500°F (816°C)—1 h, HQ-25 pct Red. Thick, at 750°F (399°C), AC+950°F (510°C)—3 h, AC; (*e*) 1500°F (816°C)—1 h, HQ-70 pct Red. Thick. at 750°F (399°C), AC+950°F (510°C)—3 h, AC.

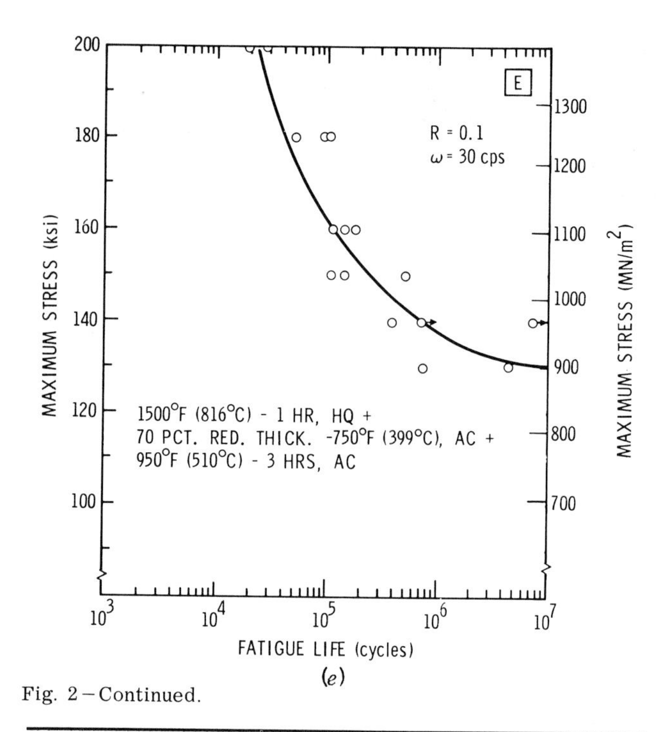

Fig. 2—Continued.

Table II. Influence of Thermomechanical Processing of the High Cycle Fatigue Resistance of 18Ni (350) Maraging Steel

Condition	Maximum Stress for 10^7 Cycle Lifetime (ksi)	(MN/m²)
Undeformed*	102	703.3
Deformed 25 pct at 75°F (24°C)	120	827.4
Deformed 70 pct at 75°F (24°C)	130	896.3
Deformed 25 pct at 750°F (399°C)	120	827.4
Deformed 70 pct at 750°F (399°C)	130	896.3

*All conditions subsequently aged for 3 h at 950°F (510°C).

Residual Stress

The X-ray residual-stress results are given in Table III; no influence of thermomechanical treatment on the residual stress level was observed. Thus changes in high-cycle fatigue resistance due to these treatments cannot be due to changes in residual stress level.

Fractography

The various thermomechanical treatments did not have a marked influence on the morphology of the fatigue fracture surface. Fatigue crack initiation was generally found to arise at an exterior corner. However, no consistent correlation was observed between the fatigue crack initiation site and cracked inclusions. In some instances, when an inclusion lay within a short distance of a sharp exterior corner, the crack initiation site was then displaced from the corner to the inclusion. Non-dispersive X-ray analysis of the inclusions associated with crack initiation indicated that they contained Ti and S. Since the analysis technique employed precluded the determination of carbon, the particles may have been Ti_2S or titanium carbosulfides as recently suggested by Baker.[25]

(a)

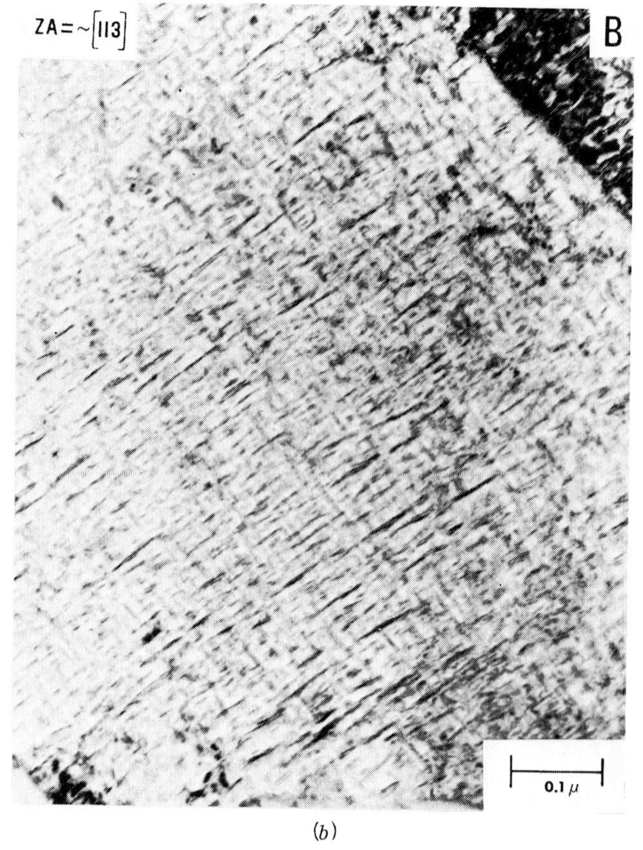

(b)

Fig. 3—Transmission electron micrographs of 18 Ni (350) maraging steel aged 3 h at 950°F (510°C). Note reverted austenite has formed at Point G.

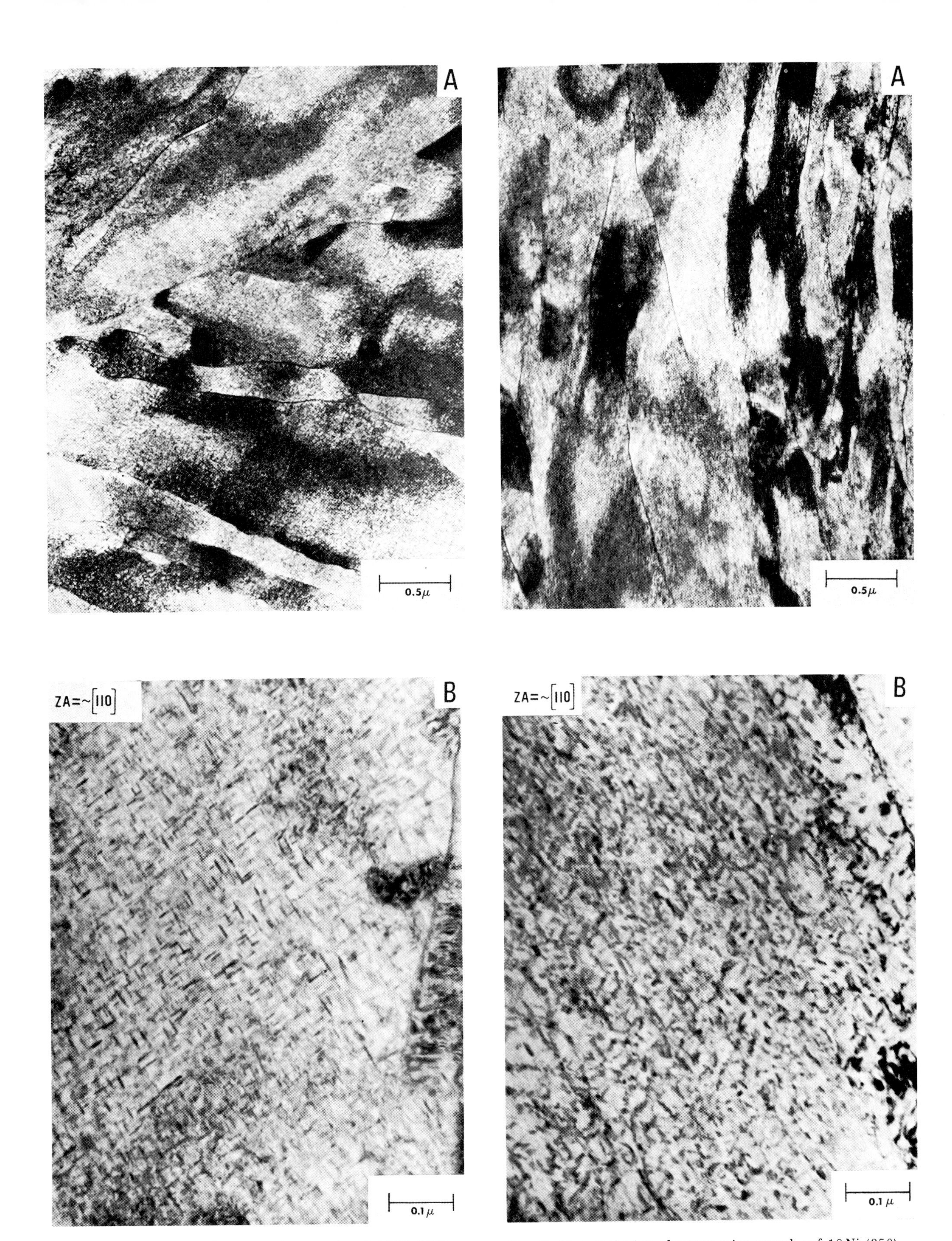

Fig. 4—Transmission electron micrographs of 18 Ni (350) maraging steel deformed 25 pct at 75°F (24°C) and aged for 3 h at 950°F (510°C).

Fig. 5—Transmission electron micrographs of 18 Ni (350) maraging steel deformed 70 pct at 75°F (24°C) and aged for 3 h at 950°F (510°C).

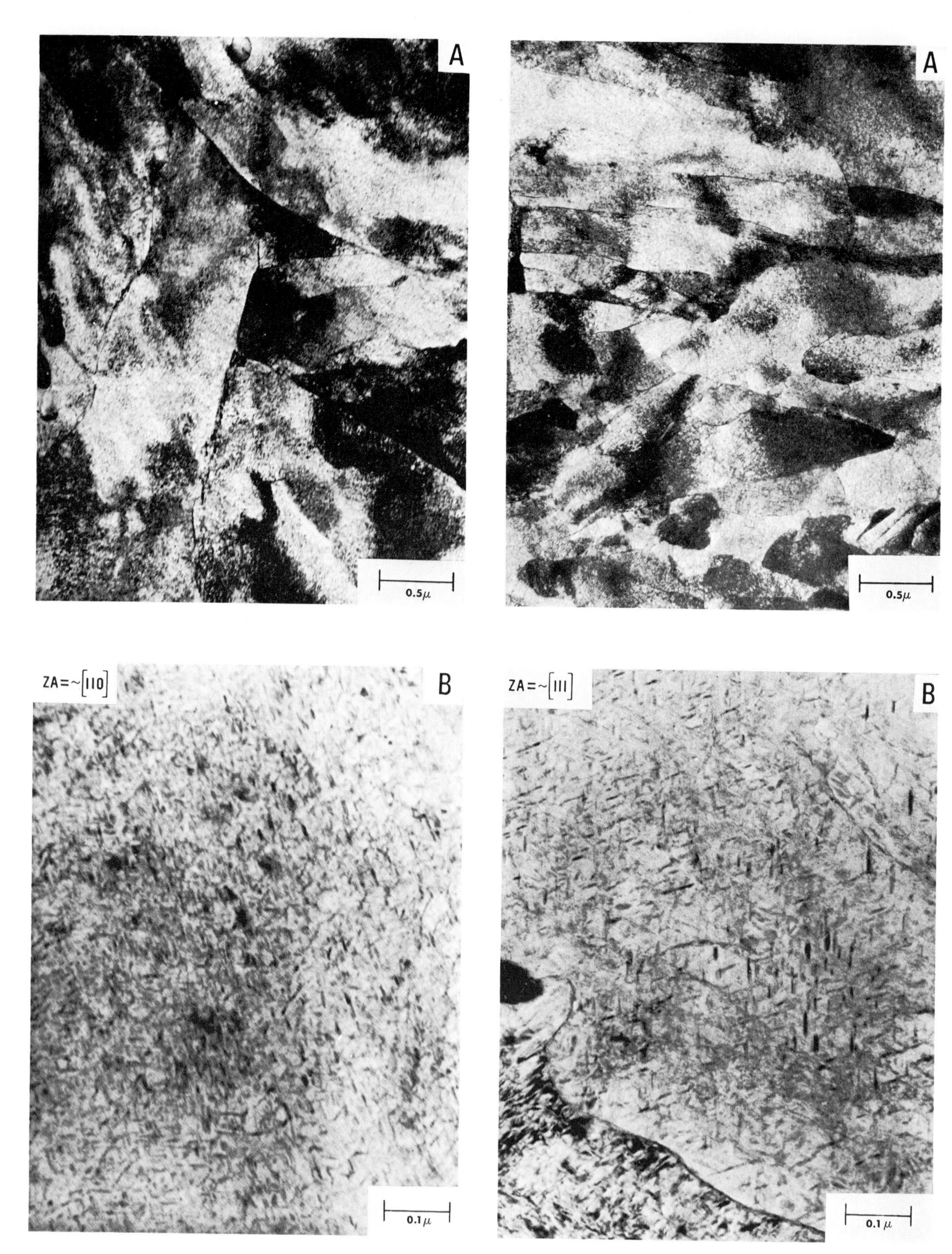

Fig. 6—Transmission electron micrographs of 18 Ni (350) maraging steel deformed 25 pct at 750°F (399°C) and aged for 3 h at 950°F (510°C).

Fig. 7—Transmission electron micrographs of 18 Ni (350) maraging steel deformed 70 pct at 750°F (399°C) and aged for 3 h at 950°F (510°C).

Table III. Influence of Thermomechanical Treatment on The Comparative Residual Stress Levels in 18Ni (350) Maraging Steel

Condition	Residual Stress* (psi)	Residual Stress* (MN/m^2)
Undeformed†	−74,282	−512.2
Deformed 25 pct at 75°F (24°C)	−63,210	−435.8
Deformed 70 pct at 75°F (24°C)	−59,633	−411.2
Deformed 25 pct at 750°F (399°C)	−72,323	−498.6
Deformed 70 pct at 750°F (399°C)	−67,303	−464.0

*Error limits ±4,000 psi (±27.6 MN/m^2).
†All conditions subsequently aged for 3 h at 950°F (510°C).

Fatigue crack propagation was characterized by regions of Stage II crack growth, as noted by the presence of fatigue striae, alternating with bursts of tensile failure. This crack growth morphology has previously been observed by Forsyth and Ryder[26] in Al alloys. Some secondary cracking parallel to the nominal tensile axis was also evident.

Texture

Typical {110} and {200} pole figures for thermomechanically processed 18 Ni (350) maraging steel are shown in Figs. 8 and 9. Deforming 25 pct at room temperature (marforming) results in a {111}⟨112⟩ texture

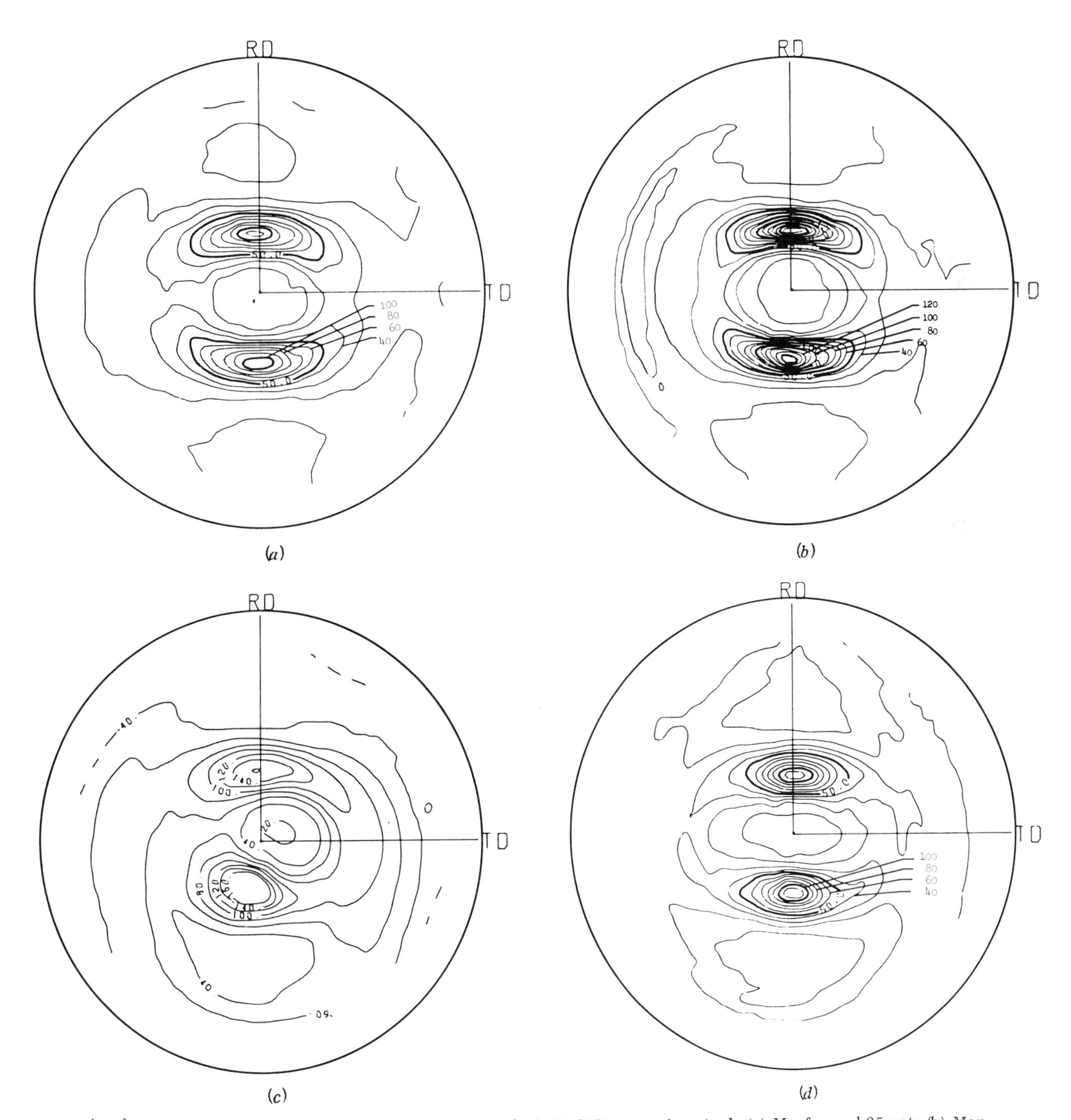

Fig. 8—{110} pole figures for thermomechanically processed 18 Ni (350) maraging steel: (a) Marformed 25 pct; (b) Marformed 70 pct; (c) Ausformed 25 pct; (d) Ausformed 70 pct.

with minor components of {100}⟨011⟩ and {112}⟨110⟩. Increasing the amount of deformation to 70 pct sharpens the {100}⟨011⟩ and {112}⟨110⟩ components at the expense of the {111}⟨112⟩ textural component.

Ausforming at 750°F (399°C) involves the ultimate formation of a strong {112}⟨110⟩ texture. Increasing the amount of deformation strengthens this textural component and eliminates the {110}⟨011⟩ component observed after 25 pct reduction. Finally a weak {111}⟨112⟩ component appears after the higher deformation.

A consideration of the textural components described above clearly shows that crystallographic texture does not significantly affect the high-cycle, low-stress fatigue behavior of maraging steel. For example, ausforming or marforming both yield the same fatigue properties while possessing vastly different textural descriptions.

DISCUSSION

Both thermomechanical treatments produce a pronounced microstructural anisotropy such that the prior austenite grains are elongated in the rolling direction and reduced through the thickness of the rolled plate. This microstructural anisotropy has a significant effect on fracture toughness,[27,28] tensile ductility,[27] and the crack propagation resistance in low-cycle fatigue.[29] However these effects are generally related to differ-

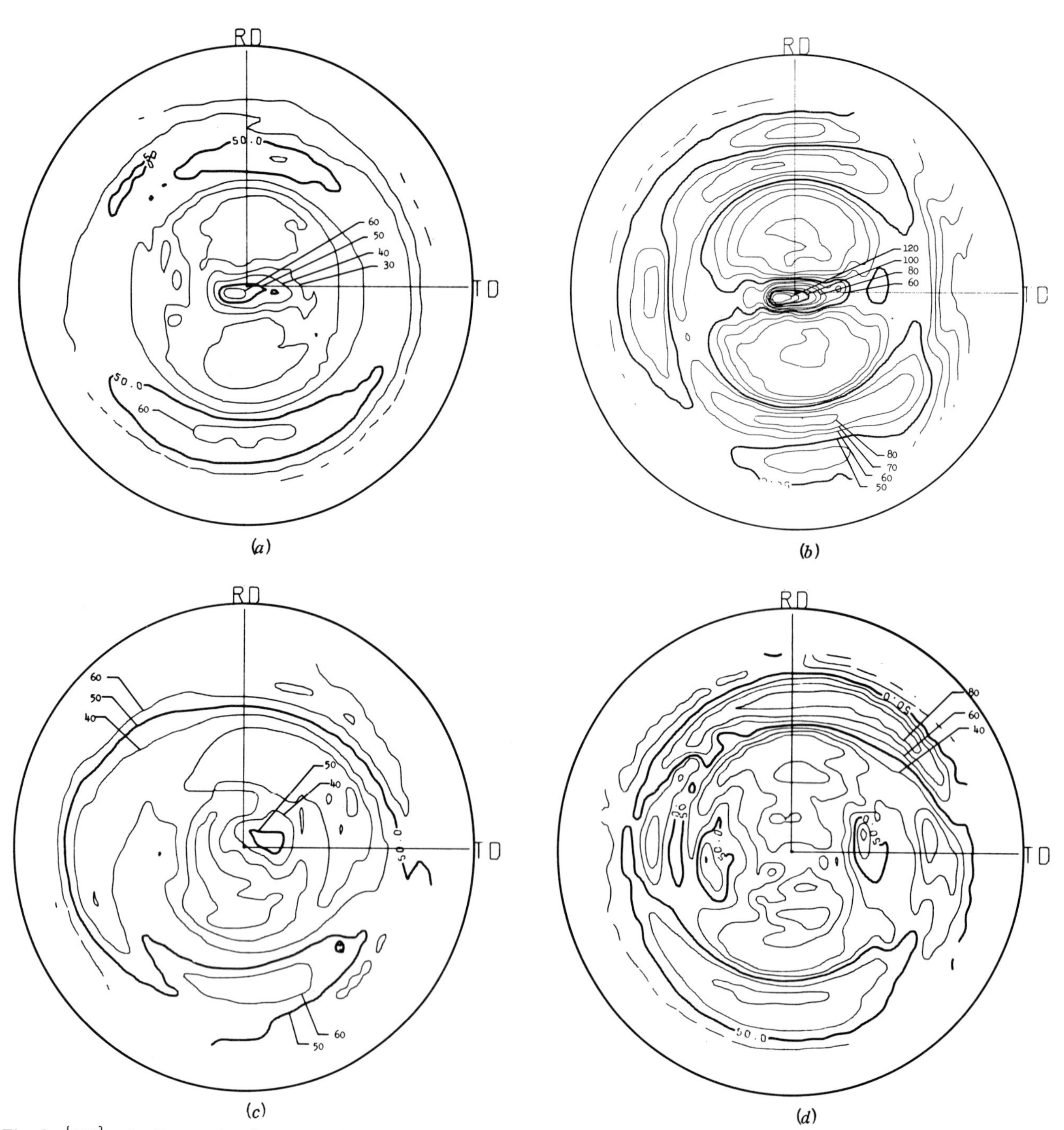

Fig. 9—{200} pole figures for thermomechanically processed 18 Ni (350) maraging steel: (*a*) Marformed 25 pct; (*b*) Marformed 75 pct; (*c*) Ausformed 25 pct; (*d*) Ausformed 70 pct.

ences between specimens taken parallel versus transverse to the prior deformation direction, whereas in this examination all specimens were taken parallel to the rolling direction. Furthermore, the grain (subgrain) shapes produced by marforming were highly elongated, while ausforming resulted in a more equiaxed substructure, the fatigue resistance after equivalent reductions was identical.

The microstructural anisotropy could play a role in the high-cycle fatigue resistance if inclusions dominated the crack nucleation event; the strain concentration effect of an oriented inclusion would be higher for transverse loading. Again, in the previous section on Fractography, it was noted that inclusions do not dominate crack initiation in 18 Ni (350) maraging steel. Thus, we can conclude that microstructural anisotropy, as such, is not the determining factor in establishing the high-cycle fatigue resistance of this steel.

Certain guidelines have recently been developed for optimizing low-stress high-cycle fatigue resistance.[30,31] Generally, material characteristics which are desirable include: a) a stable microstructure, b) no inclusions, c) a fine grain size, and d) the homogenization (dispersal) of slip.

The achievement of such a high fatigue resistance in ferrous alloys can be directly related to the classical Hall-Petch relationship:[32,33]

$$\sigma_f = \sigma_0 + Kl^{-1/2} \qquad [3]$$

where σ_f = fatigue stress, σ_0 = friction stress, K = material constant related to strength of grain boundaries, and l = grain size. Of course, these materials also display the same kind of relationship for the yield stress or the flow stress at any strain, except that the effective strength of the boundaries, *i.e.*, the constant K in Eq. [3], changes for each stress parameter. Furthermore, in fatigue, this relationship is limited to materials, *e.g.*, maraging steels, where a strong yield point and Lüders band extension are absent.[32] Generally unaged 18 Ni (350) maraging steel may be described as a lath martensite[8] where the microstructural unit that influences the strength is the block of parallel laths.[34] Upon aging at 950°F the blocks of laths disappear and the effective microstructural unit in Eq. [3] would normally revert to the prior austenite grains (which would provide little strengthening compared to precipitate and solid solution effects). However, in the case of thermomechanically treated material, a dislocation cell structure remains after aging. The presence of sharp dislocation cells (subgrains) in heavily deformed iron gives strengthening analogous to Eq. [3] although in this case the constant K refers to the cell boundary and the exponent can range from -1 for as-cold worked to $-\frac{1}{2}$ for recovered subgrains.[35-37] In either instance the retention of such a refined cell structure provides fatigue strengthening, and it is this cell structure which contributes, in part, to the increased high-cycle fatigue resistance. Furthermore, if such a dislocation substructure can inhibit the formation of extended, soft, slip bands and further prohibit large amplitude dislocation movement at relatively low stresses, then crack initiation can be retarded.[38,39] Since the dislocation structure produced by the thermomechanical treatment restricts the long-range motion of dislocations, by presenting additional barriers to such motion, crack initiation should be suppressed and fatigue resistance enhanced. Finally the high-cycle fatigue resistance should increase with decreasing cell size, *i.e.*, increasing amount of deformation, as has indeed been observed in the present investigation. This cell structure should be more effective at low-stress amplitudes in high-cycle fatigue where the mobile dislocations have difficulty in breaking through the cell boundaries, but at high-stress amplitudes, in low-cycle fatigue, these boundaries may break down and thus will not be effective in preventing the spread of deformation. Again the present results indicate that this occurs, thermomechanical processing does not significantly affect the low-cycle fatigue behavior of 18 Ni (350) maraging steel.

Wells *et al.*[31] have suggested that slip dispersal (homogenization) may be achieved by introducing finely-spaced, hard, incoherent particles into a metallic material. Such a dispersion should increase the rate of monotonic work hardening and the number of slip initiation sites[40] and thereby promote homogeneous deformation and inhibit crack initiation. The tendency for homogeneous deformation in a precipitation-hardened microstructure depends on whether the particles are cut (which gives planar inhomogeneous deformation bands) or by-passed by cross-slip (which gives homogeneous deformation). The applicability of the cutting versus by-pass mechanisms depends, all other things being equal, on the ratio of the particle radius, R, to the Burgers vector $\mathbf{b}$.[41] For $R/\mathbf{b} < 15$, the cutting mechanism is favored; while for $R/\mathbf{b} > 25$, the by-pass mechanism is anticipated. Since particle coarsening is observed following thermomechanical processing of 18 Ni (350) maraging steel, cross-slip and its accompanying fatigue improvement should be favored in the latter microstructures. Further, the most pronounced effect on fatigue should be at low deformations, *i.e.*, 25 pct reduction, where the principal changes in particle size have been observed. Finally the particle coarsening is probably accompanied by a further loss of particle-matrix coherency which will also promote cross-slip.

Although the exact mechanism by which ausforming and marforming influence the fatigue behavior of high strength steels has not been delineated, the present study does indicate that the principal microstructural effects of TMT related to changes in low-stress high-cycle fatigue are the development of a dislocation cell structure and a coarsening of precipitates, the latter tending to inhibit the formation and spread of deformation bands and promote the homogenization of deformation by cross-slip.

CONCLUSIONS

1) The low-stress high-cycle fatigue resistance of 18 Ni (350) maraging steel can be significantly improved by ausforming or marforming.

2) The stress for failure at a life of 10^7 cycles in longitudinal specimens can be raised from $\sim$ 100 ksi (689 MN/m^2) to $\sim$120 ksi (827.4 MN/m^2) with 25 pct rolling deformation and to $\sim$130 ksi (896.3 MN/m^2) with 70 pct rolling deformation prior to final aging, irrespective of the stage at which the deformation is introduced.

3) The high-cycle fatigue strength is not dominated

by the presence of inclusions. The TMT does not change the residual stress level after aging at 950°F (510°C). The crystallographic and microstructural anisotropy associated with TMT steel does not contribute to the improvement in fatigue strength.

4) The principal microstructural changes associated with both ausforming and marforming, which affect the fatigue strength, are the retention after aging of a dislocation cell structure and a slight coarsening of precipitate particles.

5) The TMT microstructures are thought to be more resistant to the formation and spread of concentrated slip bands and more likely to deform homogeneously. Both factors contribute to preventing crack initiation and lead to a higher fatigue strength in the low cyclic-stress range.

ACKNOWLEDGMENT

The authors are indebted to F. Greulich, S. Dulierre, and C. Miglionico for their technical assistance and to W. Hoover for critically reviewing the manuscript. The aid of Prof. E. Starke of the Georgia Institute of Technology in obtaining the texture pole figures is also gratefully acknowledged. This work was partially supported by the U. S. Atomic Energy Commission.

REFERENCES

1. G. W. Tuffnell, D. L. Pasquine, and J. H. Olson: *Trans. ASM,* 1966, vol. 59, p. 769.
2. R. J. Henry and R. A. Cary: *Met. Prog.,* 1969, vol. 96, no. 3, p. 127.
3. R. W. Landgraf: ASTM, STP 467, 1970, p. 3.
4. C. E. Feltner and R. Beardmore: ASTM, STP 467, 1970, p. 77.
5. F. Borik, W. M. Justusson, and V. F. Zackay: *Trans. ASM,* 1963, vol. 56, p. 327.
6. F. Ostermann: *Met. Trans.,* 1971, vol. 2, p. 2897.
7. A. Kelly and R. B. Nicholson: *Progr. Mater. Sci.,* 1963, vol. 10, no. 3, p. 149.
8. H. J. Rack and D. Kalish: *Met. Trans.,* 1971, vol. 2, p. 3011.
9. D. Kalish and H. J. Rack: *Met. Trans.,* 1971, vol. 2, p. 2665.
10. H. J. Rack and D. Kalish: *Met. Trans.,* 1972, vol. 3, p. 1012.
11. D. Kalish and H. J. Rack: *Met. Trans.,* 1972, vol. 3, p. 2289.
12. R. H. Bush: *Trans. ASM,* 1963, vol. 63, p. 885.
13. R. D. Schoone and E. A. Fischione: *Rev. Sci. Inst.,* 1966, vol. 37, p. 1351.
14. J. M. Chilton and C. J. Barton: *Trans. ASM,* 1967, vol. 69, p. 528.
15. A. L. Christenson, D. P. Koistinen, R. E. Marburger, M. Semchyshen, and W. P. Evans: *SAE Technical Report 182,* SAE Inc., New York, 1960.
16. D. P. Koistinen and R. E. Marburger: *Trans. ASM,* 1959, vol. 51, p. 537.
17. R. F. Decker, J. T. Eash, and A. J. Goldman: *Trans. ASM,* 1962, vol. 55, p. 58.
18. J. A. Roberson and A. M. Adair: *Trans. TMS-AIME,* 1969, vol. 245, p. 1937.
19. E. B. Kula and C. F. Hickey, Jr.: *Trans. TMS-AIME,* 1964, vol. 230, p. 1707.
20. C. Laird: ASTM STP 415, 1967, p. 131.
21. W. J. Plumbridge and D. A. Ryder: *Metal. Rev.* No. 136, Metals and Materials, (1969).
22. A. J. Baker and P. R. Swann: *Trans. ASM,* 1964, vol. 57, p. 1008.
23. K. Shimizu and H. Okamoto: *Trans. JIM,* 1971, vol. 12, p. 273.
24. W. A. Spitzig: *J. Materials,* 1970, vol. 5, p. 140.
25. T. J. Baker: *J. Iron Steel Inst.,* 1972, vol. 210, p. 793.
26. P. J. E. Forsyth and D. A. Ryder: *Metallurgica,* 1961, vol. 63, p. 117.
27. R. H. Bush, A. J. McEvily, and W. M. Justusson: *Trans. ASM,* 1964, vol. 57, p. 991.
28. D. Kalish and M. Cohen: *Trans. ASM,* 1969, vol. 62, p. 353.
29. R. A. Bock and W. M. Justusson: *Met. Prog.,* 1968, vol. 94, no. 6, p. 107.
30. J. C. Grosskreutz: *Met. Trans.,* 1972, vol. 3, p. 1255.
31. C. H. Wells, C. P. Sullivan, and M. Gell: ASTM, STP 495, 1971, p. 61.
32. W. L. Phillips and R. W. Armstrong: *J. Mech. Phys. Sol.,* 1969, vol. 17, p. 265.
33. M. Klesnil, M. Holzmann, P. LuKas and P. Rys: *J. Iron Steel Inst.,* (London) 1965, vol. 203, p. 47.
34. W. S. Owen, *Second International Conf. Strength Metals and Alloys,* p. 795, ASM, 1970.
35. G. Langord and M. Cohen: *Trans. ASM,* 1969, vol. 62, p. 623.
36. H. J. Rack and M. Cohen: *Mater. Sci. Eng.,* vol. 6, 1970, p. 320.
37. H. J. Rack and M. Cohen: *Frontiers in Mater. Sci.,* Marcel Dekker, N.Y., in press.
38. D. R. Wilson and J. K. Tromans: *Acta Met.,* 1970, vol. 18, p. 1197.
39. D. R. Wilson: *Phil. Mag.,* 1970, vol. 22, p. 643.
40. M. F. Ashby: *Oxide Dispersion Strengthening,* p. 143, Gordon and Breach, New York, 1968.
41. V. Gerold and H. Haberkorn: *Phys. Status Solids,* 1966, vol. 16, p. 675.

SECTION III: Mechanical Properties

Properties and Applications of Maraging Steels

R. D. WELTZIN AND C. M. BERGER

THE 18 pct Ni-maraging steels, which are in the family of iron-base alloys, are strengthened by a process of martensitic transformation, followed by age or precipitation hardening. Precipitation hardenable stainless steels are also in this group.

Maraging steels work well in electro-mechanical components where ultra-high strength is required, along with good dimensional stability during heat treatment. Several desirable properties of maraging steels are:

1) Ultra-high strength at room temperature. 2) Simple heat treatment, which results in minimum distortion. 3) Superior fracture toughness compared to quenched and tempered steel of similar strength level (*i.e.*, AISI 4340). 4) Low carbon content, which precludes decarburization problems. 5) Section size is not a factor in the hardening process. 6) Easily fabricated. 7) Good weldability.

These factors indicate that maraging steels could be used in applications such as shafts, and substituted for long, thin, carburized or nitrided parts, and components subject to impact fatigue, such as print hammers or clutches. This study determines the impact-fatigue strength of the 300 and 350 grade maraging steels, and compares their wear resistance with the wear resistance of more commonly used materials.

MATERIALS

The alloys for this study were 18 pct Ni-maraging steels aged to nominal tensile strengths of 300 ksi and 350 ksi. Impact-fatigue loading performance of these steels was compared with impact-fatigue loading performance of AISI 8620 (case hardened), Ti-6 Al-4 V, and Ti-13 V-11 Cr-3 Al. The AISI 8620 is used extensively for fatigue and wear applications, while the titanium alloys have been considered for similar uses. In addition, an aluminum alloy was tested in impact fatigue.

Wear tests were directed at obtaining the comparative wear resistance of materials commonly used for shafts or carburized components.

EXPERIMENTAL PROCEDURES

Comparative wear resistances were determined using the apparatus shown in Fig. 1. In this test, a case-hardened steel disc is abraded against the specimen under test, without lubrication, at a fixed speed of 174 rpm. The load applied resulted in an initial Hertz contact stress of 20,000 psi. Wear rate of the sample was determined by calculating the volume of the wear impression as a function of time.

Impact-fatigue behavior was evaluated using a Wiedemann-Baldwin SF-01-U Universal Testing Machine made by Wiedemann Div., Warner and Swasey Co. A special fixture was designed for this machine to allow a dynamic cyclic impact load to be applied to the test specimen. The specimen geometry used for this testing (Fig. 2) was developed by G. Koves and R. D. Weltzin ("Impact Fatigue Testing of Titanium Alloys," Journal of Materials, ASTM. September 1968).

RESULTS

Impact-Fatigue Test Results

Impact-fatigue data were recorded as applied impact stress versus the number of cycles to failure—that is, in the standard S-N curve form. Fatigue curves for annealed 300 and 350 grades of maraging steel are presented in Fig. 3 and fatigue curves for maraged stock are presented in Fig. 4. Impact-fatigue limits for both grades are 70,000 psi for the maraged material and 60,000 psi for annealed materials.

The case hardened (0.014-in. case depth) AISI 8620 steel used in fatigue loading applications has an impact-fatigue limit of 59,000 psi. The 6 Al-4 V titanium alloy with special surface treatment to improve wear

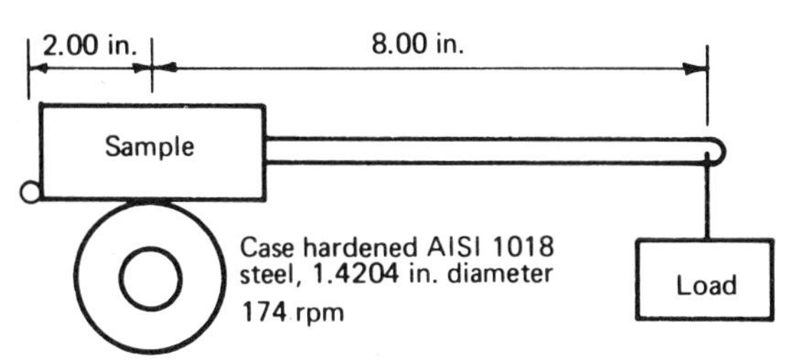

Fig. 1—Schematic of wear apparatus.

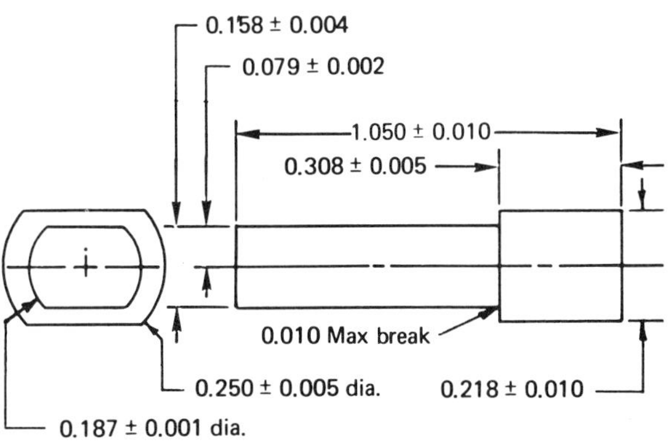

Fig. 2—Impact-fatigue specimen.

R. D. WELTZIN is an Advisory Engineer, Materials and Process Engineering, IBM General Systems Div., Rochester, Minn.; and C. M. BERGER is associated with Taussig Associates, Chicago, Illinois.

Source: *Metals Engineering Quarterly, May 1974*

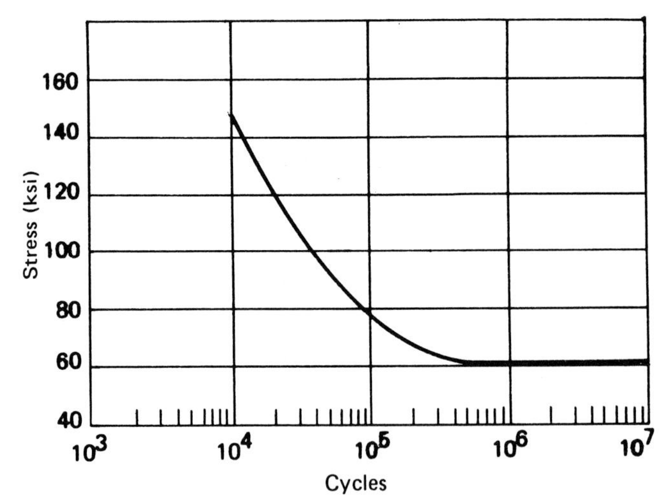

Fig. 3—Impact-fatigue S-N curve for annealed 300 and 350 maraging steel.

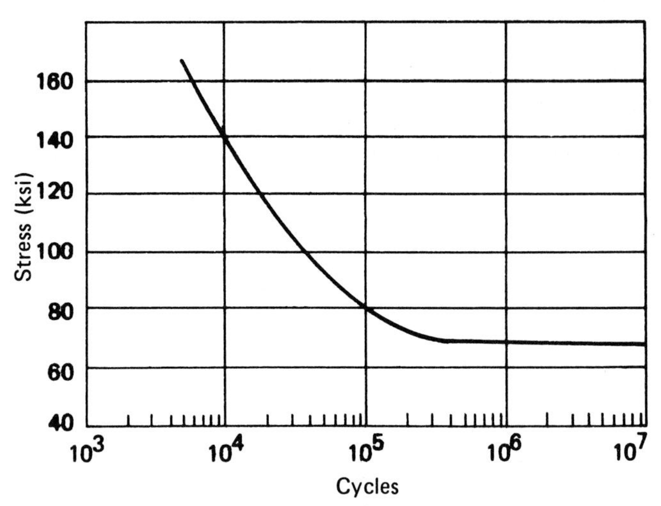

Fig. 4—Impact-fatigue S-N curve for maraged 300 and 350 steel.

resistance had a fatigue limit of 70,000 psi, while the aluminum alloy had a fatigue strength of 30,000 psi at 10^7 cycles.

Table I shows that maraged 18 pct Ni-maraging steel provides an impact-fatigue performance equivalent to any of the other materials tested. This indicates that the maraging steels could be used in applications where high static and dynamic strengths are required, and still hold close dimensional tolerances during fabrication.

Table I. Mechanical Properties of Maraging Steel and Other Alloys

Alloy	Tensile Strength (psi)	Impact Strength (ft-lb)	Impact-Fatigue Strength (psi)
300 CVM annealed	160	14.0	60,000
300 CVM (900°F for 3 h)	280	12.0	70,000
VascoMax* 350 annealed	169	–	60,000
VascoMax 350 (900°F for 10 h)	340	11.0	70,000
8620 (0.014-in. case)	–	8.3	59,000
Ti-6Al-4V	170	8.8	70,000
Ti-13V-11Cr-3Al	190	6.5	52,000
2024 T4 Aluminum	70	4.5	30,000

*Trademark of Vasco, Latrobe, Pennsylvania.

Wear Test Results

Fig. 5 presents graphically the abrasion wear test results. This is a wear evaluation of metals often used in shaft applications. Interest in shaft applications arose partially from observation of wear due to sintered bronze bushings on 416 stainless steel shafts, as reported by R. G. Bayer and J. L. Sirico (in "Wear," Vol. 9, 1966, p. 236.). This wear information is also important if maraging steels are considered as replacements for case hardened components.

Analysis of Test Results

The impact-fatigue test results indicate that the maraging steels provide good resistance to fatigue failure caused by repeated impact loading. Table I shows that aged maraging steel possesses superior impact-fatigue strength and superior fracture toughness, when compared to the case-hardened AISI 8620. This mechanical superiority (when combined with dimensional stability of the maraging steels) demonstrates that applications with impact-fatigue loading or other fatigue types of loading, should be pursued.

In several data processing machine applications, the strength considered is the fatigue strength of the material. Certain applications involving high inertial forces (due to speed requirements) suggest that the fatigue strength/density be considered. In these instances, titanium alloys could have considerable application in data processing equipment. In other applications, however, the important factor would be fatigue strength, in which maraging steels stand out as excellent.

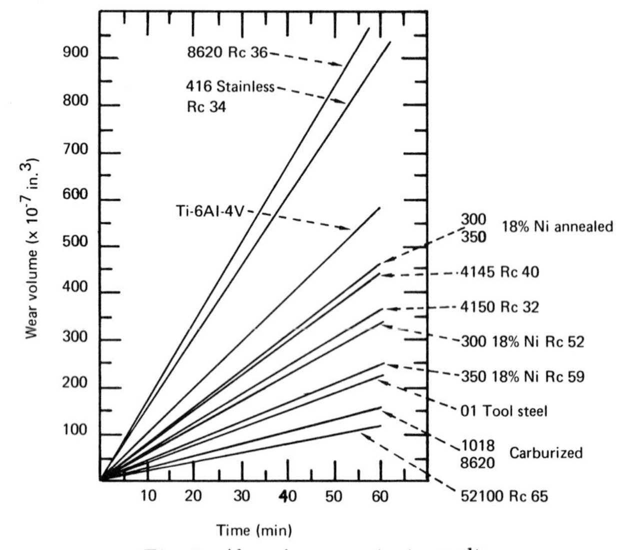

Fig. 5—Abrasion wear test results.

Table II compares several materials for fatigue strength/density and fatigue strength. (This information represents the impact-fatigue tests performed in this study.) Note that, on a fatigue strength/density basis, aluminum compares favorably. However, when considering the cross sectional area of aluminum needed to obtain the required fatigue life through re-

Table II. Impact-Fatigue Comparison of Fatigue Strength/Density and Fatigue Strength

Alloy	Tensile Strength, (psi)	Density ($lb/in.^3$)	Fatigue Strength, (psi)	Fatigue Strength/ Density ($\times 10^4$)	Tensile Strength/ Density ($\times 10^4$)
2024 T-4 aluminum	68,000	0.10	30,000	30	68
Ti-13V-11Cr-3Al	190,000	0.16	52,000	32	120
8620 case hardened	150,000	0.28	59,000	21	57
Steel, 300 maraging annealed	160,000	0.28	60,000	21	57
Ti-6Al-4V	170,000	0.16	70,000	43	105
Steel, 300 maraging	280,000	0.28	70,000	25	99
Steel, 350 maraging annealed	170,000	0.28	60,000	21	61
Steel, 350 maraging	355,000	0.28	70,000	25	127

duced stress level, the use of aluminum for most fatigue applications in data processing machines would not be recommended, primarily because of the trend toward miniaturization of components. Because of this trend, the maraging steels and other high strength steels should find extensive applications in the future. Other applications might utilize the ultra-high strength of the maraging steels (up to 350 ksi at room temperature) or their superior fracture toughness.

The study was made of the maraging steels in wear situations because little information was available on their wear resistance compared to other metals. Wear resistance of these high strength alloys is questionable because the maraged structure is a carbonless martensite.

The wear resistance of the maraging steel at RC55 is not as good as that of AISI 52100 steel at the same hardness level. This should be expected because of the difference in microstructure, which results from the difference in carbon content. Some improvements in wear resistance of the maraging steel over that of a through-hardened alloy steel, such as AISI 4145, can be noted. This, along with the wear results on the 416 stainless steel, demonstrates that maraging steels might be beneficial for certain wear applications.

SUMMARY

The properties of maraging steels clearly indicate that these steels have many potential applications in mechanical components of electro-mechanical data processing machines. Use of these steels in shafts that require good dimensional control following heat treatment should be pursued for two reasons. First, maintaining dimensions should be easier because quenching and tempering are not necessary. Second, wear data indicate that equivalent or better wear resistance is obtained from the maraging steel than from the more commonly used shaft materials.

Impact-fatigue strength of 18 pct Ni-maraging steels indicates that these steels could be used in repeated impact loading situations. The good fracture toughness, compared to that of quenched and tempered alloy steels at the same strength level, indicates possible use in high-impact, low-cycle load applications.

Finally, due to the relatively low temperature of aging, the use of the maraging steels for long, thin parts should be considered. Here, their use as a replacement for some case-hardened or nitrided components is indicated, but the potential application should be carefully studied.

An Investigation of the Fatigue Behavior of 18% Nickel Maraging Steel

G. W. TUFFNELL, D. L. PASQUINE AND J. H. OLSON

ABSTRACT. Rotating beam S – N fatigue curves were determined for bar stock from four commercially produced heats of 18% Ni maraging steel. The 10^7 cycle fatigue strength for each heat closely matched the fatigue strength of quenched and tempered alloy constructional steels of equal tensile strength.. The effect of prior annealing temperature on fatigue life in the aged condition was studied using an additional commercial heat of 18 Ni 250 maraging steel plate. Annealing at 1400 F produced maximum fatigue life; high annealing temperatures caused a reduction in fatigue life. A subsequent lower temperature annealing treatment after a high-temperature exposure partially recovered the fatigue life. "Overaging," "underaging" or simulation of a 1200 F weld heat-affected zone did not significantly affect fatigue life. Maraging after finish grinding and polishing reduced fatigue life as compared to tests made on specimens maraged prior to finishing. Shot peening treatments greatly extended fatigue life of specimens both with finish ground-maraged surfaces, and with maraged-finish ground surfaces.

SINCE THE introduction of the low carbon 18% Ni maraging steels in 1961, several studies of the influence of processing and heat treating on mechanical properties have appeared in the literature (1–8). However, relatively few fatigue data have been published (3, 4, 6–10) and the values have shown considerable scatter. There has been some evidence that such scatter is due to differing mill processing and to differing heat treating sequences. In addition, analyses of a few fatigue service failures that have come to the authors' attention have further indicated that processing history and surface treatment can indeed play a significant role in the fatigue behavior of maraging steel. The present investigation was initiated to determine the fatigue properties of several commercial heats of maraging steel heat treated according to standard practice, and to explore the dependency of these properties on several heat treating and surface treating variables. Particular emphasis was placed on annealing treatment variations. Because of the availability of comparative data and the ease of testing, the conventional rotating beam test was used.

The authors are with The International Nickel Co., Inc., Paul D. Merica Research Laboratory, Suffern, N. Y. Manuscript received July 21, 1966.

MATERIALS

The chemical analyses and mechanical properties of the five production-scale heats of maraging steel are listed in Tables 1 and 2. Although the sulfur contents for steels 2 and 4, the silicon contents for steels 3 and 4 and the manganese content for steel 5 were slightly higher than recommended, mechanical properties were not severely affected. The steels numbered 1 through 4 include air melts of 18 Ni 200 and 18 Ni 250 and consumable-electrode vacuum melts of 18 Ni 250 and 18 Ni 300. These

TABLE 1. Materials and Chemical Analyses

Steel No.	Grade*	Form	Chemical analysis, wt %†										
			Ni	Co	Mo	Al	Ti	S	P	C	Mn	Si	Fe
1	18 Ni 200 (AIR)	$\frac{5}{8}$ in. rod	18.5	8.3	3.11	0.034	0.20	0.004	0.002	0.024	0.02	0.08	Bal
2	18 Ni 250 (AIR)	$\frac{5}{8}$ in. rod	18.8	7.8	4.77	0.055	0.43	0.013	0.003	0.019	0.07	0.07	Bal
3	18 Ni 250 (CEVM)	$\frac{5}{8}$ in. rod	18.2	7.8	4.81	0.068	0.35	0.005	<0.001	0.013	0.09	0.11	Bal
4	18 Ni 300 (CEVM)	$\frac{5}{8}$ in. rod	18.7	9.4	4.72	0.067	0.56	0.011	0.003	0.016	0.06	0.11	Bal
5	18 Ni 250 (CEVM)	$\frac{1}{2}$ in. plate	18.4	7.8	4.80	0.070	0.37	0.006	0.006	0.02	0.12	0.09	Bal

* AIR = air melted, CEVM = consumable electrode vacuum melted.
† All steels contained <0.01 wt % B, Zr, O and N except steel 2 which contained 0.055 wt % O.

TABLE 2. Mechanical Properties with Standard Heat Treatment*

Steel No.	Grade	0.2% yield strength, ksi	Ultimate tensile strength, ksi	% Elong. in 1 in.	% Reduction of area	Charpy V-notch impact energy room temp, ft-lb
1	18 Ni 200	217.0	221.0	14	63	47
2	18 Ni 250	263.0	270.5	11	56	19
3	18 Ni 250	246.5	250.0	15	63	32
4	18 Ni 300	279.6	288.6	11	55	20
5	18 Ni 250	249.5	256.5	10	51	17†

* Solution anneal 1 hr at 1500 F, air cool, marage 3 hr at 900 F.
† Transverse.

steels all had very low inclusion content and were too clean to be rated for inclusions by the standard ASTM E 45 method at a magnification of $\times 100$. These four steels, in the form of mill-annealed $\frac{5}{8}$-in. diam bar stock, were used in determining the $S-N$ fatigue curves. Steel 5, an 18 Ni 250 heat in the form of mill-annealed $\frac{1}{2}$-in. plate, was used for observing the effects of annealing treatment and the other processing variables. Steel 5 also had very low inclusion content and was not rated. This steel showed a small degree of the dark banding in the aged microstructure, which has occasionally been found in other 18 Ni 250 plate material.

Experimental Procedure

To minimize effects of variation in mill annealing treatments, the bar stock of steels 1 through 4 was reannealed in the laboratory for 1 hr at 1500 F and air cooled. Smooth and notched ($K_t = 2.2$) rotating beam R. R. Moore specimens, Fig. 1, were then machined 0.025 to 0.030 in. oversize and maraged 3 hr at 900 F. Specimens were then ground to final size in the maraged condition, eliminating the slight surface oxide film formed during maraging. This is the sample preparation method commonly used for fatigue specimens of high-strength steels. A grinding technique similar to that described by Field and Kahles (9) for minimizing residual surface stress was used. As a final step, specimens were polished longitudinally with 600 emery paper to produce a 5 to 10-microin. surface finish. Except for intentional changes in annealing temperature, cooling rate, aging temperature, or surface treatment, this same method of preparation was used for specimens cut from the $\frac{1}{2}$-in. plate material of steel 5. All heat treating was carried out in air.

Specimens were cycled to failure in conventional rotating beam fatigue testing machines at a speed of 3600 or 10,000 rpm. No difference was noted between test results run at these two different speeds. In determining the $S-N$ curves, individual specimens of each steel were first tested at four or five maximum fiber stress levels ranging from 200 to 120 ksi to provide an estimate of the $S-N$ curve. Then, starting with a stress estimated to produce failure at 10^7 cycles, the ASTM staircase method of testing (11, 12) was used for the remainder of the specimens, and the mean

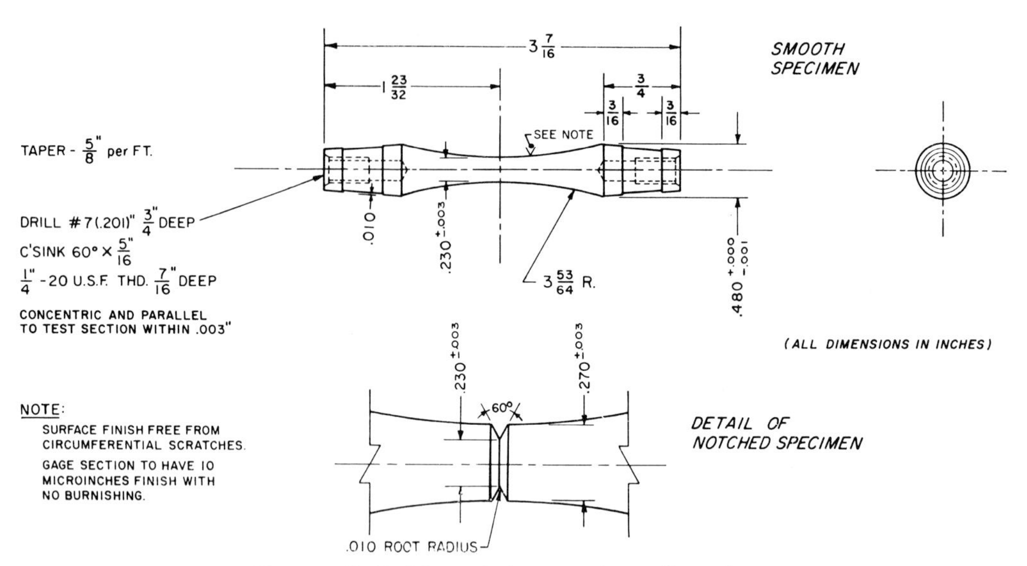

FIG. 1. R. R. Moore fatigue specimen dimensions.

fatigue strength and standard deviation (13) for a 10^7 cycle life were calculated. A few single specimens were also tested at lower stresses to estimate the stress for a life of 10^8 cycles.

For each variable studied, a batch of three specimens was prepared along with a companion tensile specimen to provide a reference strength level. The effects of annealing treatment and certain other variables were evaluated by comparing the mean life of each lot of specimens as tested at 130 ksi maximum stress. This method of testing proved quite satisfactory, giving a relatively high cycle fatigue life (failures generally between 50,000 and 2,000,000 cycles) with good discrimination between the test variables.

Effects of annealing temperature, double annealing, underaging, a simulated 1200 F weld heat affected zone, finish grinding before maraging, and shot peening were the variables investigated. Specimens were water quenched after the series of annealing temperatures to minimize possible cooling rate effects.

The simulated weld heat-affected zone was produced by quickly heating an oversize specimen at about 200 F per sec to 1200 F in a time-temperature controlled resistance heating device as described by Petersen (14). The 1200 F temperature was predetermined as being in the range for shear reversion to austenite for steel 5. Specimens in both the 1500 F annealed and the annealed and maraged prior conditions were given the 1200 F treatment then remaraged 3 hr at 90 F. Heating to the 1200 F temperature range causes about a 5 to 7% reduction of yield and tensile strength in 18% Ni maraging steels after remaraging.

The specimens comparing effects of maraging before or after finish grinding were air cooled after annealing. The specimens maraged after grinding and polishing were lightly buffed to remove any loose surface oxides. Some surface discoloration was observed to remain. The shot peening treatments were controlled to 0.005 to 0.007 Almen A2 intensity using S-70 shot. Reference 15 describes this method of measuring peening intensity.

RESULTS

S – N Curves and Computed Fatigue Strengths

S – N curves for smooth and notched specimens (annealed 1 hr at 1500 F,

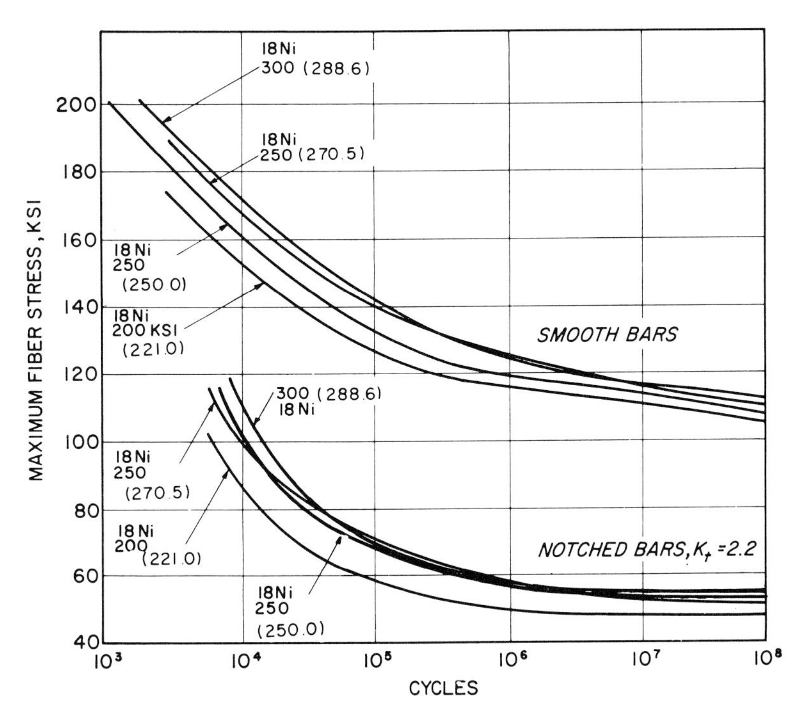

FIG. 2. Rotating beam S – N fatigue curves for commercial 18% Ni maraging steels. Ultimate tensile strength in ksi in parentheses. Confidence intervals at 10^7 cycles and survival probabilities for these data are given in Table 3 and in Fig. 3.

TABLE 3. Summary of 18% Ni Maraging Steel Rotating Beam Fatigue Properties

Steel No.	Grade	Ultimate tensile strength, ksi	Mean 10^7 cycle fatigue strength ksi*				10^7 cycle endurance ratio, 10^7 fatigue/tensile	Notch sensitivity q‡	Estimated mean 10^8 cycle strength, ksi
			Smooth bar	Standard deviation	Notched bar†	Standard deviation			
1	18 Ni 200	221.0	111.9	3.2	47.6	3.5	0.53	1.13	103
2	18 Ni 250	270.5	117	6	51.4	4.5	0.43	1.08	108
3	18 Ni 250	250.0	115.2	2.2	53.5	§	0.46	0.98	106
4	18 Ni 300	288.6	117.7	4.2	55.5	9.0	0.41	0.93	110

* Based on ASTM staircase method. No. specimens tested: Steel 1-smooth 33, notched 30; Steel 2-smooth 39, notched 20; Steel 3-smooth 29, notched 20; Steel 4-smooth 26, notched 15.
† Kt = theoretical stress concentration = 2.2.
‡ $q = Kf - 1/Kt - 1$ where Kf = ratio of smooth bar to notched bar fatigue strength at 10^7 cycles.
§ Insufficient data for calculating standard deviation.

maraged 3 hr at 900 F) are shown in Fig. 2. The curves for the smooth bars display a gradual change in slope with no well defined "knee" and no endurance limit. This shape is common for many low alloy steels hardened to these strength levels. The spread in results for the two heats of the nominal 250 ksi yield strength alloy probably reflects the difference in actual tensile strength of the two materials (270.5 and 250.0 ksi).

Table 3 lists the computed strengths for a 10^7 cycle life, the endurance ratio (fatigue/tensile), the notch sensitivity (q) and the estimated stress levels for a 10^8 cycle life. The fatigue strengths for failure at 10^7 cycles ranged from about 112 to 118 ksi, with the higher strength grades having

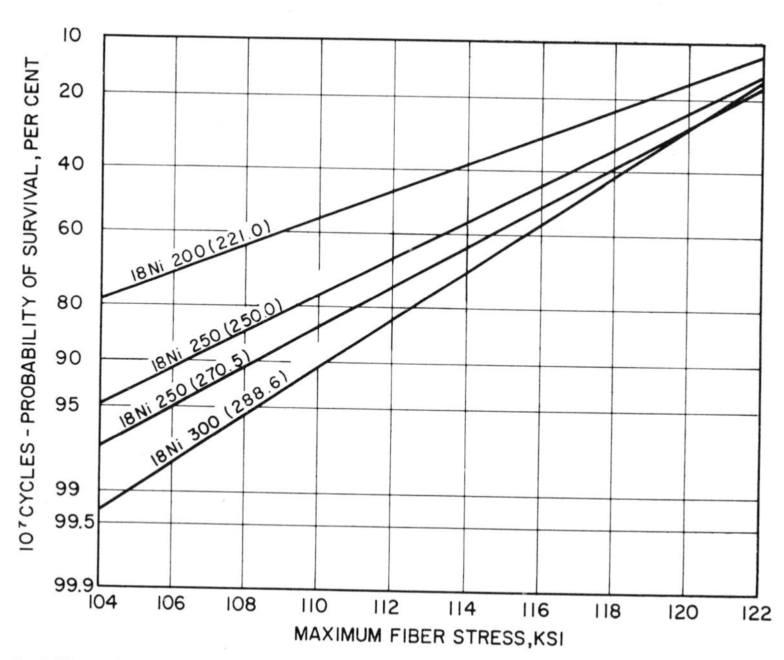

Fig. 3. Probability of surviving 10^7 cycles as a function of stress level for various heats of 18% Ni maraging steel. Measured ultimate tensile strength in parentheses.

TABLE 4. Effects of Annealing Temperature on Rotating Beam Fatigue Life of 18% Ni Maraging Steel at 130 ksi Maximum Fiber Stress (Specimens from Plate, Steel No. 5)

Annealing treatment	0.2% yield strength, ksi	Ultimate tensile strength, ksi	% elong in 1 in.	% redn. of area	Fatigue life, 1000 cycles	Mean fatigue life, 1000 cycles
1 hr 1300 F, WQ	211.3	227.4	11	55.5	929.0 220.9 489.8	546.6
1 hr 1400 F, WQ	271.0	273.1	9	50.5	1451.0 971.0 3623.3 792.1 2295.0	1826.5
1 hr 1500 F, WQ	260.8	262.9	10	56.0	667.0 1492.9 547.6	902.5
1 hr 1700 F, WQ	254.8	262.2	10	58.5	315.5 238.7 174.5	242.9
1 hr 1900 F, WQ	249.5	261.9	12	58.0	87.8 89.1 108.8	95.2
1 hr 2000 F, WQ	246.0	259.0	11	57.0	52.7 68.5 70.0	63.7
1 hr 2100 F, WQ	244.9	256.6	10	53.5	48.6 43.0 45.0	46.5
1 hr 2200 F, WQ	243.0	254.5	11	53.0	34.2 34.2 38.7	35.7

Note: All specimens maraged 3 hr at 900 F after annealing.

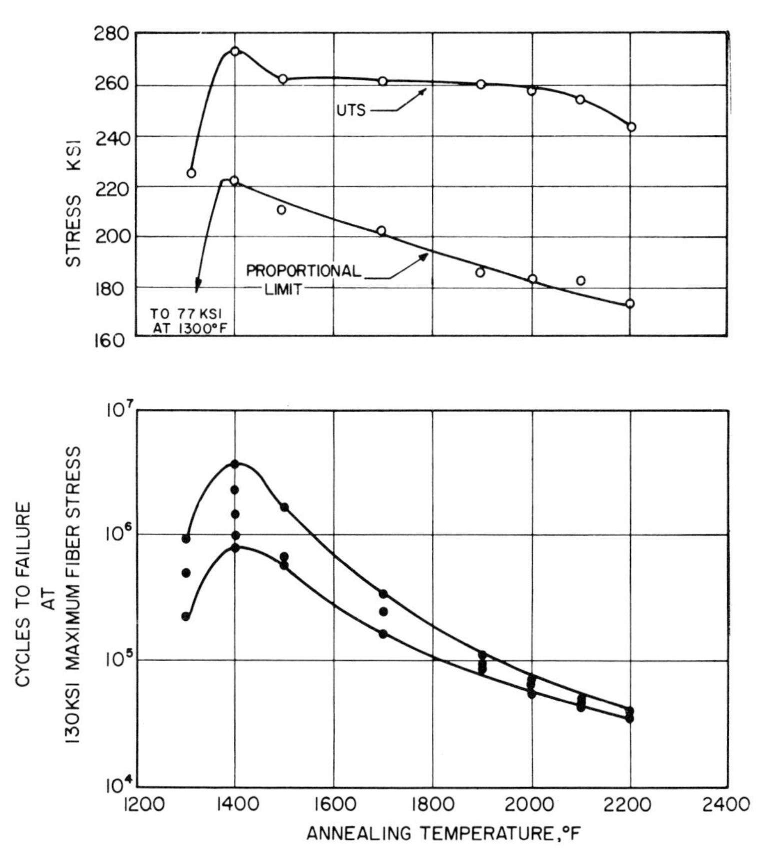

Fig. 4. Effect of annealing temperature on fatigue life, proportional limit, and ultimate tensile strength of 18 Ni 250 steel No. 5. Specimens annealed 1 hr at temperature, water quenched, maraged 3 hr at 900 F.

higher fatigue strengths. The computed stresses for lives of 10^7 cycles are based on from 15 to 39 tests of each steel.

It is also possible to use these data for constructing probability graphs as described by Finney (16). For this procedure, the per cent survival values at each stress level are used not only to calculate the estimated fatigue strength at 10^7 cycles but also the slope of the curve of per cent survival vs stress. The probability graphs for smooth bars of steels 1 through 4 are plotted in Fig. 3. The mean values for 50% survival were 111, 116, 115, and 117 ksi respectively. These values check very well with the smooth bar results calculated using the ASTM method which are listed in Table 3.

Effect of Solution Annealing Treatment

Lots of 3 to 5 fatigue blanks of steel 5 were solution annealed for 1 hr at several temperatures in the range 1300 to 2200 F and water quenched. After preparation as previously described, the specimens were all tested at 130 ksi maximum stress. The resultant fatigue life as a function of solution annealing temperature is shown in Fig. 4. Also shown are the ultimate tensile strength and proportional limit values (measured from the tensile load-extension curves) for tensile specimens annealed along with each batch of fatigue specimens. Individual fatigue values for the specimens are listed in Table 4.

Highest fatigue life was found to occur with a solution annealing temperature of

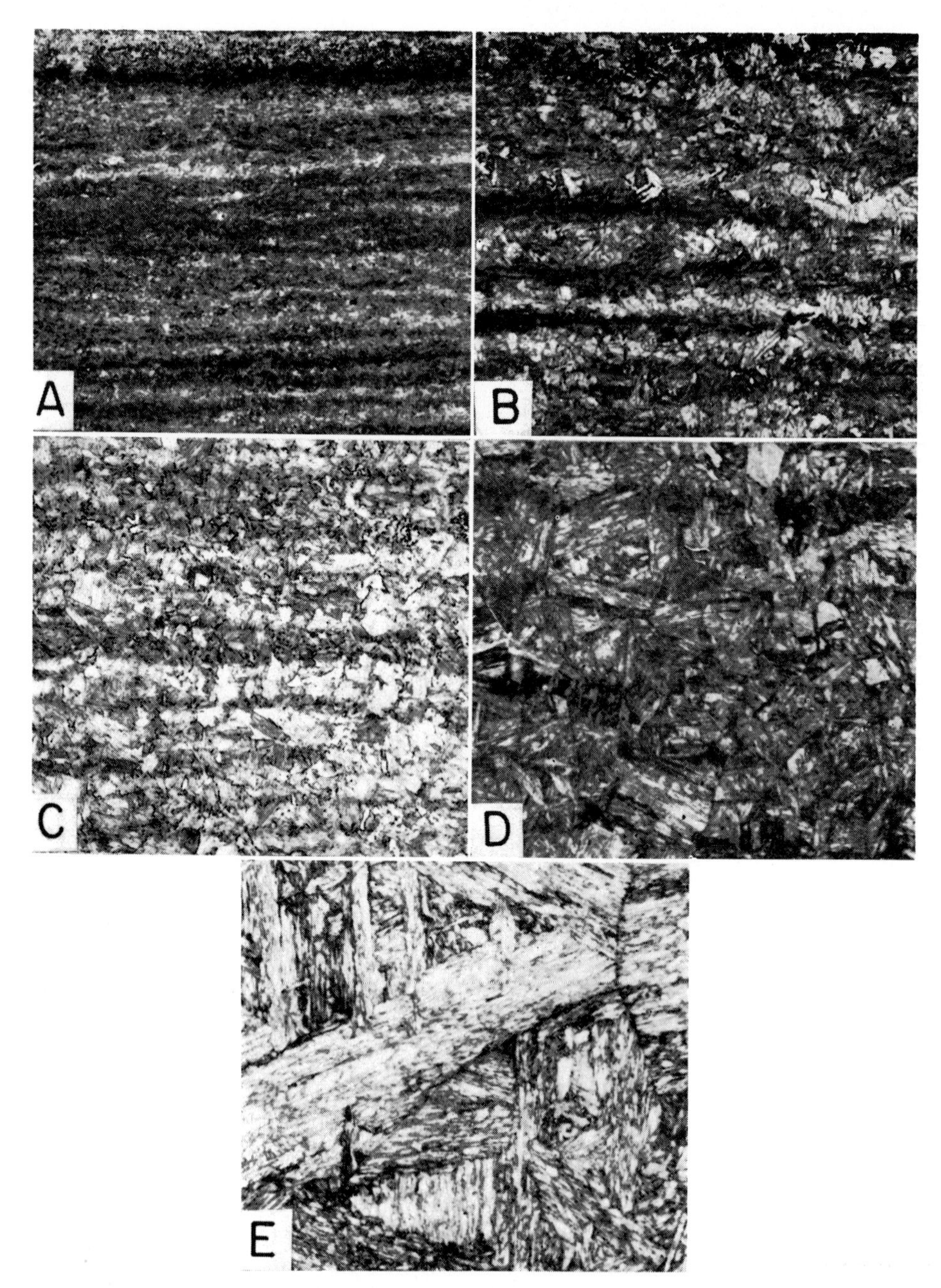

Fig. 5. Micrographs showing maraging steel structure after various annealing treatments and maraging 3 hr at 900 F. (A) 1 hr 1400 F, WQ; (B) 1 hr 1500 F, WQ; (C) 1 hr 1700 F, WQ; (D) 1 hr 1900 F, WQ; (E) 1 hr 2100 F, WQ. Etchant, modified Fry's reagent. ×250.

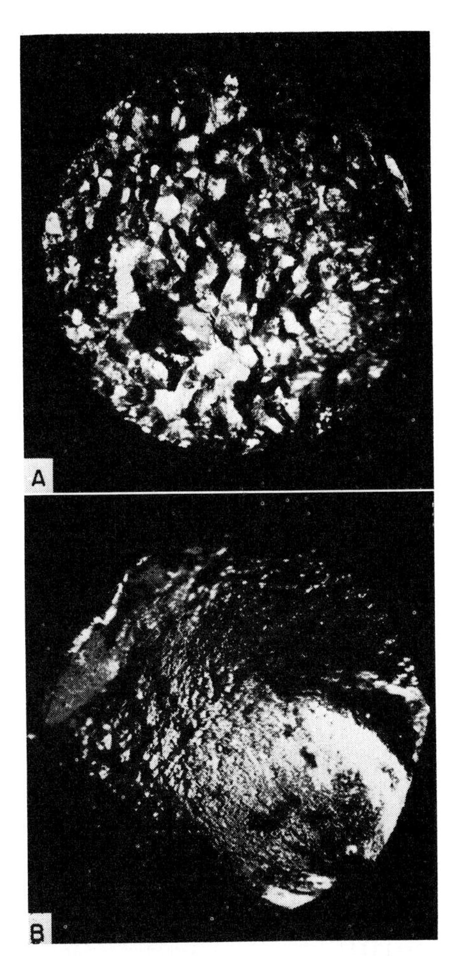

FIG. 6. Fracture surfaces of fatigue specimens. (A) Annealed 1 hr, 2200 F, maraged 3 hr, 900 F. (B) Annealed 1 hr, 1500 F, maraged 3 hr, 900 F. ×10.

1400 F. Fatigue life decreased progressively as annealing temperature was raised above 1400 to 2200 F. Fatigue life showed little or no correlation with ultimate tensile strength, since the ultimate tensile strength remained essentially constant at annealing temperatures above 1400 F. Surprisingly, in this same annealing temperature range fatigue life did correlate with the proportional limit. A 1300 F annealing treatment produced a considerably weaker structure due to the retention of some austenite, but these specimens still showed a respectable fatigue life.

Grain size remained fine (ASTM No. 7 or smaller) until an annealing temperature somewhere between 1700 and 1900 F was reached. On annealing above 1900 F, grain size increased rapidly to about ASTM No. 1 at 2100 F. Furthermore, as the solution annealing temperature was raised, the fatigue fractures became increasingly intergranular in nature. Figures 5 and 6 show examples of grain size and fracture appearance for several annealing temperatures.

Double annealing treatments in which a high temperature anneal was followed by a second lower temperature anneal were effective in restoring a considerable portion of the fatigue life lost in the first anneal. Results of three of these treatments are listed in Table 5. After the double annealing treatments, the correlation of fatigue life with proportional limit no longer held. All the double annealing treatments produced a fatigue life that was longer than would be predicted from the proportional limit.

Other Processing Treatments

Results of the overaging or underaging, the simulated reverted weld heat-affected zone, the maraging after final specimen polishing and the surface shot peening treatments are tabulated in Table 6 for triplicate specimens tested at a maximum stress of 130 ksi. Included for reference is the fatigue result for specimens given a "standard" heat treatment of annealing 1 hr at 1500 F followed by air cooling and maraging for 3 hr at 900 F. The air cooling after the anneal resulted in lower tensile strength and somewhat lower fatigue life than did water quenching (the data for Fig. 4 and Table 4).

Neither the overaging nor the underaging treatments led to a significant reduction of the fatigue properties (based on a statistical "t test" comparison at 90% confidence), despite the slightly reduced tensile strength in each case. The simulated weld heat-affected zone likewise had no damaging effect on fatigue life.

TABLE 5. Effect of Double Annealing Treatments on Rotating Beam Fatigue Life of 18% Ni Maraging Steel at 130 ksi Maximum Fiber Stress (Steel No. 5)

Annealing treatment	0.2% yield strength, ksi	Ultimate tensile strength, ksi	% elong in 1 in.	% red. of area	Fatigue life, 1000 cycles	Mean fatigue life, 1000 cycles
1 hr 1650 F, WQ	259.3*	262.5*	10*	58.0*	...	300.0*
1 hr 1650 F, AC 1 hr 1500 F, AC	257.1	265.2	11	55.5	346.0 449.0 727.0	507.3
1 hr 2000 F, WQ	246.0	259.0	11	57.0	52.7 68.5 70.0	63.7
1 hr 2000 F, WQ 1 hr 1500 F, WQ	258.1	268.1	10	52.0	160.1 143.1 85.1	129.5
1 hr 1500 F, WQ	260.8	262.9	10	56.0	667.0 1492.9 547.6	902.5
1 hr 1500 F, WQ 1 hr 1400 F, WQ	266.7	270.7	11	48.5	813.0 1341.0 1615.0	1256.3

* Estimated by interpolation, Fig. 4.
Note: All specimens maraged 3 hr at 900 F after annealing.

TABLE 6. Effect of Various Processing Treatments on Rotating Beam Fatigue Life of 18% Ni Maraging Steel at 130 ksi Maximum Fiber Stress (Steel No. 5)

Treatment	0.2% yield strength, ksi	Ultimate tensile strength, ksi	% elong.	% red. of area	Fatigue life, 1000 cycles	Mean fatigue life, 1000 cycles
Standard						
1 hr 1500 F, AC	249.5	256.5	10	51.0	410.7	
3 hr 900 F, grind					361.4 468.1	413.4
Overaging						
1 hr 1500 F, AC 3 hr 950 F, grind	233.6	246.5	14	58.0	325.5 500.2 272.9	366.2
Underaging						
1 hr 1500 F, AC 3 hr 800 F, grind	219.6	233.5	14	56.5	307.1 176.8 257.1	247.0
Simulated 1200 F Weld Heat-Affected Zone						
(a) 1 hr 1500 F, HAZ, + 3 hr 900 F, grind	...	...	..	...	189.0 508.0 590.0	429.0
(b) 1 hr 1500 F, 3 hr 900 F, HAZ, 3 hr 900 F, grind	...	...	..	...	364.0 406.5 704.0	491.5
Final Maraging						
1 hr 1500 F, AC, grind, 3 hr 900 F	251.0	259.1	11.0	55.0	135.7 131.6 160.5	142.6
Shot Peening						
(a) 1 hr 1500 F, AC, 3 hr 900 F, grind, shot peen	261.0	266.0	9	54.5	2244.8 4567.3 4699.2	3837.1
(b) 1 hr 1500 F, AC, grind, 3 hr 900 F, shot peen	...	...	..	...	705.0 4582.0 981.0	2089.3

The sequence of finish grinding and maraging operations, on the other hand, was found to affect fatigue life significantly, again based on the 90% confidence "t test." Specimens finish ground and polished directly after annealing and then given a final maraging treatment were found to have an average life of only 143,000 cycles compared to 413,000 cycles for specimens finish ground and polished after maraging. These findings are in agreement with Haynes (8).

Shot peening resulted in greatly increased fatigue life for both the maraged-last and finish ground and polished-last specimens. Roughly, the peening improved the life at 130 ksi by a factor of 10. The specimens ground then maraged showed a relatively greater improvement in life (greater than 14-fold increase) when shot peened prior to testing than did the specimens which were maraged then ground and peened (just over 9-fold increase).

Discussion

S – N Curves

In Fig. 7, the 10^7 cycle mean fatigue strengths for steels 1 through 4 are shown superimposed on a general plot of endurance limits for carbon and alloy constructional steels at various tensile strengths (redrawn from Bullens (17)). Although this curve is nearly 30 years old, the great preponderance of recently published smooth-bar rotating-beam fatigue data still falls within the indicated scatter band. Selecting examples from the more recent references at the end of this paper gives:

Steel	Ultimate tensile stress, ksi	10^7 Cycle fatigue strength, or endurance limit, ksi	Reference
Hy-Tuf	230	88.4	20
4340	150	74	20
4340	230	89.2	20
SAE 2330	150	75	20
4340	190	85	22
4340	260	98	22
4340	140	72	22
H-11	287	136.7	18

With the one exception of the H-11 reported fatigue strength, all of these data are well within the scatter band of Fig. 7. Since the maraging steels did not show an endurance limit, the 10^7 cycle fatigue strength was arbitrarily chosen for comparison. This figure refers to conventionally heat treated steels, not to certain "ausformed" steels which have shown considerable increases in fatigue strengths (18). It is seen that the fatigue strength for maraging steel at 260 ksi ultimate tensile strength is very similar to that of the quenched and tempered alloy steels. If a 10^8 cycle strength were chosen for comparison, the data points would lie about 9 ksi lower, still well within the scatter band.

Recently, a few checks of experimental higher strength maraging steels at the authors' laboratory indicate that maraging steel 10^7 cycle fatigue strength falls along the upper edge of the band of Fig. 7 extrapolated with the same curvature to more than 350,000 psi tensile strength. At these strength levels, 10^7 cycle fatigue strengths above 130 ksi have been attained. Three data points for these experimental maraging steels included in Fig. 7.

The maraging steel fatigue notch sensitivity, q, was found to be near unity, confirming findings of other investigators (3, 6). This indicates that maraging steel is fatigue notch sensitive to approximately the full extent of the notch stress concentration. Reported fatigue notch sensitivities for other alloy steels have ranged widely from about 0.5 for AISI 4340 at 230 ksi tensile strength to about unity for H-11 hot work die steel at 287 ksi tensile strength to well above unity for AISI 4340 at 150 ksi tensile strength (17, 20).

It should be noted that the fatigue results for the present investigation are for commercial material of various melt practices and that superfine surface polishing was not employed. Extremely fine surface preparation could reasonably be expected to improve the fatigue results over those reported here.

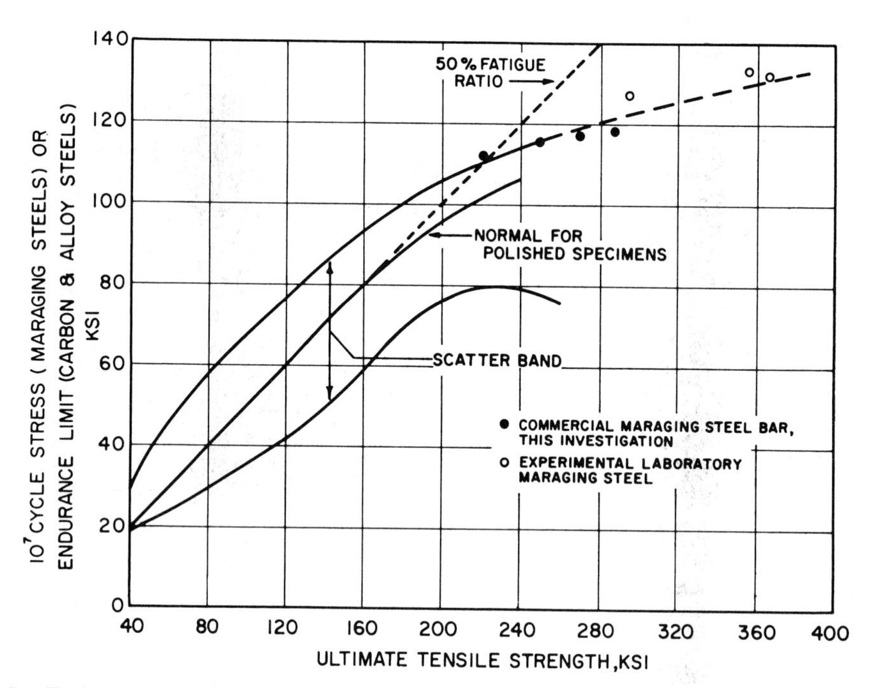

FIG. 7. Endurance limit for polished carbon and low alloy steel rotating beam fatigue specimens as a function of ultimate tensile strength from Bullens (17), and location of maraging steel mean 10^7 cycle fatigue strengths.

Processing Treatments

From this work and from other published evidence (8), the authors feel that the best fatigue specimen finishing sequence for maraging steel is to marage before finish grinding and polishing. This method parallels the preparation procedure for quenched and tempered high-strength steels which are invariably finish machined after hardening.

The high degree of scatter inherent in high-cycle fatigue data required that the comparisons among processing treatments be run at a stress that would cause failure in approximately 10^5 to 10^6 cycles. The stress of 130 ksi proved to be very satisfactory, giving a spread of almost two orders of magnitude among the various treatments while retaining relatively low scatter within each treatment group. Rigorously, the data apply only to 130 ksi, and great caution should be taken in making extrapolations from these data to higher cycle values (above 10^6 cycles). As a rough check on actual 10^7 cycle life, a few specimens of steel 5 were annealed at 1400, 1500, and 1900 F, and tested at stresses lower than 130 ksi. Approximate stresses to cause failure in 10^7 cycles were 120, 115, and 92 ksi for the 1400, 1500 and 1900 F treatments respectively. Thus the comparisons between treatments at 130 ksi, which showed longest life following an anneal at 1400 F, do reflect higher cycle fatigue behavior.

Although using only three replicate specimens leads to wide estimates of variance, statistical "t test" comparisons can still be made between the various processing treatments. The appropriate method of calculation involves an estimate of the standard deviation for each separate group of specimens after first converting the fatigue life data to log values to approximate a normal distribution. The statistical method used is outlined particularly clearly in reference 23.

Comparisons between treatments have shown that with 90% confidence (but not with 95% confidence), 1400 F annealing did indeed increase fatigue life over 1500 F annealing. Likewise, with 90% confidence, each annealing temperature above 1500 F produced a significantly lower fatigue life than did 1500 F. Also, a double annealing treatment of 2000 F + 1500 F produced a significant improvement with 90% confidence, over annealing at 2000 F alone. As a word of caution in interpreting the above, it should be noted that strict statistical randomization of the testing program was not carried out. On the other hand, the writers are not aware of any bias operating within the data.

Referring again to the 1400 F anneal, Fig. 4 shows that this heat treatment also produced a slightly higher tensile strength. The question arises, was the increased fatigue life due solely to higher strength or to other factors as well? Two things suggest that the higher strength was not wholly responsible for the increased fatigue life. First, a comparison of the 1400 F anneal mean fatigue life at 130,000 psi with Fig. 2 shows that the 1400 F data point lies substantially to the right of the "300 ksi" grade maraging steel curve even though the tensile strength of the 1400 F point is lower. Second, there is very little drop in tensile strength as annealing temperature is raised above 1500 F while fatigue life drops significantly. Hence, one would expect that the factor responsible for the clear-cut dependency of fatigue life on annealing temperature above 1500 F should also be operating to some extent at an annealing temperature of 1400 F.

Causes of the effect of annealing temperature on fatigue life are not clear. Grain size remained fine until some temperature between 1700 and 1900 F was reached; above 1900 F, grain growth was rapid. The fatigue life vs temperature curve, however, showed no discontinuity between 1700 and 1900 F which would correspond to the grain growth. Change of fracture mode roughly paralleled the grain size behavior. Below about 1700 F the fractures were transgranular, above 1700 F specimens began to show some intergranular fracture, at 2000 F fracture was predominately intergranular. Thus it is probable that above about 1700 F fatigue is influenced by grain boundary weakening or embrittlement. Below 1700 F some other effect may be operating.

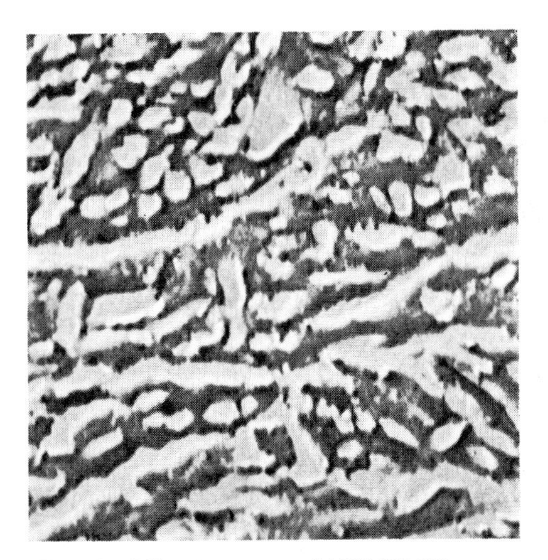

FIG. 8. Microstructure of 18% Ni 250 maraging steel "solution annealed" 1 hr at 1300 F + maraged 3 hr at 900 F. Light phase is austenite. Etchant, modified Fry's reagent. ×20,000.

The 1300 F annealing data is worthy of brief mention. This treatment is not a solution anneal in the full sense, since upon cooling to room temperature considerable stable austenite remains in the microstructure (Fig. 8). The austenite is responsible for the considerably lower tensile strength (227 ksi), and very much lower proportional limit (77 ksi). Fatigue life, however, is respectably good, higher than would be predicted for steels in general at this tensile strength. The role of the austenite in providing this apparent retardation of fatigue cracking remains to be defined. Perhaps the fine, rod-like dispersion of the phase is partly responsible.

Double annealing treatments, with the second anneal at a lower temperature than the first, always led to improved fatigue life as compared to a single anneal from the upper temperature. As an example, with a 2000 F high-temperature exposure, fatigue life was doubled by a second anneal at 1500 F. A possible explanation of this might lie in the refinement of structure during the several phase transformations which the double annealed specimens undergo. Reannealing at 1400 or 1500 F results in essentially full austenitization of the structure and retransformation to martensite upon cooling. The formation of new intragranular structure during these transformations may lessen possibly harmful prior grain boundary effects caused by the original high temperature anneal.

The innocuous effects of mild "overaging" or "underaging," or the simulation of a 1200 F (partially austenite reverted) weld head affected zone further indicate that maraging steel behaves in fatigue about as would be predicted from the effect on tensile properties.

The significant lowering of fatigue life in specimens maraged after the finish grinding and polishing operation could be the result of several factors. Although the grinding technique used should minimize residual surface stresses, it is possible that small, favorable residual compressive stresses have led to the increased life. On the other hand, maraging as a final step produces a very thin, relatively adherent oxide film on the specimens. This slight surface oxidation may serve as a weak fatigue layer or a surface roughening agent to promote early crack formation. The greatly improved fatigue life of specimens given the shot peening treatment (whether peened on a finish ground or on a ground and maraged surface) shows that maraging steel responds to peening in much the same manner as do conventional quenched and tempered steels. Using data of Lessells and Brodrick (24) for steel of equal tensile strength (260 psi), the particular shot peening treatment used in this investigation should have produced compressive stresses of about 132 ksi in the surface layer.

Closing Remarks

Work discussed in this paper involved only rotating beam tests on specimens from bar stock or from $\frac{1}{2}$-in. cross rolled plate, the specimen axis being parallel to the final rolling direction. Consequently, it should not be implied that the improvements noted for certain of the treatment variables would necessarily also apply to other types of fatigue tests or to other product forms, especially if tested in the transverse direction.

It is also possible that the 1400 F single anneal found in this investigation to give greatest fatigue life may not be applicable to all forms and conditions of maraging steel. In general, the best fatigue life appeared to be associated with a fine uniform prior austenite grain size. If a particular heat of maraging steel requires 1650 F or above for complete recrystallization to such a grain structure, then a 1400 or 1500 F secondary anneal may prove beneficial. Care should be taken to ensure that sufficient time has been allowed for adequate homogenization when annealing at 1400 F or slightly higher, otherwise, excessive amounts of austenite may be left in the final maraged structure, causing loss of tensile or fatigue strength. It should also be noted that in at least one instance annealing near 1400 F rather than the usual 1500 F has been found to reduce the fracture toughness of 18% Ni maraging steel (3).

Conclusions

Investigation of the fatigue properties of five commercial scale heats of maraging steel has shown that:

1. Rotating beam fatigue strength at 10^7 cycles for conventionally heat treated maraging steel was approximately equal to the endurance limit or 10^7 strength of low alloy steels having the same ultimate tensile strength.

2. A high annealing temperature was detrimental to fatigue life of maraging steel. A subsequent lower temperature

annealing treatment in the temperature range 1400 to 1500 F after a high temperature anneal partially recovered fatigue life.

3. An annealing temperature of about 1400 F produced maximum fatigue life, provided that the final maraged structure did not contain an excessive amount of austenite.

4. Slight "overaging," "underaging" or simulation of a 1200 F weld heat affected zone did not significantly affect fatigue life.

5. Shot peening maraging steel surfaces greatly extended fatigue life.

ACKNOWLEDGMENTS

The authors wish to thank The International Nickel Co., Inc., for permission to publish this paper. They also wish to thank R. F. Decker, S. Floreen and C. J. Novak for helpful discussions and suggestions during the experimentation, and D. T. Catchpole, C. J. Verona and R. S. Pollock for the testing of the specimens.

REFERENCES

1. R. F. Decker, J. T. Eash and A. J. Goldman, 18% Nickel Maraging Steel, ASM Trans Quart, 55 (1962) 58.
2. S. Floreen and R. F. Decker, Heat Treatment of 18% Ni Maraging Steel, ASM Trans Quart, 55 (1962) 518.
3. E. P. Gilewicz, W. A. Fragetta, V. Mehra and R. Krohn, Research on the Binary Iron-Nickel Alloys with 20 to 25 Per Cent Nickel, Technical Documentary Report No. ASD-TDR-62-996, Air Force Materials Laboratory, Research and Technology Division, Air Force Systems Command, Wright-Patterson Air Force Base, Ohio, June, 1964.
4. R. L. Jones and F. C. Nordquist, An Evaluation of High Strength Steel Forgings, Technical Documentary Report No. RTD-TDR-63-4050, Air Force Materials Laboratory, Research and Technology Division, Air Force Systems Command, Wright-Patterson Air Force Base, Ohio, May, 1964.
5. S. Floreen and G. R. Speich, Some Observations on the Strength and Toughness of Maraging Steels, ASM Trans Quart, 57 (1964) 714.
6. J. E. Campbell, F. J. Barone and D. P. Moon, The Mechanical Properties of the 18% Nickel Maraging Steels, DMIC Report No. 198, Battelle Memorial Institute, Columbus, Ohio, Feb. 24, 1964.
7. S. R. Swanson, A Survey of the Fatigue Aspects in the Application of Ultra High Strength Steels, DMIC Report No. 210, Problems in the Load Carrying Application of High-Strength Steels, Battelle Memorial Institute, Columbus, Ohio, October, 1964.
8. British Iron and Steel Institute Special Report No. 86, Metallurgical Developments in High Alloy Steels, Maraging Steels, W. Steven, 115 and Maraging Steels-Surface Protection and Surface Effects, A. G. Haynes (1964) 125.
9. M. Field and J. F. Kahles, The Surface Integrity of Machined-and-Ground High-Strength Steels, DMI Report, 210 Problems in the Load Carrying Application of High-Strength Steels, Battelle Memorial Institute, Columbus, Ohio, Oct. 26–28, 1963, 63.
10. F. Cicci, An Investigation of the Statistical Distribution of Constant Amplitude Fatigue Endurances for a Maraging Steel, Institute for Aerospace Studies, University of Toronto, UTIAS Technical Note No. 73, July, 1964.
11. American Society for Testing and Materials, A Guide for Fatigue Testing and the Statistical Analysis of Fatigue Data, ASTM Special Technical Publication No. 91-A, Second Edition (1963).
12. American Society for Testing and Materials, Manual on Fatigue Testing, Special Technical Publication No. 91 (1949).
13. W. Weibull, Fatigue Testing and Analysis of Results, Pergamon Press, New York (1961).
14. W. A. Petersen, Weld Heat-Affected Zone of 18% Nickel Maraging Steel, Welding J, 43, No. 9 (1964) Welding Research Suppl, 428-S.
15. S.A.E. Standard J442, Test Strip, Holder and Gage for Shot Peening, SAE Handbook (1964) 153.
16. D. J. Finney, Probit Analysis, Cambridge University Press (1952).
17. D. K. Bullens, Steel and Its Heat Treatment, 1, (1938) 37.
18. F. Borik, W. M. Justusson and V. F. Zackay, Fatigue Properties of an Ausformed Steel, ASM Trans Quart, 56 (1963) 327.
19. P. G. Forrest, Fatigue of Metals, Addison-Wesley Publishing Co., Reading, Mass. (1962).

20. J. W. Spretnak, M. G. Fontana and H. E. Brooks, Notched and Unnotched Tensile and Fatigue Properties of Ten Engineering Alloys at 25°C and −196°C, Trans ASM, 43 (1951) 547.
21. H. E. Frankel, J. A. Bennett and W. A. Pennington, Fatigue Properties of High Strength Steels, Trans ASM, 52 (1960) 257.
22. H. N. Cummings, F. B. Stulen and W. C. Schulte, Relations of Inclusions to the Fatigue Properties of SAE 4340 Steel Trans ASM, 49 (1957) 482.
23. M. G. Natrella, Experimental Statistics, National Bureau of Standards Handbook No. 91, U. S. Dept. of Commerce (1963).
24. J. M. Lessells and R. F. Brodrick, Shot-Peening as Protection of Surface-Damaged Propeller Blade Materials, International Conference on Fatigue of Metals, Institution of Mechanical Engineers, London and New York (1956) 7.

D. WEBBER

Fatigue results on welded 90 Tonf/in^2 maraging Steel

SYNOPSIS: Results of constant amplitude fatigue tests on $\frac{3}{16}$ in. thick plate from three manufacturers are described together with results of butt and fillet welds made by the Mig process. The results are mainly from fluctuating tension tests but some alternating results have been obtained to enable a Goodman diagram to be constructed. The effects of defects on fatigue life are discussed and the results are compared with those for lower strength steels. The addition of 1% oxygen to the argon shielding gas has also been investigated.

PORTABILITY is an important consideration in the design of military equipment and the designer of such equipment is considerably aided by the availability of high strength steel and aluminium alloys. The equipment must also be relatively robust, readily and economically constructed and be repairable in army workshops. These requirements are frequently best met by welding. However, very high strength is rarely compatible with weldability and even if it is, often involves complications in machining, heat-treatment, and control of distortion; also it is often associated with low impact and ductility properties in the vicinity of the weld.

Up to the present, these limitations have resulted in equipment having relatively highly stressed joints either being made by welding Fortiweld or Ducol 25 high strength constructional steel, or in instances where weight saving is of great importance by riveting high strength aluminium alloys.

Currently, weldable medium strength Al-Zn-Mg alloys (21 tonf/in^2 0·1% proof stress) are being used in several projects and are attractive since welds which have a heat-affected zone proof stress of about $\frac{2}{3}$ that of the plate age at room temperature.

A recent development in weldable steels is ultra-high strength 18% nickel maraging steel having a 0·2% proof stress of 90 tonf/in^2. This steel has a superior strength/weight ratio to the weldable medium strength Al-Zn-Mg alloys. It also has many other attractive properties, those published by INCO[1] being as follows:

- Ultra-high tensile strength: 95 tonf/in^2.
- Proof stress values (0·2%): 90 tonf/in^2.
- Exceptionally high toughness and impact resistance.
- Good properties at sub-zero and moderately elevated temperatures.
- Softened by a simple air cool from about 800°C to a hardness of: 200—300 DPN.
- Machined and cold formed easily.
- Hardened by simple low-temperature heat-treatment at 450—500°C.
- Quenching is unnecessary, even for the largest sections.
- Volume changes either do not occur during hardening, or are of a very low order.
- Decarburization problems do not exist.
- High resistance to stress-corrosion cracking in sea water is a characteristic of the 18% nickel maraging steel.
- Good weldability, even in the fully-hardened condition. Preheat is unnecessary, properties restored by simple low-temperature treatment.

The cost of this material is high at present (15 to 20s./lb) but, if it can be justified, it is obviously a most attractive constructional material.

A design study where weight saving was of paramount importance resulted in a project being initiated for the construction of a new military bridge in 18% nickel maraging steel. At the start of this investigation, the only fatigue information available was on smooth rotating bending specimens where the endurance limit was quoted as 42·5 tonf/in^2 at 10^7 cycles. But since welded high strength constructional steels in general have a poor fatigue resistance compared with that of the plate, the present investigation into the fatigue life of welded specimens was initiated, as this could severely limit the application of the material.

The tests are by no means complete but sufficient information is now available to indicate the likely fatigue properties of the welded material particularly at stresses above the ultimate strength of materials that have been used to date.

Test Specimens

Material. 18% nickel maraging steel of the 90 tonf/in^2 grade had not previously been made in any quantity in Britain, so orders were placed with three manufacturers, whose material will be identified as A, B and C when referring to results. Typical properties quoted by INCO[1] together with those actually obtained are given in Table 1 together with a chemical analysis. The static properties of the plate from all three manufacturers were sufficiently up to specification, although a non-standard heat-treatment was required for material A. No direct comparison has been made of percentage elongation owing to differing gauge lengths. Cold work due to hammer flattening of some of the sheets appeared to increase slightly the mechanical properties of material A.

The comparatively high strength of the material, even in the unaged condition, caused some finishing difficulties. The surface finish of material A was acceptable by constructional steel standards but it was not flat or of uniform thickness and some was rectified by hammer flattening. The finish of material B was good, but that of material C

SPECIMEN	B.S. 153 Classification
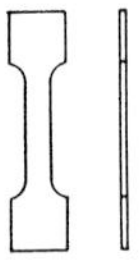 Plain Plate Ground Edges	A
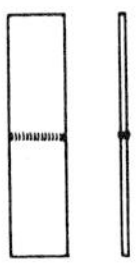 Transverse Butt Weld	E
Transverse Butt Weld Reinforcement Dressed Flush	D
Transverse Butt Weld Continuous Free of Defects Longitudinally Ground	A
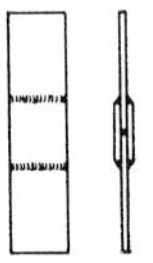 Transverse Load Carrying Fillet Weld	F
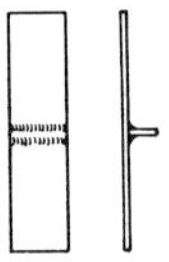 Transverse Non-load Carrying Fillet Weld	F
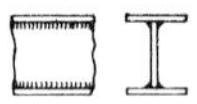 Continuous longitudinal Fillet Welds with Start Stop Positions	D

Fig. 1. Fatigue specimen configuration.

contained surface laps. The results refer only to $\frac{3}{16}$ in. thick plate since the majority of the welds in the full size structure were to be made in this thickness.

Specimen Types. Specimens were chosen to represent details that occurred in the design under consideration, plate specimens were also included as a reference. Specimen configurations tested are listed below:

Plain plate.
Transverse butt welds.
Dressed transverse butt welds.
Ground transverse butt welds.
Load carrying transverse fillet welds.
Non-load carrying transverse fillet welds.
Non-load carrying longitudinal fillet welds.

The form and classification of the specimens according to B.S. 153 "Steel Girder Bridges" fatigue clause is shown in Fig. 1.

Owing to limitations of machine load capacity it was necessary to vary the width of some specimens in order to obtain the required stresses. The widths used are listed in the tables.

Negligible distortion occurred during maraging, but in addition it was possible to "creep straighten" specimens that pulled out of line during welding, by jigging them slightly beyond the desired alignment in a stiff jig during maraging.

Welding. All specimens were welded with Mig equipment using $\frac{3}{64}$ in. diameter wire manufactured by supplier C, at indicated voltages and currents of 27—30 A and 240—280 A respectively. Gas flow was approximately 45 ft^3/h. Settings tended to the lower limit when using argon/ 1% oxygen shielding gas instead of pure argon. The composition of the welding wire was similar to that of the parent metal apart from a small increase in titanium content.

Method of Testing. Static tests were carried out in a 50-ton Denison universal testing machine, and fatigue tests in a Losenhausen U.H.S. 20 universal fatigue machine fitted with a slow cycling facility. The facility was used for load ranges above 10 tonf when the speed of testing was approximately 30 c/min. At smaller load ranges the pulsator was used and the testing speeds were 330 c/min for the first 10 minutes of a test, 500 for the next 50 min and then 660 c/min.

Pulsating tension tests were carried out at various stress ranges up to 84 tonf/in^2 with a minimum stress of 8 tonf/ in^2, which represented the approximate dead load stress in the actual design.

Alternating tension and compression tests were carried out up to a maximum stress of 48 tonf/in^2, which was the limit of the machine capacity for an acceptable specimen width, and to prevent superimposed bending stresses due to buckling. The beam specimens were stressed from 8—48 tonf/in^2 maximum fibre stress, in symmetrical three-point bending using the bending table of the machine.

Failure was taken as the formation of a fast running crack. This corresponded to complete rupture of the tension specimens or gross loss of stiffness in bending specimens.

Unless otherwise stated, all stresses quoted are based on plate area adjacent to the welds, reductions of section due to undercut have been neglected. Load carrying transverse fillet weld specimen stresses are quoted as plate stresses since for the specimen geometry used[2] failures should occur through the plate from the toe of the welds.

Results and Discussion

It is becoming common to plot fatigue results on a log stress versus log cycles to failure graph, and to fit a straight line equation to the results in the finite life region and, although straight lines could be drawn through the scatter band when the present results were plotted on this basis, they were clearly not the best fit for the large stress ranges covered in these tests. The results have, therefore, been plotted on an S-LogN basis so that the stress range as well as the maximum stress in the cycle can be plotted on the same graph. Complete results are given in the tables.

Plain Plate (Table 2. Fig. 2). Forty-three fatigue results for plate specimens taken from the three manufacturers fell within a common scatter band. Scatter, measured as the perpendicular width of the band, is fairly constant throughout the stress range, and it was possible to load cycle specimens up to the proof stress of the metal where lives of from 2,000 to 5,000 c were obtained.

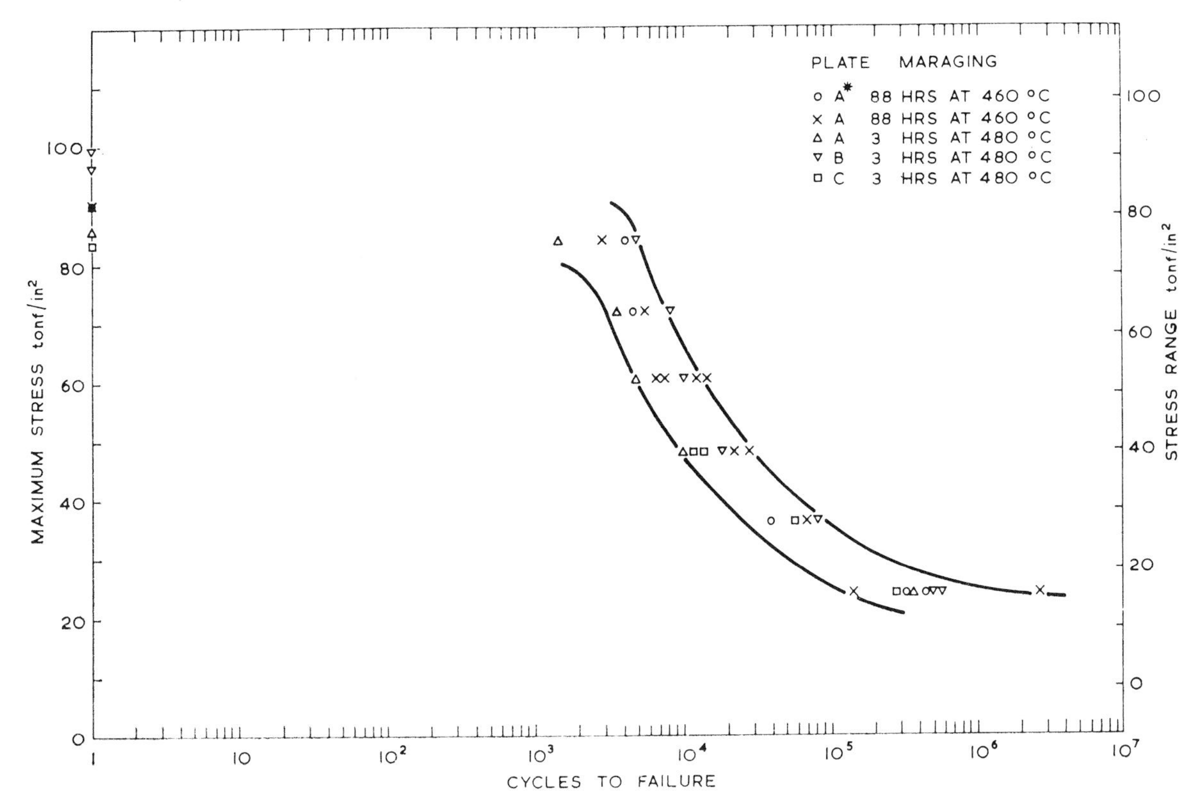

Fig. 2. Plain plate fatigue results.

The low stress end of the fatigue curve has not been well defined, but an endurance limit appears to be reached between 10^6 and 10^7 c. The life of the specimens at low stress appears to have been controlled mainly by the quality of the surface finish of the plate, while at high stresses the ultimate tensile strength appears to have a small but significant influence on life. The good mechanical properties and surface finish made material B superior overall and any benefit that might have been obtained from vacuum melting used by the manufacturers of material C was masked by a poor surface finish.

Butt Welds (*Table* 3. *Figs.* 3 *and* 5). Static tests to failure on sound welds gave results similar to those in plain plate. These are tabulated in the fatigue results as ½ c. Thirty of the 37 butt weld fatigue results are included in the scatter band drawn in Fig. 3. This band

Chemical Composition.

	A	B	C	INCO
C	0·014	0·028	0·0103	0·03 max.
Mn	0·01	0·03		0·10 ,,
Si	0·05	0·16		0·10 ,,
S	0·012	0·01	0·007	0·01 ,,
P	0·005	0·007	0·004	0·01 ,,
Ni	17·76	17·8	18·23	17—19
Ti	0·15	0·24	0·14	0·15—0·25
Co	8·40	7·98	8·62	8—9
Mo	3·27	3·30	3·00	3·0—3·5
Al	0·113	0·055	0·01	0·05—0·15
Bo	0·0035	0·003		0·003 added
Pb	0·0015	0·002		
Zr	0·02	0·010		0·02 added

Table 1. Static Properties of $\frac{3}{16}$ in. Thick 90 tonf/in² Maraging Steel Plate.

Material	0·2% Proof Stress Tonf/in²	Maximum Stress Tonf/in²	Maraging Treatment	% Elongation	Gauge Length
A*	83·6	90·2	3 h @ 480°C	9	2 in.
A	77·9	85·2	,, ,, ,, ,,	8·25	4 in.
A	79·5	85·3	,, ,, ,, ,,	7·75	4 in.
A	79·2	87·5	,, ,, ,, ,,	7·5	4 in.
A	87·2	93·2	88 ,, ,, 460°C	—	—
B	89·5	96·1	3 ,, ,, 480°C	7·5	2 in.
B	89·2	100·2	,, ,, ,, ,,	—	—
B	49·4	62·5	Unmaraged	7	2 in.
C	85·7	90·4	3 h @ 480°C	3·75	4 in.
C	86·1	89·0	,, ,, ,, ,,	5·75	4 in.
INCO(3)	85—94	89—98	,, ,, ,, ,,	14-16	$4{\cdot}5\sqrt{A}$

*Hammer flattened sheet.
(3) Maraging Steels for High Strength Application. The International Nickel Co.

Table 2. **Fatigue Properties of $\frac{3}{16}$ in. Thick 90 tonf/in^2 Maraging Steel Plate.**
Pulsating Tension. Minimum Stress in cycle 8 tonf/in^2. Specimen width 1·25 in.

Max. Stress Tonf/in^2	Cycles to Failure	Max. Stress Tonf/in^2	Cycles to Failure	Max. Stress Tonf/in^2	Cycles to Failure	Max. Stress Tonf/in^2	Cycles to Failure	Max. Stress Tonf/in^2	Cycles to Failure
Material A		Material A		Material A		Material B		Material C	
Maraged 88 h. @ 460°C		Maraged 88 h @ 460°C		Maraged 3 h @ 480°C		Maraged 3 h @ 480°C		Maraged 3 h @ 480°C	
84	4,353	84	2,992	84	1,591	84	4,780	48	14,782
72	4,810	72	6,260	72	3,693	72	8,200	48	11,828
60	10,130	60	14,479	60	4,891	60	7,233	36	56,040
60	10,479	60	12,672	60	6,907	60	10,267	24	296,410
48	14,840	48	29,040	48	9,523	48	20,520		
48	19,053	48	24,900	48	11,919	48	19,230		
36	39,880	36	64,650	36	73,080	36	86,720		
36	73,990	36	73,470	24	389,730	36	83,040		
24	349,420	24	143,680			24	578,554		
24	446,070	24	2,600,000			24	533,989		

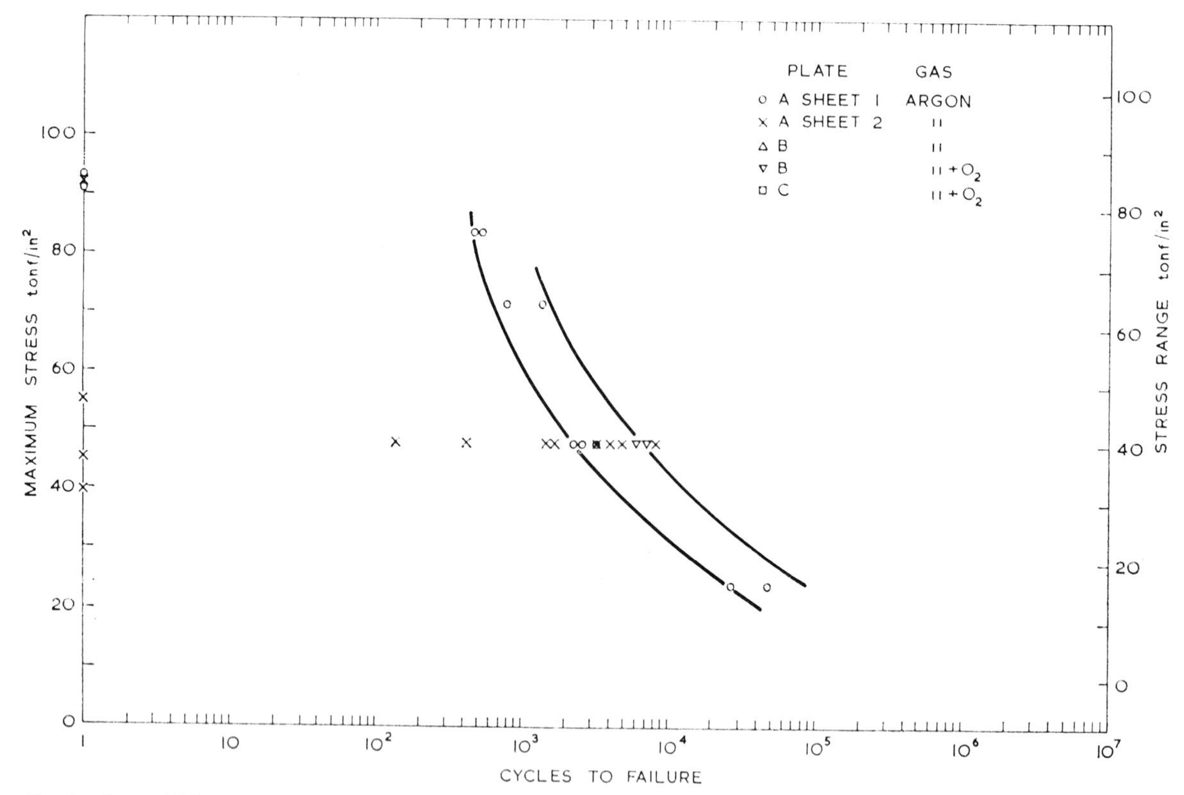

Fig. 3. Butt weld fatigue results.

is similar in width but falls lower than that of the plain plate band and does not show an endurance limit up to the limit of testing 10^7 c. All specimens that failed below the scatter band contained defects. Again, it was possible to load cycle specimens up to the 0·2% proof stress but lives were only 400—1,300 c.

Inspection of Table 3 shows that there is no significant difference between welds made in plate from different suppliers or whether argon or argon/1% oxygen was used as a shielding gas.

The former is not unexpected owing to the overriding effect of the welding process at the point of failure, but it was surprising to note that there was no significant advantage with 1% oxygen in the welding gas which gave a more stable arc much preferred by the welders. However, the difficulty of welding with argon shielding gas may become apparent when welders have to work for longer periods and under adverse conditions. INCO do not support the addition of oxygen to the shielding gas because of the formation of oxides, but in practice no significant deleterious effects that can be attributed to the presence of oxygen in the gas have been noted.

The results of alternating axial load tests have been used together with those from the fluctuating load tests in the

Table 3. Weld Results. Butt Welds.

Fluctuating Tension. Minimum Stress 8 tonf/in^2.
Argon shielding gas.

Material A Plate 1		
Max. Stress tonf/in^2	**Cycles to Failure**	**Note***
Specimen 1·25 in. wide		
Maraged 88 h @ 460°C		
84	523	T
84	477	T
72	1,384	T
72	787	T
48	2,853	T
48	2,483	T
48	3,099	T
24	48,460	T
24	27,190	T
93·2	½	T
91·5	½	T

Material B		
Max. Stress tonf/in^2	**Cycles to Failure**	**Note**
Specimen 1·25 in. wide		
Maraged 3 h @ 480°C		
48	3,570	T
48	4,800	T
48	5,490	T

Material A Plate 2		
Max. Stress tonf/in^2	**Cycles to Failure**	**Note**
Specimen 1·25 in. wide		
Maraged 88 h @ 460°C		
48	4,680	T
48	4,910	T.P.
48	4,730	T
48	3,760	T
Specimen 0·5 in. wide		
Maraged 3 h @ 480°C		
48	1,727	T.P.
48	2,546	T.P.
48	1,682	T.P.
Maraged 88 h @ 460°C		
48	140	W.C.
48	410	W.P.
48	5,990	T
48	3,670	T.P.
48	3,970	T
39·6	½	W.C.P.
45	½	W.C.
56·7	½	W.C.
92·0	½	W

*Note: T = Toe failure.
W = Weld metal failure.
P = Porosity in weld metal on plane of failure.
C = Columnar structure.

construction of a modified Goodman diagram, Fig. 5, so that lives can be predicted for other combinations of stress. Since these specimens were narrower than the fluctuating tension fatigue test specimens a few were tested in tension to see if there was a width effect but none was noted.

Dressing of the overfill or reinforcement of the weld bead, but not removing undercut, resulted in a general increase in life, but minimum results still fell near the lower limit of the scatter band, almost certainly because of the effects of residual undercut. Longitudinal grinding to remove the weld bead and undercut resulted in lives similar to those obtained for plain plate. If the stress is based on gross area of plate, the improvement, if any, was not very large because of the large percentage loss in cross-section with $\frac{3}{16}$ in. thick plate.

Fatigue failure of untreated specimens with sound welds was from the toe of the weld through the plate. Those with excessive porosity, which was clearly visible on the surface, failed through the weld metal, sometimes with a marked reduction in life, or even failed statically. Specimens cut from plate with part of the weld containing excessive porosity also frequently had one weld run of the failure face showing a marked columnar structure but they did not necessarily contain porosity, and sometimes resulted in very short lives. Fig. 6 is typical of this type of failure, the cause of which has not been established, but it is almost certainly a grain boundary effect. It is interesting also that its occurrence was also noted when argon/1% oxygen was being used as a shielding gas. Fig. 7 is a portion of an X-ray showing a concentration of porosity which has not been detrimental to fatigue life.

Grinding the specimen in the region of the weld to remove all undercut produced failure through the as-rolled plate or from occasional small porosity voids that had been exposed on the ground surface.

Fillet Welds (*Tables* 4, 5 *and* 6. *Figs*. 4 *and* 5). Load-carrying transverse fillet welds were statically as strong as the plate. This was to be expected since the geometry was such that twice the plate area was in shear or tension along the weld legs, and the throat area was 1·4 times the plate area. The majority of the fatigue results fell within a well defined, but broader scatter band of results than plain plate and butt weld results. The fatigue curve is not as steep as that for butt welds and at a stress of 8—48 tonf/in^2 the results are similar.

Some results from an initial batch of specimens fell below the scatter band. These welds, however, had considerable overfill and consequently a sharp toe radius, and there was some undercut associated with surface porosity, probably due to inadequate cleaning in the region of the weld prior to welding. These results have not been taken into consideration in plotting the scatter band shown in full lines.

The shielding gas used appeared to have no effect on the results that fall within the scatter band.

Alternating specimen results, Table 4, have been used in the construction of a modified Goodman diagram Fig. 5.

Table 3 (*continued*). **Weld Results. Butt Welds.**

Fluctuating Tension. Minimum Stress 8 tonf/in^2.
Shielding Gas Argon/1% Oxygen.

Specimens 1·25 in. wide
Maraged 3 h. @ 480°C

Material B			Material C		
Max. Stress tonf/in^2	**Cycles to Failure**	**Note**	**Max. Stress tonf/in^2**	**Cycles to Failure**	**Note**
48	7,160	T	48	8,240	T
48	7,300	T	48	2,550	T
48	7,400	T	48	3,530	T
48	6,200	T	48	2,980	T

Alternating Tests. Mean Stress zero
Argon shielding gas

Material A Plate 1			Material A Plate 2		
Max. Stress tonf/in^2	**Cycles to Failure**	**Note**	**Max. Stress tonf/in^2**	**Cycles to Failure**	**Note**
Specimen 1·25 in. wide Maraged 88 h @ 460°C			**Specimen 0·5 in. wide Maraged 88 h @ 460°C**		
±36	4,830	T	±48	490	W.C.P.
±24	8,970	T	±48	130	W.C.P.
±12	68,920	T	±48	1,680	W.P.
±12	117,950	T	±36	2,960	T
			±36	2,870	W.C.
			±36	6,710	T.P.
			±24	8,050	W.C.P.
			± 6	1,000,840	Plate
			± 6	955,510	T.P.

Table 3 (*continued*). **Weld Results. Dressed Butt Welds.**

Pulsating Tension. Minimum stress 8 tonf/in²
Argon 1% oxygen shielding gas
Maraged 88 h. @ 460°C
Specimen 1·25 in. wide

Hand dressed to remove reinforcement			
Material A			
Max. Stress Tonf/in²	Cycles to Failure	Plate No.	Note
84	1,958	3	T
72	947	4	T
72	973	4	T
48	12,726	3	T
48	7,055	3	P
48	12,520	3	T
48	3,512	4	T
48	2,480	4	T
48	3,520	4	T
48	3,700	4	T
48	2,480	4	T
48	7,920	4	T
48	2,740	4	T

Machine ground to remove reinforcement and undercut				
Material A				
Max. Stress Tonf/in²	Cycles to Failure	Plate No.	Note	Max. Stress on original area
72	434	3	P	64·0
72	7,721	3	Pl	63·6
48	16,070	3	P	37·2
48	16,510	3	P	21·9
48	12,350	3	P	40·4
48	44,800	3	Pl	41·6
48	41,710	3	Pl	39·5
48	11,530	4	P	40·9
48	500	4	P	40·2
48	34,400	4	Pl	40·2
48	38,400	4	Pl	38·0
36	85,780	3	Pl	30·0
24	412,780	3	Pl	21·8
24	815,430	3	Pl	20·1
18	6,000,000†	3		8·6
18	23,000,000†	3		8·7

Note: Pl. = Plate.
† = did not fail.

These specimens were the same width as the alternating butt weld specimens, but likewise a few tested in pulsating tension showed no significant width effect compared with the other specimens.

Results for a limited number of non-load transverse carrying fillet welds stressed in pulsating tension also fell within the scatter band of results at and below 48 tonf/in², but a result at 60 tonf/in² fell within the plate scatter band.

Longitudinal fillet welds tested as beams in fluctuating bending from 8—48 tonf/in² maximum fibre stress generally gave exceptional results falling within the scatter band of the plate. In fact, several of the failures initiated at least partly from the plate surface, otherwise the point of initiation appeared to be at the weld start/stop positions. Transverse defects reduced the lives of three of the specimens so that they fell within the scatter band for load-carrying transverse fillet welds.

The results showed that unfavourable conditions at the toe of the transverse fillet weld such as overfill and undercut could have a significant effect, also lack of root penetration could cause short life weld failures. Gross porosity also resulted in reduced lives and even low strength static failure. Three of the four non-load carrying longitudinal fillet weld results contained transverse defects as shown in Fig. 8. They appeared as cracks with a columnar structure similar to that of the butt welds, but they were not detectable by normal production X-ray, magnetic, or dye penetrant crack detecting methods. Subsequent experience has shown that ultrasonic methods using an angled probe may be applicable where there is adequate accessibility.

General. The Goodman diagram shows bands of constant life for various combinations of maximum and minimum stress. Although each life line for the welded specimen results is based on only two combinations of stress, i.e., fluctuating tension with 8 tonf/in² minimum

Table 4. Weld Results. Load Carrying Transverse Fillet Welds.

Fluctuating Tension. Minimum Stress 8 tonf/in².
Argon shielding gas.

Material A Plate 1		
Max. Stress tonf/in²	Cycles to Failure	Note
Specimen 1·25 in. wide Maraged 88 h @ 460°C		
84	15	T
84	124	T
72	360	T
72	220	T
48	2,105	T
48	1,319	T
48	2,564	T
48	990	T
48	2,480	T
24	19,730	T
24	14,770	T
18	40,090	T
12	5,059,070	T
90·1	½	T

Material A Plate 2		
Max. Stress tonf/in²	Cycles to Failure	Note
Specimen 0·5 in. wide Maraged 88 h @ 460°C		
48	2,840	T
48	3,620	T
48	3,490	T
48	2,240	T
48	3,970	T
88·5	½	
89·0	½	
Specimen 1·25 in. wide Maraged 3 h @ 480°C		
48	2,745	T
48	63	W.P.
48	2,055	T

Material B Plate 1		
Max. Stress tonf/in²	Cycles to Failure	Note
Specimen 0·85 in. wide Maraged 3 h @ 480°C		
48	6,880	T
48	10,670	T
48	8,950	T
48	9,170	T
48	11,470	T

Table 4 (*continued*). Weld Results. Transverse Load Carrying Fillet Welds.

Alternating results, Zero mean stress, Argon shielding gas		
Material A Plate 2		
Max. Stress tonf/in^2	Cycles to Failure	Note
Specimen 0·3 in. wide, Maraged 88 h @ 460°C		
±48	2,090	T
±48	550	T
±48	870	T
±48	850	T
±48	780	T
±36	2,070	T
±36	1,610	T
±36	1,620	T
±36	1,840	T
±24	8,600	T
±24	6,590	T
±12	47,260	T
±12	23,590	T
± 6	125,730	T
± 6	175,850	T

Fluctuating Tension, Minimum stress 8 tonf/in^2, Argon/1% oxygen shielding gas		
Material B Plate 1		
Max. Stress tonf/in^2	Cycles to Failure	Note
Specimen 0·85 in. wide, Maraged 3 h @ 480°C		
48	8,250	T
48	8,860	T
48	10,140	T
48	6,950	T.P.
48	$\frac{1}{2}$	P
Material C Plate 1		
Max. Stress tonf/in^2.	Cycles to Failure	Note
Specimen 0·85 in. wide, Maraged 3 h @ 480°C		
48	3,820	T
48	3,850	T
48	7,410	T
48	5,490	T

stress and alternating, it is suggested that a straight line drawn from the ultimate tensile strength point through results obtained with approximately zero minimum stress will be a conservative fit for other combinations of stress up to the alternating stress case.

The general performance of the welds follows the classification given in B.S. 153, but, in detail, non-load-carrying transverse fillet welds did not appear to be an improvement over load-carrying welds, whereas sound longitudinal fillet welds performed better than expected, and butt welds were rather disappointing. This poor performance of butt welds was considered to be due mainly to the difficulty in making a satisfactory weld with the Mig equipment and wire available at the time.

The welds giving rise to the columnar structure and transverse defects were all made in a similar period and have not occurred since. It is possible that the cause was associated with a particular batch of gas or welding wire. The analysis of these welds was not significantly different from sound welds.

Comparison with Other Results. An interesting comparison can be made between the dressed butt weld results and those for a similar weld in several materials described by Frost and Denton[3]. Extrapolating the present results to zero minimum stress on a Goodman diagram and then plotting them on the curve given by Frost and Denton, Fig. 9 shows that for mild steel and higher strength alloy steels the results fall on a common curve until the ultimate strength of the material was approached. This follows the trend noted in B.S. 153 Steel Girder Bridges[4] where identical lives are given for mild steel and high strength construction steel B.S. 968.

Conclusions

These conclusions only relate to $\frac{3}{16}$ in. thick plate, and the specimens were made by one welder using the Mig process

Fig. 4. Fillet weld fatigue results.

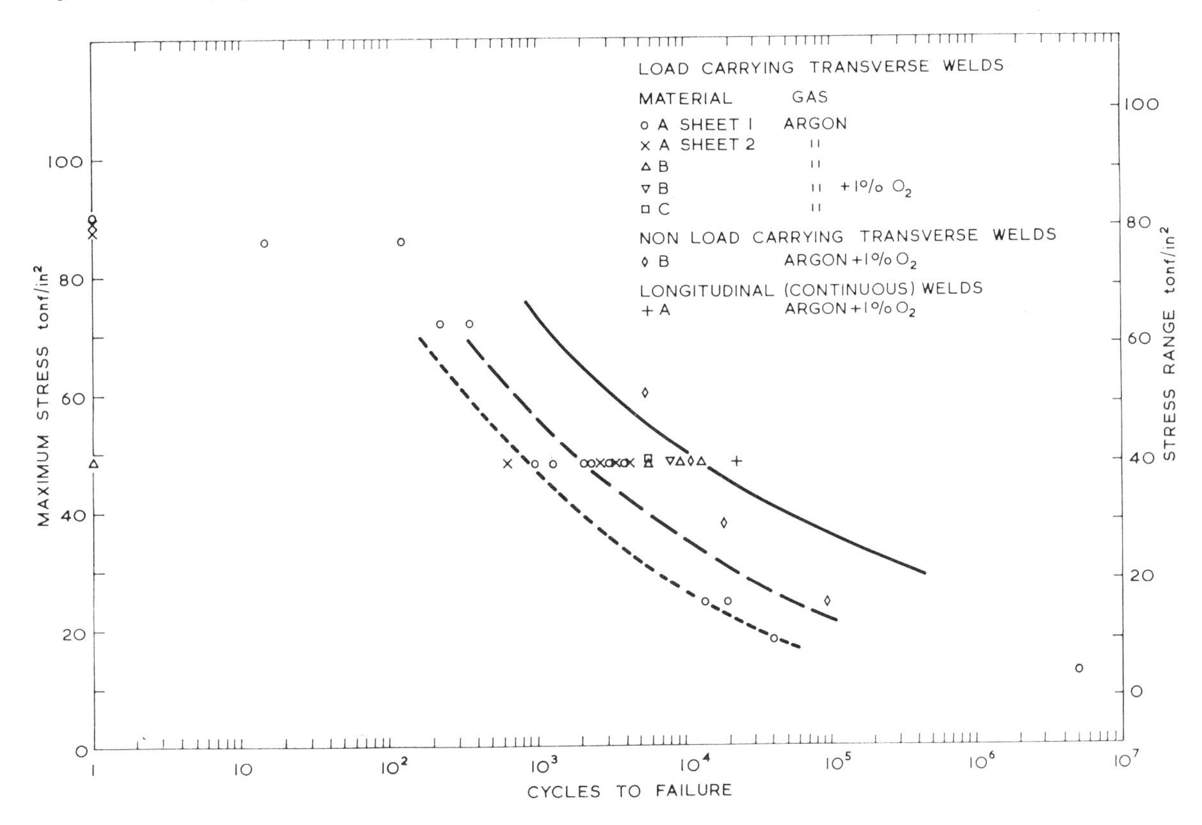

Source: *Welding and Metal Fabrication*, June 1968

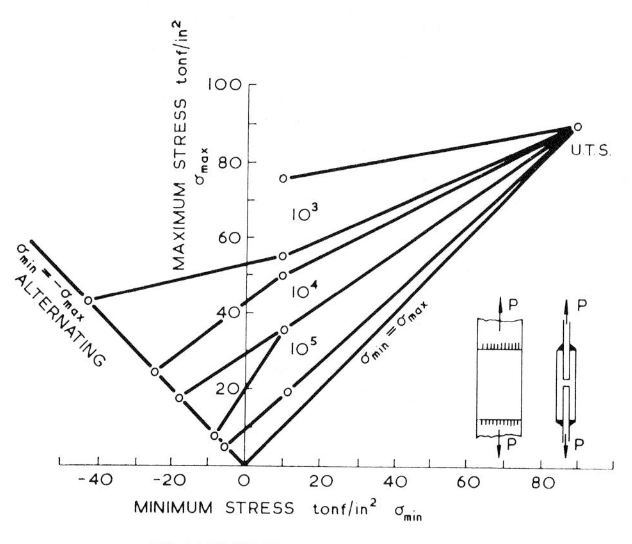

Fig. 5. Goodman diagrams.

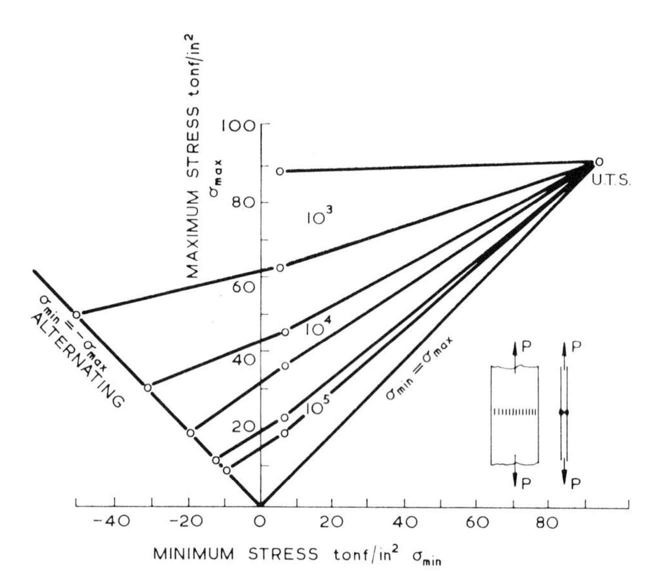

with argon or argon/1% oxygen as the shielding gas. The results could not be represented by a straight line relationship on linear or log scales.

1. All results for plain as-rolled plate specimens from three sources were similar when the stress range was from near zero minimum stress to approximately half ultimate stress. At lower stress ranges, good surface finish resulted in longer lives, while at higher stress ranges the higher ultimate strength specimens had longer lives. The endurance limit at 10^7 c was approximately 12—18 tonf/in^2 range at zero minimum stress.

2. Transverse butt welds results had similar scatter to the plate results but fell below them. There was no definite endurance limit. Dressing off the weld bead overfill without removing the undercut, increased the average fatigue life, but minimum lives remained the same. Grinding

Table 5. Weld Results. Non-Load Carrying Longitudinal Fillet Welds.

Maximum Stress 48 tonf/in^2. Minimum Stress 8 tonf/in^2. (on max. fibre stress) 5×3 in. I section made in $\frac{3}{16}$ in. plate tested in three point bending with 40 in. span.
Maraged 3 h @ 480°C
Material B

Welding Gas	Cycles to Failure	Location of Failure
Argon/1% oxygen	9,707	Transverse crack
Argon/1% "	6,796	" "
Argon/1% "	25,436	Plate
Argon/1% "	13,097	Plate/weld root
Argon	7,373	Transverse crack
Argon	14,508	Plate

Table 6. Non-Load Carrying Transverse Fillet Welds.

Fluctuating Tension Results. Minimum Stress 8 tonf/in^2.
Maraged @ 480°C. Specimen 0·85 in. wide
Material B. Argon shielding gas

Max. Stress tonf/in^2	Cycles to Failure	Note
60	5,930	T
48	9,840	T
36	18,420	T
24	93,610	T

T = Toe failure.

in the direction of loading to remove all undercut gave lives at least as good as the plate when stresses were based on the nett area remaining. The improvement was not worthwhile when stresses were calculated on gross area owing to the percentage loss of cross-section.

3. Load carrying transverse weld results were slightly inferior to the butt weld results at the lower limit of the scatter band and the best results were similar.

4. Longitudinal and transverse non-load carrying fillet welds had similar lives to the load-carrying welds at 8—48 tonf/in^2 stress range.

5. A straight line Goodman diagram from ultimate stress

Fig. 6. Butt weld failure showing columnar structure and porosity.

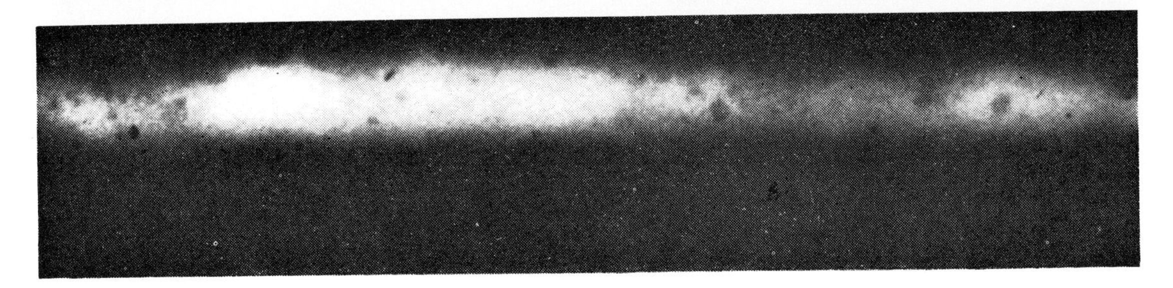

Fig. 7. Acceptable porosity in butt weld shown in an X-ray.

to pulsating tension gave a conservative representation for other stress ranges.

6. Excluding gross defects, weld profile including undercut has the largest effect on fatigue life.

7. The addition of 1% oxygen to the argon shielding gas gave a more controllable arc with no effect on the fatigue life.

8. The welds were insensitive to minor detectable defects, however other defects exist that are not readily detectable by conventional means and may shorten the fatigue life.

9. Low stress range weld fatigue results are similar to those for lower strength steels. Therefore, this steel is only suitable for short life applications or where a high level of mean stress exists.

Fig. 8. Fatigue failure in a fillet weld showing transverse weld crack.

Future Work

Much work remains to be done and it is hoped to add to the present data and obtain statistically valid data so that confidence limits can be quoted and extend the combinations of stress to obtain a more complete Goodman diagram. Differences between welders and the cumulative damage aspects must be investigated.

Other methods of welding such as manual metal-arc electrodes, pulsed Mig, Tig and electron beam all have virtues which should be evaluated together with mechanical methods of improving fatigue strength such as peening and preloading. Methods of repair also require investigation.

Acknowledgements

The author wishes to thank all of his colleagues in M.E.X.E. for their help in acquiring data and time spent in discussions during the preparation of this paper. The illustrations are Crown Copyright.

References

1. INCO. Maraging Steels for Ultra-High Strength Application.
2. D. S. Macfarlane and J. D. Harrison. Some fatigue Tests on Load carrying Transverse Fillet Welds. *Brit. Welding J.*, 1965, Vol. 12, pp. 613-623.
3. N. E. Frost and K. Denton. The Fatigue Strength of Butt-welded joints in Low Alloy Structural Steels, N.E.L. Report No. 199.
4. B.S. 153 "Steel Girder Bridges", Part 3B, Amendment No. 4.

Fig. 9. Dressed butt weld fatigue. (S/N curve[3]).

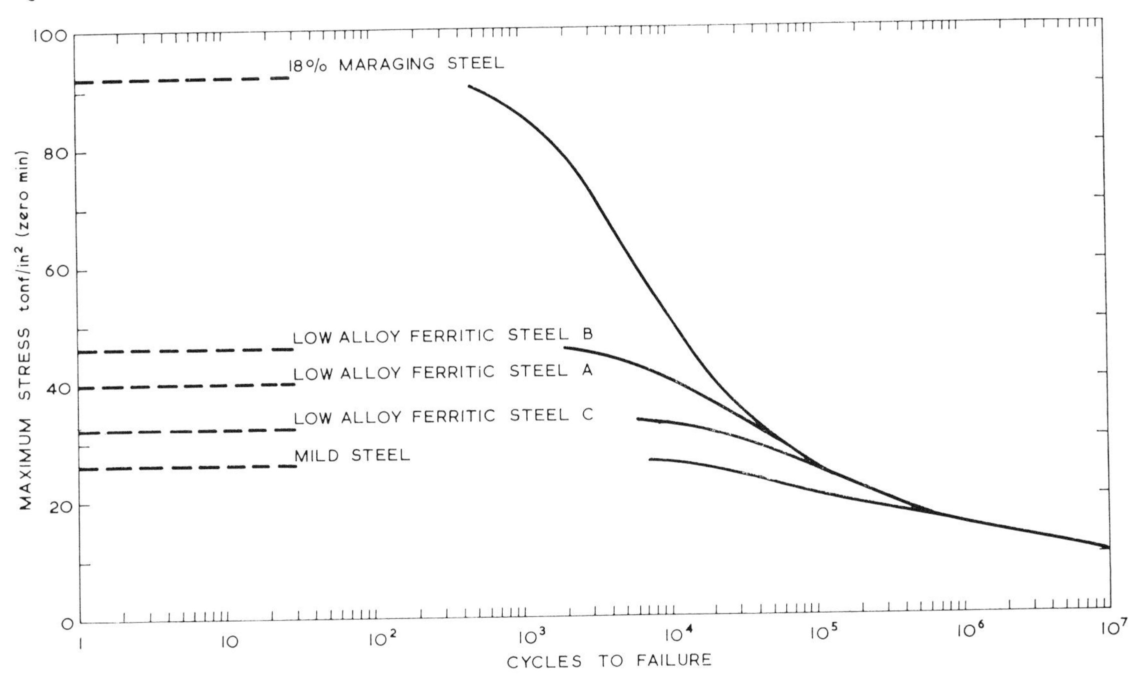

Source: *Welding and Metal Fabrication*, June 1968

*F. E. BRINE **D. WEBBER †H. G. BARON

A Note on the Effect of Maraging on the Fatigue Properties of 18% Ni Maraging Steel Plate and Weld Specimens

SUMMARY: Fatigue tests were carried out on parent metal, butt weld and fillet weld specimens of 18% Ni maraging steel in the annealed and maraged condition. Results show that: at stresses below the proof stress of annealed material the fatigue results for similar specimens were the same whether the material is tested in the annealed or maraged condition.

ONE of many virtues of maraging steel claimed by the International Nickel Co.[1], is that it can be welded in the aged condition and then re-aged to age the weld. This feature makes the material attractive for military applications as weld repairs can be made and the welds aged using local heat-treatment techniques without the need to heat-treat the whole structure.

Maraging steel also has quite good mechanical properties in the unaged condition, as is shown in Table 1.

In view of these tensile properties, it would appear possible that repair welding could be carried out in suitable instances without ageing if this was not immediately possible, and if care was taken in the subsequent use of the welded component. To investigate whether this treatment would be satisfactory under other conditions of loading, fatigue tests were carried out on plain and welded maraging steel specimens in the unaged condition.

* F. E. Brine—R.A.R.D.E. Fort Halstead.
** D. Webber—M.E.X.E. Christchurch.
†H. G. Baron—Ministry of Technology, Harefield.

Material

The steel was supplied in the 90 tonf/in^2. grade in the form of 6 ft by 2 ft by $\frac{3}{16}$ in. thick plate of the following nominal composition per cent by weight:

Nickel	18	Titanium	0·2
Cobalt	8·5	Aluminium	0·1
Molybdenum	3·2		

In the as-received annealed condition the mechanical properties were:

0·2% Proof Stress	49·4 tonf/in^2
Tensile Strength	62·5 tonf/in^2
Elongation % (on 2 in.)	10·2

Table 1. Plain Plate Specimens.

Unaged		Maraged 3 h at 480°C	
Stress Range Tons/in^2	Cycles to Failure	Stress Range Tons/in^2	Cycles to Failure
40	41,800	40	49,600
40	39,700	35	76,000
35	60,800	30	126,100
30	85,400	25	228,600
30	105,500	22	342,900
25	222,500	20	483,700
25	255,200	19	36,607,100 NB
25	278,500	19	33,109,300 NB
22	308,300	18	22,745,600 NB
20	611,300		
20	1,513,400		
20	31,000,000 NB		
19	30,000,000 NB		
18	32,000,000 NB		
17	30,000,000 NB		

Fig. 1. *18% Ni maraging steel. Annealed and maraged plain plate results.*

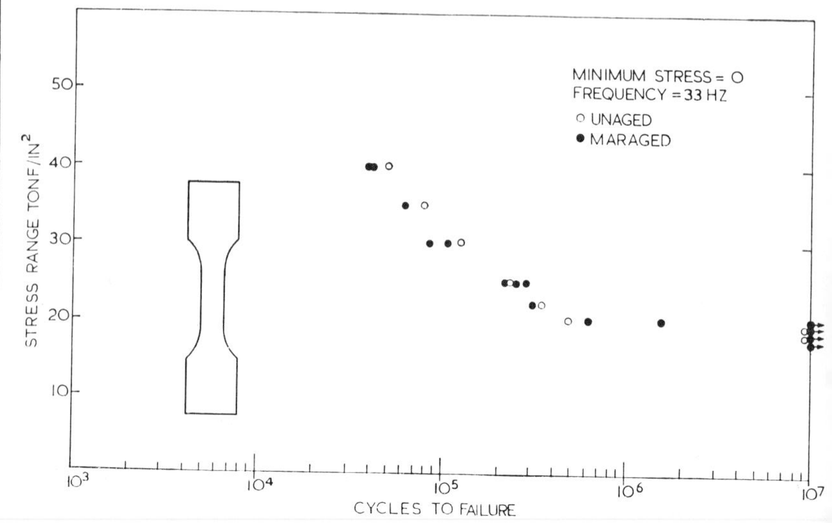

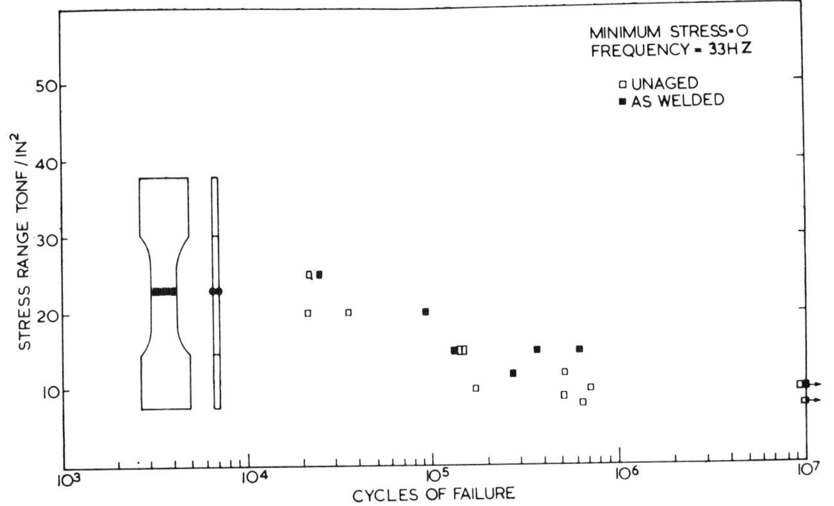

Fig. 2. 18% Ni maraging steel. As-welded and maraged butt weld results.

Table 2. Butt Weld Specimens.

Unaged		Maraged 3 h at 480°C	
Stress Range Tons/in²	Cycles to Failure	Stress Range Tons/in²	Cycles to Failure
25	24,700	20	21,000
20	91,300	20	35,200
15	127,200	15	129,200
15	610,000	15	146,700
15	360,500	12	496,000
12	276,300	11	163,400
10	13,000,000 NB	10	30,731,900
		10	687,100
		9	486,800 NB
		8	632,700
		8	20,123,200 NB

Table 3. Transverse Load-carrying Fillet Weld Specimens.

Unaged		Maraged 3 h at 480°C	
Stress Range Tons/in²	Cycles to Failure	Stress Range Tons/in²	Cycles to Failure
40	4,516	25	15,600
40	6,684	20	57,200
25	24,800	20	38,300
20	73,900	15	173,300
15	118,700	15	78,900
10	511,300	15	337,700
8	5,829,700	12	250,600
7	3,432,500	10	2,225,400
6	33,187,000 NB	10	623,700
		9	20,954,900 NB
		9	998,900
		8	27,095,900 NB

Note: Minimum Stress = zero in all tests. NB = Not broken.

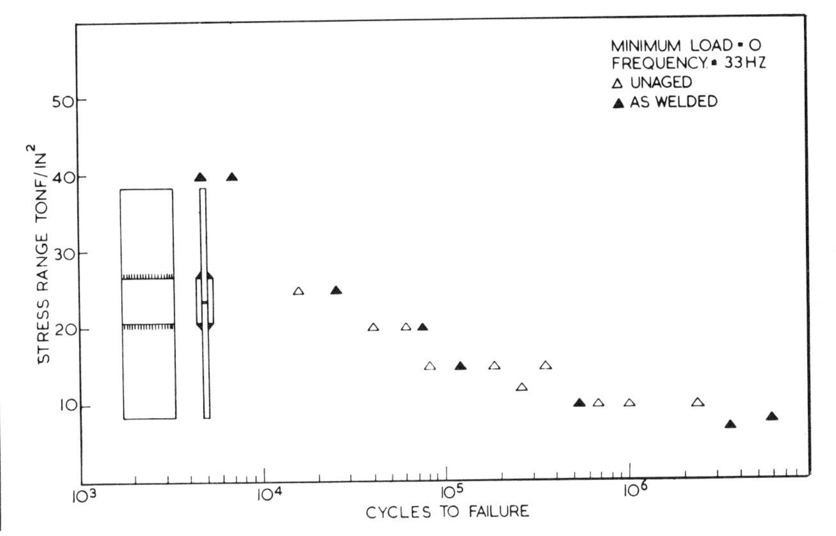

Fig. 3. 18% Ni maraging steel. As-welded and maraged fillet weld results.

After maraging for three hours at 480°C the properties were as follows:

0·2% Proof Stress	87·7 tonf/in²
Tensile Strength	89·8 tonf/in²
Elongation % (on 2 in.)	10·1

Specimens

Plain plate and transverse butt weld specimens were prepared having waisted test sections as shown in Figs. 1 and 2. Parallel-sided, load-carrying transverse fillet weld specimens were prepared as shown in Fig. 3 using cover plates cut from the same material which enabled a weld of $\frac{3}{16}$ in. leg length to be laid.

Welds were made by the metal inert gas (Mig) process with pure argon shielding gas and welding wire of substantially parent metal composition having increased titanium to allow for losses during welding. The butt welds were double-sided and were tested in the as-welded condition without any dressing.

Results and Discussion

The results are presented in the form of Tables 1-3 and S.N. curves in Figs. 1-3 for plain plate, butt welds, and fillet welds respectively. It will be seen from these that there is no significant difference between any of the results for the stress ranges examined, though it might be concluded that maraging is slightly beneficial to plate and detrimental to welds in the finite life region. A larger number of results would be required to confirm this.

This is not altogether surprising for the welds since it is often quoted[2] that the endurance limit for welds in steels is the same whatever the base material. Frost and Denton[3] have shown this particularly well for a dressed butt weld detail in a number of alloys. In these results it is only when stresses are of the order of the proof stress that there are significant differences in the curves they present.

The fact that the same results are obtained for annealed and maraged plate is interesting because it is generally found that the endurance limit of smooth laboratory specimens under alternating stress is between 0·4 and 0·6 of the tensile strength[4]. The most probable explanation lies in the stress-concentrating effect of imperfections in the as-rolled surface and perhaps an increase in notch sensitivity with increase in tensile strength.

It would be interesting to investigate this aspect of the work further. The practical implications of the present work are that at the stress levels investigated with as-rolled material, it does not matter whether the material is maraged or not.

This suggests that the field of application of high strength steels is where (a) only static stresses are encountered, (b) high levels of mean stress with a very low fluctuating stress are encountered, or (c) where a lightweight equipment having a limited fatigue life is required[5]. This situation can be improved by treatments giving rise to compressive residual stresses in regions liable to fatigue failure[6].

Conclusions

At stresses below the proof stress of the unmaraged material fatigue life of plain plate, and butt or fillet welds is not influenced by maraging.

References

1. Maraging Steels for Ultra-High Strength Applications, The International Nickel Co. (Mond) Ltd., London, 1962.
2. T. R. Gurney, Some Fatigue Tests on Fillet Welded Mild and High Tensile Steel Specimens in the As-Welded and Normalized Condition, *British Welding Journal*, Vol. 13, No. 11, 1966, pp. 648-651.
3. N. E. Frost and K. Denton, The Fatigue Strength of Butt-Welded Joints in Low-Alloy Structural Steels, NEL Report 199, 1965.
4. R. B. Heywood, Designing Against Fatigue, Chapman & Hall Ltd., London, 1963.
5. D. Webber, Fatigue Results in Welded 90 tonf/in^2. Maraging Steel, *Welding and Metal Fabrication*, Vol. 36, No. 6, 1968.
6. H. G. Baron and F. E. Brine, A note on the improvement in fatigue properties of butt welds produced by shot peening or pre-stretching, Session IX, Commonwealth Welding Conference, London, 1965.

Fatigue Crack Propagation in Martensitic and Austenitic Steels

C. BATHIAS AND R. M. PELLOUX

Fatigue crack growth rates were measured in an annealed and in an aged maraging steel and in three different austenitic steels. Microhardness measurements were used to determine the plane strain plastic zone sizes as a function of ΔK and to evaluate the cyclic flow stress of the material near the crack tip. The presence of a reversed cyclic plastic zone within the monotonic plastic zone was confirmed. The two maraging steels work soften near the tip of the crack while the three austenitic steels work harden. The fatigue crack growth rates of the maraging steels are independent of the monotonic yield stress and are typical of the growth rates of steels with a bcc crystal structure. The crack growth rates in the stainless steels are an order of magnitude lower than for maraging steels for $\Delta K < 30$ ksi $\sqrt{\text{in.}}$. The excellent fatigue crack growth resistance of austenitic stainless steels is related to the deformation induced phase transformations taking place in the plastic zone and to the low stacking fault energy of the alloys.

FATIGUE crack growth rates are usually related to the range of the elastic stress intensity factor ΔK by the empirical equation

$$\frac{dl}{dn} = C(\Delta K)^m \quad [1]$$

Recently, Hahn[1] analyzed the data available on the fatigue crack propagation rates of a wide variety of steels and concluded that the growth rates can be given by the expression:

$$\frac{dl}{dn} = 8\left(\frac{\Delta K}{E}\right)^2 \quad [2]$$

with ΔK, elastic stress intensity range in ksi $\sqrt{\text{in.}}$; E, elastic modulus in ksi; and dl/dn, growth rate in inches. Eq. [2] applies best to growth rates from 10^{-6} to 10^{-4} in./cycle. This parametric representation of the data includes steels with yield strengths from 54 to 250 ksi; however, all the steels had a bcc crystal structure, that is, they were ferritic, pearlitic, martensitic, or quenched and tempered steels. Hahn[1] concluded that at a given ΔK the fatigue crack growth rate in steels is insensitive to yield strength, composition and ductility.

Barsom[2] working with HY-80, HY-130, 10 Ni-Cr-Mo-Co, and 12 Ni-5 Cr-3 Mo steels obtained the following empirical relation for the crack growth rate:

$$\frac{dl}{dn} = 0.66 \times 10^{-8} (\Delta K)^{2.25} \quad [3]$$

Barsom[2] also concluded that Eq. [3] fitted the fatigue crack growth rate data of 19 high yield strength steels.

Throop[3] reported that for K_{max} below the range of 30 to 40 ksi $\sqrt{\text{in.}}$, substantial differences in mechanical properties of 4340 steels tempered at different temperatures have little influence on the crack growth rate. For K_{max} above 40 ksi $\sqrt{\text{in.}}$, the crack growth rate was related to the index $(EYK_{IC})^{-1}$ or $(EU\epsilon_F)^{-4}$ [E is Young's modulus, Y and U are yield and ultimate tensile strength, ϵ_F is the true strain at fracture, and K_{IC} is the plane strain fracture toughness.] Throop[3] reviewed 50 values of the exponent m in Eq. [1] for low, medium, and high strength steels and found an average value of $m = 3.42$ with m ranging from a low of 2.2 to a high of 6.7. In general it appears that m depends upon the range of growth rate measurements over which it is taken; it varies from 2 if it is averaged from 10^{-6} to 10^{-4} in./cycle to 3 if it is averaged from 10^{-6} to 10^{-3} or 10^{-2} in./cycle.

It has been shown[7] that under constant cyclic load range, each fatigue striation observed on the fracture surface corresponds to one load cycle. The spacing of the fatigue striations is a measure of the microscopic or local rate of advance of the crack front. Following some extensive measurements, Bates[4] found the following relationship for a wide variety of steels:

$$\frac{dl}{dn} \text{ (striation spacing)} = 5.4\left(\frac{\Delta K}{E}\right)^{2.1}$$

Bates[4] found that for $\Delta K < 25$ ksi $\sqrt{\text{in.}}$ the average striation spacing is larger than the macroscopic growth rate. For $\Delta K > 25$ ksi $\sqrt{\text{in.}}$ the two growth rates are equal.

In a similar study with 4340 and 18 Ni maraging steels Miller[5] also found that the striation spacing is larger than the macroscopic growth rate. The average values of m for these steels were:

macroscopic growth rate $m = 3.13$
microscopic growth rate $m = 2.15$

In all cases however it can be said that for quenched and tempered steels the striation spacing is never more than twice the macroscopic growth rate at a given ΔK level. In this work we shall show that this is not the case for the austenitic steels.

The fact that the fatigue crack growth rate of steels is independent of the monotonic yield strengths appears to contradict a theoretical crack growth model which relates the growth rate dl/dn to the crack tip opening displacement (CTOD). McClintock[6] proposed such a crack extension model by assuming on the basis of slip line field analysis, that the growth rate dl/dn was equal to half the cyclic CTOD of the material.

$$\frac{dl}{dn} = \tfrac{1}{2}\text{CTOD cyclic} = \frac{0.25\,(\Delta K)^2}{2YE} \quad [5]$$

C. BATHIAS is Staff Engineer, Aerospatiale, Paris, France. R. M. PELLOUX is Associate Professor, Massachusetts Institute of Technology, Cambridge, Mass. 02139.
Manuscript submitted March 20, 1972.

The second part of this equation is given by Rice.[8] (Y is the 0.2 pct offset cyclic flow stress of the material near the crack tip.) The correlation between the growth rate and the CTOD has been confirmed experimentally by electron fractography of fatigue fractures of programmed load tests. There is a one to one correlation between the load amplitude and the striation spacing.[7] Since in steels the fatigue crack growth rate is found to be independent of the yield stress, Eq. [5] will be meaningful only if we assume that the cyclic flow stresses, Y, approach the same limit for all steels regardless of their initial monotonic yield stress. Cyclic deformation within the plastic zone is best understood by considering a small process zone of size ρ which is cyclically deformed under N cycles of increasing plastic strain amplitudes as it approaches the crack tip. The total number of plastic strain cycles N can be estimated from the ratio of the plastic zone size to the crack growth rate. For most engineering materials this ratio is on the order of 10^2 to 10^3 over a wide range of growth rates. For mode I of crack opening the distribution of plastic strains within the plastic zone, R, is not known, but we can assume with McClintock[16] that by analogy to mode III of crack tip opening the strains vary as $1/r$ where r is the distance to the crack tip. Thus, fatigue hardening or softening of the plastic zone is essentially a low cycle fatigue process under increasing plastic strain amplitudes. Near the crack tip, the cyclic flow stress of different steels may reach a limit which is independent of the original monotonic yield strength. The resulting fatigue crack growth rate will also be independent of the monotonic yield stress.

In the work reported here, we have used microhardness measurements to determine plastic zone sizes as a function of ΔK and to evaluate the cyclic flow stress of the material near the fatigue crack tip. Fatigue crack growth rates, striation spacings, and plastic zone sizes were measured in a maraging steel and in three austenitic steels. Our measurements confirm the presence of a reversed cyclic plastic zone within the monotonic plastic zone as suggested by Rice.[8] The fatigue crack growth rates of the austenitic steels are much lower than for the maraging steels. The difference in growth rates is attributed to the slip character and work hardening behavior of the austenitic steels.

I) MATERIALS AND TEST PROCEDURES

The chemical analyses and the mechanical properties of the steels used in this research are given in Tables I and II.

Heat Treatment

The maraging steels were austenitized for 1 h at 1500°F and air cooled. Maraging steel "M" was tested in the annealed condition. Maraging steel "MA" was given a three hour aging treatment at 900°F. The stainless steels (16-13) and (24-20) were quenched in water from a homogenization anneal of 1 h at 2000°F. The grain sizes of the stainless steels were 100 μ for (24-20) and 200 μ for (16-13). A commercially available grade of 304 stainless steel was used. The (16-13) and (24-20) stainless steels deform extensively by mechanical twinning when the tensile plastic strains are larger than 10 pct.[9,11] The 304 stainless undergoes a martensitic transformation during plastic deformation.[10]

Mechanical Testing

Double cantilever beam specimens were used for the fatigue crack propagation tests. The specimen dimensions were 3.75 in. wide by 3.625 in. high by $\frac{3}{8}$ in. thick. The original notch length was 1.650 in. The specimen geometry was designed to give a slow rate of increase of ΔK with increasing crack length in order to have steady state plastic zone sizes over finite crack length intervals. All tests were performed at room temperature, at 10 cps, on a hydraulic fatigue machine. Crack lengths were measured with an absolute accuracy of 0.001 in. The minimum load was kept at 250 lb. The stress intensity range was obtained from the following equation:

$$\Delta K = \left[\Delta\sigma\sqrt{a}\,(29,6 - 185.5\left(\frac{a}{w}\right) + 6.557\left(\frac{a}{w}\right)^2 - 1017\left(\frac{a}{w}\right)^3 + 638.9\left(\frac{a}{w}\right)^4\right]$$

with a, crack length and w = 3 in. An airtight plexiglass cell was used for crack propagation tests in an inert argon environment (less than 200 ppm of water vapor).

Metallography

A portable electropolishing technique was used to prepare the outside surface of the DCB specimen for crack length measurements and the different polished sections used for microhardness measurements. The electropolishing conditions are given in Table III. A

Table I. Chemical Analysis

Designation	C	Cr	Ni	Mo	Co	Ti	Al	Si	Mn
16-13	0.02	16.4	13.1	–	–	–	–	0.56	1.35
24-20	0.15	23.5	19.8	–	–	–	–	1.76	0.87
304	0.012	18.4	8.7	–	–	–	–	0.63	1.26
Maraging	0.014	–	18.76	4.81	8.82	0.66	0.09	0.02	0.05

Table II. Mechanical Properties

	0.2 Pct Yield	UTS	Microhardness, Vickers
16-13	28	73	145
24-20	40	95	180
304	35	85	–
M, maraging steel annealed	140	145	350
MA, maraging steel aged 900°F	260	280	610

Table III. Electropolishing Solution and Electroplating Solution

Electropolishing Solution

Perchloric Acid 9 Pct–Acetic Acid 45 Pct–Butylcellosolve 46 Pct
30 V–40°F–2 to 3 Min

Electroplating Solution

Ferrous Chloride 500 g/1000 ml of water
Manganese Chloride 5 g/1000 ml of water
Temperature 200°F current density 180 to 280 amp/sq ft
Total time per specimen = 5 h

pure iron layer of 0.20 in. was electroplated on one half of the fracture surfaces to protect the edges of the fracture in order to allow hardness measurements to be taken very close to the fracture plane. The plated specimen was then sectioned with a diamond cut-off wheel along a plane parallel to the surface but at mid-thickness. The sectioned surface was carefully ground, mechanically polished and electropolished for the microhardness measurements. Polishing of the section is very critical in order to avoid any surface distortion which will lead to a scatter in microhardness measurements.

The microhardness measurements were made with a Microdurimet Leitz with a hardness load of 25 g except for the aged maraging steel (MA) which required a 50 g load. The measurements were taken at different crack lengths in a direction normal to the plane of fracture.

The microscopic growth rates were obtained by measuring the average spacing of fatigue striations in a direction parallel to the macroscopic crack growth direction. All the fractography work was done by scanning electronmicroscopy at 0 deg tilt of the specimen.

II) RESULTS

Fatigue Crack Growth Rates

Fig. 1 gives the crack growth rates for the maraging steels and the austenitic steels. Each curve represents the data of at least three different tests. The error in the crack growth rate measurements is less than ± 15 pct. The crack growth rates are exactly the same for the two maraging steels although the different heat treatments gave a ratio of yield stress of $\frac{260}{140} = 1.86$. The slope is 2.5 which is in close agreement with a slope of 2.88 reported by Miller.[5] In order to evaluate the influence of room humidity on the growth rate, a test was run in dry argon with the aged maraging steel. As shown in Fig. 1, the slope does not change markedly but the growth rate curve in dry argon is shifted downward. The difference in growth rates between air and dry argon is on the order of a factor of 2.

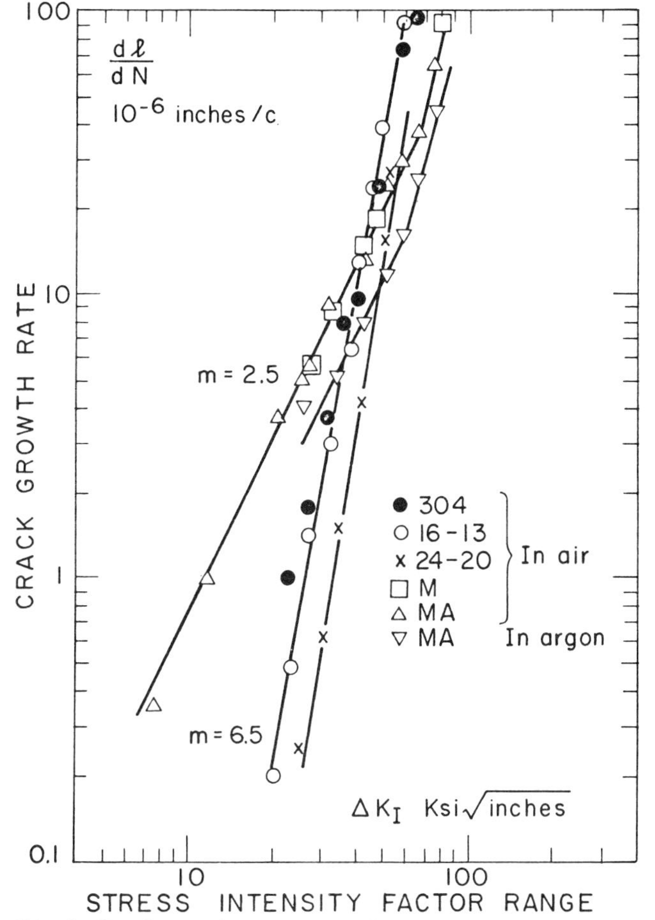

Fig. 1—Crack growth rates vs ΔK for maraging and stainless steels.

The slopes of the $(dl/dn, \Delta K)$ curve for the austenitic steels are approximately 6.5 and the growth rates in the austenitic steels are markedly lower than in the ferritic alloys below $\Delta K = 30$ ksi $\sqrt{\text{in.}}$. For growth rates above 2×10^{-5} in./cycle, the crack growth rate in the austenitic is larger than in the ferritic steels. This cross-over is unaccounted for.

Measurements of striation spacings for the different alloys are reported in Figs. 2, 3, and 4. Each point represents the average of five measurements. Striation spacing measurements with the SEM were limited to growth rates larger than 5×10^{-6} in./cycle. All the SEM measurements were made in the center section of the fatigue specimens and the crack lengths were measured accurately with the stage vernier of the SEM. Typical SEM micrographs of the striations are shown in Figs. 5 and 6.

For the maraging steels the striation spacings and the macroscopic rates do not differ by more than a factor of 2 at 30 ksi $\sqrt{\text{in.}}$. For the austenitic steels the difference between the striation spacings and the macroscopic growth rates is very large. It reaches two orders of magnitude at 20 ksi $\sqrt{\text{in.}}$. This result indicates that the process of fatigue crack propagation in austenitic steels is very localized and the crack front does not advance uniformly along the whole front. It is interesting to note that the two curves of striation spacings vs. ΔK for the maraging and for the austenitic steels fall within the same scatter band, Fig. 4, indicating that the microscopic growth rates at a given ΔK are quite similar although the overall growth rates are different. At large growth rates, the striation spacings are smaller than the macroscopic growth rate because of the extensive amount of crack propagation per cycle due to tearing by ductile fracture (void formation).

Measurements of Plastic Zone Sizes

Different attempts have been made to measure plastic zone sizes at the tip of fatigue cracks. The different techniques of transmission microscopy,[12] X-ray microbeam[13] or Moiré fringes[14] have been only partially successful. However, recently the Fe- 3 Si etching technique was used by Hahn[15] to decorate dislocations and reveal plastic deformation in fine grained Fe-3 Si. The etching technique revealed the presence of a reversed flow or cyclic plastic zone within the monotonic plastic zone as predicted by Rice.[8] Fig. 7 reproduces the sketch used by Rice[8] to explain the presence of the two plastic zones. The reversed plastic zone OA is one-fourth the size of the monotonic plastic zone size OB which can be calculated from fracture mechanics. Within the region AB, the stress range $\Delta\sigma$ is elastic and the cyclic plastic strain range is zero. The monotonic plastic strain increases from zero at B to ϵ_{PA} at A. ϵ_{PA} for mode I of crack opening can be calculated by analogy to the plastic strain distribution in mode III, that is, ϵ (total) at a distance r from the tip of the crack is equal to ϵ_y (yield strain) XR/r. In Fig. 7, we have: $OB/OA = R/r = 4$ and ϵ (total) at $A = 4\epsilon$ yield. Consequently the monotonic plastic strain at A is quite small since it is only three times the yield strain of the material. Although the cyclic plastic strain range is zero from B to A, the large plastic stress ranges in the region AB result in cyclic plastic microstrains which will work harden or work soften the material. In the reverse plastic zone, OA, the stress range is $2Y$ and the cyclic plastic strain range increases from zero at A to infinity near the crack tip O. In summary, we should expect the boundaries of the two plastic zones to be associated with two well defined transitions in the work hardened state of the material.

Since the etching techniques are time consuming and difficult to use with maraging and stainless steels, the plastic zone sizes were determined by microhardness measurements. However, etching experiments were conducted successfully in the two types of steels and the sizes of the etched plastic zones were equivalent to the sizes measured by microhardness. The scatter of the measurements is on the order of ± 5 Vickers units. The hardness indentations were taken along

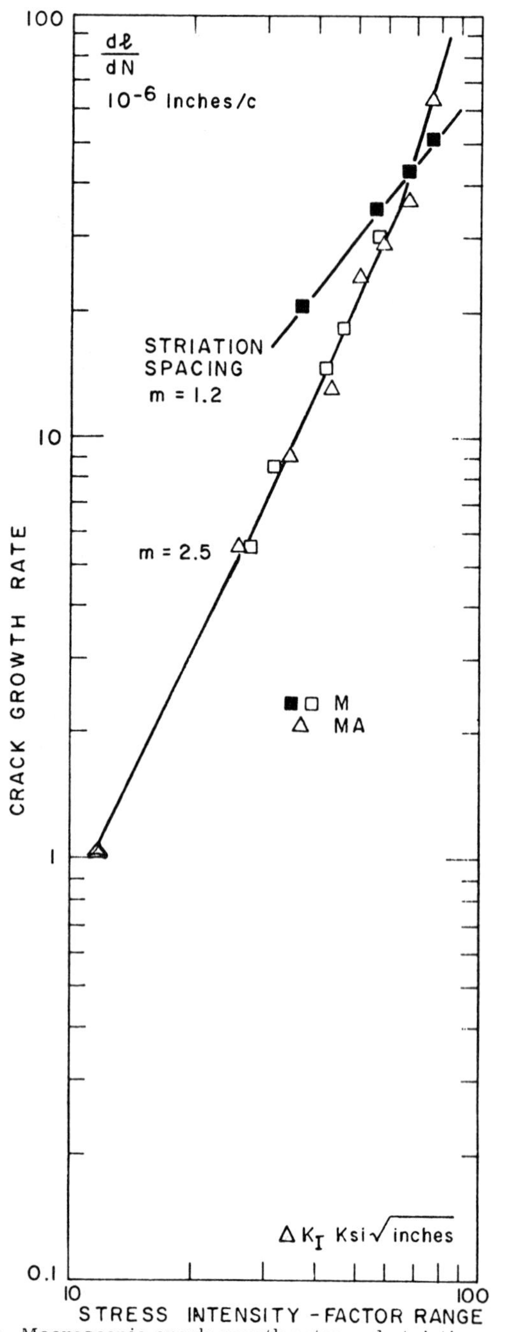

Fig. 2—Macroscopic crack growth rates and striation spacing for maraging steels.

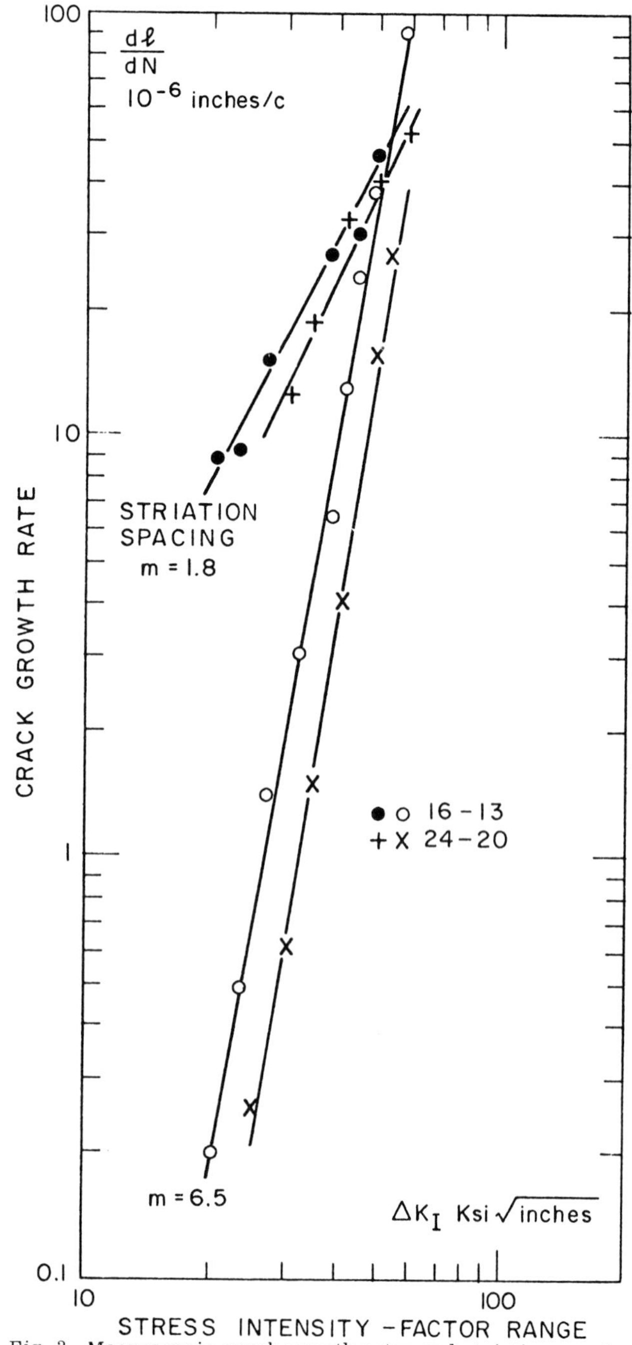

Fig. 3—Macroscopic crack growth rates and striation spacing for stainless steels.

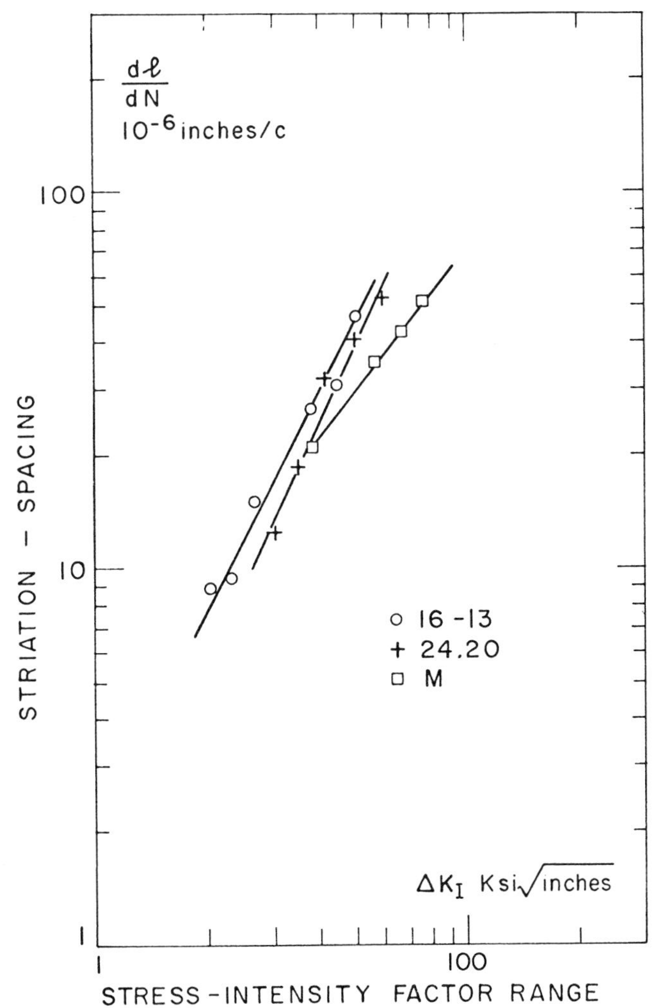

Fig. 4—Striation spacing vs ΔK for maraging and stainless steels.

directions normal to the fracture plane at crack length intervals such as to cover a range of ΔK values from 20 to 75 ksi $\sqrt{\text{in.}}$. Figs. 8 and 9 show four typical plots of the hardness readings as a function of distance from the fracture surface. Following the nomenclature proposed by Hahn,[15] the sizes of the monotonic and cyclic plastic zones are referred to as r''_x, r''_y and r'_x, r'_y, respectively, with $r'_y/r'_x = r''_y/r''_x = 4$. Fig. 10 shows the butterfly appearance of the plastic zones at the crack tip. Each hardness curve presents two plateaus which define clearly the plastic zone sizes r''_y and r'_y. The transition between the undeformed material and the monotonic plastic zone r''_y corresponds to an average change of 20 Vickers units. The austenitic steels are work hardening near the fatigue crack tip and the two maraging steels are work softening. It was quite surprising to see that even the annealed maraging steel would work soften. The transition between the monotonic and the cyclic zone is usually well defined except when r'_y is too small for a meaningful hardness indentation. The uniform hardness plateau of the monotonic plastic zone is not accounted for since the small plastic strain amplitudes should increase from B to A, Fig. 10. The hardness plateaus are probably due to the butterfly wing shape of the plastic zone. An element away from the crack surface will undergo a large number of cycles of small plastic strains which may be equivalent for cyclic hardening or cyclic softening

Fig. 5—Fatigue fracture surface of the annealed maraging steel (ΔK = 50 ksi $\sqrt{\text{in.}}$).

to a few cycles of large plastic strains near the crack surface. For a given material the saturation hardness measurements within the monotonic plastic zones are the same regardless of the level of ΔK. This result is in agreement with Hahn's result[15] that the cumulative plastic strain is less than 10 pct in the monotonic zone.

Microhardness increases or decreases sharply within the reversed zone r'_y depending upon the type of steels. It was difficult to measure an exact limit of the microhardness near or at the fracture surfaces. However, limit hardness measurements were obtained for the larger r'_y zones which correspond to large

Fig. 6—Fatigue fracture surface of the austenitic steel 25 Cr-20 Ni (ΔK = 45 ksi $\sqrt{\text{in.}}$).

ΔK values. Table V summarizes the maximum hardness readings reported near the fracture surfaces. There does not appear to be a common hardness limit near the fracture surfaces of the different steels although the values are quite close except for the aged maraging.

The four to five plastic zone sizes r''_y measured (on each alloy) are reported in Fig. 11. The slope of each curve is close enough to 2 to state that the data fits the

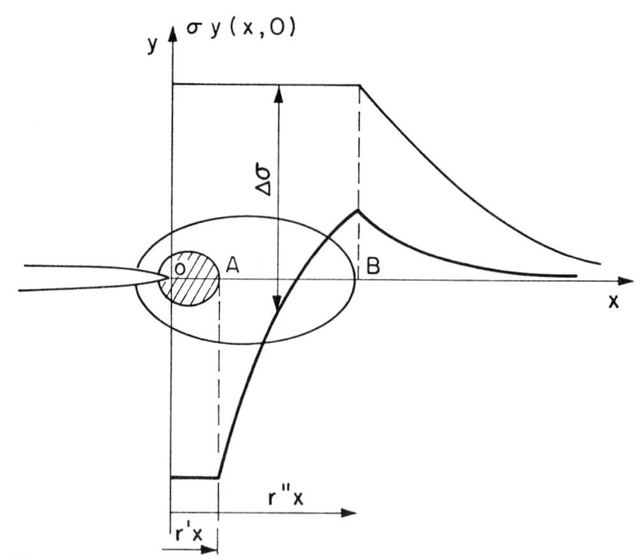

Fig. 7—Schematic representation of the monotonic and cyclic plastic zones.

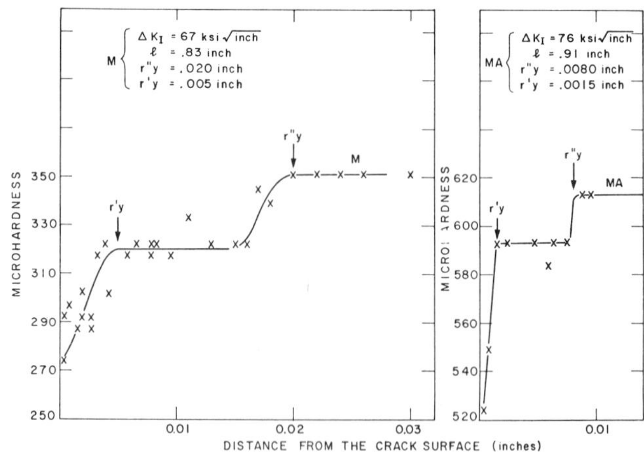

Fig. 8—Plastic zone size determination by microhardness in two maraging steels.

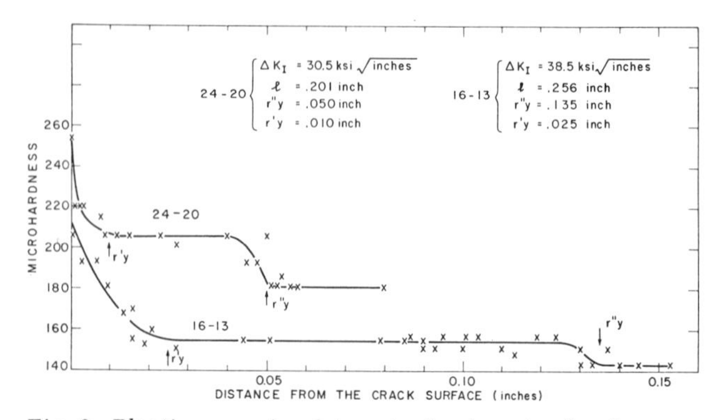

Fig. 9—Plastic zone size determination by microhardness in austenitic steels.

Table IV. Plastic Zone Sizes Measured at the Tip of Fatigue Cracks

Irwin	monotonic zone theoretical, plane stress	$r = (1/2\pi)(K^2/Y)$
Rice	monotonic zone theoretical, plane strain	$r = 0.157\,(K/Y)^2$
Yokobori	monotonic zone, plane stress	$r = 0.17\,(\Delta K/Y)^2$
Liu	monotonic zone, plane stress	$r = 1.30\,(\Delta K/Y)^2$
Hahn	monotonic, plane strain	$r = 0.13\,(K/Y)^2$
Pelloux	montonic, plane strain	$r = 0.05$ to $0.10\,(\Delta K/Y)^2$

Table V. Microhardness Readings Near the Fracture Surfaces

Material	Core Hardness	ΔK ksi $\sqrt{\text{in}}$	Hardness Near Fracture Surface	Equivalent Flow Stress ksi
Maraging M	350	76	270	126
Maraging MA	610	76	550	264
Austenitic 16-13	145	45	260	121
Austenitic 24-20	180	70	320	151

general equation given by fracture mechanics

$$r''_y = C\left(\frac{\Delta K}{Y}\right)^2 \qquad [6]$$

with the following values of C:

Austenitic steels $C = 0.045$ to 0.055
Maraging steels $C = 0.060$ to 0.10

The dependence of r''_y upon the monotonic flow stress Y shown in Fig. 12 for $\Delta K = 40$ ksi $\sqrt{\text{in.}}$ also agrees with Eq. [6]. In all cases, the size of the reversed plastic zone r'_y was on the order of $\frac{1}{4}$ to $\frac{1}{5} r''_y$, as predicted by Rice.[8] For the maraging steels r'_y was difficult or impossible to measure by microhardness for $\Delta K < 60$ ksi $\sqrt{\text{in.}}$.

III) DISCUSSION

Plastic Zone Sizes

Table IV gives different parametric expressions for the plastic zone sizes measured experimentally at the tip of fatigue cracks by Yokobori,[13] Liu,[14] Hahn,[15] and in this work. Yokobori[13] and Liu[14] were not able to determine the reversed plastic flow zones or at least their results do not indicate such zones. In this work as well as in Hahn's work, the cyclic plastic zone sizes r'_y were measured and found to be about $\frac{1}{4}$ of the size of the monotonic plastic zones which are given in Table IV.

The monotonic plastic zone sizes measured by microhardness are very close to the sizes predicted by the theoretical equation:

$$r''_y = 0.10\left(\frac{\Delta K}{Y}\right)^2 \qquad [5]$$

This is due to the fact that microhardness measurements can detect local work hardening due to plastic strains as small as 1 pct. A calibration of the hardness values against a cyclic saturation stress will be required to relate hardness readings to a quantitative estimate of the plastic strain distribution within the plastic zone.

One main drawback of the hardness technique is in the size of the indentation (10 μ) which limits the re-

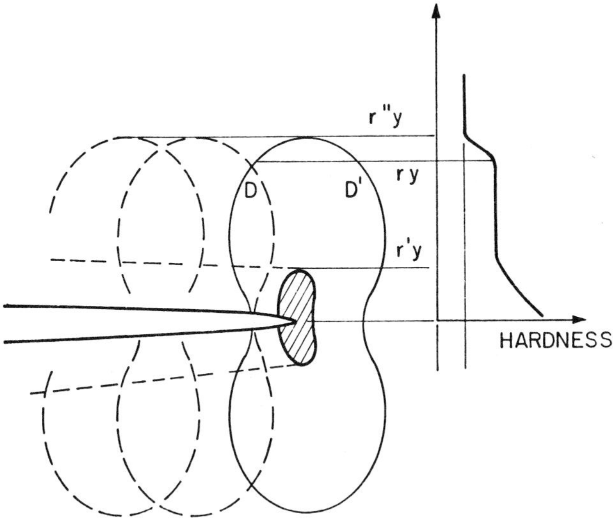

Fig. 10—Schematic representation of the monotonic and cyclic plastic zones.

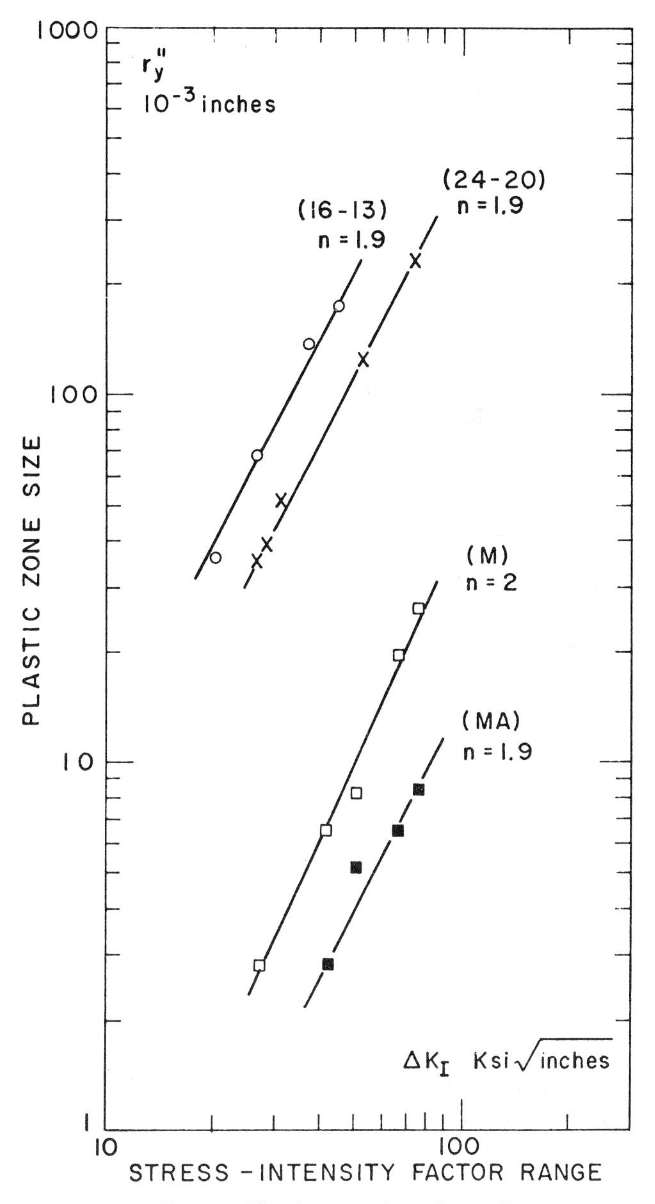

Fig. 11—Plastic zone size r''_y vs ΔK.

versed plastic zone sizes to be measured to 50 to 80 μ. Consequently very small plastic zone sizes can only be measured by means of an etching technique. Bathias[9] has used this technique extensively with stainless steels and found that combining microhardness measurements and etching is one of the best approaches to plastic zone size measurements.

Fatigue Crack Growth Rates

The marked difference between the fatigue crack growth rates in austenitic and ferritic steels below 10^{-5} in./cycle is quite striking. However, this difference can best be accounted for by the fact that the microscopic growth rates of the two types of steels are about the same in the whole range of ΔK investigated. From this observation it is concluded that crack growth in the austenitic steels is a non-uniform and very localized microscopic process along the crack front. At a given time the crack advances only along a fraction of the crack front with the rest of the front remaining "dormant". This mechanism of crack growth can be explained by the fact that austenitic steels have the following characteristics:

—high work hardening coefficient due to phase transformation during plastic deformation

—planar slip character

—crystallographic fracture path and irregular fatigue crack front

The phase transformations induced by plastic deformation in the Fe-Ni-Cr-C alloys have been reviewed by Lecroisey.[11] The fcc to bcc (α phase) and fcc to hcp (ϵ phase) transformations within the M_S and M_D temperature range depend upon the alloy chemistry and the amount of plastic deformation. The Fe-18 Cr-8 Ni (304 stainless) deforms readily to α martensite. The Fe-16 Cr-13 Ni alloy will form ϵ martensite within a network of twins.[11] The third alloy Fe-24 Cr-20 Ni deforms by twinning. Bathias[9] found by etching that these phase transformations are confined to the reversed plastic zone where the plastic strain amplitudes are large. Within the monotonic plastic zone the plastic deformation is associated with dislocations motion and multiplication. Since the crack growth rates are about the same for the three alloys the exact nature of the phase transformation within the plastic zone does not seem to be critical in controlling the growth rate.

Another explanation for the slow growth rate of the austenitic steels involves the high work hardening coefficient associated with the phase transformations. A high work hardening coefficient will result in a lower plastic strain range for a given total strain range and will lead to a better accomodation of the applied strains within the plastic zone. The discontinuous crack growth rate may be due to the extensive plastic strain accommodation of each grain at the tip of the crack. The accommodation of cyclic strains is enhanced by the reversibility of slip and the extensive microtwinning associated with a planar slip character. This reversibility corresponds to a large Bauschinger strain and again to a lower plastic strain range for a given total strain range. The importance of the planar slip character is clearly demonstrated in Fig. 6. In each grain the crack front is strongly crystallographic and the orientation of the fatigue striations changes from grain to grain. This chaotic but crystallographic appearance of the fracture

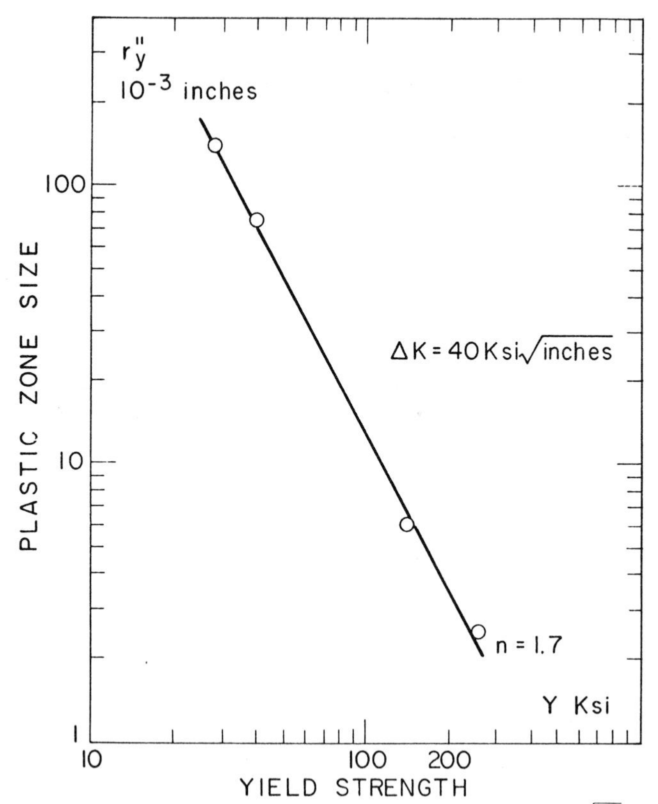

Fig. 12—Plastic zone size r''_y vs Y for ΔK = 40 ksi $\sqrt{\text{in.}}$

surface may in part account for the discontinuous microscopic process of fatigue crack growth in the austenitic steels. In each grain the fracture plane proceeds from a nucleating or focal point which seems to correspond to a local reinitiation site along the grain boundary. It appears that the crack front is arrested at the grain boundary before proceeding to the next grain. This discontinuous process of crack propagation from grain to grain leads to the formation of a wavy crack front with a wavelength corresponding to the grain size. The crack front advances by tunnelling ahead locally by the length of a grain and the crack is then arrested at the grain boundary of the next grain. The role of the phase transformation and of the planar slip character is to provide extensive accommodating strains in the grain in front of the crack during crack arrest.

We also have to account for the unique macroscopic fatigue crack growth rate for all bcc steels and for the fact that the relationship between striation spacing and ΔK is about the same for fcc and bcc steels. Microhardness measurements have failed to show a unique saturation cyclic flow stress in the reversed plastic zone for all the steels tested. If this stress existed it would be quite low since the annealed maraging steel did work soften markedly from a monotonic yield stress of 140 ksi. An estimated saturation cyclic stress can be obtained by assuming that the striation spacing is on the order of one half the CTOD (see McClintock, Ref. 6),

$$\frac{\text{CTOD}}{\text{cyclic}} = 2\,\frac{dl}{dn} = \frac{0.25\,(\Delta K)^2}{EY} \qquad [5]$$

By fitting the striation spacing data of Fig. 4 to Eq. [5], we obtain a value of the cyclic flow stress of 270 ksi. This value of Y is much larger than the equivalent saturation flow stresses measured by microhardness near the crack tip. Table V shows that the measured cyclic flow stresses are below 120 ksi except for the aged maraging alloy. (In this latter case the reversed plastic zone is too small to allow a valid measurement of the microhardness near the fracture surface.) The discrepancy between Eq. [5] and the experimental results are corrected easily if we assume that the crack growth rate is only a fraction of the CTOD.

The practical implications of the results of this work are quite interesting when one considers the difference in growth rates between austenitic and ferritic steels below 10^{-5} in./cycle. At a given ΔK level the austenitic steels offer an excellent resistance to fatigue crack growth as compared to the quenched and tempered steels including maraging steels. This difference is not due to the fcc crystal structure but to the deformation induced phase transformation and to the low stacking faults of the austenitic alloys. In a corrosive environment such as sea water the difference would be even more pronounced. However, the tensile yield and fracture strengths of the 304, (16-13) and (24-20) stainless steels are quite low which means that their endurance limits will also be low. It will be worthwhile to see if high strength austenitic steels can combine a high endurance limit with an excellent resistance to crack growth.

IV) CONCLUSIONS

1) Fatigue crack growth rates of steels with a fcc structure are an order of magnitude lower than for steels with a bcc structure for $\Delta K < 30$ ksi $\sqrt{\text{in.}}$. For $\Delta K > 40$ ksi $\sqrt{\text{in.}}$ and growth rates larger than 10^{-5} in./cycle the growth rates fall approximately in the same scatter band.

2) The fatigue crack growth rates of maraging steels are independent of the monotonic yield stress and are typical of the growth rates of steels with a bcc crystal structure.

3) The excellent fatigue crack growth resistance of austenitic stainless steels is related to their low stacking fault energy and their planar slip character.

4) Monotonic and cyclic plastic zone sizes can be measured at the tip of fatigue cracks by means of microhardness measurements.

5) Maraging steels in the annealed or aged condition work soften at the tip of a fatigue crack.

6) The austenitic steels 16 Cr-13 Ni-Fe, 24 Cr-20 Ni-Fe, and 304 stainless steel work harden in the plastic zone at the tip of a fatigue crack.

ACKNOWLEDGMENTS

This work was supported by Air Force Materials Laboratories under a corrosion fatigue research contract number F33615-70-C-1785.

REFERENCES

1. G. T. Hahn, R. C. Hoagland, and A. R. Rosenfield: AF 33615-70-C-1630, Battelle Memorial Institute, Columbus, Ohio, August, 1971.
2. J. M. Barsom, E. J. Imhof, and S. T. Rolfe: ARL-B-23103, U. S. Steel Corp., Monroeville, Pennsylvania, 1968.
3. J. F. Throop and G. A. Miller: *Amer. Soc. Test. Mater., Spec. Tech. Publ.* 467, p. 154, American Society for Testing and Materials, 1969.
4. R. C. Bates and W. G. Clark: 68-1D7-RPAFC-P1, Westinghouse Research Laboratories, Pittsburgh, Pennsylvania, 1968.
5. G. A. Miller: *Trans. ASM,* 1969, vol. 62, p. 651.
6. F. A. McClintock: *Fracture of Solids,* p. 65, John Wiley, New York, 1963.
7. R. M. Pelloux and J. C. McMillan: *Amer. Soc. Test. Mater., Spec. Tech. Publ.* 415, p. 505, American Society for Testing and Materials, 1966.
8. J. R. Rice: *Amer. Soc. Test. Mater., Spec. Tech. Publ.* 415, p. 247, American Society for Testing and Materials, 1966.
9. C. Bathias: *Mem. Sci. Rev. Met.,* April, 1971, p. 233.
10. C. Bathias: Thesis, University Poitiers, Poitiers, France, 1972.
11. F. Lecroisey and A. Pineau: *Met. Trans.,* 1972, vol. 3, p. 387.
12. M. Klesnil and P. Lukas: *Proc. of the Second Int. Conf. on Fracture,* p. 725, Chapman and Hall, London, 1969.
13. T. Yokobori, K. Sato, and Y. Yamaguchi: *Reports of Research Institute for Strength and Fracture of Metals,* vol. 5, p. 49, Tokohu University, 1970.
14. H. W. Liu and N. Iino: *Proc. of the Second Int. Conf. on Fracture,* p. 812, Chapman and Hall, London, 1969.
15. G. T. Hahn, R. G. Hoagland, and A. R. Rosenfield: *Met. Trans.,* 1972, vol. 3, pp. 1189-1202.
16. F. A. McClintock: *Fracture of Solids,* p. 65, Interscience Publishers, New York, 1963.

THE FRACTURE BEHAVIOUR OF MARAGING STEEL IN THIN SECTIONS*

K. Brown† and P. D. Smith†

SYNOPSIS: In structural sections that are sufficiently thin, the apparent fracture toughness may be higher than the plane-strain fracture toughness.

A study was undertaken to determine the most suitable of several models to predict the failure of pressurized thin-wall maraging steel tubes containing partial thickness axial flaws, and to determine a suitable fracture toughness parameter for use in the selected model. The best prediction was obtained with the ASME Boiler & Pressure Vessel Code when using fracture toughness values determined on test samples of the same thickness but of different geometries.

The effects of sheet thickness and notch sharpness on the measured fracture toughness were determined. The results suggest that the use of some plane-stress fracture toughness tests will lead to overly optimistic estimates of the design parameter for fracture toughness.

1. INTRODUCTION

In structures of sufficient cross section that plane-strain fracture mechanics are applicable, the fracture toughness, K_{1c}, can be used as a design parameter, and to establish the structure life and the frequency of non-destructive testing inspections. However since the majority of steel structures do not utilize sections sufficient to develop plane-strain conditions, elasto-plastic fracture mechanics must be used.

In the studies made by the ASTM E24 Committee [1, 2] to establish the range of testing conditions over which valid plane-strain fracture toughness data could be obtained, it was noted that in many cases, high values of fracture toughness were measured on sections thinner than those which can develop plane-strain conditions. In most cases this was an experimental artefact which resulted from the use of invalid data in the calculation of K_{1c} [3]. However it appeared that data obtained for a material in a particular section could be applied to design for that material in the same section. This conclusion has been subject to dispute [4] but if it is correct, there are potentially large savings in utilizing the higher toughness of thinner sections.

The failure of pressurised thin-wall cylinders containing axial defects has been treated by Irwin [5], Hahn et al. [6] and the pressure vessel and pipeline codes [7, 8]. By using published values of the plane-strain fracture toughness or fracture toughness measures based on empirical correlations with Charpy impact data, many workers have been able to use these treatments to predict accurately the failure pressures of cylinders of a variety of materials which have machined or fatigue cracked artificial flaws [6] [9]. The predictions have been successful even though the wall sections have usually been too thin for plane-strain fracture toughness to apply, and the defects have often been blunt in comparison with the fatigue cracks in samples used to generate the fracture toughness data.

This study was designed to obtain data for the prediction of the failure of pressurized thin-wall tubes containing sharp axial flaws, from experimental burst tests on tubes containing blunt spark-machined flaws. The data thus obtained were fitted to four models of the failure of axially flawed pressurised cylinders, to assess which model best predicted the failure conditions of the tubes. Fracture toughness data for this assessment were determined on centre-notched tensile and three-point bend test specimens containing spark machined notches. To indicate the effect of notch sharpness, further bend tests were conducted on samples containing fatigue cracks. These tests yielded crack growth resistance curves, ('R' curves), which were then used in the selected model to predict the behaviour of tubes containing sharp axial flaws.

2. EXPERIMENTAL

(a) *Tube Studies*

Three tubes of maraging steel, 50 mm diameter, 100 mm long, were extruded at 1200°C from commercial Grade 300, 2070 MPa (300,000 psi) tensile strength billet. These were solution heat treated for 1 hour at 816°

* Manuscript received 12th May 1977.

† Australian Atomic Energy Commission, Research Establishment, Lucas Heights, NSW.

C, machined inside and out to produce a uniform wall thickness and remove oxide scale and extrusion lubricant, and then aged at 482° C for 3 hours.

Artificial axial flaws of the lengths and depths indicated in Table 1 were spark machined at approximately the mid-section of two of the tubes, using 0.13 mm thick copper sheet as a machining electrode. Although fatigue crack defects would have been desirable a shortage of tubes and the experimental difficulties associated with growing a partial thickness crack in a thin wall made the use of fatigue cracks impractical.

The tubes were pressurized to failure at ambient temperature (~ 25°C) using a soluble oil and water mixture, and taking care that no air was entrained in the tube. During pressurization the tubes were held in sliding 'O'-ring seals so that no axial stress component was present (Fig. 1). A pressure gauge with a maximum load indicator was used to record the burst pressure.

The hoop stress at failure, σ_H, was calculated from the burst pressure, P, using the formula,

$$\sigma_H = PR^*/t \qquad (1)$$

where R is the tube radius and t the wall thickness. The flow stress of the material was taken directly as the failure stress of the unflawed tube, Tube 1.

The hoop stress at failure was determined for varying crack lengths using the four treatments described.

(i) The Hahn, et al. [6] treatment for tough and relatively ductile material with relatively short cracks, in which the hoop stress at failure, σ_H, of a cylindrical tube containing an axial flaw, and the flow stress, $\bar{\sigma}$, are related by

$$\sigma_H = \bar{\sigma}/M \qquad (2)$$

where M is the Folias bulging factor [10] given by

$$M = (1 + 1.61\ C^2/R^*t)^{\frac{1}{2}} \qquad (3)$$

where C is the half flaw length
and R* is the tube radius.

(ii) The Hahn, et al. [6] treatment for low to medium toughness vessels in which the fracture toughness of the material governs the failure stress. The failure stress σ_H is given by

$$\sigma_H = [K_{1c}/(\pi C\phi_3)^{\frac{1}{2}}]M \qquad (4)$$

where

$$\phi_3 = \ln [\sec (\pi M\sigma_H/2\bar{\sigma}]^2/[\pi M\sigma_H/2\bar{\sigma}]^2 \qquad (5)$$

This 'brittle' solution applies for short cracks where $(K_{1c}/\sigma_y)^2 < 7C$; σ_y is the yield stress of the material.

To calculate the failure stresses in the presence of partial thickness flaws, the Duffy et al. [9] extension of the solutions for through-wall flaws was applied to both ductile and brittle Hahn et al. treatments, i.e. to (i) and (ii) above.

The failure stress for a partial thickness flaw, σ_H, was related to that of a through-thickness flaw of the same length by

$$\sigma_H/\sigma_c = (A_o - A)/[A_o - A\sigma^*/\sigma_c] \qquad (6)$$

where σ_c is the failure stress for an unflawed tube,
σ^* is the failure stress for a tube containing a through-wall defect of the same length as the surface defect,
A_o = Ct, is the area of the through wall defect,
A is the area of the partial thickness defect.

(iii) A simplified treatment of Irwin [5] which strictly applies only to a flawed flat plate and does not contain a bulging factor. The fracture toughness is related to the failure

TABLE 1. Experimental and Calculated Burst Stresses for Tubes Containing Spark Machined Flaws

Tube No.	Wall Thickness (mm)	Internal Radius (mm)	Flaw Length 2C (mm)	Flaw Depth (mm)	Burst Pressure MPa	Failure Hoop Stress MPa	K ($MPa\ m^{\frac{1}{2}}$) Calculated Hahn brittle Solution (6)	Calculated Irwin Solution (5)	Calculated ASME Solution (7)	Present Exptl. Det. (approx.) (see Fig. 8)
1	0.60	24.83	—	—	50.0	2070	—	—	—	80-105
2	1.250	24.15	3.0	0.32	103.5	1998	203	70	83	
3	1.265	24.20	8.0	0.55	68.2	1305	77	60	97	

stress by the simplified equation

$$1.1\ \sigma_H \sqrt{(\pi a/Q)} \quad (7)$$

where 'a' is the flaw depth and 'Q' is flaw shape parameter given by

$$\phi^2 - 0.212\ (\sigma/\bar{\sigma})^2 \quad (8)$$

in which ϕ is a shape factor.

(iv) The ASME Boiler & Pressure Vessel Code (7) method for K_1 determination. This is an extension of the Irwin treatment and in the absence of bending stresses the stress intensity is given by

$$K_1 = \sigma_m M_m \sqrt{\pi} \sqrt{a/Q}$$

in which

a is the flaw depth,
σ_m is the membrane or hoop stress,
M_m is a correction factor for membrane stresses,
Q is a flaw shape parameter which is determined from the crack depth, a, and length, 2c.

The predictions of the Hahn, et al. [6] ductile model (i) were compared with the experimental results by plotting the predicted hoop stress at failure against crack length. The other models were assessed by calculating the values of fracture toughness that gave the experimental burst stresses in the model and comparing these with values determined in fracture toughness test samples containing spark machined notches.

(b) *Fracture Toughness Measurements*

Most fracture toughness measurements were made on spark machined or fatigue cracked test specimens whose geometry is not in accordance with recommended ASTM test procedures [2]. To determine the effect of thickness, the sides of test specimens were machined away (Fig. 2). According to the ASTM standards, specimens machined to less than ~ 3 mm were sub-thickness.

The specimens were solution heat-treated under the same conditions used for the tubes, and then fatigue cracked by three-point bending, or spark machined to produce a crack initiator. The samples were then aged. Attempts to produce a satisfactory fatigue crack after ageing were unsuccessful because of the softness of the loading system and the high strength of the aged material. In other respects ASTM procedures were followed.

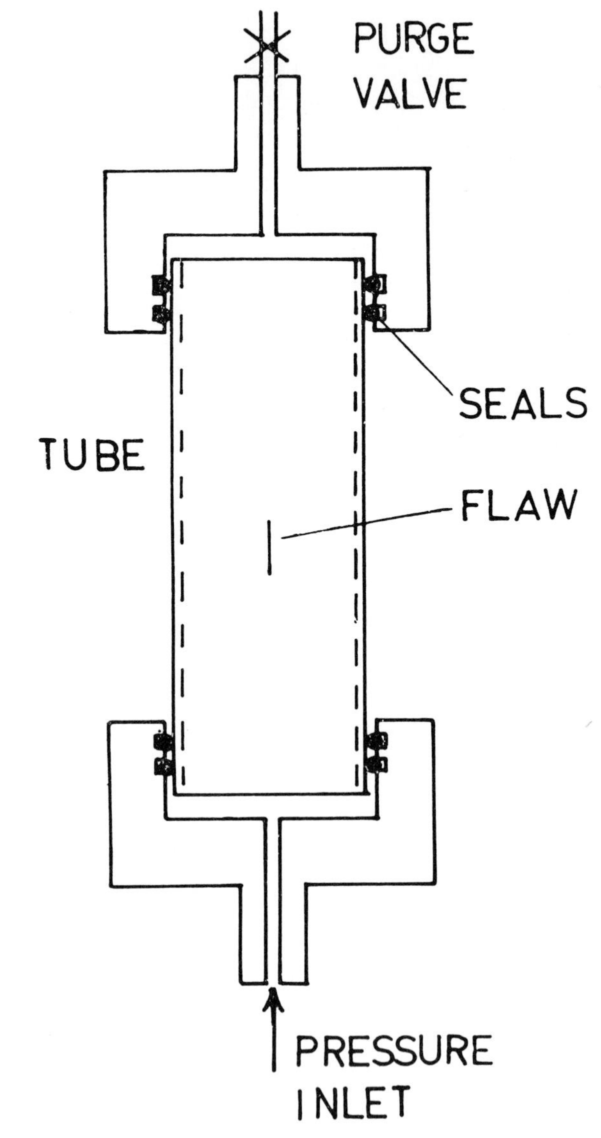

Fig. 1. Schematic diagram of pressure testing rig.

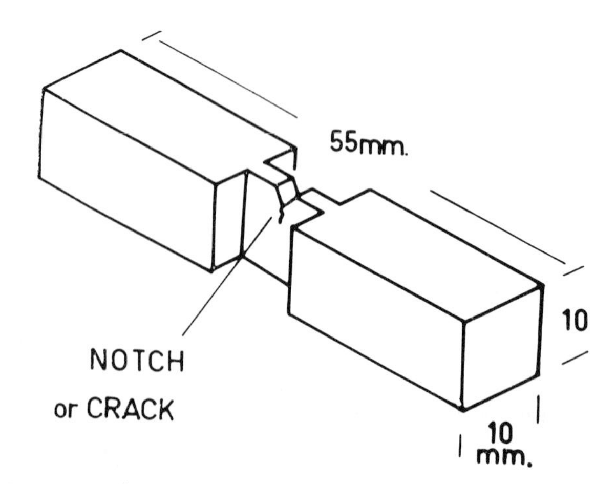

Fig. 2. Three-point bend fracture toughness test specimen.

Crack growth resistance curves ('R' curves) were constructed from the load versus crack mouth opening curves for three point bending, using the methods of Jones and Brown [3] for determining the crack length. The value of the crack growth resistance (K_R) at the maximum test load was then plotted against specimen thickness for samples of less than 5 mm thickness.

Further tests were conducted on grade 300 maraging steel sheet tensile specimens, 1.65 mm thick, containing a central transverse slot which was spark machined completely through the specimen after ageing, (Fig. 3). By assuming a crack length equal to the machined crack length, (i.e. that no slow crack growth preceded failure) a fracture toughness parameter for the material was calculated from the maximum tensile stress before catastrophic failure. The corrections for specimen geometry were those used by Orange [11] in the basic Irwin equation

$$K_c = Y\sigma\sqrt{c + r}$$

where σ = failure stress (MPa)
c = half crack length (metres)
K_c = critical stress intensity factor (MPa $m^{\frac{1}{2}}$)
Y = calibration factor

$$Y = (1 - 0.025\,\lambda^2 + 0.06\,\lambda^4)\left(\pi \text{ secant } \frac{\lambda\pi}{2}\right)^{\frac{1}{2}}$$

$\lambda = 2a/w$
w = specimen width
r = plastic zone correction factor
$r = (K/\sigma_{ys})^2/n\pi$
$n = 2.83$

When necessary the specimen thickness was reduced after ageing by machining both specimen surfaces as shown in Fig. 3.

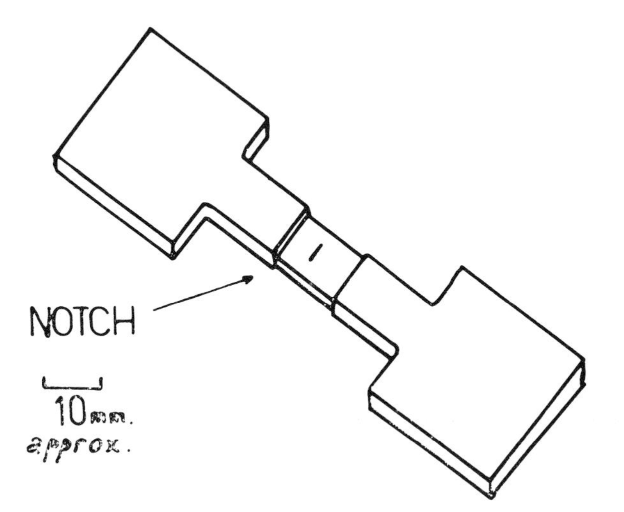

Fig. 3. Centre-notched tensile test specimen.

(c) *'R' Curve Techniques*

Crack growth resistance curves, 'R' curves, describe the variation in the fracture resistance, K_R, of the *material* as a crack grows. They are usually plotted as K_R versus crack length, a, or crack growth increment, Δa. In materials of the thicknesses and type of those used in this study, K_R generally increases with crack length because of an increase in the plastic zone size at the tip of the growing crack.

The variation in stress intensity, K, at the tip of a crack in a *structure* may be plotted at a particular applied stress on the same axes as the 'R' curve of the *material.* At any particular stress and crack length, the stability of the crack can then be assessed. If the stress intensity, K, applied to the crack tip, exceeds the crack growth resistance, K_R, then growth occurs. If the crack growth sufficiently increases K_R, such that $K_R = K$, growth ceases. The point at which a stress intensity/crack length curve is tangent to the 'R' curve defines the stress, fracture resistance and crack length at unstable failure.

These parameters were determined graphically for the pressurized tubes by superimposing plots of stress intensity against crack length, for tubes of appropriate geometry, on 'R' curves for three-point bend test samples of approximately the same thickness.

For this construction, 'R' curves for specimens 1 mm thick were used to approximate the behaviour of the 1.25 mm wall thickness tubes; however because of the similarity of the curves for 0.5, 1 and 2 mm samples, this approximation is believed to be within experimental accuracy.

We believed that crack growth would occur initially in the thickness direction (increasing 'a') and that when it had penetrated the wall, it would then grow lengthwise (increasing 'c').

The stress intensity versus crack depth 'a' curves were determined using the methods presented in the ASME Boiler & Pressure Vessel Codes (7) described in (iv) above.

After the crack had penetrated the wall, the curves for increasing crack length 'c', were determined from the Orange (11) version of the Irwin equation given earlier in (b). The same crack growth resistance curve was considered to apply for growth in both thickness and axial directions.

3. RESULTS

All the maraging steel tubes tested burst catastrophically, (Fig. 4). In tubes 2 and 3 the fracture grew through the wall from the spark machined flaws then propagated axially over the entire length of the tubes. At the tube ends, which were restrained by the grips, it travelled circumferentially. With the exception of a small area of flat fracture at the root of the flaw (Fig. 5) the failure was under plane-stress conditions with a typical ductile shear fractography.

The burst stresses and the fracture toughness values calculated by either the Hahn et al. [6], Irwin [5] or ASME code [7] treatments are shown for each tube in Fig. 6 or Table 1.

The fracture resistance curves, ('R' curves) for three-point bend test specimens containing fatigue cracks are shown in Fig. 7. The fracture toughness calculated by ASTM methods for thicknesses greater than 5 mm, or by 'R' curve methods for thinner specimens, is plotted against specimen thickness in the composite curve in Fig. 8. The plane-strain fracture toughness agrees well with values presented in the literature [12] despite the non-

Fig. 4(a) Flawed maraging steel tube No. 2 after failure.
(b) Flawed maraging steel tube No. 3 after failure.

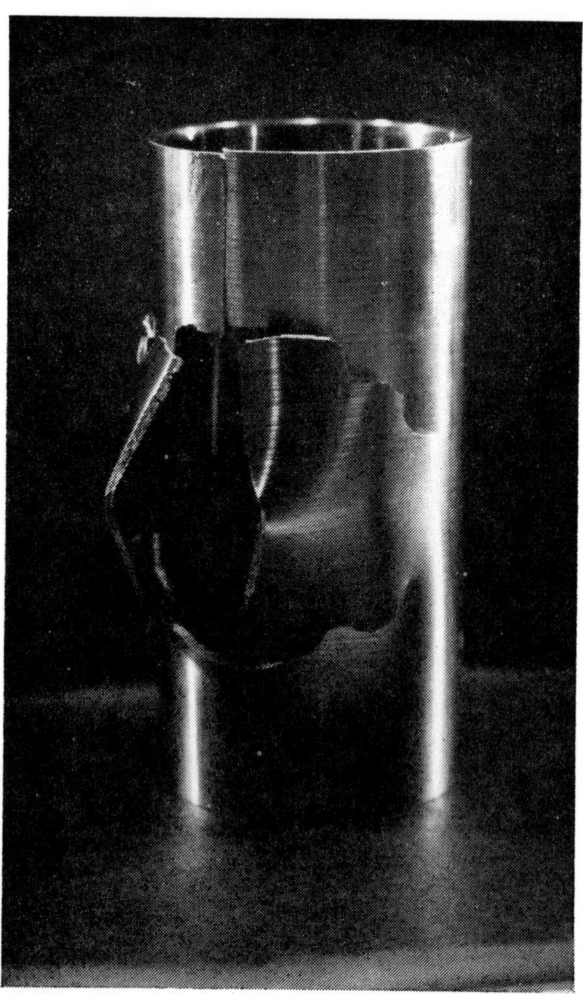

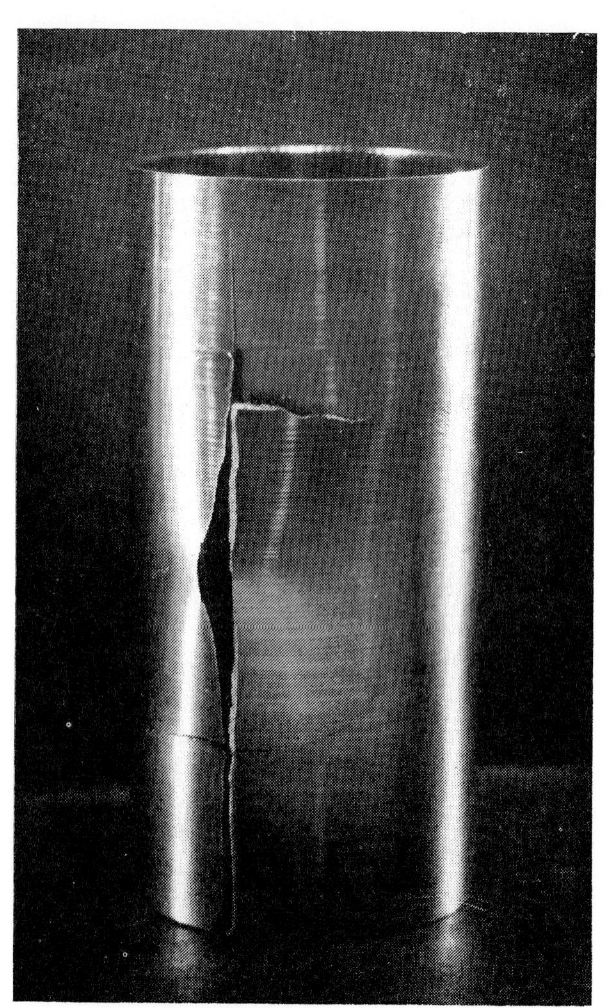

standard specimen geometry and the fact that fatigue cracking was carried out before final ageing.

A maximum fracture toughness is suggested for sections ~ 1 mm thick; however note that the stress intensity at which crack growth can first be detected (Fig. 8) is close to the plane-strain fracture toughness, and appears independent of sample thickness. Similar results have been reported for Grade 190 maraging steel [13].

The fractography of all of these samples, in the range 1 to 10 mm, was similar. Crack

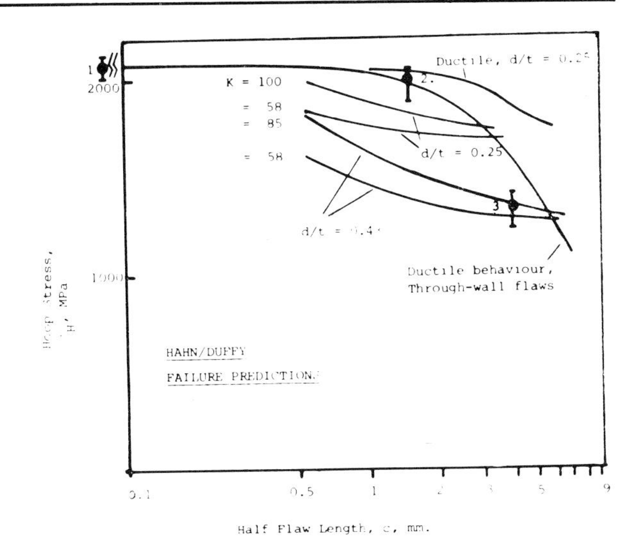

Fig. 6. Plots of the failure hoop stress for tubes of wall thickness 't' showing the predictions made using the Hahn et al. models[6], and the Duffy et al.[9] treatment for partial thickness flaws of depth 'd'. The curve describing the stability of through-wall flaws in ductile material is also shown, together with the actual failure conditions for each tube.

Fig. 5. Fracture surfaces in the vicinity of the artificial defect in Tubes 2 and 3 (Tube 3 top). The areas of plane-strain and plane-stress failure are shown in b.

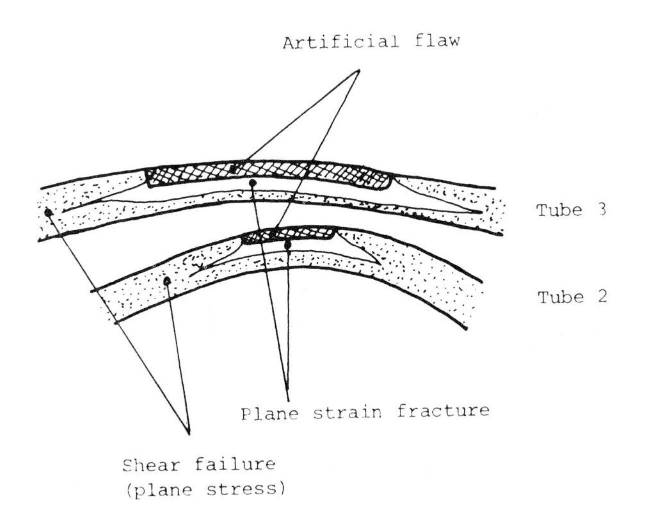

initiation occurred under predominantly plane-strain conditions; however during growth, shear lips were formed, and these increased in width up to 0.4 mm after ~ 1 mm of crack growth. The width of the fully developed shear lips was approximately constant in all samples and consequently the relative proportion of shear failure increased with decreasing thickness. In those samples less than 1 mm thick, the fully developed fracture occurred entirely by shear. In the transition region at the notch root a triangular region of plane-strain failure remained even in the thinnest samples tested.

The results obtained for the sheet tensile specimens with spark-machined flaws also suggested a maximum fracture resistance at about 1 mm thickness (Fig. 8); however the actual values were significantly higher than those measured on the fatigue-cracked bend specimens. Unlike the tensile specimens, three-point bend samples containing spark-machined flaws instead of fatigue cracks, showed no significant sensitivity to specimen thickness (Fig. 8).

The fractography of the sheet tensile specimens was similar to that of the three-point bend samples except that no plane-strain region persisted in the fully developed fracture, even in specimens up to 1.65 mm thick. It seems likely that this behaviour resulted from the lower restraint of this type of specimen in the sheet thickness direction.

The pressurized tubes and all test samples having a spark-machined notch, contained a small shear lip, ~ 0.05 mm wide, at the root of the notch (Fig. 9). This was not present in any fatigue-cracked sample.

4. DISCUSSION

It is clear that the ASME Code (7) provides the best prediction of the design fracture toughness for this material in this section thickness, Table 1. The values of fracture toughness determined for tubes with spark-machined flaws fall within the experimental scatter of values determined for the same flaws in centre-notched tensile and three-point bend-test specimens (Fig. 8).

The Hahn et al. [6] 'ductile' treatment provides a satisfactory prediction of the burst stress of tube 2, but it is clearly inappropriate for tube 3 (Fig. 6) as its failure stress is even lower than that predicted for through wall flaws of the same length. The Hahn et al. 'brittle' model yielded an excessively high toughness for tube 2, and a slightly low value for tube 3. Excessively low fracture toughness values were indicated by the simplified Irwin treatment [5].

The behaviour of the tube containing axial defects may be explained graphically by using 'R' curves as described in (c) above. Schematic fracture resistance curves for samples containing a blunt spark machined notch, (Fig. 10), and experimentally determined 'R' curves for samples containing a sharp fatigue crack (Fig. 11) are shown. The blunt notch requires a higher stress (i.e. higher stress intensity, K) for crack initiation, but as the crack grows through the tube wall, the fracture resistance will fall towards that of a growing sharp crack. This can be seen in the fractography of these samples, where the small shear lip precedes the region of plane strain fracture (Fig. 9).

Unless the loading is very 'hard' the stress intensity at the crack tip will remain supercritical and unstable fracture will occur. Because the loading in the pressurized tubes is not sufficiently hard, unstable fracture in the thickness direction follows initiation. Once the defect grows almost through the wall, the stress intensity for longitudinal growth is

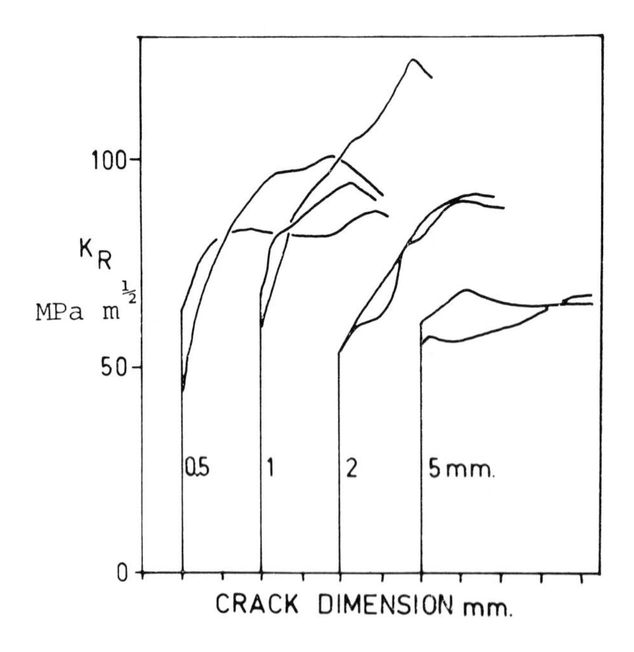

Fig. 7. Fracture resistance curves ('R' curves) determined experimentally for duplicate bend-test specimens of the thicknesses indicated.

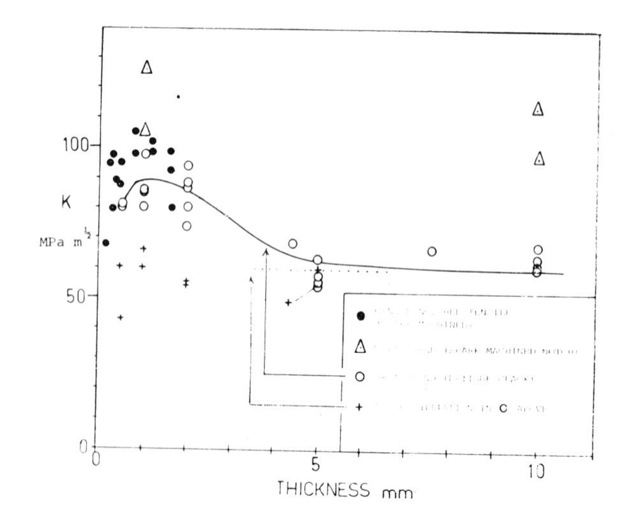

Fig. 8. Plot of experimentally determined fracture toughness versus sample thickness.

Fig. 9. Plane-strain fracture region in a centre notched tensile specimen 1.65 mm in thickness. Note the shear lip at the root of the spark machined notch. Magnification X35.

clearly supercritical (c.f. broken curves in Fig. 10) and unstable fracture continues. Thus the appropriate fracture toughness parameter to be used in failure prediction is that measured at the point of initiation.

It is now possible to examine the behaviour of the tubes, had sharp defects been present. In this case the fracture resistance rises following crack initiation, however the relative softness of the loading on the crack causes the stress intensity to increase at so fast a rate that instability is reached after only very small crack growth increments (c.f. solid curves in Fig. 11). This growth is insufficient for significant increases in fracture resistance to be achieved, and the effective design fracture toughness parameter approximates the value at crack initiation. Again, when the partial thickness crack penetrates the wall, the stress intensity for longitudinal growth is supercritical (c.f. broken curves).

For fatigue cracked tubes of geometry identical to tubes 2 and 3, this analysis suggests approximately 20% lower burst stresses (i.e. 1600 MPa and 1050 MPa respectively) because of the sharper defect. However it can

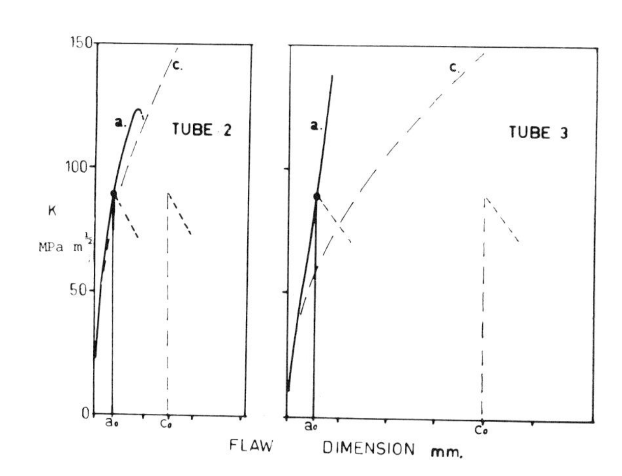

Fig. 10. Stress intensity versus crack length curves for the two tubes. Schematic 'R' curves for crack growth from a spark machined notch are superimposed at a_0, the initial flaw depth (solid curve), and C_0, the initial half-length of the flaw (broken curve). Curve (a) (solid curve) refers to the flaw growth in the thickness direction and (c) — (broken curve) to the growth of a through-wall flaw in the axial direction. Both curves are calculated for the actual failure stress of each tube.

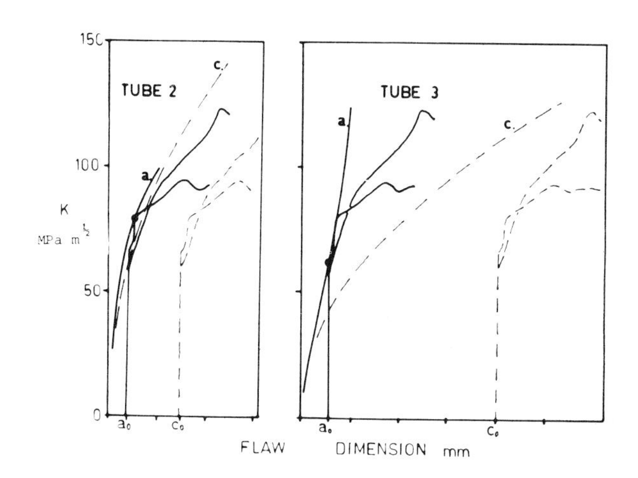

Fig. 11. Plots showing the failure criteria for tubes containing sharp axial flaws. The plots are as in Fig. 10, except that duplicate 'R' curves determined experimentally for fatigue cracked samples are shown, and maximum failure stresses of 1600 MPa for a tube of identical geometry to tube 2, and 1050 MPa for tube 3, are predicted.

be seen that the estimate depends on relatively small differences in the shape of the experimentally determined 'R' curves, e.g. note the double tangency for tube 3. The toughness at initiation is shown experimentally to be close to the plane-strain fracture toughness. This is supported by the fractography of the fatigue-cracked specimens, where the initial crack growth is under predominantly plane-strain conditions, and there is no small shear lip.

In a hard loading structure, where significant crack growth and consequent increases in fracture resistance can precede unstable failure, a higher design fracture toughness can be used. This is exemplified by the three-point bend-test specimens used to generate the data. Here the appropriate toughness parameter to describe the onset of unstable fracture at thicknesses around 1 mm is ~ 50% higher than the plane-strain fracture toughness.

5. CONCLUSION

The most appropriate model to predict the failure of pressurized thin-wall maraging steel tubes containing axial flaws is that used in the ASME Boiler & Pressure Vessel Codes (7).

The plane-strain fracture toughness is the most appropriate and conservative parameter for design in thin-wall structures of grade 300 maraging steel. Higher experimentally determined toughness values may be used for failure prediction only if the compliance of the structure and loading permit sufficient stable crack growth before unstable fracture. This may be determined by using crack growth resistance curves ('R' curves).

ACKNOWLEDGEMENTS

The authors are grateful to Mr. A. Ridal and Mr. S. Spain for making the tubes used in this study. The assistance of Mr. B. Zybenko and Mr. C. Duke is gratefully acknowledged. Thanks are due also to Dr. P. M. Kelly for numerous helpful discussions.

REFERENCES

1. W. F. Brown and J. E. Strawley, *'Plane Strain Crack Toughness Testing of High Strength Metallic Materials'*, 1966 ASTM STP 410.
2. *'Plane Strain Fracture Toughness Testing of Metallic Materials'* ASTM Standards, E399-74, '1974 Annual Book of ASTM Standards', ASTM, Philadelphia 1974.
3. M. H. Jones and W. F. Brown in *'Review of Developments in Plane Strain Fracture Toughness Testing'*, 1970 ASTM STP 463, 63.
4. J. G. Kaufman in *'Review of Developments in Plain Strain Fracture Toughness Testing'*, 1970 ASTM STP 463, 95.
5. G. R. Irwin, *J. Appl. Mech.* 1962, **29**, 651.
6. G. T. Hahn, M. Sarrate and A. R. Rosenfield, *Int. J. Fract. Mech.*, 1969, **5**, 187.
7. ASME Boiler & Pressure Vessel Code, Section II, *'In Service Inspection of Nuclear Power Plant Components'*, 1974, Article A3000, 117.
8. USA Standards Institute, USAS B-31 *'Code for Pressure Piping'*, 1968B, ASME, N.Y.
9. A. R. Duffy, A. R. Eiber and W. A. Maxey, *'Recent Work on Flaw Behaviour in Pressure Vessels'*, Symposium on Fracture Toughness Concepts, UKAEA, April 1969.
10. E. S. Folias, *Int. J. Fract. Mech.*, 1965, **1**, 104.
11. T. W. Orange in *'Fracture Analysis'*, 1973, ASTM STP 560, 122.
12. Inco Europe Ltd., *'18 per cent Nickel Maraging Steels, Engineering Properties'*, 1976, Pub. 4419.
13. S. T. Rolfe and S. R. Novak in *'Review of Developments in Plane Strain Fracture Toughness Testing'*, 1970, ASTM STP 463, 92.

Fracture Toughness Study of a Grade 300 Maraging Steel Weld Joint

Plane strain fracture toughness is determined directly from welded joints while weld simulation specimens add correlative data

BY J. G. BLAUEL, H. R. SMITH AND G. SCHULZE

ABSTRACT. The plane strain fracture toughness (K_{Ic}) of a gas tungsten-arc (GTA) welded joint prepared from 10 mm (0.39 in.) thick, grade 300, maraging steel was determined. Transverse three-point bend specimens, 10 × 10 × 55 mm (0.39 × 0.39 × 2.17 in.), were prepared with notches and precracks preferentially located in the weld and the heat-affected zone (HAZ), the outline of which had been revealed by prior polishing and etching. All critical K_I values less than about 270 kp mm$^{-3/2}$* (76 ksi-in.$^{1/2}$) satisfied ASTM validity criteria for K_{Ic}; these included determinations for all microstructures except one that was characterized by a very high stable austenite content.

The effect of microstructural and property variations characteristic of the HAZ on the determination of K_{Ic} was studied using "weld simulation" specimens thermally treated in a Smit weld simulator. Results were correlated with those obtained from actual welds on the basis of equivalent microstructures, mechanical properties, and thermal treatments. From this study, a tendency towards embrittlement due to short time exposure at 750 C was detected. It was also noted that K_{Ic} appeared to be independent of grain size.

Typical K_{Ic} values of the weld joint were as follows: for the base material: 250 kp mm$^{-3/2}$ (71 ksi-in.$^{1/2}$) for the high austenite zone: greater than 290 kp mm$^{-3/2}$ (82 ksi-in.$^{1/2}$) for the transformed zone: 230 kp mm$^{-3/2}$ (65 ksi-in.$^{1/2}$) and for the weld: 190 kp mm$^{-3/2}$ (54 ksi-in.$^{1/2}$).

**kp = kilopond, a unit of force exactly equal to the kilogramforce (kgf), is frequently used in Europe but is relatively unknown in the United States. In this paper, all expressions involving the kilopond have been converted to, and are followed by, U.S. customary units in parentheses — ed.*

J. G. BLAUEL and H. R. SMITH are with Institut für Festkorpermechanik der Fraunhofer Gesellschaft e.V., Freiburg, and G. SCHULZE is with Schweisstechnische Lehr-und Versuchsanstalt, Berlin, West Germany.

Introduction

There is a continuing demand for ultra high strength materials for application in weight critical structures. For example, most commercial and military aircraft use these materials in landing gears and other secondary structure. In most cases the designs require that the materials be heat treated to a strength level near that which is the maximum obtainable, a condition which unfortunately is also quite brittle. The designer is allowed to take a calculated risk concerning the possibility of brittle fracture because the failure of a secondary structure is not necessarily followed by catastrophic failure of the aircraft.

However, these risks can be minimized by a thorough understanding of the nature of the brittle fracture characteristics of the material — especially under welding conditions. In this regard, the tools provided by fracture mechanics can be utilized.

A number of workers have reported the results of fracture toughness testing of weld joints (Refs. 1-5). However it should be stated that extreme analytical and experimental difficulties in obtaining valid and meaningful results are introduced by the variation in microstructural properties and the operation of residual stresses resulting from the complex temperature profiles attending the multipass welding. A solution for the material part of this problem can be approached through augmenting the information derived from weld joint tests by use of a weld simulator (Refs. 6,7). In such an apparatus, localized regions of the heat-affected zone can be synthesized in cross sectional areas large enough to yield valid test results.

One of the more attractive weldable (Refs. 7,8,9), ultra high strength materials available today is maraging steel. It can be fabricated in a relatively soft solution annealed condition to final dimensions, and then aged to full hardness. The aging process involves the precipitation of intermetallic compounds in an essentially carbon free matrix at relatively low temperatures (450-500 C) with no accompanying phase transforma-

Fig. 1 — Microstructure of grade 300 maraging steel after hot-rolling, solution annealing at 820 C for 4 h, and aging at 480 C for 4 h. An extremely nonuniform structure due to incomplete recrystallization is indicated. (Electrolytic etch with 10% Cr_2O_3; X100, not reduced)

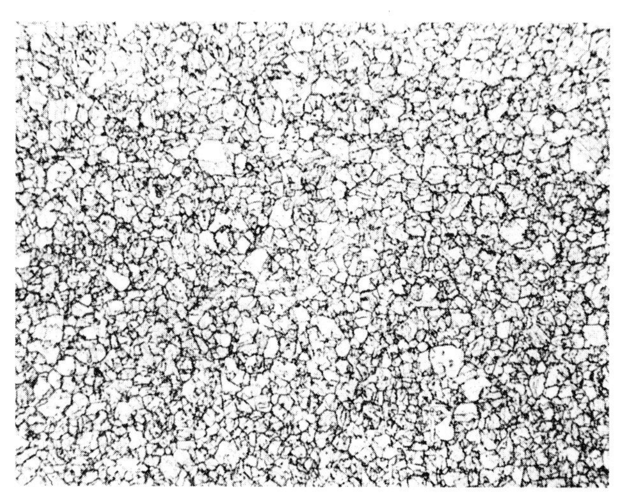

Fig. 2 — The material shown in Fig. 1 after an additional solution annealing treatment at 900 C for 4 h. Recrystallization is complete. (Etch and magnification same as Fig. 1)

Table 1 — Composition of Base and Filler Metals for X2 NiCoMo 18 9 5, wt%

	C	Si	Mn	P	S	Al	Co	Mo	Ni	Ti
Base metal	0.004	0.02	0.02	0.005	0.004	0.085	9.05	4.94	17.8	0.79
Filler metal	0.004	0.02	0.02	0.006	0.006	0.120	8.80	4.80	18.2	0.98

Table 2 — Mechanical Properties and Toughness of X2 NiCoMo 18 9 5 material

Ultimate tensile strength, σ_2 , kp/mm² (ksi)	216	(303)
Yield strength, σ_y , kp/mm² (ksi)	210	(299)
Fracture toughness, K_{Ic} , kp - mm $^{3/2}$ (ksi-in. $^{1/2}$	246	(69)
Vickers hardness, HV 50, kp/mm² (ksi)	610	(867)

tions. These features minimize the problems of dimensional stability, distortion, surface oxidation and decarburization commonly associated with quenched and tempered steels.

Depending on its composition, maraging steel can be aged to strength levels ranging from 140 to 240 kp/mm² (200 to 340 ksi) at toughness levels high in comparison to corresponding steels. For this reason maraging steel in the 190-210 kp/mm² (270-300 ksi) yield strength range was chosen as the research material of this program.

Material, Apparatus and Procedure

Material

The material was X2 NiCoMo 18 9 5 steel supplied by Stahlwerke Südwestfalen, Geisweid, and identified by their tradename as HFX 760. This material corresponds to American grade 300 18% Ni maraging steel. It was first prepared as a vacuum induction melted and electron-beam re-melted 260 mm (10.24 in.) round block, which was subsequently hot rolled to a 10 mm thick plate and solution annealed at 820 C for ½ h. When aged at 480 C for 4 h, the material's microstructure appeared as in Fig. 1. This was considered to be too nonuniform to permit sensitive measurements of the grain size in the HAZ of a welded specimen, and consequently the material was resolution annealed at 900 C for ½ h. This resulted in the fine grained uniform microstructure shown in Fig. 2, which was then considered satisfactory for subsequent investigations.

The chemical composition is shown in Table 1, while mechanical properties and toughness of the reannealed material in the aged condition are shown in Table 2. From this point on, whatever prior processing (i.e., welding or simulation) was involved, specimens were subsequently aged at 480 C for 4 h.

Heat Treatment

At temperatures higher than 750 C, 18% nickel maraging steel (nominally 18% Ni, 9% Co, 5% Mo) exists as a face centered cubic, single phase austenite. On cooling, the austenite transforms by shear to body centered cubic martensite, irrespective of cooling rate. The low carbon Fe-Ni-martensite so formed is soft and ductile. On reheating in the 450-550 C range the alloy is hardened by the precipitation of intermetallic phases as Ni_3Mo and Ni_xTi_y (Refs. 10, 13). Optimum hardness is obtained by aging from 3 to 8 h in this range. However, after prolonged exposure, the alloy overages by a process involving reversion of the martensite to austenite, probably in the regions immediately surrounding the intermetallic compounds (Ref. 11). At higher temperatures (but still below 750 C), there is a tendency for the elements Ni and Ti to partition to the extent that a Ni-rich phase is retained on cooling to room temperature. It has been reported that the tendency to form stable austenite is strongest at 650 C (Refs. 12, 13), and at this temperature the required times are very short (of the order of seconds) (Refs. 14, 15). Furthermore, heating rate, hold time, and the number of overlaying cycles play important parts in the process (Ref. 15). These points are important in understanding the microstructure of a HAZ that must experience brief exposures to such conditions.

Welding

It has been demonstrated that the arc welding methods employing inert, protective gas produce sound joints of superior strength and toughness in maraging steel. Of these, the gas tungsten-arc welding (GTAW) process is easiest to control with respect to critical welding parameters.

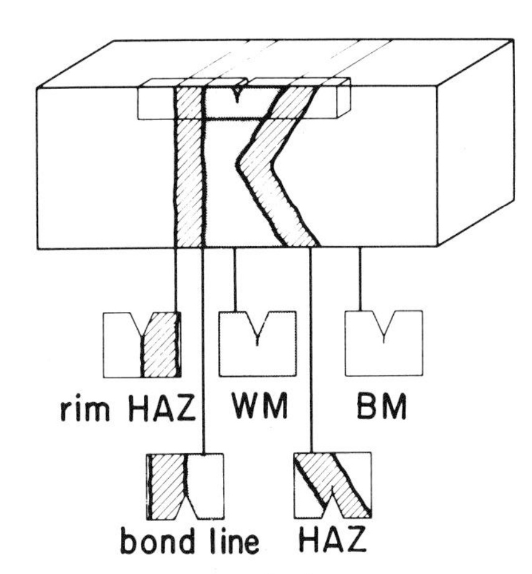

Fig. 3 — K configuration weld joint showing specimen and notch locations. BM = base metal, WM = weld metal zone, HAZ = heat-affected zone

Table 3 — Parameters for GTA Welding — by Passes

Parameter	Pass number 1	2	3	4	5	6
Current, A	195	210	210	230	210	210
Voltage, V	13	15	15	15	15	15
Welding speed, cm/sec,	0.18	0.18	0.13	0.18	0.18	0.09
in./min	4.2	4.2	3.1	4.2	4.2	2.1
Energy input, kJ/cm,	14.1	17.6	24.2	19.2	17.6	35
kJ/in.	35.8	44.7	61.5	48.8	44.7	88.9

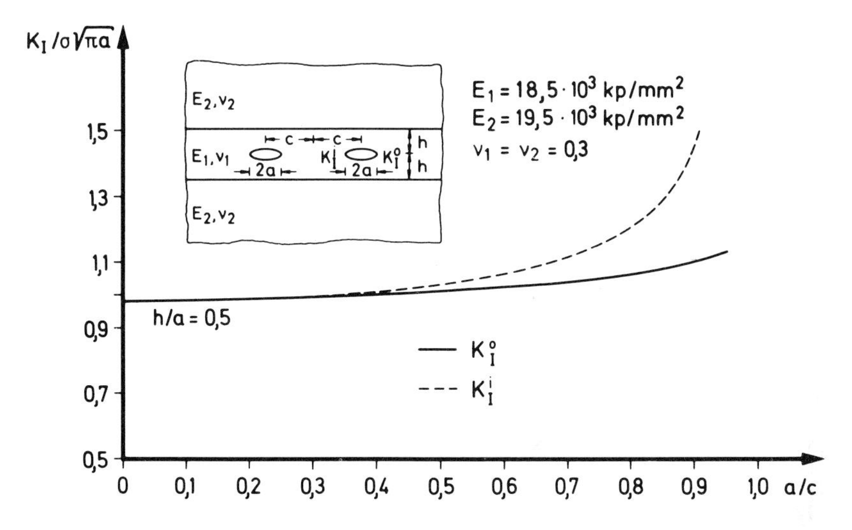

Fig. 4 — Stress intensity factors for coplanar cracks located in a layered medium (Ref. 19)

Accordingly, the GTAW process was chosen to weld the K-configuration joints shown in Fig. 3. The welding was parallel to the rolling direction. The HAZ that forms on the straight, perpendicular wall of the joint has itself straight boundaries. This facilitates the location of the notch and precrack in subsequently fabricated test specimens. The plates were hand welded with multiple passes using the parameters shown in Table 3. The filler metal (2 mm or 0.079 in. diam) composition is shown in Table 1. Radiographically acceptable welds were obtained when exceptional measures were taken to maintain cleanliness from pass to pass. (Ref. 16).

Weld Simulation

A Smit weld simulator (Ref. 17) was employed to synthesize localized HAZ microstructures in the cross section of test specimens by controlled resistance heating and water cooling. Reference cycles were obtained by running a bead on the surface of a plate, and recording the temperature as a function of time at various selected distances from the weld fusion line. These reference cycles, in turn, provided a basis for constructing simple treatments that would simulate the effect imposed by single pass welding. Peak temperatures of 650, 700, 750, 800, and 1350 C were selected on the basis of microstructural analysis as those best suited to provide a critical survey of the strength and toughness behavior of the HAZ.

The temperature history for each simulation specimen was controlled by a thermocouple spot welded to the surface at the middle of its length. Although there was a considerable gradient from end to end, the temperatures in the surface-to-surface planes were uniform. The phase transformations occuring in each treatment were measured *in situ* by a dilatometer.

Fracture Mechanics in Relation to the Weld Joint

In an elastic, isotropic, and homogenous medium containing a crack, the stress intensity factor K for mode I loading is given by:

$$K_I = \sigma (\pi a)^{1/2} Y \quad (1)$$

where σ is the gross section stress, a is the crack length, and Y is a factor which accounts for the boundary influences associated with a particular configuration and loading mode (in this program, a three point bend system was used).

In performing a fracture toughness test, a specimen containing a well defined fatigue precrack is loaded to failure. According to fracture mechanics, the criterion for failure is described in terms of a critical stress intensity level with respect to the material. This is defined as the material's fracture toughness. When constraint to deformation at the crack tip is severe enough to approximate closely the plane strain state of stress, the critical stress intensity factor is called K_{Ic}. The procedure in determining a valid value for K_{Ic} has been specified by ASTM (Ref. 18). Under the conditions of this specification, the value so determined is independent of geometry, and consequently, it is desirable whenever possible to evaluate toughness in terms of K_{Ic}.

When the principles of fracture mechanics are applied to the study of a weld joint, the degree of microstructural anisotropy and the state of residual stress in the joint must be considered.

The effect of anisotropy on the stress distribution at the crack tip can be examined by referring to simple cases which can be handled analytically. In Fig. 4, for example, the stress intensity solutions for a system for two coplanar cracks are shown contained in a layered medium in which the elastic constants are typical for different zones of the weld joint (Ref. 19). The stress intensity factor K_I^i pertains to the inner tips, while K_I^o applies to the outer tips of the cracks. When the cracks lie far apart (in Fig. 4 this corresponds to small values of a/c) there is no interaction of the stress fields. K and K_I^o become identical, and the representation is that of a single crack. K_I for the layered medium is normalized by $\sigma(\pi a)^{1/2}$, the value of K for a homogeneous medium, to permit direct observation of the anisotropic influence. It can be seen that for the given difference in elastic constants of the var-

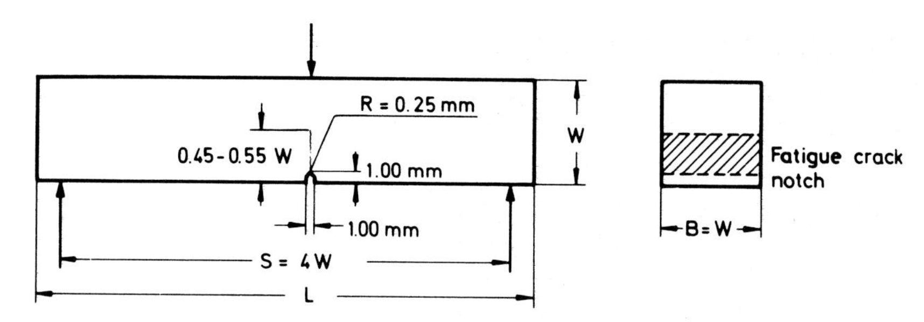

Fig. 5 — Three point bend specimens for K_{Ic} determination

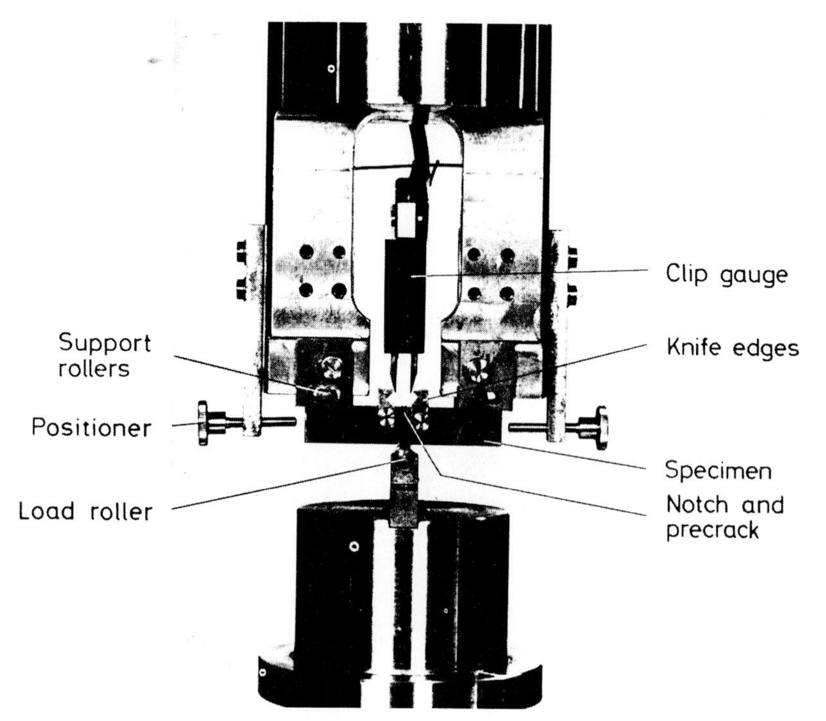

Fig. 6 — Three point bend testing fixtures

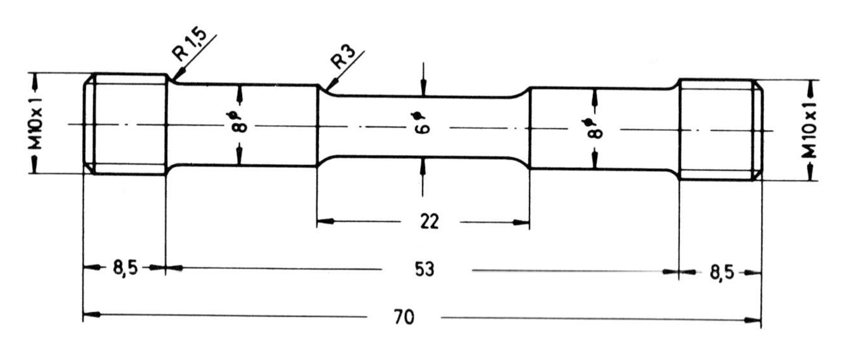

Fig. 7 — Round bar tensile specimen with double reduced section

ious microstructures represented in the HAZ of maraging steel, the ratio of $K_I / \sigma (\pi a)^{1/2}$ is only about 2 or 3% less than one. It is concluded from this and further analytical results (Ref. 19) that the linear elastic stress analysis for an isotropic and homogeneous medium is still appropriate.

When determining critical stress intensity levels (K_{Ic}) as material values, the influence of the anisotropy and inhomogenity on the formation of the plastic zone at the crack tip must also be considered. The size of the plastic zone influences the fracture toughness, i.e., the resistance against instable fracture, and that, in turn, can be defined as a function of the ratio of the elastic stress intensity factor K_{Ic} to an effective local yield strength which takes into account the differing plastic deformation behavior of the various microstructures and their interaction. In this program representative values of σ_Y for local yield strength (0.2% offset in uniaxial tension) were determined from round-bar tension experiments and used to check the validity criteria for K_{Ic} measurements (see also later).

With respect to residual stress it has been shown that the following observations are pertinent to maraging steel welds (Ref. 20):

(1) The residual stress is dominant in the longitudinal direction (i.e. parallel to the weld); however, transverse residual stress at a much lower level also is developed.

(2) The residual stress is compressive along the weld centerline, balanced by tensile stresses in the HAZ on each side.

All the specimens are taken transverse by cutting at right angles to the weld length. Their cracks are extended in a direction through the thickness of the plate and in a plane parallel to the weld length (Fig. 3). Then only the minor transverse residual stresses can superpose on the external applied stress, and they will add or subtract according to their distribution indicated in (Ref. 2). However, their influence was not apparent when driving the fatigue crack through different zones of the weld. It is generally known that cutting the weld length in segments, as is done in preparing the small transverse test specimens, should lower the residual stress level.

In summary, it appeared that the influence of residual stress on the results of the program was insignificant.

Evaluation of Strength, Toughness and Microstructure

To evaluate K_{Ic}, three point bend specimens as depicted in Fig. 5 were chosen. A preliminary program indicated that specimens taken transverse to the rolling direction with precracks directed through the thickness gave rise to the lowest fracture toughness values. All subsequent bend specimens were fabricated in this manner. Generally, the specimens were prepared to final dimensions (with chevron or straight notch) when the material was in the soft solution-annealed condition. They were then aged and fatigue precracked prior to testing. In the case of the weld specimens, however, it was necessary to first prepare the specimens to rough dimensions only. Then, one of the side surfaces of each was polished and etched to reveal the weldment and HAZ. The position of the notch was located as desired, and then the preparation was continued as for the unwelded specimens.

The specimen size (i.e. width and thickness) was limited by what could be treated in the Smit simulator. Thus, the width-thickness dimensions for all specimens were 10 × 10 mm with the precrack length being 5 mm. The load fixtures were arranged so that specimens of all lengths could be tested; however, the major span length for each was 40 mm (1.58 in.). A clip gauge was designed to fit into knife edges mounted on the specimen adjacent to the notch. The assembled apparatus is shown in Fig. 6. Load-displacement curves obtained from the tests were used to calculate K_{Ic} in accordance with the specified procedure (Ref. 18).

The restrictions on specimen dimensions governing the determination of K_{Ic} are given by:

$$a, B \geq 2.5 (K_{Ic} / \sigma_Y)^2 \qquad (2)$$

where a is crack length, B is thickness, and σ_Y is the "effective" material's yield strength. With a limiting crack length of 5 mm, the maximum valid K_{Ic} that could be measured was about 270 kp-mm$^{-3/2}$ (76 ksi-in.$^{1/2}$) when the lower limiting yield strength was 190 kp/mm^2 (270 ksi).

Tensile tests were performed using the specimen shown in Fig. 7. In consideration of the simulation treated specimens, the design was arranged such that the 690 C isotherm was always located in the lower-stressed shoulder of the reduced section. This is important due to the tendency for soft austenite to form and stabilize at this temperature. It was necessary to use longitudinal specimens to determine the strength of the weldment and the transformed zone immediately adjacent to it in an actual weld. Vickers hardness was also measured for each zone of interest.

The most important microstructural features were considered to be the prior austenitic grain size and the percentage of retained austenite. These were measured on the carefully ground and polished fracture surfaces of broken bend specimens. Electrolytic etching with 10% Cr_2O_3 revealed the prior austenitic grain outline quite clearly if the structure had first been age hardened. Ordinary etching with V2A Beize (50 ml HCl, 5 ml HNO_3, 0.15 ml Sparbeize inhibitor) developed the austenitic-martensitic microstructure and it was found that overlaying the second treatment on the first produced a microstructure in which all the salient features could be seen. Additionally, retained austenite content was measured by x-ray diffraction (Ref. 21). Grain size was determined by the intercept method in the prescribed manner (Ref. 22).

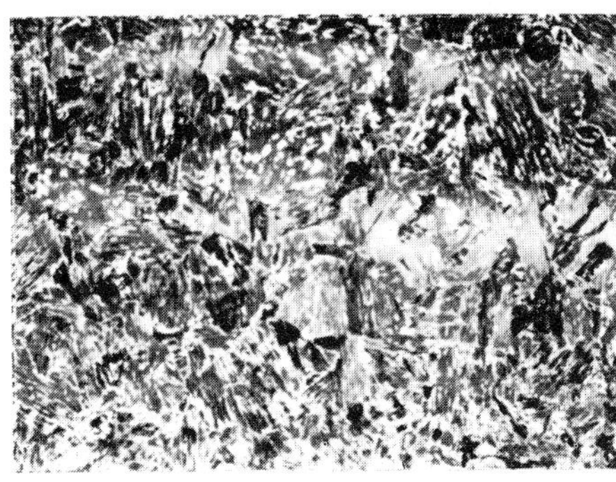

GTA weld, dark etching zone

K_{Ic} or K_Q = 246–337 Kp/mm$^{3/2}$
σ_Y = 189 Kp/mm^2
Hv_{50} = 516–619 Kp/mm^2
γ_S = 2–55%
GS = 22 μ

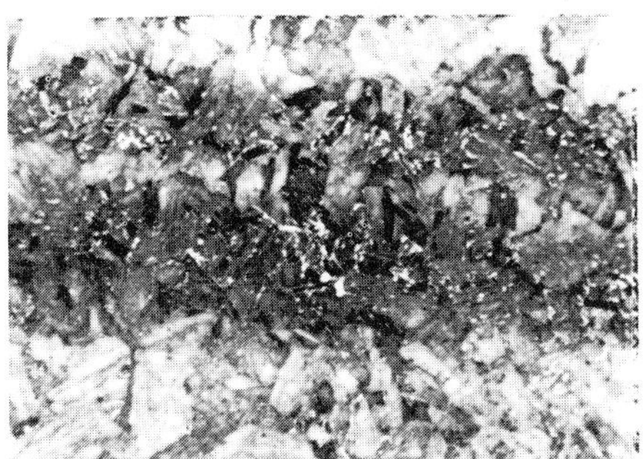

Simulated structure, ϑ_S = 650°C

K_Q = 300 ± 10 Kp/mm$^{3/2}$
σ_Y = 186 Kp/mm^2
Hv_{50} = 536 Kp/mm^2
γ_S = 12%
GS = 18 μ

Simulated structure, ϑ_S = 700°C

K_Q = 323 ± 30 Kp/mm$^{3/2}$
σ_Y = 182 Kp/mm^2
Hv_{50} = 540 Kp/mm^2
γ_S = 21 %
GS = 20 μ

Fig. 8 — Microstructures and properties of simulated structures compared with the dark etching zone of a GTA weld. (Etch: electrolytic with 10% Cr_2O_3 + V2A Beize; X500, reduced 35%)

ϑ_s = peak temperature of simulation cycle
K_Q = fracture toughness not fulfilling all ASTM specifications
σ_Y = yield strength
Hv50 = Vickers hardness, load 50 kp
γ_s = weight percent of retained austenite
GS = prior austenitic grain size

Results and Discussion

The weld joint was analyzed in terms of three areas of interest: (1) the weld metal zone, (2) the transformed zone, and (3) the dark etching zone. The microstructures and properties of these areas are compared with those of the simulation treated material in Figs. 8,9. The weld metal zone of the joint is seen in Fig. 12.

The dark etching zone is characterized by high stable austenite (γ_s) content distributed in a thin band at the "cold" edge of the transformed zone. Such a distribution influences the accuracy of the determination, and the percentages reported must be considered as approximate. The γ_s gradient was severe, varying from 0 to 50% in about 3 mm. The 650 and 700 C simulation treatments developed structures which corresponded with the weld's dark etching zone. However, the γ_s contents were uniform in and around the central cross section of the specimen, and the overall levels were lower at 12 and 21% respectively. The higher level coming from the 700 C treatment is reasonable since the dilatometer information for all treatments showed a sharp phase change at 690 C.

The area between the dark etching and weld metal zones consisted of transformed and recrystallized martensite containing virtually no retained austenite. The prior austenitic grain size varied from about 24 μ at the edge of the dark band to about 50 μ near the fusion line. The 750 and 800 C simulation treatments produced microstructures comparable to those seen in the recrystallized zone near the dark band, with the grain size being 24 and 26 μ respectively. Some γ_s in the amounts of 8% for the 750 C treatment and 5% for the 800 C treatment were measured, in contrast to the recrystallized zone. The 1350 C treatment developed a structure similar to the HAZ near the

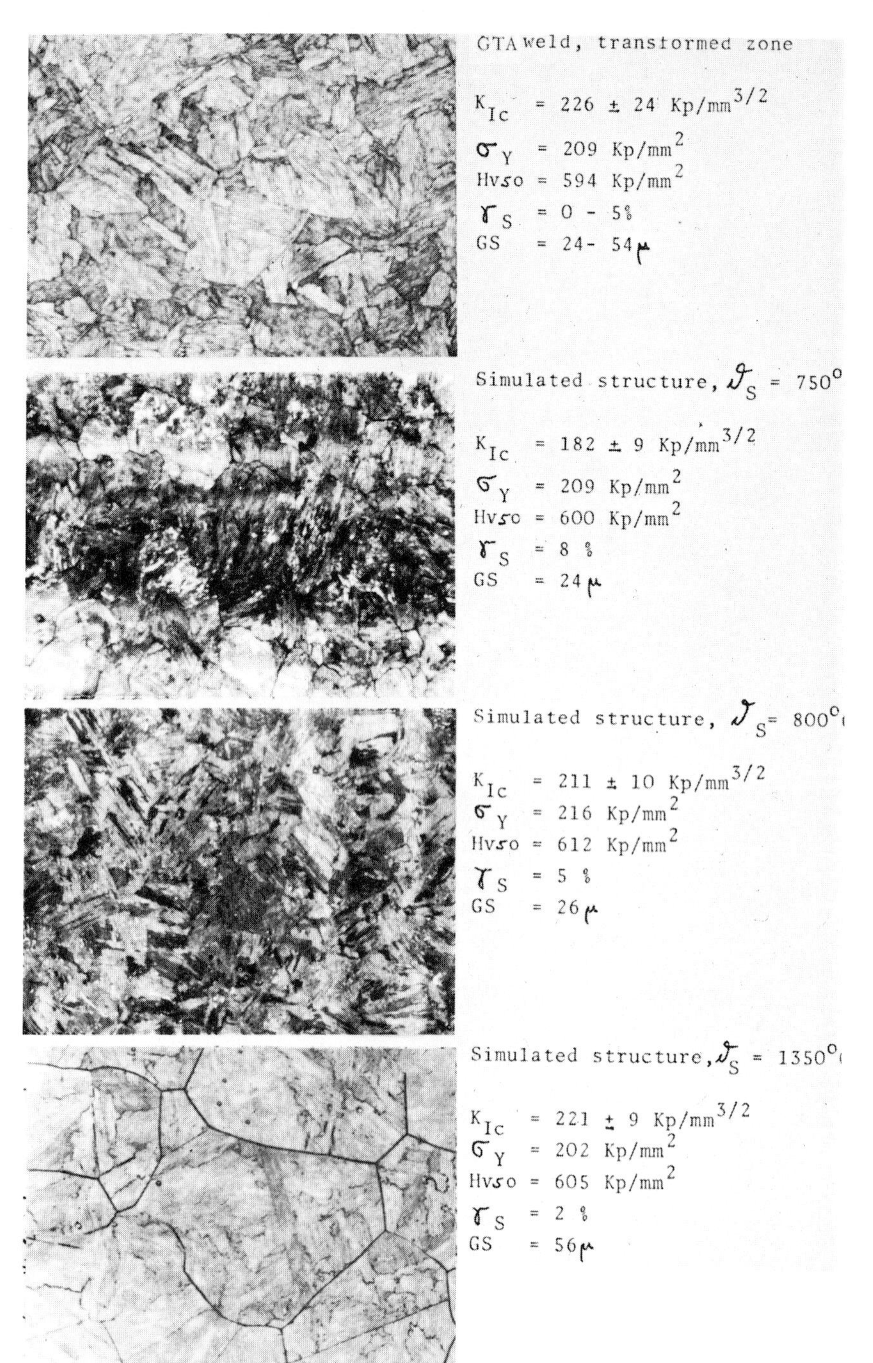

Fig. 9 — Microstructures and properties of simulated structures compared with the transformed zone of a GTA weld. (X500, reduced 35%)

bond line, the grain size being 56 μ and the γ_s content less than 2%. The weld metal zone of the joint consisted of dendrites of transformed martensite containing a dispersion of γ_s pools.

The plane strain fracture toughnesses reported for all structures except those containing high γ_s contents are valid in accordance with ASTM specifications, and are identified as K_{Ic} in Figs. 8-11. Values reported as K_Q in these figures are considered to be invalid only because the crack lengths were slightly too short, and they may be treated as good estimates of K_{Ic}.

The number of tests was 5 as a minimum for uniform structures and between 10 and 20 for the weld structures zones. In the figures mean values are given together with their standard deviation.

Figure 12 shows a typical example of a load-displacement curve of a fracture mechanics test together with views of the fracture surface and the corresponding microstructure.

Concerning the dark etching part of the HAZ, it was very difficult to locate precracks from specimen to specimen in zones of equal γ_s content due to the steep concentration gradient noted above. This gave rise to scattered K_Q results, a detraction that was not observed in the simulation specimens. This serves to emphasize the importance of the weld simulation approach as an adjunct to direct testing of a weld joint.

It can be seen from Figs. 8-11 that of all the structures represented in the weld joint, that of the weld metal zone is the most brittle, in spite of its low yield strength. However, the weld metal was largely ignored because its properties can be controlled by the filler metal composition. The dark etching zone developed fracture toughnesses varying from 246 to 337 kp-mm$^{-3/2}$ (69-95 ksi-in$^{1/2}$). The simulation results indicate that γ_s contents as low as 12% can elevate the fracture toughness significantly, which is in direct contrast to the weld metal zone results. It has been reported that the morphology of γ_s in maraging steel exerts an important influence on fracture toughness (Ref. 4). The point could not be pursued in this program because the martensite matrix of the weld metal was harder than that of the dark etching zone probably as a result of the higher Ti content due to the filler metal composition.

It is interesting to note that although the precrack location in specimens coming from the recrystallized part of the HAZ was random with respect to grain size, the K_{Ic} scatter was negligible. A detrimental influence of increasing grain size on K_{Ic} is not apparent, an observation supported by the results of the 750, 800 and 1350 C simulation treatments. This may be due to the fact that the overlying martensitic structure was uniform in size irrespective of the prior austenitic grain size.

A metallurgical effect of as yet unknown origin appears to be influencing the toughness of the 750 C treatment.

It can be seen in Fig. 10 that the K_{Ic} value is significantly lower than those of the other treatments and that of the base metal.

Conclusions

The following conclusions can be drawn regarding the fracture mechanics analysis of a GTA welded joint of a grade 300 maraging steel:

1. The plane strain fracture toughness K_{Ic} can be determined directly from the weld joint; however, weld simulation specimens offer an extremely valuable supplement to any such testing program.

2. Concentrations of retained austenite ranging to 50% are found in a narrow region of the HAZ, apparently

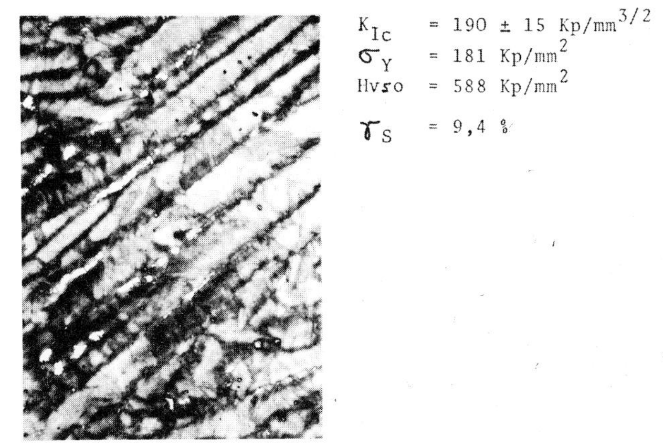

Fig. 10 — Weld metal zone microstructure and properties (Etch: electrolytic with 10% Cr_2O_3 + V2A Beize; X500, reduced 18%)

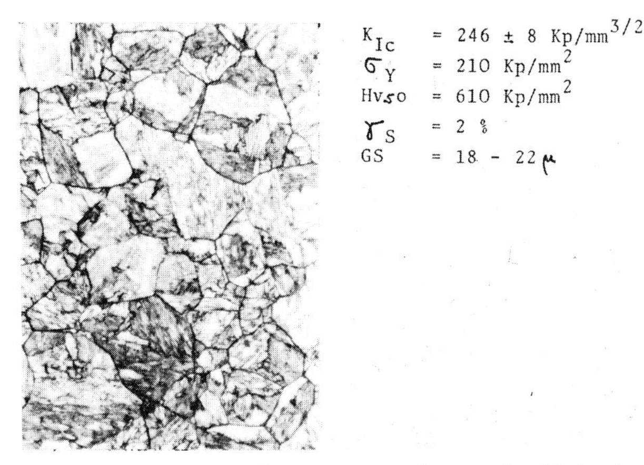

Fig. 11 — Base metal microstructure and properties (Etch: electrolytic with 10% Cr_2O_3 + V2A Beize; X500, reduced 18%)

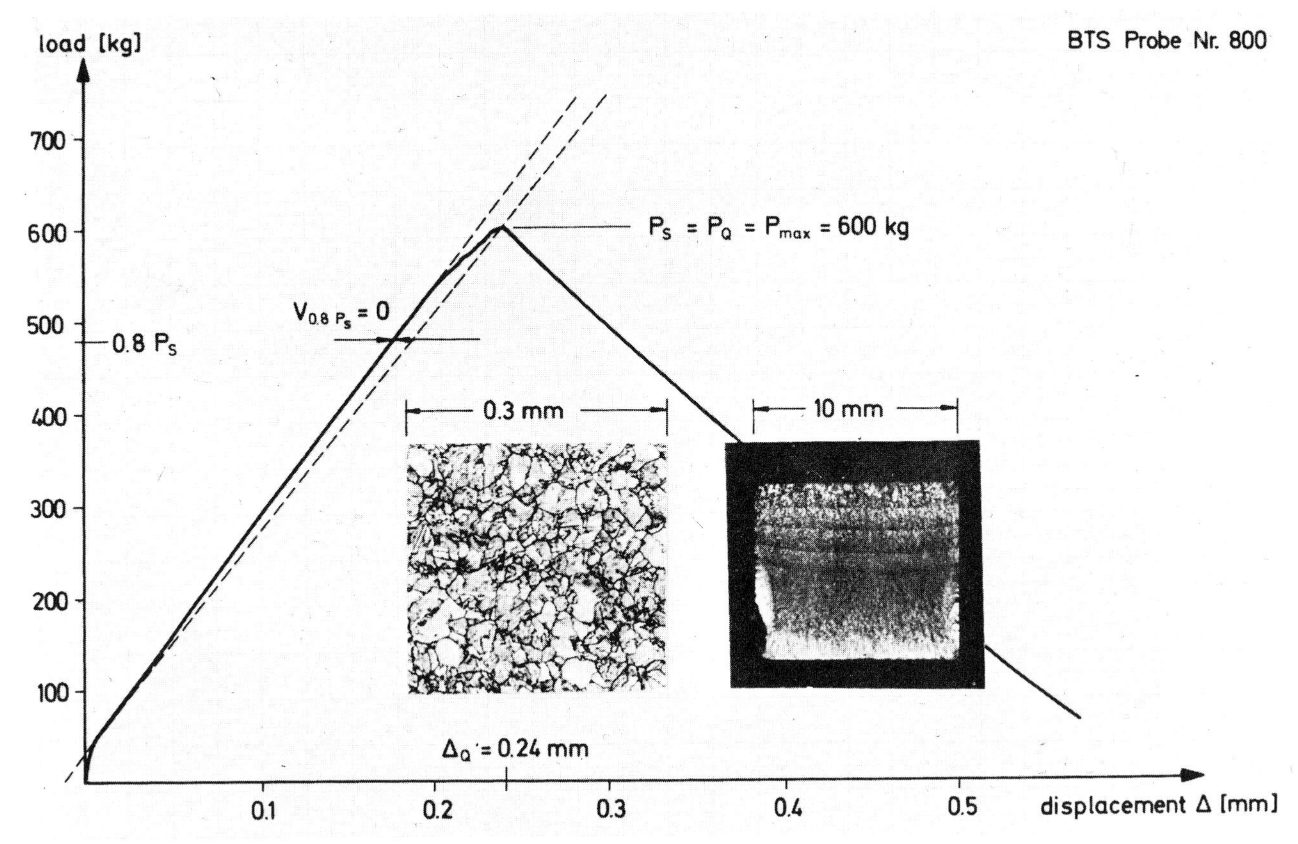

Fig. 12 — Load-displacement curve of a fracture mechanics test of the transformed structure in the HAZ of a GTA weld, together with the microstructure in the region of the fatigue crack tip and a macrograph of the crack surface

as a result of repeated very short time exposure to temperatures between 650 and 700 C. The fracture toughness K_{Ic} in this zone is elevated significently, but the attendent lowering of yield strength is not serious.

3. Prior austenitic grain size in the recrystallized portion of the HAZ does not influence K_{Ic}.

4. A possible embrittling effect is associated with planes of the HAZ experiencing short time exposure to 750 C.

Acknowledgement

The work for this program was performed under the auspices of the Deutscher Verband fur Schweisstechnik (DVS). We wish to thank the Arbeitsgemeinschaft Industrieller Forschungsvereinigungen (AIF) for financial support.

References

1. Bailey, N. "Weldability and Toughness of Maraging Steel", *Metal Construction British Welding Journal,* Vol. 3, pp 1-5, 1971.

2. Cottrell, C. L. M. "Fracture Toughness Concepts Applied to the Zones of Welded Joints", *British Welding Journal,* Vol. 15, 262-267, 1968.

3. Kies, J. A., Smith, H. L., Romine, H. "Fracture Toughness and Critical Defect Sizes With Welds of 18% Nickel Maraging Steels", *Metals Engineering Quarterly,* Vol. 6, 37-47, 1966.

4. Salmon-Cox, P. H., Birkle, A. J., Reisdorf, B. G., Pellisier, G. E. "An Investigation of the Mechanical Properties and Microstructure of 18 Ni (250) Maraging Steel Weldments", *Transactions, ASM Quarterly,* Vol. 60, 1967.

5. Weiss, B. Z., Steffens, H. D., Seifert, K. "Fracture Toughness of the Heat Affected Zone in 14 Cr Mo V 69 Steel and 18 Ni Maraging Steel", *Welding Journal,* Vol. 51, 9, (Sept.) 1972, Research Suppl., pp. 449-s to 456-s.

6. Pepe, J. J., Savage, W. F., "The Weld Heat Affected Zone of the 18 Ni Maraging Steels", *Welding Journal,* Vol. 49, no. 12 (Dec.) 1970 Research Suppl., pp. 454-s to 553-s.

7. Lang, H. J., Kenyon, N. "Welding of

Maraging Steels" Welding Research Council Bulletin 159, 1971.

8. Canonico, D. A., "Gas Metal Arc Welding of 18% Nickel Maraging Steel", *Welding Journal,* Vol. 43, no. 10 (Oct.) 1964, Research Suppl., pp. 433-s to 442-s.

9. Adams, C. M. Jr., Travis, R. E., "Welding of 18% Ni-Co-Mo Maraging Alloys", *Welding Journal,* Vol. 43, no. 5 (May) 1964, Research Suppl., pp. 193-s to 197-s.

10. Baker, A. J., Swann, P. R. "The Hardening Mechanism in Maraging Steels", *ASM Transactions,* Vol. 57, pp. 1008-1011, 1964.

11. Reisdorf, B. G., Baker, A. J. "The Kinetics and Mechanisms of the Strengthening of Maraging Steels", Technical Report AFML - TR - 65-390, 1965.

12. Peters, D. T. "A Study of Austenite Reversion during Aging of Maraging Steels", *ASM Transactions,* Vol. 61, pp. 62-74, 1968.

13. Floreen, S. F. "The Physical Metallurgy of Maraging Steels", *Metallurgical Review,* Vol. 13, pp. 115-128, 1968.

14. Petersen, A., "Weld Heat Affected Zone of 18% Nickel Maraging Steel", *Welding Journal,* Vol. 43, no. 9 (Sept.) 1964, Research Suppl., pp. 428-s to 432-s.

15. Goldberg, A., O'Connor, D. G., "Influence of Heating Rate on the Transformation of an 18% Nickel Maraging Steel", *Nature,* Vol. 213, pp. 170-171, 1967.

16. Blauel, J. G., Smith, H. R., Schulze, G., "Messung der Riss-zahigkeit an Schweissverbindungen und thermisch simulierten Gefügen aus einem martensitaushärtbaren Stahl", Report no. 4/73, Institut für Festkörpermechanik, Freiburg, 1973.

17. Adrichem, T. I., Kas, I. "Calculation, Measurement and Simulation of Weld Thermal Cycles", Company Report of Smit-Weld, N.V. Nijmwegen, Holland, 1969.

18. ASTM-Designation E 399-70 T, "Tentative Method of Test for Plane Strain Fracture Toughness of Metallic Materials", *ASTM Book of Standards,* Part 31, 1970.

19. Ratwani, M., "Wechselwirkung von Rissen", Report 5/72, Institut fur Festkörpermechanik, Freiburg, 1972.

20. Adams, Jr. C. M., Corrigan, D. A., "Mechanical and Metallurgical Behavior of Restrained Welds in Submarine Steels", MIT Final Report on Contract No. NObs - 92077, 1966.

21. Averbach, B. L., Cohen, M., "X-ray Determination of Retained Austenite by Integrated Intensities", *Transaction of AIME,* Vol. 176, pp. 401-414, 1948.

22. ASTM-Designation E 112-4, "Average Grain Size of Metals", *ASTM Book of Standards,* Part 31, 1969.

A Metallographic Investigation of the Factors Affecting the Notch Toughness of Maraging Steels

A. J. BIRKLE, D. S. DABKOWSKI, J. P. PAULINA AND L. F. PORTER

ABSTRACT. A new metallographic technique for investigating the factors affecting the toughness of steels is that of extraction fractography, a technique made possible by the extreme depth of focus of the electron microscope. This paper describes the results of using this technique in an investigation of nine maraging steels with widely varying notch toughness (25 to 70 ft-lb Charpy V-notch energy at 190 ksi yield strength). In conjunction with extraction fractography, light microscopy, thin-film electron microscopy, microprobe analysis, and electron- and x-ray diffraction analyses were used to help identify the causes for the difference in toughness among the nine steels. The results of the investigation indicate that by examining extraction fractographs with metallographic, x-ray- and electron-diffraction, and microprobe techniques it is possible to isolate the major factors responsible for low notch toughness in maraging steels. Thin-film transmission electron microscopy and analysis of insoluble residue, while generally substantiating the results obtained by analysis of the fractographs, are less definite and therefore less useful than the fractographic studies. Zirconium nitride and titanium sulfide particles were found on the fracture surfaces of the steels having the lowest notch toughness. Titanium carbide and nitride particles were found on the fracture surfaces of several of the steels, and aluminum nitride particles were found on the fracture surface of one of the steels. The fracture surfaces of the two high-toughness steels were remarkably free of extracted particles. The maximum notch toughness attainable at a yield strength of 190 ksi in the steels studied was about 70 ft-lb Charpy V-notch energy at 80 F.

AT A GIVEN strength level, several factors may influence the notch toughness of steels. Some of these are: 1) grain size; 2) microcleanliness; 3) grain-boundary precipitates; 4) second-phase and general precipitates; and 5) embrittlement of grain boundaries or matrix without the presence of resolvable precipitates.

Although most of these factors can be examined individually with careful metallographic techniques, they frequently interact in a complex fashion so that it is difficult to determine which is the controlling factor in a given case, even after an extensive, careful metallographic study.

A new technique (1–5) that gives promise of providing a rapid and convenient method for determining the critical factors influencing the notch toughness of steels is the extraction-fractograph technique made possible by the extreme depth of focus of the electron microscope. With this technique,

The authors are associated with the United States Steel Corp., Applied Research Laboratory, Monroeville, Pa. Manuscript received April 29, 1965.

the nature of a freshly fractured surface can be examined and the size, shape, distribution, and identification of intermetallic and nonmetallic particles on the fracture surface can be determined. Because only the path of the fracture is examined, it is possible to determine the critical factors influencing toughness without evaluating each of the above-mentioned factors seaparately.

This paper is presented to illustrate the use of the fractographic technique, in conjunction with additional techniques suggested by the fractographs, to gain a more precise understanding of the mechanisms and constituents that are related to toughness variations in a series of nine experimental maraging steels of widely different notch toughness. Eight of the steels of about the same strength level (183 to 198 ksi yield strength) ranged in notch toughness from 25 to 68 ft lb Charpy V-notch energy at 80 F. The ninth steel, although of considerably lower strength (159 ksi yield strength), exhibited exceptional toughness (108 ft-lb Charpy V-notch energy at 80 F). In addition to extraction fractography, light microscopy, thin-film electron microscopy, microprobe analysis, electron- and x-ray diffraction analysis, and special analyses of insoluble residue were conducted on selected samples.

Materials and Experimental Work

Chemical Composition

The chemical composition of the nine steels examined is given in Table 1. Six of the steels are of the 18Ni – 8Co – Mo type and three are of the 12Ni – 5Cr – Mo type. The steels are designated alphabetically in order of decreasing notch toughness.

Melting and Processing Procedures

Both small laboratory heats and large-size production heats were examined; thus, heat sizes of 25 lb, 100 lb, 1000 lb, 20 tons, and 80 tons are represented. The 20 and 80-ton production heats were made by double-slag (lime-alumina) electric furnace practice, the 1000-lb heat was vacuum-consumable-electrode remelted, and the 25 and 100-lb heats were made by vacuum-induction furnace melting.

Heat Treatment

With the exception of Steel B (material supplied by the Naval Research Laboratory) (6) and Steel D, the plates from the 18Ni – 8Co – Mo steel heats were all solution annealed at 1700 F for 1 hr per in. of section (1 hr minimum), water-quenched, and aged at 900 F for either 3 or 5 hr. Steel B was solution-annealed at

TABLE 1. Chemical Composition
(Check analysis)

Steel	Composition, %												
	C	Mn	P	S	Si	Ni	Cr	Mo	Co	Ti	Zr	Al*	N
A	0.004	0.029	0.002	0.003	0.036	17.68	ND†	1.93	8.13	0.13	0.003	0.043	0.004
B‡	0.008	0.003	0.003	0.004	<0.01	18.40	0.002	3.24	8.83	0.17	0.005	0.004	0.001
C	0.023	0.088	0.004	0.008	0.094	12.10	5.21	2.86	ND	0.24	0.010	0.38	0.009
D	0.008	0.10	0.003	0.007	0.056	17.40	ND	3.50	7.99	0.12	0.017	0.33	0.006
E	0.026	0.044	0.003	0.010	0.062	12.10	4.47	2.97	ND	0.20	0.003	0.38	0.009
F§	0.016	0.090	0.002	0.006	0.055	17.55	ND	3.17	8.09	ND	0.029	0.22	0.003
G	0.032	0.032	0.005	0.012	0.022	18.29	ND	2.83	8.09	0.14	ND	0.18	0.007
H	0.024	0.044	0.007	0.013	0.042	11.74	5.38	3.18	ND	0.30	0.001	0.47	0.008
I	0.027	0.082	0.003	0.006	0.049	17.64	ND	3.01	8.17	ND	0.032	0.24	0.003

* *Acid soluble.*
† *Not determined.*
‡ *This heat also contained 0.065 W.*
§ *This heat also contained 0.10 V.*
Note: All the steel contained 0.002 or less oxygen.

1500 F, air cooled, and aged at 900 F for 3 hr. Steel D was solution annealed at 2200 F, water quenched, and aged at 900 F for 3 hr. In addition, for x-ray diffraction studies, samples from Steel D were given a special embrittling treatment, which consisted of quenching the samples from the 2200 F solution annealing temperature to 1400 F and holding at 1400 F for 6 hr. The samples were then water quenched and aged at 900 F for 3 hr.

The 12N – 5Cr – Mo steels were solution annealed at 1500 F for 1 hr per in. of section (1 hr minimum), water quenched, and aged for 3, 5, or 30 hr at 900 F.

Mechanical Tests

Tests specimens were machined from heat treated plates. Duplicate 0.252-in. diam tension test specimens were tested at room temperature, and duplicate Charpy V-notch specimens were tested at 80 F. All specimens were longitudinally oriented. Charpy V-notch specimens were machined with the notch normal to the plate surface.

Metallographic Studies

As stated in the introduction, the principal technique used to examine the steels was that of extraction fractography. The technique consisted of examining the frac-

TABLE 2. Processing History

Steel	Description	Heat treatment
A	1000-lb vacuum consumable-electrode remelted heat, 18Ni – 8Co – 2Mo steel, 1-in. thick cross-rolled plate.	1700 F—1 hr, WQ*, 900 F—3 hr
B	25-lb vacuum-melted heat, 18Ni – 9Co – 3Mo steel, high-purity vacuum-melted raw materials, hammer forged at 2200 to 1500 F into 0.6 by 3-in. section bar stock.	1500 F—1 hr, AC,† 900 F—3 hr
C	Production 20-ton heat, 12Ni – 5.2Cr – 2.9Mo steel, ½-in. thick cross-rolled plate.	1500 F—1 hr, WQ, 900 F—3 hr
D	Laboratory vacuum-carbon-deoxidized 100-lb heat, 18Ni – 8Co – 3.5Mo steel, ½-in. thick cross-rolled plate.	2200 F—1 hr, WQ, 900 F—3 hr
E	Production 20-ton heat, 12Ni – 4.5Cr – 3Mo steel, 2-in. thick cross-rolled plate.	1500 F—2 hr, WQ, 900 F—30 hr
F	Laboratory vacuum-carbon-deoxidized 100-lb heat, 18Ni – 8Co – 3.2Mo – 0.1V steel, ½-in. thick cross-rolled plate.	1700 F—1 hr, WQ, 900 F—5 hr
G	Production 80-ton heat, 18Ni – 8Co – 2.8Mo steel, ½-in. thick cross-rolled plate.	1700 F—1 hr, WQ, 900 F—5 hr
H	Production 20-ton heat, 12Ni – 5Cr – 3.2Mo steel, 3-in. thick cross-rolled plate.	1500 F—3 hr, WQ, 900 F—5 hr
I	Laboratory vacuum-carbon-deoxidized 100-lb heat, 18Ni – 8Co – 3Mo steel, ½-in. thick cross-rolled plate.	1700 F—1 hr, WQ, 900 F—5 hr

* *Water-quenched.*
† *Air-cooled.*

TABLE 3. Grain Size and Mechanical Properties

Steel	ASTM grain size	Yield (0.2% offset), ksi	Tensile strength, ksi	Elongation in 1-in., %	Reduction of area, %	Impact properties* Charpy V-notch energy absorption at 80 F, ft-lb
A	8	159	167	16.0	72.2	108
B	10	190†	199	16.0	68.1	68
C	8	190	193	13.5	61.3	45
D	2	198	209	14.0	62.0	35
E	5½	183	190	16.5	65.7	41
F	6	187	201	15.0	61.3	31
G	9	187	195	13.5	62.0	28
H	4	190	197	13.0	55.2	25
I	6	184	200	15.5	61.8	27

* *Average of duplicate longitudinal specimens except where noted.*
† *Results obtained from Reference 1.*

ture surface of freshly broken Charpy V-notch impact test specimens by using the carbon-replication technique, which is capable of showing the nature of the fracture surface and extracting any nonmetallic or intermetallic particles that may be on the surface.

The procedure for preparing extraction fractographs consisted of placing a freshly broken Charpy V-notch impact test specimen within a vacuum evaporator and depositing a heavy film of carbon on the fracture surface. The shadowed specimen was then immersed in a 1% solution of bromine in methyl alcohol for 1 min to loosen the particles to be extracted. Then a $1\frac{1}{2}$-hr soak in ethyl alcohol was employed to remove all traces of the bromine-methyl-alcohol solution. At this stage, the carbon film still adhered tightly to the specimens; therefore, it was necessary to dissolve the fracture surface by electropolishing. The specimen was the anode, and a stainless steel cathode was used. An electropolish for 5 sec in a perchloric-acetic-acid solution (100 ml acetic acid and 70 ml 65% perchloric acid) loosened the carbon film so that the extraction replica could be floated off the metal surface in ethyl alcohol, put on a copper grid, dried, and examined in an electron microscope. This procedure permitted the examination of the contours of the fracture surface and the shape and distribution of the extracted precipitates and inclusions. Precipitates and inclusions isolated on extraction fractographs may be identified by selected-area electron-diffraction, x-ray diffraction, and microprobe techniques.

The following additional metallographic techniques were also used in this study: 1) light metallographic studies were made on longitudinal specimens etched with a 1% solution of bromine in methyl alcohol and Vilella's reagent, and 2) thin-film specimens for transmission microscopy were prepared by slicing 0.030-in. thick wafers from the samples and sanding the wafers on a series of increasingly fine emery papers to 0.005-in. thickness. These pieces were subsequently thinned by the Bollman technique (7).

Special Chemical Analysis

Another method used in an attempt to determine the nature of the nonmetallic inclusions present was quantitative and qualitative analysis of insoluble residue. For this analysis, the matrix of the steel was dissolved away in an electrolytic cell, leaving a residue of particles which were then collected on a millipore filter. The residue was examined by differential-thermal analysis (DTA) (8), x-ray diffraction, and wet chemical analysis.

Results and Discussion

Mechanical Properties

The processing history and the mechanical properties of the steels are summarized in Tables 2 and 3, respectively. The yield strength – toughness relation for the nine steels is shown in Fig. 1. It

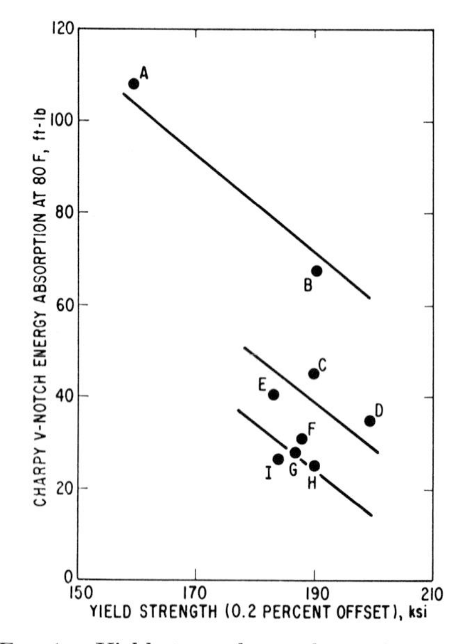

Fig. 1. Yield strength–notch toughness relation for the steels investigated. Solid black lines indicate the generally observed slope of the strength-toughness relation for maraging steels of this strength level, and letters refer to the steel designation.

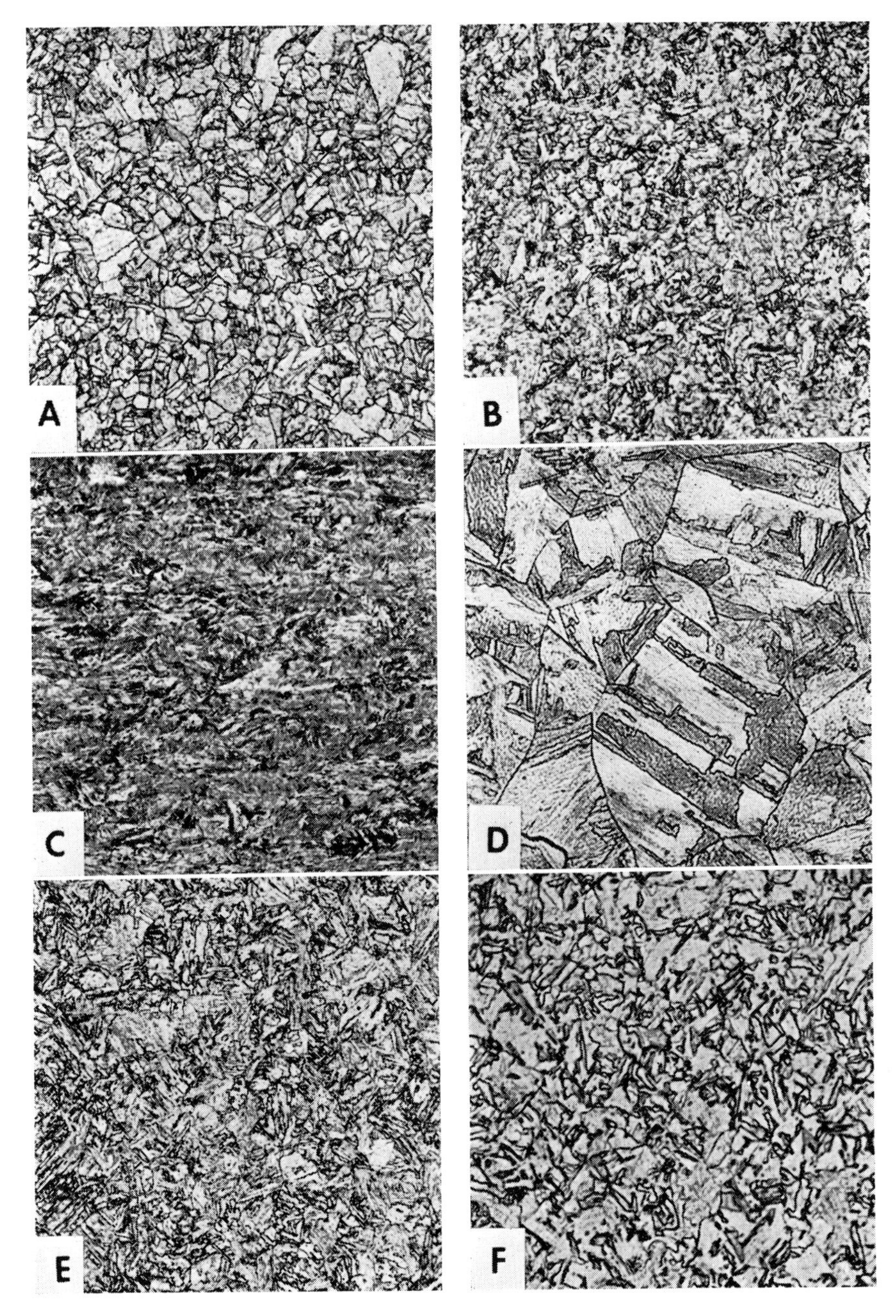

FIG. 2. Microstructure of steels. Solution annealed and aged microstructure shown except for Steel F (solution annealed only). (A) Steel A, (B) Steel B, (C) Steel C, (D) Steel D, (E) Steel E, and (F) Steel F. Bromine-methyl alcohol and a Vilella's reagent etch; ×100.

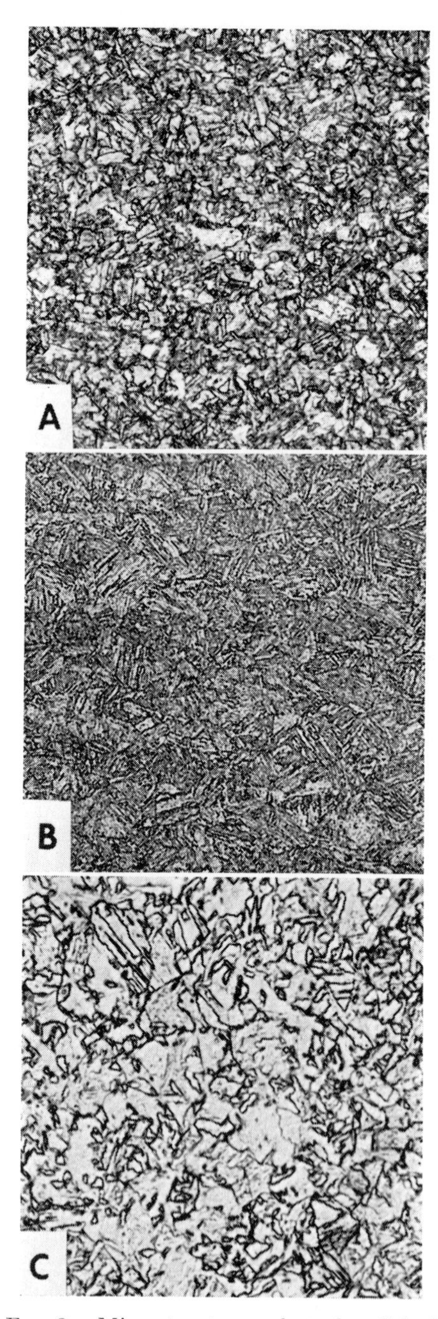

FIG. 3. Microstructures of steels. Solution annealed and aged microstructure shown except for Steel I (solution annealed only). (A) Steel G, (B) Steel H, and (C) Steel I. Bromine-methyl alcohol and a Vilella's reagent etch; X100.

will be noted that, with the exception of Steel A (material supplied by The International Nickel Co.), the yield strengths of all of the steels lie in the range 183 to 198 ksi. On the basis of the expected decrease in toughness with increasing strength, the steels may be grouped into three toughness levels as follows:

Steel designation	Nominal toughness at 190 ksi yield strength, Charpy V-notch energy absorption at 80 F, ft-lb
A, B	70
C, D, E	40
F, G, H, I	25

These steels thus represent a series with toughness varying by a factor of three and should provide an ideal group of samples for the intended study.

Light Metallography

To provide an indication of the variations in microstructure among the nine steels in this study, light micrographs representative of the microstructure of each of the steels at $\times 100$ are shown in Fig. 2 and 3. It will be observed that the microstructures generally consist of well-outlined equiaxed grains with a very fine acicular structure within the grains.

As indicated in Table 3, the grain size varies from ASTM grain size No. 10 to 2. Steels rolled or forged to light gages ($\frac{1}{2}$-in. thick) at low finishing temperatures (below 1500 F) develop distorted grains during the later stages of rolling. If the solution annealing temperature is low (1500 F), the distortion is not completely removed and partially unrecrystallized, fine-grained microstructures such as are shown for Steels B and C are obtained. If the finishing temperatures are higher (above 1600 F), as occurs in rolling heavy sections, the grains recrystallize during rolling and the grain size of the plates may be coarse even though the solution annealing temperature is low (1500 F) (Steels E and H). Solution annealing at 1700 F generally results in complete recrystallization of the distorted grain

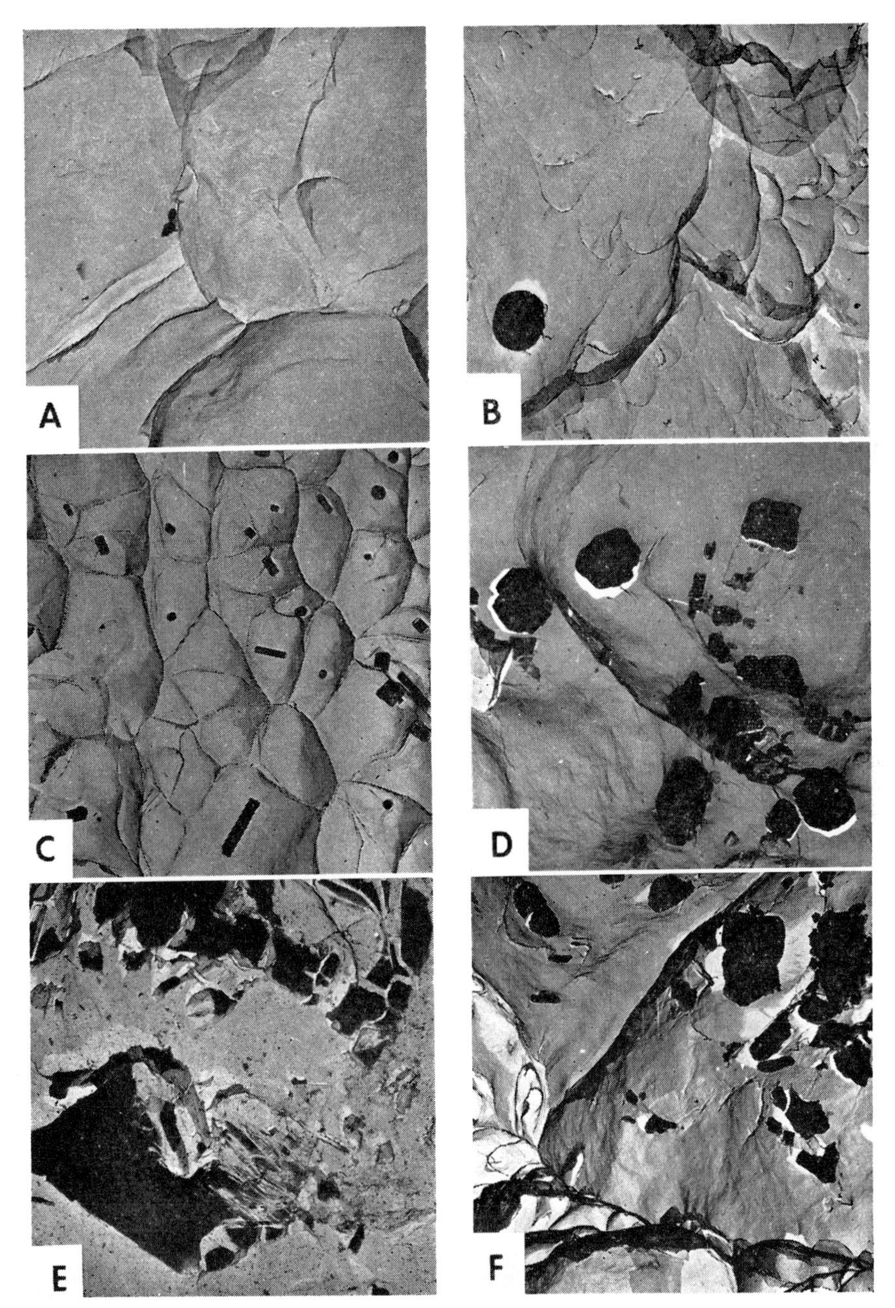

FIG. 4. Extraction fractographs from broken Charpy V-notch specimens. (A) Steel A, (B) Steel B, (C) Steel C, (D) Steel D, (E) Steel E, and (F) Steel F; ×4000.

structure produced in light sections at low finishing temperatures and may or may not result in some grain growth (Steel A, F, G, and I). Solution annealing at 2200 F produced a very coarse-grained structure (Steel D).

Comparison of grain size with notch toughness, Table 3, shows that the three steels with the best notch toughness are fine-grained (Steels A, B, C, ASTM grain size No. 8 to 10), but the steel with the coarsest grain (Steel D, ASTM grain size No. 2) has better notch toughness than many finer grained steels and one of the finest grained steels (Steel G, ASTM grain size No. 9) has low notch toughness. This emphasizes the difficulties involved in using conventional metallographic techniques to study the factors influencing notch toughness. Although grain size probably influences the toughness of maraging steels in the expected manner (the coarser the grain size, the lower the toughness), the effect is apparently being obscured in some cases by more important factors.

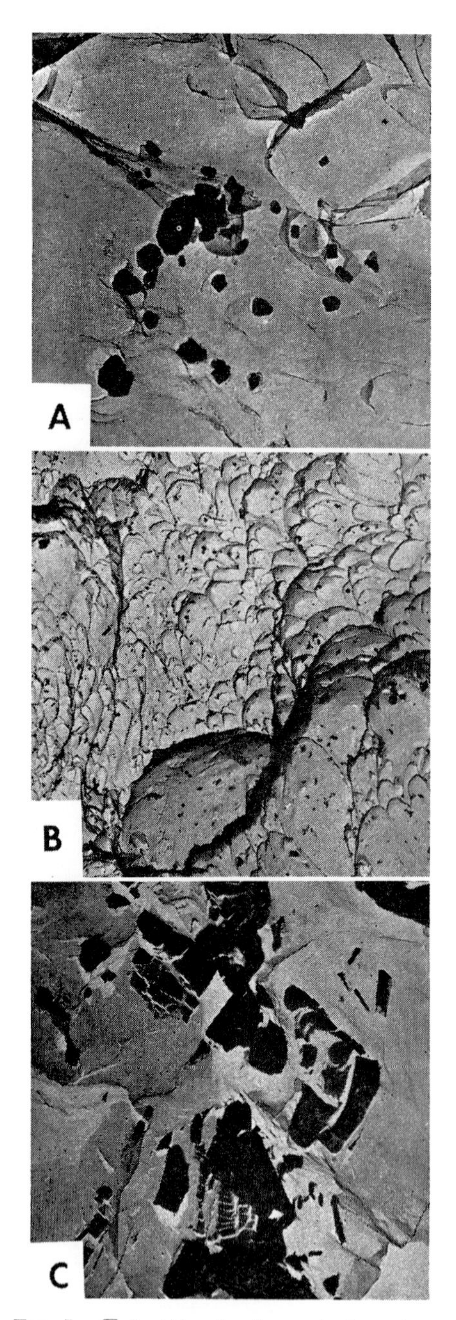

Fig. 5. Extraction fractographs from broken Charpy V-notch specimens. (A) Steel G, (B) Steel H, and (C) Steel I; ×4000.

Extraction Fractography

When there is no cleavage or intergranular failure, failure is believed to occur by the formation, growth, and coalescence of voids to develop a shear fracture surface. Such voids are often initiated by intermetallic and nonmetallic particles in the steel (2, 4). For a given strength steel, large nonmetallic and intermetallic particles are believed to permit voids to form at lower levels of triaxial stress than small particles. Likewise, for particles of a given size, a steel having few particles is expected to exhibit toughness superior to that of a steel having many particles.

Typical extraction fractographs of each of the steels are shown in Fig. 4 and 5. Several different forms of fracture surface and extracted particles (generally opaque) are present. The fractographs show very limited areas of the fracture surface and are not, therefore, necessarily representative of the entire surface. Careful examination of the extraction replicas with the electron microscope indicated that the extraction fractographs of Steels A and B, the two high-toughness steels, were notably free of extracted particles. The fractographs of Steels C through I showed heavy concentrations of extracted particles in some areas and very light concentrations in other areas. For Steels C through I there was no obvious correlation between the number of extracted particles observed and the notch toughness of the steels.

Examination of the fractographs led to several generalizations concerning the fracture surfaces.

When the steel was free of particles that could initiate the formation of voids (Steel A), fracture resulted in the formation of 10-micron diam ductile dimples.

Large dimples were also formed when large particles initiated fracture (Steels B, D, E, and I). Very small, finely divided particles resulted in much smaller dimples (0.1-micron diam) as in Steel H. Thus the size of the fracture facets or dimples in itself is not a good indication of the toughness of a particular steel. Poor toughness can result from large particles which initiate large fracture dimples or from a fine dispersion of very small particles which initiate many small fracture dimples. It may be that the volume of the material deformed, that is the depth of the dimples in addition to their size, determines the toughness of the steel. If this is true, a study of the fracture surface in three dimensions would be required to obtain a correlation between volume of metal deformed and toughness of the steel.

Fractographs from Steel E show a very flat fracture surface almost free of any dimple formation. Flat surfaces are believed to be indicative of areas of brittle grain-boundary fracture.

Several different types of nonmetallic and intermetallic particles are seen in the fractographs. Perhaps most prevalent are 1) the large, generally rectangular-shaped particles in the fractographs of steels D, E, and I. Fractographs of Steels B, D, F, and G show 2) very opaque particles that instead of being rectangular are more rounded and irregular in form. The particles in the fractograph of Steel C and some of the particles in the fractograph of Steel G are 3) small, opaque-elongated rectangles (rod-like). The fractograph of Steel H is markedly different from the others. This fractograph shows 4) a very finely divided dispersion of small, irregularly shaped, opaque particles.

The reduced toughness of Steels C through I as compared with Steels A and B is undoubtedly associated with the nonmetallic and intermetallic particles observed on the fracture surfaces. As has been indicated, the relative toughness of Steels C through I could not be readily correlated with the number of nonmetallic and intermetallic particles on the fracture surface, but the fractographs did yield invaluable information concerning the identity of the particles responsible for the lower toughness of the steels. This information was obtained from additional examinations of the fractographs with x-ray diffraction, electron-diffraction, microprobe, and special chemical tests.

X-Ray Diffraction Analysis

Table 4 shows the results of x-ray diffraction analysis of extraction replicas taken from the fracture surfaces of Steels C, D, E, F, H, and I. Good correspondence between the "d" values obtained from the extraction replicas and the known "d" values of the suspected compounds indicates that particles of the suspected compounds are present on the fracture surface of the steel. The data shown in Table 3 may be summarized as follows:

Steel	Identity of particles on the fracture surface
C	AlN and possibly Al_2O_3
D (water quenched from 2200 F)	TiN
D (held at 1400 F)	ZrN and TiC
E	TiC
F	ZrN and probably TiS
H	Ti_2S and probably TiC and TiN
I	ZrN

TABLE 4. Comparison of Known "d" Values for Suspected Compounds with "d" Values Obtained by X-Ray Diffraction Analysis of the Extraction Fractographs From Several Steels

d(A)*	Steels examined, "d" values*						
	C	D‡	D§	E	F	H	I
AlN							
2.70	2.68	...	...	...	...	...	...
2.49	2.49	...	2.50	...	...	2.49	...
2.37	2.37	...	...	...	...	...	...
1.83	1.83	...	...	...	1.85	1.85	...
1.56	1.55	...	...	1.55	...	1.54	...
1.41	1.41	...	1.39	...	1.40	...	...
1.35	1.34	1.36	1.37	...	...	...	...
1.32	1.32	...	1.33	...	1.32	...	1.33
1.30	1.30	1.29	1.30	1.30	...	1.30	...
ZrC							
2.70	2.68	...	...	...	...	...	...
2.34	...	...	...	...	...	...	...
1.65	...	...	...	...	...	...	...
1.41	1.41	...	1.39	...	1.40	...	1.33
1.35	1.34	...	1.33	...	...	...	...
1.17	...	...	...	...	...	...	...
ZrN							
2.64	...	...	2.65	...	2.66	...	2.64
2.29	...	...	2.29	2.27	2.30	...	2.31
1.62	1.60	1.61	1.61	...	1.62	1.60	1.62
1.38	1.39	1.36	1.39	...	1.38	1.38	1.38
1.32	1.32	...	1.33	1.30	1.32	1.30	1.33
TiC							
2.51	2.49	...	2.50	...	...	2.49	...
2.18	...	...	2.16	2.17	...	...	...
1.53	1.53	1.52	1.52	1.55	...	1.54	...
1.31	1.30	1.29	1.30	1.30	1.32	1.30	1.33
1.25	...	1.24	1.25	1.24	...	1.24	...

d(A)*	Steels examined, "d" values*						
	C	D‡	D§	E	F	H	I
Al_2O_3							
3.48	...	...	...	...	3.46		
2.55	2.55						
2.38	2.37	...	2.40				
2.09	2.10						
1.74							
1.60	1.60	1.61	1.61	1.59	1.62	1.60	1.62
1.40	1.41	...	...	...	1.40		
1.37	1.37	1.36	1.38	...	1.38	1.38	1.38
1.24	...	...	1.25	...	1.22	1.24	
TiO_2 (Rutile)							
3.25	...	...	...	...	3.27		
2.49	2.49	...	2.50	...	...	2.49	
2.19	...	...	...	...	...	2.21	
1.69	...	...	...	...	1.69		
1.62	1.60	1.61	1.61	...	1.62	1.60	
1.36	1.37	1.36	1.38	...	1.38	1.38	
TiO_2 (Anatase)							
3.51							
2.38	2.37	...	2.40				
1.89							
1.70	...	...	...	...	1.69	...	
1.67							
1.48	...	...	1.49	...	1.46		
1.26	...	1.24	1.25	1.24	...	1.24	

TiN								
2.44	...	2.46	...	2.45	...	...	...	
2.12	2.10	2.14	...	...	...	2.13	...	
1.50	...	1.52	1.52	...	...	1.51	...	
1.28	1.30	1.29	1.30	1.30	...	1.30	...	
1.22	...	1.24	...	1.24	1.22	1.22	...	
TiS								
3.19	...	...	3.19	...	3.18	...	3.19	
2.86	...	...	2.86	...	2.84	...	2.84	
2.61	...	2.60	...	...	2.62	...	...	
2.13	...	2.14	...	...	...	2.13	...	
1.71	...	...	...	...	1.69	...	...	
{1.65 1.61}	1.60	1.61	1.61	...	1.62	1.60	1.62	
1.48	...	...	1.49	...	1.46	...	...	
1.43	1.41	...	...	...	...	...	...	
1.39	1.39	...	1.39	...	1.40	1.38	1.38	
1.30	1.30	1.29	1.30	1.30	1.32	1.30	...	
Ti_2S								
2.77	2.79	...	...	2.79	...	2.77		
2.69	2.68							
2.22	...	...	...	...	2.22	2.21		
1.96								
1.86	...	...	...	1.86	1.85	1.85		
1.74								
1.60	1.60	1.61	1.61	1.59	1.62	1.60	1.62	
1.54	1.55	1.52	1.52	1.55	...	1.54		
1.38	1.39	1.36	1.39	...	1.38	1.38	1.38	
1.30	1.30	1.29	1.30	1.30	1.32	1.30		
1.21	...	...	...	...	1.22	1.22		

* *"d" values of high intensity lines taken from ASTM powder-index numbers shown below:*

Compound	*Index No.*	*Compound*	*Index No.*
AlN	8–262	Tis	12–534
ZrC	11–110	Ti_2S	11–664
ZrN	2–0956	Al_2O_3	10–173
TiC	6–0614	TiO_2 (Rutile)	4–0551
TiN	6–0642	TiO_2 (Anatase)	4–0477

† *For each suspected compound only the "d" values obtained from the indicated x-ray diffraction photographs that were within ±0.02 were listed.*
‡ *Austenitized at 2200 F, water-quenched, and aged at 900 F for 3 hr.*
§ *Austenitized at 2200 F, cooled to 1400 F and held for 6 hr to embrittle, then water-quenched and aged at 900 F for 3 hr.*
Note: When good correlation was found to exist between the "d" spacings of the unknown diffraction pattern and the known "d" spacings of the suspected compounds, good correlation was generally found between the relative intensities of the unknown and known diffraction lines.

Comparison of the x-ray diffraction results with the fractographs (Fig. 4) led to the conclusion that the small, rod-shaped particles shown in the fractograph of Steel C (Fig. 4C) are aluminum-nitride particles. This evidence was further substantiated by selected-area electron diffraction of the rod-shaped particles. The relative freedom of the fracture surface of this steel from such particles as TiC, TiN, TiS, and ZrN probably explains why the toughness of the steel is superior to that of all of the steels except Steels A and B.

The x-ray diffraction results obtained for Steel D are particularly informative. When this steel was water-quenched from 2200 F, the particles on the fracture surface were identified as titanium nitride and the Charpy V-notch energy was 35 ft-lb at 80 F. When the steel was cooled from 2200 to 1400 F and held at 1400 F for 6 hr, the toughness dropped to a Charpy V-notch energy of 23 ft-lb at 80 F, and the particles on the fracture surface were identified as zirconium nitride and titanium carbide. Thus it appears that ZrN is more stable than TiN and, after long times at 1400 F, ZrN forms and the titanium is then free to react with the carbon in the steel. It is not known whether the ZrN and TiC are more deleterious to the notch-toughness than TiN or whether the decreased toughness is due to the larger amount of precipitate present. Evidence to be presented later does indicate that the precipitate formed by long-time holding at 1400 F or by slow cooling from the hot-rolling temperature is present as a film on the austenitic grain boundaries. A film-like grain-boundary precipitate would be expected to be particularly detrimental.

The x-ray diffraction analysis of Steel E indicated that titanium carbide was present. The fractograph of Steel E, like that of Steel D, showed a predominance of rectangular-shaped particles. Thus, it appears that rectangular-shaped particles seen on the fracture surface may be either titanium carbide or nitride.

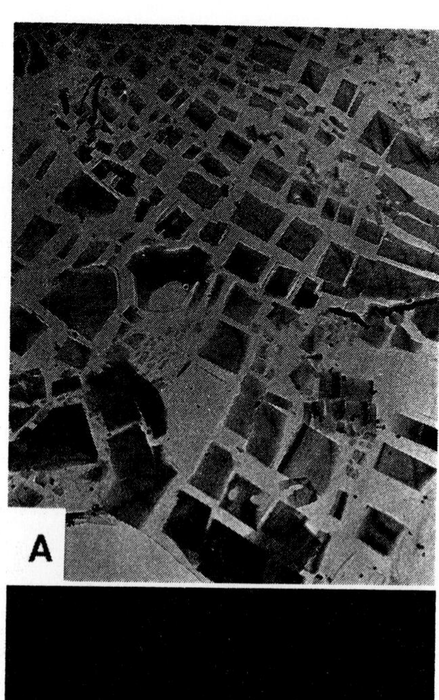

FIG. 6. (A) Extraction fractograph and (B) selected-area electron-diffraction pattern of a typical thin film of particles found on the fracture surface of Steel G. ×20,000.

The x-ray diffraction analysis of the fracture replicas from the three steels that contained a zirconium addition, Steels D (held at 1400 F) F, and I, showed that all three steels had large amounts of zirconium nitride on the fracture surface. Another recently conducted study (9) has shown that 0.03% Zr in 12% Ni maraging steels causes a marked reduction in notch toughness (about 30 ft-lb Charpy V-notch energy at 70 F) from that which would have been obtained if no zirconium was present.

There is definite evidence of titanium sulfide (TiS) on the fracture surface of steel H, and TiS is probably present on the fracture surface of Steel F. Although the chemical analysis of Steel F shows that the sulfur content in the steel is not abnormally high (0.006%), Steel H had the highest sulfur content of any of the steels examined (0.013%). The extraction fractographs of these steels (particularly of Steel H) indicate that the sulfides are present on the fracture surface as very fine, irregularly shaped particles. As will be discussed later, the presence of titanium sulfide in Steel H was also confirmed by microprobe techniques.

Selected-Area Electron Diffraction

In addition to x-ray diffraction, selected-area electron diffraction is sometimes helpful in identifying the composition of the particles on an extraction replica of the fracture surface. This technique has the advantage that the particles being analyzed can be selected; and if the electron-diffraction analysis is successful, identification of the particular particle or group of particles is assured. However, in order for the electron-diffraction analysis to be successful, the particles should be thin enough to transmit the diffracted electron beam, and it is necessary to obtain a good measurement of the electron-diffraction pattern. The latter requirement is particularly difficult. Large particles are usually single crystals, and several isolated diffraction spots rather than diffraction rings are often obtained. Under these conditions, most of the diffraction-ring spacings can only be obtained by compiling the diffraction patterns from several differently oriented particles of the same type.

When small, thin particles in many different orientations are closely grouped, it is possible to obtain a fairly complete diffraction-ring pattern in a single diffraction

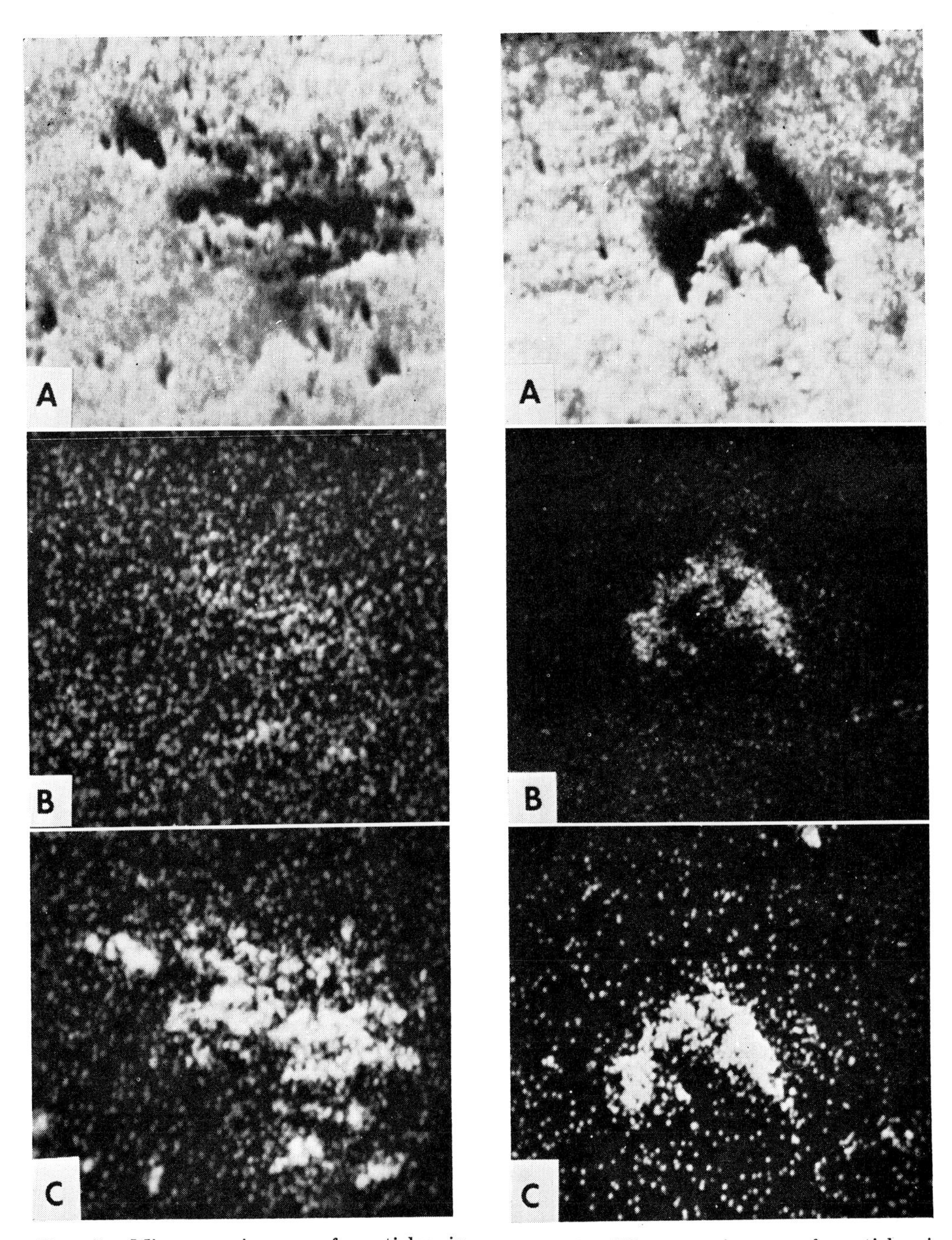

Fig. 7. Microscan images of particles in extraction fractographs from Steel E. (A) Electron image, (B) x-ray scan image, sulfur radiation, and (C) x-ray scan image, titanium radiation; ×270.

Fig. 8. Microscan images of particles in extraction fractographs from Steel H. (A) Electron image, (B) x-ray scan image, sulfur radiation, and (C) x-ray scan image, titanium radiation; ×270.

photograph. Under these conditions, analysis of the pattern is simplified and identification of the composition of the particles is more certain. Figure 6 illustrates such a condition. Here a large area of the fracture surface of Steel G was found to contain a regular array of rectangular particles (Fig. 6A). These particles are believed to have formed during slow-cooling from the hot rolling temperature because they have not been observed in samples of maraging steel that were water-quenched after hot rolling. The flat nature of the fracture surface indicates that the film of particles probably lies on an austenite grain boundary. The selected-area electron-diffraction photograph shown in Fig. 6B was obtained from the area shown in Fig. 6A, and the analysis of the diffraction pattern indicated that the film was probably titanium carbide.

Microprobe Examination

The extraction replicas of the fracture surface can also be examined by microprobe. The Cambridge microprobe analyzer has proven particularly effective for these studies because microscan images

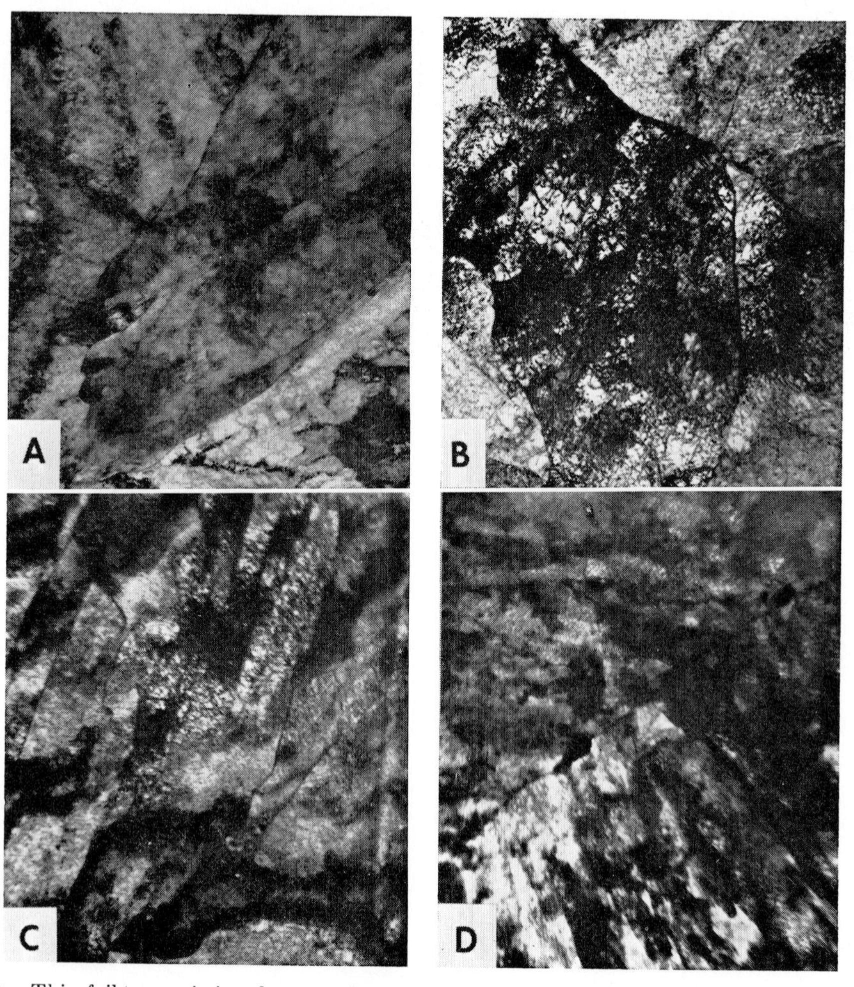

Fig. 9. Thin-foil transmission electron micrographs from (A) Steel A, (B) Steel B, (C) Steel C, and (D) Steel D. ×20,000.

of the x-ray fluorescence of suspected elements can be directly compared with the electron image. Figure 7 shows electron (Fig. 7A) and microscan (Fig. 7B and C) images of the extracted particle from the fracture surface of Steel E. Because chemical analysis showed that the steel contained 0.010% S, it was expected that sulfides would be found on the fracture surface; but, as discussed earlier, x-ray diffraction studies indicated that only titanium carbide was present. The microscan image for sulfur, Fig. 7B, shows only a very faint indication that sulfur is present in the extracted particles. The general spotting over the entire photograph is due to the electronic "noise" in the system that has been accentuated by the long exposure required to obtain any indication of sulfur. The low concentration of sulfur in the extracted particles is indicated by the low contrast between the brighter areas of the photograph and the background.

A microscan image for titanium confirms that that extracted particles are titanium rich. Unfortunately, the Cambridge microprobe analyzer is unable to check for elements of low atomic weight such as carbon, nitrogen, and oxygen, so it is not possible to determine whether the particles are titanium carbides, nitrides, or oxides.

Figure 8 shows the electron and microscan images of extracted particles from the fracture surface of Steel H. In agreement with the x-ray diffraction analysis, the microprobe examination shows the presence of titanium sulfide (the large central mass of extracted particles in Fig. 8A) and several areas of titanium-rich sulfur-free particles (presumably titanium carbides and nitrides) near the bottom and sides of the photograph (compare Fig. 8C with 8B).

Transmission-Electron Microscopy

In Fig. 9, thin-film transmission-electron micrographs of Steels A and B (Fig. 9A and B), the two high-toughness steels, are compared with those of Steel D (Fig. 9C and D) to see if there is an obvious difference in the dislocation structure or grain boundaries that might account for the lower toughness of Steel D.

All three of the steels have a high dislocation density. The dislocation density of Steel B appears to be somewhat lower than that of Steel A or Steel D and the substructure of Steel B appears to be somewhat coarser than that of the other two steels. The grain boundaries and subboundaries of all three of the steels are quite distinct. This is probably an indication that precipitate exists on the boundaries. In Fig. 9D the presence of isolated particles of precipitate is readily observed along the grain boundary that runs diagonally across the photomicrograph.

Analysis of Insoluble Residue

A chemical analysis of the insoluble residue left after dissolving the matrix of Steels A and G is shown in Table 5. About three times as much titanium was found in the residue of the low-toughness steel (Steel G) as in the high-toughness steel (Steel A). X-ray diffraction and differential-thermal analysis indicated that the titanium was present as titanium carbide or nitride. Zirconium carbide was

TABLE 5. Results of a Quantitative and Qualitative Analysis of Inclusions in Steel A and G

Steel	Composition, % Ti	Mo	Ni	Co	Al*
Composition, Percentages of Elements Present as Nonmetallic Inclusions					
A	0.02	0.001	0.001	0.001	0.003
G	0.06	0.04	0.002	0.001	0.005

Qualitative Analysis of the Inclusions by X-Ray Diffraction and Differential Thermal Analysis†

Steel	
A	The major constituents could not be identified by x-ray diffraction and differential-thermal-analysis techniques. TiN and Ti (C, N) were identified as a minor constituent by differential thermal analysis.
G	The major constituent was Ti (C, N) based on x-ray diffraction and differential-thermal-analysis techniques. ZrC was identified by x-ray diffraction.

* *Present as the oxide.*
† *It was not possible to identify any compounds of Mo, Ni, and Co.*

also identified by x-ray diffraction as a consitituent in the residue of Steel G.

Summary

The examination and analysis of extraction fractographs by metallographic, x-ray- and electron-diffraction, and microprobe techniques has been shown to be a powerful method of isolating the major factors responsible for low notch-toughness in maraging steels. Thin-film transmission electron microscopy and analysis of insoluble residue, while generally substantiating the results obtained by analysis of the fractographs, are less definite and therefore less useful than the fractographic studies. These studies, in conjunction with the examination of the steels by light microscopy, have led to the following conclusions regarding the factors influencing the notch toughness of maraging steels:

1. Although fine-grained maraging steels would be expected to have higher notch toughness than coarse-grained steels, the effect of grain size is apparently obscured by other factors in some of the steels examined in this study.
2. The fracture surfaces of high toughness heats are remarkably free of extracted particles.
3. Except for the two steels with high toughness, it was not possible to correlate the relative amounts of extracted particles on the fracture surfaces with the notch-toughness of the steels.
4. The extraction replicas from the fracture surfaces of the steels that had the lowest notch toughness contained particles of zirconium nitride or titanium sulfide.
5. Titanium carbide and nitride particles were found on the extraction replicas from several of the steels.
6. Zirconium nitride and titanium carbide and nitride were found to exist in the form of rectangular particles. Slow-cooling from the hot rolling temperature appeared to produce very thin films of titanium carbide or nitride particles, probably at austenite grain boundaries.
7. Aluminum nitride occurred as small, rod-shaped particles; and titanium sulfide was present as very small, irregularly shaped particles.

Acknowledgment

The foregoing paper describes one phase in a study of feasibility of developing a high-toughness alloy steel weldment with a yield strength in the range of 180 to 210 ksi. The majority of the investigation described was carried out under Bureau of Ships Contract No. NObs-88540, which is under the technical supervision of T. J. Griffin, Head, Metals Section, Hull Division, Bureau of Ships. The support, encouragement, and advice of the various Navy activities associated with the project are gratefully acknowledged.

This paper is published for information only and does not necessarily represent the recommendations or conclusions of the Bureau of Ships.

The authors also wish to thank R. F. Decker of the International Nickel Co. for supplying samples of Steel A and W. S. Pellini of the Naval Research Laboratory for supplying samples of Steel B. The assistance of our associates W. R. Bandi and P. A. Stoll for conducting special chemical and x-ray diffraction analysis and of J. A. Gula of the Edgar C. Bain Laboratory for Fundamental Research, U. S. Steel Corp., for aid in the microprobe studies is gratefully acknowledged.

REFERENCES

1. W. R. Warke and A. R. Elsea, Electron Microscopic Fractography, Battelle Memorial Institute, DMIC Memorandum 161 (1962).
2. C. D. Beachem, An Electron Fractographic Study of Mechanism of Ductile Rupture in Metals, U. S. Naval Research Laboratory, NRL Report 5871 (1962).
3. W. R. Warke, Some Observations on the Electron-Microscopic Fractography of Embrittled Steels, Battelle Memorial Institute, DMIC Memorandum 187 (1964).
4. C. D. Beachem and R. M. N. Pelloux, Electron Fractography—A Tool for the Study of Micro-Mechanics of Fracturing Proc-

esses, Symposium on Fracture Toughness Testing and Its Application, ASTM (1964).

5. C. D. Beachem, Electron Fractographic Studies of Mechanical Fracture Processes in Metals, Trans ASME Journal of Basic Engineering, Paper No. 64-Met-12.
6. J. E. Srawley and T. C. Lupton, Tensile and Impact Properties of Some Maraging Steel Compositions, U.S. Naval Research Laboratory, NRL Report 5916 (1963).
7. P. M. Kelley and J. Nutting, Techniques for the Direct Examination of Metals by Transmission in the Electron Microscope, J Inst Metals, 87 (1959) 385.
8. W. R. Bandi, H. S. Karp, W. A. Straub and L. M. Melnick, Determination of Nonmetallic Compounds in Steel—I, Talanta, 11 (1964) 1327.
9. E. P. Sadowski, Twelve Percent Ni Maraging Steels, ASM Woodside Panel (1963).

SECTION IV:
Cryogenic and Elevated-Temperature Properties

Low-Temperature Elastic Properties of a 300-Grade Maraging Steel

H. M. LEDBETTER AND D. T. READ

Elastic properties of an annealed 300-grade maraging steel (18 Ni, 9 Co, 5 Mo pct by weight) were studied between room temperature and liquid-helium temperature. Longitudinal and transverse ultrasonic velocities were determined by a pulse method. The reported elastic constants are: longitudinal modulus, shear modulus, Young's modulus, bulk modulus, and Poisson's ratio. Except for the bulk modulus, the room-temperature elastic constants are all lower than those of iron; and their temperature dependencies are regular in the studied temperature region.

ELASTIC properties of a 300-grade maraging steel (18 Ni, 9 Co, 5 Mo) were studied between room temperature and liquid-helium temperature. Elastic constants were determined by measuring the velocities of both longitudinal and transverse ultrasonic waves in a polycrystalline aggregate. The following constants are reported here: C_ℓ = longitudinal modulus, G = shear modulus, B = bulk modulus, E = Young's modulus, and ν = Poisson's ratio.

Maraging steels exhibit some of the best available combinations of high strength and high toughness at room temperature.[1] Some alloy steels with high nickel contents also have good toughness at cryogenic temperatures. However, the usual heat treatment of maraging steels would probably embrittle them at these temperatures. Thus, maraging steels would probably be used at low temperatures in their annealed conditions. An annealed 300-grade maraging steel was studied (300 indicates a yield stress of approximately 300×10^6 psi). This alloy has a body-centered-cubic crystal structure, and it is ferromagnetic at room temperature. Other grades of annealed maraging steels are expected to be similar elastically because of their similar chemical compositions.

Elastic constants are interesting for two principal reasons. First, elastic constants are related directly to interatomic forces; thus, they are connected with a variety of solid-state phenomena, including maximum attainable strengths, phase stabilities, and lattice specific heats. Second, elastic constants are essential design parameters; the elastic constants must be known to compute deflections due to applied loads or stresses due to temperature changes of constrained components.

While some data on the temperature dependence of the Young's modulus of a 250-grade maraging steel are available,[2] data on the other elastic constants (shear modulus, bulk modulus, Poisson's ratio) and on other grades of maraging steel have not been reported. Furthermore, because of the many elastic-constant anomalies exhibited by iron-nickel alloys (both those with higher and lower nickel contents than maraging steels), and because of the complicated chemistry of maraging steels, their elastic behavior cannot be estimated confidently by usual predictive methods. Therefore, detailed experimental studies of the elastic behavior of these alloys are required.

H. M. LEDBETTER is Research Metallurgist and D. T. READ is a Physicist and NRC-NBS Postdoctoral Research Associate, 1975-76, with the Cryogenics Division, National Bureau of Standards, Boulder, CO 80302.

Manuscript submitted December 31, 1976.

EXPERIMENTAL

Material

The studied material was obtained from a commercial source in the form of a $10 \times 23 \times 108$ cm ingot. The chemical (mill) analysis of the material by weight is: 18.41 Ni, 9.27 Co, 4.95 Mo, 0.002 C, 0.12 Al, 0.01 Mn, 0.006 P, 0.003 S, 0.01 Si, 0.59 Ti, balance Fe. The Rockwell B hardness is 105. The mass density is 8.188 g/cm^3 as determined by Archimedes's method using distilled water as a standard. The microstructure was an ill-defined Widmanstätten type with a ferrite particle size of about 0.02 mm. Original austenite grain boundaries were invisible.

Room-Temperature Sound Velocities

Room-temperature sound velocities were measured by a pulse technique.[3] Briefly, a quartz piezoelectric transducer with a fundamental resonance frequency of 10 MHz was cemented with phenyl salycilate (salol) to one end of a specimen having flat and parallel faces. The specimen in this case was a $1.25 \times 1.25 \times 1.6$ cm parallelpiped with opposite faces ground flat and parallel within 2.5 μm. Ultrasonic pulses about one μs in duration were sent into the specimen by electrically exciting the transducer. The pulses propagated through the specimen, reflected from the end, and propagated back. The echoes were detected by the transducer and displayed on an oscilloscope. The sound velocity was computed by

$$v = 2\ell/t \qquad [1]$$

where ℓ is the specimen length and t is the transit (round-trip) time. On the oscilloscope, t was the time between subsequent echoes. The oscilloscope was calibrated against a precision time-mark generator. An x-cut quartz transducer was used for longitudinal waves, and an ac-cut quartz transducer was used for transverse waves.

Low-Temperature Sound Velocities

Low-temperature measurements of the sound velocities were made by a pulse-superposition technique.[4] Briefly, the repetition rate of the pulse was increased so that each pulse coincided with the second echo of the preceding pulse. Since the excitation voltage was large compared with the echo voltages, the oscilloscope display consisted of alternating pulses of excitation voltages and "echo" voltages where the "echo" voltage represented the sum of all odd-numbered echoes of the nonsuperimposed case. Because of interference effects, the envelope of the summed odd-numbered echoes is highly sensitive to small changes in the ultrasonic velocity that are caused in this case by cooling.

The transducer-specimen bonding material was a stopcock grease. Temperatures were monitored with a chromel-constantan thermocouple placed near the specimen. Cooling rates were about 2 K/min. The specimen holder was described previously.[5] Cooling was achieved by lowering the specimen-holder assembly stepwise into the ullage of a helium dewar. No thermal contraction corrections were made. For this alloy, the maximum thermal contraction correction to the elastic constant, which applies at T = 0 K, is 0.3 pct. No correction was made for the transducer-cement-coupling phase shift; the McSkimin[6] analysis gives a correction of less than 0.5 pct in the velocity, assuming a maximum phase shift of π. The transit time was corrected to allow for the thickness of the transducer; this correction is approximately one cycle at 10 MHz; thus, the observed longitudinal-wave and transverse-wave transit times were reduced about 1.0 and 0.5 pct, respectively.

RESULTS

Room-temperature sound velocities are given in Table I. Velocities were measured in three orthogonal directions for three orthogonal polarizations. The experimental uncertainties in these velocities are estimated to be 0.5 pct. The directional invariance of the sound velocities shows that texture is insignificant in this material.

Longitudinal and shear elastic constants were computed from the velocities according to

$$C_\ell = \rho v_\ell^2 \quad [2]$$

and

$$G = C_t = \rho v_t^2 \quad [3]$$

where ρ is the mass density. The uncertainty of these constants is estimated to be 1 pct.

The temperature variations of C_ℓ and C_t are shown in Figs. 1 and 2. Curves in those figures are least-squares fits of the data to the Varshni[7] relationship:

$$C = C^0 - \frac{s}{\exp(t/T) - 1} \quad [4]$$

where C is C_ℓ or C_t, C^0, s, and t are adjustable parameters, and T is temperature. The value of C at T = 0 K is C^0, and $-s/t$ is the high-temperature limit of the temperature derivative dC/dT. By invoking an Einstein oscillator model, it can be shown that t is the Einstein temperature. Parameters C^0, s, and t are

Table I. Room Temperature Acoustic Wave Velocities in Units of 10^6 cm/s

Propagation Direction	v_l	v_{t_1}	v_{t_2}
x	0.5673	0.2882	0.2879
y	0.5648	0.2881	0.2865
z	0.5589	0.2848	0.2841

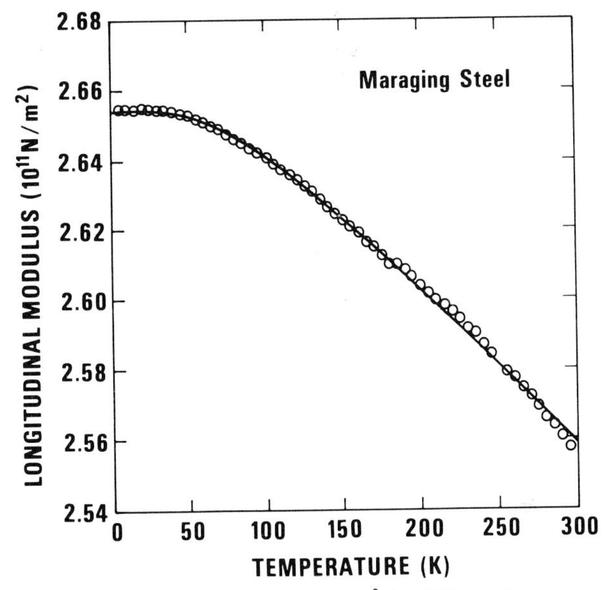

Fig. 1—Longitudinal modulus $C_\ell = \rho v_\ell^2$ for 300-grade maraging steel as a function of temperature.

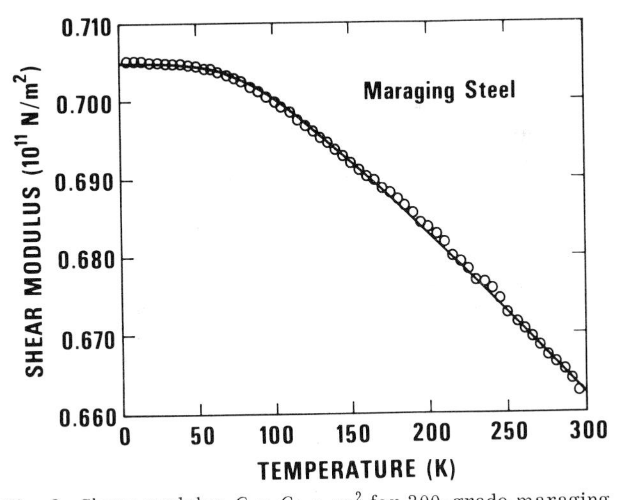

Fig. 2—Shear modulus $G = C_t = \rho v_t^2$ for 300-grade maraging steel.

given in Table II. Average differences between measured and curve values are 0.04 and 0.05 pct for the longitudinal and transverse cases, respectively.

Temperature variations of Young's modulus, the bulk modulus, and Poisson's ratio are shown in Figs. 3 to 5. These elastic constants were computed from Eq. [4] and the parameters in Table II using the relationships:

$$B = C_\ell - (4/3)G \quad [5]$$

$$E = 9GB/(G + 3B) \quad [6]$$

and

$$\nu = (E/2G) - 1. \quad [7]$$

The experimental uncertainties of these elastic constants are estimated to be ±1 pct. Values of these elastic constants at selected temperatures are given in Table III, and temperature coefficients of the elastic constants are given in Table IV.

DISCUSSION

The present study is discussed in two parts: the room-temperature elastic constants and the changes of the elastic constants due to cooling to liquid-helium temperature.

As shown in Table III, the Young's and shear moduli of the steel are about twenty pct lower than those of unalloyed iron, the bulk modulus is about the same, and Poisson's ratio is about ten pct higher. Usually, the elastic-stiffness constants of alloys are lower than those of their base metal. Two of the three principal alloying elements in the steel—nickel[11] and molybdenum[12]—lower the elastic stiffnesses of iron, while cobalt[11] raises them. Nickel has the largest effect of the three. Thus, it is appropriate to compare the elastic constants of the steel with those of the corresponding iron-nickel alloy, and this is done in Table III. Except for the bulk modulus, the elastic constants of the steel are very close to those of an iron-18 nickel alloy. Cobalt and molybdenum seem to have very little effect on Young's modulus, the shear modulus, and Poisson's ratio, that is, on elastic constants that are determined either entirely or largely by the resistance of a material to shear deformations.

Table II. Parameters Determined from the Varshni Equation

	Longitudinal	Transverse
C°, 10^{11} N/m^2	2.654	0.7047
s, 10^{11} N/m^2	0.1036	0.05959
t, K	221.9	264.4

Table III. Elastic Constants of a 300-Grade Maraging Steel at Selected Temperatures and Those of Iron and Iron-18 Nickel at Room Temperature in Units of 10^{11} N/m^2 Except ν, Which is Dimensionless

Material	Temperature, K	E	G	B	ν
Maraging steel, 300-grade	300	1.756	0.663	1.676	0.325
	200	1.806	0.683	1.692	0.322
	100	1.848	0.700	1.708	0.320
	0	1.859	0.705	1.714	0.319
Iron*	300	2.140	0.831	1.680	0.288
Iron-18 Ni†	300	1.710	0.653	1.463	0.309

*Ref. 8.
†Ref. 9, 10.

Table IV. Temperature Coefficients $(1/C)(dC/dT)$ at Room Temperature in Units of 10^{-4} K^{-1}

Material	E	G	B	ν
Maraging steel	−2.92	−3.18	−0.98	1.04
Unalloyed iron*	−2.11	−2.31	−0.90	0.95

*Computed from data in Ref. 8 using a Voigt-Reuss-Hill arithmetic average.

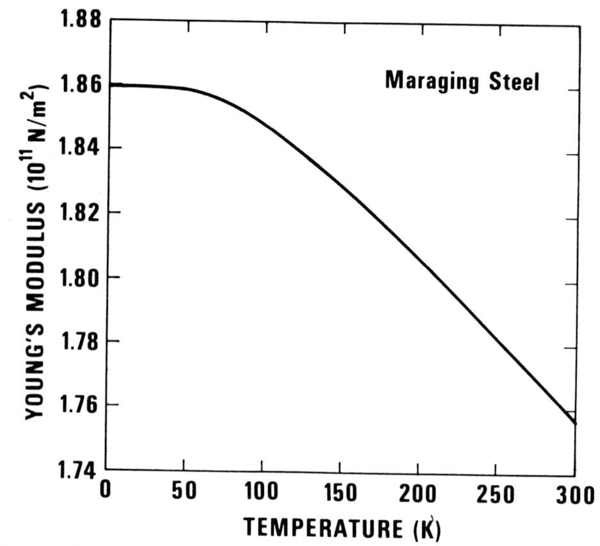

Fig. 3—Young's modulus for 300-grade maraging steel.

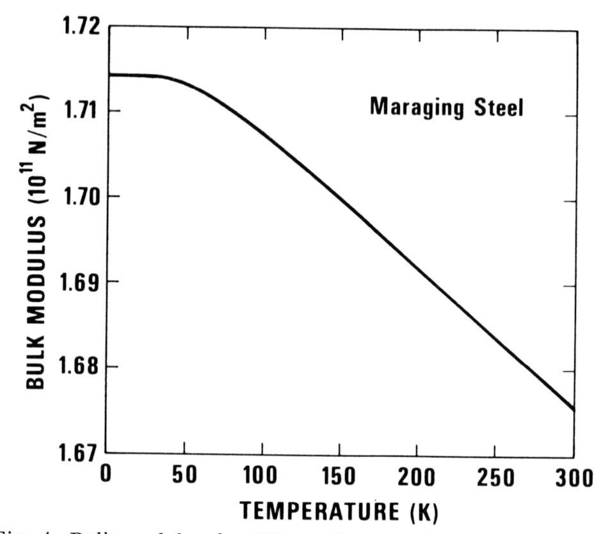

Fig. 4—Bulk modulus for 300-grade maraging steel.

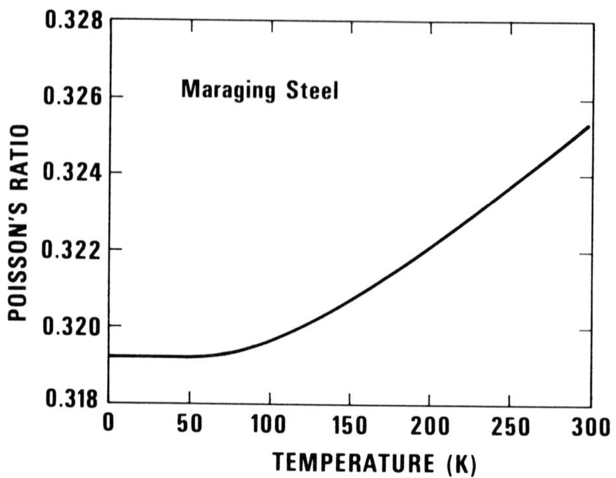

Fig. 5—Poisson's ratio for 300-grade maraging steel.

The bulk modulus is determined by the resistance of a material to dilatational deformations, and it is surprising that the bulk moduli of the steel and of unalloyed iron are the same. There appears to be no simple explanation for the unexpectedly high bulk modulus of the maraging steel. The usual effects of alloying are that the bulk modulus changes approximately in the same way as Young's modulus and the shear modulus. However, there is no *a priori* requirement that the bulk and shear moduli change similarly, and the present case is an interesting departure from usual behavior. Possible explanations for this effect may lie in the high bulk modulus of molybdenum, which is fifty pct higher than that of iron. Of all the elastic constants, the bulk modulus would be expected to come closest to a Vegard-law type behavior[13] because it is the most structure-independent elastic constant, depending much more on atomic volume than on other factors. Also, the bulk modulus is the elastic constant least affected by the usual magnetic energies that contribute to the elastic constants of iron-base alloys.[14]

As shown in the figures, the temperature behavior of the elastic constants of maraging steel is quite regular; that is, qualitatively the same as that exhibited by almost all simple metals and alloys. The salient features of regular behavior are: zero slope at zero temperature, as required by the third law of thermodynamics, continuous decrease with increasing temperature, consistent with the softening of interatomic bonding forces due to increased thermal vibrations; and a linear slope at high temperatures.

The regular temperature behavior of the maraging steel is unexpected. While no low-temperature elastic constants have been reported previously for body-centered-cubic iron alloys with so high a nickel content, some data exist for lower nickel content alloys. Weston, Naimon, and Ledbetter[10] reported the low-temperature elastic constants of iron-nickel alloys containing 3.5, 5, 6, and 9 pct nickel. These alloys show regular behavior above about 40 K, but are anomalous at lower temperatures. The anomalies are due, presumably, to magnetic transitions. But the maraging steel shows no such behavior. It is well known that iron-nickel alloys containing more than about 30 pct nickel are face-centered cubic and have very large anomalies in their elastic constants between room temperature and liquid-helium temperature. These anomalies are also due to magnetic effects.[15]

The maraging steel has considerably higher temperature coefficients of Young's modulus and the shear modulus than unalloyed iron, as shown in Table IV. This suggests that the steel has a lower Debye temperature than iron, consistent with the Debye temperatures reported for bcc iron-nickel alloys by Weston, Naimon, and Ledbetter.[10] Practically, this means the elastic stiffness of the steel, upon cooling, increases more than in the case of iron. Simple computations show that the Debye temperature of the maraging steel is 421 K, while iron has a Debye temperature of 472 K.

As expected from theory, the Einstein temperature of 247 K computed from the data in Table II is lower than the Debye temperature. In this case, the Einstein temperature is rather inaccurate because it is determined from an approximate model and from the temperature dependence of the elastic constants. The Debye temperature is computed directly from the elastic constants.

CONCLUSIONS

In the present study, several conclusions were reached concerning the elastic properties of annealed 300-grade maraging steel:

1) The Young's modulus, shear modulus, and Poisson's ratio are all about 15 pct lower than the corresponding elastic constants of unalloyed iron. The bulk modulus is the same as that of iron.

2) Except for the bulk modulus, all the elastic constants are essentially the same as those of an iron-18 nickel alloy, implying that the effects of cobalt and molybdenum on the elastic constants are effectively canceled.

3) All the elastic constants show regular behavior over the entire studied temperature range, 4 to 300 K, indicating the absence of magnetic transitions that have been reported at low temperatures in other body-centered-cubic iron-nickel alloys.

4) Temperature coefficients of Young's modulus and the shear modulus are about 40 pct higher than those of unalloyed iron. This is consistent with a lower Debye temperature for the steel.

ACKNOWLEDGMENT

This study was supported by the Advanced Research Projects Agency of the U.S. Department of Defense.

REFERENCES

1. A. Maynée, J. M. Drapier, J. Dumont, D. Coutsouradis, and L. Habraken: *Cobalt-Containing High-Strength Steels*, p. 50, Centre D'Information Du Cobalt, Brussels, 1974.
2. F. R. Schwartzberg, S. H. Osgood, and R. Herzog: *Cryogenic Materials Data Handbook*, U.S. Air Force Systems Command (Tech. Doc. Rpt. No. AFML-TDR-64-280), 1968.
3. R. T. Beyer and S. V. Letcher: *Physical Ultrasonics*, pp. 79-87, Academic Press, New York, N.Y., 1969.
4. H. J. McSkimin: *J. Acoust. Soc. Amer.*, 1961, vol. 33, pp. 12-16.
5. E. R. Naimon, W. F. Weston, and H. M. Ledbetter: *Cryogenics*, 1974, vol. 14, pp. 246-49.
6. H. J. McSkimin: *IRE Trans. Ultrason. Eng.*, 1957, vol. 5, pp. 25-43.
7. Y. P. Varshni: *Phys. Rev.*, 1970, vol. 2, pp. 3952-55.
8. J. A. Rayne and B. S. Chandrasekhar: *Phys. Rev.*, 1961, vol. 122, pp. 1714-16.
9. H. M. Ledbetter and R. P. Reed: *J. Phys. Chem. Ref. Data*, vol. 2, pp. 531-618, 1974.
10. W. F. Weston, E. R. Naimon, and H. M. Ledbetter: *ASTM STP 579*, pp. 397-420, Amer. Soc. Test. Mater., Philadelphia, Pa., 1975.
11. W. C. Leslie: *Met. Trans.*, 1972, vol. 3, pp. 5-26.
12. S. Takeuchi: *J. Phys. Soc. Jap.*, 1969, vol. 27, pp. 929-40.
13. N. F. Mott: *Rep. Prog. Phys.*, 1962, vol. 25, pp. 218-43.
14. W. F. Brown: *Phys. Rev.*, 1936, vol. 50, pp. 1165-72.
15. G. Hausch: *Phys. Status Solidi*, 1973, vol. 15, pp. 501-10.

N. KENYON
E. P. SADOWSKI
P. P. HYDREAN

International Nickel Company, Inc.,
Paul D. Merica Research Laboratory,
Sterling Forest, Suffern, N. Y.

Elevated Temperature Properties of Maraging Steel Plates and Welds

The creep rupture behavior, and the effects of elevated temperature exposure in air and hydrogen on the subsequent room temperature properties of a 12 percent Ni-5 percent Cr-3 percent Mo maraging steel are described. Tests have been made on several heats of plate and on gas tungsten-arc, gas metal-arc, and electroslag welds. On the basis of the results obtained, maraging steels offer promise as high-strength steels for service at elevated temperatures.

Introduction

THE attractive properties of maraging steels have been investigated mainly for components that operate at normal atmospheric temperatures, although maraging steel dies are performing successfully at elevated temperatures. The combination of high strength and ease of fabrication that maraging steels offer could also be useful for higher temperature structural applications. Elevated temperature properties, however, have received comparatively little attention. Published creep data are limited and appear to be restricted to relatively short-time tests on the 18 percent nickel steels [1, 2, 3].[1]

In the work described here, tests have been made up to 10,000 hours on the 12Ni-5Cr-3Mo grade, which has a nominal room temperature yield strength of 180 ksi [4]. The elevated temperature strength, the creep rupture behavior, and the effects of exposure at elevated temperature on subsequent room temperature properties have been examined. Plates from several heats, and a variety of welds were included in the investigation.

Experimental Procedure

Plates Examined. The compositions, methods of melting, and plate thicknesses of the four commercial heats tested as plate are shown in Table 1. Heat IV was made with greater than normal amounts of the hardening elements aluminum and titanium to see if a high-hardener heat, in the overaged condition, would perform better at elevated temperatures than the standard composition.

Heat Treatment of Plates. The plates were heat treated as follows:

Heat I – 1500 deg F/1 hr + 900 deg F/3 hr
Heat II – 1600 deg F/1 hr + 900 deg F/3 hr
Heat III – 1700 deg F/1 hr + 1400 deg F/1 hr + 900 deg F/3 hr
Heat IV – 1700 deg F/1 hr + 1400 deg F/1 hr + 1000 deg F/3 hr

The treatment used for heat I is usually considered to be the standard heat treatment. The double annealing treatment used for heats III and IV is a later development. Double-solution-annealed material has in some instances exhibited a better combination of strength and toughness at room temperature than the same material given a single anneal.

The customary aging temperature for maraging steels is 900 deg F. A higher aging temperature (1000 deg F) was used for the high hardener heat IV to overage it to the desired strength.

Details of the Welds. Samples from the following welds have been tested:

(*a*) A 2-in-thick gas metal-arc (MIG) weld made in a few passes with comparatively heavy beads. This has been called the high-deposition-rate (HDR) MIG weld.

(*b*) A 2-in-thick MIG weld made in many passes with fine stringer beads. This is the low-deposition-rate (LDR) MIG weld.

(*c*) An electroslag weld made with one pass in 2-in-thick plate. This weld was made as part of an experimental program to determine the suitability of the electroslag process for welding maraging steels. The process is not being used industrially to weld steels of this strength level.

(*d*) A 1-in-thick gas tungsten-arc (TIG) weld. The compositions of the plates and filler wires used to make the welds, and details of the welding conditions are given in Tables 1 and 2, respectively. The MIG and TIG welds were made with heats V and VII and the electroslag weld with heats VI and VIII. All the welds were made with the welding direction parallel to the final rolling direction of the plate.

The plates were in the fully heat treated condition prior to welding, and blanks from the welds were aged before testing. The postweld aging treatment used for the MIG and TIG welds was 900 deg F for 3 hours. The electroslag weld was aged at 900 deg F for 24 hours.

Testing Procedure

Tensile Tests. Unless stated otherwise, tensile specimens (1.0 × 0.252-in-dia gauge length) were machined transverse to the final

[1] Numbers in brackets designate References at end of paper.

Contributed by the Metals Engineering Division for publication (without presentation) in the JOURNAL OF BASIC ENGINEERING. Manuscript received at ASME Headquarters, November 30, 1970. Paper No. 71-Met-E.

Table 1 The compositions of the plates and filler wires

Heat No.	Melting Process	Plate Section Size	Ni	Cr	Mo	Al	Ti	C	Si	Mn	S	P
		HEATS TESTED AS PLATE										
I	Air	2"	12.1	4.5	3.0	.38	.20	.03	.06	.04	.010	.003
II	Air	2"	12.1	5.0	3.2	.27	.23	.02	.10	.06	.007	.002
III	Vacuum	1"	12.0	5.0	3.1	.19	.21	<.003	.03	.05	.004	--
IV	Vacuum	1"	12.1	4.9	3.1	.66	.32	.017	.03	.04	.006	<.001
		MATERIALS USED TO MAKE THE WELDS										
V	Air	1" and 2"	12.2	4.8	3.4	.23	.21	.011	.15	.05	.005	.005
VI	Vacuum	2"	12.1	4.9	2.9	.23	.21	.002	.05	.02	.005	.003
VII	Vacuum	Filler Wire	12.3	4.9	3.4	.26	.30	.006	.05	.02	.003	.001
VIII	Vacuum	Filler Wire	12.0	5.2	2.4	.10	.43	.004	.02	.01	.004	.002

Composition, Wt. % (columns Ni through P)

Table 2 Welding conditions

Weld	Joint Geometry	Diameter of Filler Wire (ins)	No. of Passes	Volts	Amps	Travel Speed (in/min)	Wire Feed (in/min)	Shielding
TIG	1" Double U	1/16	22	10	220	4	20	35 cfh Argon
LDR MIG	2" Double U	1/16	35	30	300	20	200	50 cfh Argon
HDR MIG	2" Double U	1/16	18	30	300	10	200	50 cfh Argon
Electroslag	2" Square Butt	1/8	1	40	600	--	--	Proprietary Flux

rolling direction in the plates, and transverse to the direction of welding in the weldments. Tests were made at room temperature, 650 F, 750 F, 850 F, and 900 deg F.

Impact Toughness Tests. Charpy V-notch impact specimens were machined in the same direction as the tensile samples and were notched through the plate thickness, or through the weld thickness. Tests were made only at room temperature.

Creep Rupture Tests. Creep rupture tests were made at 900 deg F on samples from the electroslag weld, the high-deposition-rate MIG weld and two heats of plate. The samples were taken transverse to the plate rolling direction, and transverse to the direction of welding. The gauge length (0.252 in. dia) was 1 in. for the electroslag weld and plate samples, but was reduced to $^3/_4$ in. for the MIG weldment to insure that the gauge length consisted of weld metal only. The electroslag welds were wide enough to make this reduction in gauge length unnecessary.

Exposure Tests at Elevated Temperatures. Aged plate specimens were exposed in creep furnaces for up to 10,000 hours at 650 F and 850 deg F. Some were exposed without stress, others were stressed to $^1/_4$ of the elevated temperature tensile strength. A relatively large specimen with a 5-in. gauge length was used. At the end of the exposure, the gauge section was machined into a standard tensile specimen and a Hounsfield impact specimen, both of which were tested at room temperature. Since the results of these tests proved to be the same whether the samples were stressed or not during exposure, subsequent welded sample blanks were exposed unstressed. Weld exposures were for 5000 hours at 650 F and 850 deg F. After exposure, standard Charpy and tensile samples were machined and tested at room temperature.

A few plate samples (heat II) were exposed for up to 1000 hours in a 1000 psi hydrogen atmosphere at 750 F and 850 deg F. The strength and toughness were subsequently measured at room temperature.

The austenite contents of some of the samples exposed at elevated temperatures were measured by X-ray diffraction.

Results

Strength and Toughness at Room Temperature. The room tem-

Table 3 The strength and toughness of the plates and welds (1) at room temperature

Material	0.2% Y.S. (ksi)	U.T.S. (ksi)	Elong. (%)	R.A. (%)	Impact Toughness (CVN) (Ft-Lbs)
Heat I	187.2	190.4	15	58	39
Heat II	189.2	194.3	16	63	39
Heat III	180.1	183.9	15	65	82
Heat IV	190.7	195.6	14	65	41
TIG Weld	172.3	177.0	14	65	97
MIG Weld (LDR)	171.1	176.9	13	49	50
MIG Weld (HDR)	170.4	176.8	10	41	44
Electroslag Weld	156.4	168.2	13	46	29

(1)Transverse samples from the weldments failed in the weld.

Table 4 Tensile strength at elevated temperature (1)

Material	0.2% Y.S. (ksi)	U.T.S. (ksi)	Elong. (%)	R.A. (%)
	At 650°F			
Heat I	146.4	155.7	14	57
TIG Weld	136.3	149.4	12	58
MIG Weld (LDR)	140.8	147.1	11	44
MIG Weld (HDR)	137.7	145.5	10	42
Electroslag Weld	129.7	140.7	10	39
	At 750°F			
Heat I	144.1	153.4	14	61
	At 850°F			
Heat I	122.2	137.3	17	64
TIG Weld	119.2	132.5	16	62
MIG Weld (LDR)	125.7	133.3	9	35
MIG Weld (HDR)	118.5	128.9	8	32
Electroslag Weld	116.5	126.5	11	49
	At 900°F			
Heat III	116.4	127.9	19	70
Heat IV	128.5	143.4	20	72

(1)Transverse samples from the weldments failed in the weld.

perature strength and impact toughness of base plates and weldments are shown in Table 3. The tensile strengths of the plates are in the range 185 to 195 ksi, while the toughness varies from about 40 ft-lb to 80 ft-lb. The toughness of heat III is better than that of heats I and II, presumably because heat III was lower in residual elements and tested in 1 in. section, whereas heats I and II were air melted and tested as 2 in. sections. Heat III also had the benefit of the double-annealing treatment. The difference in the toughness of heats III and IV, both of which were vacuum melted and tested as 1 in. plate, reflects the difference in the alloying approach. Pronounced overaging of a high-hardener heat (in this case, heat IV) in order to lower its strength generally does not result in optimum toughess.

The tensile strengths of the TIG and MIG welds represent 90 to 95 percent joint efficiency. The lower tensile strength of the electroslag weld (168 ksi versus 177 ksi for the MIG and TIG welds) results from a loss in aluminum and titanium that occurred during the electroslag welding process.

The toughness of the welds is in the expected ranking order; TIG is tougher than MIG which in turn is tougher than electroslag. The results for the two MIG welds demonstrate that the better toughness is obtained when fine stringer beads are used.

The impact toughness of these TIG and MIG welds at 97 and 44–50 ft-lb, respectively, are usually high. More typical values would be 60 ft-lb for the TIG and 40 ft-lb for the MIG welds [5].

Tensile Strength at Elevated Temperatures. As the test temperature increases the strength decreases (Table 4), and at approximately the same rate for both plate and welds (Fig. 1). The plate is the strongest and the electroslag weld the weakest of the samples at all test temperatures. At 850 deg F, the strengths are 70–75 percent of the room temperature strengths.

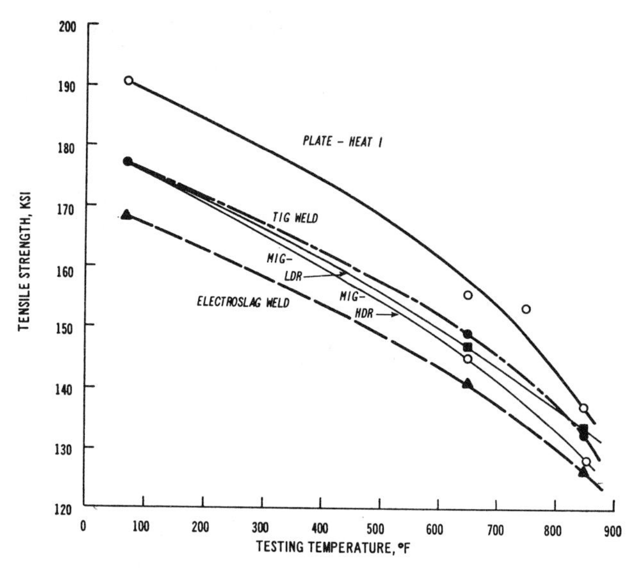

Fig. 1 Elevated temperature tensile strength of plate and welds

Creep Rupture Properties of Plates and Weldments. The results of creep rupture tests at 900 deg F on heat III, heat IV, the high deposition rate MIG weld, and the electroslag weld are plotted in Figs. 2 through 5.

At a given stress, the high-hardener heat IV has longer rupture life than heat III (Fig. 2), and a lower minimum creep rate (Fig. 3). The high-hardener heat was also stronger in short-time ten-

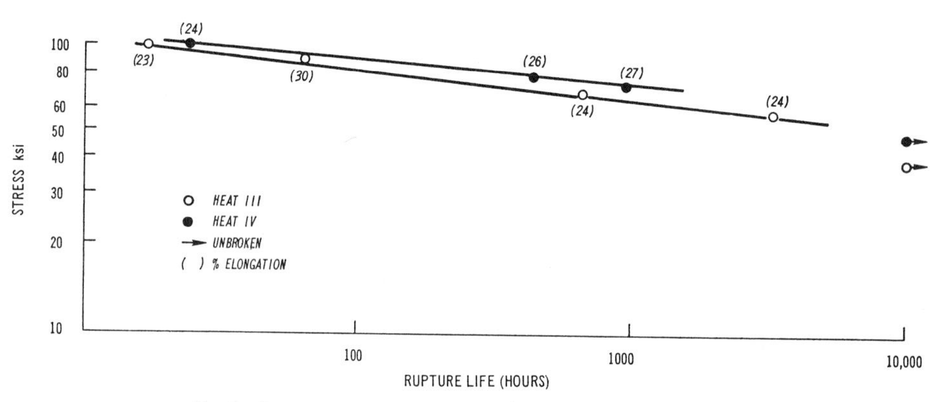

Fig. 2 Stress versus rupture life at 900 deg F for heats III and IV

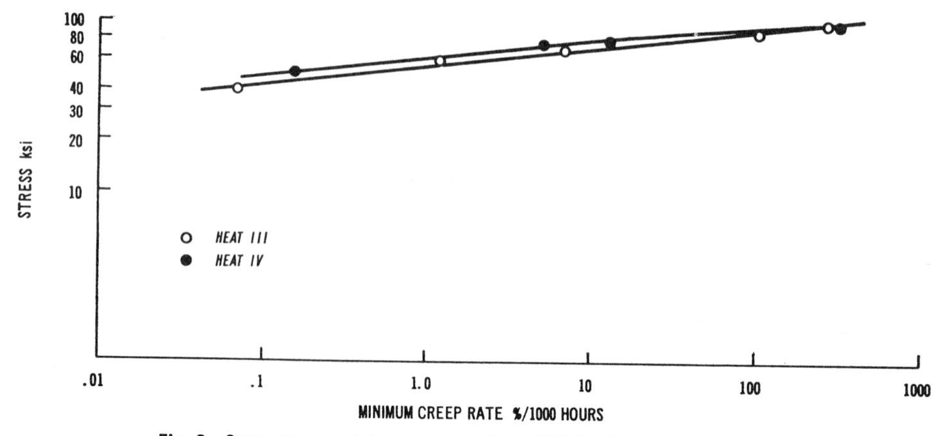

Fig. 3 Stress versus minimum creep rate at 900 deg F for heats III and IV

Table 5 Stress-time relationships for deformations of 0.5, 1.0, and 2.0 percent at 900 deg F

	Stress in ksi for a Creep Strain of										
	0.5%			1.0%				2.0%			
Heat No.	10 Hrs	100 Hrs	1000 Hrs	10 Hrs	100 Hrs	1000 Hrs	10,000 Hrs	10 Hrs	100 Hrs	1000 Hrs	10,000 Hrs
Heat III	78	54	(29)	86	66	46	(30)	96	74	(59)	--
Heat IV	79	55	(32)	88	70	(44)	--	95	78	59	(42)

	Stress in ksi for a Total Strain of										
	0.5%			1.0%				2.0%			
	10 Hrs	100 Hrs	1000 Hrs	10 Hrs	100 Hrs	1000 Hrs	10,000 Hrs	10 Hrs	100 Hrs	1000 Hrs	10,000 Hrs
Heat III	62	38	--	78	62	(38)	--	92	72	56	--
Heat IV	--	--	--	79	60	(41)	--	90	76	56	(41)

( )Denotes extrapolated value.

NOTE: Creep strain does not include strain obtained on initial loading.

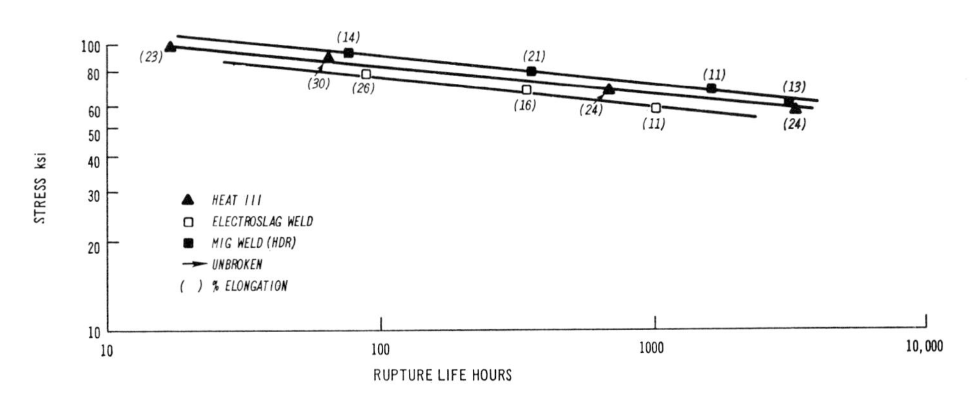

Fig. 4 Stress versus rupture life for plate (Heat III) and welds at 900 deg F

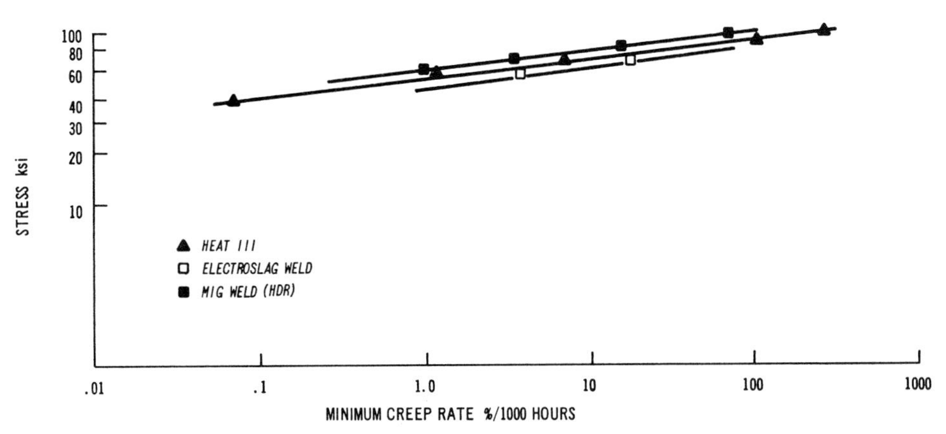

Fig. 5 Stress versus minimum creep rate for plate (heat III) and welds at 900 deg F

Table 6 The effect of long-time exposures at 650 deg F and 850 deg F on the room temperature properties of heat I

Sample	Temp. (°F)	Time (Hrs)	Load (ksi)	0.2% Y.S. (ksi)	U.T.S. (ksi)	Elong. (%)	R.A. (%)	Impact Toughness CVN, Ft-Lbs: Actual	Impact Toughness CVN, Ft-Lbs: Est. from Hounsfield	Hounsfield (Ft-Lbs)	Austenite (%)
Heat I	NO EXPOSURE			187.2	190.4	15	58	39	--	16.5	0
	650	5,000	39	205.1	208.1	13	58	--	33	13.3	0
	650	5,000	0	205.1	208.1	13	57	--	36	14.4	0
	650	10,000	39	205.6	208.3	15	56	--	35	14.1	6.7
	650	10,000	0	208.3	209.7	14	56	--	31	12.4	6.3
	850	5,000	34	172.3	192.1	16	61	--	41	16.4	NA
	850	5,000	0	174.3	192.1	16	61	--	43	17.1	NA
	850	10,000	34	170.2	189.9	16	59	--	43	17.1	19.0
	850	10,000	0	171.0	189.1	16	60	--	40	16.0	19.2

sile tests at 900 deg F (Table 4). The creep data for heats III and IV are tabulated as stress-time relationships for given strains in Table 5.

When the creep properties of the welds are compared with those of heat III, the MIG weld is slightly stronger than the plate while the strength of the electroslag weld is somewhat less (Figs. 4 and 5). The slopes of the curves are approximately the same for welds and plate.

Results of Exposures at Elevated Temperatures. After 5000 hours at 650 deg F, the room temperature tensile strength of heat I increases by approximately 20 ksi and the impact toughness decreases (Table 6). These properties change very little with further exposure to 10,000 hours, although the austenite content of the steel increases to approximately 6 percent.

The toughness and tensile strength are comparatively unchanged by exposure at 850 deg F, but the yield strength decreases approximately 15 ksi. After 10,000 hours at 850 deg F the steel contains 19 percent austenite. For comparison, after 10,000 hours at 900 deg F, the unbroken creep samples of heats III and IV contain 23 and 21 percent austenite, respectively.

The reaction of welds to prolonged exposure at elevated temperature is similar to the plate behavior, but the changes in properties, at least for the MIG and TIG welds, are more marked. The tensile strengths of the welds increase substantially after 5000 hours at 650 deg F and the toughnesses decrease (Table 7).

After the 850 deg F exposure the increase in the strengths of the welds is relatively small while the toughnesses are approximately half their original values. The microstructure has altered and austenite is visible (Fig. 6).

Effect of Exposure in Hydrogen at Elevated Temperature. After 500 hours at 750 deg F in a pressurized hydrogen atmosphere, the tensile strength of longitudinal specimens from heat II increases nearly 30 ksi and the impact toughness decreases approximately 30 ft-lbs (Table 8). With exposure at 850 deg F, the strength again increases after 500 hours but after 1000 hours the strength starts to decrease as overaging takes place. The decrease in strength is accompanied by an increase in toughness.

Discussion

When the elevated temperature tensile strengths of plates and welds are calculated as a percentage of their respective room temperature strengths, the results for all the samples are very

Table 7 The room temperature strength and toughness of the welds after 5000 hr at 650 deg F and 850 deg F (1)

Sample	0.2% Y.S. (ksi)	U.T.S. (ksi)	Elong. (%)	R.A. (%)	Impact Toughness (CVN) (Ft-Lbs)
	BEFORE EXPOSURE				
TIG	172.3	177.0	14	65	97
MIG (LDR)	171.1	176.9	13	49	50
MIG (HDR)	170.4	176.8	10	41	44
Electroslag	156.4	168.2	13	46	29
	AFTER 5000 HOURS AT 650°F				
TIG	210.4	216.4	12	53	23
MIG (LDR)	220.3	225.4	4	17	15
MIG (HDR)	221.0	224.0	6	24	10
Electroslag	179.4	190.4	9	46	17
	AFTER 5000 HOURS AT 850°F				
TIG	174.0	193.0	14	57	45
MIG (LDR)	175.7	193.8	7	15	27
MIG (HDR)	180.7	195.8	6	15	24
Electroslag	159.6	176.7	10	40	18

(1) All samples failed in the weld metal.

Table 8 Effect of exposure at elevated temperature in a 1000 psi hydrogen environment

Sample	Temp. °F	Time Hrs.	0.2% Y.S. (ksi)	U.T.S. (ksi)	Elong. (%)	R.A. (%)	Impact Toughness (CVN) (Ft-Lbs)
Heat II	NO EXPOSURE		185.4	189.0	14	65	65
	750	500	214.0	221.0	10	47	37
	850	500	225.2	229.4	11	47	28
	850	1000	208.3	222.0	12	47	32

NOTE: These samples were tested parallel to the final rolling direction.

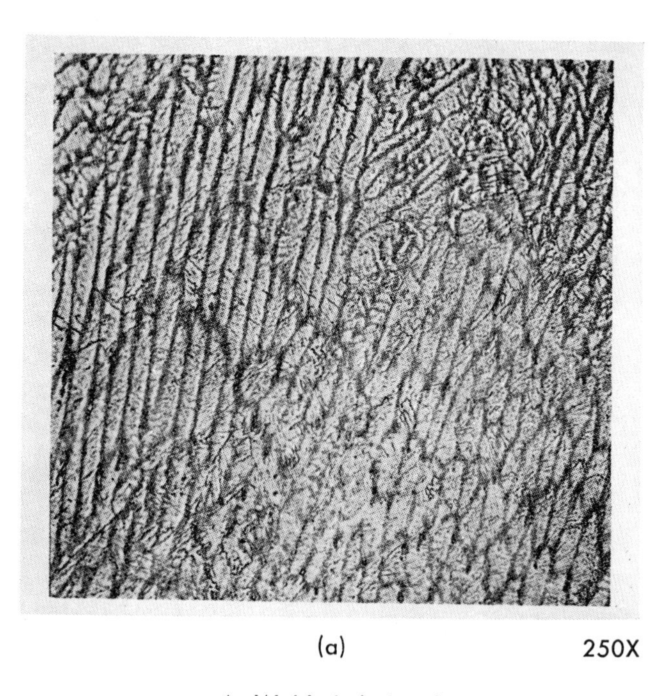

(a) 250X

As-Welded + Aged

(b) 250X

After 5000 Hours at 850°F

Fig. 6 Structure of low deposition rate MIG weld

similar (Table 9). The values compare well with those shown in Fig. 7 for a variety of 18 percent Ni steels [6].

Welds and plates with similar room temperature strengths also have similar creep rupture strengths, at least at 900 deg F. The stress to cause a minimum creep rate of 0.01 percent/1000 hours is approximately 30 ksi for heat III and 35 ksi for heat IV. These values are very close to $^1/_4$ of the 900 deg F tensile strengths of the two heats.

The fact that the plots of stress vs. rupture time are straight lines indicates that no damaging structural change occurred in the time period studied. Since there was no large decrease in rupture ductility with time, embrittlement does not seem to be occurring.

The increase in strength after exposure at 650 F and 850 deg F is caused by age-hardening of the steel. This reaction was not affected by the application of stresses equal to one quarter of the tensile strength (Table 6). Higher stresses are said to enhance the age-hardening [3]. The results in Table 10 also give an indication of this. The increase in room temperature strength after exposure for 5060 hours at 650 deg F under a stress equal to the yield stress was greater than would be expected for this steel exposed without stress.

Table 10 Effect of exposure at elevated temperature under a stress equal to the yield stress

COMPOSITION OF STEEL

Ni	Cr	Mo	Al	Ti	C	Si	Mn	S	P
11.9	5.0	3.0	.31	.12	.024	.09	.04	.003	.007

ORIGINAL HEAT TREATMENT

1500°F for 1 hr, air cool + 900°F for 3 hrs, air cool

EXPOSURE CONDITIONS

5060 hours at 650°F under a stress of 134,000 psi

EFFECT OF EXPOSURE

	0.2% Y.S. (ksi)	U.T.S. (ksi)	Elong. (%)	R.A. (%)
Before Exposure	189.1	192.2	15.0	62
After Exposure	224.7	226.7	16.5	69

Table 9 Elevated temperature tensile strengths of plates and welds expressed as a percentage of their tensile strengths at room temperature

Sample	650°F	750°F	850°F	900°F
Heat I	82	81	72	--
Heat III	--	--	--	69
Heat IV	--	--	--	73
TIG Weld	84	--	75	--
MIG (LDR) Weld	83	--	75	--
MIG (HDR) Weld	82	--	73	--
Electroslag Weld	84	--	75	--

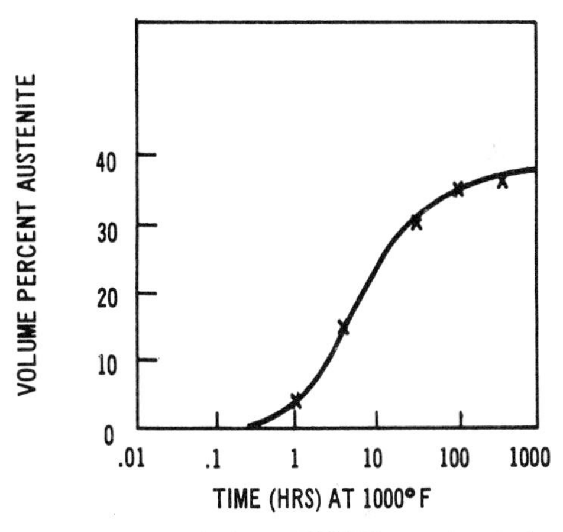

Fig. 8 Formation of austenite in an 18Ni 250 maraging steel exposed a 1000 deg F (reference [7])

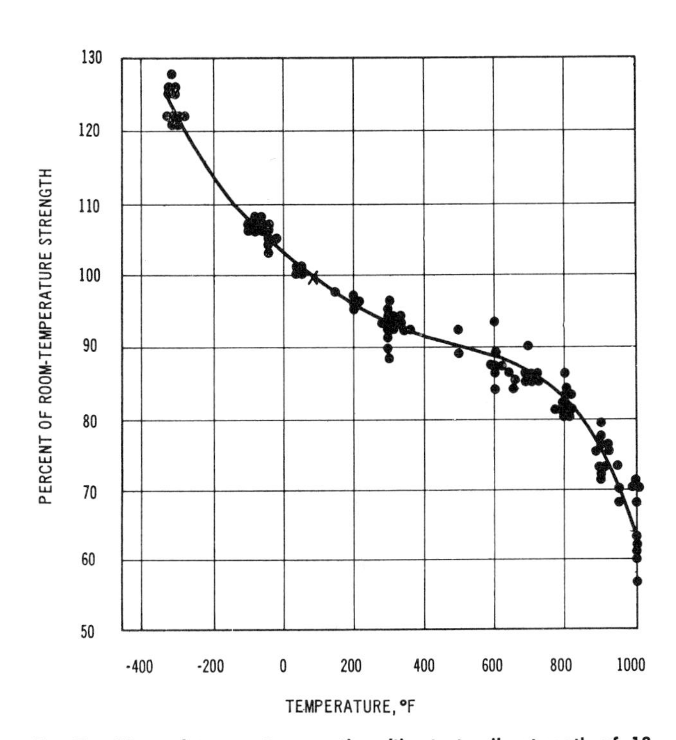

Fig. 7 Effect of temperature on the ultimate tensile strength of 18 percent nickel maraging steels aged at 900 deg F for 3 hr

The aging response of heat I can be seen from the results in Table 6. Strength increased with exposure for 5000 hours at 650 deg F, but there was no further change with exposure to 10,000 hours. The reactions that occurred in the period between 5000 and 10,000 hours are open to interpretation since the shapes of the aging curves are not known exactly. A balance might have been established between the aging and overaging tendencies of the steel, with the result that there was little change in strength with time. Or, more likely perhaps, the strength could have reached a maximum between 5000 and 10,000 hours, and by 10,000 hours had started to decline.

At 850 deg F, a maximum strength occurred in a relatively short time. The results for heat II for example indicate that the maximum occurred between 500 and 1000 hours (Table 8). Nevertheless, after 10,000 hours at 850 deg F, the tensile strength of heat I is almost exactly the same as that of the unexposed material. This is in spite of the fact that the yield strength had decreased and the steel contained 19 percent austenite.

The overaging of a maraging steel occurs through a combination of overaging and the formation of a soft austenite that is stable at room temperature [7]. The relative importance of these two factors depends on the aging or exposure temperature. At the higher temperatures the formation of reverted austenite assumes greater importance. Evidence of this is the fact that there is three times as much austenite in heat I after 10,000 hours at 850 deg F than after 10,000 hours at 650 deg F.

At the higher temperatures at which enough austenite can form to influence the properties, the rate of formation of austenite obviously becomes important. Fig. 8 shows the increase in austenite content with time for an 18 percent Ni steel exposed at 1000 deg F [7]. The amounts of austenite do not apply to this discussion, they are for a higher nickel steel exposed at a higher temperature; but the shape of the curve is relevant. After an incubation period, the formation of austenite is comparatively rapid and then lessens as it approaches equilibrium. The room temperature strengths of heat I specimens exposed for 10,000 hours at 850 deg F were only very slightly different from those of samples exposed for 5000 hours. Further changes taking place with exposure beyond 10,000 hours might also be expected to be small.

Although the discussion so far has concentrated on the plates examined, the same trends also apply to the welds. After 5000 hours at 650 deg F, the weld strengths are probably close to their maximum, while after 5000 hours at 850 deg F the maximum strength has been passed and the results fall on the overaged portion of the curve.

As expected, the decrease in toughness observed after some of the exposures was accompanied by an increase in strength. In those cases where the strength after exposure was similar to the original strength, the toughness was also unchanged (Table 6).

These effects of the elevated temperature exposures were not altered by the presence of a hydrogen atmosphere. Maraging steels would be expected to be resistant to high temperature hydrogen attack because of their very low carbon contents and the presence of carbide stabilizers such as chromium, molybdenum, and titanium. They have shown themselves to be quite resistant to low temperature hydrogen embrittlement [8].

Summary

In quenched and tempered alloy steels, increases in tensile strength often result in poorer fabricability and weldability; for example, as the strength increases the alloys frequently become more susceptible to weldment cold cracking. The age-hardening maraging steels offer an effective way to increase strength while retaining good fabricability, which makes them attractive as ultra-high-strength steels for use at normal temperatures. On the basis of the results shown here, maraging steels offer similar promise as high-strength steels for service at elevated temperatures.

Conclusions

1 Maraging steel plates and weldments exhibit high strengths at elevated temperatures. Tensile strengths at 850 deg F are 75 percent of the room temperature strengths.

2 Plates and weldments have similar creep strengths. At 900 deg F, the stresses to produce a minimum creep rate of 0.01 percent/1000 hours are approximately equal to one quarter of the tensile strengths.

3 During exposures at elevated temperatures, the steels first age-harden then overage. Austenite forms on prolonged exposure and the amount increases with increasing temperature. However, the room temperature tensile strengths remain unchanged after exposure for 10,000 hours at 850 deg F in spite of the formation of 19 percent austenite.

4 The 12Ni-5Cr-3Mo maraging steel is insensitive to high temperature hydrogen attack. Elevated temperature exposure in a pressurized hydrogen atmosphere has the same effect as exposure in air.

References

1 Floreen, S., and Decker, R. F., "Maraging Steel for 1000° F Service," *ASM Trans. Quarterly*, Vol. 56, 1963, pp. 403–411.

2 Hoenie, A. F., "Determination of Mechanical Property Design Values for 18Ni-Co-Mo 250 and 300 Grade Maraging Steels," AFML-TR-65-197, North American Aviation, July 1965.

3 Martin, G., "Maraging Steels Provide Ultra-High-Strength," *Materials in Design Engineering*, Dec. 1964, pp. 104–107.

4 Sadowski, E. P., "12%Ni Maraging Steel," *Metals Engineering Quarterly*, Vol. 5, Feb. 1965, pp. 56–64.

5 Lang, F. H., "Welding of 12%Ni Maraging Steel," *Welding Journal*, Vol. 47, 1968, pp. 25-s to 34-s.

6 "The Mechanical Properties of the 18 Percent Nickel Maraging Steels," DMIC Report 198, 1964.

7 Peters, D. T., "A Study of Austenite Reversion During Aging of Maraging Steels," *Trans. ASM*, Vol. 61, 1968, pp. 62–73.

8 Gray, H. R., and Troiano, A. R., "How Hydrogen Affects Maraging Steel," *Metal Progress*, Vol. 85, 1964, pp. 75–78.

SECTION V:
Stress-Corrosion Cracking Behavior

Stress Corrosion Cracking in 18% Ni (250) Maraging Steel*

B. C. SYRETT*

Abstract

The susceptibility of 18% Ni (250) maraging steel to stress corrosion cracking in 3.5% NaCl solution has been investigated. Metallographic and fractographic examinations show distinct differences between specimens broken under free corrosion conditions and those broken under hydrogen charging conditions. It is proposed that stress corrosion cracking occurs by anodic path dissolution at potentials more noble than about -600 mV (SCE), that hydrogen embrittlement cracking occurs at potentials more active than about -800 mV, and that a mixture of anodic path dissolution and hydrogen embrittlement cracking occurs in the intervening potential range of -600 mV to -800 mV. Stress corrosion cracking in freely corroding maraging steel (potential $\approx$ -580 mV) is thought to occur by anodic path dissolution.

It is now well established that high strength steels under tensile stress are susceptible to catastrophic cracking failures in aqueous solutions under a wide variety of conditions.

The susceptibility of maraging steels to this environmentally induced cracking is dependent on such parameters as pH[1,2] and NaCl content[3] of the corrodent, electrode potential[4,5] and testing temperature.[6] Although there has been considerable interest in these steels in recent years because of their possible application in the fabrication of rocket casings and specialized marine vessels, the cracking mechanism operative under free corrosion conditions is still debated. In approximately neutral NaCl solutions, for instance, some workers[6] suggest that stress corrosion cracking (SCC) occurs by anodic path dissolution while others[7,8] propose a hydrogen embrittlement cracking mechanism.

The aim of the present work is to study the SCC of 18% Ni (250 grade) maraging steel in 3.5% NaCl solution and to characterize the cracking mechanisms operating at various electrode potentials. This steel is one of several which have been evaluated recently[9] in our laboratory for possible application in advanced marine vessels; it has already been used in the foils system of the Canadian FHE 400 hydrofoil craft.

Experimental Procedure

Cantilever Test

The susceptibility of 18% Ni maraging steel to SCC was determined using a cantilever testing rig similar to that used by Brown and Beachem.[10] This rig was adapted so that either constant-load or rising-load tests could be performed.

*Submitted for publication June, 1970.

*Corrosion Section, Physical Metallurgy Div., Dept. of Energy, Mines and Resources, Ottawa, Canada. Present address: International Nickel Co., Paul D. Merica Research Laboratory, Sterling Forest, Suffern, N. Y.

A general description, including the heat treatment, of the 18% Ni (250 grade) maraging steels used in this work is given in Table 1, and the chemical compositions are shown in Table 2. The standard cantilever test specimens were both top and side notched (Figure 1), but a few were prepared without side notches. Specimens were machined in the solution annealed condition, heat treated, then precracked in the top notch by fatiguing, using a Krouse Reverse Bend Plate Fatigue Testing Machine operating at a rate of 1725 cycles/min. The peak tensile stress was adjusted so that a crack about 0.020-inch deep was developed in 20,000 cycles. Occasionally, the depth of the fatigue precrack was as small as 0.013-inch or as large as 0.050-inch, but some preliminary work suggested that within this range, the precracking rate had little, if any, effect on SCC properties. The major axis of the specimen was in the rolling direction, and the depth (D) was that of the original rolled plate. The final depth, after machining, was constant for specimens taken from the same plate, but varied from plate to plate in the range 0.440-inch to 0.500-inch. The thickness (B) was always 0.375-inch. Unless otherwise stated, the thin heat treatment oxide was not removed from the specimen surface prior to SCC tests.

Whenever feasible, specimens of the same size (cut

TABLE 1 – General Description of the 18 Ni (250) Maraging Steels

Source (Plates 1, 2, and 3): Vanadium-Alloy Steel Co.

As-received condition (Plates 1, 2, and 3): Consumable arc vacuum melted 0.5-inch plate in solution annealed condition.

Final specimen thickness after machining: Plate 1, 0.440-inch; Plate 2, 0.500-inch; Plate 3, 0.482-inch.

Additional heat treatment (Plates 1, 2, and 3): 3 hr at 482 C (900 F), air cooled.

Note: Specimens machined from Plates 1 and 2 have top and side notches, while those from Plate 3 have top notches only.

TABLE 2 – Chemical Composition[1] of the 18 Ni (250) Maraging Steels, Weight Percent

Plate No.	C	Si	Mn	S	P	Al	Ti	B	Mo	Co	Ni	Zr	Ca
1	0.017	0.04	0.04	0.008	0.004	0.06	0.38	0.003	4.58	7.50	18.78	0.01	0.005
2,3	0.018	0.04	0.04	0.008	0.004	0.06	0.40	0.003	4.46	7.92	19.22	0.01	0.003

(1)Analysis at the Department of Energy, Mines and Resources.

from the same plate) were used to study the effect of a given variable on SCC, but where this was not possible, care was taken to differentiate the results.

During SCC tests, the precracked region of the specimen was surrounded by 250 ml of 3.5% NaCl solution (pH 6.0 to 6.5) contained in a plastic vessel. Build-up of corrosion products was minimized by flushing the vessel continuously with fresh salt solution at a rate of no less than 250 ml/h. The electrochemical potential of freely corroding specimens could be recorded continuously, and, when desired, the potential could also be potentiostatically controlled to within ±1 mV. All potentials quoted in this work are with reference to the saturated calomel electrode (SCE).

In constant load tests, and to a more limited extent in rising load tests, the progress of SCC could be monitored by means of a linear variable displacement transducer positioned to measure the deflection of the cantilever beam as cracking proceeded; the output from the transducer was fed into a potentiometric chart recorder.

Stress Intensity Factor

The opening mode stress intensity factor, K_I, applied to a cantilever test specimen is a function not only of the applied load, but also of the crack depth. In a constant load SCC test, the load remains constant but the crack length, and thus K_I, increases as SCC takes place, until the critical stress intensity for terminal fracture is reached; the magnitude of the initial stress intensity factor, termed K_{Ii} by Brown and Beachem,[10] depends on the precrack depth (depth of notch plus fatigue crack), "a", which may be measured after fracture.

$$K_{Ii} = \left(\frac{B}{B_N}\right)^{1/2} \frac{\beta M}{BD^{3/2}} \qquad (1)$$

The units of K_{Ii} are ksi$\sqrt{}$inch; M is the bending moment on the notch (i.e., the arm length x load at the end of the arm) and β is a function of a/D,

$$\beta = 4.12\sqrt{\frac{1}{(1 - a/D)^3} - (1 - a/D)^3} \qquad (2)$$

All other variables are defined in Figure 1. The term

$$\left(\frac{B}{B_N}\right)^{1/2}$$

does not appear in the equation used by Brown and Beachem, and is the Freed-Krafft correction for the side notching.[11]

For a particular alloy in a selected heat treatment condition, a characteristic stress intensity, K_{ISCC}, has been defined by Brown,[12] above which SCC will definitely occur in a freely corroding specimen.

In rising load tests, a nominal stress intensity K^*_{Ii}, is computed from values of the initial crack depth ("a") and the load at fracture. The loading rate is calculated in terms of the increase in nominal stress intensity per minute; the loading vessel and cantilever beam alone are heavy enough to impart an initial stress intensity of about 25 ksi$\sqrt{}$inch to the cantilever specimen, so the loading rate is calculated by subtracting this initial stress intensity from K^*_{Ii} and dividing by the testing time.

Some rising load tests were conducted on dry specimens in order to obtain a base-line stress intensity value; because SCC does not occur during loading in these tests, K^*_{Ii} can be considered to approximate the fracture toughness, K_{Ic}. In fact, the dimensions of the specimens used in this work did not meet the requirements recommended by the ASTM Committee E-24[13] for a strictly valid plane-strain K_{Ic} determination. However, the error is

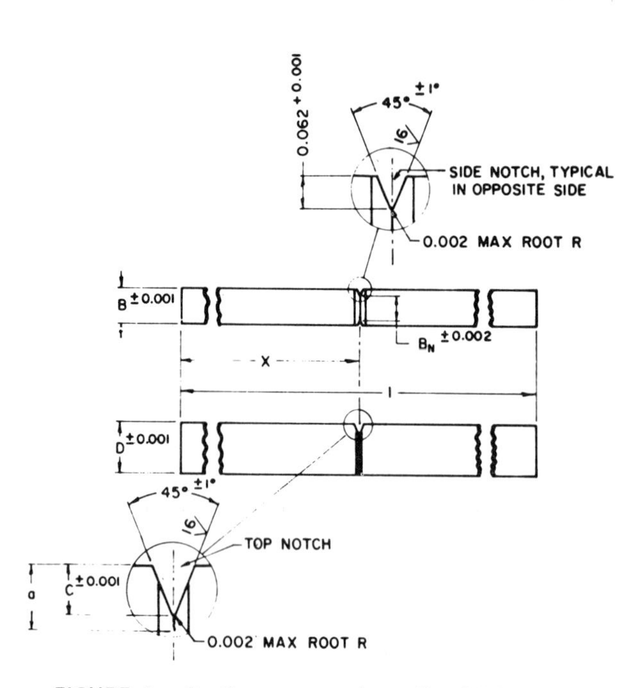

FIGURE 1 – Cantilever test specimen. X = 3.0-3.2 inch; l = 6.0-6.4 inch; D = 0.440-0.500 inch = specimen depth; B = 0.375 inch = specimen thickness; B_N = 0.250 inch = specimen thickness in plane of notches; C = 0.113 inch = depth of machined notch; and *a* = notch depth plus fatigue crack depth.

thought to be small, especially in standard specimens where the side notching would tend to induce plane-strain conditions. Therefore, the subscript I is retained here, if for no other reason than to signify that the crack grows in the tensile opening mode, mode I, as opposed to the sliding modes II and III.

Estimation of the pH of the Corrodent Within an Advancing Crack

The suggestion that the pH of the corrodent within an advancing crack is different from the pH of the bulk solution[7,8] prompted attempts to measure this pH directly, using a method similar to that described elsewhere.[14] The method involves the use of filter paper strips which have been impregnated with a pH indicating solution, and dried. A propagating stress corrosion crack is stopped short of failure, and the specimen is immersed in liquid nitrogen to freeze the liquid within the crack; the stress corrosion crack is then extended mechanically to failure to expose the frozen solution on the crack surfaces. As soon as the solution has thawed, the indicating paper is used to absorb the corrodent and the pH is estimated from the subsequent color indications.

This procedure was adopted in the present work but, in addition, a second method was employed which was considered easier and probably just as accurate: after removing the specimen from the salt solution, the sides were quickly wiped dry, so that only the corrodent within the crack remained. The indicating paper was then held against the edge of the crack to absorb part of the small volume of corrodent within, and the pH was estimated from the subsequent color change in the paper. The first method was considered no more effective than the second in detecting possible pH-gradients within the crack, because preliminary tests showed that, in order to absorb sufficient corrodent to give a valid color indication in the paper, it was necessary to take corrodent from regions close to, as well as at, the crack tip.

The wedge opening loading specimens used in these tests were cut from the broken ends of cantilever test specimens, such that the precrack lay in the rolling plane (Figure 2). The wedge, forced into the slot using a vise, was made from the same maraging steel, but to minimize any complicating side reactions, the wedge was kept above the waterline throughout the test period.

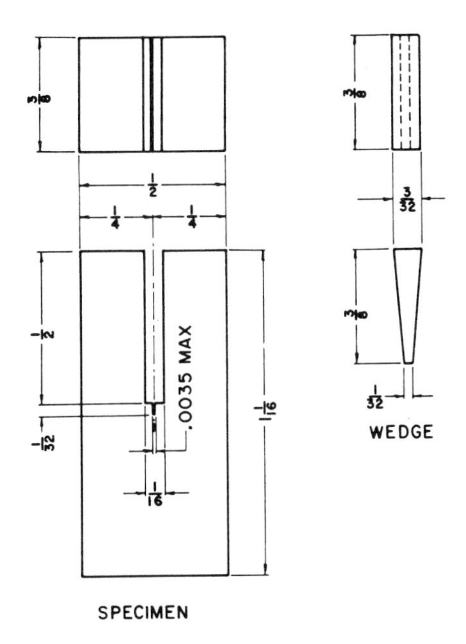

FIGURE 2 — Wedge opening-loading specimen.

Results

Biefer and Garrison[15] have shown that when cantilever specimens of 18% Ni (250) maraging steel are polarized to zinc potentials in 3.5% NaCl solution, the values of K^*_{Ii} obtained are dependent upon the loading rate chosen. They demonstrated that at some loading rates, the computed value of K^*_{Ii} could give a misleading measure of SCC susceptibility. Thus, it seemed prudent to investigate the possibility of similar trends in specimens tested under more noble conditions; in this way, a suitable loading rate could then be selected for subsequent rising load tests in which electrochemical potential was to be the variable. Free corrosion conditions were chosen for this study, because of the practical significance, and because of the experimental simplicity. The maraging steel specimens were cut from Plates 1 and 2. As a basis for comparison, some of these specimens were broken in air at a rapid loading rate (>5000 psi√inch/min): duplicate tests gave K_{Ic} values of 69.9 ±0.5 ksi√inch (Plate 1) and 87.7 ±1.9 ksi√inch (Plate 2). It is not clear why two maraging steel plates of nominally the same composition and heat treatment should have such largely different K_{Ic} values, but other workers have reported this sort of variation in mechanical properties in different batches of 18% Ni maraging steel (for example, see Reference 4). In these specimens tested in air, the presence of side notches effectively suppressed "shear lip" formation, resulting in flat fractures in the plane of the notches.

So that all the results might be compared directly, the loading rate has been plotted as a function of K^*_{Ii}/K_{Ic} in Figure 3. The points are seen to fall on two quite distinct curves: one shows a continuous decrease in K^*_{Ii}/K_{Ic} with decreasing loading rate, and the other, perhaps surprisingly, shows an initial increase in K^*_{Ii}/K_{Ic} to values greater than unity, followed by a fairly rapid decrease at lower loading rates. At some loading rate between 5 and 10 psi√inch/min (time to failure >80 hr), the two curves appear to meet.

An examination of the various specimens showed that there were two possible types of SCC, one type being associated with each curve. The first type occurred in the plane of the precrack and notches: specimens cracking in this manner are represented by the continuously decreasing K^*_{Ii}/K_{Ic} curve. The second type of SCC occurred along well-defined planes inclined at about 60° to, and on either side of, the precrack. This path of cracking has been termed "Type 2" by Carter,[16] and cracking in the plane of the precrack has been termed "Type 1"; this terminology is adopted in the present paper.

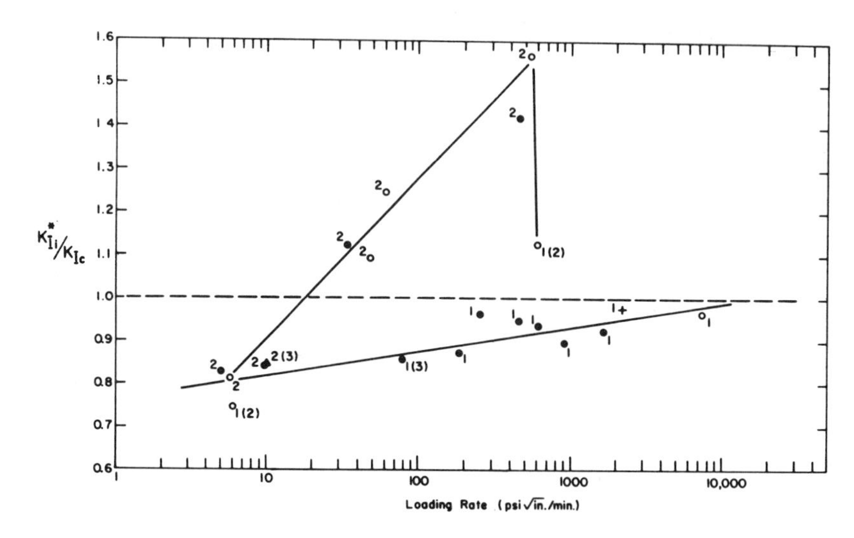

FIGURE 3 – Effect of loading rate on K^*_{Ii}; free corrosion conditions. o Plate 1: K_{Ic} = 69.9 ±0.5 ksi√inch. + Plate 1: free corrosion for 100 hours at zero load, followed by drying and fracture in air. • Plate 2: K_{Ic} = 87.7 ± 1.9 ksi√inch. ▲ Plate 2: heat treatment oxide removed prior to testing. 1, 2, 3–Respectively, Types 1, 2, and 3 cracking over most of the crack front. Brackets, around a number signify this type of cracking over a small section of the crack front.

Metallographic examination shows that at loading rates of about 10 psi√inch/min or below, all specimens suffer at least some Type 2 cracking (Figure 4). When Type 1 and Type 2 cracking co-existed, they usually did so at different points along the precrack front; there was seldom evidence of both types of cracking co-existing in the same vertical longitudinal section. Occasionally, a third crack type, "Type 3", was encountered, which ran for short distances at 90° to the precrack plane. After extensive metallography, it was clear that Type 3 cracking was usually of a purely mechanical nature, and followed stringers of inclusions and carbonitrides in the rolling plane. These inherently weak planes were also responsible for Type 3 cracking in some specimens tested in air. Very occasionally, Type 3 cracks were observed to follow prior austenite grain boundaries. The crack type leading to failure is marked next to each point in Figure 3.

The cantilever beam deflection was recorded as a function of time during some rising load tests on freely corroding specimens and also during some tests performed in air. During the latter, the deflection was a linear function of time up to fracture but in the SCC tests, slow crack growth was reflected in a departure from linearity and an increase in the rate of beam deflection. At rapid loading rates, it was difficult to state with certainty when slow crack growth commenced, but in tests at lower loading rates, there was a sufficient departure from the linear trace to show that crack growth had clearly occurred long before final fracture (Figure 5). It was apparent that even at the slowest loading rates, SCC took place at stress intensities significantly lower than K^*_{Ii}. Therefore, unless the loading rates were so low as to be impracticable, K^*_{Ii} would not be expected to have a value as low as K_{ISCC} at free corrosion potentials.

Figure 5 also shows the free corrosion potential vs time curve for the same specimen. This curve is typical of all rising load and constant load test specimens, and is, in fact, similar, except at the point of fracture, to the curves obtained for unstressed samples. At fracture, a negative

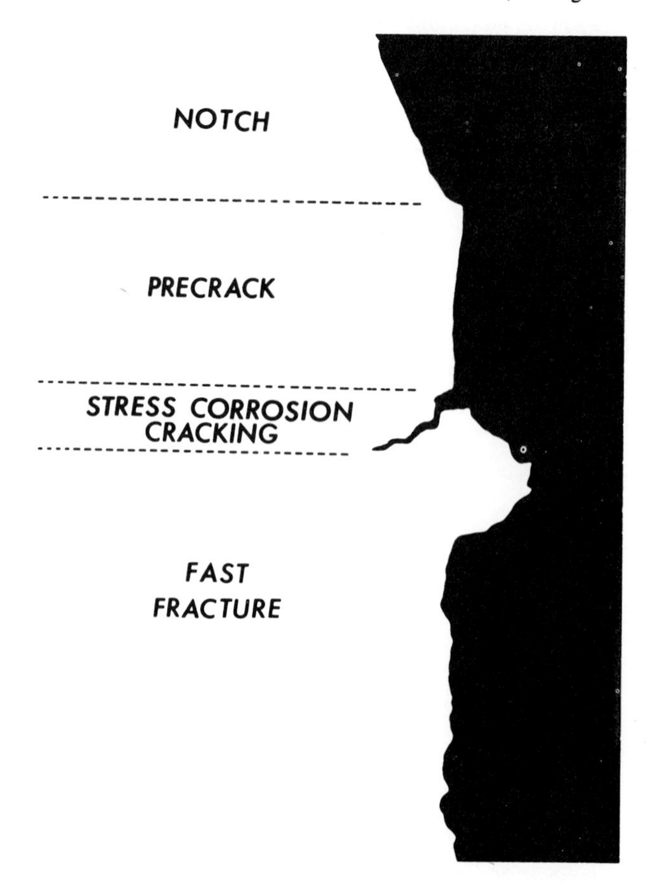

FIGURE 4 – Type 2 cracking in a freely corroding specimen loaded at 9.9 psi√inch/min.

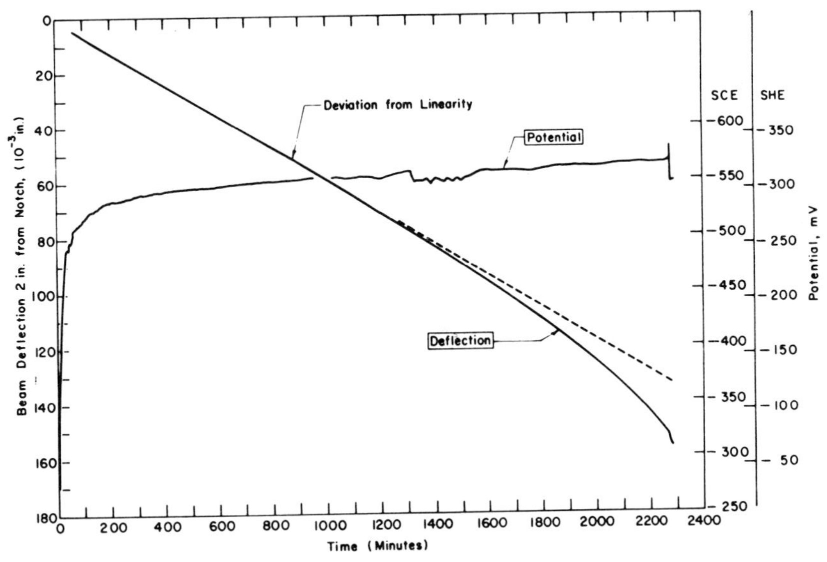

FIGURE 5 — Cantilever beam deflection and potential recordings for a freely corroding specimen loaded at 34.2 psi√inch/min.

(more active) jump in potential was invariably recorded, sometimes as high as 40 mV, and this was followed by a rapid rise to more noble potentials. The prefracture value was always regained within two minutes, but after this characteristic fall and rise in potential, the behavior was less reproducible.

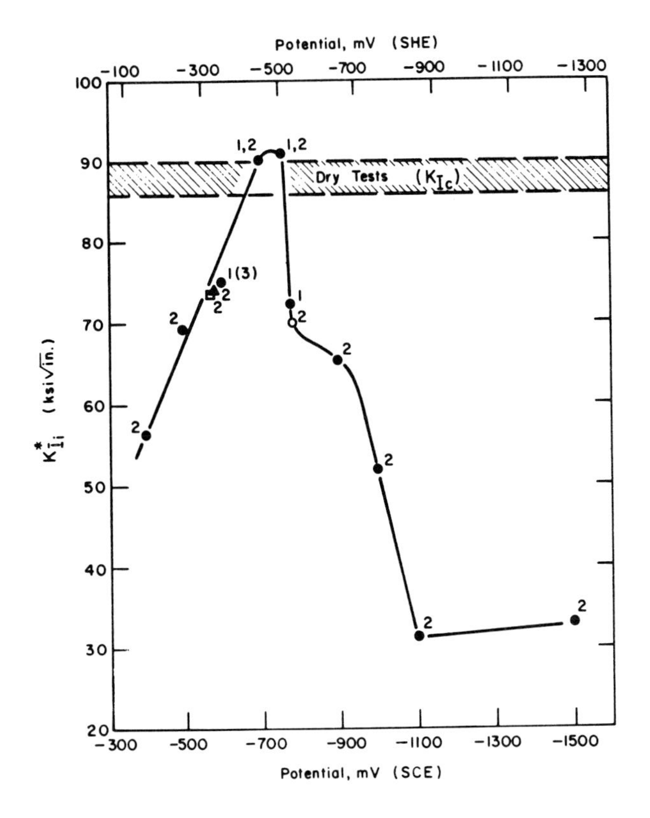

FIGURE 6 — Effect of potential on the K^*_{Ii} values. • Potential controlled by a potentiostat. o Specimen coupled to aluminum alloy 5083. □ Free corrosion. △ Free corrosion, heat treatment oxide removed prior to testing. 1, 2, 3—Respectively Types 1, 2, and 3 cracking over most of the crack front. Brackets around a number signify this type of cracking over a small section of the crack front.

In order to establish whether the heat treatment oxide on the specimen surface was having any effect on the cracking process, two specimens were tested at a loading rate of 10 psi√inch/min, one with the oxide removed with 2/0-grade emery paper, and the other in the usual oxide-coated condition. The potential-time curves were shown to be almost identical, the maximum difference between the two curves being 17 mV; and the K^*_{Ii} values differed by only 0.45 ksi√inch (about 0.6%). It was considered reasonable to assume, therefore, that the heat treatment oxide played little, if any, part in the cracking process.

One unstressed specimen was allowed to corrode freely in the 3.5% NaCl solution for 100 hours, before it was dried and broken in air. The K^*_{Ii} value calculated for this specimen was only a little below the mean K_{Ic} value (Figure 3), so the prior free corrosion probably had no effect on the mechanical properties.

Effect of Potential on K^*_{Ii}

It has been shown that K^*_{Ii} values are fairly reproducible when freely corroding 18% Ni maraging steel is loaded at rates below 10 psi√inch/min; furthermore, previous work[15] has shown that when this steel is polarized to -1060 mV, K^*_{Ii} values are relatively insensitive to loading rate, at rates below 10 psi√inch/min. Consequently, it was considered reasonable to study the effect of electrochemical potential on K^*_{Ii} at a loading rate of 9.2 (±1.9) psi√inch/min.

Specimens were cut from Plate 2 for this investigation, and the results are shown in Figure 6. Although evidence is later presented that at a potential of about -1100 mV, $K^*_{Ii} \approx K_{ISCC}$, in general this is not the case. In Figure 6, a high K^*_{Ii} value may indicate a high K_{ISCC} value, a low SCC rate, a long induction period, or a combination of the three; this rising load test, therefore, provides a convenient method of quickly assessing the resistance of a metal to SCC under a variety of testing conditions, without having to stipulate the controlling parameter.

The potential of most specimens was controlled potentiostatically, but one was coupled to an aluminum alloy 5083 "sacrificial anode", which polarized the steel to an average potential of -782 mV (+11, -19 mV), and two specimens were allowed to corrode freely without any potential control. Although a comparison of freely corroding specimens with those under externally imposed potentials is not strictly valid, these results have been included in Figure 6 to facilitate a discussion of the SCC mechanism operative in freely corroding maraging steel. The freely corroding specimens, also discussed in the last section, were similar except that the heat treatment scale on one had been removed with 2/0 grade emery paper. In both cases, the free corrosion potential dropped fairly rapidly to more active values during the first 10 hours, but during the subsequent 73 hours before failure, it changed by less than 40 mV. For the purposes of Figure 6, therefore, the average potential in the 10 to 83-hour period was computed for each specimen.

The maximum resistance to SCC occurred at a potential of -750 mV; at this potential, the K^*_{Ii} value lay in the range of K_{Ic} values for this plate. However, despite the high K^*_{Ii} value, metallographic examination showed that the specimen had suffered some SCC.

At potentials more noble than -600 mV and more active than -800 mV, Type 2 cracking was dominant, but in the intermediate range of -600 mV to -800 mV, both Types 1 and 2 cracking occurred. The crack type leading to failure is marked next to each point in Figure 6.

Specimens which were polarized to potentials of -400 mV and -500 mV rapidly developed a black oxide film.

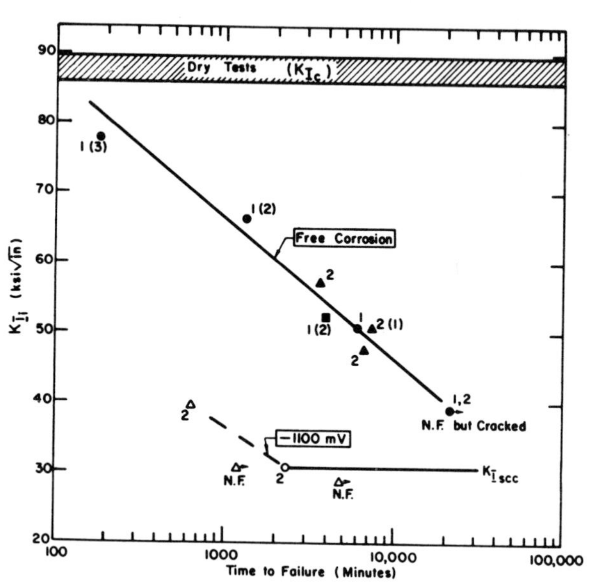

FIGURE 7 — Results of constant load tests, under (a) free corrosion conditions; (b) hydrogen charging conditions (polarized to -1100 mV). ●■▲ Free corrosion. ○△ Polarized to -1100 mV. ●○ Standard specimens (Plate 2) in 3.5% NaCl solution. ▲△ Specimens without side notches (Plate 3) in 3.5% NaCl solution. ■ Specimen without side notches (Plate 3) in 3.5% NaCl solution containing 4 mg/liter $NaAsO_2$. N.F. No failure during time indicated. 1, 2, 3—Respectively Types 1, 2, and 3 cracking over most of the crack front. Brackets around a number signify this type of cracking over a small section of the crack front.

Electron probe analysis of this film indicated that iron, nickel, and cobalt were present in much lower quantities than in the parent metal. It was deduced that these elements dissolve preferentially in the 3.5% NaCl solution, leaving behind an oxide seemingly enriched in molybdenum, titanium, silicon and manganese.

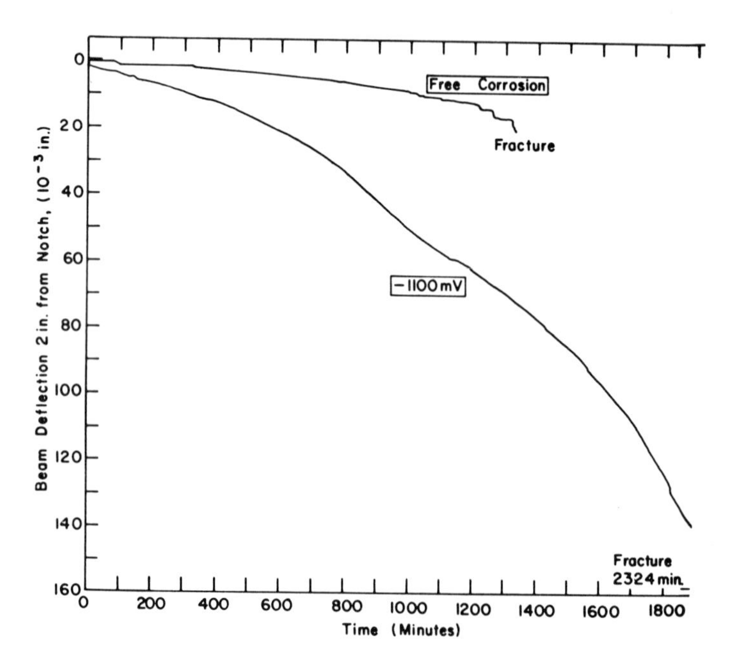

FIGURE 8 — Beam deflection vs time curves for specimens tested under (a) free corrosion conditions (K_{Ii} = 66.3 ksi√inch), and (b) hydrogen charging conditions (-1100 mV; K_{Ii} = 30.9 ksi√inch).

Constant Load Tests

Specimens for this series of constant load tests were cut from Plates 2 and 3: it will be recalled that the standard side notches were absent in specimens taken from Plate 3 and that the final thickness was fractionally smaller than in those specimens taken from Plate 2. However, Plates 2 and 3 were originally both part of a larger single plate, and can therefore, be expected to exhibit identical properties.

Most of these specimens were allowed to corrode freely in the salt solution under the influence of a constant load. Figure 7 shows how the time-to-failure was very dependent on the initial stress intensity, K_{Ii}, but it also illustrates the independence of the results to side notching. A few specimens were also tested under hydrogen charging conditions at -1100 mV; although results are sparse, there is little doubt that K_{ISCC} is about 31 ksi$\sqrt{}$inch at this potential. It is significant that in the rising load tests, reported in the last section, the specimen polarized to -1100 mV fractured at a K^*_{Ii} value of 31.4 ksi$\sqrt{}$inch; the similarity between this value and the suggested K_{ISCC} would seem to be evidence of a very rapid crack propagation rate at stress intensities above K_{ISCC}. On comparing the curves in Figure 7, it might appear that the SCC propagation rate at -1100 mV is from 30 to 35 times the rate at free corrosion potentials. In fact, the cantilever beam deflection vs time curves recorded during some of these constant load tests, provide evidence that the difference between these two SCC rates is not quite so marked. It was found that, under free corrosion conditions, there was an apparent incubation period when little or no deflection was recorded: this period ranged from 0.6-hour at high K_{Ii} values to a few days at low K_{Ii} values. Specimens polarized to -1100 mV showed no incubation period, even when K_{Ii} was only slightly greater than K_{ISCC}, and slow crack growth commenced immediately. The beam deflection vs time curves in Figure 8 reflect the progress of SCC both in a specimen polarized to -1100 mV and in a freely corroding specimen. Even though K_{Ii} for the latter specimen was over twice that for the hydrogen charged specimen, the rate of SCC is seen to be lower under free corrosion conditions; also the absence of an incubation period at -1100 mV contrasts the 90-minute period of constant beam deflection for the free corrosion specimen. The small jumps in beam deflection, recorded for the freely corroding specimen, were of some interest: careful observation of the edges of the advancing crack made it clear that the jumps were associated with the rapid ductile tearing of "islands" of metal which remained after the main crack front had undermined these areas.

When one of the freely corroding specimens under constant load (K_{Ii} = 39.2 ksi$\sqrt{}$inch) showed little sign of cracking after 15 days of exposure, the test was ended. Subsequent metallographic examination, however, showed that the specimen had been cracking and, given time, would probably have broken; it can be deduced, therefore, that the value of K_{ISCC} lies below 40 ksi$\sqrt{}$inch.

The suggestion[7,8] that hydrogen embrittlement cracking occurs in the high strength steels even under free corrosion conditions, prompted a constant load test in which the regular supply of salt solution was replaced by 3.5% NaCl solution containing 4 mg $NaAsO_2$ per liter (pH = 6.5). It is well established that arsenite or arsenate additions to an aqueous solution act as a "cathodic poison" for the combination of hydrogen atoms to molecular hydrogen. The cathodic reaction includes a discharge step, in which hydrogen ions are reduced to hydrogen atoms and adsorbed on the steel surface, and a combination step, in which two hydrogen atoms combine to form a molecule of hydrogen. However, before the atoms of hydrogen combine to form molecules, some are absorbed by the metal. Since the arsenite addition hinders the combination reaction, the ultimate result is an increase in both the quantity of atomic hydrogen on the metal surface, and the quantity of hydrogen entering the metal. If a hydrogen embrittlement cracking mechanism is operating under free corrosion conditions, therefore, the addition of sodium arsenite ($NaAsO_2$) to the corrodent can be expected to promote cracking and reduce the time to failure. In fact, the arsenite addition had little, if any, effect on the cracking rate (Figure 7).

The crack type leading to failure is marked next to each point in Figure 7. Both Type 1 and Type 2 cracking occurred in the freely corroding steel, but when they co-existed, they usually did so at different points along the precrack front. Occasionally, however, both types of cracking co-existed in the same vertical longitudinal section, Type 1 cracking initiating at the precrack tip, and propagating some distance before Type 2 cracking commenced.

Fine Structure of the Stress Corrosion Cracks

Optical metallography showed that all specimens cracked partly along prior austenite grain boundaries and partly across these grains. The relative proportion of each varied a little, depending on such variables as polarizing potential and applied load (or loading rate), but generally, they were present in about equal proportions. In specimens failing by Type 2 SCC, the proportion of intergranular attack was observed to be slightly higher than in specimens cracking in the precrack plane. This behavior was probably promoted because the grain structure was elongated in the rolling direction (perpendicular to the precrack plane); during Type 2 SCC, there would be greater opportunity for grain boundary attack than during Type 1 cracking.

Although the major crack path was the same for specimens cracking under free corrosion or hydrogen charging conditions, the micrographs show obvious differences. In Figure 9, which is a section through a specimen broken at -1100 mV, there is evidence of some transgranular cleavage, but, more significantly, there is no sign of any attack on the crack sides. In the freely corroding and anodically polarized specimens, on the other hand, the major cracks were characterized in many instances by a side attack (Figure 10). These hairline cracks appeared only on the sides of main cracks which had limited access to the bulk corrodent. Once a crack had "yawned", side attack of this type ceased and, when general corrosion rates were high as they were at -500 mV, previously formed hairline cracks tended to be

FIGURE 9 – Stress corrosion cracking in 18% Ni maraging steel, polarized to -1100 mV (SCE). 500X

FIGURE 10 – Stress corrosion cracking in 18% Ni maraging steel, polarized to -500 mV (SCE). 500X

obliterated by general corrosion. In no instance was this hairline attack observed to occur on the outer machined surfaces of the specimen, nor was its formation dependent on the action of stress, because an unstressed freely corroding specimen exhibited this type of attack on the precrack surfaces. Free corrosion specimens which failed in less than one hour showed little or no evidence of side attack; thus, all the available evidence suggests that the composition or pH of the corrodent within the advancing crack is changing sufficiently with time to cause secondary attack on the crack sides.

Fractographic Examination

Many of the fracture surfaces of the broken specimens were examined in the scanning electron microscope. Just as SCC susceptibility was affected by potential, so also were the fractographic features. Hydrogen embrittlement cracking was characterized by an abundance of quasi-cleavage faces (Figure 11), whereas specimens broken at -400 mV showed a roughened or pitted surface without any sign of cleavage (Figure 12). Free corrosion specimens were quite similar to those tested at -400 mV, and, like all specimens, showed an increase in the proportion of intergranular attack when Type 2 cracking had occurred, thus confirming the metallographic observations.

Although the characteristic feature of hydrogen embrittlement cracking, quasi-cleavage, was not apparent in free corrosion or anodically polarized specimens, there remains the possibility that, in these specimens, hydrogen embrittlement cracking occurred initially but was followed by general corrosion and pitting, which obliterated any evidence of quasi-cleavage. However, even at the tip of the SCC region where general corrosion would be a minimum, careful examination failed to provide any evidence of hydrogen embrittlement cracking.

In the SCC zone of the fracture surface, isolated regions of "dimpling", normally associated with ductile rupture, were sometimes observed. Such areas were probably caused by the ductile tearing of "islands" of metal which remained after the main stress corrosion crack front had advanced further into the metal.

Solution Chemistry Within the Crack

Preliminary tests[17] showed that many indicators were unsuitable for estimating the pH of the corrodent within an advancing stress corrosion crack; but methyl orange paper and bromophenol blue paper were considered acceptable if they were prepared using respectively 0.01% and 0.02% aqueous solutions of the indicators.

Although many tests were performed on specimens which had been freely corroding, there was never any indication that the pH of the corrodent within the crack dropped as low as 3.8, which was the minimum value quoted by Brown[7] for this maraging steel-salt water system. In the majority of tests, the pH was greater than (or possibly equal to) about 4.5, this value being the upper limit of the useful indicating range of the papers used. Very occasionally, there was reason to believe that the pH had dropped as low as 4.2, but such results were few and hardly convincing.

Polarizing the specimen to -400 mV had no noticeable effect on the pH of the corrodent within the crack: again the estimated pH was greater than about 4.5 in most cases.

Because of the similarities between a stress corrosion crack and a crevice, it was considered appropriate to

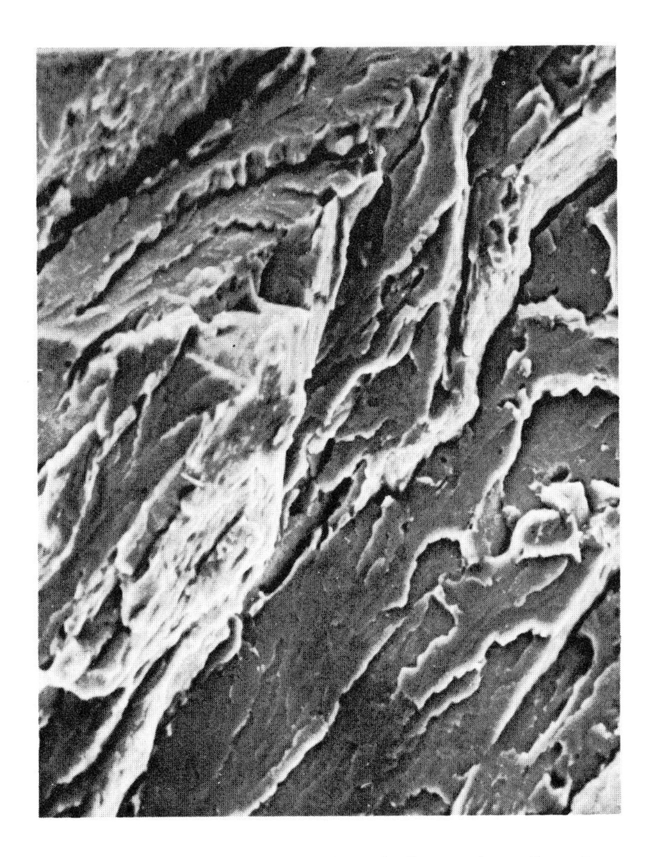

FIGURE 11 — Quasi-cleavage in a hydrogen charged specimen (polarized to -1100 mV). 1200 X

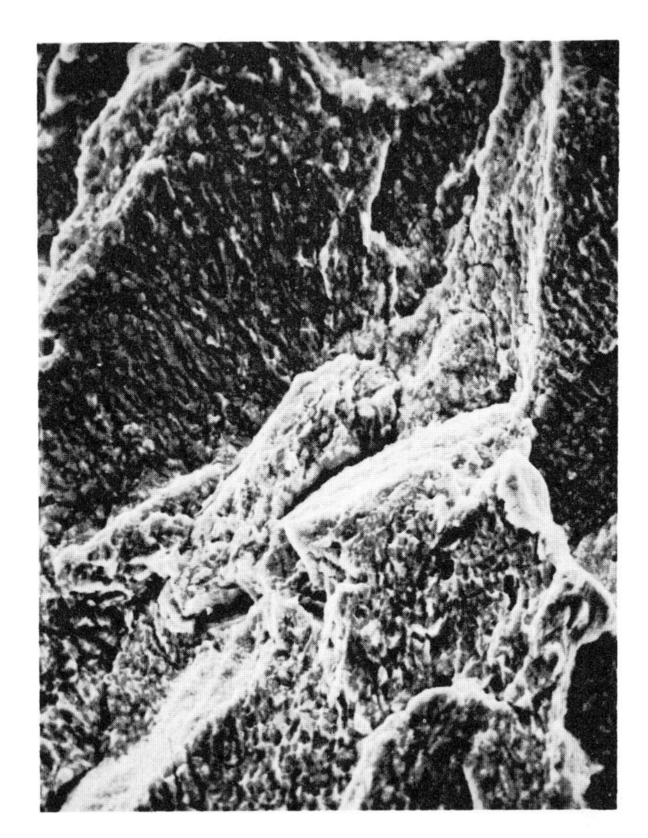

FIGURE 12 — Pitted fracture surface of a specimen polarized to -400 mV. 1200X

study the changes which might occur in the solution within a crevice. Two maraging steel bars, separated by two 0.001-inch thick Teflon spacers, were held in this configuration by rubber bands in 3.5% NaCl solution. After exposure times of up to 100 hours, the pH of the corrodent within the crevice was estimated using the bromophenol blue and methyl orange indicating papers. There was complete agreement between these results and those taken in the SCC tests, the pH rarely indicating below 4.5 even when the steel was held at a potential of -400 mV.

Discussion

Figure 6 shows that by polarizing the specimen to slightly more positive (noble) potentials, the susceptibility to cracking increases; the converse is true when the specimen is polarized slightly in the active direction. Such behavior is considered by some[5,18,19] to be evidence of SCC by anodic path dissolution. The classical interpretation is that, if cracking results from anodic dissolution at the crack tip, the application of a cathodic current would be expected to suppress this reaction and decrease the susceptibility to cracking; conversely, the application of an anodic current would be expected to promote dissolution and increase the rate of cracking. On the other hand, if hydrogen is produced at cathodic sites during the corrosion reaction, and hydrogen embrittlement cracking is promoted in the metal, the application of a cathodic current could be expected to generate more hydrogen and shorten the time to failure. Thus, Figure 6 would be interpreted as providing evidence of anodic path dissolution at potentials more noble than -750 mV and hydrogen embrittlement cracking at potentials more active than -750 mV.

In contrast to this proposal, it has been suggested[7,8] that the type of response to potential shown in Figure 6 does not preclude the possibility of hydrogen embrittlement cracking at free corrosion potentials. It is proposed that the conditions existing at the specimen surface need not be the same as conditions within the crack: hydrolysis reactions occurring within an advancing crack may locally reduce the pH of the corrodent. The reversible potential for hydrogen ion reduction is dependent on pH:

$$E_o = -0.2415 - 0.0591 \text{ pH volts (SCE)} \qquad (3)$$

Thus, at the specimen surface, where the free corrosion potential is approximately -580 mV and the salt solution has a pH of 6.5, hydrogen ion reduction would be impossible, even if overvoltage effects could be ignored. However, if the corrodent within the confines of the crack were sufficiently acidic, hydrogen ion reduction becomes quite feasible. A decreasing susceptibility to cracking with increasing cathodic polarization must depend on factors which influence the rate of hydrogen ion reduction. At a given potential, this rate will be a function not only of the pH of the corrodent within the crack, but also of the difference between this pH and the critical pH given by Equation (3). Because hydrolysis reactions are repressed as the potential is forced to more active values, the pH of the corrodent will tend to increase, though not as high as the critical pH.[7] The net effect would be a decrease in the rate of hydrogen ion reduction.

Of course, the potential will eventually reach a value active enough for hydrogen ion reduction to become possible on the specimen surface, and the classical explanation, discussed above, describes why the cracking susceptibility would subsequently increase with increasing cathodic polarization.

In the present work, therefore, there can be little doubt that hydrogen embrittlement cracking occurs at potentials more active than -800 mV. At more noble potentials, the cracking mechanism is not so clear-cut. The pH of the corrodent within the crack could not be measured accurately, because the useful indicating range of the papers was so limited. However, there is little doubt that, under free corrosion conditions and in specimens polarized to -400 mV, the pH of the corrodent within the crack is greater than, or equal to, 4.5. This value is the upper limit of the useful indicating range of the papers used. Assuming a pH of 4.5, and neglecting overvoltage effects, hydrogen ion reduction would be possible at -507 mV. Thus, in specimens polarized to -400 mV, hydrogen embrittlement cracking would be impossible if the potential within the crack was also -400 mV. However, France and Greene[20] have demonstrated that large potential gradients may exist in crevices during potentiostatic anodic polarization of stainless steels. If the same is true for 18% Ni maraging steel, the potential at the crack tip may be much more active than the -400 mV impressed at the specimen surface. Unfortunately, the present work does not resolve the latter possibility.

The secondary attack observed on the sides of stress corrosion cracks, which had propagated under free corrosion conditions, could be evidence of a pH drop within the confines of the crack. However, if the pH is no less than 4.5 in a specimen polarized to -400 mV, under free corrosion conditions the pH is likely to be significantly higher. Also, any overvoltage effects will tend to decrease the likelihood of hydrogen ion reduction at free corrosion potentials.

At potentials near the free corrosion potential, any changes in conditions within the crack which would stimulate hydrogen ion reduction, would also promote anodic dissolution. Thus, even if the pH did decrease and the potential was active enough to allow hydrogen ion discharge in this region, there is little reason to believe that hydrogen embrittlement cracking occurred in preference to anodic path dissolution. In fact, the differences in the fractographic and metallographic features would support the proposal that anodic path dissolution occurred at the free corrosion potential while hydrogen embrittlement cracking occurred at potentials more active than -800 mV. Furthermore, the single test performed in 3.5% NaCl solution containing $NaAsO_2$ suggested that hydrogen embrittlement processes were minimal, or entirely absent, at free corrosion potentials.

Therefore, all of the available evidence points to an anodic path dissolution mechanism at potentials more noble than about -600 mV, and to a hydrogen embrittlement cracking mechanism at potentials more active than about -800 mV. It seems reasonable to assume that, at intermediate potentials, anodic path dissolution and hydrogen embrittlement will each play a role in slow crack growth, the relative importance of hydrogen embrittlement increasing as the potential becomes more active.

The potential vs time curve in Figure 5 gives no further clue as to the cracking mechanism operative under free corrosion conditions. However, the potential changes occurring at the point of fracture are worthy of mention. The initial potential jump in the active direction was probably caused by the sudden exposure of the clean, oxide free fracture surface to the NaCl solution; subsequent rises to more noble potentials, followed by a reversal of this trend, were likely due to the rapid oxidation of the fracture surface to form a partially protective film (similar in nature to an air-formed film) and the subsequent breakdown by general corrosion processes. In support of these suggestions is the observation that the potential of maraging steel in 3.5% NaCl solution jumps rapidly to more active potentials when the surface film in contact with the corrodent is removed with emery paper. The potential returns equally as quickly to its original value when abrasion of the surface is halted; subsequent slower changes to more active potentials are similar to those shown in Figure 5. The potential jump induced under these conditions may be over 50 mV, and it is in the direction opposite to any potential changes which would be caused by mechanical disturbance of the corrodent.

It has been shown that SCC may occur in three directions: in the precrack plane (Type 1 cracking) or along planes inclined at about 60° to, and on either side of the precrack plane (Type 2 cracking). When Type 2 cracking initiates, it has the effect of blunting the sharp precrack: this results in a higher load being required for the onset of fast fracture. Thus, the curve associated with Type 2 cracking in Figure 3 becomes quite understandable. At a loading rate of about 500 psi√inch/min (time-to-failure = approximately 200 min) the "blunting effect" is at a maximum and K^*_{Ii} may exceed 1.5 K_{Ic}. At slower loading rates, the blunting effect can still be appreciable, but the crack has time to propagate further across the specimen and the load to cause terminal fracture is increasingly lowered as the loading rate is decreased. At higher loading rates, above 500 psi√inch/min, there appears to be insufficient time (< 100 min) for Type 2 SCC to occur to any appreciable extent, and the small amount of SCC which does occur is confined to Type 1. Above 10,000 psi√inch/min (time-to-failure < 8 min), corrosion processes play no part, and fracture is purely mechanical.

Although the subject is still debated, there is good reason to believe that SCC actually occurs along one or more of the three "zero isoclinic surfaces"; these surfaces are characterized by zero shear, and separation may occur across them by displacements entirely in the direction of loading. Theoretical and experimental support for this proposal is given elsewhere in some detail.[21]

Conclusions

It is proposed that 18% Ni maraging steel suffers SCC

by anodic path dissolution at potentials more noble than about -600 mV; at potentials more active than about -800 mV, it appears that hydrogen embrittlement cracking is dominant; and, in the intervening potential range of -600 mV to -800 mV, it is thought that anodic path dissolution and hydrogen embrittlement cracking both play a part in the cracking mechanism. The potential of freely corroding maraging steel is more noble than -600 mV, so SCC by anodic path dissolution can be expected. These conclusions are based on the observed response to polarization, the influence of additions of arsenic to the corrodent, metallographic and fractographic examinations, and estimates of the pH of the corrodent within the advancing crack.

Acknowledgments

The author is indebted to G. J. Biefer and L. P. Trudeau for their helpful and stimulating discussions, and also J. G. Garrison and K. Pickwick for their assistance with the experimental work.

References

1. J. A. S. Green and E. G. Haney. *Corrosion,* **23,** 5 (1967).
2. H. P. Leckie. Effect of Environment on Stress Induced Failure of High-Strength Maraging Steels. *Proceedings of the Conference on Fundamental Aspects of Stress Corrosion Cracking,* p. 411, R. W. Staehle, A. J. Forty, and D. van Rooyen, Editors, National Association of Corrosion Engineers, Houston, 1969.
3. J. A. S. Green and E. G. Haney. "A Stress Corrosion Test for Foil and Strip." Presented at ASTM Conference, Atlantic City, N. J., 1966.
4. S. W. Dean and H. R. Copson. *Corrosion,* **21,** 95 (1965).
5. R. N. Parkins and E. G. Haney. *Trans. AIME,* **242,** 1943 (1968).
6. A. Rubin. Report 2914, USAF Research Contract DA-04-495-ORD-3069, 1964.
7. B. F. Brown. "On the Electrochemistry of Stress Corrosion Cracking of High Strength Steels", presented at the Fourth International Congress on Metallic Corrosion, Amsterdam, Netherlands, 1969.
8. H. P. Leckie and A. W. Loginow. *Corrosion,* **24,** 291 (1968).
9. G. J. Biefer and J. G. Garrison. "Stress Corrosion Cracking Tests on Some High Strength Steels, Using the USNRL Cantilever Method", Mines Branch Technical Bulletin TB 114, Department of Energy, Mines and Resources, Ottawa, Canada, 1969.
10. B. F. Brown and C. D. Beachem. *Corr. Sci.,* **5,** 745 (1965).
11. C. N. Freed and J. M. Krafft. *J. Materials,* **1,** 770 (1966).
12. B. F. Brown. *Met. Rev.,* **13,** 171 (1968).
13. ASTM Standards, Part 31, 1099 (1969) May.
14. B. F. Brown, C. T. Fujii, and E. P. Dahlberg. *J. Electrochem. Soc.,* **116,** 218 (1969).
15. G. J. Biefer and J. G. Garrison. "Cantilever Stress Corrosion Cracking Tests on 18% Nickel Maraging Steel: Effect of Loading Rate and Cathodic Pretreatment", Physical Metallurgy Division Internal Report PM-R-68-5, Department of Energy, Mines and Resources, Ottawa, Canada, 1968.
16. C. S. Carter. Boeing Document D6-19770, 1967 (available from Defense Documentation Center).
17. B. C. Syrett and J. G. Garrison. "The Limitations of Measuring pH by Colour Indicators", Mines Branch Technical Bulletin TB117, Department of Energy, Mines and Resources, Ottawa, Canada, 1969.
18. B. F. Brown. Report of NRL Progress, 40 (1958) May.
19. P. C. Hughes. *J. Iron and Steel Inst.,* **204,** 385 (1966).
20. W. D. France and N. D. Greene. *Corrosion,* **24,** 247 (1968).
21. B. C. Syrett and L. P. Trudeau. "Stress Corrosion Cracking Along Zero Isoclinics", *Corrosion,* **27,** 216 (1971).

Stress Corrosion Cracking of 18 Pct Ni Maraging Steel in Acidified Sodium Chloride Solution

R. N. Parkins and Elwood G. Haney

Stress corrosion cracking of two heats of 18 pct Ni maraging steel in rod form immersed in an aqueous solution of 0.6N NaCl at pH 2.2 has been studied on unnotched specimens stressed in a hard tensile machine. Austenitizing temperature in the range 1830° to 1400° F has been shown to have a marked influence on the propensity to crack, the lowest austenitizing temperature producing the greatest resistance to failure. In the most susceptible conditions, the cracks followed the original austenite grain boundaries; but when the steels were heat treated to improve their resistance to stress corrosion, the cracks became appreciably less branched and showed significant tendencies to become transgranular. Electron metallography of the steels indicated the presence of small particles, possibly of titanium carbide, along the prior austenite grain boundaries and these particles were more readily detectable in the structures that were most susceptible to cracking. Crack propagation rates, which appeared to be dependent upon applied stress and structure, were usually in the region of 0.5 mm per hr and may, therefore, be explained on the basis of a purely electrochemical mechanism. However, there is some evidence from fractography that crack extension may be assisted by mechanical processes. Anodic stimulation reduced the time to fracture, although cathodic currents of small magnitudes delayed cracking; further increase in cathodic current resulted in a sharp drop in fracture time, possibly due to the onset of hydrogen embrittlement.

THE use of the high strength maraging steels, with their attractive fracture toughness characteristics, is restricted because of their susceptibility to stress corrosion cracking in chloride solutions. Although this limitation has resulted in investigations of the stress corrosion susceptibilities of these steels, there have been few systematic studies aimed at defining the various parameters that determine the level of susceptibility. It is the case that the usual tests have been performed with the object of defining some stress or time limit, on unnotched or precracked specimens, within which failure was not observed,[1] but while such results may be of some use in design considerations, they are necessarily concerned only with the steels as they currently exist and not with their improvement to render them more resistant to stress corrosion failure. This omission may be considered unfortunate because the indications are that stress corrosion in maraging steels shows dependence on structure in following an intergranular path, and since experience with other systems of intergranular stress corrosion cracking is that susceptibility may be varied by modifying heat treatments, a similar effect may be expected with maraging steels. It is sometimes from such observations that a fuller understanding of the mechanism of stress corrosion crack propagation begins to emerge, leading in time to the development of more resistant grades of material. The present work was undertaken to study only one aspect of the influence of heat treatment upon the cracking propensities of the 18 pct Ni maraging steel, namely the effect of austenitizing temperature, although certain ancillary measurements and experiments have been undertaken.

ELWOOD G. HANEY, Member AIME, is Fellow, Mellon Institute, Carnegie-Mellon University, Pittsburgh, Pa. R. N. PARKINS, formerly visting Fellow, Mellon Institute, is Reader, Department of Metallurgy, University of Newcastle upon Tyne.

Manuscript submitted October 31, 1967. IMD

EXPERIMENTAL TECHNIQUES

Most of the measurements were made on a steel, A, having the analysis shown below, although a few results were obtained on a steel, B, having a slightly different composition.

Both steels were supplied in the austenitized condition, A as 3/8-in-diam rod and B as 1/2-in.-diam rod.

Cylindrical tensile test pieces were machined from the rods; the overal length was 2 1/2 in., the gage length 1 in. and the diameter 0.128 to 0.136 in. The stress corrosion tests were carried out with the specimens strained in tension in a hard beam testing machine, the necessary total strain being applied to the specimen over a period of about 30 sec, after which the moving crosshead was locked in position and the load allowed to relax as crack propagation proceeded; the load relaxation was recorded.

The load was applied after the specimen had been brought into contact with the corrosive solution, the latter being contained in a polyethylene dish having a central hole through which the specimen passed, leakage being prevented by the application of a film of rubber cement. The specimen was in contact with the solution for over half of its gage length and the solution was exposed to the air during testing. The solution was prepared from distilled and deionized water to which NaCl was added, 0.6N, and the pH adjusted to 2.2 by HCl additions. The composition of the solution

	A	*B*
C (wt pct)	0.019	0.02
Si	0.08	0.10
Mn	0.07	0.10
S	0.007	0.006
P	0.006	0.005
Mo	4.80	4.86
Co	9.17	9.35
Ni	18.85	18.60
Al	0.10	0.13
Ti	0.59	0.54
B	0.003	0.005
Zr	0.011	0.015
Ca	0.05 (added)	0.05 (added)

was chosen to coincide with the most aggressive of the range of acid chloride solutions tested by Green and Haney.[2]

Tests in which anodic or cathodic currents were applied during stress corrosion were made in a cell that contained a platinum electrode, in the form of a wire wound over a Perspex former some 2 in. diam. In these tests the specimen gage length was increased to 2 in. with 1 1/2 in. immersed in the solution and the whole was contained in a glass tube sealed at the bottom with a rubber gasket through which the specimen passed. Current was supplied from a filtered constant current dc power source. In some experiments in which electrode potential measurements were made, the specimen was electrically insulated from the grips of the testing machine with mica strips and the potential measurements were made with reference to a saturated calomel electrode via a string bridge, the potential difference being indicated on a voltmeter.

Specimen preparation involved sealing the machined test pieces in a helium atmosphere in vycor tubes. Austenitizing was for 1 hr at the appropriate temperature and was followed by air cooling and subsequent aging at 900° F for 3 hr. After heat treatment, the specimens were ground longitudinally with emery paper to a 3/0 finish to remove the circumferential machining marks. They were degreased immediately prior to testing.

For optical microscopy, an etching solution containing 25 cc HNO_3, 50 cc HCl, 150 cc H_2O, and 1 g $CuCl_2$

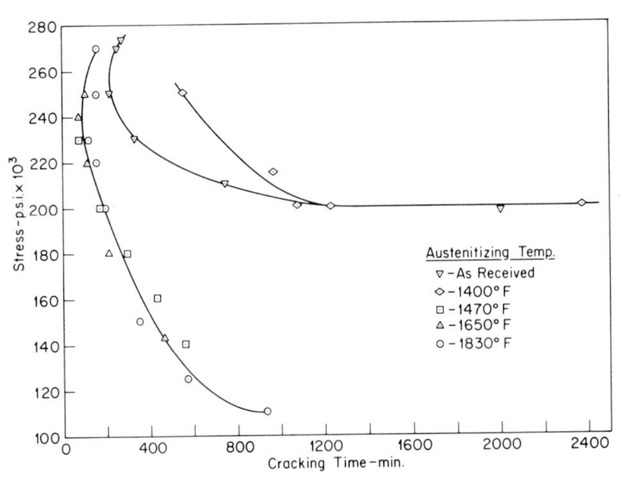

Fig. 1—Effects of stress and austenitizing temperature on the time to total fracture for steel A.

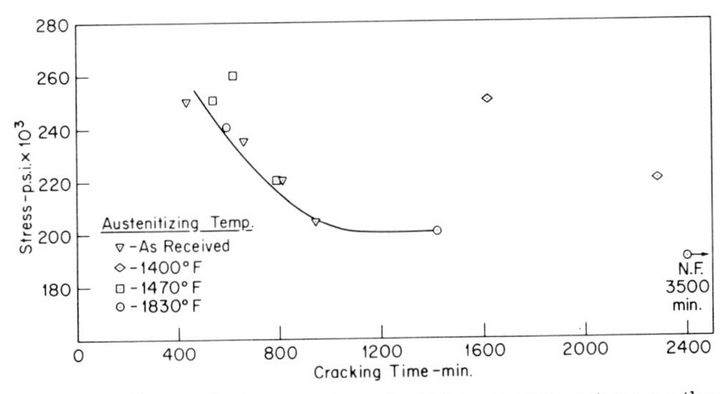

Fig. 2—Effects of stress and austenitizing temperature on the time to total fracture for steel B.

was found satisfactory. Foils for electron microscopy were prepared by grinding samples to about 0.008 in. thickness followed by reduction to 0.002 in. thick in 50 pct HCl at 10 amp. Final preparation was by electrochemical polishing in a solution containing 450 g Na_2CrO_4 to 3 liters glacial acetic acid at 0.56 amp per sq cm and 70 v. Polishing was continued until the first indication of perforation of the foil. Replicas were taken from the fracture surfaces of some specimens for examination in the electron microscope. By discarding the first few plastic replicas, clean samples relatively free of corrosion products or other debris were obtained. The clean plastic replicas were coated with carbon, the plastic dissolved and the carbon shadowed with platinum at 45 deg to provide the sample to be examined.

RESULTS

Effects of Austenitizing Temperature and Composition. Fig. 1 shows the time to total fracture as a function of initial stress for specimens of steel A austenitized at various temperatures. The curves take the characteristic form in suggesting a threshold stress below which failure will not occur. The most interesting feature however is the markedly increased susceptibility of the material to cracking following austenitizing at temperatures of 1470° F and above, although the actual value of the austenitizing temperature above 1470° F is without effect. It is noteworthy that, not only is the time for fracture at a given stress considerably increased when austenitizing is carried out at 1400° F, but the threshold stress is increased by a factor of about two. These differences are not obviously related to variations in the mechanical properties of the steel resulting from differences in heat treatment, since the mechanical properties showed only small, possibily insignificant, changes with austenitizing temperature, as the following results indicate.

Fig. 2 shows a similar trend in the results for steel B, the advantages of keeping the austenitizing temperature at as low a value as possible again being apparent. With this steel, however, it would appear that the threshold stress could not be raised much by a reduction in austenitizing temperature, since even with the highest temperatures the steel had a relatively high threshold stress and was appreciably less susceptible to stress corrosion than steel A.

Since different austenitizing temperatures would result in different cooling rates as specimens were air cooled, a few tests were carried out in which the cooling rate was intentionally exaggerated. Specimens of steel A were water quenched or furnace cooled from 1650° F and stress corrosion tested at 230,000 psi. Fracture times fell in the range 70 to 100 min, in good

Steel	Treatment	0.2 pct YS, ksi	UTS ksi	R.A. pct
A	As Rec. + 900°F.	260	281	50
A	1470° + 900°F.	252	278	55
A	1650° + 900°F.	250	276	55
A	1830° + 900°F	258	272	50
B	As Rec. + 900°F	250	284	55
B	1400° + 900°F	252	282	50

agreement with those obtained on air-cooled specimens. These results were not affected by cooling in liquid nitrogen prior to aging.

Load Relaxation and Crack Profiles. The relaxation of the load during a stress corrosion test did not follow the same pattern in all tests; Fig. 3 shows typical examples of relaxation curves. Where the initial stress was below the yield stress the load remained constant for a period of time depending, among other factors, upon the initial stress. With initial stresses in excess of the yield stress, creep resulted in a diminishing rate of relaxation in the early stages of the test, as is apparent from Curve 1 in Fig. 3. In the later stages of a test, the load relaxed at an increasing rate but to an extent that varied considerably from specimen to specimen. Thus, in Fig. 3, Curve 1 shows a relatively long relaxation curve and Curve 2 one that is short in the sense that the load drop prior to sudden fracture was less marked. This difference in behavior was related to the number of cracks that developed in a specimen and hence to the susceptibility of the sample to stress corrosion, marked load relaxation being observed when a specimen developed many cracks and little relaxation occurring when one or a few cracks were present. Samples of steel A treated at 1830° or at 1650°F and tested at high stresses invariably exhibited many branched cracks and long load relaxation curves. Steel A austenitized at 1470°F or below and all samples of steel B developed fewer than six cracks when stress corroded and exhibited proportionately less load relaxation.

The specimens used in the tests in which relatively large anodic or cathodic currents were applied were exceptional in that although they invariably developed many cracks, ~ 40, they always fractured after relatively small amounts of load relaxation, as is apparent from Curves 3 and 4 in Fig. 3. This was possibly due to their greater length and the higher level of elastic energy in the specimens for otherwise comparable conditions.

In addition to variations in the number of cracks from specimen to specimen, there were differences in the crack shape. With specimens that were allowed to proceed to complete failure the shape of the stress corrosion crack at fracture was readily apparent from the differing reflectivities of the oxide covered stress corrosion fracture and the bright surface exposed by the final mechanical fracture. Some tests were stopped before complete failure and crack profiles determined either by sectioning and metallographic examination at appropriate intervals, 0.005 in. or by raising the stress to propagate the stress corrosion crack to complete fracture. The crack profiles followed the general pattern of behavior observed in austenitic stainless steels[3] and low carbon steels,[4] in that in the early stages cracks had thumbnail profiles but later spread more rapidly at the periphery of the specimen than the center. With fine cracks, *i.e.*, those that had not extended sufficiently to cause the macroscopic yield strength to be exceeded and so had not yawned, peripheral spreading was apparent when the crack had reached a depth of about 0.5 mm, but when the crack was open the thumbnail profile was apparent to radial depths of almost 1.0 mm. Typical profiles are apparent in Fig. 4. These observations are consistent with an explanation involving a limiting crack depth due to ohmic drop down the crack, which may be expected to be less down an open than a fine crack, hence the greater radial depth with open cracks. Cracks usually branched at some small distance below the surface, Fig. 10, and again the usual explanation, that they follow paths, in the macroscopic sense, normal to the maximum resolved tensile stress, would appear valid.

The size of stress corrosion cracks, reflecting their time and stress dependence, and the number of cracks in specimens were found to be related through the load immediately prior to complete failure. Fig. 5 shows that the area of stress corrosion crack associated with the exposed fracture surface is dependent upon the fracture load and that the latter was greatest for specimens containing only one crack and smallest for specimens containing many cracks, specimens containing a few cracks, 2 to 6, occupying an intermediate position. The result is independent of the structure or composition of the steel within the limits studied in this investigation. The obvious explanation of Fig. 5 is that it is related to the amount of load relaxation and hence, as already indicated, to the number of cracks developed. Load relaxation will result from plastic deformation of the uncracked material beyond a crack and will be additive for all the cracks in a single specimen. It is likely, therefore, that load re-

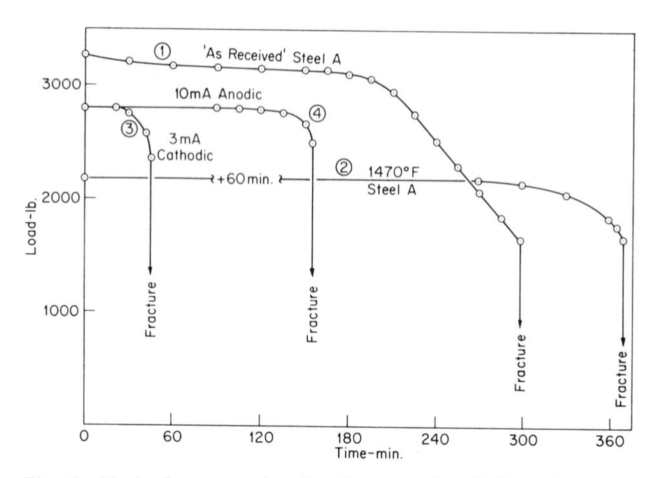

Fig. 3—Typical curves showing the ways in which the loads on specimens relaxed during crack propagation.

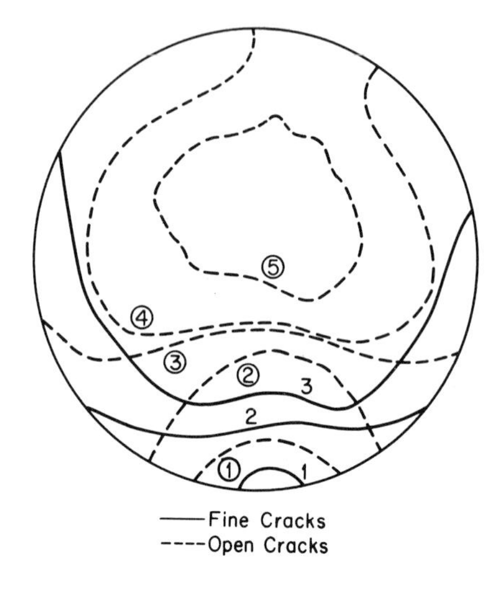

Fig. 4—Showing crack profiles at various stages of propagation.

laxation will be more marked in specimens containing a number of cracks than in those containing only one. In the latter case, a small stress corrosion crack will be able to propagate a relatively brittle mechanical failure because the applied load remains high, whereas with the marked load relaxations associated with the presence of many stress corrosion cracks the latter must propagate much further before one of them becomes large enough to create the stress conditions at a relatively small load for sudden fracture.

The results shown in Fig. 5 have interesting implications to stress corrosion tests of the constant strain type in general. Most stress corrosion tests are of the constant strain type, since these usually more realistically produce the stress conditions resulting in service failures, and normally require only simple jigs, and the measure that is most frequently taken as an indication of the susceptibility to stress corrosion is the time to total failure under given stress conditions. It is apparent from Fig. 5, however, that a specimen in a highly susceptible condition, resulting in a large number of cracks, may take significantly longer to proceed to total failure once load relaxation has begun than a specimen in a less susceptible condition developing only one or a few cracks. The time to total fracture will, of course, include the time that elapses to the beginning of load relaxation and this may counteract the trend just mentioned; but where the time to load relaxation is short, the total time to fracture may be largely dependent upon the time during which the load is relaxing. This appears to be so at the highest stresses employed in the present tests, since Figs. 1 and 2 show a tendency for the times to fracture to increase somewhat as the stress is increased in the highest ranges.

Crack Propagation Rates. Measurements of the rates of stress corrosion crack propagation are useful in helping to decide whether or not crack propagation is entirely dependent upon an electrochemical mechanism and for this reason some such measurements were made in the present work. It is usual to estimate the average rate of propagation from measurements of maximum crack depth and total time of exposure but it is clear that the rate may vary considerably during the various states of a test. In the present work an incubation period was detectable during which crack propagation in the strict sense did not occur. No systematic attempt was made to measure the incubation period over a large range of stress levels or for different heat treatments, but from the few measurements that were made, it was apparent that this period could constitute a significant part of the total time to fracture. Thus,

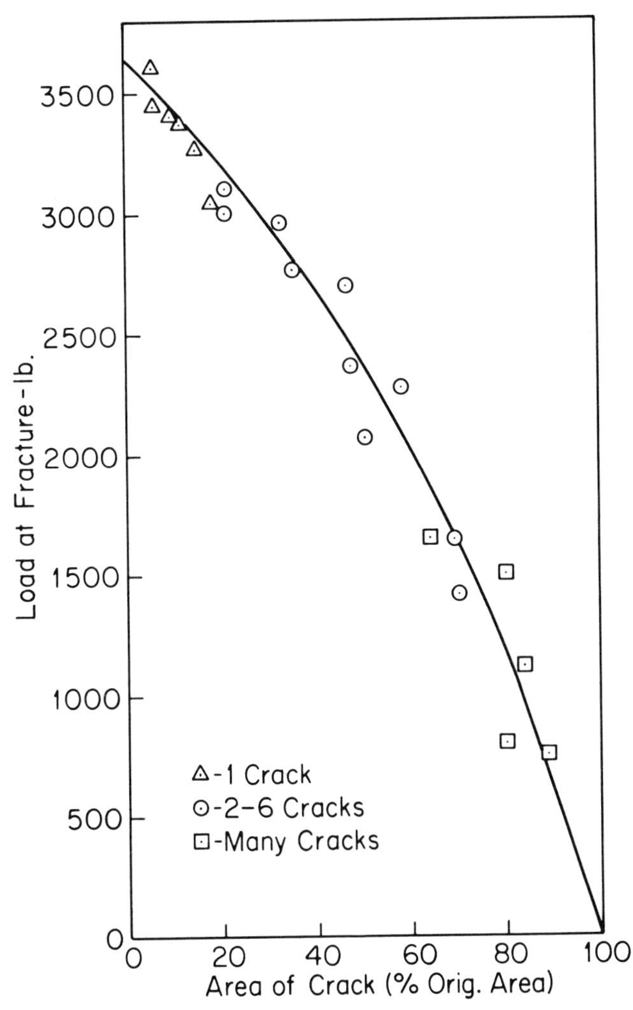

Fig. 5—The variation of load at the instant of final fracture as influenced by the area of the largest stress corrosion crack and the number of cracks.

Fig. 6—Showing surface fissures and the development of stress corrosion cracks therefrom. Magnification 970 times.

in a sample of steel A, austenitized at 1830°F and tested at an initial stress of 150,000 psi, cracks were not detected after 90-min contact with the solution, although they were present after 160 min and total failure occurred in 350 min. Similarly, steel B austenitized at 1470° F and tested at 230,000 psi, exhibited no cracks after 240 min, although total failure occurred in a similar test after 650 min.

This incubation period was not one in which no reactions occurred, however, since microscopical examination of samples invariably revealed some surface fissures whether or not crack propagation had occurred. These fissures tended to occur in groups, were particularly prevalent near the water line, and stress corrosion cracks proper appeared to be initiated from the tips of such features. Fig. 6 shows typical examples of fissuring, with stress corrosion cracks, in various stages of advancement, clearly associated with the fissures. There was no obvious structural dependence of the location of fissures, but their major axes were normal to the surface and they were more numerous and deeper the higher the stress applied to the specimen in which they formed. Their prevalence and size was also found to be variable in the specimens to which anodic or cathodic currents were applied and Fig. 7 shows this variation; at higher cathodic currents, >1 ma per sq in., fissures were not detected. In a period of about 1 hr, the fissures typically extended to depths of about 0.004 mm in moderately stressed specimens, increasing to 0.015 mm in anodically stimulated, 1 ma per sq in., conditions.

Since these fissures were found in all specimens and they formed in the early stages of a test, it seems reasonable to associate their formation with the incubation period, together with any period involved in the undermining of surface oxide films. The development of stress corrosion cracks from the tips of fissures may depend upon the establishment of some precise chemical condition within the confines of the fissures and whether or not cracks form may be largely determined by whether or not the stress level is sufficient to prevent the reactions being stifled by blockage of the fissures. The prevalence of fissures near the water line, presumably associated with oxygen depolarization, sometimes resulted in failure of specimens in such regions. Although relatively few specimens fractured at the water line, and when this did occur the result was disregarded, the tendency was most marked in those tested at stresses in the region of the threshold stress.

The measurements that were made to determine crack propagation rates were insufficiently reproducible to determine any effects of stress level or heat treatment in anything other than a qualitative sense. Of all the results obtained, 0.5 mm per hr was the fastest and 0.15 mm per hr the slowest rate observed, with an average rate of about 0.3 mm per hr. The effect of ignoring the incubation period, the variability of which may well have been the cause of the apparent variability in crack propagation rates, will be to produce somewhat slower rates or propagation than in fact obtained but, for reasons discussed below, this should not involve an error that would increase the rate by more than a factor of about two. There appeared to be some stress dependence of the crack propagation rate in that results on steel A in the "As Received" condition were as follows: at 260,000 psi — 0.4 mm per hr, at 230,000 psi — 0.25 mm per hr, and at 200,000 psi — 0.15 mm per hr. Similarly, steel A austenitized at 1830° F gave a crack propagation rate of 0.4 mm per hr at 250,000 psi whereas the same steel austenitized at 1400°F gave a crack propagation rate of 0.15 mm per hr.

The various figures quoted above refer to the propagation of both fine and open cracks, at various stresses and in material having had different austenitizing treatments, so that it is not surprising that the results show scatter, more especially because of the inclusion of the incubation time in the total time used for measuring the propagation rate. Some of these difficulties, especially the last one, may be surmounted and a maximum crack propagation rate obtained from the load relaxation curves for specimens containing a single crack, if the assumption is made that load relaxation is directly related to the spreading of the crack. For specimens containing a single crack, this seems a reasonable assumption and is supported by the observation that the load relaxation varied quadratically with time, except in the minute or two before complete failure. With such specimens, the stress corrosion crack retained its thumbnail shape to complete fracture and its area would therefore increase as the square of the radius, to a first approximation. If the linear rate of propagation remains constant, then the area would be expected to increase with the square of the time. Consequently, once the crack is spreading under conditions

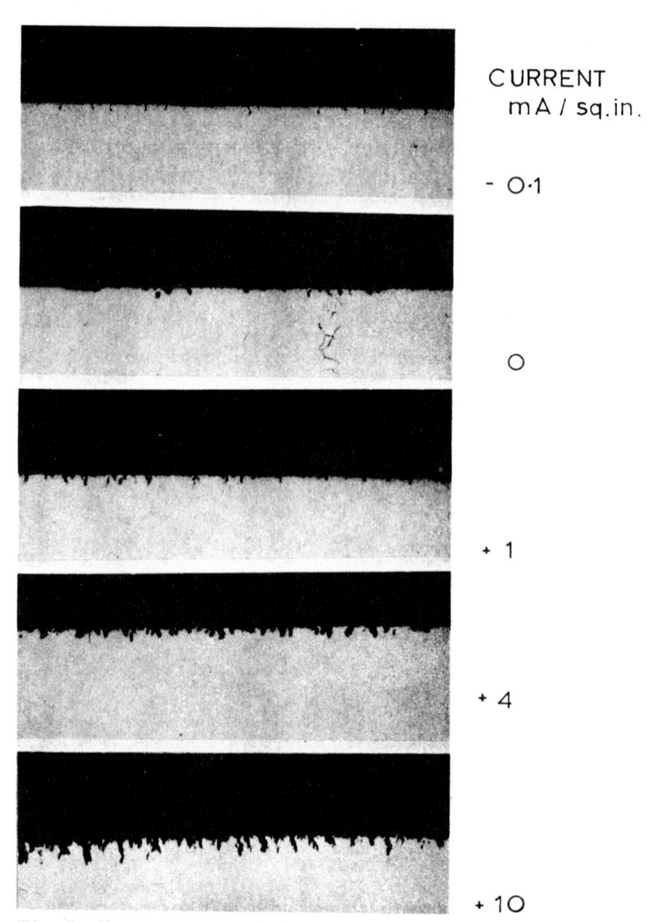

Fig. 7—Showing the variation in surface fissuring with applied current. Magnification 95 times.

where it is accompanied by load relaxation, the latter would be expected to fall as a quadratic function of time, as indeed was observed. Furthermore, taking load relaxation to occur until the stress on the uncracked portion reaches some constant value, *i.e.*, neglecting work hardening, it is easily shown that the change in area Δa, of the stress corrosion crack is directly proportional to the change in load, Δl, *i.e.*,

$$\Delta a = \frac{\Delta l}{\sigma_{y.s.}}$$

where $\sigma_{y.s.}$ is the effective yield stress. The crack propagation rates calculated from the load relaxation curves in these circumstances were as follows:

Steel A 'As Received'	0.6 mm per hr
Steel A Austenitized 1830°F	0.8 mm per hr
Steel B 'As Received'	0.5 mm per hr

It seems reasonable to consider these as the maximum rates of stress corrosion crack propagation relevant to the system studied in this investigation, since they were observed under the most extreme conditions, and refer to the propagation of an existing crack without an incubation period. Since they are not vastly different from the values quoted earlier, averaging 0.3 mm per hr, and which would be expected to be in error because of the inclusion of the incubation period in the time used in calculating the crack propagation rate, it would appear justifiable to suggest that the crack propagation rates in general lay in the range 0.3 to 0.8 mm per hr. Such rates are fairly typical of stress corrosion crack propagation rates[5] and, on the basis of a purely electrochemical dissolution process, would require a current density at the crack tip of about 0.2 to 0.5 amp per sq cm to maintain this rate of advancement. Such current densities are of an order that is not frequently met in corrosion reactions, due to the incidence of polarization effects, and could be interpreted as evidence of crack propagation involving a purely mechanical stage thereby reducing the current requirements by reduction of the electrochemical stage in the propagation. However, there is evidence[6, 7] that localized corrosion rates, especially in chloride and in acid solutions, may be significantly higher than bulk corrosion rates, in which cases the need for invoking a mechanical stage in crack propagation is unnecessary.

Electrochemical Measurements. Measurement of the electrode potential of a specimen undergoing a stress corrosion test followed a typical pattern apparent from the curves shown in Fig. 8. The steady potential that the specimen was allowed to reach prior to the application of the load showed evidence of shift immediately the load was applied, the potential falling to more negative values as the load relaxed due to creep. The beginning of marked load relaxation, associated with the formation of an open crack, was accompanied by a more marked change in potential to more negative values, as is readily apparent from a comparison of the curves in Fig. 8. The potential began to approach a steady value in the later stages of the test, but complete fracture produced a further sharp change in potential, that recovered almost as sharply. The results appear consistent with an explanation involving film rupture exposing unfilmed metal, especially in view of the marked changes in potential when the load began to relax rapidly and when the fracture process was suddenly completed. No sharp fluctuations in potential were observed over very short periods of time, other than at complete separation.

The application of anodic or cathodic currents to specimens undergoing stress corrosion tests is useful in deciding whether failure of a stressed specimen in a corrosive medium is due to hydrogen embrittlement or stress corrosion cracking.[8] Fig. 9 shows that, for the system reported herein, anodic polarization reduces the cracking time while small cathodic currents prolong the life very significantly, behavior that is characteristic of a classical stress corrosion system. Relatively high cathodic currents reduce the time to failure and indicate that this system is one in which hydrogen embrittlement can be induced; but that at normal potentials, failure is by stress corrosion cracking. This was supported by metallographic examination and also by observations of the load relaxation curves obtained in experiments on polarized specimens. Thus, cathodically polarized specimens, 3 to 10 ma per sq in., were associated with very significantly faster load relaxation rates than were observed under any other conditions; and, by reversing the polarity of a specimen undergoing load relaxation, the rate of relaxation could be varied,[9] almost instantaneously, by an order of magnitude, as the current was

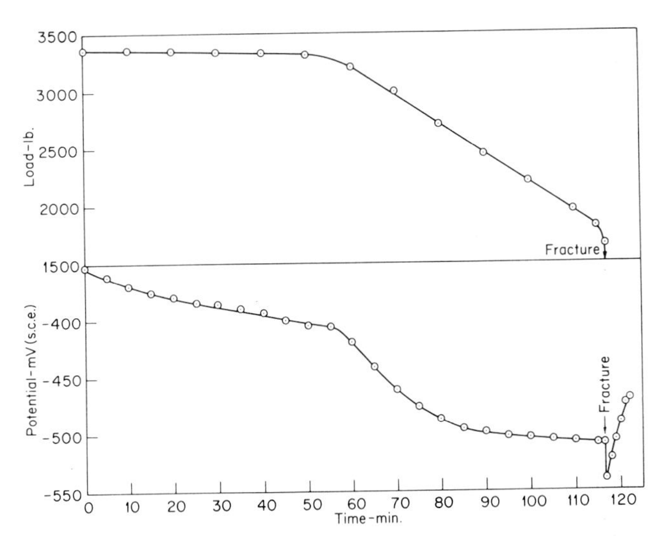

Fig. 8—Electrode potential/time and load/time curves, indicating that marked load changes are reflected in potential changes.

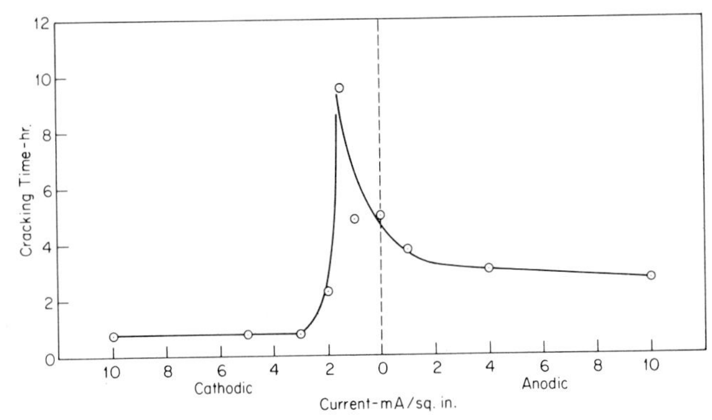

Fig. 9—Effects of applied currents on the time to failure of steel A austenitized at 1830°F and tested at 230,000 psi.

changed from +10 to −10 ma per sq in.

Metallography. Optical microscopy revealed two distinctly different types of cracking in specimens fractured in conventional stress corrosion tests. In circumstances of high susceptibility to stress corrosion, *e.g.*, resulting from a high austenitizing temperature in the case of steel A, the cracking was typically intergranular, with numerous branches, Fig. 12. Where cracking times were relatively long, *e.g.*, with steel A austenitized at relatively low temperatures and tested at stresses in the vicinity of the threshold and in almost all tests on steel B, the cracking was not so obviously structure dependent and branching was confined to short, thick, arms. The differences between the two types of cracking were also apparent on examination at much lower magnifications. Fig. 10 shows the marked branching usually observed with the fine intergranular cracks detected in highly susceptible material and Fig. 11 the broad cracks associated with less susceptible conditions and suggestive, even at this magnification, of little structural dependence. Short transgranular cracks were apparent developing from intergranular cracks in some of the specimens in highly susceptible conditions as shown in Fig. 12. Specimens subjected to cathodic polarization, >1.5 ma per sq in., during stress corrosion differed from all others in that, while the cracks were predominantly intergranular, significant amounts of straight transgranular cracking, suggestive of cleavage cracks, were apparent, Fig. 13. Similar features were detected in unstressed specimens cathodically charged at 5 ma per sq in. for 16 hr before being allowed to undergo delayed fracture in a constant strain test in air.

Optical microscopy was of little use in examining the details of the structure of these steels and electron microscopy of foils was relied upon for this purpose. However, optical microscopy did reveal that etching of the austenitic grain boundaries occurred more readily the higher the austenitizing temperature and that the austenitic grain size of steel A varied according to the heat-treatment temperature as follows:

Austenitizing Temperature	No. of grains per sq in. at X100
'As Received'	1100
1400°F	840
1470°F	370
1650°F	450
1830°F	128

Electron metallography on thin foils indicated some differences between samples removed from specimens of differing susceptibility to stress corrosion. Thus, apart from the usual matrix containing numerous fine precipitate particles in a highly dislocated structure, and which was observed in all specimens, the original austenite grain boundaries were lighter and more clearly delineated the more susceptible the material to cracking. Moreover, there was a tendency for the boundaries in highly susceptible material to contain more of a relatively coarse precipitate, Fig. 14. The latter has not yet been positively identified but there are some indications that it may be a carbide and it certainly is different from the fine precipitate present within the grains. Other features noticed in foils were tendencies towards shallow grooving at the austenite grain boundaries and for the periphery of a perforation to be coincident with austenite boundaries, suggestive of attack upon the latter to the point where a piece of metal became detached.

Fractography. Replicas from fracture surfaces showed essentially two types of area. Fig. 15 is typical of one of these, exhibiting flat and undulating areas sometimes broken by sharply defined steps and by frills or fringes. Exceptionally repetitively stepped areas were observed, but a commonly observed effect was that shown in Fig 16 in which the fringes have a spacing very similar to that of the short transgranular cracks shown in the optical micrograph of Fig. 12.

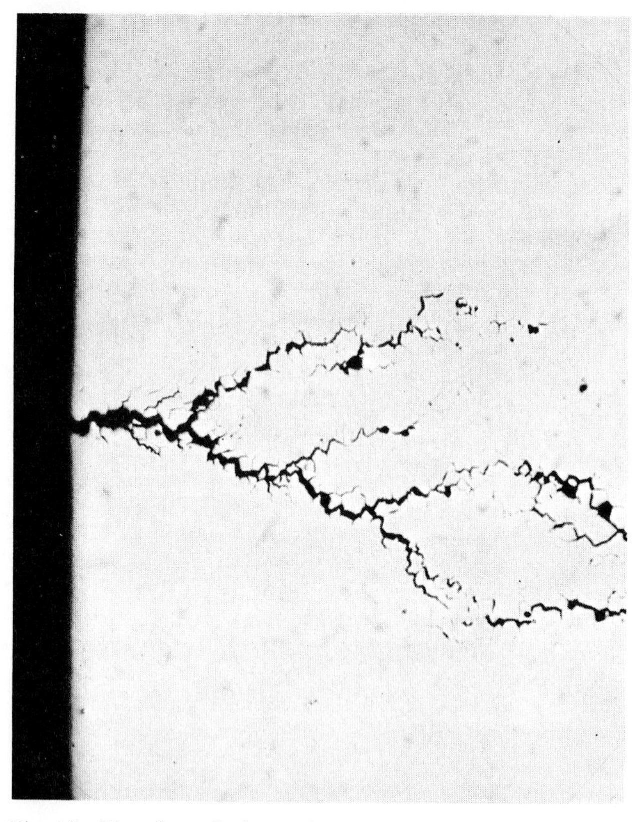

Fig. 10—Fine, branched, cracks exhibited by highly susceptible material. Magnification 65 times.

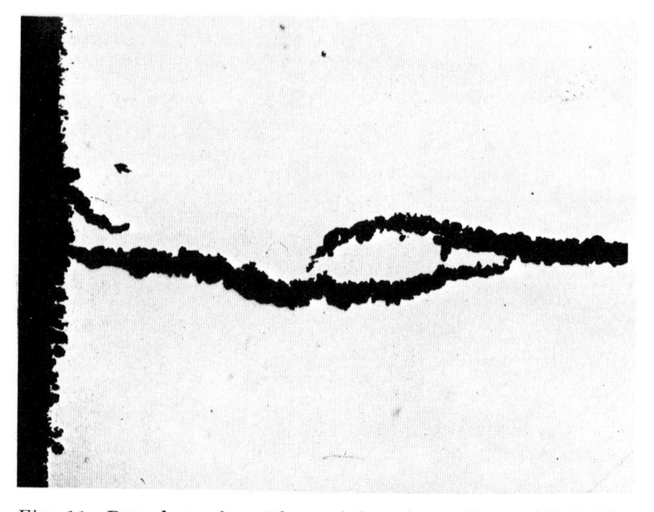

Fig. 11—Broad cracks with much less branching exhibited by more resistant material. Magnification 71 times.

The other type of area found on fracture surfaces showed extensive dimpling, although the latter did not always have the same shape, size, or clarity of delineation from specimen to specimen. Examples are shown in Figs. 17 and 18; Fig. 17 approaches most closely, perhaps, the classical dimpling usually associated with ductile rupture; Fig. 18 shows the directions of fracture to be opposite on different sides of the boundary located near the center of the photograph and also shows an extensive area that is almost featureless. Some of the dimpled areas were more undulating and less sharply defined than those in Fig. 17, while others exhibited small zones within which steps were apparent, and others showed fringes. Dimpling was widely observed on samples where the fracture surface comprised about 90 pct stress corrosion fracture and 10 pct mechanical failure, and it is difficult to imagine that all of the areas showing dimples should be associated with the surface suddenly exposed on completion of fracture.

No river patterns of the type associated with cleavage were observed, although exceptionally small areas resembling quasi-cleavage were detected. Specimens fractured during or after cathodic charging at 5 ma per sq in. or higher only exceptionally showed dimpling, although they did show fringes, frequently forming a network or aligned in parallel rows.

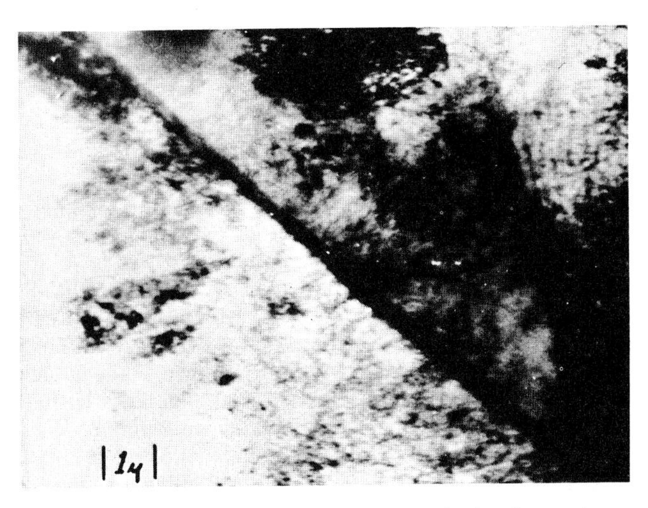

Fig. 14—Transmission electron micrograph showing grain boundary precipitate.

Fig. 12—Short transgranular cracks from intergranular cracks in highly susceptible material. Magnification 1170 times.

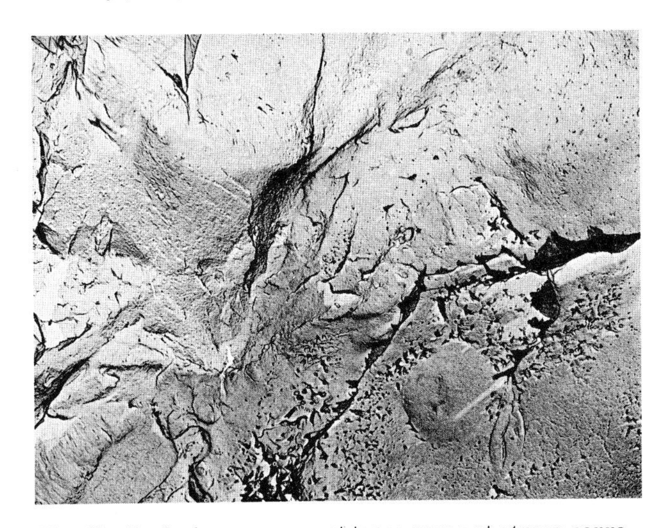

Fig. 15—Typical appearance of large areas of stress corrosion fracture surface, showing steps and fringes. Magnification 7125 times.

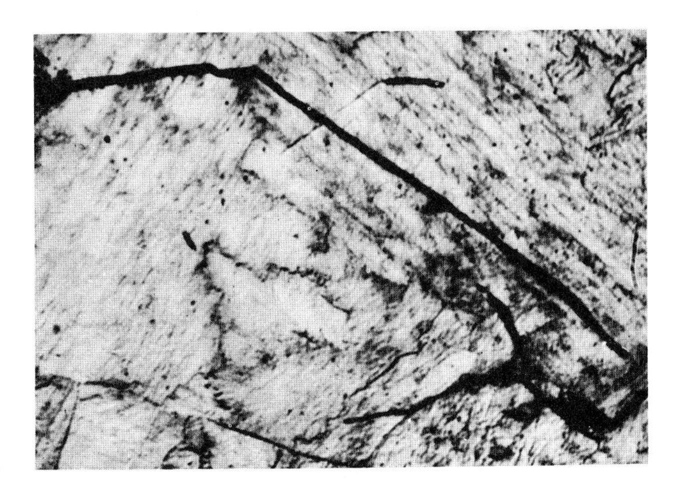

Fig. 13—Straight transgranular cracks in specimen cathodically polarized at 5 ma per sq in. Magnification 1170 times.

Fig. 16—Fringes indicative of small secondary cracks from major fracture. Magnification 7125 times.

DISCUSSION

Notwithstanding the relatively low pH (2.2) of the solution with which the present work has been concerned, and the fact that small gas bubbles were frequently detected on the surfaces of specimens before the latter cracked, it is considered that the failures observed, except when relatively high cathodic currents were applied, were instances of stress corrosion cracking rather than of hydrogen embrittlement. The evidence in support of this statement is that cracking could be very appreciably delayed by cathodic treatment at intermediate current densities, but that failure could be accelerated by anodic stimulation or by the application of relatively high cathodic current, in extension of some results quoted by Dean and Copson.[9] Similarly, significant differences were apparent in the crack slopes and paths taken by specimen fractured at high cathodic current densities as compared with specimens not so treated. Thus, in the former case, straight transgranular cracks were detectable crossing grains, Fig. 13, whereas in other cases, the cracks had the appearance shown in Figs. 11 or 12.

In the freely corroding condition, the initiation of stress corrosion cracking proper was delayed by an incubation period during which the air-formed film was disrupted and short, structurally independent, fissures developed. The incubation period was very short with highly susceptible specimens tested at the highest stresses and relatively long with less susceptible conditions, the effect of stress probably being related to its influence on mechanical rupture of the surface film, as suggested by the increased electrochemical activity reflected in the potential changes that accompanied load changes on the specimen, Fig. 8. Anodic stimulation also reduced the incubation period and, in addition, increased the frequency with which surface fissures were formed. On the other hand, increasingly large cathodic currents diminished the number of fissures and, indeed, none were detected at current densities above 1.0 ma per sq in. It appears significant that specimens to which relatively small cathodic currents were applied, 1.5 ma per sq in. or less, very few cracks were detected in the fractured specimens, as compared with the very large number, ~ 40, in all other specimens tested in this series and to which Fig. 9 refers. Additionally, this would appear to support the suggestion that the mechanism of cracking was different according to whether the conditions of testing lay to one side or the other of the peak on Fig. 9.

The fissures formed in the incubation period do not all continue to propagate into true stress corrosion cracks, as is apparent from Figs. 6 and 7. The conditions that determine whether or not a surface fissure continues to propagate into what may be termed a crack, are probably concerned with the stress and/or electrochemical conditions at the tip of the fissure. Thus, the fact that extensive cracking was only experienced when extensive fissuring was detected, suggests that the latter are playing some positive role and that cracks are unlikely to propagate, in the solution used in the present work, from a plane surface. This could be due to the necessity for some geometrical feature within which some critical species, or concentration thereof, can collect, without stifling by blockage of the fissure with corrosion products. (A black film, presumably of magnetite, was noticed on the surfaces of cracked specimens, especially in the vicinity of cracks; with anodic stimulation this film formed within minutes of the start of the experiment and covered the whole of the exposed surface.) Stress will presumably assist in maintaining a continuous path between the bulk of the solution and that located at the crack tip.

In material having the austenitic grain boundaries in a susceptible condition, penetration from the tips of fissures was sustained, and recognizable stress corrosion cracks formed, presumably as the result of local electrochemical action resulting from the peculiarities of the structure or composition of the grain boundary region. The majority of fissures remain dormant, however, possibly because once a few have propagated cracks, the stress conditions at the tips of the majority are no longer conducive to continued reaction. In specimens of relatively high resistance to stress corrosion, and especially if these were tested at stresses approaching the threshold value, the propagation of a fissure into a crack was markedly less dependent upon the positions of austenite grain boundaries. It is true that cracks of the types shown in Fig. 11 were sometimes seen to be associated with the austenite grain boundaries when viewed at an appropriate magnification; but, in general, the paths that they followed were at least as much transgranular as they were intergranular and appeared to be dictated by stress conditions rather than by structure. In this

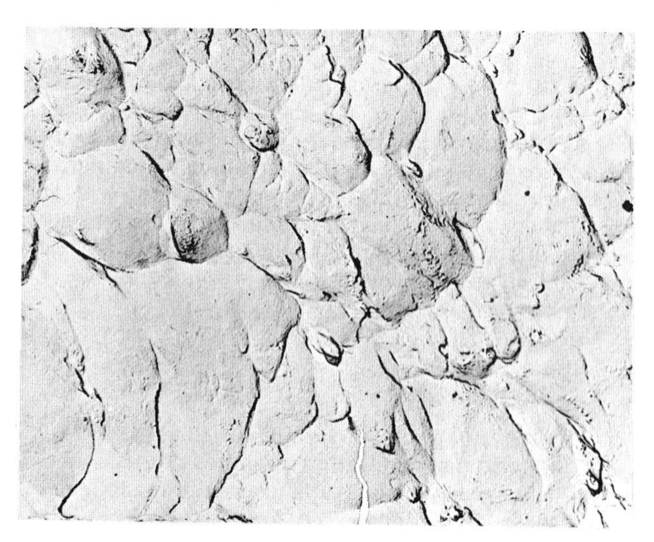

Fig. 17—Clearly delineated dimples on fracture surface. Magnification 7125 times.

Fig. 18—Dimples indicative of fracture extending in opposite directions on opposite sides of boundary. Magnification 5450 times.

latter respect, they resembled the surface fissures that formed in the, so-called, incubation period; and it may be the case that cracks such as these should be considered as forming by essentially the same mechanism as that which results in fissures. It seems significant that the curve shown in Fig. 2 for steel B is virtually coincident with that in Fig. 1 for steel A austenitized at 1400° F, and that it was only with the specimens to which those curves refer that cracking of the type shown in Fig. 11 was observed.

Irrespective of whether the crack was propagating along the original austenite grain boundaries or without much reference to structural features, it continued as a fine crack until the stress on the uncracked material reached the macroscopic yield stress and the crack opened or yawned. The fastest rate of crack propagation was observed during this latter period and corresponded to a linear rate of 0.8 mm per hr, although the average rate for all measurements was about half of this value. To maintain such a crack propagation rate by purely electrochemical processes would require a current density at the crack tip of about 0.5 amp per sq cm, and while such values are not commonly encountered in electrochemical processes, it has been shown that they can occur if depolarization is assisted by plastic straining[6] or by stimulation of the cathodic reaction by appropriate location of efficient cathodic sites.[7] Whether or not either of these circumstances obtained in the present work is unknown, but both appear feasible in the present context and it may be unnecessary to explain the crack propagation rates observed by recourse to arguments involving short bursts of mechanical rupture alternating with periods of relatively slow electrochemical crack propagation. On the other hand, the examination of fracture surfaces has revealed evidence of steps and of dimples, both of which are suggestive of either an intermittent process or one in which the overall propagation is assisted by some mechanical tearing. The load relaxation curves during crack propagation at the yield stress were not smooth, but whether this was due to some stick-slip behavior in the pen recorder or was indeed indicative of intermittent crack propagation is uncertain, since no special precautions were taken to examine for small, but sharp, load drops in the specimens. The weight of the evidence does seem to suggest that crack propagation was assisted by purely mechanical processes and these possibly played an increasing part as the stress corrosion crack extended until in the final stage they were entirely responsible for sudden completion of failure.

Although the microstructural evidence indicates that two different mechanisms of crack propagation may have been operative in the system studied, the least resistance to stress corrosion cracking was associated with cracking that followed the austenite grain boundaries. Moreover, it is apparent that the austenitizing temperature may markedly influence the likelihood of attack in such regions. This effect of austenitizing temperature does not appear to be simply related to grain size but rather to some effect that occurs at all temperatures above about 1460° F. The electron microscopy showed particles of a precipitate, possibly of titanium carbide, in the prior austenite grain boundaries, and these particles were larger and in greater number in the highly susceptible material than in more resistant samples. Whether or not this structural difference is responsible for the difference in cracking propensity is unknown at the present time, but if the precipitate is of TiC, it is possible that this is effective either by acting as points of intense cathodic activity or by inducing physical or chemical changes in the adjacent material that renders the latter more susceptible to dissolution. The differences in composition of the two steels employed in the present investigation give no obvious clue as to the nature of the precipitate or the reason for the difference in cracking susceptibility of the two steels, since the compositions are too complex for the small differences to have obvious significance. This is perhaps not surprising in view of the fact that the results of Dean and Copson[9] covering a very much more extensive range of steels show no significant trends with composition, except possibly when the nickel content is increased from about 18 to 25 pct. However, it would appear that the matter should continue to be pursued since there is little doubt that in the most susceptible condition for cracking there is marked structural dependence of the cracking and that this can be related to the properties of the grain boundary regions as influenced by heat treatment.

CONCLUSIONS

1) Two types of cracking have been found in stress corroded 18 pct Ni maraging steel samples, one associated with a relatively high threshold stress, ~ 200,000 psi, and the other with a threshold at about half this value.

2) The more susceptible condition is associated with crack propagation along the prior austenite grain boundaries and is markedly dependent upon the temperature at which austenitizing is performed. Electron microscopy has indicated that these effects may be related to the formation of a carbide precipitate at the boundaries. The indications are that this type of cracking may be controlled through heat treatment or, possibly, composition.

3) The markedly less structure dependent cracking associated with the high threshold stress may not be capable of control by these means, but its occurrence may be restricted to solutions similar to that used in the present work.

4) Measured crack propagation rates are of the order of those that may be explained on a purely electrochemical basis, although there is some evidence that propagation may be intermittent and involve states of mechanical advancement.

5) The cracking is amenable to electrochemical control and in such a manner as to suggest that in freely corroding circumstances is due to stress corrosion and not hydrogen embrittlement.

ACKNOWLEDGMENTS

This work was performed while one of us, RNP, held a visiting fellowship at Mellon Institute. The assistance of numerous colleagues is gratefully made, especially to Dr. G. Pollard and to Mr. G. Goldberg. The support of the work by the Scaife Family of Pittsburgh is gratefully acknowledged.

REFERENCES

[1] A. Rubin: Aerojet General Corp., rept. no. 2914, August, 1964.

[2] J. A. S. Green and E. G. Haney: Preprint for Symposium on Stress Corrosion Testing during ASTM Annual Meeting, June, 1966, Atlantic City.

[3] J. G. Hines and R. W. Hugill: in *Physical Metallurgy of Stress Corrosion Fracture* edited by T. N. Rhodin, p. 193, Interscience Publishers, 1959.

[4] M. Henthorne and R. N. Parkins: *Corrosion Sci.,* 1966, vol. 6, p. 357.

[5] R. N. Parkins: *Met. Rev.,* 1964, vol. 9, p. 201.

[6] T. P. Hoar and J. M. West: *Proc. Roy. Soc.,* 1962, vol. 268, Ser. A, p. 304.

[7] M. Henthorne and R. N. Parkins: *Br. Corros. J.,* Sept., 1967, vol. 2,

[8] E. H. Phelps and R. B. Mears: *First International Congress on Metallic Corrosion,* 1961, Butterworth and Co., p. 157.

[9] H. P. Leckie: Preprint for Conference on Fundamental Aspects of Stress Corrosion Cracking, Columbus, Ohio, Sept., 1967. To be published in Proceedings of Conference.

[10] S. W. Dean and H. R. Copson: *Corrosion,* 1965, vol. 21, p. 95.

Stress-Corrosion Cracking Behavior of an 18 Pct Ni Maraging Steel

A. J. STAVROS AND H. W. PAXTON

Stress-corrosion cracking of an 18 pct Ni maraging steel in aqueous solutions was studied using precracked cantilever beam specimens. By appropriate heat treatments, six different structures having the same yield strength were obtained. Although significantly different plane strain fracture toughness values (K_{Ic}) resulted, it was found that the threshold plane strain stress intensity (K_{Iscc}) was the same for all structures. K_{Iscc} had the same value in 3 pct NaCl at various pH values, in 1N H_2SO_4, and in distilled water. Specimens tested in 3 pct NaCl under both anodic and cathodic applied potentials also exhibited this same K_{Iscc} value. Fractographic inspection of the crack surfaces revealed no apparent differences due to changes in solution, pH, or applied potential. The crack path was intergranular in all cases. However, specimens austenitized at 1500°F exhibited crack branching, whereas in specimens austenized at much higher temperatures branching no longer occurred. Aging time and temperature seemed to change only the time to failure. The mechanism most consistent with all observations appears to be hydrogen cracking.

THE stress-corrosion cracking behavior of high-strength maraging steel has not been investigated extensively. Failure of maraging or other high-strength martensitic steels in aqueous solutions does not seem to depend on the presence of a specific ion, as is the case, for example, in the austenitic stainless steels. The only safe generalization that can be made to date about the cracking behavior of maraging steels is that higher yield strength tends to increase the susceptibility to cracking.

Many of the previous investigations into the stress-corrosion behavior of maraging steels are summarized in the reviews by Matteoli and Soyga,[1] Phelps,[2] and Kennedy and Whittaker.[3] Even though the heat treatment of maraging steel affects its electrochemical characteristics, almost all these previous investigations were concerned with determining the effects of either electrochemical variables or metallurgical variables but not both.

The investigation described below was therefore designed to study the effects of both types of variables simultaneously. By working with a number of different microstructures having identical yield strength, it was possible to consider the effects produced on cracking behavior through variations of austenite grain size, amount of austenite, precipitate size and distribution, and fracture toughness. Electrochemical variables were studied simultaneously by varying the test solution, pH, applied potential, and atmosphere above the solution.

Since the initiation of a crack usually occurs at a rate many orders of magnitude slower than the propagation rate, any test involving smooth specimens is essentially a measure of the resistance of the material to crack initiation, *e.g.*, by pitting. To avoid the long and poorly reproducible initiation period of the stress-corrosion process, a sharp artificial crack was incorporated into the test specimen.

Another important feature of precracked tests is that the concepts of linear elastic fracture mechanics can be applied to describe mathematically the effects of geometry and stress on conditions at the crack tip. Brown[4] has recently reviewed the application of fracture mechanics to stress-corrosion testing; all symbols used in this paper will have the same definitions as presented in Brown's review.

EXPERIMENTAL PROCEDURES

The steel investigated was an 18 pct Ni (300 grade) maraging type purchased in the form of hot-rolled and straightened bar. All experiments were performed using material from a single heat. The composition, as determined by the manufacturer, is given in Table I.

Heat treating was done in neutral salt baths. The two austenitizing treatments examined were a) an anneal of 2 hr at 1500°F and b) an anneal of 1 hr at 2300°F followed by direct quenching to 1700°F and holding for 4 hr. Three aging treatments, each of which hardened the previously austenitized steel to R_c 52-53, were selected: 1) 800°F for 10 hr, 2) 900°F for $3\frac{1}{2}$ hr, and 3) 900°F for 100 hr.

The stress-corrosion test specimens were cantilever beams approximately 1/2 by 5/8 in. in cross section and 12 in. long. A V-notch 0.22 in. deep with a root radius of less than 0.01 in. was machined $5\frac{1}{2}$ in. from one end. The specimens were also side-grooved on each side to a depth of 5 pct of the width (approximately 0.025 in.). After heat treatment the specimens were precracked on a constant load fatigue machine. The procedures outlined above for obtaining precracked cantilever beam specimens all conform to published recommendations for insuring plane strain conditions at the crack tip.[5]

Testing was done in distilled water, 3 pct NaCl of various pH values, and 1N H_2SO_4. Reagent grade chemicals and distilled water were used to prepare the salt and acid solutions. Each solution was placed in a

A. J. STAVROS, formerly with the Department of Metallurgical and Materials Science, Carnegie-Mellon University, Pittsburgh, Pa., is at Homer Research Laboratories, Bethlehem Steel Corp., Bethlehem, Pa. H. W. PAXTON is Head, Department of Metallurgy and Materials Science, Carnegie-Mellon University.

Manuscript submitted April 6, 1970.

Table I. Composition of Maraging Alloy

Element	C	Mn	P	S	Ni	Co	Mo	Ti	Al	Si	Zr	Fe
Wt pct	0.005	0.01	0.004	0.006	18.23	8.99	5.08	0.67	0.08	0.01	0.027	Bal

sealed $\frac{1}{2}$ pt polyethylene container surrounding the crack. Provision in these cells was made for a gas inlet and outlet tube and occasionally for a Luggin probe and an auxiliary platinum electrode.

The procedure for testing the precracked specimens involved: 1) adding solution to the cell, 2) bubbling hydrogen or oxygen gas into the solution for 15 to 60 min, 3) bolting a 30 in. lever arm onto the free end of the specimen, and 4) adding the desired load to the end of the lever arm. In controlled potential tests, the potential was applied potentiostatically at least 10 min before the load was applied.

An approximate indication of crack growth was obtained by monitoring the deflection of the cantilever with a dial gage. Time to failure was recorded when the falling lever arm tripped a microswitch.

Data obtained from these precracked cantilever tests are plotted as initial stress intensity, K_{Ii}, vs the time to failure, t_f. K_{Ii} can be calculated from any of several expressions which Rolfe *et al.*[6] have shown can be simplified to the form illustrated in Fig. 1. The various treatments differ slightly in the value given the $f(a/w)$ term.

To establish the threshold stress intensity, K_{Iscc}, the step loading technique was employed. If no crack growth was evident after 10,000 min the load was increased slightly. It has previously been shown that this technique does not alter the K_{Iscc} value from that obtained by direct loading.[7]

The toughness of the steel was also determined from cantilever beams. Several specimens from each heat treatment were loaded to failure in air. The stress intensity at fracture, K_{Ix}, gives an approximation of the plane strain stress intensity (a measure of fracture toughness), K_{Ic}, which is normally determined by more elaborate methods.

After a stress-corrosion failure the fractured surfaces of many specimens were inspected directly with a scanning electron microscope. In some instances when the surfaces were less rough, plastic-carbon replicas were examined in a transmission electron microscope.

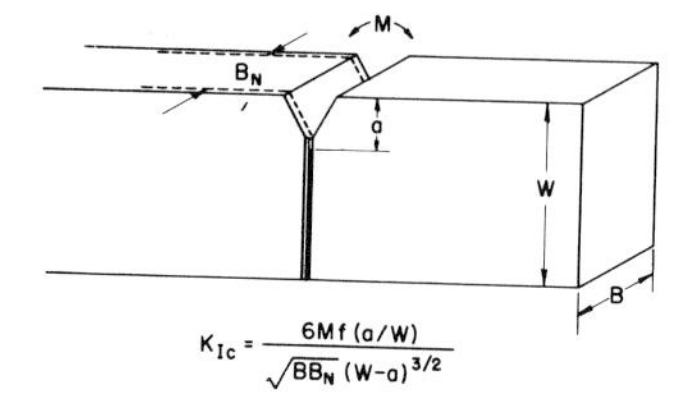

Fig. 1—Cantilever beam specimen.

Table II. Effect of Heat Treatment on K_{Ix}

Heat Treatment	K_{Ix}, ksi$\sqrt{\text{in.}}$	Average
1500°F/2 hr + 800°F/10 hr	122, 115.5 122	119.9
1500°F/2 hr + 900°F/3½ hr	102.5, 92, 112, 90.5	99.2
1500°F/2 hr + 900°F/100 hr	68, 71.5	69.8
2300°F/1 hr + 1700°F/4 hr + 800°F/10 hr	58.5, 61.5, 55.6	58.5
2300°F/1 hr + 1700°F/4 hr + 900°F/3½ hr	53.5, 59.5 60.0	57.5
2300°F/1 hr + 1700°F/4 hr + 900°F/100 hr	52.3, 51.5, 54.2	52.6

RESULTS

Microstructural Variations

Microstructures produced by the six different heat treatments were documented by standard techniques. The prior austenite grain size was determined by comparing optical micrographs with standard ASTM grids. Austenitization at 1500°F resulted in a grain size of ASTM no. 9 while austenitizing at 2300° + 1700°F yielded a grain size of ASTM no. 0. A substantial variation in size between individual grains was noted within each specimen, especially after the high temperature anneal.

The fine structure resulting from the heat treatments was observed by thin film electron microscopy. Both austenitizing treatments produced a typical martensitic structure composed of laths with a high density of dislocations.[8]

Aging at 800°F for 10 hr did not change the appearance of this structure to any extent. The dislocation density seemed slightly lower, but no precipitates were visible at 60,000X (Kinetic studies by other investigators indicate the presence of two types of molybdenum-precipitate below 850°F[9]).

Aging 3 hr at 900°F resulted in two types of precipitate in the martensite matrix. These precipitates have been previously identified as Ni_3Mo and σ - FeTi.[9] Aging for 100 hr at 900°F results in larger precipitates and approximately 20 pct reverted austenite.

When the higher temperature austenization treatment was employed large grain boundary precipitates were observed in addition to the usual matrix precipitates. These large precipitates are believed to be Ti(C,N).[10]

Fracture Toughness Variations

Table II presents the K_{Ix} values obtained for each of the six heat treatments selected for investigation. It is evident that austenitizing at higher temperatures for longer times results in a more brittle structure. Aging at 800°F results in better toughness than aging at 900°F, and overaging gives a lower toughness relative to underaging. The toughness variations, however, are not related to the stress-corrosion cracking behavior.

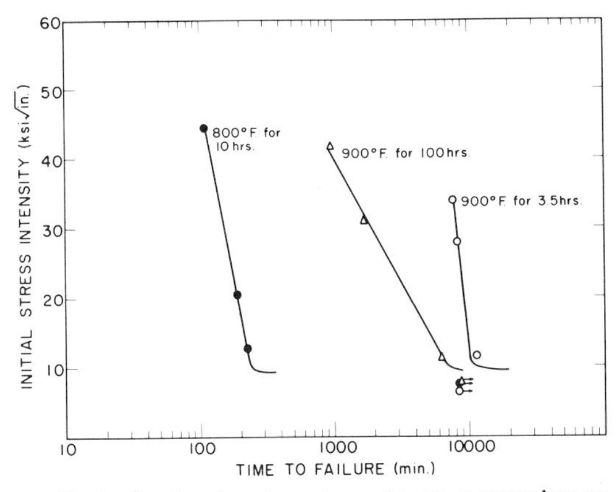

Fig. 2—Effect of aging treatment on stress-corrosion susceptibility in 3 pct NaCl at pH 6.3 of 18 pct Ni maraging steel austenitized 2 hr at 1500°F.

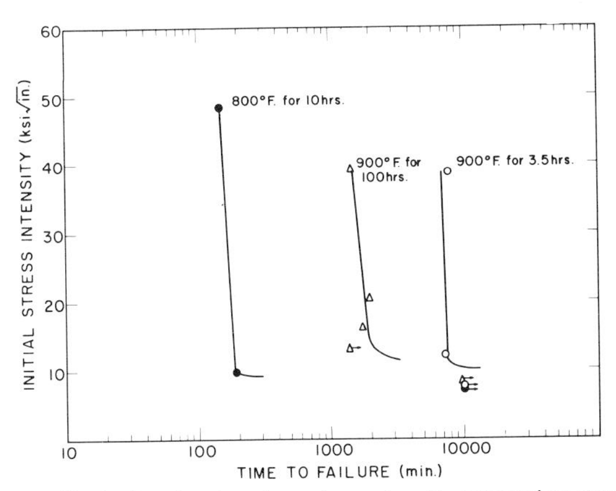

Fig. 3—Effect of aging treatment on stress-corrosion susceptibility in 3 pct NaCl at pH 1.7 of 18 pct Ni maraging steel austenitized 2 hr at 1500°F.

Cracking Behavior

Sodium Chloride Solutions

Fig. 2 illustrates the cracking behavior in deaerated 3 pct NaCl at pH 6.3 of maraging steel annealed at 1500°F and aged. The time to failure appears relatively insensitive to stress intensity but very dependent on aging treatment. K_{Iscc} has a value of approximately 10 to 15 ksi$\sqrt{\text{in}}$. for all three aging treatments. When the pH of the salt solution is changed to 1.7 the same type of behavior is observed, Fig. 3. K_{Iscc} is still 10 to 15 ksi$\sqrt{\text{in}}$.

Figs. 4 and 5 illustrate the cracking behavior in deaerated 3 pct NaCl of pH 6.3 and 1.7, respectively, of maraging steel annealed at 2300° + 1700°F and aged. Time to failure appears much more dependent on stress intensity and appreciably less so on aging treatment. K_{Iscc}, however, still has a value of 10 to 15 ksi$\sqrt{\text{in}}$.

The mode of cracking in all cases was intergranular with respect to the prior austenite grains. Steel annealed at 1500°F exhibited crack branching whenever the stress intensity at the crack tip exceeded approximately 31 ksi$\sqrt{\text{in}}$., whereas annealing at the higher temperatures never resulted in branching.

Branching was in the form of two divergent cracks, approximately 45 deg (measured at the side surface) to the plane of the fatigue crack. These two cracks continued to branch further until the structure shown in Fig. 6 resulted.

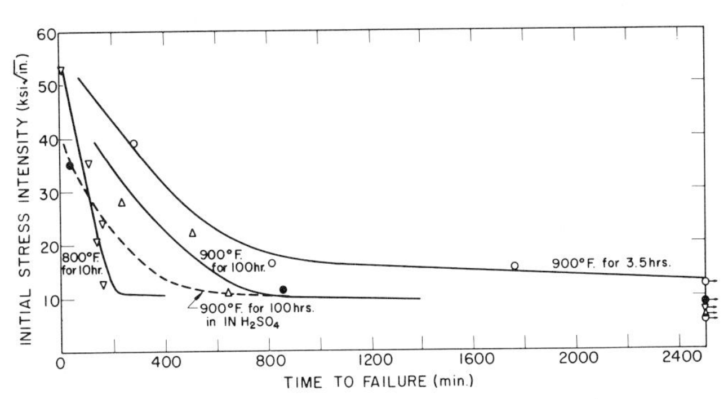

Fig. 4—Effect of aging treatment on stress-corrosion susceptibility in 3 pct NaCl at pH 6.3 of 18 pct Ni maraging steel austenitized 1 hr at 2300°F + 4 hr at 1700°F.

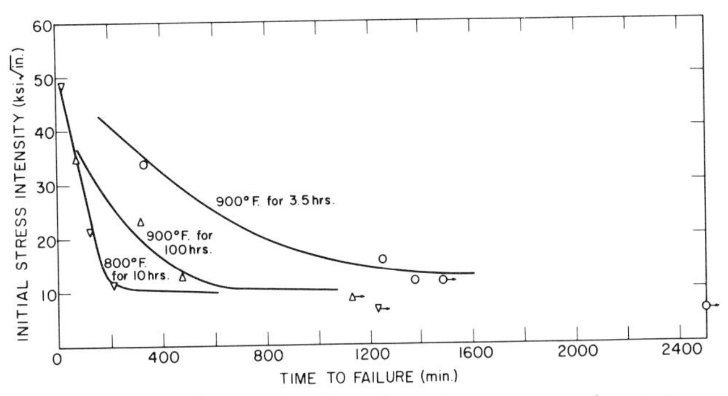

Fig. 5—Effect of aging treatment on stress-corrosion susceptibility in 3 pct NaCl at pH 1.7 of 18 pct Ni maraging steel austenitized 1 hr at 2300°F.

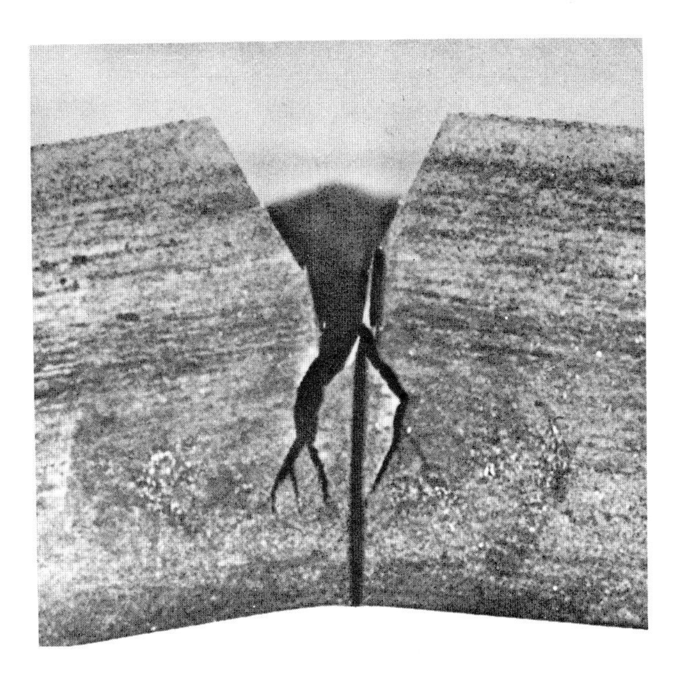

Fig. 6—Extensive crack branching.

Sulfuric Acid Solutions

Several experiments were conducted in H_2-saturated 1N H_2SO_4. The stress intensity-failure time plots were similar to those obtained in 3 pct NaCl for material given equivalent heat treatments, Figs. 4 and 7. Branching was also observed in specimens given the 1500°F annealing treatment.

Distilled Water

Deaerated distilled water was also used as a test environment. The resulting stress intensity-failure time curves were similar to those obtained in the salt and acid solutions. Fig. 8 presents a comparison of cracking behaviors in distilled water, sulfuric acid, and sodium chloride for a single heat treatment. The stress intensity dependence and the threshold stress intensity appear unaffected by environment.

Electrode Potential Variations

Most previous investigations on stress-corrosion cracking of martensitic steels have relied on electrochemical concepts to suggest the controlling mechanism. The effect of potential on cracking behavior in 3 pct NaCl of pH 6.3 for one heat treatment is shown in Fig. 9. All applied potentials lower t_f with respect to the unpolarized condition at all K_{Ii}. Thus, the elec-

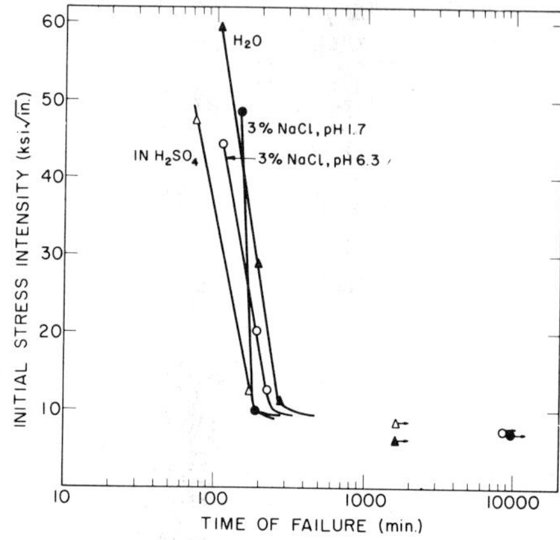

Fig. 8—Effect of environment of stress-corrosion susceptibility of 18 pct Ni maraging steel austenitized 2 hr at 1500°F and aged 10 hr at 800°F.

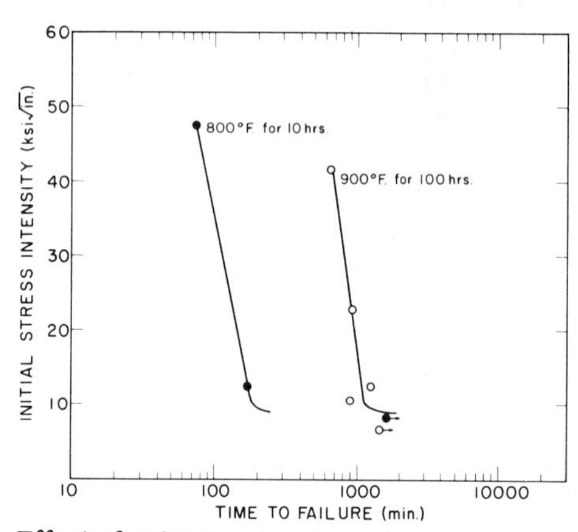

Fig. 7—Effect of aging treatment on stress-corrosion susceptibility in 1N H_2SO_4 of 18 pct Ni maraging steel austenitized 2 hr at 1500°F.

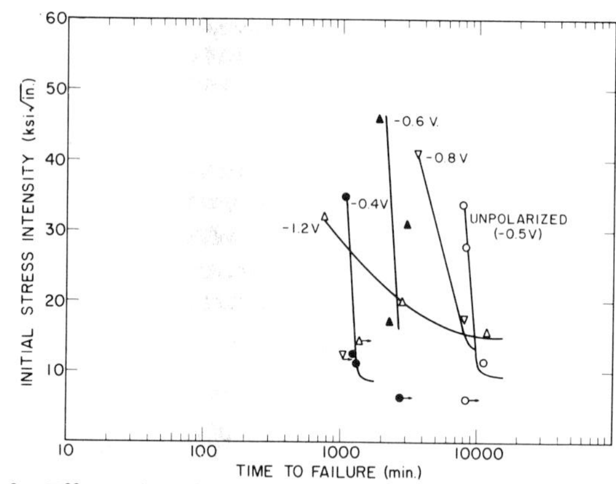

Fig. 9—Effect of applied potential on stress-corrosion susceptibility of 18 pct Ni maraging steel austenitized 2 hr at 1500°F and aged $3\frac{1}{2}$ hr at 900°F.

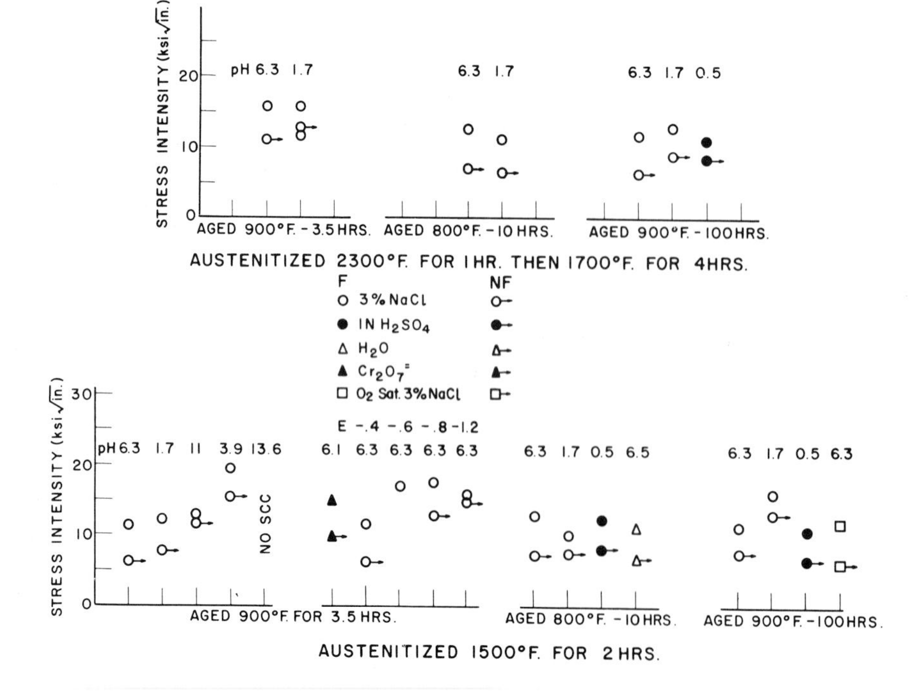

Fig. 10—K_{Iscc} of 18 pct Ni maraging steel in various heat treated conditions and environments.

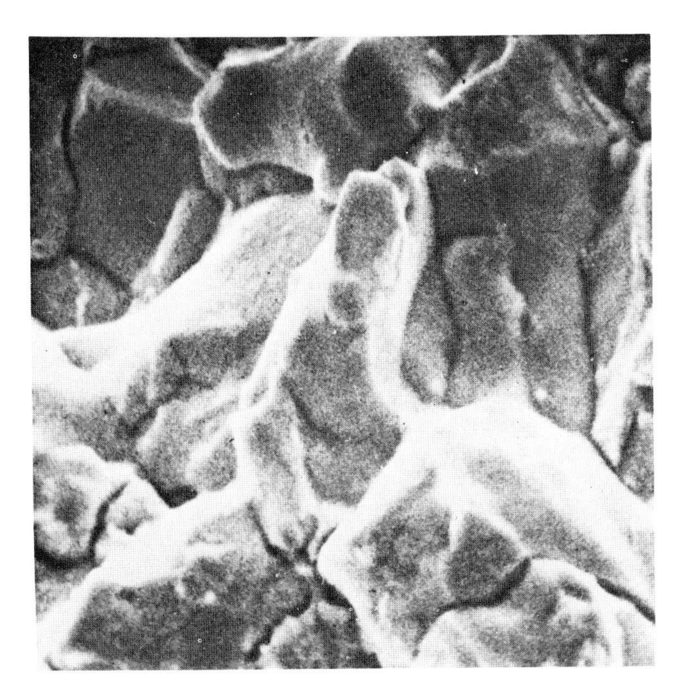

Fig. 11—Typical stress-corrosion topography of 18 pct Ni maraging steel austenitized 2 hr at 1500°F. Scanning electron microscopy. Magnification 1875 times.

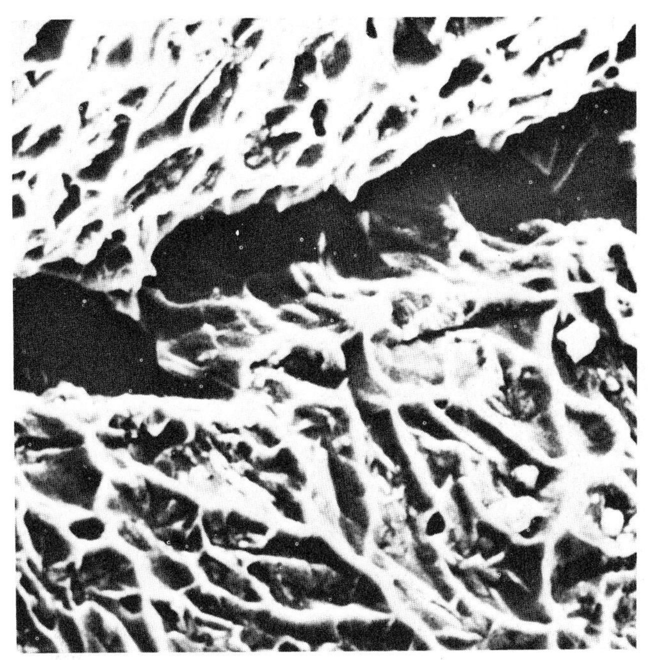

Fig. 12—Typical stress-corrosion topography of 18 pct Ni maraging steel austenitized 1 hr at 2300°F + 4 hr at 1700°F. Scanning electron microscopy. Magnification 1875 times.

trochemical means of distinguishing between possible mechanisms[11] does not yield valid results. K_{Iscc}, however, does not appear to be measurably affected by applied potential.

Further Tests

Since the threshold stress intensity for this maraging steel appears invariant on the basis of the above results, further tests were conducted to determine K_{Iscc} in other solutions; a) in 3 pct NaCl at other pH values, b) in oxygen-saturated 3 pct NaCl, and c) in 3 pct NaCl + 1.5 pct $Na_2Cr_2O_7$ at pH 6.1. Fig. 10 presents the threshold stress intensities obtained for all tests conducted. The value is always about 10 to 15 ksi$\sqrt{in}$.

Fractography

Inspection by scanning electron microscopy of the surfaces of fractured stress-corrosion specimens annealed at 1500°F revealed that in all cases failure is intergranular, the individual grain facets being rather featureless. Many secondary cracks or crevices were always evident. The surface appearance did not seem to depend on aging treatment, test solution, pH, or applied potential and always appeared as in Fig. 11.

Austenitizing at the higher temperatures (2300° + 1700°F) did change the surface appearance, however, as illustrated in Fig. 12. Cracking was still intergranular with many secondary cracks present, but the individual grain facets were no longer smooth. They appeared to have many tear ridges, similar to those attributed to quasicleavage.[12]

The severe roughness of the stress-corrosion fracture surfaces made preparation of replicas very difficult, especially for material annealed at the higher temperatures. Replicas obtained from material annealed at 1500°F showed features characteristic of stress-corrosion fractures in high strength martensitic steels, Fig. 13. Although a thorough examination of all specimens was not conducted, Fig. 13 is typical of all tests run on 1500°F annealed specimens.

DISCUSSION

The main experimental observations from this study are:

1) K_{Iscc} was found to be constant throughout all tests and has a value of 10 to 15 ksi$\sqrt{in}$.
2) Low temperature aging (800°F) results in shorter failure times than high temperature aging (900°F). At 900°F underaging ($3\frac{1}{2}$ hr) is more beneficial than overaging (100 hr).
3) Time to failure decreased with both anodic and cathodic polarization.
4) Cracking occurred intergranularly in all cases. The grain facets were either smooth and featureless or covered with tear ridges, depending only on whether the austenitizing treatment was at 1500°F or at 2300° + 1700°F.

These observations are most consistently explained as resulting from a hydrogen cracking mechanism. Since the theories of hydrogen cracking depend on the amount of hydrogen present in critical areas and not on its source, the exact nature of the environment should have little influence as long as the necessary amount of hydrogen can enter the steel. The similar stress intensity-failure time curves observed in sulfuric acid, sodium chloride, and distilled water is evidence for this statement provided hydrogen is actually available in each of these environments.

Immersion of maraging steel into sulfuric acid results in the evolution of hydrogen on the steel surface and hence probable entry of some hydrogen into the matrix.

Fig. 13—Typical stress-corrosion electron fractograph of 18 pct Ni maraging steel austenitized 2 hr at 1500°F. Magnification 3400 times.

Recent evidence also indicates that hydrogen can enter the metal at the crack tip in sodium chloride solutions. Brown[13] has determined the solution chemistry near the tip of a propagating crack in several steels, including a maraging steel, tested in sodium chloride at various initial pH values. In all cases the pH at the crack tip was 3.6 to 3.8 and the ferrous ion concentration about one molar. If similar behavior in distilled water is assumed, then inspection of the Pourbaix diagram for the iron-water system indicates that the thermodynamic requirements for the generation of hydrogen are met within the crack. Thus all test solutions provided a source of hydrogen.

The effect that aging treatment had on failure time must be accounted for by assuming that the resulting structures affect either the hydrogen entry rate or the hydrogen diffusion rate. If the tests of material annealed at 1500°F are considered, it is seen that time to failure is changed more by varying aging treatment than by varying test solution or applied potential. Since these latter two variables cannot change the diffusion rate of hydrogen in the matrix it appears that aging influences the time to failure by affecting the hydrogen diffusion rate. This situation is consistent with the known fact that structural changes affect the trapping of hydrogen in maraging steels.[14]

The electrochemical means of distinguishing between anodic dissolution stress-corrosion and hydrogen cracking depends on the imposition of applied potentials. If the hydrogen generated by the cathodic reaction of a corrosion process is the cause of cracking then the application of a cathodic potential should increase the amount of hydrogen and shorten time to failure. Conversely, application of an anodic potential should decrease the available hydrogen and lengthen time to failure. If, on the other hand, anodic dissolution is responsible for cracking the application of a cathodic potential should lengthen time to failure, whereas, an applied anodic potential should shorten time to failure (if attack remains localized).

The various types of curves which may be encountered in use of the polarization method have been summarized by Bhatt and Phelps.[11] The experimentally observed decrease in failure time with both anodic and cathodic applied potentials is interpreted as indicating an anodic dissolution mechanism at anodic potentials and a hydrogen cracking mechanism at cathodic potentials. No direct information on the operating mechanism under open-circuit conditions is obtained in this type of experiment.

A change of mechanism with potential does not appear consistent when K_{Iscc} is independent of potential. Recent literature has suggested that hydrogen may be generated even on specimens anodically polarized.[15,16] If this occurred, K_{Iscc} should remain unaffected by potential even though the time to failure decreases with both anodic and cathodic applied potentials.

The assumption of hydrogen generation at anodic potentials may not be necessary if we consider that the potential at the crack tip may be quite different from the controlled bulk potential. The crack is very narrow (it can be estimated from fracture mechanics as[17] about 0.0002 in.), which suggests, in accordance with theoretical treatments,[18,19] that much of the crack tip may be relatively unaffected by the value of the potential outside the crack.

If the crack tip potential is unchanged then the solution chemistry in this region will always be suitable for hydrogen entry into the metal. A hydrogen mechanism could then operate regardless of the applied bulk potential and K_{Iscc} should remain independent of potential.

The shortened time to failure observed when a potential was applied is probably a reflection of the fact that a portion of the hydrogen responsible for cracking may have entered the steel in regions easily influenced by the applied potential (*e.g.* the crack edges or the upper crack surfaces). Since the diffusion coefficient of hydrogen is high (10^{-6} - 10^{-7} sq in per sec) in martensite, transport of hydrogen to the crack tip will not be limiting. The precise relationship of failure time to applied potential will be then determined by the way in which the applied potential affects the generation and entry of hydrogen in these exposed regions. Any applied potential which induces hydrogen to enter the metal will be detrimental, regardless of whether it is anodic or cathodic.

Additional support for a single mechanism causing cracking is provided by the similarity of the fracture surfaces under all conditions. The differences in appearance between grain facets annealed at 1500°F and those annealed at 2300° + 1700°F may be attributed to the preferential precipitation of Ti (C, N) at the prior austenite grain boundaries when annealing takes place at the higher temperatures. However, these carbonitrides, which are detrimental to fracture toughness, do not have any obvious effect on hydrogen cracking as is evident by the uniform K_{Iscc} value for all heat treatments.

Crack branching during stress-corrosion crack growth has recently been reviewed by Carter.[20] He found that a critical stress intensity was required before branching occurs. In this investigation the critical stress intensity was $31\,ksi\sqrt{in.}$ The existence of a "branching threshold" may be sought in structural terms. At $K_I = 31\,ksi\sqrt{in.}$ the plastic zone diameter at the crack tip[21] is nearly the same size as the prior austenite grains, ASTM 9. A larger plastic zone diameter might result in partial transgranular fracture which could require more energy than does the intergranular mode.

SUMMARY

This investigation has shown that the heat treatment given a maraging steel, through its effect on the resulting structure, can have a pronounced influence on stress-corrosion cracking behavior. Annealing and aging treatments were varied in a manner designed to keep the yield strength constant. The threshold stress intensity was found to be independent of heat treatment, test solution, pH value, and applied potential. Such behavior suggests that a single mechanism—hydrogen cracking—is responsible for stress-corrosion.

Although the threshold stress intensity was invariant, the time to failure could be changed slightly by varying the test solution, to a somewhat greater extent by varying the pH or applied potential, and by two orders of magnitude by varying the aging treatment. The hydrogen cracking mechanism was shown to be consistent with these failure time changes. Aging treatments were very effective in affecting failure times because the diffusion coefficient of hydrogen in martensite can be changed by the number and type of trapping sites which are dependent on aging treatment.

Fractographic analysis also indicated that a single mechanism probably causes cracking in all cases, since the fracture surfaces did not change in appearance as the test conditions were varied.

Another influence of structure on stress-corrosion behavior is illustrated by the phenomenon of crack branching. Branching was found to occur when the crack-tip plastic zone interacted with some structural unit which appears to be related to the prior austenite grain size. Fine-grained material which is produced by lower austenitizing temperatures branches extensively thus increasing the time to failure relative to the nonbranching case.

ACKNOWLEDGMENTS

This research was supported by the Advanced Research Projects Agency of the Department of Defense and was monitored by U.S. Naval Research Laboratory under Contract No. Nonr-760(31). The authors are appreciative of many stimulating discussions with associates interested in the problems of stress-corrosion. A. J. Stavros was an International Nickel Co. Fellow during part of this work and gratefully acknowledges this support.

REFERENCES

1. L. Matteoli and T. Songa: in Stress Corrosion Cracking in Aircraft Structural Materials, AGARD Proceedings No. 18, 1967.
2. E. H. Phelps: in Fundamental Aspects of Stress Corrosion Cracking, National Association of Corrosion Engineers, Houston, Texas, 1969.
3. J. W. Kennedy and J. H. Whittaker: *Corrosion Sci.,* 1968, vol. 8, p. 359.
4. B. F. Brown: *Met. Rev.,* 1969, vol. 13, p. 171.
5. W. F. Brown and J. E. Srawley: Am. Soc. Testing Mater., Spec. Tech. Publ. No. 410, 1967.
6. S. T. Rolfe, S. R. Novak, and J. H. Gross: Paper presented at Am. Soc. Testing Mater. Meeting, Atlantic City, June, 1966.
7. R. P. M. Procter and H. W. Paxton: *Trans. Quart. ASM,* 1969, vol. 62, p. 989.
8. J. M. Chilton and C. J. Barton: *Trans. Quart. ASM,* 1967, vol. 60, p. 518.
9. S. Floreen: *Met. Rev.,* 1969, vol. 13, p. 115.
10. C. J. Barton, B. G. Reisdorf, P. H. Salmon Cox, J. M. Chilton, and C. E. Oskin, Jr.: Air Force Materials Lab., AFML-TR-67-34, March, 1967.
11. H. J. Bhatt and E. H. Phelps: *Proceedings of Third International Congress on Metallic Corrosion,* vol. 2. p. 285, Swets and Zeitlinger, Amsterdam, 1969.
12. C. D. Beachem and R. M. N. Pelloux: in Am. Soc. Testing Mater., Spec. Tech. Publ. No. 381, 1965.
13. B. F. Brown: extended abstract of paper presented at Fourth Intern. Congress on Metallic Corrosion, Sept., 1969.
14. H. R. Gray: Ph.D. Thesis, Case Institute of Technology, Cleveland, Ohio, 1967, (available University Microfilms, Order No. 67-16074).
15. J. H. Shively *et al.*: *Corrosion,* 1967, vol. 23, p. 215.
16. W. D. Benjamin and E. A. Steigerwald: Air Force Materials Lab., p. 519, AFML-TR-68-80, April, 1968.
17. B. F. Brown and C. D. Beachem: *Corrosion Sci.,* 1965, vol. 5, p. 745.
18. W. D. France and N. D. Greene: *Corrosion,* 1968, vol. 24, p. 247.
19. E. M. Gutman: *Proceedings of Third International Congress on Metallic Corrosion,* vol. 2, p. 377, Swets and Zeitlinger, Amsterdam, 1969.
20. C. S. Carter: Doc. D6-23871, The Boeing Co., Renton, Wash., March, 1969, (available CFSTI, AD687725).
21. G. R. Irwin: *Eng. Fract. Mech.,* 1968, vol. 1, p. 241.

Fracture Toughness and Stress Corrosion Characteristics of a High Strength Maraging Steel

C. S. CARTER

The effect of heat treatment on the tensile, fracture toughness, and stress corrosion properties of a high strength maraging steel (nominal composition 16.3 Ni-12.87 Co-4.98 Mo-0.78 Ti) is described. A maximum ultimate tensile strength of 323 ksi, combined with a fracture toughness K_{Ic} of 62 ksi $\sqrt{in}$, was achieved. This strength level appears to be the maximum which can be achieved in maraging type steels without decreasing the crack tolerance below that of currently used high strength low alloy steels. Reversion to austenite did not improve either the fracture toughness or stress corrosion resistance relative to completely martensitic microstructures with equivalent strength.

SINCE their introduction in 1960 the 18 pct Ni maraging steels have been increasingly used for a variety of aerospace and other applications. The commercially available 200, 250, and 300 grades contain 17 to 19* Ni, 7 to 9 Co, 3 to 5 Mo and titanium within the range 0.15 to 0.80 depending upon the strength-grade required.[1] The highest strength which can be guaranteed for the 300 grade in heavy section form is 280 ksi. While higher strength grades have been developed,[2,3] the fracture toughness of these alloys has limited their structural application. The purpose of this study was to explore the feasibility of increasing the minimum tensile strength of 300 grade maraging steel by modifying the alloy content while maintaining an acceptable level of fracture toughness. In addition the effect of heat treatment on fracture toughness and stress corrosion was investigated.

*Alloy content is given as wt pct.

ALLOY CONTENT

Previous studies have shown that titanium has a potent hardening effect on 18 pct Ni maraging steel (approximately 90 ksi per 1 pct addition) but has an adverse effect on fracture toughness.[1] Therefore, the titanium content in this program was restricted to the upper end of the 300 grade range; i.e., 0.8 pct. To achieve the desired strength the cobalt content was increased from 8.5 to 9.5 pct to 12 to 13 pct, thereby enhancing the age-hardening response.[4] To maintain the M_S temperature at a similar level to the 300 grade (300 to 350°F) the nickel content was decreased by 2 pct.

A 200 lb. heat was vacuum induction melted by Carpenter Technology Corporation to this analysis and then remelted under vacuum into an 8 in. diam ingot. The ingot was press forged to 5 in. sq, then to $2\frac{3}{4}$ in. sq, and finally to 1 in. sq. Chemical analysis of the heat is shown in Table I, the composition being essentially that aimed for.

C. S. CARTER is Senior Engineer, Commercial Airplane Group, The Boeing Company, Renton, Wash.

Manuscript submitted October 26, 1970.

EXPERIMENTAL PROCEDURE

Tension specimens 0.25 in. diam were machined from the 1 in. sq bar. After heat treatment the specimens were tensile tested at room temperature at a strain rate of 0.005 in. per in. per min through to yield and 0.02 in. per in. per min to fracture.

Single edge notch specimens (0.5 in. thick, 1.0 in. wide, 5.0 in. long) were also machined from the bar and fatigue precracked after heat treatment. These were loaded in 3 point bending according to the ASTM recommended practice for plane strain fracture toughness testing,[5] and the results analyzed according to this procedure.

Standard size (0.394 in. sq) Charpy specimens were used to determine the stress corrosion properties. These were fatigue precracked after heat treatment; the maximum stress intensity imposed during fatigue cracking was estimated to be less than 30 ksi $\sqrt{in.}$ The specimens were subsequently loaded in cantilever bending to selected initial stress intensity levels using a hydraulic loading system. The stress intensity was calculated according to the relationship given by Brown Brown and Srawley[6] for pure bending. A supply of $3\frac{1}{2}$ pct aqueous sodium chloride was dripped into the notch, delivery being commenced just prior to load application. Specimens were exposed until either failure occurred or a minimum period of 100 hr elapsed. If fracture did not occur the specimens were rapidly broken open in laboratory air and macroscopically examined for evidence of crack growth.

Austenite contents were measured by X-ray diffraction using the technique described by Lingren.[7]

RESULTS

Effect of Solution Annealing

The effect of double solution annealing (1 hr + 1 hr) was investigated, followed by an aging treatment of

Table I. Chemical Composition, Wt Pct

Ni	Co	Mo	Ti	C	Mn	Si	P	S	B	O, ppm	N, ppm
16.30	12.87	4.98	0.78	0.004	0.01	0.01	0.005	0.003	0.0018	14	10

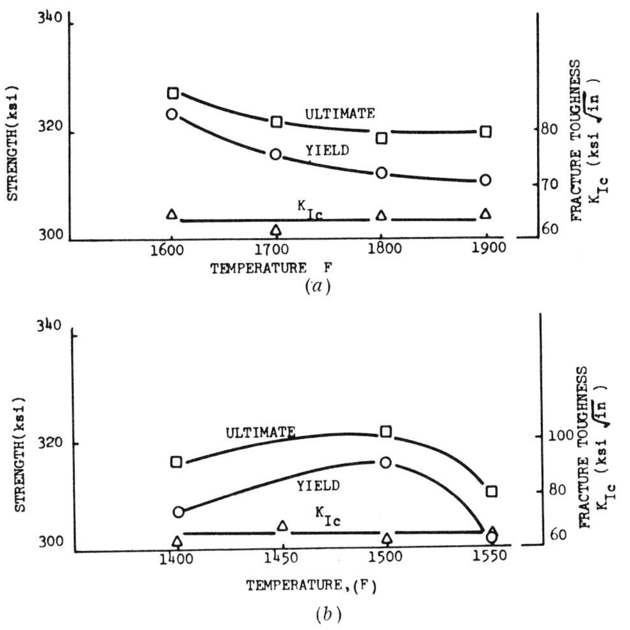

Fig. 1—Effect of solution treatment (aged 900°/8 hr). (*a*) Effect of first solution annealing temperature (second solution anneal at 1500°F); (*b*) effect of second solution annealing temperature (first solution anneal at 1700°F).

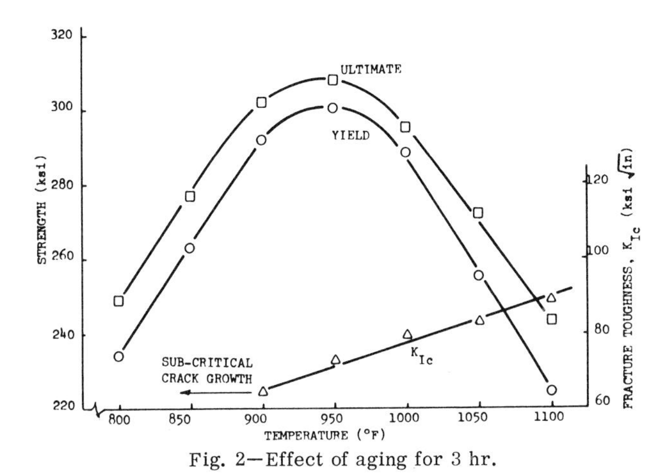

Fig. 2—Effect of aging for 3 hr.

900° F for 8 hr. Results are shown in Fig. 1. Increasing the first annealing temperature slightly reduced the tensile properties but had no effect on the plane strain fracture toughness K_{Ic}. Optimum properties were achieved with a second solution annealing temperature of 1500°F; other temperatures decreased the tensile strength properties although there was no effect on fracture toughness. In view of these results all further tests were conducted after double solution annealing: 1700° F (1 hr) + 1500°F (1 hr).

Effect of Aging Treatment

TENSILE AND FRACTURE TOUGHNESS PROPERTIES

The effect of aging temperature and time was initially examined using single specimens for both fracture toughness and tensile properties. Results of aging at temperatures within the range 800° to 1100°F for

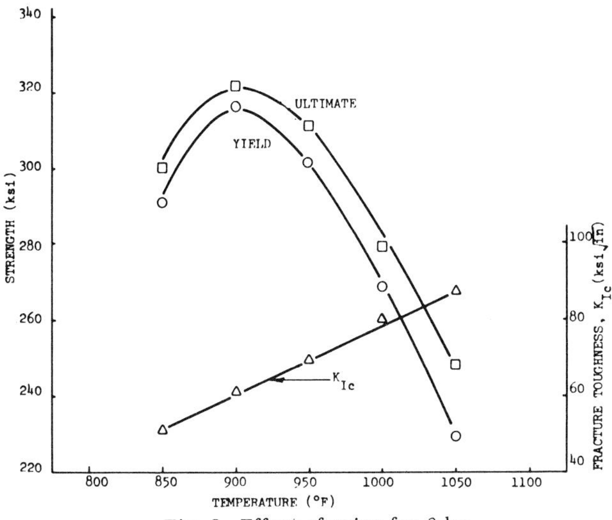

Fig. 3—Effect of aging for 8 hr.

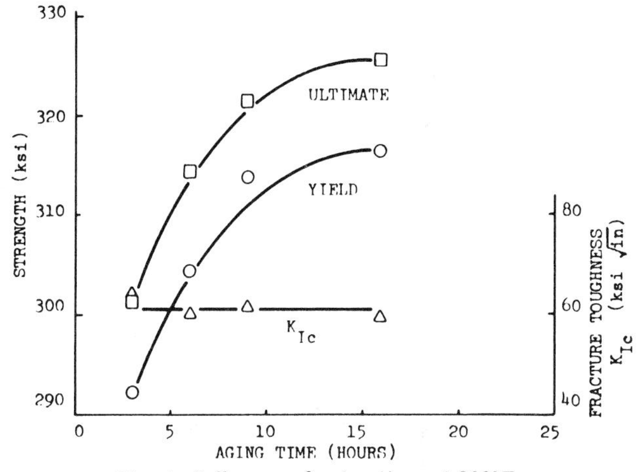

Fig. 4—Influence of aging time at 900°F.

periods of 3 and 8 hr are shown in Figs. 2 and 3, respectively. Tensile strength response was typical of age-hardening type materials. For the 3 hr aging period a peak tensile strength of 307 ksi was obtained after heat treatment at 950°F. After aging at 900°F for 8 hr the strength increased to 318 ksi. Extending the aging period at 900°F to 16 hr increased the tensile strength by 5 ksi, Fig. 4. These results were confirmed by triplicate tests on specimens aged at 900°F for 8 and 16 hr, Table II.

Subcritical crack growth occurred during the rising load fracture toughness tests on specimens aged at 800°F (3 and 8 hr) and 850°F (3 hr). Cracking occurred in an integranular manner and extensive branching occurred on a macroscopic scale. Because the branch cracks dissipate the applied load, a K_{Ic} value (for a single crack) could not be determined.[8] Similar behavior was reported for 350 grade maraging steel heat treated and tested in an identical manner.[3] This was attributed to a stress corrosion--hydrogen embrittlement crack growth mechanism, laboratory air being the corrosive mechanism. A similar mechanism would appear to explain the subcritical crack growth discussed above. This phenomenon was not observed when specimens were aged at temperatures of 900°F and higher.

Table II. Mechanical and Fracture Toughness Properties

Aging Treatment	Yield Strength, ksi	Ultimate Strength, ksi	Reduction of Area, pct	Fracture Toughness K_{Ic} ksi $\sqrt{in.}$
900°F for 8 hr	315.8	321.8	N.M.	61.5
	309.6	312.6	60	63.0
	316.3	320.4	60	63.5
	(313.9)	(318.3)	(60)	(62.7)
900°F for 16 hr.	316.2	325.4	N.M.	59.1
	317.0	321.6	55	62.2
	320.8	322.8	57	65.7
	(318.0)	(323.8)	(56)	(62.3)

Note: 1 NM—Not Measured.
2 ( )—Indicates Average.

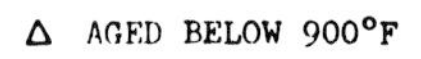

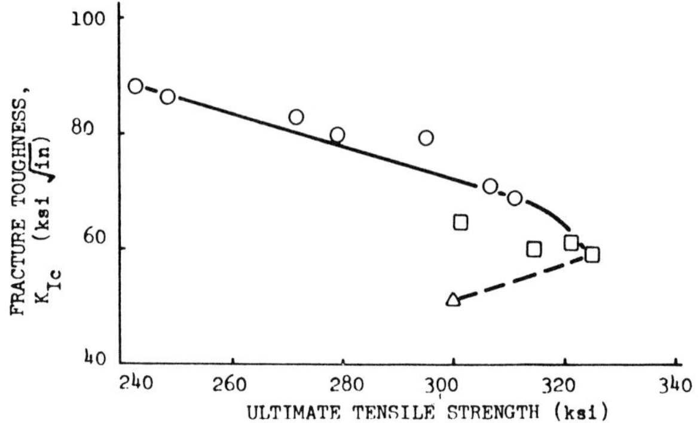

Fig. 5—Relationship between tensile strength and toughness.

As shown in Figs. 2 and 3 the fracture toughness increased with aging temperature. Aging at 850°F (8 hr) led to lower fracture toughness than at 900°F despite the 20 ksi greater tensile strength at the higher aging temperature. There was no significant effect of aging time at 900°F on the fracture toughness although the tensile strength was increased by 25 ksi when the time was increased from 3 to 16 hr, Fig. 4. This was confirmed by additional fracture tests on specimens aged for 8 and 16 hr, Table II. The relationship between strength and toughness is shown in Fig. 5. Aging at temperatures above 900°F gave the best combination of strength and toughness although maximum strength necessitated aging at 900°F.

The relationship between mechanical properties and retained austenite content is shown in Fig. 6. It is apparent that the decrease in strength following aging at temperatures exceeding 900°F can be associated with austenite reversion.

Fracture toughness increased with austenite content but the combination of strength and toughness was no better than can be achieved with other grades of maraging steel having a completely martensitic structure and fully age hardened to comparable strength. For example, the fracture toughness of 250 grade maraging steel is 80 to 100 ksi $\sqrt{in.}$ combined with a tensile strength of approximately 260 ksi.[1] Thus, the deliber-

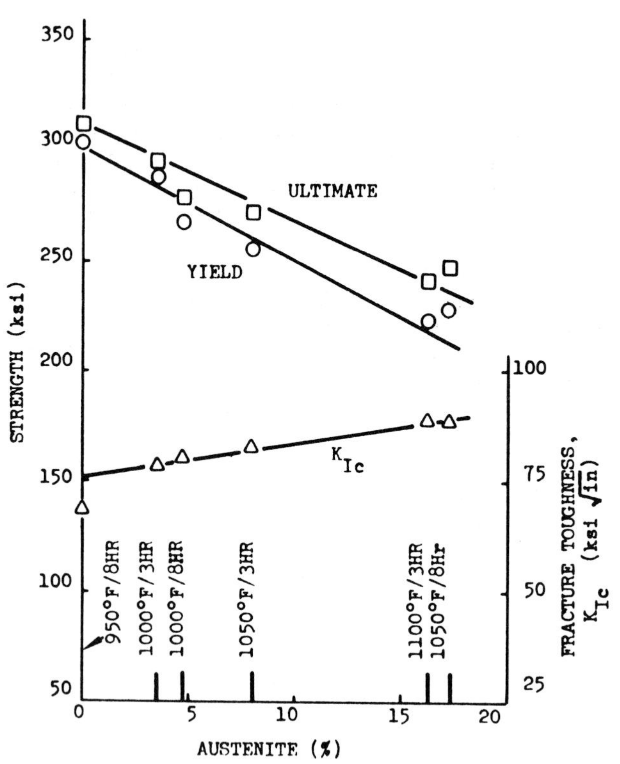

Fig. 6—Relationship between austenite content and properties.

ate reversion to austenite is not beneficial with regard to strength/toughness.

To investigate the influence of a duplex microstructure of martensite and unaged martensite on strength/toughness the following heat treatment study was performed. Single edge notch specimens were quenched from a solution annealing temperature of 1500°F to selected temperatures which encompassed the M_s to M_f temperature range. This provided a structure of martensite, and untransformed austenite. Specimens were held for 2 hr and then reheated to 900°F and held for 8 hr, to age the martensite. Air cooling to room temperature transformed the residual austenite to martensite (unaged). The fracture toughness values obtained and the corresponding hardness values are shown in Fig. 7. An increase in toughness was only obtained at the expense of marked decrease in hardness and hence in strength. The strength/toughness combination obtained is no better than can be achieved from the commercial grades of maraging steel aged to peak strength in a conventional way.

STRESS CORROSION PROPERTIES

Stress corrosion curves of initial stress intensity K_{Ii} vs time to failure are shown in Fig. 8 for four aging treatments. The threshold stress intensity K_{Iscc} below which stress corrosion cracking did not occur was approximately 10 ksi $\sqrt{in.}$ for all heat treatment conditions except for the 850°F for 3 hr treatment which gave a K_{Iscc} of less than 10 ksi $\sqrt{in.}$ However, the time to failure at a given initial stress intensity level increased with aging temperature.

The stress corrosion curves show that the time to failure was essentially independent of initial stress intensity for K_{Ii} levels exceeding a value approximately 15 ksi $\sqrt{in.}$ for the 850°F and 900°F aging treatments. This value increased to 20 to 25 ksi$\sqrt{in.}$ at higher aging temperatures. Previous studies have shown that this time independent behavior can be correlated with an essentially constant rate of crack growth.[8] Also, extensive crack branching was observed in all specimens; a constant crack velocity is an essential requirement for the occurrence of this phenomenon. Since visual observation showed that the period for crack initiation was small compared to the total time to failure, an estimate of the stress corrosion crack velocity (for the stress intensity range from 15 or 20 ksi$\sqrt{in.}$ to K_{Ic}) could be obtained by dividing the stress corrosion crack length by the time to failure. Velocities obtained in this way are presented in Table III, and summarized in Fig. 9. Velocity decreased by almost two orders of magnitude as the aging temperature increased from 850°F to 1050°F. This can be contrasted with the K_{Iscc} parameter which was essentially independent of heat treatment. Therefore the increase in failure time with aging temperature noted above is due to the accompanying increase in K_{Ic}, Figs. 2 and 3, which dictates larger critical crack lengths for mechanical fracture, and the decrease in crack velocity.

The velocity data is compared with that reported for other grades of 18 pct maraging steels in the stress intensity independent regime of growth in Fig. 10. It appears that there is an approximately linear relationship between the logarithm of crack velocity and yield strength when the aging temperature is not less than 900°F. It is interesting to note that the velocity estimated for the 1050°F for 3 hr age condition containing 18 pct austenite is similar to that for a completely martensitic 250 grade maraging steel (aged 900°F for 3 hr) at a comparable strength level. On the other hand, the K_{Iscc} associated with this 250 grade steel was 40 ksi $\sqrt{in.}$ compared to 10 ksi $\sqrt{in.}$ for the overaged steel. Thus, reverted austenite can be seen to have no beneficial effect on stress corrosion resistance when strength level is taken into account.

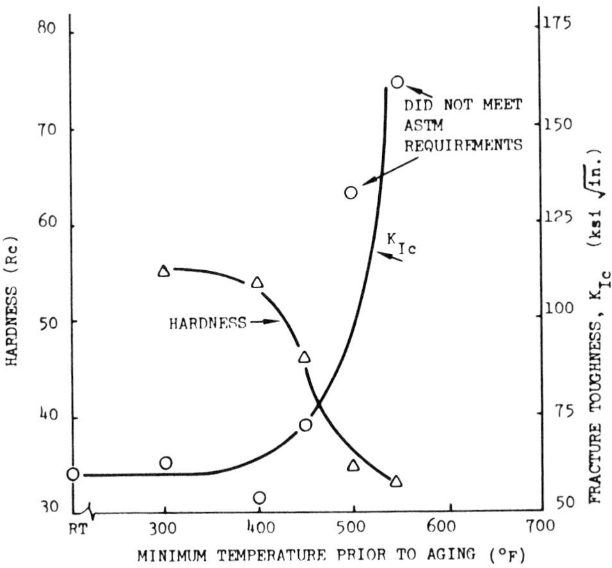

Fig. 7—Effect of interrupted quenching on properties.

Table III. Stress Corrosion Crack Velocity Data

Aging Treatment	Initial Stress Intensity, ksi $\sqrt{in.}$	Crack Velocity, in. per min
850°F for 3 hr	14.2	1.2 X 10^{-3}
	25.0	1.6 X 10^{-3}
	40.0	1.5 X 10^{-3}
		(1.4 X 10^{-3})
900°F for 8 hr	25.0	1.2 X 10^{-4}
	35.0	5.4 X 10^{-4}
	50.0	2.2 X 10^{-4}
		(2.9 X 10^{-4})
900°F for 16 hr	14.7	2.5 X 10^{-4}
	23.5	2.8 X 10^{-4}
		(2.7 X 10^{-4})
1050°F for 3 hr	25.0	8.8 X 10^{-5}
	38.7	6.3 X 10^{-5}
	58.3	5.7 X 10^{-5}
	80.0	6.3 X 10^{-5}
		(6.9 X 10^{-5})

Note: ( ) Indicates Average.

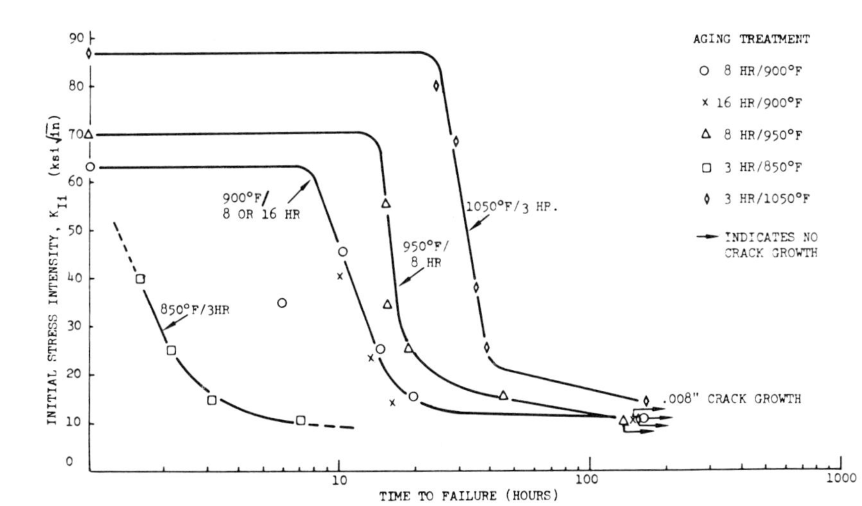

Fig. 8—Effect of aging on stress corrosion properties.

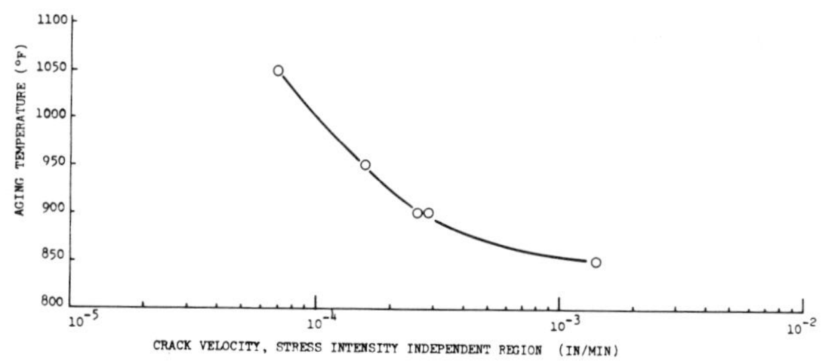

Fig. 9—Effect of aging temperature on stress corrosion crack velocity.

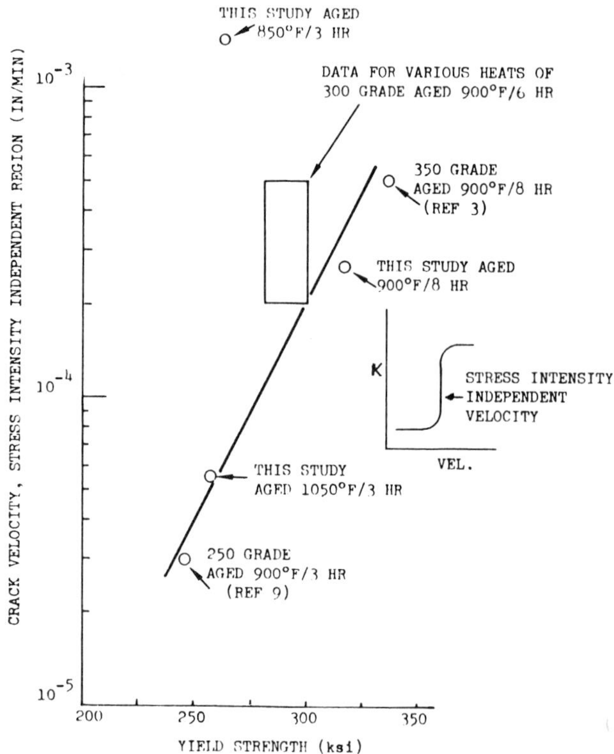

Fig. 10—Relationship between yield strength and stress corrosion crack velocity.

DISCUSSION

The relationship between strength and fracture toughness for various maraging steels with strength levels exceeding 300 ksi is shown in Fig. 11. The major question to be answered is what is the minimum toughness level required. Estimates of this can be made by means of fracture mechanics. Using the maximum stress to be applied in service and inspection capability for crack detection the minimum toughness required to avoid brittle fracture can be estimated. Alternatively, we can consider maintaining the critical flaw size at the same level as in currently used high-strength low alloy steels. The conditions for brittle fracture can be written as:

$$K_{Ic} = \sigma\sqrt{\pi c}\, S \qquad [1]$$

where

σ = Applied stress

C = Critical crack length

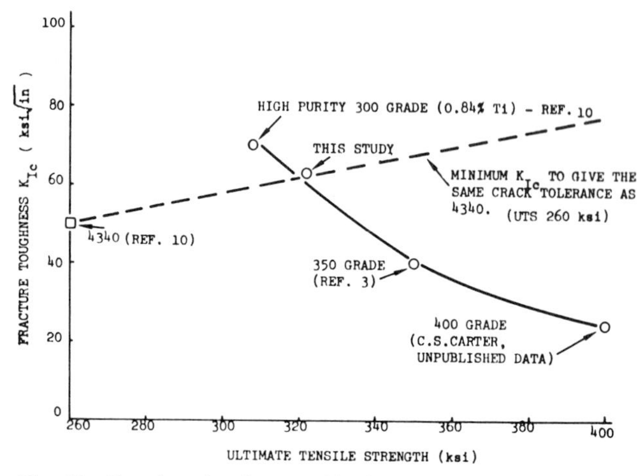

Fig. 11—Fracture toughness of high strength maraging steels.

S = A factor depending on crack shape and loading conditions

Since design stresses are frequently based on the ultimate tensile strength (σ_u) the applied stress can be expressed as

$$\sigma = \sigma_u D \qquad [2]$$

where D is a design factor.

Thus Eq. [1] can be written as:

$$K_{Ic} = \sigma_u \sqrt{\pi c}\, DS \qquad [3]$$

It can be seen from Eq. [3] that the fracture toughness must increase in direct proportion to the tensile strength to maintain the same critical crack size. The alloy 4340 has been widely used for airplane landing gear and other airframe applications at the 260 to 280 ksi strength level. At this strength level the plane strain fracture toughness is approximately 50 ksi $\sqrt{\text{in.}}$[10] Using the above approach the toughness required to maintain the critical flaw size at the same level as in 4340 can be determined. This minimum toughness level is compared with the maraging steel data in Fig. 11 and shows that the critical flaw size in the experimental maraging steel described in this report is maintained at the same level as in the 4340 steel in spite of the 60 ksi increase in tensile strength level. However, the crack tolerance of the higher strength maraging grades (350 and 400) is lower than that of 4340. Thus, 320 ksi appears to be the maximum tensile strength which can be achieved in the maraging system if crack resistance is to be maintained at least at the same level as in low alloy steels.

Stress corrosion crack growth in low alloy high strength steels is stress intensity independent over a wide range of stress intensity in most heat treatment conditions.[11] Such velocities for a number of these steels are compared with the maraging steel data in Fig. 12. At equivalent strength levels the stress corrosion crack velocity in maraging steel is slower than in the low alloy steels. Comparison with the 4340 steel data shows an order of magnitude difference. Wei and Landes[16] have suggested that fatigue crack growth in an aggressive environment can be estimated by adding the sustained load influence of the environment to the fatigue crack growth rate in an inert en-

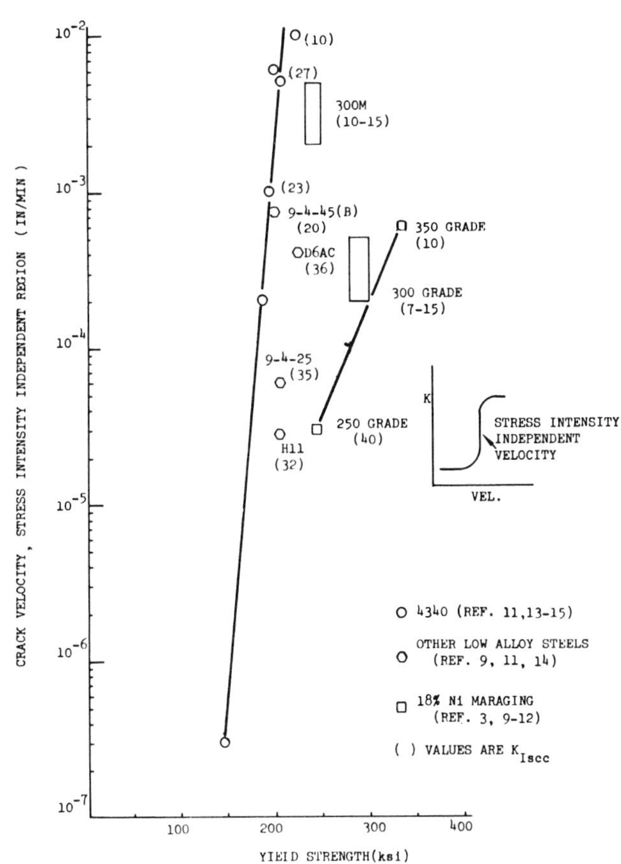

Fig. 12—Comparison of stress corrosion crack velocities in maraging and low alloy steels.

vironment on a cyclic basis. The fatigue crack growth rate characteristics of high strength low alloy and 18 Ni (300) maraging steels are reported to be similar in an inert environment.[17] Therefore, the Wei–Landes hypothesis combined with the data shown in Fig. 11 suggests that fatigue crack growth resistance of maraging steels in a sodium chloride environment will be significantly better than in low alloy steels of similar strength at low cyclic frequencies.

CONCLUSIONS

1) An ultimate tensile strength level of 320 ksi can be achieved in a maraging steel containing 16.3 Ni-12.87 Co-4.98 Mo-0.78 Ti. The associated plane strain fracture toughness K_{Ic} was 62 ksi $\sqrt{\text{in.}}$ It is shown that this combination of strength and toughness provides a similar resistance to brittle fracture as 4340 steel heat treated to a 260 ksi tensile strength level. Increasing the strength of maraging steels to higher levels results in a lower crack tolerance than in 4340.

2) Although reverted austenite reduced the strength, the fracture toughness and stress corrosion resistance were no greater than other grades of 18 pct Ni maraging steel in the fully martensitic condition at comparable strength levels.

3) A duplex microstructure of aged and unaged martensite did not offer any benefit with respect to strength/toughness.

4) Increasing the aging temperature from 850° to 1050°F significantly decreased the stress corrosion crack velocity in the region of stress intensity independent crack growth but had an insignificant effect on the threshold stress intensity K_{Iscc}.

5) At equivalent strength levels the stress corrosion crack velocity in the stress intensity independent region is approximately an order of magnitude less in maraging steels than in 4340 type steels.

REFERENCES

1. A. M. Hall and C. J. Slunder: The Metallurgy, Behavior, and Application of the 18-Percent Nickel Maraging Steels, NASA Report SP-5051, 1968.
2. G. W. Tufnell and R. F. Cairns: *ASM Trans. Quart.*, 1968, vol. 61, p. 798.
3. C. S. Carter: *Met. Trans.*, 1970, vol. 1, p. 1551.
4. S. Floreen: Met. Rev., vol. 126, Metals and Materials, September 1968, p. 115.
5. *Proposed Method of Test for Plane Strain Fracture Toughness of Metallic Materials, ASTM Stand.*, May,1969, Pt. 31, p. 1099.
6. W. F. Brown and J. E. Srawley: *Plane Strain Crack Toughness Testing* STP 410, ASTM 1967.
7. R. Lingren: *Metal Progr.*, April 1965, p. 102.
8. C. S. Carter: *Stress Corrosion Crack Branching in High Strength Steels*, presented at the National Symposium on Fracture Mechanics, Lehigh Univeristy, August 1969, to be published in *J. Eng. Fract. Mech.*
9. S. Mostovoy, H. R. Smith, R. G. Lingwall, and E. J. Ripling: *A Note on Stress Corrosion Cracking Rates*, presented at the National Symposium on Fracture Mechanics, Lehigh University, August 1969, to be published in *J. Eng. Fract. Mech.*
10. C. S. Carter: Evaluation of a High Purity 18% Ni (300) Maraging Steel Forging Air Force Mater. Laboratory Rept. No. AFML-TR-70-139, June 1969.
11. C. S. Carter: *Observations on the Stress Corrosion Crack Propagation Characteristics of High Strength Steels*, Boeing Document D6-25274, 1970.
12. C. S. Carter: *The Tensile, Fracture Toughness, and Stress Corrosion Properties of Vacuum Melted (300) Maraging Steel*, Boeing Document D6-23888, 1969.
13. V. J. Colangelo and M. S. Ferguson: *Corrosion*, 1970, vol. 25, p. 509.
14. C. S. Carter: *Corrosion*, 1969, vol. 25, p. 423.
15. J. P. Gallagher: *Corrosion Fatigue Crack Growth Behavior Above and Below K_{Iscc}*, Naval Res. Lab. Rept. No. 7064, May 1970.
16. R. P. Wei and J. D. Landes: *Mater. Res. Stand.*, 9, no. 7, 1969, p. 25.

SECTION VI:
Environmental Cracking Behavior

Environmental Cracking of 18 Ni 200 Maraging Steel★

J. B. GILMOUR*

Abstract

The susceptibility of 18 Ni 200 maraging steel (1380 MPa YS) to environmental cracking in 3.5% NaCl solution has been investigated. This alloy has been found to be susceptible to cracking due to hydrogen embrittlement, with $K_{ISCC} < 26$ MPa·m$^{1/2}$ at -1.10V_{SCE}. Deoxygenation of the salt solution has been shown to greatly decrease the time to failure when the alloy is subjected to the mildly hydrogen embrittling conditions given by coupling to cadmium.

Preliminary screening tests carried out by others in this laboratory showed that 18 Ni 200 maraging steel was more resistant to environmental cracking (EC)[(1)] than 18 Ni 250 maraging steel.[1-3] The latter alloy had developed a large number of environmental cracks when used in an advanced marine vessel, and the 18 Ni 200 was considered as a possible replacement. Therefore, an extensive test program was undertaken to evaluate more precisely the EC properties of the 200 grade maraging steel. Specimens used in this program were of the precracked cantilever type and were tested under static applied load.

★Submitted for publication June, 1976.

*Corrosion Section, Physical Metallurgy Research Laboratories, CANMET, Department of Energy, Mines, and Resources, Ottawa, Canada.

(1)Environmental cracking is used as a general term and includes hydrogen embrittlement cracking and stress corrosion cracking.

Experimental Procedure

All the specimens for this investigation were machined from a single 12.7 mm thick plate of 18 Ni 200 maraging steel produced by Cameron Iron Works Inc., Houston, Texas. The steel was vacuum consumable-electrode arc-melted, hot rolled, and solution annealed at 815 C for 1 hour. Metallographic examination showed significant alloy banding. The manufacturer's heat analysis is given in Table 1.

The standard cantilever test bars (Figure 1) were cut from this plate, most having the length of the bar in the rolling direction and the crack propagation in the thickness direction (L-S crack

TABLE 1 – Analysis of 18 Ni 200 Maraging Steel

Element	Wt%
Ni	18.81
Co	8.57
Mo	3.24
Ti	0.19
C	0.01
Si	0.01
Mn	0.05
S	0.007
Al	0.07
B	0.004
Zr	0.004

orientation[4]). A few bars were prepared with an L-T crack orientation. The machined bars were aged for 3 hours at 480 C in air and were air cooled. This resulted in a hardness of Rc 41.5, corresponding to a yield strength of approximately 1380 MPa. The thin oxide film formed during heat treatment was not removed before testing. Syrett[2] has shown that this oxide does not affect the EC characteristics of the 18 Ni 250 maraging steel. The cantilever bars were fatigue precracked in the notch using a Krouse reverse-bend fatigue machine. A precrack approximately 0.4 mm long was formed in 30,000 to 50,000 cycles.

A detailed description of the experimental procedures and equipment used in this investigation has been given by Syrett[2] and by Syrett and Biefer.[3] All the tests were performed at room temperature in a 3.5% NaCl solution replenished at a rate of 4 liters per day. The test specimens passed through holes cut in a 250 ml plastic bottle and silicone sealing compound was used to caulk the joints. In most tests, the corrosion vessel was open to the atmosphere and the salt solution assumed to be saturated with 0.2 atmospheres of oxygen. In other tests, referred to as deoxygenated, nitrogen which had been passed over copper turnings at 400 C was bubbled through the salt solution in a closed corrosion vessel and also through the salt solution reservoir.

Cantilever beam deflections were recorded either continuously, using a linear voltage displacement transducer coupled to a chart recorder, or intermittently, using a dial gauge.

Electrochemical potentials were measured with a saturated calomel reference electrode via an agar salt bridge, and all potentials quoted are with respect to the saturated calomel electrode. Some test specimens were potentiostatically controlled using a platinum counter electrode in the corrosion vessel. In addition, cadmium anodes were used to polarize two series of test specimens. Periodic pH measurements were made during the experiments.

Stress intensity values were calculated using the following equations given by Brown and Beachem,[5] with the Freed-Krafft correction for side notches.[6]

$$K_{Ii} = \left(\frac{B}{B_N}\right)^{1/2} \frac{\beta M}{BD^{1.5}}$$

$$\beta = 4.12 \left(\frac{1}{(1\text{-}a/D)^3} - (1\text{-}a/D)^3\right)^{1/2}$$

B, B_N, and D are the specimen dimensions shown in Figure 1, and a is the total depth of machined notch and fatigue precrack, measured after specimen failure or termination of the test and sectioning of the specimen. M is the total bending moment at the notch. Most stress intensities do not conform to ASTM requirements for plane strain[7] fracture toughness testing because the total defect size (notch plus precrack) and specimen width are not greater than $2.5(K/\sigma_{ys})^2$.

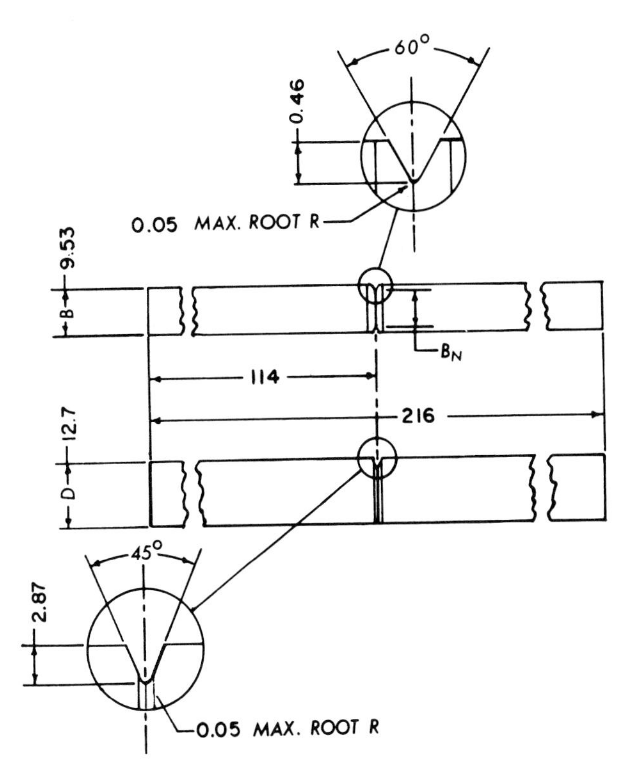

FIGURE 1 – The top and side notched cantilever beam specimen. All dimensions in mm.

Experimental Results

A tabulation of the experimental results is given in Table 2. The results are divided into groups of tests, having similar environmental conditions, and the data given for the first specimen of each group are representative of all tests in that group.

Dry Tests

To determine the approximate fracture toughness of the steel, several specimens were rapidly loaded in air. For the L-S orientation, 137 and 132 $MPa{\cdot}m^{1/2}$ were obtained, and for the L-T orientation, 129 $MPa{\cdot}m^{1/2}$. Biefer and Garrison[1] obtained 134 and 125 $MPa{\cdot}m^{1/2}$ for L-S specimens tested under similar conditions.

Free Corrosion Conditions

No specimens failed when tested under free corrosion conditions. The most severe conditions were a total test period of 1170 hours at an initial stress intensity of 113 $MPa{\cdot}m^{1/2}$. However, during all free corrosion tests in air saturated solution (specimens 4 to 7), a slow continuous deflection of the beam was noted. Metallographic examination of a section through the precrack area (Figure 2) showed that this steady deflection was due to corrosion and microcracking of the maraging steel along the sides and particularly at the tip of the precrack and resulted in blunting of the precrack. In these specimens tested with an L-S crack orientation, there was a tendency for this corrosion and cracking to follow the banded structure of the steel. One specimen, having an L-T orientation (not listed in Table 1) and tested in a similar manner, showed similar crack blunting due to corrosion and microcracking. However, in this case, there was no preferential attack in the rolling plane.

One free corrosion test was performed in a deoxygenated solution, (specimen 8, Table 2) at an initial stress intensity of 93 $MPa{\cdot}m^{1/2}$. No deflection of the beam was noted in 620 hours. The load was then increased to a stress intensity of 113 $MPa{\cdot}m^{1/2}$ for an additional 70 hours. Again, no deflection of the beam was noted, and the test was terminated. The specimen was sectioned in the precrack area and is shown in Figure 3. The precrack was blunted primarily by the application of the load, and there appear to be two very small environmental cracks starting at the bottom of the precrack.

Anodic Conditions

Two specimens potentiostatically held at potentials of -0.45 and

TABLE 2 – Summary of Test Results

Specimen No.	Test Conditions		Potential $V_{(SCE)}$	pH	K_{Ii} $MPa \cdot m^{1/2}$	Time to Failure (hours)
1	Dry		–	–	137	
2					133	
3[(1)]					129	
4	FC	air sat.	-0.59	6-7	113	NF (1170)
5					92	NF (1000)
6					80	NF (1000)
7					55	NF (1020)
8[(1)]	FC	de-ox.	-0.62	6.0	93	NF[(2)] (620)
9	PC	air sat.	-0.45	9.0	93	NF (380)
10			-0.35	9.0	93	204
11	PC	air sat.	-1.10	6-7	113	9.2
12					101	19.1
13					96	21.5
14					77	35.9
15					75	11.5
16					60	24
17					60	23
18					57	23
19					45	24.4
20					26	38.5
21					15	NF (405)
22	PC	air sat.	-1.40	9.0	96	7.8
23	PC	air sat.	-0.90	7.0	110	612
24	Cd	air sat.	-0.75	8-9	115	427
25[(1)]					93	1551
26					87	2554
27	Cd	de-ox.	-0.80	8-9	112	72
28[(1)]					90	128
29					65	256
30					55	360
31					37	936

(1) L-T crack orientation, all others L-S
(2) K_I increased to 113 $MPa \cdot m^{1/2}$ for an additional 70 hours before terminating test.
FC = Free corrosion
PC = Potentiostatically controlled
Cd = Coupled to cadmium
NF = No failure

$-0.35V_{(SCE)}$ at 93 $MPa \cdot m^{1/2}$ exhibited no environmental cracking, although the $-0.35V_{(SCE)}$ specimen did fail after 204 hours. However, this failure was the result of overload of the specimen after reduction of the cross section by anodic dissolution.

Cathodic Conditions

A series of tests were carried out at a potential of $-1.10V_{(SCE)}$, a potential only slightly more active than the value provided by cathodic protection systems using zinc sacrificial anodes. The times to failure are given in Table 2 and plotted in Figure 4. At a stress intensity as low as 26 $MPa \cdot m^{1/2}$, a specimen failed in less than 40 hours, while at 15 $MPa \cdot m^{1/2}$ no crack propagation as measured by beam deflection had occurred in 400 hours, and the test was terminated. Specimens loaded to a K_{Ii} of greater than approximately 60 $MPa \cdot m^{1/2}$ exhibited a well defined incubation period before the start of crack propagation, whereas those loaded to lower stress intensities did not.

An indication of the variation in crack growth rate with stress intensity may be obtained from the initial beam deflection rate for each specimen. These data, plotted in Figure 5, were obtained by measuring the slope of the deflection vs time record for the first few hours of well defined crack growth after the incubation period. Although there was crack branching in all the specimens, measurement of the beam deflection rate for the start of crack propagation yields data proportional to crack growth rate for the initial environmental extension of single fatigue precrack, the only point at which a meaningful stress intensity can be calculated. The curve in Figure 5 has the same functional relationship as that presented by Carter[8] for other high strength steels.

Several of the specimens of this series were removed from the test apparatus after the crack had propagated for several hours but before final fast fracture. Metallographic sectioning showed that there was extensive crack branching (Figure 6), which was most pronounced at the center of the specimen. However, the crack front had advanced further down the side of the specimen than it had in the center, an effect which had been noticed in other high strength materials under hydrogen charging conditions.[9]

A single specimen (number 22) was tested while polarized to $-1.40V_{(SCE)}$ at 96 $MPa \cdot m^{1/2}$. This specimen failed in only 7.8 hours and exhibited an incubation period of 6.9 hours. The initial beam

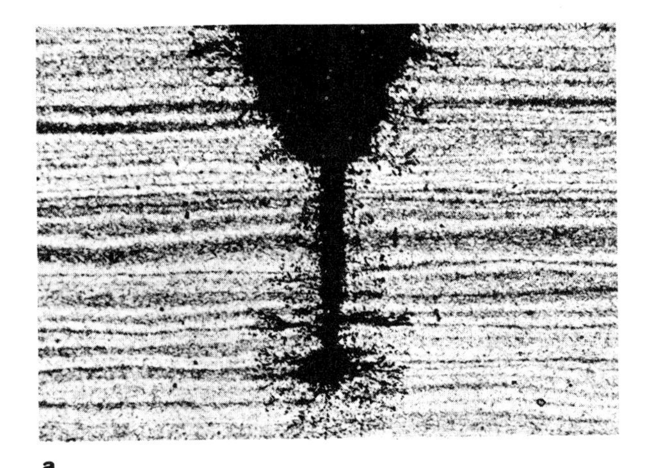

a

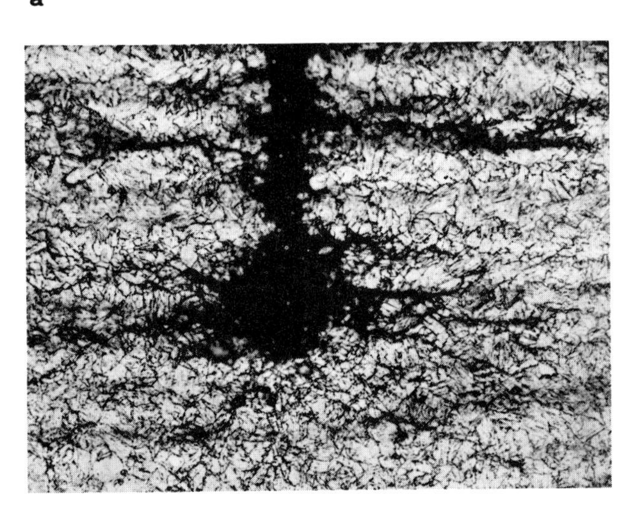

b

FIGURE 2 – The precrack area of Specimen 6 after free corrosion in air saturated 3.5% NaCl solution for 1000 hours, K_{Ii} = 80 MPa·m$^{1/2}$. 2% nital etch. (a) 60X, and (b) 325X.

FIGURE 3 – Micrograph of the precrack tip of Specimen 8 after free corrosion in deoxygenated 3.5% NaCl solution for 620 hours at a stress intensity of 93 MPa·m$^{1/2}$ and a further 70 hours at 113 MPa·m$^{1/2}$. Nital etch. 500X

deflection rate was approximately four times the rate for a specimen polarized to -1.10V$_{(SCE)}$ at a similar stress intensity.

Several experiments were carried out with specimens potentiostatically polarized to potentials between -0.70 and -1.0V$_{(SCE)}$. The results of these tests were not conclusive due to the experimental difficulties in carrying out long term potentiostatic tests. With the exception of Specimen 23, which failed in 612 hours after being polarized to -0.90V$_{(SCE)}$ at an initial stress intensity of 110 MPa·m$^{1/2}$, the results are not included in Table 2 and will not be discussed in detail. However, one characteristic was apparent from these tests. In all cases, no beam deflection was detected for most of the exposure time. Specimen 23, which failed at 612 hours, had not started to crack as measured by beam deflection at 596 hours. Thus, there would appear to be a long incubation period before the advent of crack propagation. This effect was found for other test conditions described below.

Cadmium Coupled Specimens

Sacrificial cadmium anodes were used to polarize specimens to potentials between the free corrosion condition where resistance to EC was high, and -1.10V$_{(SCE)}$, where failure was rapid. Two series of tests were carried out, one in air saturated solutions and one in deoxygenated solutions. The results of these tests are shown in Figure 7 and the data tabulated in Table 2. The presence of air in the system results in a corrosion potential of -0.75V$_{(SCE)}$ compared with a value of -0.80V$_{(SCE)}$ in the deoxygenated system. The times to failure in the air saturated tests were extremely long (over 2500 hours), and at lower stress intensities the times to failure would be even longer. Thus, no effort was made to find the threshold stress

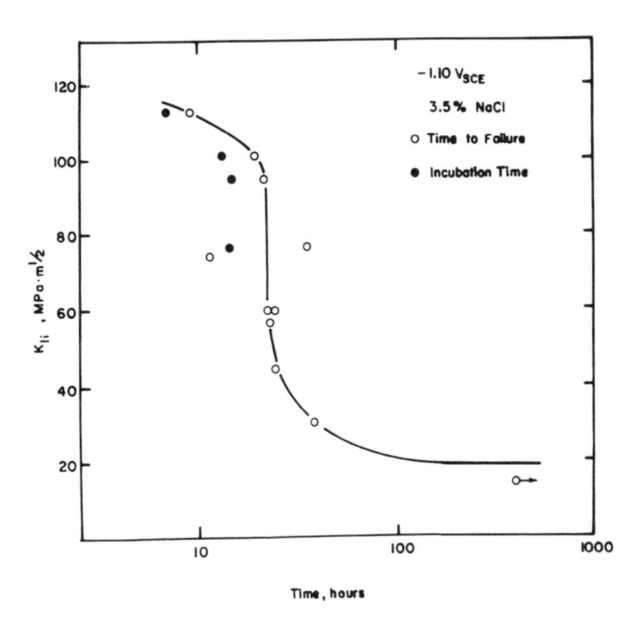

FIGURE 4 – Total time to failure for precracked specimens of 18 Ni 200 maraging steel polarized to -1.10V$_{(SCE)}$ in 3.5% NaCl solution. Incubation time for specimens having a well defined incubation period in excess of 3 hours is also shown.

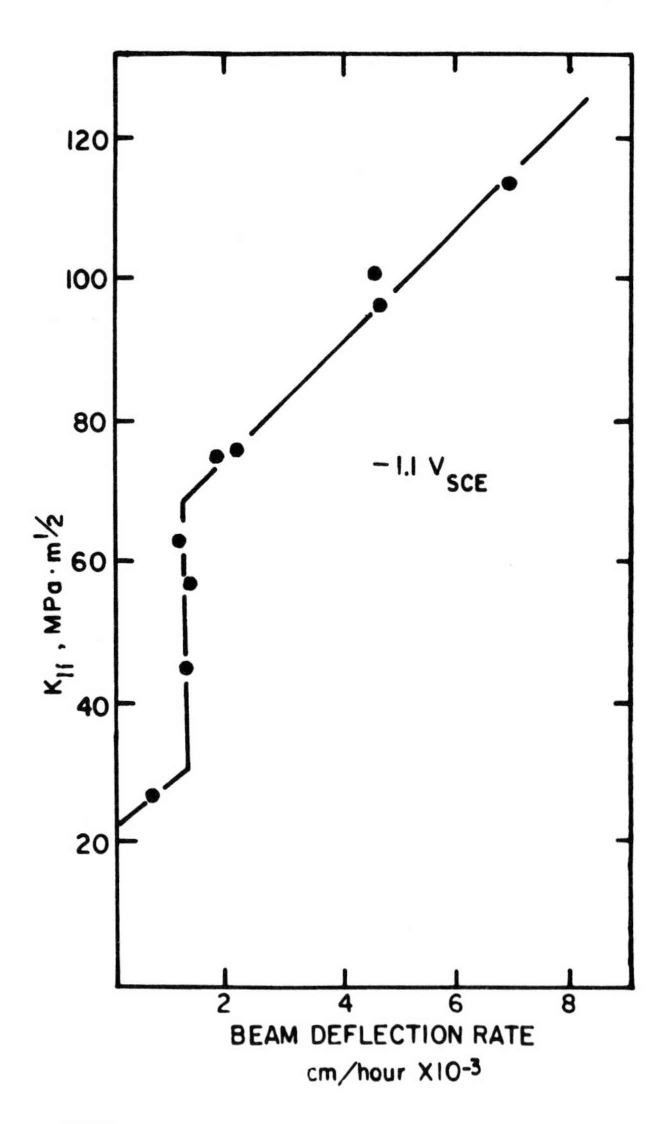

FIGURE 5 — Initial beam deflection rate 5 cm from the notch for 18 Ni 200 maraging steel polarized to -1.1V$_{(SCE)}$ in 3.5% NaCl solution.

FIGURE 6 — Typical branched cracks toward the center of Specimen 19 polarized to -1.10V$_{(SCE)}$ at K_{Ii} = 45 MPa·m$^{1/2}$ in 3.5% NaCl solution.

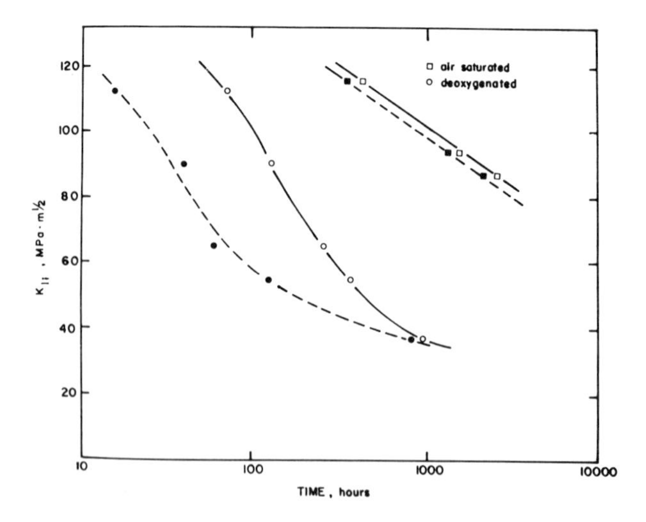

FIGURE 7 — The effect of oxygen on the time to failure (solid lines) for precracked specimens of 18 Ni 200 maraging steel coupled to cadmium in 3.5% NaCl solution.

intensity (K_{ISCC}) under these conditions. The incubation times for the specimens in the air saturated solution consisted of almost all the time to failure and are also shown in Figure 7.

In the deoxygenated salt solution, incubation times and times to failure were much shorter, as shown in Figure 7. However, particularly at the higher loads, the incubation period was not as well defined as it was in the tests in air saturated solutions, and some crack propagation may have taken place at even earlier times than those shown. It will be noted that the specimen loaded to 37 MPa·m$^{1/2}$ did fail in 936 hours, and thus the threshold stress intensity (K_{ISCC}) is somewhat lower than this value under these conditions.

Once cracking began in air saturated solution, the crack propagation rate did not appear to be substantially different from that observed in the deoxygenated solution.

Corrosion Rates and Polarization Measurements

To help interpret the EC test results, cathodic polarization curves were determined for the maraging steel in the neutral salt solution used in the EC tests. Standard techniques were used,[10] and the instrumentation included a Wenking potentiostat with the potential stepping device set for one 10 mV step per minute. The samples of maraging steel used were cut from the ends of the cantilever bars and spot welded to a steel rod. The steel rod was passed through the hollow electrode holder, and the junction of the holder and specimen sealed with epoxy. Air or prepurified nitrogen was bubbled through the solution in the closed polarization vessel to produce aerated or deoxygenated conditions, respectively. The specimen was allowed to corrode freely in the solution for 3 hours before the potential scan was started. The cathodic polarization curves obtained are shown in Figure 8.

Figure 9 shows potential time curves for maraging steel freely corroding in both air saturated and deoxygenated neutral 3.5% NaCl solution. A steady state corrosion potential of approximately -0.65V$_{(SCE)}$ was reached in the deoxygenated solution in only a few hours, whereas a period of up to 30 hours was required to reach

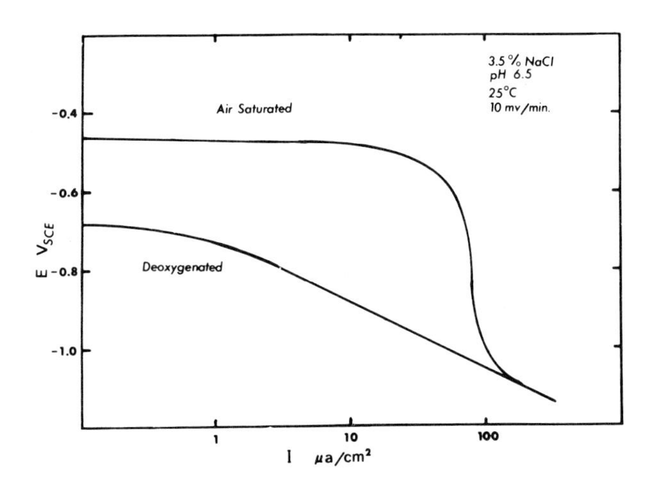

FIGURE 8 — Cathodic polarization curves for 18 Ni 200 maraging steel in air saturated and deaerated 3.5% NaCl solution, pH 6.5 at 25 C one 10 mV step per minute.

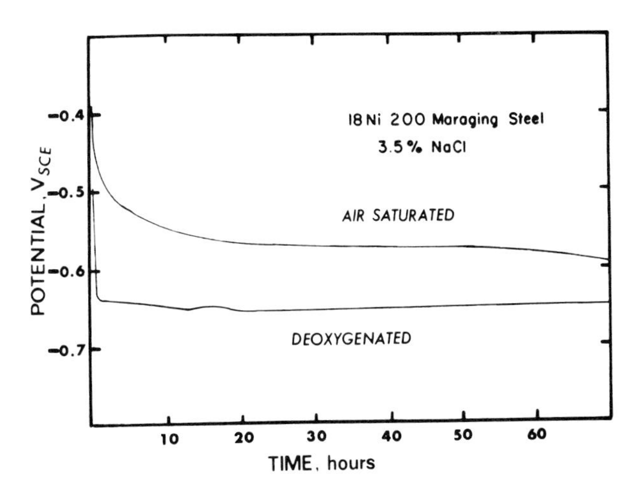

FIGURE 9 — The time-potential relationship for 18 Ni 200 maraging steel freely corroding in neutral 3.5% NaCl solution at room temperature.

-0.59V$_{(SCE)}$ in the air saturated solution. Weight loss measurements on small specimens immersed in deoxygenated neutral 3.5% NaCl solution gave a negligible corrosion rate of 0.2 mdd, whereas the specimen in the aerated solution corroded at a rate of 15 mdd.

SEM Fractography

A limited amount of SEM fractography was carried out on specimens tested under a variety of conditions, including high and low K_{Ii} at -1.10V$_{(SCE)}$, and in both aerated and deoxygenated solutions coupled to cadmium. All environmental crack surfaces examined exhibited both intergranular and transgranular (quasi-cleavage) crack paths. There was some indication that more intergranular cracking took place in the initial environmental extension of the precrack at the lower K_{Ii} levels, but because of the branched nature of the crack, this was difficult to confirm. When the crack has branched beyond the nominal plastic zone size, no stress intensity may be calculated for the tip of any one crack, and thus correlation of the fracture path with stress intensity is not possible.

Discussion

All the results presented have demonstrated that 18 Ni 200 maraging steel is susceptible to environmental cracking due to hydrogen embrittlement. The SEM examinations of EC surfaces showed the typical quasicleavage and intergranular fracture associated with hydrogen embrittlement.

In contrast to the results of other investigations, no specimens failed under free corrosion conditions. Bodner and Carson[11] found K_{ISCC} = 42 MPa·m$^{1/2}$ under free corrosion conditions in flowing sea water at about 12 C for 12.7 x 12.7 mm specimens cut from the same plate as the specimens reported here. Kenyon, Kirk, and Van Rooyen[12] determined K_{ISCC} = 103 MPa·m$^{1/2}$ for 18 Ni 200 maraging steel in 3.5% NaCl solution. This difference may be caused by the variation in test procedures including specimen preparation, precrack method, and heat treatment variables from one laboratory to another. Dautovitch and Floreen[13] have reported a wide variation in EC properties of maraging steels with small changes in residual alloying elements and in processing variables.

In high strength ferrous alloys under free corrosion conditions, the local corrosion reaction occurring within the restricted geometry of the precrack leads to the development of potential-pH conditions below the hydrogen line on a Pourbaix diagram.[14] The cathodic reaction in the precrack is then the hydrogen ion reduction, and the resulting hydrogen is believed to enter the alloy and lead to environmental cracking. However, in the present series of tests, environmental cracking leading to failure does not occur for several possible reasons. First, the precrack in these specimens is relatively shallow, and thus the corrosion reaction occurring in the precrack is not as occluded as it might be with a deeper precrack (and wider specimen). Second, the specimen size precludes the attainment of plane strain conditions at the precrack tip, and it has been shown[15] that the susceptibility to environmental cracking of maraging steel is greatest for plane strain conditions. Third, the corrosion and microcracking within the precrack blunt the precrack, thereby reducing the stress intensity. Once the crack has been blunted, it is unlikely for a failure to occur by environmental cracking. Sandoz[16] found crack blunting by corrosion at the tip of the precrack in other high strength steels.

It is of interest to note that the "microcracking", which took place under free corrosion conditions, will occur in the absence of a high stress. In Figure 2a, this cracking is visible on the sides of the machined notch where the stress is low. Syrett[2] has reported a similar phenomenon in 18 Ni 250 maraging steel. This micro-cracking appears different from the environmental cracking observed under cathodic conditions and may be a type of intergranular corrosion.

In deoxygenated 3.5% NaCl solution the corrosion rate, and thus the hydrogen production rate, is very low. However, one might speculate that if a stressed precracked specimen was left for a sufficiently long time, failure would result. The slow buildup of hydrogen in the metal at the crack tip should eventually reach a level at which cracking would commence. The two small cracks in Figure 3 may be the start of environmental cracks as no similar feature was found in specimens dry loaded and sectioned.

The experiments performed at -1.10V$_{(SCE)}$ show the extreme susceptibility of 18 Ni 200 maraging steel to hydrogen embrittlement cracking. The threshold stress intensity is less than 26 MPa·m$^{1/2}$. The crack growth kinetics at -1.10V$_{(SCE)}$ show a region where the growth rate is independent of stress intensity. The functional relationship is similar to that reported for a variety of high strength materials.[8,16]

Qualitatively, the overall crack growth rate under conditions of hydrogen charging appears to be dependent upon the rate of hydrogen production or arrival, probably at the crack tip. Specimen 22, polarized to -1.40V$_{(SCE)}$ at a stress intensity of 96 MPa·m$^{1/2}$, exhibited a total time of crack growth before failure of only 0.9 hours, whereas specimen 13, at -1.10V$_{(SCE)}$ and the same initial stress intensity, had a crack propagation time of over 6 hours. Also, the initial beam deflection rate at -1.40V$_{(SCE)}$ was nearly four times as great as at -1.10V$_{(SCE)}$.

Metallographic examination of the environmental cracks formed at -1.10V$_{(SCE)}$ did not reveal any internal or subsurface cracks. Repeated grinding and polishing showed that cracks which, in a particular plane, did not appear to be connected to the original crack, were in fact connected on some other plane.

The most interesting results are those obtained with the specimens coupled to cadmium in air saturated and deoxygenated solutions. It has been shown that some maraging steels have maximum resistance to environmental cracking at potentials slightly more active than the free corrosion potential.[15] At these potentials in oxygen containing solutions, the main cathodic reaction is oxygen reduction. Corrosion is minimized or eliminated with little or no hydrogen introduced into the alloy, resulting in an overall enhancement of EC resistance.

The cathodic polarization curves (Figure 8) show that oxygen reduction is the main cathodic reaction in neutral salt solution between the free corrosion potential and -0.90$V_{(SCE)}$. However, the cathodic reduction of hydrogen ion also occurs in the portion of this potential range below the reversible hydrogen potential, and the observed polarization curve is the sum of the 2 cathodic reactions: (1) hydrogen ion reduction, and (2) oxygen reduction. Thus, the effect of oxygen in delaying EC initiation must involve the kinetics of the entry of hydrogen into the metal. Barth and Troiano[17] have proposed that hydrogen ions may be discharged on the surface of the absorbed oxygen layer, and under these conditions the entry rate of hydrogen into the metal is reduced. The results of the present work would support the concept of a reduced rate of hydrogen entry into the maraging steel in aerated salt solution. The extremely long incubation periods observed in air saturated solution while coupled to cadmium are presumed to reflect the very slow rate of hydrogen entry under these conditions.

Under the pH-potential conditions of the specimen when coupled to cadmium in the air saturated salt solution, the equilibrium hydrogen pressure is close to one atmosphere. However, in the deoxygenated solution the only accessible cathodic reaction is the hydrogen ion discharge and because of the somewhat more active potential the equilibrium hydrogen pressure will be much higher. There will be no surface coverage of oxygen to restrict the entry of hydrogen and cracking may initiate much earlier. Of course, the potential-pH conditions measured in the bulk environment do not necessarily apply in the precrack.[14] However, in these experiments the $Cd(OH)_2$ (the corrosion product of the cadmium anode) tends to buffer the solution to a basic pH and this effect will probably extend into the precrack.

The extremely long incubation periods before the onset of crack propagation of specimens coupled to cadmium in oxygen containing solution illustrate the danger in using an arbitrary time limit for such experiments. These results also suggest that the use of mild cathodic protection to reduce environmental cracking may lead to unexpected failure if a deoxygenated solution develops in cracks, crevices, and undercoatings, etc. In addition, crack propagation may begin after prolonged exposure to the environment as a result of the slow buildup of hydrogen. When cracking initiates, the rate of propagation appears to be independent of the oxygen content of the bulk environment.

Conclusions

The maraging steel studied has been found to be susceptible to environmental cracking due to hydrogen embrittlement. There is no evidence of crack propagation by an anodic dissolution mechanism. Under free corrosion conditions in air saturated 3.5% NaCl solution, the tendency of the very highly stressed precrack tip to corrode preferentially results in crack blunting which eliminates further environmental cracking. However, others have shown that this material will fail by environmental cracking under free corrosion conditions. Mild cathodic protection using cadmium anodes results in long incubation periods before the start of crack propagation in air saturated solutions. In deoxygenated solutions cathodic protection by cadmium is detrimental, and K_{ISCC} for these conditions is less than 37 MPa·m$^{1/2}$. At a potential of -1.10$V_{(SCE)}$, the threshold value is less than 26 MPa·m$^{1/2}$. The dry fracture toughness was approximately 130 MPa·m$^{1/2}$.

These results all apply to one plate of maraging steel which exhibited highly undesirable alloy banding. Floreen and Dautovitch[13] have reported that a great variation in properties has been found between various heats of maraging steel. Nevertheless, structures fabricated from this alloy for use in a marine environment and subjected to cathodic protection would undoubtedly develop environmental cracks.

Acknowledgments

The author is indebted to G. J. Biefer and R. J. Brigham for their helpful discussions, and to J. G. Garrison and A. Blouin for their assistance with the experimental work.

References

1. G. J. Biefer and J. G. Garrison. NATO, AGARD CP98. Specialists Meeting on Stress Corrosion Testing Methods (1971).
2. B. C. Syrett. Corrosion, Vol. 27, p. 270 (1971).
3. B. C. Syrett and G. J. Biefer. The Rising-Load Cantilever Test: A Rapid Test for Determining the Resistance of High Strength Materials to Environmental Cracking, Mines Branch Research Report R227, Department of Energy, Mines and Resources, Ottawa, Canada (1970).
4. R. J. Goode. Materials Research and Standards, Vol. 12, p. 31 (1972).
5. B. F. Brown and C. D. Beachem. Corr. Sci., Vol. 5, p. 795 (1965).
6. C. N. Freed and J. M. Krafft. J. Materials, Vol. 1, p. 770 (1966).
7. ASTM Standards, Part 31, p. 919 (1972).
8. C. S. Carter. Eng. Fracture Mechanics, Vol. 3, p. 1 (1971).
9. B. C. Syrett and L. P. Trudeau. Corrosion, Vol. 27, p. 216 (1971).
10. N. D. Green. Experimental Electrode Kinetics. Rensselaer Polytechnic Inst., Troy, New York (1965).
11. M. Bodner and J. A. H. Carson. Private Communication.
12. N. Kenyon, W. W. Kirk, and D. Van Rooyen. Corrosion, Vol. 27, p. 320 (1971).
13. D. P. Dautovitch and S. Floreen. The Stress Corrosion and Hydrogen Embrittlement Behavior of Maraging Steels. Presented at International Conference on Stress Corrosion Cracking and Hydrogen Embrittlement of Iron Base Alloys, Firminy, France, June, 1973.
14. B. F. Brown. ARPA Coupling Program on Stress Corrosion Cracking. NRL Report 7160 (1970).
15. H. W. Hayden and S. Floreen. Corrosion, Vol. 27, p. 429 (1971).
16. G. Sandoz. The Resistance of Some High Strength Steels to Slow Crack Growth in Salt Water. NRL Memorandum Report 2454 (1972).
17. C. F. Barth and A. R. Troiano. Corrosion, Vol. 28, p. 259 (1972).

Hydrogen Induced Delayed Cracking in Maraging Steel

By T. Boniszewski, PH.D.

The susceptibility of 18% Ni–Co–Mo maraging steel to delayed hydrogen cracking was examined using the constant load rupture test. The steel was in the microstructural condition to be found in the weld heat-affected zone immediately after welding, i.e., austenitised at very high temperature and quenched. The hydrogen was dissolved in the specimens during austenitisation. In this condition the maraging steel showed low susceptibility to hydrogen embrittlement, and behaved similarly to 9% Ni steel. However, the metallurgical reasons for the low susceptibility of the maraging steel to hydrogen embrittlement were different from those applying in the case of 9% Ni steel. After ageing, the tensile properties of the coarse grained, weld HAZ type, microstructure of the maraging steel were similar to those obtained after the standard heat treatment.

Introduction

The possibility of welding 18% Ni–Co–Mo maraging steel in a plate form necessitated the evaluation of the susceptibility of its heat-affected zone (HAZ) microstructure to delayed cracking, in the presence of dissolved hydrogen. Previous investigations[1,2] carried out at BWRA have shown that there is a correlation between hydrogen-induced cracking in the weld HAZ and hydrogen-induced fracture of notched specimens subjected to a constant load. These constant load rupture (CLR) tests must be carried out on specimens whose microstructure is representative of that in weld HAZ. Such CLR tests have been used in the present investigation.

The susceptibility of 18% Ni–Co–Mo maraging steel to hydrogen embrittlement has already been reported in the literature,[3–6] but the tests were carried out in the fully heat treated condition, *i.e.*, solution annealed and aged, and the hydrogen was introduced into the specimens by a cathodic reaction at the surface. It was found that maraging steel was less susceptible to hydrogen-induced delayed fracture than medium carbon low alloy steels of similar strength. Whilst these findings are useful in assessing the behaviour of the maraging steels after pickling and electroplating, and in corrosive environments, they are not applicable to the as-quenched, *i.e.*, solution treated microstructure obtaining in the HAZ.

Report B2/WSS/54/64 of the British Welding Research Association, circulated to members in February, 1965.

Dr. Boniszewski is a Senior Scientific Officer in the Association's Metallurgical Laboratory (Ferrous). 1008

Table I

Chemical composition of 18% Ni–Co–Mo maraging steel examined (cast CCTN)

Element	*Wt., %*
Carbon	0·015
Nickel	18·1
Cobalt	7·3
Molybdenum	4·4
Chromium	0·03
Titanium	0·52
Aluminium	0·03
Manganese	0·03
Silicon	0·1
Sulphur	0·007
Phosphorus	0·007
Zirconium	~0·02
Calcium	<0·005
Boron	<0·002

Table II

Mechanical properties reported by the manufacturer for 18% Ni–Co–Mo maraging steel examined (cast CCTN)

Property	*Annealed 820°C. and aged at:*			
	390°C./3 hr	*420°C./3 hr*	*450°C./3 hr*	*480°C./3 hr*
Limit of Proportionality, tons/sq.in.	62	67	66	82
0·1% Proof Stress, tons/sq.in.	77	84·5	93	103
0·2% Proof Stress, tons/sq.in.	80·5	89	98	107
UTS, tons/sq.in.	91	97	105	113·5
NTS (0·3 in. bar, Kt = 10–12) tons/sq.in.	153	149	164	168·5
NTS/UTS ratio	1·68	1·53	1·56	1·49
Elongation, $4\sqrt{Ao}$, %	15·7	15·7	15·7	12·4
Reduction in Area, %	46·2	48·6	50·8	49·2
Charpy V-notch at 20°C., ft-lb	27·5, 28	23·5, 23	21·5, 20	19·5, 20
Impact Value, ft-lb −78°C.	18·5, 19	17·5, 18	17·5, 19	17·5, 17
Impact Value, ft-lb −196°C.	13·5, 13	11, 11	10, 10	8·5, 10

In the present investigation, the notched tensile specimens of 18% Ni–Co–Mo maraging steel were tested after a thermal treatment that would simulate the conditions in the HAZ, as far as the experimental technique would permit. Hydrogen was dissolved in the specimens at high temperatures, in the austenite range, and retained by fast cooling comparable to that which may occur in the HAZ of a relatively thick plate. The specimens were tested in tension and also under constant load, and in the latter tests their time to fracture was measured at various nominal stresses. Hydrogen free specimens, treated thermally in a similar way, but in argon, were tested for comparison.

Experimental Details

Material and specimens

The 18% Ni–Co–Mo steel examined was obtained from The International Nickel Co. (Mond), in the form of ¾ in. diameter annealed bar. The steel was produced in the induction furnace as a 2,300 lb commercial cast which was air melted. The composition of the steel is given in Table I. The mechanical properties of this melt, as reported by the manufacturer, are given in Table II.

The tensile specimens, plain and notched, used in the present investigation are shown in Fig. 1. They were machined from the bar, two specimens side by side.

Thermal treatment and hydrogen charging

To examine the properties of 18% Ni–Co–Mo maraging steel in a microstructural condition similar to that obtaining in the HAZ, the tensile specimens were given the following heat treatments:

Series A: Austenitising 1050°C./20 min and quenching.
Series B: Austenitising 1100°C./15 min and quenching.
Series C: Austenitising 1250°C./15 min, cooling in the furnace (10 min) to 1100°C. 15 min and quenching.

The specimens were placed in the furnace at the peak temperatures. The specimens to be eventually aged were austenitised in an argon atmosphere, and were cooled in air. The notched specimens to be

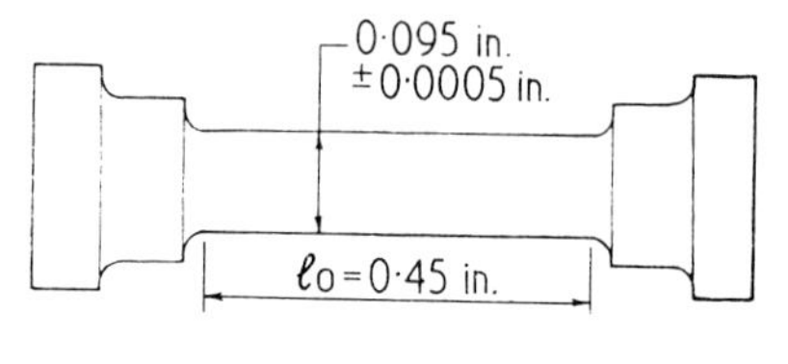

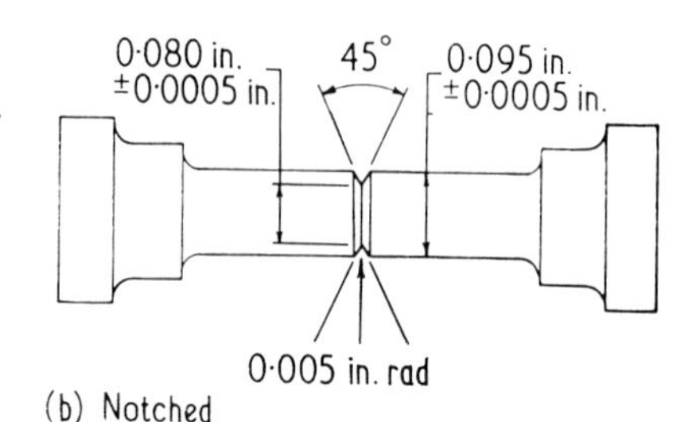

1—*Specimens for* (a) *plain tensile, and* (b) *notched tensile and constant load rupture tests. Other dimensions as for Hounsfield No. 11 tensile test piece*

(a)

(b)

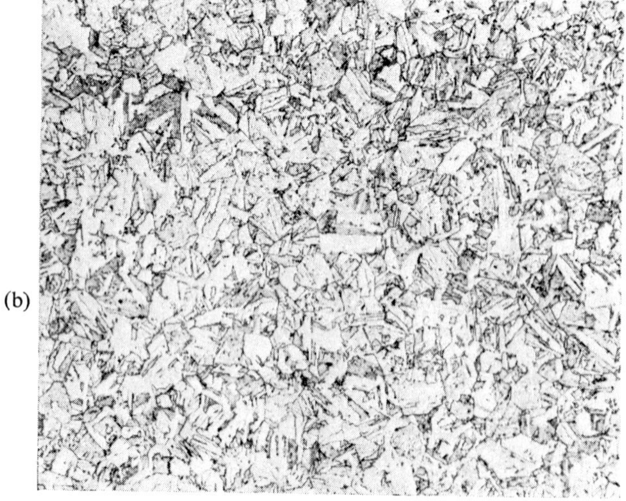

(c)

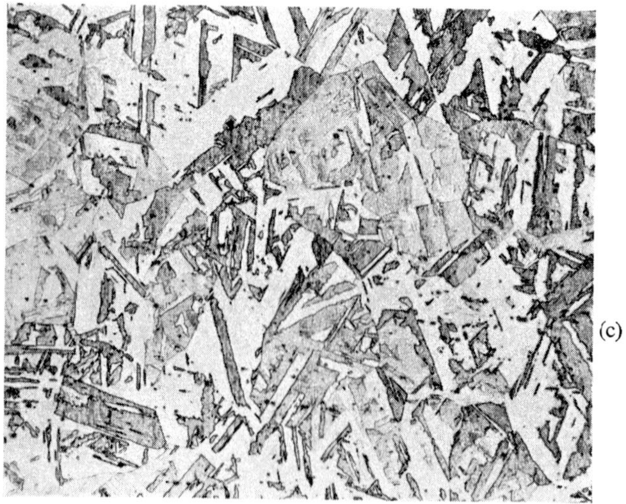

2—*General microstructure and grain size of the three series of specimens. Etched in acid ferric chloride* × 100
(a) *Series A: 1050°C./20 min*
(b) *Series B: 1100°C./15 min*
(c) *Series C: 1250°C./5 min → 1100°C./15 min*

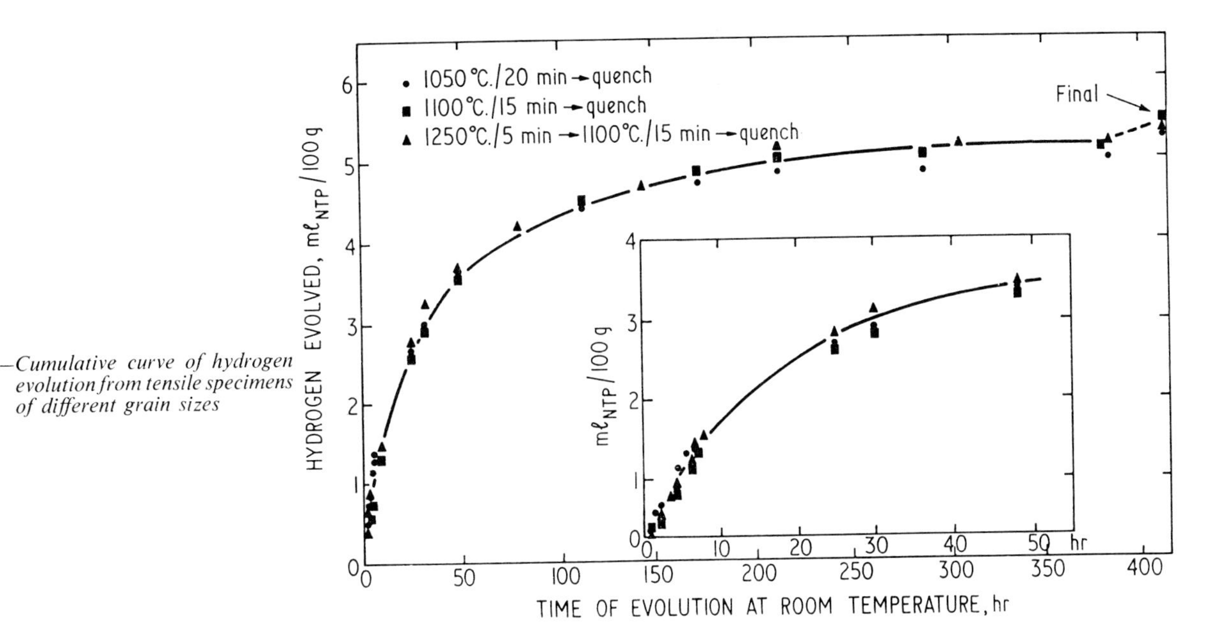

3—*Cumulative curve of hydrogen evolution from tensile specimens of different grain sizes*

Table III

Size of prior austenite grains of the three series of specimens tested

Series	*Grain Diameter,* μ	*ASTM No.*
A	50	No. 6
B	75	No. 4
C	250	No. 1

subjected to hydrogen delayed cracking, in the as-quenched condition, were quenched with the rate corresponding to thermal severity No. 24, *i.e.*, obtaining in the HAZ of the trithermal CTS test, 2 in. top plate on 2 in. bottom plate, welded without preheating. The quenching was done with argon blast. It was necessary to apply this high quenching rate to retain hydrogen in the specimens. The constant load rupture (CLR) specimens were austenitised in hydrogen atmosphere at about 1 atm. H_2. Some specimens tested as-quenched, for comparison, were austenitised in argon atmosphere.

The three different austenitising conditions produced grain sizes shown in Fig. 2 (*a–c*). The diameters of the prior austenite grains and their respective ASTM numbers are given in Table III.

The hydrogen content of notched specimens of all the three series was the same, *i.e.*, about 5·25 ml/100 g measured at S.T.P. There was practically no difference between Series A and B in spite of the difference of 50°C. in the final austenitising temperature. Hydrogen evolution from the CLR specimens is shown in Fig. 3. It can be represented by a single curve for all the three series of specimens.

Testing

Tensile tests were carried out on a 5 ton Instron machine at a cross-head speed of 0·05 cm/min. Constant load rupture (CLR) tests were carried out on ¾ ton Denison creep testing machines.

Results

Tensile properties of simulated HAZ microstructures after full heat treatment

Plain and notched tensile specimens were austenitised at high temperatures, cooled in air and aged at 480°C. for 3 hr. Their properties are given in Table IV together with the properties of the specimens which received the standard heat treatment. It can be seen that the increase in the austenitising temperature up to 1250°C. and the coarsening of the prior austenite grain had practically no effect on the tensile properties. The present results are in agreement with the results of Floreen and Decker.[7]

Table IV

Tensile properties after standard heat treatment compared with those of coarse grain specimens representing the HAZ in fully aged condition

Property	*820°C./1 hr 430°C./3 hr*		*1050°C./20 min 480°C./3 hr*		*1100°C./15 min 480°C./3 hr*		*1250°C./5 min 1100°C./15 min 480°C./3 hr*	
0·1% PS, tons/sq.in.	103·2	104·7	98·3	104·5	101·6	98·5	104·3	105·9
0·2% PS, tons/sq.in.	104·7	106·7	100·3	105·9	104·5	100·7	107·2	108·0
UTS, tons/sq.in.	105·3	110·2	103·8	106·6	108·0	104·7	110·4	111·5
Elong., %	6·4	8·2	8·8	7·8	9·2	9·0	13·9	12·8
Red. in Area, %	54·1	51·2	54·5	51·1	54·8	52·4	53·3	45·1
NYS, tons/sq.in.	127·7	130·4	130·7	126·3	125·3	126·4	126·8	126·2
NTS, tons/sq.in.	142·8	143·1	143·0	139·3	141·1	139·6	141·5	141·0
NTS/UTS ratio	1·33		1·34		1·32		1·26	

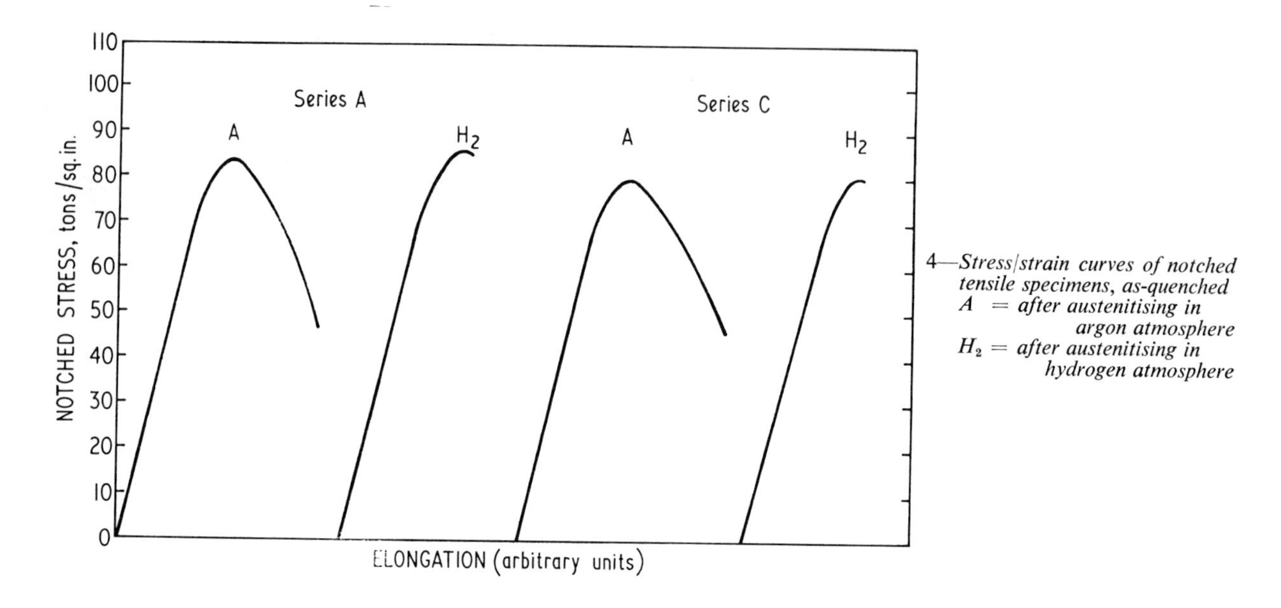

4—Stress/strain curves of notched tensile specimens, as-quenched
A = after austenitising in argon atmosphere
H_2 = after austenitising in hydrogen atmosphere

Constant load rupture tests of simulated HAZ microstructures

The notched tensile specimens treated in hydrogen and argon were first tested in tension to determine the notched tensile and notched yield* strengths. Hydrogen had no effect on the NYS and the NTS, but the notch ductility of the hydrogen-charged specimens was reduced considerably, as can be seen in Fig. 4.

With the value of the NTS established, the hydrogen-charged notched specimens were subjected to constant load. The rupture characteristics of the three series, with different grain sizes, are shown in Fig. 5. The specimens with the finest grain size had the lowest index of susceptibility to delayed hydrogen cracking, and the specimens with the coarsest grain

* The notched yield strength was determined from Instron recording charts taking the load at which the first deflection from a straight line was observed.

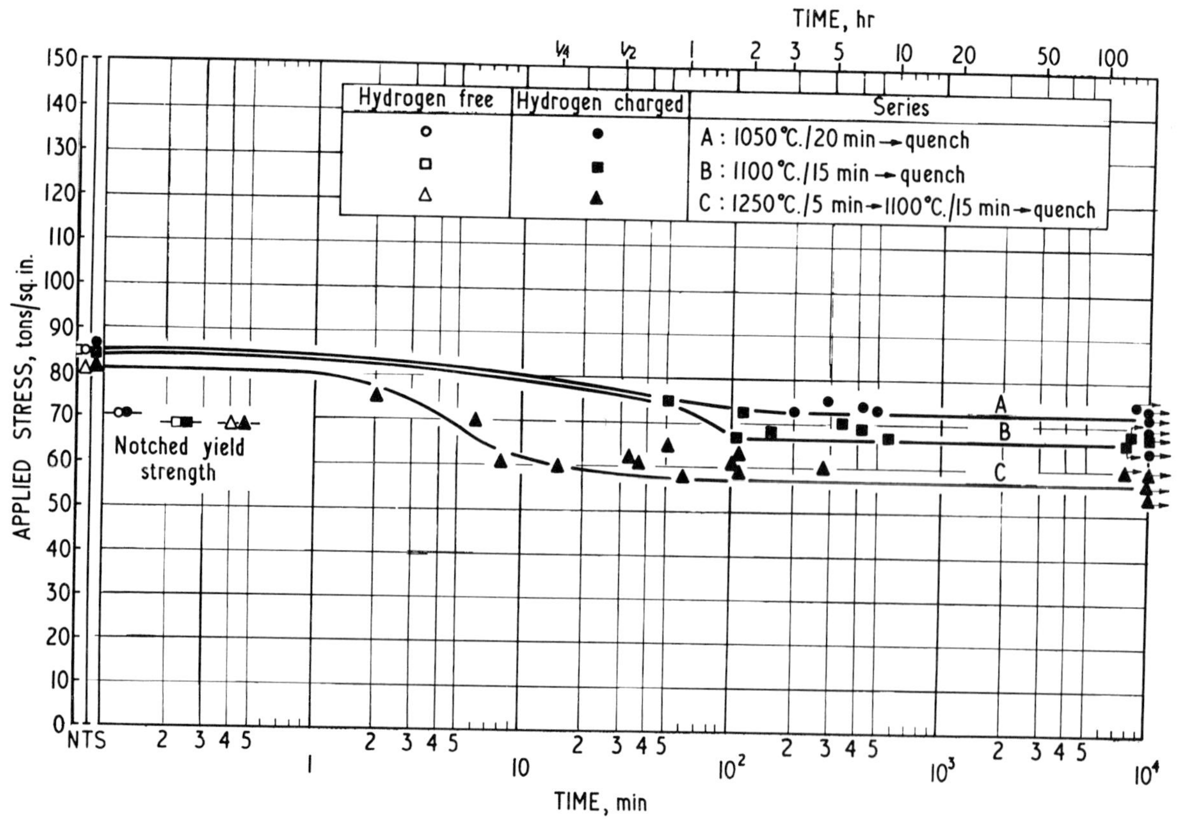

5—Constant load rupture curves of 18% Ni–Co–Mo maraging steel in the as-quenched condition

6—*As-quenched microstructure of 18% Ni–Co–Mo steel. Etched in acid ferric chloride* × 500
(a) *Series A: 1050°C./20 min*
(b) *Series B: 1100°C./15 min*
(c) *Series C: 1250°C./5 min → 1100°C./15 min*

size had the highest index of susceptibility to delayed hydrogen cracking (Table V).

It is interesting to notice that the lower critical stress (LCS) of Series A is somewhat above the measured NYS. This indicates that a considerable amount of plastic deformation was necessary to initiate hydrogen cracking in the fine grained specimens. With the increase in the grain size, the LCS dropped below the stress at which plastic deformation could be detected in the notched specimens by means of the Instron tensile machine.

Metallography of simulated HAZ microstructure and the mode of hydrogen cracking

To reveal the general solution treated microstructure of the HAZ, 18% Ni–Co–Mo steel can be etched either in acid ferric chloride solution or in 20% Nital (for composition see Appendix A). The first solution gives a pronounced contrast between differently oriented crystals, by producing finely spaced surface etch pits of various shapes, depending on the crystal orientation.

Figures 6(*a–c*) show the general microstructure of the solution-treated and quenched CLR specimens with the three different grain sizes. The appearance of the microstructure is complex:

(i) There is a general network of the prior austenite grains
(ii) Within the prior austenite grains one can see large massive crystals of different orientations. Some of these crystals may be thermal twins of the prior austenite and others may be massive ferrite grains
(iii) Within the certain massive crystals one can see fine elongated laths, aligned in the same direction. This effect can be seen more clearly in Figs. 7*b* and *c*. Where the micro-section is normal to the longest axis of laths, the sheaves of laths give a characteristic cellular appearance (Fig. 7*a*).

Table V
The indices of susceptibility to hydrogen embrittlement of 18% Ni–Co–Mo steel as quenched

Series	*Heat Treatment*	$\frac{(NTS)_A - (LCS)_{H_2}}{(NTS)_A}$
A	1050°C./20 min→quench	0·13
B	1100°C./15 min→quench	0·21
C	1250°C./5 min→1100°C./15 min→quench	0·28

The examination of surface replicas in the electron microscope did not reveal any further details in the as-quenched microstructure of the maraging steel.

An attempt was made to examine thin foils from the gauge length of the CLR specimens. The thin areas examined were not large enough to permit the determination of the crystallographic orientation of the individual ferrite crystals within the sheaves. However, it was found that the dislocation density of the as-quenched maraging steel was extremely high (Fig. 8). The dislocations appeared very irregular and diffuse and were arranged in dense clouds. Yamashita and Taneda[8] have shown that such cloudy dislocation configurations may result from the allotropic transformation during quenching of iron, whilst slow cooling does not produce cloudy dislocation structure.

To obtain additional information on the type of austenite transformation products in the 18% Ni–Co–Mo steel, specimens with polished surfaces were subjected to various heat treatments in hydrogen

Source: *British Welding Journal*, Nov 1965

(a)

(b)

(c)

7—*As-quenched microstructure of 18% Ni–Co–Mo steel. Etched in acid ferric chloride* × 1000
(a) *Series A: 1050°C./20 min*
(b) *Series B: 1100°C./15 min*
(c) *Series C: 1250°C./5 min → 1100°C./15 min*

atmosphere. The hydrogen atmosphere was used in this case to protect the polished surfaces from oxidation. After the heat treatments the morphology of the polished surfaces was examined in the microscope.

Figure 9 shows a typical surface appearance of a specimen austenitised at 820°C. for 1 hr and cooled slowly, to simulate air cooling which is used after the standard solution annealing of maraging steel. The pre-polished surface showed well pronounced rumpling. The needle-like surface relief is of interest here, as opposed to the thermally etched prior austenite grain boundaries. This surface relief is the result of surface upheavals associated with the homogeneous shear or 'shape' deformation accompanying the martensitic transformation in steels[9-12]. The surface

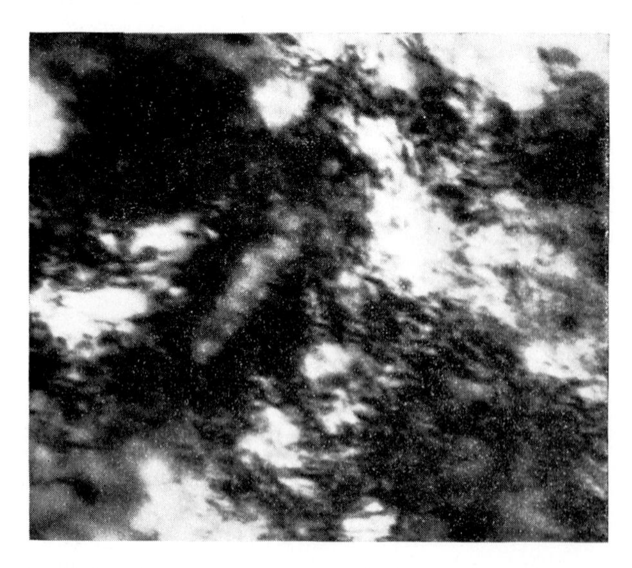

8—*Dislocation structure in as-quenched maraging steel microstructure comparable with that of the weld HAZ*
Thin foil × 160,000

9—*Pre-polished surface of 18% Ni–Co–Mo maraging steel after 820°C./1 hr in H_2 and slow cooling (equivalent to air cooling)*
Oblique illumination × 1000

10—*Pre-polished surface of 18% Ni–Co–Mo maraging steel after 1050°C./20 min in H_2 and argon blast quench (simulated HAZ cooling)* Oblique illumination × 1000

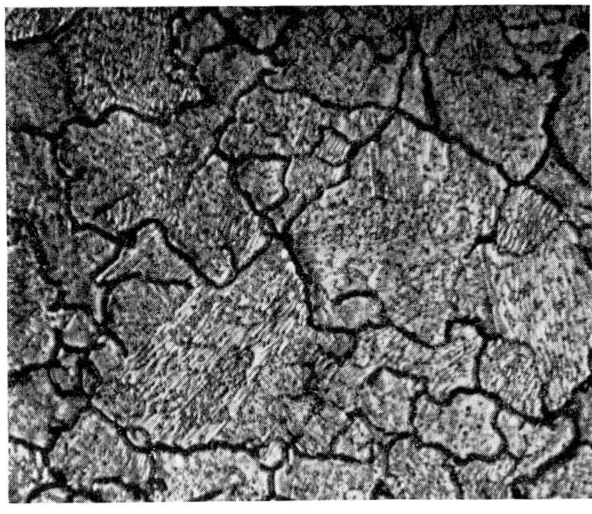

11—*Pre-polished surface of 18% Ni–Co–Mo maraging steel after 1100°C./15 min in H_2 and argon blast quench (simulated HAZ cooling)* Oblique illumination × 1000

12—*As-quenched maraging steel (Series A) etched in half-strength acid ferric chloride* Oblique illumination × 1000

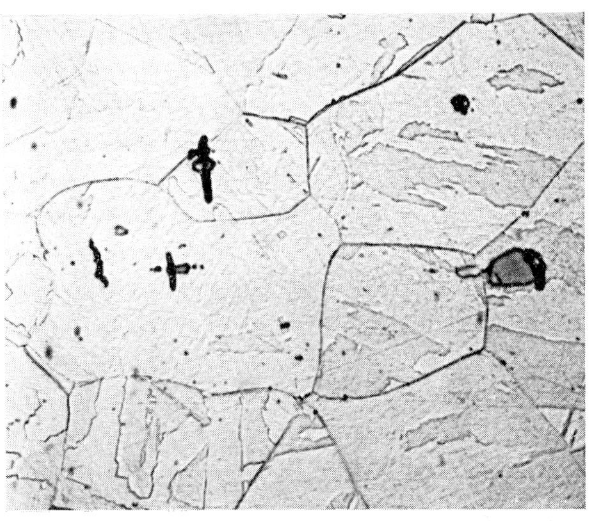

13—*Crack nucleation at non-metallic inclusions in a CLR specimen of as-quenched 18% Ni–Co–Mo steel (Series A). Etched in 20% Nital* × 1000

shears illustrated in Fig. 9 indicate that on cooling from the standard austenitising temperature (820°C.), martensitic transformation takes place in 18% Ni–Co–Mo maraging steel.

Figure 10 shows a typical surface appearance of a specimen austenitised at 1050°C. for 20 min and quenched in argon blast to simulate weld HAZ cooling in thick plate. After this heat treatment, the number of surface shear markings decreased considerably. If present, only one to three surface shears were observed within the individual prior austenitic grains.

Figure 11 shows a typical surface appearance of a specimen austenitised at 1100°C. for 15 min and quenched in a way similar to the preceding specimen. After quenching from 1100°C., and higher austenitising temperatures, the martensitic surface shears disappeared altogether. The prior austenite grains became divided into 'massive' areas, showing, within their interior, a fine striated texture. This 'surface' microstructure appears to correlate with the internal microstructures revealed by ordinary metallographic methods (compare Fig. 11 with Figs. 6*a* and *b*, and Figs. 7*a* and *b*).

Occasionally a rather unusual effect was observed when the as-quenched maraging steel with the coarsest grain size was etched in acid ferric chloride of half the strength of that given in Appendix A. After prolonged etching a large cellular pattern* appeared in some massive crystals (Fig. 12). The exact nature of this pattern is not known to the authors, but it is possible that it represents the regions of segregation that had occurred during the solidification of the ingot.

* This large cellular pattern should not be confused with that shown in Fig. 7*a*.

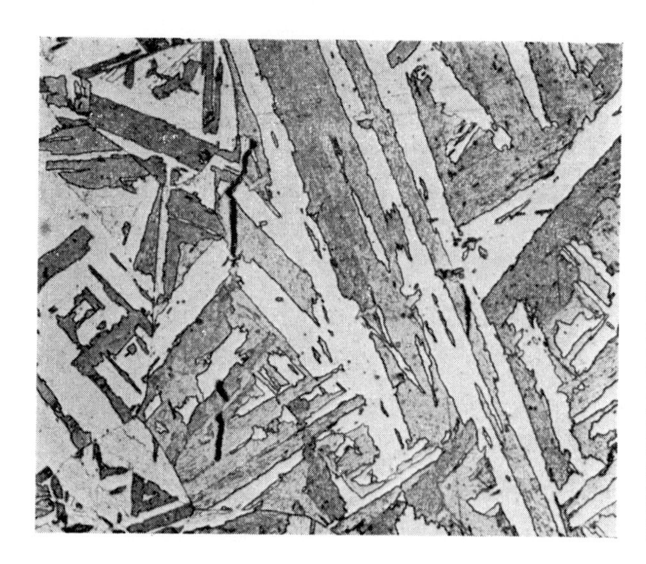

14—*Transcrystalline cracks in a CLR specimen of as-quenched 18% Ni–Co–Mo steel (Series C). Etched in acid ferric chloride* × 250

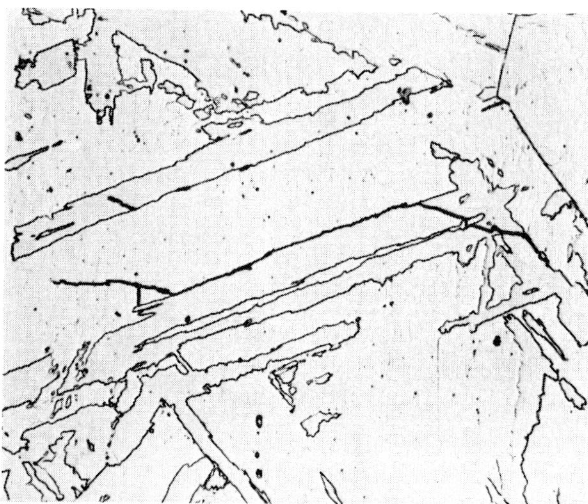

15—*Transcrystalline cracks in a CLR specimen of as-quenched 18% Ni–Co–Mo steel (Series C). Etched in 20% Nital* × 500

To study the mode of hydrogen cracking in the as-quenched 18% Ni–Co–Mo maraging steel, the fractured CLR specimens and those which remained unbroken after all the hydrogen had evolved from them, were examined. Adjacent to the fracture face, a number of subsidiary cracks were observed in the notch root area of the broken specimens. In the fine grained specimens (Series A) a large majority of the subsidiary cracks formed at non-metallic inclusions (titanium carbonitrides) and propagated in a transcrystalline manner (Fig. 13). The mode of crack initiation at non-metallic inclusions is consistent with the LCS of Series A being above the NYS, because some appreciable plastic deformation is necessary to initiate such cracks. In Series A no microcracks were found in the CLR specimens which remained unbroken.

In medium and coarse grain specimens (Series B and C) the subsidiary cracks were usually transcrystalline, and they did not seem to be associated with non-metallic inclusions (Fig. 14). It appeared that some transcrystalline cracks in the CLR specimens of Series B and C might begin at the boundaries of the massive crystals (Fig. 15). They either ran along the boundaries or projected from them.

(a)

(b)

16—*Unbroken CLR specimen (Series C) with microcracks in the plastic zones. Etched in 25% Nital*
(a) *General view of the notch area* × 65
(b) *Area of crack ringed in* (a) × 350

The initiation of the microcracks in the CLR specimens of Series B and C was also clearly associated with plastic deformation. The microcracks were found most readily along the plastic zones of the notched specimens and particularly in the centre of the gauge length where two plastic zones met at the top of an imaginary cone (*see* Fig. 15 in Ref. 2).

The examination of the CLR specimens that remained unbroken (stressed below the LCS), after all the hydrogen had evolved from them, revealed the presence of microcracks in Series B and C, but there were no cracks in Series A. The cracks were also transcrystalline and they were preferentially located in the plastic zones and zone intersections. However, there were no microcracks immediately under the root of the notch, in the area of so-called triaxial tension. Figure 16 shows an example of a specimen with the microcracks in the plastic zones.

Discussion

Consideration of microstructure

Notched tensile specimens of 18% Ni-Co-Mo steel, as-quenched and having a microstructure similar to that to be found in the weld HAZ, appeared to have a relatively low susceptibility to delayed hydrogen cracking under constant load. The susceptibility to delayed hydrogen cracking increased with the prior austenite grain size, *i.e.*, the peak austenitising temperature.

The finest grain size specimens (Series A), representing the HAZ microstructure to be expected in welding of sheet material with low heat inputs, had an index of susceptibility to hydrogen embrittlement as low as that of vacuum melted 9% Ni steels with carbon content not exceeding 0·15% C.[13] The specimens with coarser grains (Series B and C) had indices of resistance to hydrogen embrittlement comparable with those of vacuum melted 0·02% C–9% Ni steel or air melted 0·1% C–9% Ni steel.[2,13]

This similarity between 18% Ni–Co–Mo maraging steel and 9% Ni steels can be taken only as far as the general end result is concerned, and it cannot be rationalised in the same terms. The two steels differ considerably in their content of alloying elements, and the maraging steel is practically carbon free. Whilst the microstructures in the CLR specimens of 9% Ni steels (0·1–0·2% C) were unambiguously martensitic,[2,13] the microstructures in such specimens of the 18% Ni–Co–Mo maraging steel cannot be regarded as truly martensitic, on the basis of the evidence obtained in the present investigation.

After austenitising at the standard heat treatment temperature (820°C.), the maraging steel appears to transform to martensite, as indicated by the surface shears (Fig. 9). This observation is in agreement with the observations of Baker and Swan[14] who reported that, in maraging steels, martensite is formed on cooling. It is also in agreement with the results of Cheney,[15] who used hot stage microscopy and observed surface shears in 18% Ni–Co–Mo maraging steel cooled from 760°C. In Cheney's photographs, the alpha phase formed was unmistakably martensitic.

The simulated weld HAZ heat treatment of the CLR specimens produced blocks of alpha phase within the prior austenite grains. Such microstructure has been described by Gilbert and Owen[16] as 'massive' transformation product. When this microstructure is etched in Nital, only the massive blocks are revealed as in Fig. 15 (*see* also Fig. 5 in Ref. 16). Etching of this microstructure in the acid ferric chloride (AFC) solution (*see* Appendix A) revealed small elongated crystals within the massive blocks (Figs. 6 and 7). Similar microstructure was also observed on the prepolished surfaces (Fig. 11). The morphology of these fine alpha crystals, aligned in the blocks, appears to resemble bainitic colonies observed in the steels studied previously.[2]

It is possible that the austenitising temperature may affect the type of transformation products obtained in 18% Ni–Co–Mo maraging steel on cooling. It is known that in medium carbon, plain and low alloy steels, and in 0·4% C–24% Ni–Fe alloy, the Ms temperature is raised by increasing the austenitising temperature.[17-20] The increase in austenitising temperature of 18% Ni–Co–Mo steel may cause a change from the martensitic transformation, *i.e.*, accompanied by surface shears, to another type of transformation in which small alpha phase crystals are aligned in massive blocks.

There is some evidence in the literature that Fe–Ni alloys [16,12] may transform either to martensite or to massive alpha phase, depending on composition and cooling rate. At Ni contents below 6%, the transformation is always massive. At Ni concentration above 18%, the transformation is always martensitic. The alloys with 6–15% Ni may transform to either of the two products, at two distinctly different temperatures, depending on the cooling rate. The 18% Ni–Co–Mo steel may belong to this transitional group of alloys. If in this steel, the increase in austenitising temperature can raise the transformation temperature, this could explain the morphology of the transformation products obtained in the CLR specimens heat treated to simulate HAZ conditions.

Sometimes high values are obtained for the fracture toughness of the HAZ of maraging steel, sometimes the values are relatively low.[21] The reason for this behaviour is unknown, but it is not impossible that it may be due to the difference in the morphology of the transformation products obtained in different cases. One would expect the martensitic microstructure to have good toughness, whilst the massive alpha phase (resembling bainitic colonies) would have lower toughness.

Hydrogen embrittlement

18% Ni–Co–Mo maraging steel has superior resistance to hydrogen embrittlement, when compared with medium carbon low alloy, high strength steels, in the as-quenched condition. The medium carbon low alloy steels examined by Watkinson *et al*[1] had a UTS of the order of 120 tons/sq.in. and an NTS of the order of 140 tons/sq.in. after quenching. These steels could only withstand 10–30 tons/sq.in. under constant load when hydrogen was dissolved in them. Their high susceptibility to hydrogen induced cracking was associated with twinned martensite in their microstructure. The 18% Ni–Co–Mo steel had an NTS of the order of 85 tons/sq.in. in the as-quenched condition. Under constant load it could withstand stresses higher than 65 tons/sq.in. in the presence of dissolved hydrogen.

The maraging steel examined had a better resistance to hydrogen embrittlement than the low carbon alloy steels examined previously[2] and containing bainite in their microstructure. On the other hand, the mechanism of crack initiation in the maraging steel (particularly in Series B and C) suggested that the microstructure of this steel is akin to that of the bainitic alloy steels.

To understand the anomalous behaviour of the 18% Ni–Co–Mo maraging steel one should take into account:

(i) Its high content of alloying elements
(ii) Extremely high dislocation density observed in this steel in the as-quenched condition
(iii) A very slow rate of hydrogen evolution, which is indicative of the slow rate of bulk diffusion.

The comparison of hydrogen evolution curves of the maraging steel (Fig. 3) with those of the steels examined previously shows that, after 20 hr for example, the CLR specimens of the maraging steel lost 50% of their hydrogen content while those of the other steels lost almost all their hydrogen content.[2,13]

Factors (i) and (ii) have probably a very strong effect on factor (iii). The slow rate of hydrogen diffusion in the maraging steel may also explain why dormant microcracks were found in the unbroken CLR specimens. No such cracks were ever observed in the steels examined previously.[2,13]

The mechanism of crack initiation in the maraging steel examined might well be similar to that in the bainitic steels studied previously.[2] The difference in cracking behaviour would then lie in the different rate of crack propagation which could be affected by the factors discussed above.

It is possible that the high nickel content of the carbon-free alpha phase in maraging steel is important in lowering its susceptibility to cracking. Nunes and Larson[22] have shown that Ni increases the work hardening capacity and decreases the flow-stress/temperature dependence of ferrite. These two factors indicate that nickel increases the toughness of ferrite. Since the modes of brittle cleavage and hydrogen cracking are the same, it is possible that nickel may improve the resistance of the alpha phase ferrite in maraging steel to the propagation of hydrogen cracking.

It is interesting to notice that the high austenitising temperatures typical of the weld heat-affected zone had no detrimental effect on the tensile properties of the maraging steel after ageing. The coarse grain specimens were neither weaker nor had lower ductility parameters than those heat treated under the standard conditions.

Conclusions

(1) 18% Ni–Co–Mo maraging steel, containing dissolved hydrogen and quenched from high temperatures typical of the weld HAZ, has a low susceptibility to delayed hydrogen cracking under constant load.

(2) The as-quenched, simulated HAZ, microstructure is morphologically rather bainitic than martensitic and the microcracking is initiated in a transcrystalline manner, except in the fine grained specimens, where it initiates at non-metallic inclusions.

(3) After ageing, the coarse HAZ type microstructure has the same tensile properties as the microstructure of the maraging steel heat treated under the standard conditions.

Acknowledgments

The author wishes to thank Messrs. L. W. M. Nex and G. T. Hall for their help with the experimental work.

Appendix A

Etching solutions for 18% Ni–Co–Mo maraging steel

(1) Acid ferric chloride:
10 g ferric chloride
100 ml hydrochloric acid
100 ml distilled water.
Etching time: about 20 seconds.

(2) 15–25% solution of HNO_3 in alcohol for the solution annealed (quenched) microstructure.
Etching time: 30 sec–2 min depending on the concentration.

(3) 2% solution of HNO_3 in alcohol for the aged microstructure.

REFERENCES

1. F. Watkinson *et al*: *Brit. Welding J.*, 1963, vol. 10, pp. 54–62.
2. T. Boniszewski *et al*: *Brit. Welding J.*, 1965, vol. 12, pp. 14–36.
3. W. Steven: "Metallurgical developments in high-alloy steels", Special Report 86, Iron and Steel Institute, 1964 pp. 115–124.
4. A. G. Haynes: ibid, pp. 125–133.
5. L. R. Scharfstein: *J. Iron Steel Inst.*, 1964, vol. 202, pp. 158–159.
6. H. R. Gray and A. R. Troiano: *Met. Prog.*, 1964, vol. 85, pp. 75–78.
7. S. Floreen and R. F. Decker: *ASM Trans.*, 1962, vol. 55, pp. 518–530.
8. T. Yamashita and Y. Taneda: *J. Phys. Soc., Japan*, 1962, vol. 17, pp. 527–531.
9. A. B. Greninger and A. R. Troiano: *Trans. AIME*, 1940, vol. 140, pp. 307–331.
10. A. B. Greninger and A. R. Troiano: ibid, 1949, vol. 185, pp. 590–598.
11. R. B. G. Yeo: *ASM Trans.*, 1964, vol. 57, pp. 48–61.
12. E. A. Wilson: "Metallurgical developments in high-alloy steels", Special Report 86, Iron and Steel Institute, 1964, pp. 155–157.
13. T. Boniszewski and R. G. Baker: *Brit. Welding J.*, 1965, vol. 12, pp. 349–362.
14. A. J. Baker and R. P. Swan: 21st Ann. Meet. Electr. Micr. Soc. Am., Aug., 1963.
15. D. M. Cheney: *Met. Prog.*, 1964, vol. 85, pp. 92–93.
16. A. Gilbert and W. S. Owen: *Acta Met.*, 1962, vol. 10, pp. 45–54.
17. M. R. Meyerson and S. J. Rosenberg: *Trans. ASM*, 1954, vol. 46, pp. 1225–1256.
18. M. G. H. Wells: *J. Iron Steel Inst.*, 1961, vol. 198, pp. 165–179.
19. M. G. H. Wells and D. R. F. West: *J. Iron Steel Inst.*, 1962, vol. 199, p. 472.
20. A. S. Sastri and D. R. F. West: "The effect of austenitising conditions on the kinetics of martensite formation", S & T Memo 4/64, Min. Aviation, May, 1964.
21. *Reports of Progress of Welding Research Council*, 1964, vol. 19, p. 32.
22. J. Nunes and F. R. Larson: *Trans. AIME*, 1963, vol. 227, pp. 1369–1377.

SECTION VII:
Castings

Cast Maraging Steel

Cast maraging steel with high strength and toughness should allow use of the castings at higher strength-density ratios than now used on low alloy steel castings.

by E. P. Sadowski and R. F. Decker, Research Laboratory of The International Nickel Co., Inc., Bayonne, N. J.

Abstract

A high strength cast maraging steel possessing an excellent combination of ductility, notch tensile and impact properties was developed using a statistical approach. Use of the Box-Wilson method of experimentation established the composition limits of the alloy with only 41 experimental heats. The results point to the usefulness of statistics in alloy development and yielded a cast alloy with many advantages such as deep hardening, good machinability and little distortion or dimensional change during a simple heat treatment.

Introduction

The excellent combination of properties of the recently developed maraging steels in wrought form[1] prompted an investigation for a cast version of the alloy. It was believed that the industrial and defense needs for a high strength cast alloy with high toughness could be filled by a maraging steel. The other advantages inherent in the maraging steels, such as simple heat treatment, through hardening in large section sizes, dimensional stability on aging, good machinability, good stress-corrosion cracking resistance and good weldability were further incentive for the development of a cast alloy.

Targets were set up for the development program. These were room temperature properties of 240,000 psi yield strength (0.2 per cent offset), 5 per cent elongation, 10 per cent reduction of area, 15 ft-lb Charpy V-notch impact energy and notched tensile strength/tensile strength >1. Furthermore, it was desirable to have castability and fluidity comparable with cast low alloy steels and inherent weldability sufficient to permit repair welding without undue difficulty.

The amenability of the cobalt-molybdenum hardened maraging steel to conventional atmospheric casting practices made this a logical choice as an alloy base. Since the variables included four base elements, iron-nickel-cobalt-molybdenum, and several possible supplemental hardeners, simplification and acceleration were sought by centering the alloy study around a statistical design, the Box-Wilson method.[2,3,4]

Material and Melting Practice

All test materials were from 30 lb experimental heats which were induction melted under an argon atmosphere using magnesia crucibles. The heats were poured at 2850 to 2900 F into keel blocks (Fig. 1) in dry sand molds. All test samples were obtained from the keel block legs.

Forty-one heats were evaluated. The overall composition range (by weight per cent) investigated was:

C	0.006 — 0.050
Ni	14.9 — 17.8
Co	6.6 — 14.1
Mo	4.28 — 4.79
Ti	0.13 — 0.83
Cb	0.0 — 0.22
V	0.0 — 0.51
Cu	0.0 — 2.25
Al	0.03 — 0.31
Fe	bal.

Source: *Modern Castings*, Vol 43, No. 1 (Feb 1963)

The compositions of all heats investigated are listed with the properties of each heat later.

Electrolytic iron, nickel, cobalt and copper were used, whereas molybdenum, columbium and vanadium were added as ferro alloys. In four heats, molybdenum chips were used instead of the ferro alloy. The listed components, along with carbon in the form of BB7 graphite, made up the base charge. After melt down, 0.05 per cent calcium in the form of Ca Si was added for desulfurization. Aluminum rod and titanium sponge were then added, followed by 0.02 per cent zirconium (as Ni Zr) and 0.003 per cent boron (as Fe B).

Material Evaluation

Room temperature tensile (0.357 in. diameter, 1.4 in. gage length, shoulder type), notch tensile (0.300 in. major diameter, 0.212 in. minor diameter, 0.0005 in. notch root radius, $K_t = 12$), and Charpy V-notch impact energy were determined at room temperature on each heat.

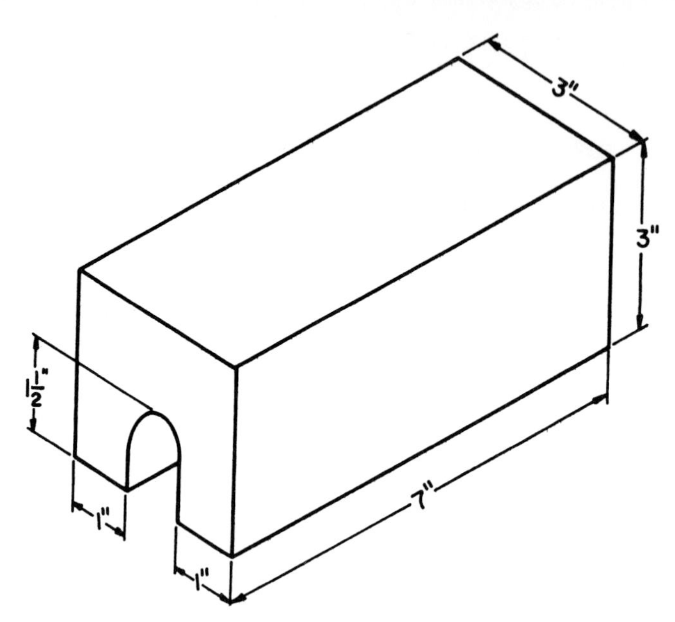

Fig. 1. Dimensions of cast keel blocks.

Homogenization Effect at 2200 F on Tensile Properties

The need for homogenization of the cast material prior to aging was found in the first series, as shown in the tensile data given in Table 1. A homogenization temperature of 2200 F was selected for the early compositional studies; this increased ductility in

TABLE 1. Chemical Composition and Mechanical Properties of Cast Maraging Steel — First Series

Heat No.	C	Ni	Mo	Co	Ti	Cb	V	Al	Cu	Hardness, Rc As-Cast	Hardness, Rc H+A	Condition[A]	YS (0.2% offset), 1000 psi	TS, 1000 psi	% Elong. in 1.4 in.	% RA	CVN Impact Energy, ft-lb	NTS, 1000 psi	NTS/TS
1	0.028	16.0	4.55	6.57	0.24	—	—	0.10	—	27.0	43.5	C+A	212.0	214.0	5.0	10.0	—	—	—
												H+A	204.4	215.0	12.0	46.5	20.2	321.5	1.49
2	0.015	17.3	4.60	9.90	0.13	—	—	0.16	—	30.5	46.5	H+A	217.3	233.3	8.0	34.0	11.2	318.4	1.46
3	0.015	16.2	4.55	7.00	0.42	—	—	0.24	1.98	29.5	52.0	C+A	241.2	255.8	3.0	4.0	—	—	—
												H+A	240.7	257.8	6.5	21.5	11.0	291.5	1.21
4	0.020	16.4	4.55	9.80	0.39	—	—	0.07	2.07	29.5	52.0	C+A	240.0	262.0	1.5	3.5	—	—	—
												H+A	256.0	270.7	4.5	15.0	9.7	308.5	1.20
5	0.017	17.3	4.55	7.02	0.14	0.15	—	0.25	2.25	31.5	48.0	H+A	224.3	242.3	3.0	9.5	11.5	317.4	1.41
6	0.050	16.4	4.57	9.55	0.20	0.21	—	0.09	2.10	31.0	54.0	C+A	—	233.7	Nil	Nil	—	—	—
												H+A	257.0	275.0	2.0	1.0	4.2	215.3	0.83
7	0.023	15.8	4.65	6.75	0.36	0.16	—	0.10	—	29.0	49.0	C+A	193.9	217.8	3.5	7.5	—	—	—
												H+A	208.0	220.5	8.0	26.5	11.0	301.1	1.44
8	0.020	17.0	4.44	10.05	0.42	0.17	—	0.28	—	32.0	49.0	H+A	248.2	263.3	7.0	23.0	8.5	340.9	1.37
9	0.018	16.0	4.55	6.85	0.20	—	0.45	0.10	2.05	28.5	50.5	C+A	238.0	248.5	0.5	4.0	—	—	—
												H+A	233.6	250.5	7.0 in 1 in.	19.5	9.7	320.8	1.37
10	0.018	16.6	4.55	9.60	0.20	—	0.49	0.30	2.05	30.5	53.5	C+A	264.2	265.2	1.5	2.0	—	—	—
												H+A	277.5	288.0	3.0	6.5	4.5	229.6	0.82
11	0.016	16.0	4.60	6.84	0.44	—	0.47	0.25	—	31.0	50.0	C+A	214.9	220.4	1.5	3.5	—	—	—
												H+A	226.5	245.3	7.0 in 1 in.	20.0	9.7	330.5	1.45
12	0.017	16.2	4.67	9.71	0.42	—	0.49	0.12	—	30.5	51.5	C+A	231.1	253.5	1.5	Nil	—	—	—
												H+A	244.5	260.0	4.5	12.0	8.0	334.3	1.36
13	0.024	16.1	4.68	6.80	0.20	0.19	0.49	0.26	—	30.5	50.0	C+A	208.2	236.1	0.5	2.0	—	—	—
												H+A	226.7	244.1	4.5	8.5	7.0	279.0	1.23
14	0.018	16.1	4.60	9.56	0.20	0.22	0.50	0.12	—	31.0	52.0	C+A	230.5	247.0	1.5	1.5	—	—	—
												H+A	248.3	261.7	3.0	6.5	5.5	286.5	1.15
15	0.014	16.2	4.65	6.70	0.38	0.17	0.51	0.10	2.10	30.0	51.0	C+A	222.8	253.5	Nil	0.5	—	—	—
												H+A	233.3	248.8	5.0	16.0	8.2	294.0	1.26
16	0.016	16.7	4.65	9.70	0.46	0.15	0.47	0.31	2.20	31.0	55.0	C+A	251.3	261.8	0.5	2.0	—	—	—
												H+A	277.6	291.3	3.0	5.0	5.0	228.2	0.82
17	0.023	16.0	4.60	9.70	0.26	—	—	0.08	—	30.0	51.5	C+A	227.9	248.0	1.5	3.0	—	—	—
												H+A	237.3	249.8	5.0	11.0	9.0	341.9	1.44
18	0.035	16.7	4.55	9.70	0.41	0.20	—	0.11	2.05	29.5	52.0	C+A	236.0	259.1	2.0	1.5	—	—	—
												H+A	251.7	267.7	5.0	14.5	6.5	283.1	1.12
19	0.017	15.0	4.35	6.40	0.20	0.20	0.47	0.08	1.85	30.0	48.0	C+A	215.5	228.8	3.0	2.0	—	—	—
												H+A	217.0	234.1	4.5	11.0	7.5	289.2	1.33

Note:

(A) Heat Treatment: (C+A): As-Cast plus aged 3 hr at 900 F
(H+A): Homogenized 4 hr at 2200 F, air cooled plus aged 3 hr at 900 F

TABLE 2. Effect of 0.1 Per Cent Element Added on Base Composition

Element	Hardness, Rc	Y.S. (0.2% offset), psi	T.S., psi	Elong., % in 4D	RA, %	CVN Impact Energy, ft-lb	Notched Tensile Strength, psi	Element Varied, %	Range, %
Co	+0.10	+ 950	+ 900	−0.08	−0.27	−0.13	− 810	1.50	7.0 -10.0
Ti	+0.49	+1910	+1950	+0.13	+0.29	−0.11	+5850	0.15	0.20- 0.50
Cb	+0.48	+1140	+1320	−0.88	−3.96	−1.16	−9640	0.12	0- 0.25
V	+0.14	+2800	+2800	−0.34	−2.08	−0.74	−2800	0.25	0- 0.50
Al	+1.18	+3360	+3910	−0.26	−0.94	−0.51	−2910	0.10	0.10- 0.30
Cu	+0.14	+1100	+1140	−0.12	−0.52	−0.10	−1800	1.00	0- 2.00
Avg.	50.6	239,000	254,200	5.5	16.9	9.1	294,800		

Average values correspond to the average of the 16 heats in the analyses and pertain to a base composition of:

	C	Ni	Mo	Co	Ti	Cb	V	Al	Cu
Added	0.05	17.0	4.50	8.50	0.35	0.12	0.25	0.20	1.00
Analyses	0.02	16.4	4.58	8.40	0.30	0.09	0.22	0.18	1.05

The above effects correspond to a heat treatment of 4 hr at 2200 F + 3 hr at 900 F.

every case and, in most cases, increased yield and tensile strengths. Based on these results all subsequent heats were homogenized prior to aging.

Statistically Designed First Series

The compositions of the first 16 heats listed in Table 1 were chosen to study the major maraging element, cobalt, plus several supplemental elements uncovered in earlier work, titanium, vanadium, columbium, aluminum and copper. Heat Numbers 17, 18 and 19 listed in Table 1 did not enter into the statistical analyses.

The Box-Wilson method of experimentation was followed. Essentially this procedure is used to determine the direction of steepest ascent up the response surface.

The chemistry range of the experimental alloys surveyed statistically was:

C	0.014 — 0.050
Ni	15.8 — 17.3
Co	6.6 — 9.8
Mo	4.45 — 4.68
Ti	0.13 — 0.46
Cb	0.0 — 0.22
V	0.0 — 0.51
Cu	0.0 — 2.25
Al	0.07 — 0.30
Fe	bal.

The statistical analysis was made on the basis of percentage of each element added to a 17.0 per cent nickel, 8.5 per cent cobalt, 4.5 per cent molybdenum and 0.03 per cent carbon base. A factorial design of ¼ x 2,[6] 6 factors, 16 observations was chosen. The amount of each variant added was – Co – 7 or 10 per cent, Ti – 0.20 or 0.50 per cent, Cb – none or 0.25 per cent, V – none or 0.50 per cent, Al – 0.10 or 0.30 per cent and Cu – none or 2.00 per cent.

Five of the heats in the first series (3, 4, 8, 12 and 18) met the target requirements of 240,000 psi yield strength (0.2 per cent offset), 5 per cent elongation, 10 per cent elongation and NTS/TS >1. The range of impact strength on these heats was from 6.5 to 11 ft-lb CVN. Three other experimental compositions, Heat Numbers 9, 15 and 17, had yield strengths between 233,000 and 237,000 psi and also met the ductility requirements.

The results of the statistical analyses are shown in Table 2. The latter results were obtained by a Yates analysis[5] of the experimental data. The relative effectiveness of each of the elements given in Table 2 refers to the expected effect of 0.1 per cent element added, on the average mechanical properties of the base composition given in Table 2. As the table indicates, all the variants were effective in increasing the hardness, yield strength and tensile strength. Second, the table indicates that all elements except titanium reduced the notch tensile strength and ductility. All elements were deleterious to impact strength.

Titanium Effect on Mechanical Properties of the Statistically Established Base Composition

Since the statistical study indicated that, of the six variants investigated, titanium was the most promising supplementary hardener, a new series with increasing titanium was melted. Heats 20 through 24 (Table 3) show the effect of higher titanium (0.29 to 0.83 per cent) on a 17 per cent nickel, 8.5 per cent cobalt, 4.5 per cent molybdenum base. The composition limits of this series were:

C	0.012 — 0.015
Ni	16.5 — 17.0
Co	7.60 — 8.55
Mo	4.40 — 4.50
Ti	0.29 — 0.83
Al	0.13 — 0.16
Fe	bal.

The tensile test results (Table 3, Fig. 2) of homogenized and aged samples showed an increase in yield strength from 211,000 psi for the 0.29 per cent titanium heat to 247,000 psi for the highest titanium level (0.83 per cent). Ductility and impact strength decreased at the higher titanium levels (0.56 per cent and above) and notch tensile strength decreased with increased titanium content (Fig. 3). However, NTS/TS was still high, the range being 1.15 to 1.50.

Notch tensile strength varied from 341,000 psi at 0.29 per cent titanium to 303,000 psi at 0.83 per cent titanium. Statistically, from the first series it was predicted that the notch tensile strength would increase with increasing titanium content. Apparently, the upper titanium level of the first series (0.50 per cent added) was close to optimum and the original calculations could not foresee going through a maximum.

The excellent correlation between the properties obtained on heat 20 and calculated values from the Box-Wilson series is notable (Table 3). This adds confidence in the use of this research technique.

Although heat 23 met most of the target requirements, composition variables other than titanium were investigated in an effort to obtain a better combination of properties.

High Cobalt and Titanium Effect

Heats 1 and 17 and the statistical analyses (Table 2) had pointed to beneficial effects of cobalt up to the 10 per cent level. The possible effects of some interactions were computed from the Box-Wilson series, and are given in Table 4. The cobalt-titanium interaction appeared to be most effective on notch properties. To follow this lead and to establish the effects of higher cobalt at the 0.2 and 0.5 per cent titanium levels, heats 25 through 29 were melted and tested. The composition range of the five heats was:

C	0.011 — 0.018
Ni	16.5 — 17.4
Co	9.5 — 14.1
Mo	4.43 — 4.55
Ti	0.16 — 0.44
Al	0.084 — 0.10
Fe	bal.

Yield strength (Table 5) increased with increasing cobalt content at a specific titanium level (heats 25, 26 and 29 at low titanium and heats 27 and 28 at high titanium). The notch tensile strength reached a maximum at a specific cobalt level dependent upon the titanium content (Fig. 4). The 14.1 per cent cobalt heat 29 had low notch tensile strength. At lower cobalt levels the notch tensile strength varied from 288,000 psi to 351,000 psi with NTS/TS of 1.06 to 1.43. The best combination of properties in this series was obtained on heat 27.

Defining a Composition Range

Upon establishment of the cobalt and titanium ranges for optimum notch properties (Fig. 4), nickel and molybdenum were varied over a feasible commerical composition range. Heats 30 through 37 (Table 6) were an attempt to bracket this chemistry range for commercial use. The range of composition for this series was:

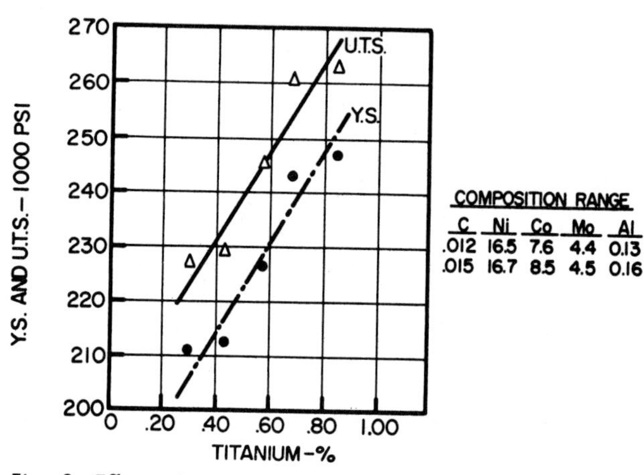

Fig. 2. Effect of titanium on smooth bar tensile properties of statistically derived base composition.

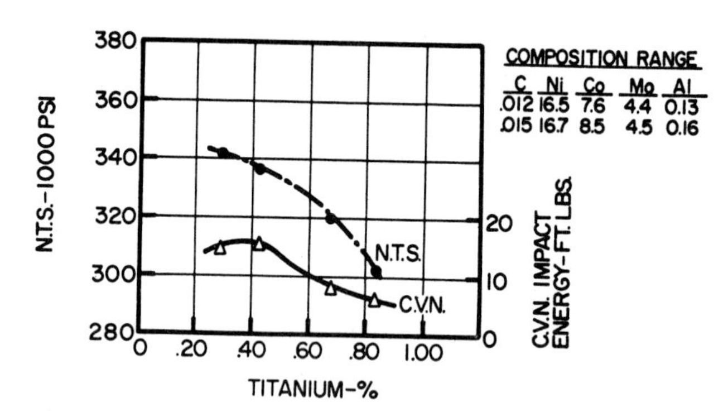

Fig. 3. Effect of titanium on notch properties of statistically derived base composition.

C	0.007 — 0.019
Ni	14.9 — 17.8
Co	9.25 — 10.60
Mo	4.28 — 4.79
Ti	0.19 — 0.43
Al	0.08 — 0.10
Fe	bal.

Comparison of heats 30 and 37 and heat 31 with 32 reveals the effect of nickel on mechanical properties on both the high and low sides of the composition limits of this series. With molybdenum, titanium and cobalt at the low level increasing nickel content from 14.9 (heat 30) to 17.2 per cent (heat 37) resulted in increased yield strength at the rate of

TABLE 3. Effect of Titanium on Mechanical Properties of Statistically Established Base

Heat No.	Composition, Wt. % (Fe Bal)						Hardness (Rc)		Condition[a]	YS (0.2% offset), 1000 psi	TS 1000, psi	Elong., % in 1.4 in.	RA, %	CVN Impact Energy, ft-lb	NTS, 1000 psi	NTS/TS
	C	Ni	Mo	Co	Ti	Al	As-Cast	H+A								
Calculated values after correction of average values of Table 2 for removal of Cu, V and Cb							—	48.1	H+A	220.0	234.2	8.6	31.9	13.3	294.8	1.41
Above corrected for lower Co obtained on heat 20							—	47.2	H+A	211.0	226.1	9.3	34.3	14.5	338.6	1.50
20	0.012	16.6	4.43	7.60	0.29	0.14	33.5	47.0	H+A	211.0	226.6	10.0	42.0	15.0	341.4	1.50
21	0.012	16.7	4.48	8.30	0.42	0.13	33.0	47.5	H+A	212.5	229.5	10.5	43.0	15.8	338.5	1.47
22	0.012	16.7	4.40	8.55	0.56	0.15	30.5	50.5	H+A	226.5	245.2	3.0	6.0	8.0	Poor Casting	
23	0.012	16.5	4.50	8.13	0.67	0.16	31.5	52.0	H+A	242.7	260.8	6.5	27.0	7.8	319.8	1.23
24	0.015	16.7	4.48	8.25	0.83	0.16	35.0	51.5	H+A	247.2	262.9	3.5	11.0	6.2	302.9	1.15

Note:

(a) H+A: Homogenized 4 hr at 2200 F + aged 3 hr at 900 F.

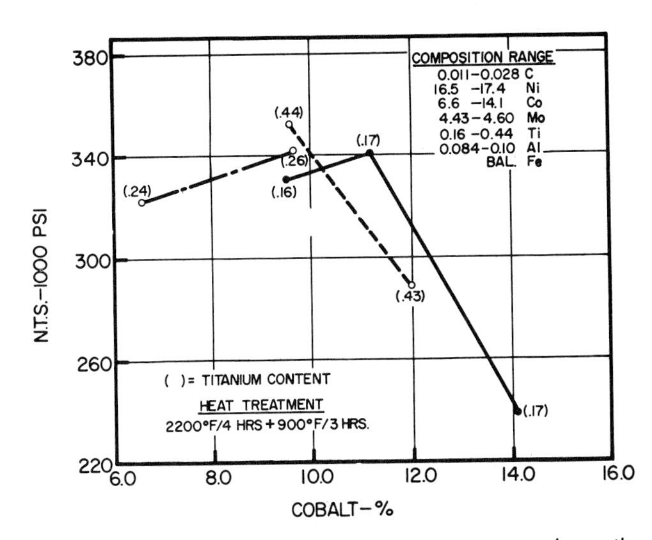

Fig. 4. Relationship of cobalt and titanium to notch tensile properties of cast maraging steel.

1270 psi 10.1 per cent and increased notch tensile strength at a rate of 1150 psi/0.1 per cent.

At the high side of the series, heats 31 and 32, increasing the nickel content resulted in increased yield strength and notch tensile strength at the rates of 960 and 300 psi/0.1 per cent, respectively. Increase in nickel content also resulted in a moderate decrease in ductility and CVN impact energy at both the high and low sides of the chemistry range.

Heats 33 and 34 and heats 35 and 36 show the effects of molybdenum. Increase of this element increased yield strength and decreased notch tensile strength.

Based on all the previously discussed experimental alloys, the composition range given in the accompanying data sheet (Table 7) is suggested. The mechanical properties of 8 heats approximating the suggested composition limits are given in Table 8. Heats 31, 35 and 38 indicate the properties which were obtained when the nickel, cobalt and titanium were on the high side of the chemistry range. The properties were:

0.2% YS, 1000 psi	244-248
Elong., % in 4D	5.5-11.5
RA, %	27-34
CVN Impact energy, ft-lb	10.5-14.5
TS, 1000 psi	258-261
NTS, 1000 psi	313-354
NTS/TS	1.20-1.37

TABLE 4. Effects of Interactions on Mechanical Properties of Cast Maraging Steel

Interaction	NTS 1000, psi	YS 1000, psi	Elong., %	RA, %	CVN Impact Energy, ft-lb
CoTi, CbAl	+20.4	+0.4	+0.2	+0.5	+1.2
CoCb, TiAl	+ 5.9	+3.0	+0.4	+0.9	+0.4
CoV, AlCu	+ 1.6	+1.7	−0.1	−0.2	+0.2
TiV, CbCu	+ 0.1	−3.4	+0.1	+1.0	+0.7
CbV, TiCu	− 3.9	−1.0	+0.3	−2.1	+0.7
VAl, CoCu	+ 0.1	+2.4	—	—0.8	—0.5

Since only a fractional replicate was used in the Box-Wilson series, the statistical analyses do not establish which of the two interactions listed for each effect is responsible. Examination of the test data pointed to the presence of Co and Ti as being considerably more effective than Cb and Al.

TABLE 5. Effect of Cobalt and Titanium on Mechanical Properties of Cast Maraging Steel

Heat No.	Composition, Wt. % (Fe Bal) C	Ni	Mo	Co	Ti	Al	Hardness (Rc) As-Cast	H+A	YS (0.2% offset), 1000 psi	TS, 1000 psi	Elong.,% in 1.4 in.	RA, %	CVN Impact Energy, ft-lb	NTS, 1000 psi	NTS/TS
25	0.014	16.7	4.43	9.53	0.16	0.09	34.5	49.5	227.5	241.5	6.5	26.5	10.0	329.6	1.37
26	0.014	16.8	4.48	11.20	0.17	0.09	34.5	51.5	232.5	251.7	5.5	17.5	8.8	339.6	1.35
27	0.011	16.5	4.52	9.50	0.44	0.08	35.0	50.0	230.2	246.2	9.5	32.5	10.8	351.3	1.43
28	0.014	17.1	4.55	12.00	0.43	0.09	35.0	53.0	257.5	272.0	3.5	4.5	5.2	287.9	1.06
29	0.018	17.4	4.50	14.10	0.17	0.10	37.0	54.5	265.5	282.6	3.0	6.0	6.8	239.3	0.84

Note:
H+A: Homogenized 4 hr at 2200 F + 3 hr at 900 F.

TABLE 6. Effect of Varying Nickel, Cobalt, Molybdenum and Titanium on Mechanical Properties of Cast Maraging Steel

Heat No.	Composition, Wt. % (Fe Bal) C	Ni	Mo	Co	Ti	Al	Hardness (Rc) As-Cast	H+A	YS (0.2% offset), 1000 psi	TS, 1000 psi	Elong.,% in 1.4 in.	RA, %	CVN Impact Energy, ft-lb	NTS, 1000 psi	NTS/TS
30	0.007	14.9	4.31	9.32	0.19	0.08	31.0	43.0	190.5	206.0	10.5	44.0	16.8	305.4	1.48
31	0.007	17.4	4.63	10.50	0.39	0.10	32.0	50.5	244.5	259.6	11.5	28.0	12.5	350.1	1.35
32	0.008	15.3	4.63	10.60	0.38	0.09	33.5	47.5	224.3	238.9	8.5	37.0	15.8	343.9	1.44
33	0.011	15.1	4.72	9.30	0.39	0.08	34.0	47.5	215.2	220.9	defective	4.0	13.5	331.0	1.50
34	0.009	15.1	4.41	9.25	0.43	0.10	33.5	45.5	203.8	219.0	9.0	43.0	17.0	339.0	1.55
35	0.012	16.9	4.79	10.60	0.20	0.10	32.0	51.5	247.7	261.4	5.5	27.5	10.5	313.0	1.20
36	0.019	17.8	4.38	10.60	0.19	0.09	33.0	49.0	243.7	259.8	3.5	13.0	11.0	330.0	1.27
37	0.016	17.2	4.28	9.42	0.20	0.08	32.5	46.0	219.7	235.1	6.5	26.0	12.5	331.9	1.41

Note:
H+A: Homogenized 4 hr at 2200 F + 3 hr at 900 F.

Source: *Modern Castings*, Vol 43, No. 1 (Feb 1963)

TABLE 7. Composition Range and Mechanical Properties of a Cast Maraging Steel

	heat treated 4 hr/2200 F plus 3 hr/900 F
Ni	16 -17.5
Co	9.5-11.0
Mo	4.4- 4.8
Ti	0.15- 0.45
C	0.03 max.
Mn	0.1 max.
Si	0.1 max.
S	0.01 max.
P	0.01 max.
Al	0.05- 0.15
B	0.003 added
Zr	0.02 added
Ca	0.05 added
Fe	bal.
Yield strength, 0.2 per cent offset, psi	228,000-248,000
Ultimate strength, psi	242,000-261,000
Elongation, %	5-11
Reduction of area, %	11-34
Notched tensile, 0.3 in. bar, $K_t = 12$, psi	313,000-354,000
CVN impact energy, room temp., ft-lb	9-15

The preceding properties were obtained with a heat treatment of 4 hr at 2200 F plus 3 hr at 900 F.

Homogenization Time and Temperature Effect on Mechanical Properties

Six heats in the prescribed composition range, heats 31, 35 and 36, and three additional heats 39, 40, and 41 were tested after being homogenized at temperatures between 1800 and 2200 F for 4 to 16 hr. The results are given in Table 9. Lowering the homogenization temperature below 2100 F generally resulted in lowering the reduction of area, notch tensile strength and impact strength. However, NTS/TS was still at least 1.1, except in one case where a value of 1.0 was obtained with the 1800 F homogenization temperature (heat 31).

Increasing the time of homogenization at 1800 and 1900 F from 4 hr to either 8 or 16 hr resulted in some recovery of the impact and notch tensile properties obtained at the higher homogenization temperatures but full recovery was never realized.

The effect of homogenization temperature on mechanical properties is shown graphically in Figs. 5 through 8. The latter plots are for the results obtained on heats 39, 40 and 41. In all cases, optimum ductility was obtained with the 2100 F treatment. Generally, optimum notch tensile and impact properties were also obtained with the latter temperature. The range of properties obtained with the 2100 F homogenization temperature is listed in Table 10.

It should be noted that heats 39, 40 and 41 were

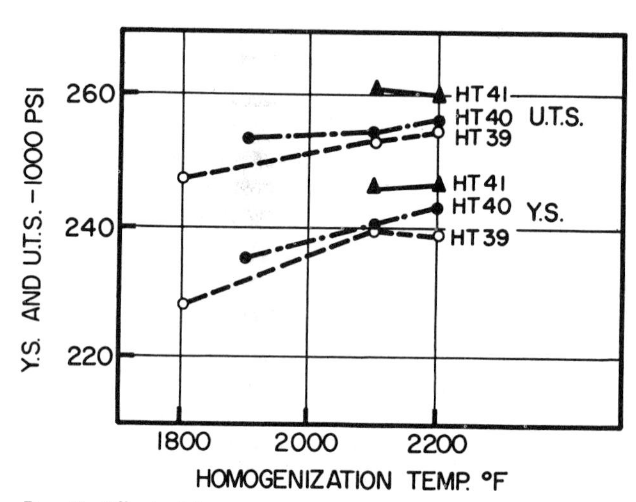

Fig. 5. Effect of homogenization temperature on yield and tensile strength (4 hr treatment).

TABLE 8. Heats Upon Which Cast Maraging Steel Composition Range Is Defined[A]

Heat No.[B]	C	Ni	Mo	Co	Ti	Al	Mn	Si	B	Zr	P	S	Hardness (Rc) As-Cast	Hardness (Rc) H+A	Y.S. (0.2% offset), 1000 psi	T.S., 1000 psi	Elong., % in 1.4 in.	RA, %	CVN Impact Energy, ft-lb	NTS, 1000 psi	NTS/TS
17	0.023	16.0	4.60	9.70	0.26	0.08	0.06	0.11	0.0036	<0.01	—	—	30.0	51.5	237.3	249.8	5.0	11.0	9.0	341.9	1.44
25	0.014	16.7	4.43	9.53	0.16	0.088	0.048	0.062	0.0032	<0.01	—	—		49.5	227.5	241.5	6.5	26.5	10.0	329.6	1.37
26	0.014	16.8	4.48	11.2	0.17	0.092	0.056	0.10	0.0032	<0.01	—	—		51.5	232.5	251.7	5.5	17.5	8.8	339.6	1.35
27	0.011	16.5	4.52	9.50	0.44	0.084	0.052	0.075	0.0028	<0.01	—	—	29.5	50.0	230.2	246.2	9.5	32.5	10.8	351.3	1.43
31	0.007	17.4	4.63	10.50	0.39	0.10	<0.04	0.055	0.0033	<0.01	.005	.007		50.5	244.5	259.6	11.5	28.0	12.5	350.1	1.35
35	0.012	16.9	4.79	10.60	0.38	0.090	<0.04	0.065	0.0030	<0.01	.004	.006		52.0	247.7	261.4	5.5	27.5	10.5	313.0	1.20
36	0.019	17.8	4.38	10.60	0.19	0.087	<0.04	0.071	0.0032	<0.01	.004	.005		49.0	243.7	259.8	3.5	13.0	11.0	330.0	1.27
38	0.006	17.3	4.63	11.38	0.27	0.03	<0.02	<0.02	0.0028	<0.01	.002	.007		50.5	245.3	257.9	8.0	34.0	14.5	354.1	1.37

Note:

(A) Heat treatment — (H+A) — Homogenized 4 hr/2200 F + aged 3 hr/900 F

(B) Heat 38 was made using Mo chips — all other heats had Fe Mo in the charge

TABLE 9. Effect of Time and Temperature of Homogenization on Mechanical Properties of Cast Maraging Steel[A]

Heat No.[B]	C	Ni	Mo	Co	Ti	Al	Mn	Si	B	Zr	P	S	Homogenization Time, hr	Homogenization Temp, F	Y.S. (0.2% off-set), 1000 psi	T.S., 1000 psi	Elongation, %	RA, %	CVN Impact Energy, ft-lb	NTS, 1000 psi	NTS/TS
31	0.007	17.4	4.63	10.50	0.39	0.10	<0.04	0.055	0.0033	<0.01	0.005	0.007	4	2200	244.5	259.6	11.5	28.0	12.5	350.1	1.35
													4	2000	248.5	264.5	5.0[C]	20.0	9.0	318.2	1.20
													4	1800	244.9	283.5	9.0	21.0	7.5	284.1	1.00
35	0.012	16.9	4.79	10.60	0.38	0.090	<0.04	0.065	0.0030	<0.01	0.004	0.006	4	2200	247.7	261.4	5.5	27.5	10.5	313.0	1.20
													4	2000	248.2	262.8	5.5	21.0	9.8	304.5	1.16
													4	1800	237.3	254.4	5.0	15.5	5.3	276.4	1.09
36	0.019	17.8	4.38	10.60	0.19	0.087	<0.04	0.071	0.0032	<0.01	0.004	0.005	4	2200	243.7	259.8	3.5	13.0	11.0	330.0	1.27
													4	2100	238.5	254.5	5.0	15.5	—	338.1	1.33
													4	2000	244.3	259.2	3.5	10.0	8.5	300.3	1.16
													4	1800	235.1	250.0	3.0	5.0	6.5	296.3	1.18
39[B]	0.012	16.92	4.72	10.40	0.38	0.054	<0.05	<0.04	0.0020	<0.01	0.003	0.005	4	2200	238.9	254.5	9.0	39.0	11.8	349.8	1.37
													4	2100	239.4	253.5	11.0	39.5	17.0	361.3	1.43
													4	1800	228.3	247.5	7.0	26.5	8.5	331.5	1.34
													8	1800	233.9	250.5	9.0	29.5	12.5	345.1	1.37
40[B]	0.018	16.90	4.53	10.45	0.38	0.066	<0.05	<0.04	0.0021	<0.01	0.003	0.007	4	2200	243.0	256.5	8.0	27.5	14.5	361.5	1.43
													4	2100	238.9	254.5	10.0	35.5	13.0	354.2	1.39
													4	1900	235.4	253.5	7.0	24.0	10.0	342.1	1.35
													16	1800	234.8	251.5	10.0	38.0	11.0	328.9	1.31
41[B]	0.014	16.90	4.72	10.30	0.41	0.078	<0.05	<0.04	0.0022	<0.01	0.002	0.005	4	2200	246.0	260.0	5.0	15.5	10.5	339.0	1.30
													4	2100	245.5	260.5	8.0	31.0	13.5	355.5	1.36
													8	1900	239.9	257.0	5.0	13.5	8.5	325.3	1.26

Composition, Wt % (Fe Bal); Mechanical Properties.

Note:
(A) All specimens aged 3 hr at 900 F after homogenization.
(B) Heats 39, 40 and 41 were made using molybdenum chips rather than Fe Mo.
(C) Small defect.

made with molybdenum chips instead of ferromolybdenum in the melt charge. This may account, at least in part, for the high notch properties of these heats even when the lower homogenization temperatures are used.

Homogenization Treatment Effect on Microstructure

Typical microstructures with and without homogenization at 2100 F are shown in Fig. 9. A relatively coarse structure containing <2 per cent retained austenite was observed in the "cast and aged" condition (Figs. 9a and 9b). The homogenized material was completely transformed to a body centered cubic phase having a somewhat finer structure (Figs. 9c and 9d).

Homogenization Treatment Effect on Segregation

The effects of homogenization on the removal of segregation of the major alloying elements were determined. Scans of 12.5 mm specimens were obtained with a Heinrich Miniature Probe attachment on a GEXRD-3 unit. A one mm aperture was used. The

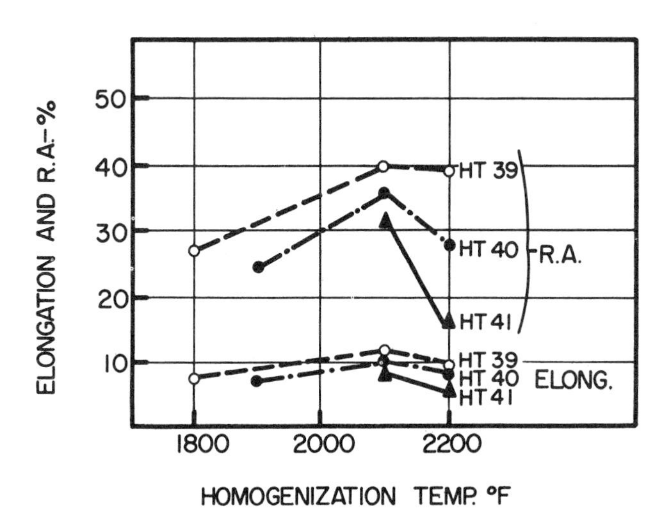

Fig. 6. Effect of homogenization temperature on ductility (4 hr treatment).

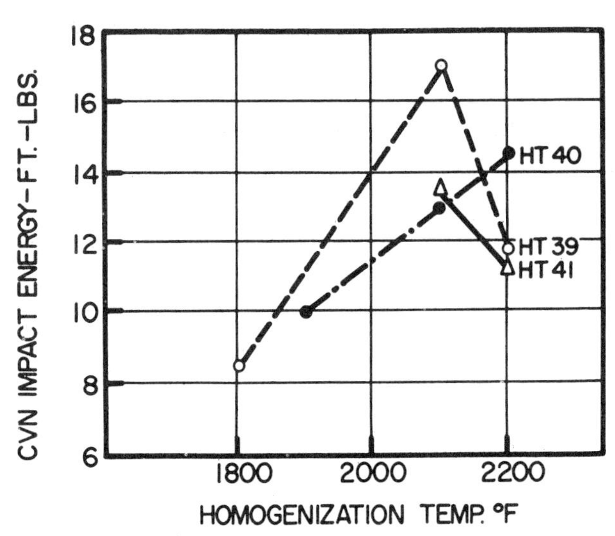

Fig. 7. Effect of homogenization temperature on impact energy (4 hr treatment).

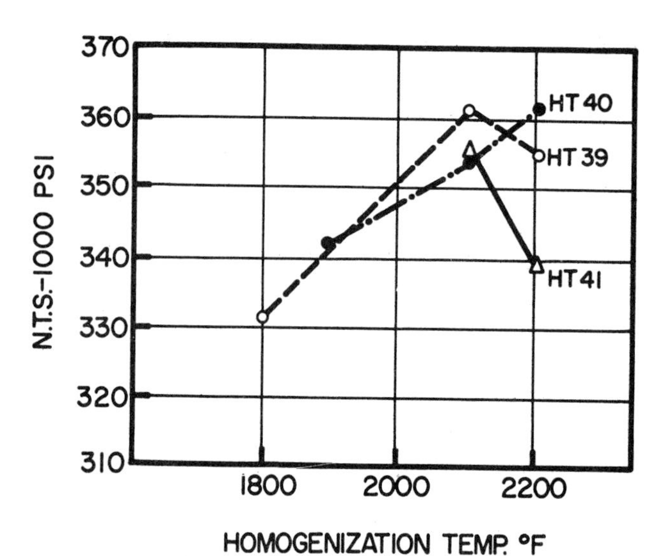

Fig. 8. Effect of homogenization temperature on notch tensile strength (4 hr treatment).

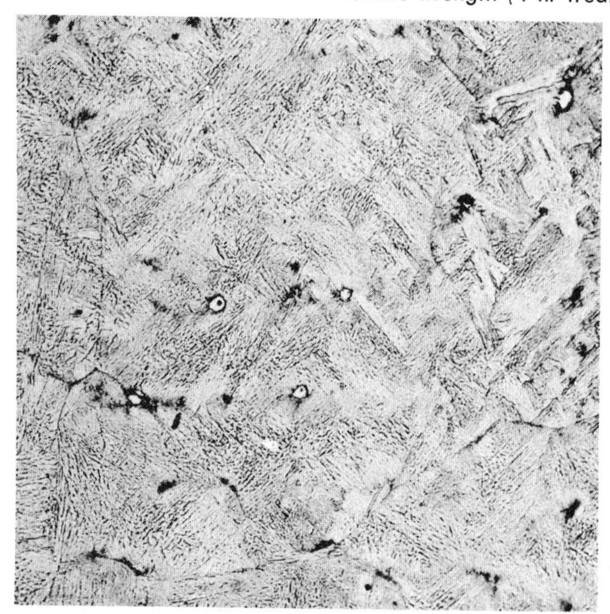

Fig. 9a. As-cast plus aged. Fry's reagent etch. 100 ×.

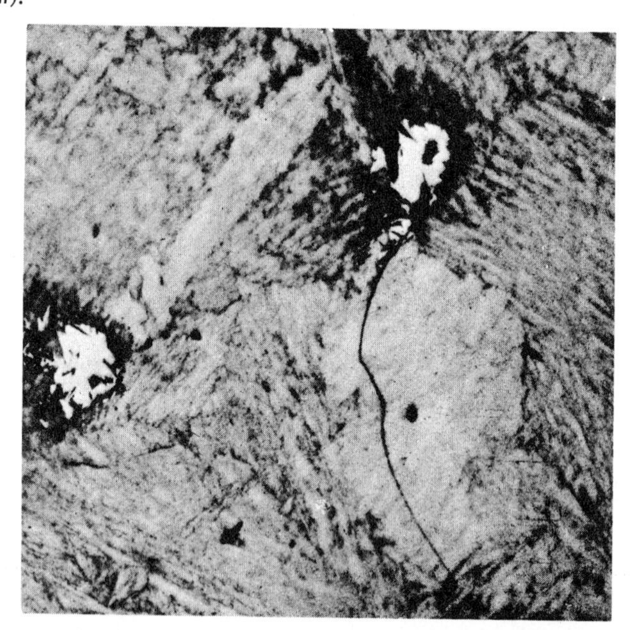

Fig. 9b. As-cast plus aged. Fry's reagent etch. 500 ×.

Fig. 9c. 4 hr/2100 F, AC plus age. Fry's reagent etch. 100 ×.

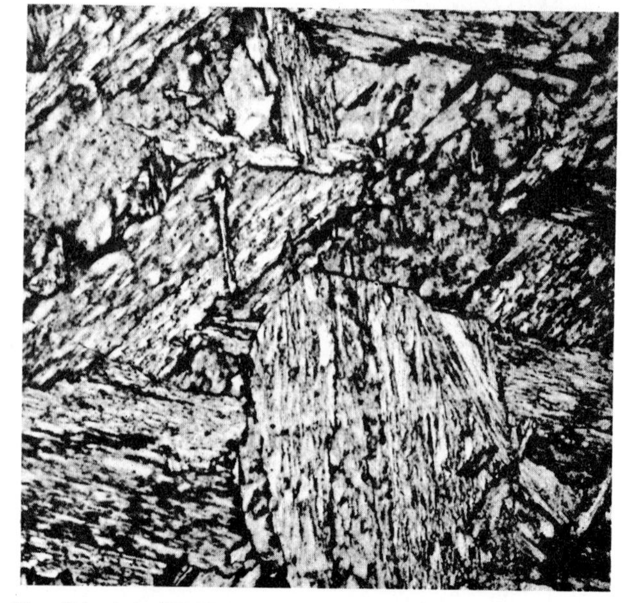

Fig. 9d. 4 hr/2100 F, AC plus age. Fry's reagent etch. 500 ×.

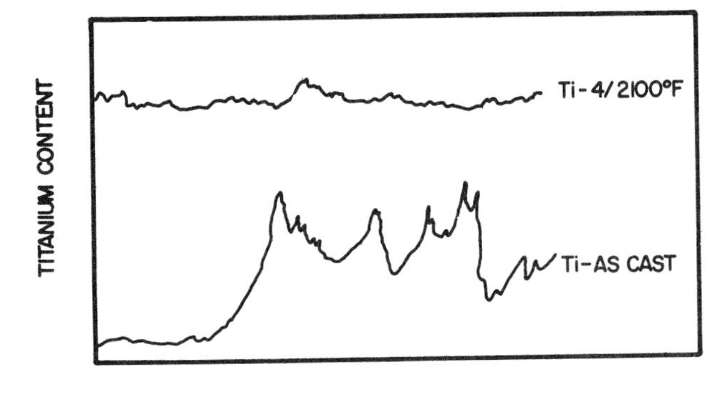

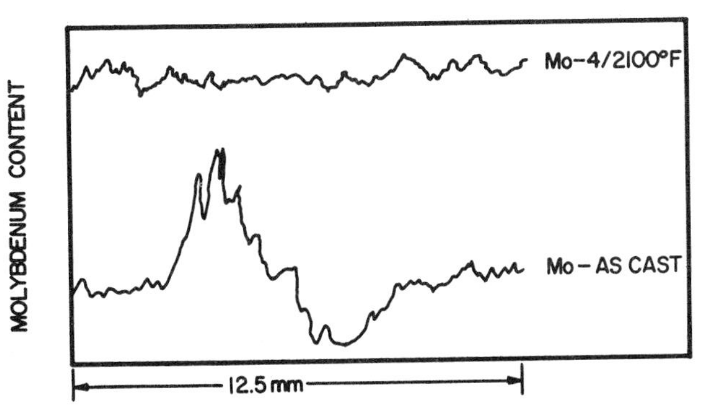

Fig. 10. Effect of homogenization at 2100 F on segregation of titanium and molybdenum.

TABLE 10. Range of Mechanical Properties of Cast Maraging Steel Obtained Using a 2100 F Homogenization Treatment[a]

Yield strength, 0.2%, psi	238,000 - 244,000
Ultimate tensile strength, psi	254,000 - 260,000
Elongation, %	5 - 11
Reduction of area, %	16 - 40
Notch tensile strength (K_t = 12), psi	338,000 - 361,000
CVN impact energy, room temp., ft-lb	13.0 - 17.0
NTS/TS	1.33 - 1.43

Notes:
(a) Homogenized 4 hr, air cooled, then aged 3 hr at 900 F.
(b) Based on four heats — three of which contained molybdenum chips in the base charge rather than ferro molybdenum.

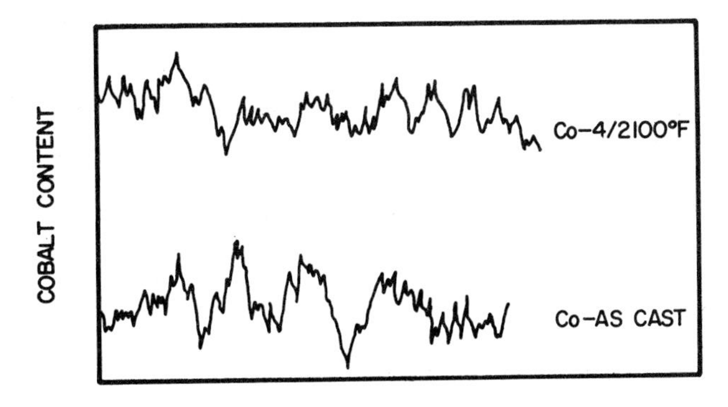

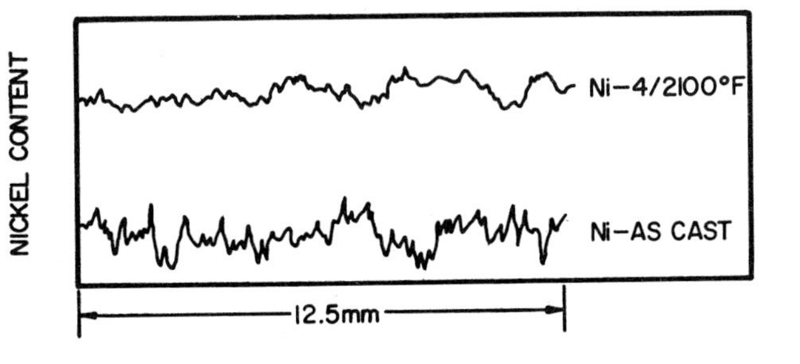

Fig. 11. Effect of homogenization at 2100 F on segregation of nickel and cobalt.

scan before and after heat treatment was taken over the same area on the same specimen. The results are shown in Figs. 10 and 11, and should be analyzed qualitatively rather than quantitatively.

The segregation of molybdenum and titanium was quite severe, whereas the nickel and cobalt were rather well distributed. However, the homogenization treatment (2100 F) resulted in removal of the severe segregation. Scans were also made on material homogenized at 2000 and 2200 F, and similar results were observed. It would appear that removal of the segregation is a major influence in the higher ductility and notch strength of the cast and homogenized alloy.

Comparison of Cast Maraging Steel and Other Ultra High Strength Cast Steels

The room temperature mechanical properties of the cast maraging steel and the properties reported for other high strength cast steels are given in Table 11. The maraging steel at an equivalent or higher yield strength appears to have superior reduction of area, Charpy V-notch impact energy and NTS/TS. The 234,000 yield strength obtained by Schaffer, et al,[8] on modified A.I.S.I. 4340 + silicon is the best of many heats, and was obtained using high purity charge material and carefully controlled thermal gradients during solidification.

It should be noted that the values listed in Table 11 for 8740, 4340, 4330 and H-11 steels are reported to have been obtained in commercial practice by careful control of the melt from the initial charge to pouring. Electrolytic iron and sponge iron briquettes are used instead of scrap and castings are made in ceramic molds.

Fluidity

Two fluidity spirals were poured from one heat of each of – cast maraging steel, 300M and A.I.S.I. 4340.

TABLE 11. Comparison of Mechanical Properties of Ultra High Strength Cast Steels

Material	Heat Treatment	Mechanical Properties					
		Y.S. (0.2% Offset), 1000 psi	TS, 1000 psi	Elong., %	RA, %	CVN Impact Energy, ft-lb	NTS/TS
Maraging steel	2200 F, 4 hr, AC plus 900 F, 3 hr, AC	228-248	241-261	5-11	11-34	9-14.5	1.20-1.37
	2100 F, 4 hr, AC plus 900 F, 3 hr, AC	238-244	253-260	5-11	16-40	13-17	1.33-1.43
300 M (6) (7)	2250 F, 4 hr, AC plus 1650 F, one hr, FC to 1350 F, OQ, temper 6 + 6 hr, 600 F	(0.1%) 221.3	257.5	5.8	10.6	10.5	ND
	Same as above except OQ from 1650 F	226.8 219.3	235.8 234.2	1.5 1.5	4.0 4.5	10.0	1.18 1.09
	1750 F, one hr, AC plus 1600 F, one hr, OQ plus 6 + 6 hr, 600 F	216-220	256-268	6-8	12.0	10.0	ND
Mod. 4340 + high silicon (8) (see text of report)	2200 F, 8 hr, AC plus 1750 F, 8 hr, AC plus 1225 F, 12 hr, FC + 1600 F, 5 hr, FC to 1400 F, OQ, plus 7 + 7 hr, 400 F	(Air) 234.0 (Vac.) 247.0	290.0 290.0	5.9 8.6	9.1 20.3	7.5 (−40 F) 10.4 (−40 F)	ND ND
4340 + high (7) silicon	1700 F, one hr, AC plus 1600 F, one hr, OQ plus 600 F, 4 hr, AC — impact specimens, 2 + 2 hr, 600 F	236.1 253.6	270.0 289.5	2.0 3.7	8.7 6.7	10.5 10.0	ND ND
Mod. 4340 + (7) high silicon	1700 F, one hr, AC plus 1600 F, one hr, OQ plus 2 + 2 hr, 600 F	240.0 241.1	269.0 268.0	6.5 2.0	18.0 5.5	10.0 —	1.08 1.03
8740 (9) (10)	Not given, but said to consist of homogenization, austenization, quench and temper.	230.0	285.0	4.5	—	—	—
4340 (9) (10)		215.0	280.0	6.0	—	—	—
4330 (9) (10)		200.0	250.0	9.0	—	—	—
H-11 (9) (10)		205.0	260.0	9.5	—	—	—
UHS-260 (10) (experimental)	Same as above	222.0	261.0	4.0	—	—	—

( ) — Numbers in parentheses refer to source data given at end of paper.
ND — Not determined.

The average length of the spirals (spiral channel dimensions — 7/16-in. width, 3/8-in. overall height with 0.15 in. radius) obtained from a pouring temperature of 2850 F were — maraging steel — 27 in., 300M — 25 in., A.I.S.I. 4340 - 20 in. Based on these results, the fluidity of the cast maraging steel is considered to have equivalent or better fluidity than that of low alloy steels.

Weldability

Semi-automatic tungsten arc surface bead tests were performed on heat 38. Tests were made on samples with and without homogenization at 2200 F. No surface cracking was observed in either sample. Examination of cross-sections of the samples showed the homogenized samples to be free from cracking, whereas the as-cast material had a tendency towards underbead cracking in this test.

Conclusions

A cast maraging steel offers an attractive combination of strength and toughness. This alloy has a low carbon iron - 17 per cent nickel martensitic base. A capability for maraging is provided by 10 per cent cobalt, 4.5 per cent molybdenum and 0.4 per cent titanium.

The alloy appears to have castability and fluidity suitable for air melting and casting and adequate weldability for repair and welding to other components.

This alloy should exhibit the features already demonstrated in wrought 18 per cent Ni maraging steel viz — (a) simple heat treatment without quench, (b) deep hardening, (c) good machinability, (d) low distortion and dimensional changes on hardening, (e) retention of strength and toughness up to 1000 F, (f) scaling resistance in air at 1000 F and (g) lack of transition behavior at low temperatures.

These unique features should allow use of the castings at higher strength/density ratios than now used on low alloy steel castings. Furthermore, designs can now be made which in the past were impossible with conventional steels.

Acknowledgment

The authors wish to express their appreciation to the management of The International Nickel Co., Inc., for permission to publish this paper.

References

1. Decker, R. F., Goldman, A. J., Eash, J. T., "18% Ni Maraging Steels," *Transactions Quarterly*, A.S.M. p. 58 (March 1962).
2. G. E. P. Box and K. B. Wilson, "On the Experimental Attainment of Optimum Conditions," *Journal of the Royal Statistical Society* B, 13 (1951).
3. G. E. P. Box, "The Exploration and Exploitation of Response Surfaces," *Biometrics, vol. 10*, no. 16 (March 1954).
4. G. E. P. Box and J. S. Hunter, "Multifactor Designs for Exploring Response Surfaces," *Annals of Mathematical Statistics, vol. 27* (1956).
5. Bennett, C. A., and Franklin, N. L. *Statistical Analyses in Chemistry and the Chemical Industries,* John Wiley & Sons, p. 499 (1954).
6. "300 M Ultra High Strength Steel," Pamphlet No. 258. p. 6-7, Published by The International Nickel Co.
7. Unpublished The International Nickel Co. work.
8. P. S. Schaffer, P. J. Ahearn and M. C. Flemings, "Vacuum Induction Melting High Strength Steels," AFS TRANSACTIONS, *vol. 68*, p. 551 (1960).
9. "Castings Tackle the Muscle Jobs," *Steel, vol. 144*, p. 84 (Aug. 24, 1959).
10. "A Review of Certain Ferrous Castings Applications in Aircraft and Missiles," DMIC Report no. 120, OTS PB151077 (Dec. 1959).

SECTION VIII:
Welding

18% Nickel Maraging Steel Metallurgy and Its Effects on Welding

R. J. Knoth and F. H. Lang

Weldability and weld metal properties of the 18% Ni 250 grade maraging steel were studied to determine the suitability of this material for weld fabrication. Welds produced by the TIG process resulted in plane strain fracture toughness in the weld and in the heat affected zone that compared with that of the base plate. The MIG and short-arc welds developed lower weld toughness in the weld metal. Strength properties were considered excellent for each of the welding methods. While weld filler metals could be adjusted to vary the yield strength by about 70,000 psi, weld ductility and toughness were sacrificed as strength increased. Welding was followed by a 3-hr aging treatment at 900 F to harden the weld metal and the heat affected zone.

THE 18% NI MARAGING steels have good weldability characteristics. These characteristics can be best described in relation to the physical metallurgy, mechanical properties and the chemical composition of the steels under discussion. Optimum weld metal properties are readily obtainable. Each of these topics will be discussed.

Physical Metallurgical Characteristics

Table 1 shows typical properties obtainable in three 18% Ni maraging steels whose compositions are balanced to give nominal yield strengths of 200, 250, and 280 ksi. Notice the excellent impact resistance of these alloys at the strength levels obtained. The 250 ksi material has received most commercial attention to date and is the alloy referred to most frequently in this paper.

Dilatometry curves provide a simple way to look at the physical metallurgical characteristics that are pertinent to the weldability of an alloy. Figure 1 schematically shows the dimensional changes of 18% Ni maraging steel as it is cycled through the normal heat treatment schedule. Upon air cooling to room temperature, the material transforms to a soft ductile martensite. This transformation is not dependent on cooling rate, so the ordinary aspects of hardenability associated with carbon hardened steels do not apply. This low-carbon martensite is age hardenable by reheating to about 900 F for 3 hr. The increment of hardening at this stage is dependent upon the content of Ti, Al, Co, Mo.

The very small change in length occurring during the aging treatment is important because it accounts for the excellent dimensional stability of the steel during fabrication heat treatment.

It is possible to analyze how the heat of welding affects this material. If the aged martensite is heated beyond the aging temperature range, transformation to austenite begins, as indicated by the deviation from linearity in the curve in Fig. 2. Near the range of 1200 F the austenite forms

R. J. Knoth is associated with The International Nickel Co., Inc., New York, N.Y. F. H. Lang is with the P. D. Merica Research Laboratory, The International Nickel Co., Inc., Sterling Forest, N.Y.
This paper was presented October 5, 1965 at the ASM sponsored Fracture Toughness session held at Birmingham, Ala. at the Fall meeting of the American Welding Society.

Table 1. Compositions and Properties of Three Grades of 18% Ni Maraging Steel

Nominal Chemical Composition

Grade	Ni	Co	Mo	Ti	Fe
18Ni 200	18	8.5	3	0.2	bal
18Ni 250	18	8	5	0.4	bal
18Ni 280	18	9	5	0.6	bal

	Annealed Properties			Maraged (900 F/3 hr) Properties				
Grade	Y.S., ksi	U.T.S., ksi	Elong., %	Y.S., ksi	U.T.S., ksi	Elong., %	R.A., %	CVN ft/lb
18Ni 200	117	145	17	200	215	13	51	40
18Ni 250	116	146	19	250	260	11	47	25
18Ni 280	115	147	17	280	285	9	40	15

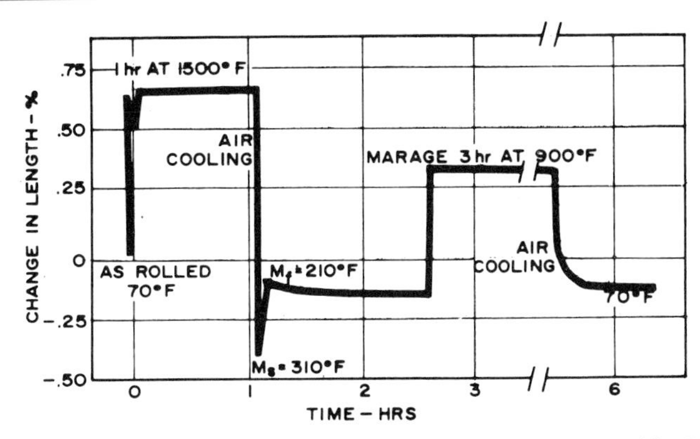

Fig. 1. Dilatometry behavior of 18% Ni maraging steel cycled through normal heat treatment.

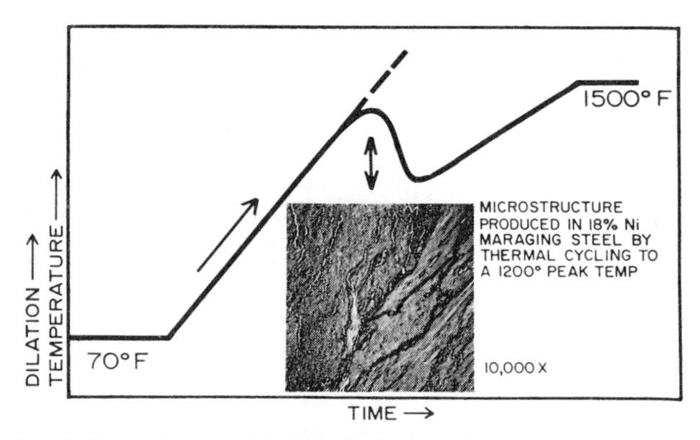

Fig. 2. Tracing of dilatometer record showing heating transformation of martensite → austenite.

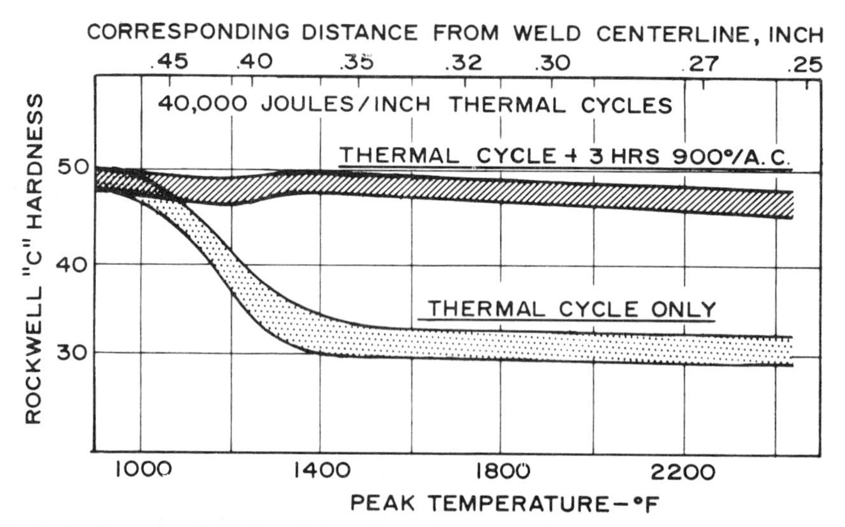

Fig. 3. Hardness of synthetically cycled specimens. Lower curve shows hardness after thermal cycling and upper after thermal cycling and aging.

by diffusion. If the material is cooled to room temperature from this region, a small amount of this austenite remains as a stable phase. Its distribution is very fine and the phase itself is so small that it can only be observed by electron microscopy (as shown in Fig. 2). Austenite formed by heating beyond 1200 F is developed by a shear transformation of martensite, indicated by the sharp change in the curve of Fig. 2. Upon cooling to room temperature, this austenite transforms to a soft, ductile martensite.

Figure 3 shows properties of a synthetically produced weld heat-affected-zone in the 18 Ni 250-grade maraging steel. The as-welded heat-affected-zone of material, welded in the maraged condition, shows gradual softening as the peak temperature attained increases above 1000 F. A uniform hardness is observed in material that has been annealed and has transformed back to martensite. A post-weld aging treatment at 900 F for 3 hr restores practically the entire heat-affected-zone to the original hardness and strength level. Only two small areas do not attain original hardness. These are the regions near the 1200 F peak temperature, where some stable austenite has formed, and the region experiencing temperatures above about 1800 F, where grain coarsening has occurred. With regard to toughness, there is no degradation in any region of the HAZ. The area heated to 1200 F, where austenite has been stabilized, actually shows improved toughness.

The effects of weld thermal cycles on properties of material welded in the annealed condition are shown in Fig. 4. The region experiencing peak temperatures of around 900 F ages slightly during the welding process. The regions above about 1200 F behave in a manner similar to that previously discussed. Again, a post-weld aging at 900 F for 3 hr provides full hardness in the heat-affected-zone, except for slight softening in the 1200 F region and in the grain-coarsened region. The fact that maraging steels soften, rather than harden, in the heat-affected-zone eliminates the need for pre-heat and post-heat treatments and accounts for the marked resistance to cracking in maraging steel weldments.

The particular characteristics that make maraging steel desirable from a weld fabrication viewpoint are summarized as follows:

1. The materials provide high notch-toughness along with high strength.
2. Forming and machining can be performed on soft annealed material. In addition, the maraging steels have reasonably good machinability in the aged condition.
3. Simplified welding procedures can be used with no pre-heat or post-heat, although, of course, the joints are aged after welding to develop maximum strength.
4. Weld crack sensitivity is minimized since the heat affected zone is softened by the heat of welding. In contrast, high carbon steels develop hard, brittle heat affected zones which are crack sensitive unless stringent pre-heat and post-heat requirements are maintained.
5. Hardenability aspects commonly associated with carbon steels do not apply.
6. The maraging heat treatment is simple and involves little dimensional change. Quenching is not required and local post-weld aging is possible.
7. There is no property degradation of practical significance in the heat-affected-zone after the post-weld aging treatment.

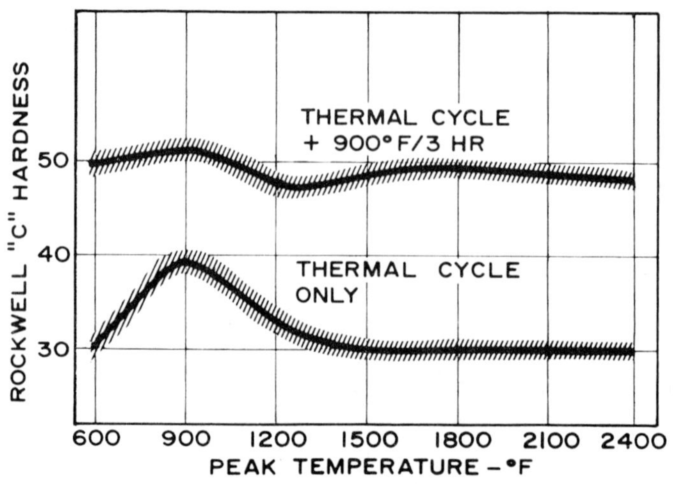

Fig. 4. Hardness of specimens synthetically cycled in the annealed condition.

Table 2. Weld Properties Obtained in 18% Ni Maraging Steel

Weld (Aged 900 F, 3 hr)	Weld thickness, in.	0.2 Y.S., ksi	U.T.S., ksi	Elong., %	CVN ft/lb
TIG	0.062	235	246	3.5	—
MIG	½	220	235	7	13
Short-Arc	½	228	245	4	11

Filler: 18Ni 8Co 4.5Mo 0.50Ti 0.20Al
Base: 18Ni 250 ksi maraging steel

Table 3. Effect of Filler Composition on Properties of 18% Ni Maraging Steel Welds

Filler					Weld, in.	Maraged Properties*				
Ni	Co	Mo	Ti	Al		Y.S., ksi	U.T.S., ksi	Elong., %	R.A., %	CVN ft/lb
17	7.3	2.0	0.7	0.05	½ TIG	201	213	10	47	24
18	8	4.5	0.5	0.2	0.062 TIG	235	246	3.5	18	—
18	9.5	4.5	0.7	0.2	0.062 TIG	271	276	2.5	10	—

* Aged 900 F/3 hr.

Weld Properties

Weld filler metals similar to the base composition produce sound inert-gas-shielded arc welds with excellent combinations of strength and toughness. A simple post-weld aging treatment at about 900 F for 3 hr is used to harden the weld metal and HAZ. For example, Table 2 shows typical properties obtained in TIG, MIG and short-arc welds using the same filler wire composition.

Weld-metal hardener content can be adjusted to obtain lower or higher weld strength. Table 3 shows three filler metal compositions that provide yield strengths in the range 200 to 270 ksi. Two items should be noted at this point. First, weld ductility and toughness are sacrificed as strength is increased; and second, the increase in strength obtained with increased hardener content is not a limitless process. The richer alloy compositions become more prone to microsegregation on freezing and, upon aging, the high alloy areas may revert to austenite. These large pools of austenite, although inherently tough, are low in strength and serve to initiate fracture in the hard martensite matrix. Also, this reverted austenite does not respond to aging and the maximum strength level obtainable reaches a plateau.

Aerospace and hydrospace design calculations are often based upon fracture toughness criteria which indicate the ability of a material to resist the propagation of fracture in the presence of a flaw. Figure 5 shows plane strain fracture toughness values obtained in 250 ksi weldments made with the various processes shown (1). With the TIG process, toughness in the weld and in the HAZ compared well with base plate toughness. In MIG and short-arc, weld-metal toughness was lower. Ranges of typical K_{IC} values of welds of various yield strengths are shown in Fig. 6.

Weld properties obtained with the covered electrode process are shown in Table 4. Although the strength properties are typical and consistent with those obtained by the inert-gas-shielded processes, deposit toughness is not as satisfactory. Furthermore, there has been less experience with the covered electrode process, and consequently less confidence generated in its use.

18% Ni maraging steels have been electron-beam welded experimentally with relatively good success. Table 5 shows weld properties reported for ¼-in. thick joints (2). The stringent controls associated with the use of this process, however, have restricted its use in commercial applications.

The high strength and toughness of the maraging steels and their ease of fabrication have led to their consideration for industrial pressure vessel applications. A program was recently sponsored at Lehigh University to determine how 18% Ni steel weldments behave in the presence of low-cycle loading in the plastic strain range as might occur in pressure vessels. Lehigh cantilever bar specimens of maraged 18% Ni 250 base material and MIG welds were tested under constant-deflection reversed bending at 200 cpm. The specimen blanks, ¾ by 2-½ by 18 in., were butt welded parallel to and coincident with the axis of flexure. Although the standard Le-

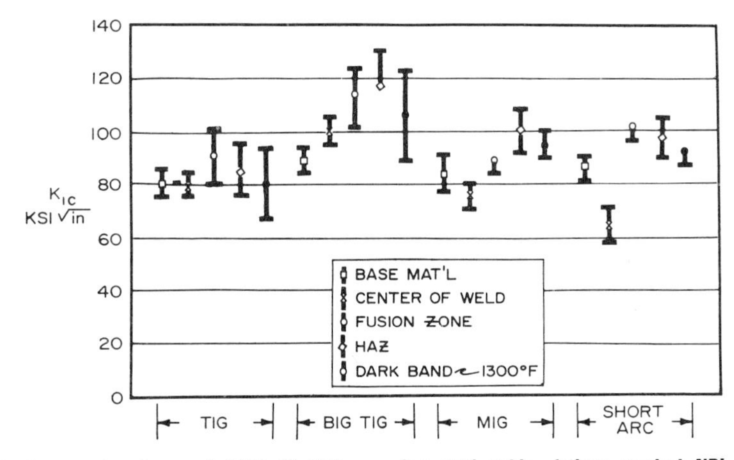

Fig. 5. Fracture toughness of 18% Ni 250 maraging steel welds—fatigue cracked NRL 3-point loaded specimens.

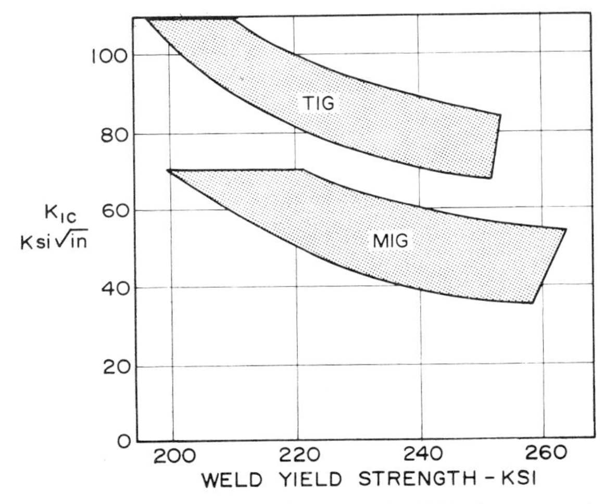

Fig. 6. Variation of K_{IC} values with yield strength in 18% Ni maraging steel welds.

Table 4. Properties of Covered Electrode Welds in 18% Ni 250 KSI Maraging Steel

Yield strength	220/230 ksi
Tensile Strength	230/240 ksi
Elongation	5/10%
Reduction of Area	20/40%
CVN	5/15 ft/lb

Table 5. Electron Beam Welds in 18% Ni 250 Ksi Maraging Steel*

Specimen	Heat treatment	Y.S., ksi	Properties U.T.S., ksi	Elong. (in 2 inches) %
Base	Aged 925 F/3.8 hr	264	267	8
Weld	As Welded	243	243	1.5
Weld	Aged 925 F/3.8 hr	266	267	2.5

* Composition: 18.0Ni, 7.7Co, 4.6Mo, 0.50Ti, 0.08Al, 0.03C.

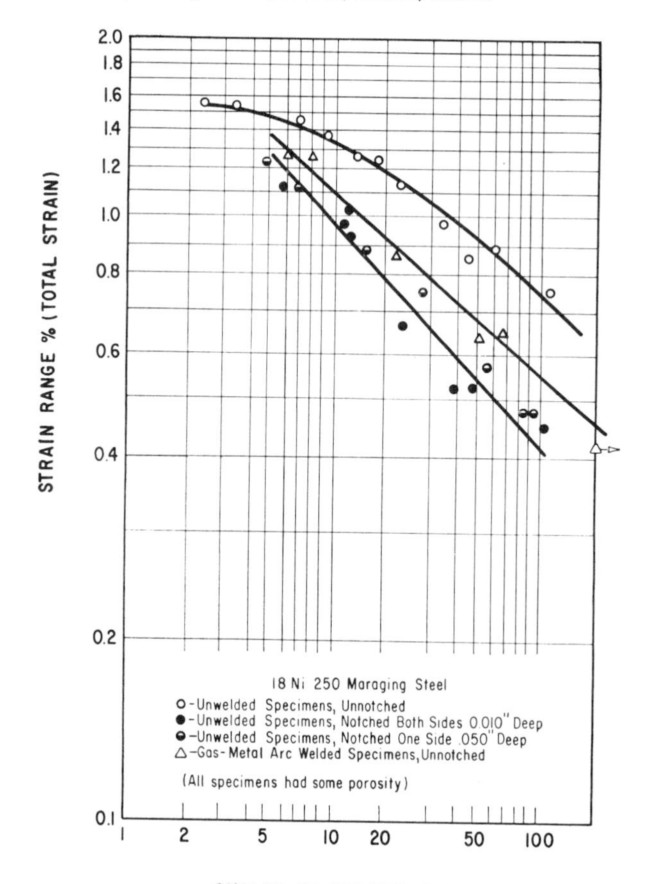

Fig. 7. Results of low cycle fatigue tests on 18% Ni 250 base material and welds.

high specimen is welded perpendicular to the axis of flexure (3), it was felt that this modification, having the flexure concentrated in all-weld-metal, was a more severe test of the weldment. Tests were also run on the same base plate heat in the unnotched and notched condition. The notched bars had standard Charpy V-notches machined in, either 0.050 in. deep on one side or 0.010 in. deep on both sides.

The results obtained are shown in Fig. 7. The curve through the weld data points, while somewhat below that of the unnotched base plate, is clearly above that of the notched base plate. As an aid in interpreting the log-log plot of the data, the notched bar and welded bar lives were calculated as a percentage of the unnotched bar life for a number of total strain range values. Notched bar life averaged 27% of the unnotched bar life while the welded bar life averaged 40% of that of the unnotched bar.

The need to make joints between dissimilar metals is often encountered in commercial applications. The 18% Ni maraging steels can be joined to a variety of dissimilar metals using either a general purpose, high nickel alloy welding electrode, such as INCO-WELD*A electrode, or a maraging filler metal. Typical properties resulting when maraging steel was joined to mild steel, 4340 steel and Type 304 stainless steel are shown in Table 6.

Control Guide Lines

When considering guide lines for obtaining optimum weld properties, the first item, and one of the most important, is the composition of filler wire. Table 7 shows the effects of excessive or insufficient levels of the major alloying elements in the filler metal as well as in the base material.

Dropping below about 17% nickel results in weld metal of reduced toughness. Also weld-crack sensitivity is increased. Excessive amounts of nickel promote the formation of reverted austenite on aging. As mentioned above, appreciable quantities of

* Registered Trademark, The International Nickel Company, Inc.

Table 6. Properties of Weld Joints Between 18% Ni 250 Maraging Steel and Various Dissimilar Metals

Metal joined	Filler metal	Weld type	As-welded Y.S., ksi	As-welded CVN, ft-lb	Maraged 900 F/3 hr Y.S., ksi	U.T.S., ksi	Elong., %	Bk.***	CVN ft-lb
Mild steel	All-purpose*	Cov. elect.	57	80	47	68	16	B	97
4340	All-purpose	Cov. elect.	56	80	61	100	12	W	69
Type 304	All-purpose	Cov. elect.	51	95	54	91	36	B	97
Mild steel	18% Ni**	MIG	63	25	50	67	5	B	16
4340	18% Ni	MIG	147	16	170	180	4	B	8
Type 304	18% Ni	MIG	62	28	61	90	32	B	14

* 73% Ni, 14% Cr, 8% Fe, Cb, Mo, Mn
** 18% Ni, 8% Co, 4.5% Mo, 0.50 Ti, 0.10 Al, bal Fe
*** Bk = Break, W = Weld, B = Base Plate

Table 7. Effect of Alloying Elements on 18% Nickel Maraging Steel

Element	Amount present	In parent material	In welds
Nickel	Too low	Lowers strength	Lowers strength and resistance to cracking
	Too high	Promotes austenite lowering strength and toughness	Segregated austenite pools reduce strength and toughness
Cobalt or molybdenum	Too low	Lowers strength	Lowers strength and resistance to hot cracking
	Too high	Raises strength but lowers notch properties	Lowers both yield and notch strength
Titanium	Too low	Lowers strength	Increases porosity
	Too high	Lowers notch properties	Promotes segregated stable austenite forms low melting point sulfides

Table 8. Role of Excessive Residual Elements

Elements	Effects
Carbon	Ties up titanium and molybdenum to reduce strength.
Sulfur	Particularly in welds, low melting point titanium sulfides increase hot shortness and lower ductility and toughness.
Aluminum	Increases strength but lowers toughness.
Silicon and manganese	Lowers impact properties, silicon increases hot-crack sensitivity.
Boron and zirconium	Improves toughness *in plate*, increases crack sensitivity and reduces toughness *in welds*.
Hydrogen	Welds with more than 3 ppm are crack sensitive.
Oxygen	Additions to shielding gas reduce impact strength of welds.

this austenite not only lower achievable strength levels but, because of their formation in pools at grain boundaries, also contribute to low ductility and toughness.

Cobalt and molybdenum serve as principal hardening elements and are controlled to achieve maximum strength-toughness combinations. It has been shown that reduced molybdenum in welds can result in increased hot-crack sensitivity. On the other hand, excessive molybdenum favors increased amounts of austenite reversion with its accompanying ill effects.

Titanium is a very potent supplemental hardener (about 10 ksi/0.1%) and is also necessary in welds to assure thorough puddle deoxidation. When not present in sufficient amounts, aging response will be lowered. It has been demonstrated that a minimum titanium level of about 0.3% is necessary to control weld porosity in the gas-shielded arc processes. On the other hand, excessive titanium increases strength but at the expense of ductility. Segregated titanium compounds have been found associated with pools of reverted austenite. The affinity of titanium for sulfur promotes formation of low melting titanium sulfides which are potential contributors to weld hot-crack sensitivity.

The residual elements necessarily present in maraging steels also have important effects on properties, Table 8.

Wire quality has been found to be an extremely important factor in its effect on weld properties. Freedom from seams, scale, oxides and lubricant is a "must". Vacuum annealing and de-gassing of wire is recommended to assure freedom from surface contamination.

The general theme to be emphasized throughout the welding of maraging steel is cleanliness. The proper shielding of welds by inert gas is extremely important. In fact, there is evidence that increased shielding, such as a trailing shield will promote better weld properties. Unless such auxiliary shielding is used, travel speeds should not exceed about 10-in. per min to ensure freedom from porosity. The frequently used expedient of adding oxygen to the shielding gas to improve arc stability must be avoided because of the deleterious effects of oxygen on weld properties.

The final important control factor is the post-weld aging treatment. Aging at 900 F for 3 hr provides satisfactory properties. However, in critical applications other combinations of temperature and time can sometimes be found which are more beneficial with a particular fabricator's techniques and facilities. Thus, it is recommended that test welds be used, if necessary, to derive the optimum aging treatment for critical applications.

Summary

In summary, the following conclusions can be made: The 18% Ni maraging steels are attractive from a fabrication viewpoint because of their unique inherent characteristics which make them especially compatible to welding. These characteristics include minimum distortion, high resistance to cracking, absence of hardenability restrictions, and little change in HAZ properties. Excellent weld metal properties are obtainable in terms of strength and impact toughness. Also, weld plane-strain fracture toughness and plastic fatigue behavior are reasonably satisfactory. The alloy is also compatible to joining to dissimilar metals. To fully exploit these advantages in commercial applications, the above discussed guide lines of weld cleanliness and process control must be faithfully followed.

REFERENCES

1. Plane Strain Fracture Toughness of 18 Ni (250) and 18 Ni (200) Maraging Welded Steel Plate, H. E. Romine, U. S. Naval Weapons Laboratory NWL Report No. 1959 (January 1965).
2. Welding 18% Ni Maraging Steel by Electron Beam, M. L. Kohn and C. D. Schaper, Metal Progress (May 1964) 93.
3. The Effect of Composition and Microstructure on the Low Cycle Fatigue Strength of Structural Steels, R. D. Stout and A. W. Pense, ASME Metals Engineering Conference, May 1964, Paper No. 64 - Met. 9.

The Welding of 12% Nickel Maraging Steels

A yield strength and toughness combination of at least 160 ksi and 50 ft-lb is obtained in gas-shielded arc welds

BY R. J. KNOTH AND W. A. PETERSEN

ABSTRACT. This paper outlines research into the weldability of maraging steel compositions which exhibit a high level of toughness in the 160–180 ksi yield strength range.

Using the gas-shielded arc processes, experimental filler metals were developed. These produced sound welds with strength and toughness approaching that of the base metal. No preheat was necessary, and a short postweld maraging treatment provided yield strengths in excess of 160 ksi in combination with at least 50 ft–lb in the Charpy V-notch test.

A detailed study of the weld heat-affected zone showed that, after the postweld maraging treatment, there was essentially no deterioration in either the strength or toughness of the heat-affected base metal.

Introduction

Research outlining a series of maraging compositions that provided excellent toughness in the 160–180 ksi yield strength range has been reported.[1] Like the higher strength 18% Ni maraging steels,[2, 3] these 12% nickel alloys responded to a combination of martensite formation and age hardening. Table 1 shows the composition and properties of two alloys which were used as base metals in a study of the general weldability of the 12% nickel maraging alloys.

This paper describes the findings of these welding studies. The subject matter is presented in two parts with the first section describing the development of filler metals having properties approaching those of the base metals. Results with both the gas metal-arc and the gas tungsten-arc processes are discussed. The second part of the paper describes the results of a comprehensive study of the weld heat-affected zone.

R. J. KNOTH and W. A. PETERSEN are associated with the Research Laboratory of The International Nickel Co., Sterling Forest, Suffern, N. Y.

Paper presented at the AWS National Fall Meeting held in San Francisco, Calif., during Oct. 5–8, 1964.

Filler Metal Development

The best possibility of developing a filler metal with properties approaching those of 12% Ni base metal compositions shown in Table 1 appeared to lie in filler metals of approximately matching compositions. Most welds were made by the manual gas tungsten-arc process since filler metals of various compositions could be obtained quickly by using a glass-tube extraction process. Initial work was concerned with investigating the crack sensitivity of matching compositions, and composition modifications to balance strength and toughness favorably were then explored. Based on the initial work, gas metal-arc filler metals also were prepared and evaluated.

Gas Tungsten-Arc Process

Evaluation of X-Weld Soundness

Preparation of Materials and Test Procedure. Based on the known effects of the five major alloying elements in the plate materials, several filler metal compositions were prepared for evaluation. These had the following composition ranges: Ni—9.0 to 13.0%; Cr—3.5 to 5.3%; Mo—1.0 to 3.0%; Al—0.04 to 0.19%; Ti—0.15 to 0.35%.

Filler metal rods were obtained from 30 lb melts using a glass-tube extraction process as described by Witherell and Fragetta.[4] To ensure low gas content in the rods, the melts were made in an induction furnace under an argon blanket atmosphere. The rods were used uncoated and were fed manually to make X-weld crack test specimens of the type shown in Fig. 1. The usefulness of this type weld in detecting weld crack sensitivity has been described elsewhere.[5]

Conventional argon - shielded welding techniques were employed using fully aged 12% Ni steel for base material. Weld passes were deposited in the V-grooves, alternating from one side to the other with each successive pass. No preheat was used, and the interbead temperature was held below 250° F. After completing the 3 in. joints, five transverse cross sections were prepared, etched and examined for defects.

Test Results. No cracking or porosity was observed in any of the X - welds. Adequate deoxidation was obtained even when as little as 0.04% aluminum was used in conjunction with 0.15% titanium in the filler metal.

Effect of Composition on Weld Strength and Toughness

Preparation of Materials and Test Procedure. Several series of filler metals of the type studied in the X-weld tests were prepared for evaluation in 1/2 in. butt welds. These compositions had systematic variations in nickel, chromium and molybdenum as well as in the major hardening elements, aluminum and titanium. As before, the filler metal rods were obtained from 30 lb air induction melts with the glass-tube extraction process.

These rods were used to join 1/2 x 3 x 5 in. plates (along the 5 in. length) of fully aged 12% nickel steel. The plates were beveled to a single V-groove with a 60 deg included angle and had about a 5/32 in. root spacing. During welding, the plates were firmly held on the welding platen with three heavy duty C-clamps on each side. This restraint kept warpage across the released finished joint to less than 1 deg.

The welded plates were sectioned into 1/2 in. transverse slices, polished, etched and examined for defects. Smooth transverse tensile bars (0.252 in. diam) and standard Charpy V-notch specimens were machined as shown in Fig. 2. In addi-

Table 1—The Composition and Mechanical Properties of Heats of 12% Ni Maraging Steel Tested

Alloy	Heat size	Melting atmosphere	Composition, wt-% Ni	Cr	Mo	Al	Ti	C	S	P	B	Zr	Si	Mn	Plate thickness, in.	Heat treatment	Room temperature properties in transverse direction[a] U.T.S., ksi	Y.S. ksi	Elong., %	R.A., %	Rc	CVN, ft-lb
1	2000 lb	CEVM[b]	12.1	4.98	3.17	0.22	0.18	0.02	0.005	0.002	0.005	0.01	0.06	0.06	1/2	Annealed + aged[d]	174.4	169.0	16.4	72	40	75
2	20 tons	Air	12.0	5.10	2.90	0.35	0.27	0.02	0.008	0.005	[c]	[c]	0.09	0.07	1/2	Annealed + aged[d]	189.0	184.0	15.0	67	41	45

[a] U.T.S.—ultimate tensile strength; Y.S.—yield strength (0.2%); Elong.—elongation; R.A.—reduction in area; CVN—Charpy impact.
[b] Consumable electrode vacuum remelt.
[c] Not analyzed.
[d] Annealed 1500° F/1 hr air cooled; maraged 900° F/3 hr air cooled.

Table 2—Effect of Filler Metal Composition on the Strength and Toughness of Gas Tungsten-Arc Butt Welds[a] in 12% Ni Maraging Steel

Weld no.	Filler metal[c] composition, wt-% Ni	Cr	Mo	Al	Ti	Treatment	Type test	Properties[b] U.T.S., ksi	Y.S. (0.2%), ksi	Elong., %	R.A., %	Bk[e]	CVN, ft-lb 70° F	CVN, ft-lb −40° F
Titanium comparison														
14	12.7	3.57	2.88	0.15	0.35	Aged 900° F/3 hr	Transverse	167	163	13	60	W	60.0	..
15	12.3	3.42	2.69	0.17	0.44	Aged 900° F/3 hr	Transverse	177	173	13	54	W	52.3	..
Aluminum comparison														
16	11.4	5.0	2.20	0.08	0.28	Aged 900° F/3 hr	Transverse	162	155	12	60	W	60	..
17	12.7	5.25	1.96	0.24	0.37	Aged 900° F/3 hr	All weld metal	179	173	14	59	...	57.5	..
Nickel series														
14	12.7	3.57	2.88	0.15	0.35	Aged 900° F/3 hr	Transverse	167	163	13	60	W	60.0	..
18	13.6	3.56	2.75	0.08	0.28	Aged 900° F/3 hr	Transverse	167	162	12	60	W	51.3	..
19	14.4	3.48	2.69	0.13	0.35	Aged 900° F/3 hr	Transverse	167	165	7	60	B	46.0	..
20	15.2	3.48	2.74	0.12	0.31	Aged 900° F/3 hr	Transverse	168	164	7	59	B	49.3	..
21	15.4	3.39	2.66	0.20	0.34	Aged 900° F/3 hr	Transverse	167	164	9	60	B	38.8	..
22	16.2	3.36	2.63	0.18	0.33	Aged 900° F/3 hr	Transverse	168	164	11	59	B	38.5	..
Molybdenum series														
23	12.6	5.24	2.45	0.24	0.32	Aged 900° F/3 hr	Transverse	168	163	9	59	B	49.8	..
24	12.6	5.33	2.92	0.17	0.37	Aged 900° F/3 hr	Transverse	170	166	11	58	B	48.3	..
25	12.5	5.18	3.34	0.13	0.31	Aged 900° F/3 hr	Transverse	168	164	12	57	B	52.8	..
26	12.5	5.18	3.34	0.13	0.31	Aged 900° F/3 hr	All weld metal	177	168	15	64	...	...	..
27	12.4	5.11	3.79	0.18	0.35	Aged 900° F/3 hr	Transverse	184	178	13	58	W	44.5	..
Chromium series														
28	12.7	1.25	2.87	0.16	0.45	Aged 900° F/3 hr	Transverse	165	159	13	61	W	62.3	..
29	12.5	2.39	2.83	0.09	0.39	Aged 900° F/3 hr	Transverse	162	157	13	60	W	70.7	..
30	12.1	4.48	2.71	0.15	0.37	Aged 900° F/3 hr	Transverse	166	161	13	59	WB	60.8	..
31	11.9	5.64	2.65	0.25	0.50	Aged 900° F/3 hr	Transverse	166	164	12	60	B	53.5	..
32	11.9	5.64	2.65	0.25	0.50	Aged 900° F/3 hr	All weld metal	186	177	4[d]	19[d]	...	...	..

Table 2—continued on next page

Source: *Welding Research Supplement*, Jan 1965

Table 2—Continued

Weld no.	Filler metal[c] composition, wt-% Ni	Cr	Mo	Al	Ti	Treatment	Type test	Properties[b] U.T.S., ksi	Y.S. (0.2%), ksi	Elong., %	R.A., %	Bk[e]	CVN, ft-lb 70° F	CVN, ft-lb −40° F
Low hardener series														
33	9.0	5.17	2.30	0.08	0.32	Aged 900° F/3 hr	Transverse	160	152	12	59	W	52.5	..
34	10.0	(5.10)[f]	(2.20)	0.06	0.24	Aged 900° F/3 hr	Transverse	155	148	13	60	W	71.7	..
35	12.3	5.30	1.00	0.06	0.24	Aged 900° F/3 hr	Transverse	163	157	12	58	W	60.5	47
36	12.2	(5.20)	1.50	0.04	0.15	Aged 900° F/3 hr	Transverse	156	149	12	60	W	64.5	60
As-welded properties														
17	12.7	5.25	1.96	0.24	0.37	As-welded	Transverse	141	127	15	71	W	93	..
28	12.7	1.25	2.87	0.16	0.45	As-welded	Transverse	135	122	15	69	W	122	..
30	12.1	4.48	2.71	0.15	0.37	As-welded	Transverse	140	128	15	68	W	95	..
Series aged at 800° F														
19	14.4	3.48	2.69	0.13	0.35	Aged 800° F/1 hr	Transverse	167	161	14	61	W	46	..
23	12.6	5.24	2.45	0.24	0.32	Aged 800° F/1 hr	Transverse	169	163	16	60	W	55.5	..
29	12.5	2.39	2.83	0.09	0.39	Aged 800° F/1 hr	Transverse	155	150	13	66	W	93	..

[a] Welds made on aged plate (alloy no. 1 in Table 1.)
[b] U.T.S.—ultimate tensile strength; Y.S.—yield strength; Elong.—elongation; R.A.—reduction in area; CVN—Charpy impact.
[c] Filler rods vacuum-extracted from 30 lb air induction melts.
[d] Defect in fracture.
[e] Location of tensile test break: W—weld; B—base metal.
[f] Approximate analysis in parenthesis.

tion, all-weld metal 0.252 in. diam tensile bars were obtained from some of the welds. The machined specimens were aged at 900° F for 3 hr and tested at room temperature. Several Charpy bars were also broken at −40° F. Some as-welded properties were determined and a few specimens were tested after only abbreviated aging treatments.

Test Results. Table 2 shows the filler metal compositions studied and the test results obtained. Strength and toughness were most strongly affected by aluminum and titanium variations. Yield strength increased and toughness decreased with an increase in either of these elements (e.g., compare welds 14 and 15, and welds 16 and 17). In general, increases in chromium, molybdenum or nickel also decreased toughness and increased yield strength but the effects of these elements were less pronounced than those of aluminum or titanium and the effects of chromium, molybdenum and nickel in the series shown were sometimes masked by the concurrent changes in aluminum and titanium.

In fully aged welds, strength-toughness combinations as high as 177 ksi yield strength and 53.5 ft-lb were obtained (welds 31 and 32). Yield strengths in excess of 160 ksi with at least 50 ft-lb in the Charpy V-notch test were obtained with all compositions ranging as follows: Ni—12.1 to 13.6%; Cr—3.4 to 5.2%; Mo—2.7 to 3.3%; Al—0.10 to 0.20%; Ti—0.35 to 0.46%. The very limited data indicated only a minor degradation in toughness values at temperatures down to −40° F. This characteristic is not unexpected in view of the known behavior of the 18% nickel grades.

The as-welded yield strengths were not strongly dependent on composition within the limits studied, and were generally in the 120–130 ksi range. Some attractive combinations of weld strength and toughness were obtained with only a 1 hr age at 800° F; however, the response to this treatment was quite dependent on composition with the higher hardener compositions responding more strongly.

Gas Metal-Arc Process

Effect of Composition on Weld Strength and Toughness

Material Preparation and Test Procedure. Laboratory 40 lb vacuum-induction melts of several compositions that gave combinations of high strength and toughness with the gas tungsten-arc process were processed into 0.062

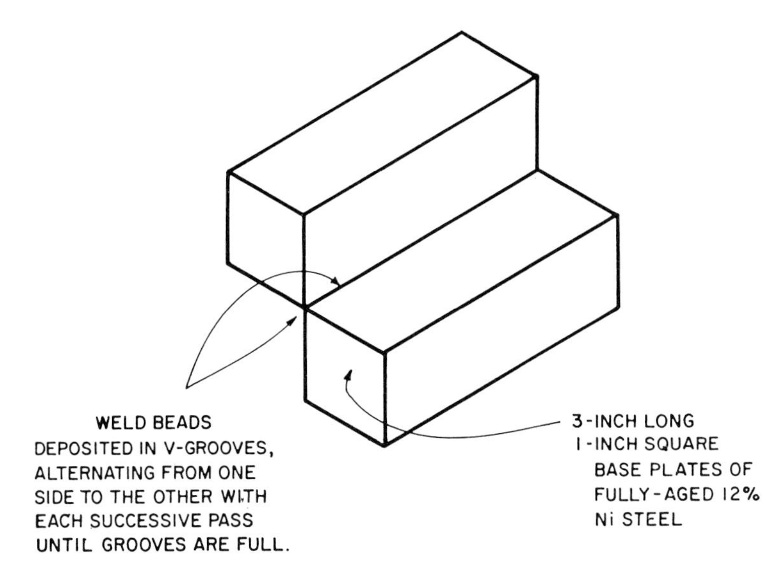

Fig. 1—X-weld crack test specimen used to evaluate filler metal crack sensitivity

in. diam wires. In addition, wires containing combinations of either high titanium-low aluminum or low titanium-high aluminum were produced for evaluation. Four-pass restrained butt welds in $^1/_2$ in. thick aged plate were made using these filler metals and pure argon shielding. No preheat was employed, and the interpass temperature was kept to a maximum of 250° F.

In an effort to reduce grain size in the weld deposit, a number of additional $^1/_2$ in. welds were made with faster travel speed and an 8 pass procedure. In one of these welds (no. 45 in Table 3) the influence of a small amount of oxygen in the shielding gas was investigated with regard to its effect on operability.

In a further effort to reduce deposit grain size, two of the 0.062 in. diam wires were drawn down to 0.035 in. diam and used to make $^1/_2$ in. welds by both the conventional gas metal-arc and the short circuiting arc processes.

All-weld-metal tensile bars and transverse tensile and Charpy bars obtained from these welds were tested at room temperature after aging at either 900° F for 3 hr or at 850° F for 6 hr. In some cases, as-welded properties and properties after abbreviated aging also were determined.

Test Results. The compositions and test results are shown in Table 3. The strengths obtained were comparable to those developed by gas tungsten-arc welds, but toughness was lower. The high titanium-low aluminum compositions gave high yield strengths (above about 160 ksi) and improved toughness. In contrast, the high aluminum-low titanium compositions gave lower toughness values for the strengths obtained.

Welds 44–47 made with the 8-pass procedure exhibited properties comparable to those made with the 4-pass procedure using the same wires. It is interesting to note that one of these 8-pass welds (no. 47) aged for only 1 hr at 800° F exhibited a strength-toughness combination of 165 ksi and 47 ft-lb.

Although the operability with pure argon for shielding was satisfactory, the addition of oxygen did increase arc stability. The oxygen, however, had a deleterious effect on the toughness of the aged deposits (compare welds 44 and 45).

Short circuiting arc process results were equivalent to those of the standard gas metal-arc process with 0.062 in. diam filler metal (compare welds 48 and 41 and welds 49 and 42). One of the best combinations of properties (162 ksi yield strength and 52 ft-lb) was obtained with the gas metal-arc process using 0.035 in. diam filler metal and a 850° F for 6 hr postweld aging treatment (weld 50).

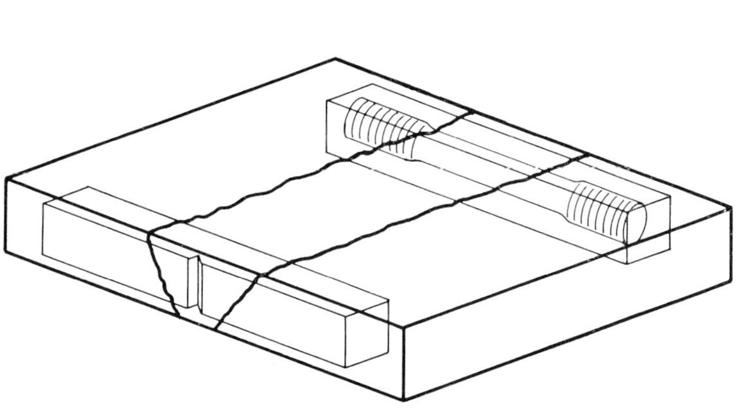

Fig. 2—Location of transverse tensile specimens and Charpy V-notch impact specimens cut from $^1/_2$ in. butt welds

Summary of Filler Metal Development

Filler metal development work to date has shown that sound welds were produced in 12% Ni maraging steel under heavy restraint. No preheat was needed, and properties approaching those of the base metal were obtained with the gas-shielded arc processes. The achievement of these attractive combinations of weld strength and toughness is gratifying, but is not believed to represent an end-point in filler metal development for the 12% Ni maraging steels.

Heat-Affected Zone

In general, the characteristics of the heat-affected zone of the 12% Ni maraging alloys were found to be quite similar to those of the 18% Ni maraging steel.[6] There was no deterioration in toughness throughout the heat-affected zone, and minor softening only occurred in two regions experiencing peak temperatures in the vicinity of 1200 and 2400° F.

Experimental Procedure

The effects of weld thermal cycling on the properties of the heat-affected zone were studied with the aid of a high-speed time-temperature controller.[7] This device allowed the rapid heating and cooling of finite volumes of metal through thermal sequences duplicating those to be expected at various positions within the weld heat-affected zone.

The thermal cycles chosen for study were equivalent to those resulting from a top-pass weld on $^1/_2$ in. thick maraging steel plate welded with $^3/_{16}$ in. diam covered electrodes at an energy input of 40,000 joules/in.[5] Transverse Charpy and tensile blanks were cut from solution treated and aged 1 in. thick 12% Ni maraging steel

Source: *Welding Research Supplement*, Jan 1965

plate—alloy 1 in Table 1. After cycling to peak temperatures ranging from 1000 to 2400° F, the test specimens were reaged at 900° F for 3 hr, machined to standard configuration and tested.

Test Results

Hardness. There was no appreciable change in the hardness of the reaged heat-affected zone as a result of thermal cycling—Table 4. The hardness of the originally solution treated and aged plate was reduced from Rc 38 to 27.5 in the weld heat-affected zone, but reaging after simulated welding restored the hardness of the heat-affected zone and base metal to Rc 38.5–40. The region which had been heated to about 1200° F had a hardness at the lower end of this range.

Strength. Slight strength reduction was observed in two distinct regions of the cycled and reaged heat-affected zone—Table 4. The first, which was heated to 1200° F, showed a loss of about 5% in yield strength and 7% in ultimate strength. This region coincided with the area of lowered hardness mentioned previously. The ultimate tensile strength of the coarse grained or the high peak temperature region was unaffected by thermal cycling; however, a 5% loss in yield strength was encountered in the samples cycled to peak temperatures greater than 2000° F. No significant changes were observed in elongation or reduction of area throughout the heat-affected zone.

Although the losses in the simulated single pass weld heat-affected zone were small, a brief survey was made of the effects of a second simulated weld pass on the strength of a previously coarse grained region. Material cycled to 2400° F and then recycled to 1000–2400° F peak temperatures to approximate this condition, showed that the yield strength of a considerable portion of the coarse grained region was restored to the original base metal values. Thus, in the coarse grained regions of an actual multipass weldment, no additional reduction in strength over that found in a single pass weld would be anticipated. The region heated to approximately 1200° F would remain slightly softened unless recycled into the austenitization temperature range. This zone would be about 5% weaker than the surrounding base metal and heat-affected zone.

Toughness. The toughness of the heat-affected zone showed no sign of deterioration as a result of thermal cycling—Table 4. All the broken Charpy specimens exhibited considerable lateral expansion and no evidence of brittle fracture. A marked increase in toughness was found in specimens cycled to a peak temperature of about 1200° F. This increase was associated with the formation of austenite.

Table 3—Effect of Filler Metal Composition on the Strength and Toughness of Gas Metal-Arc Butt Welds[a] in 12% Ni Maraging Steel

Weld no.	Filler metal composition,[b] wt-% Ni	Cr	Mo	Al	Ti	Heat treatment	Type test	Properties[c] U.T.S., ksi	Y.S. (0.2%), ksi	Elong., %	R.A., %	Bk[d]	CVN, ft-lb
1/2 in. weld—0.062 in. filler metal—4 pass procedure													
37	12.7	3.55	2.79	0.12	0.32	Aged 900° F/3 hr	Transverse	175	170	12.2	61	B	33.2
38	11.9	3.51	2.90	0.16	0.40	Aged 900° F/3 hr	Transverse	173	168	14.2	63	B	24.8
39	11.7	3.70	2.98	0.31	<0.01	Aged 900° F/3 hr	Transverse	163	157	3.8	10	W	31.2
40	11.9	3.85	3.04	0.49	<.01	Aged 900° F/3 hr	Transverse	174	166	12.3	50	W	22.2
41	11.7	3.73	3.04	<0.02	0.27	Aged 900° F/3 hr	Transverse	170	162	11.2	52	W	44.3
42	11.5	3.03	3.03	<0.02	0.47	Aged 900° F/3 hr	Transverse	175	167	11.8	59	B	37.5
43	13.4	5.00	2.00	0.05	0.45	Aged 900° F/3 hr	Transverse	175	169	12.0	56	W	44.7
43	13.4	5.00	2.00	0.05	0.45	Aged 850° F/6 hr	Transverse	180	173	11.0	52	W	43.0
1/2 in. weld—0.062 in. filler metal—8 pass procedure													
44	Same as no. 38					As-Welded	Transverse	140	122	16	72	W	...
44	Same as no. 38					Aged 900° F/3 hr	Transverse	174	166	13.0	55	W	32.5
45	Same as no. 38[e]					As-Welded	Transverse	147	134	14.0	60	W	...
45	Same as no. 38					Aged 900° F/3 hr	Transverse	...	...	...	..	..	22.7
46	Same as no. 41					Aged 900° F/3 hr	Transverse	167	164	6	27	W	44
46	Same as no. 41					Aged 800° F/1 hr	Transverse	162	159	6	25	W	47
47	Same as no. 42					Aged 900° F/3 hr	Transverse	177	171	3	13	W	36
47	Same as no. 42					Aged 800° F/1 hr	Transverse	170	165	10	50	W	47
1/2 in. weld—0.035 in. filler metal—short-circuiting arc													
48	Same as no. 41					Aged 900° F/3 hr	Transverse	164	160	10	47	W	40.2
48	Same as no. 41					Aged 850° F/6 hr	Transverse	167	161	9	31	W	35
49	Same as no. 42					Aged 900° F/3 hr	Transverse	176	171	8	30	W	42.5
49	Same as no. 42					Aged 850° F/6 hr	Transverse	182	175	12	51	W	31
1/2 in. weld—0.035 in. filler metal—gas metal-arc													
50	Same as no. 41					Aged 900° F/3 hr	Transverse	163	159	12	54	W	52.5
50	Same as no. 41					Aged 850° F/6 hr	Transverse	166	162	6	28	W	52
51	Same as no. 42					Aged 900° F/3 hr	Transverse	171	166	7	24	W	39.5
51	Same as no. 42					Aged 850° F/6 hr	Transverse	176	171	13	56	W	40.7

[a] Weld nos. 37–45 made on aged plate (alloy no. 1 in Table 1); Weld nos. 45–52 made on aged plate (alloy no. 2 in Table 1).
[b] Filler metals vacuum melted.
[c] U.T.S.—ultimate tensile strength; Y.S.—yield strength; Elong.—elongation; R.A.—reduction area; CVN—Charpy impact.
[d] Location of break: W—weld; B—base metal.
[e] About 4% O_2 added to shielding gas.

Table 4—Properties of the Synthetically Produced Weld Heat-Affected Zone in 12% Ni Maraging Steel

Weld thermal cycle peak temperature, °F	Hardness, Rc After cycling	Hardness, Rc Aged[a]	Tensile properties[b] Y.S. (0.2%), ksi	U.T.S., ksi	Elong. in 0.5 in., %	R.A., %	CVN[a,b] at 70° F, ft-lb	% austenite
Base metal	38	40	169.0	174.4	16.4	72.2	75	None
1000	37	40	169.0	174.4	17.2	74.3	72	None
1100	34	39.5	169.0	174.4	16.5	72.0	72	None
1200	30	38.5	161.0	162.0	16.4	75.2	88	1.7
1250	28.5	39	164.0	167.0	16.4	70.2	103	1.6
1300	27	40	166.0	171.0	16.5	70.5	80	None
1400	27	40	169.0	175.0	16.4	71.5	79	<1.0
1600	27	40	172.0	178.0	16.4	71.5	77	None
1800	27.5	40	169.0	179.0	16.4	71.2	75	None
2000	27.5	40	161.0	178.0	17.3	72.5	82	None
2200	27.5	40	160.0	176.0	16.4	71.8	88	None
2400	27.5	40	160.0	174.0	16.3	70.8	86	None

[a] Specimens aged at 900° F for 3 hr after thermal cycling.
[b] Y.S.—tensile strength; U.T.S.—ultimate tensile strength; Elong.—elongation; R.A.—reduction in area; CVN—Charpy impact.

Fig. 3—Microstructure produced in 12% Ni maraging steel by thermal cycling to a 1200° F peak temperature and aging at 900° F for 3 hr. Austenite and ferrite precipitated along prior austenite grain boundaries and at the intersection of martensite platelets as a result of thermal cycling are clearly visible. Replica electron micrograph. Modified Fry's etchant. X 10,000 (reduced 50% on reproduction)

Austenite Reversion. The increased toughness and loss of strength within the region of the heat-affected zone cycled to a peak temperature of about 1200° F was associated with the decomposition of a small amount of martensite. The austenite and ferrite so formed did not respond to the aging treatment as did the martensitic matrix, and a slight strength-loss resulted. Similarly, the presence of alloy-rich austenite led to the observed increase in toughness.

Austenite and ferrite formed as discrete particles within the martensitic matrix and were too fine to be observed with the light microscope. The electron microscope revealed the distribution of these phases (Fig. 3) in a sample that had been thermally cycled to the 1200° F peak temperature and contained 1.7% austenite. The austenite and ferrite precipitated along prior austenite grain boundaries and at the intersection of martensite platelets.

Conclusion

1. Twelve percent nickel maraging compositions can be welded successfully with the conventional gas-shielded arc processes. No preheat or postheat precautions are necessary, and weld properties approaching those of the base metal can be obtained with a simple postweld aging treatment.

2. Experimental filler metals were developed which provided yield strengths in excess of 160 ksi, and as high as 177 ksi, with at least 50 ft-lb in the Charpy V-notch test.

3. There was little degeneration of the properties of the weld joint due to the effects of weld thermal cycling. Only minor softening (about 5% decrease in yield strength) occurred in the region experiencing 1200° F peak temperatures, but this was accompanied by a marked increase in toughness.

References

1. Sadowski, E. P., "12% Nickel Maraging Steel," presented at ASM Annual Meeting, October 1963, Cleveland, Ohio.
2. Decker, R. F., Eash, J. T., and Goldman, A. J., "18% Nickel Maraging Steel," *Trans. ASM*, **55** (1), 58–76 (1962).
3. Floreen, S., and Decker, R. F., "Heat Treatment of 18% Ni Maraging Steel," *Ibid.*, **55** (3), 518–530 (1962).
4. Witherell, C. E., and Fragetta, W. A., "Weldability of 18% Nickel Steel," WELDING JOURNAL, **41** (11), Research Suppl., 481-s to 487-s (1962).
5. U. S. Pat. 2,422,489, description of "X-Weld Crack Test."
6. Petersen, W. A., "Weld Heat-Affected Zone of 18% Nickel Maraging Steel," WELDING JOURNAL, **43** (9), Research Suppl., 428-s to 432-s (1964).
7. Nippes, E. F., and Savage, W. F., "Development of Specimen Simulating Weld Heat-Affected Zones," *Ibid.*, **28** (11), Research Suppl., 534-s to 546-s (1949).

Submerged-Arc Welding of 18% Nickel Maraging Steel

A development program results in the definition of suitable welding materials and optimum parameters for utilization of inherent high-deposition rate characteristics

BY F. D. DUFFEY AND W. SUTAR

ABSTRACT. In the past five years, technological advances in the field of engineering materials have been remarkable. An impressive accomplishment occurred in 1960, when the International Nickel Co. announced the development of a group of nickel-alloy low-carbon martensitic steels which possess certain unique features:

1. Excellent ductility at very high strength levels.
2. Good strength-to-weight ratio.
3. Attractive fracture toughness at high strength.
4. Good fabricability and weldability.
5. Full hardening response by simple heat treatment.

For these reasons, the materials have attracted wide interest among engineers and fabricators engaged in the construction of high-strength structures.

This paper deals with the results of an extensive development program designed to establish the suitability of the submerged-arc process for joining unlimited thicknesses of the 18Ni - 7Co - 5Mo member of the maraging "family." Preliminary testing utilizing commercially available materials indicated that this process exhibited excellent potential for depositing very sound welds, which responded fully to the optimum base material age-hardening treatment. Further investigations resulted in the delineation of optimum filler metal chemistry and welding composition type, the establishment of welding parameters for material thicknesses up to 4 in. and a very thorough evaluation of all factors associated with welded joints in this alloy.

Unlike most high-strength steels, the 18Ni - 7Co - 5Mo alloy appeared to be *relatively* insensitive to the effects of heat-input during welding. Thus, major emphasis in the program centered on utilizing to the fullest extent the inherent high-deposition rate potential of the submerged-arc process.

Joints in plate up to $1^1/_4$ in. thick were executed in two passes, with heavier thicknesses being joined by welding a single pass with the joint axis in the vertical fixed position. It has been found that regardless of the technique applied, the deposits exhibit excellent soundness, with yield strengths of 230 to 235 ksi, ultimate tensile strengths of 238 to 245 ksi, elongations of 4 to 8%, and reductions of area of 20 to 35%. Such welds exhibit considerably less fracture toughness than is observed for base plate however, as measured by the shallow, partial-thickness-crack test configuration.

A corollary benefit has been derived from the program, inasmuch as a very considerable body of data concerning the metallurgical characteristics of the alloy system has been generated, contributing to an excellent understanding of material behavior.

F. D. DUFFEY is Assistant Welding Engineer and W. SUTAR was Project Engineer in the Welding Dept., Newport News Shipbuilding and Dry Dock Co., Newport News, Va.

Paper presented at the AWS 45th Annual Meeting held in Detroit, Mich., during May 4-8, 1964.

Introduction

Approximately three years ago, the authors' company became interested in the fabrication of very large, solid propellant rocket motor cases for use in near-future space exploration programs.

In cooperation with major aerospace manufacturers, an intensive evaluation program was initiated to study suitable manufacturing techniques, economic factors, inspection techniques, facilities suitability, and all related factors. Of prime importance to the ultimate success of hardware resulting from such a program was a very careful selection of the material used for construction. Accordingly, heavy emphasis was placed on this facet of the program, and a thorough screening of candidate materials was undertaken, with the following basic criteria governing material selection:

1. The material should be fabricable by "state-of-the-art" techniques, and particularly, must be amenable to heavy equipment construction practices.
2. A material possessing a very high strength, coupled with adequate ductility and fracture toughness was sought, to take advantage of the highest strength-to-weight ratio compatible with other requirements.
3. Acceptance by the material of environmental conditions to be encountered during manufacture, fabrication, testing and end use was obviously required.

With the above goals in mind, the choice soon narrowed to a selection of one or more high strength steels. In this category, AISI 4340 heat-treated to the 180–220 ksi yield strength level appeared a logical choice, since a considerable amount of experience had been gained nationally in its use. In addition, because the data available at that time indicated that the new 18Ni - 7Co - 5Mo maraging alloy possessed a number of unique features which made it highly attractive for consideration in the proposed program, a decision was made to evaluate both materials in the initial investigations. Following the generation of sufficient data, further study of the less attractive material would be dropped and full program effort devoted to the more promising alloy.

It is beyond the scope of this report to record in detail the comparative data derived from the preliminary studies. Suffice to say that, due mainly to the more exotic joining and heat-treating procedures required for AISI 4340, it was fairly rapidly relegated to the "less attractive" category, with the 18% Ni maraging alloy emerging as being potentially most capable of fulfilling program requirements.

Evaluation Program Scope–18% Ni

Once the decision has been

made to concentrate on the 18Ni–7Co–5 Mo alloy, a development program was initiated to study primarily:

1. Base material properties
2. Compatibility with fabrication techniques
3. Weldability

Since the size of the contemplated hardware was quite large, it was apparent that plate thicknesses much greater than anything heretofore used in solid propellent motor cases would be involved. Preliminary design studies indicated that 1/2 in. to approximately 4 in. would be the thickness range. It was also evident that the yield strength required would be in the area of 230 to 250 ksi, and that welds closely matching the base material strength required would have to be made in these thicknesses, since hardware size made adoption of the "roll-and-weld" concept mandatory.

Base Material Properties

Initial development program material procurement centered on two 3/4 x 110 x 240 in. plates, rolled from a 22 ton air-melt heat produced by U. S. Steel Corp. Subsequently, portions of other heats, all produced by air-melt practice and mostly of a chemistry designed to afford a nominal 250 ksi yield strength, were obtained. Table 1 shows chemistries of the materials involved; the ability of the steel manufacturers to produce this alloy within fairly close chemistry limits is evident.

Although the base material properties of all plates listed above were studied, the most complete evaluations were conducted using the above mentioned 3/4 in. material, Heat No. X13371, and since this is typical of those shown, most of the data presented will pertain to this heat. Furthermore, since this paper is concerned primarily with welding, and because detailed studies of the alloy system are reported in numerous papers,[1-4] a minimum of detail will be given on this facet of the program to afford a basis for comparison in assessment of welded joint properties.

Aging Response

Standard R-1 round bar tensile specimens (Fed. Test Method Std. No. 151) were machined from the X-13371 plates and were aged at 875, 900, 925 and 1000° F for times ranging from 1 to 15 hr. From the data obtained, the aging response curves shown in Fig. 1 were compiled, enabling the following conclusions to be drawn:

1. For aging times of 1/2 to 6 hr, the material is quite sensitive to the temperature employed.
2. Aging temperatures above 925° F should be avoided, inasmuch as failure to fully respond as well as a tendency to "overage" are evident.
3. For the temperature range from 875 to 925° F, response is essentially uniform, with full strength being attained slowly for the lower and much more rapidly for the higher temperature.
4. Prolonged aging times in the 875 to 925° F range produce no noteworthy decrease in tensile strength.

Heat treatment for all preliminary base material studies was conducted in atmospheric electric furnaces in the laboratory. A subsequent section of the report will show that completely satisfactory aging of large maraging welded assemblies can be accomplished in suitably controlled production furnaces fired by fossil fuel.

Uniaxial Mechanical Properties

Figure 2 illustrates the results of studies of the mechanical properties of a representative number of the materials involved in the program. Noteworthy are the lower average strength values for the X53013 material and the larger than average spread of values for the 4 in. material—Heat X14819. While there is no readily explainable reason for the first anomaly, the latter is due to the fact that the properties range shown for the 4 in. material encompasses data representative of test specimen locations at the surface, quarter-point and center of the material.

It has been found that the temperature used in solution-annealing of the 18% Ni maraging alloy has a potent effect on the properties which will be obtained upon subsequent aging. For a material with a given chemical analysis, solution-annealing at 1500° F will result in significantly higher post-aging strengths than will accrue when a temperature of 1650 to 1700° F is employed. Conversely, when a heat is high in the precipitation hardening elements, Ti and Al, a higher solution-annealing temperature must be employed in order to attain a given strength level than would be utilized for a material which is "leaner" in these two important elements. It is generally possible therefore, given a group of heats with slight but significant variations in total Al + Ti, to attain a fairly constant post-aging strength level by judicious selection of the solution-annealing temperature for each heat. It should be cautioned, however, that temperatures in excess of 1700° F appear to be harmful, due to a tendency toward grain growth and the formation of deleterious grain-boundary precipitates.

Fracture Toughness

The resistance of a high strength material to the propagation of surface or internal defects under applied stress is vitally important in the manufacture of high strength, low safety margin structures. An extensive investigation of the fracture toughness of the program materials has been conducted,

Table 1—18% Ni Maraging (250 ksi grade) Plates Utilized in Evaluation Program

Material thickness, in.	Source heat no.	Element weight, %									
		C	Ni	Co	Mo	Ti	Mn	Si	P	S	Al	Cu
1/2	U. S. Steel X53013	.02	17.48	8.01	4.77	.47	.02	.06	.005	.012	.08	.12
1/2	Bethlehem[a]	.03	18.35	7.97	4.70	.45	.10	.20	.004	.003	.13	.07
3/4	U. S. Steel X13371	.02	17.83	7.41	4.70	.46	.04	.08	.004	.009	.11	.05
3/4	U. S. Steel X14359	.02	17.73	7.40	4.80	.39	.03	.09	.005	.005	.07	.11
1 1/4	U. S. Steel X14636	.03	18.37	8.49	4.70	.42	.06	.10	.005	.010	.13	...
2	U. S. Steel X14637	.02	18.30	8.32	5.03	.42	.05	.05	.005	.008	.05	.10
4	U. S. Steel X14819	.01	17.55	8.79	4.85	.55	.08	.03	.006	.009	.08	.03
3 3/4	U. S. Steel X53201	.02	17.90	7.30	4.87	.39	.05	.07	.005	.004	.12	...

[a] Heat number not available.

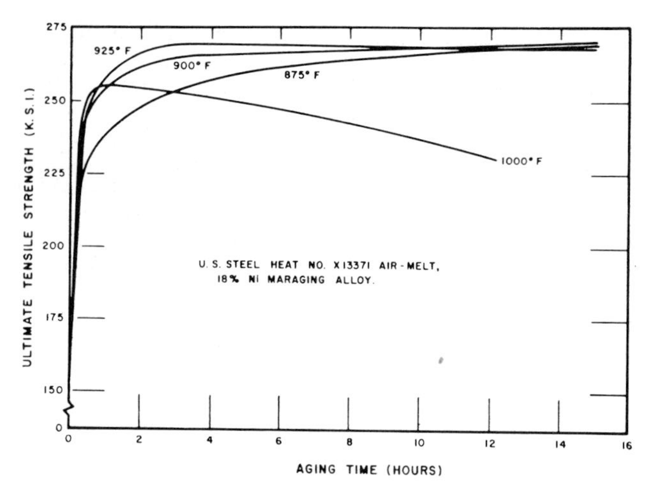

Fig. 1—Tensile strength of base material as a function of aging cycle

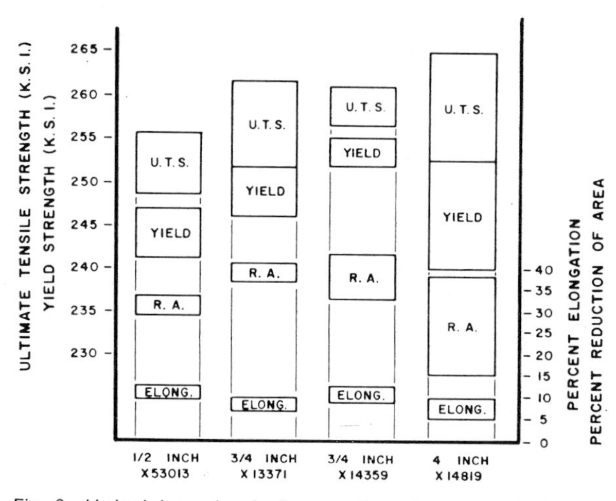

Fig. 2—Uniaxial mechanical properties of representative base material. Heat treatment: (1) solution annealed at 1500° F–1 hr/in.; (2) aged at 900° F–3 hr

largely by means of the shallow, partial-thickness surface crack, uniaxially tensile loaded configuration. It is not the authors' intent to enter into a detailed description of the mechanics of the test method, there being excellent descriptions in pertinent literature.[5,6] Figure 3 illustrates the results obtained in testing Heat X13371 in this manner, using both the nominal 3 hr aging treatment as well as an extended aging at 900° F. Similar data have been compiled for all other materials used in the program, and with allowance for minor experimental scatter, all exhibit very similar fracture toughness characteristics. The following general conclusions have been reached.

1. Compared to other high-strength alloys, 18% Ni maraging steel exhibits attractive fracture toughness at strength levels at which the other alloys are seriously impaired in this respect.

2. Surface cracks up to approximately 0.250 in. long do not affect the fracture stress of the 18% Ni alloy, aged to give ultimate strengths of approximately 260 ksi.

3. Extended aging, for times up to 12 hr, lowers the critical flaw length somewhat, due mainly to the resultant increase in strength.

4. State-of-the-art nondestructive testing techniques are capable of discerning flaws considerably smaller than the size found to be critical at the 260 ksi strength level.

Many additional factors germane to suitability of the material for use in the contemplated motor case program were studied. Susceptibility to stress-corrosion was found to be negligible. The alloy lent itself to conventional fabricating techniques such as cutting, machining, forming, etc. Tests indicated that subassemblies could be constructed in the solution-annealed condition and subsequently aged in existing furnace facilities. Distortion during this operation was found to be slight, and predictable. It was felt that aged subassemblies could be successfully welded, with local zone heating of the resulting joint employed to bring that area up to full strength.

Welding Development Program

Since the motor case fabrication program was predicted on use of the roll-and-weld concept, a thorough understanding of the weldability of the material was mandatory. At the initiation of the program, relatively little was known in this area, but one technical paper[7] indicated that the 18% Ni maraging alloy was potentially weldable by most conventional processes. Also, due to the metallurgical characteristics of the alloy system, high heat inputs appeared to be possible, since the heat-affected zones were relatively soft and ductile due to localized solution annealing. Thus it was felt that the economic advantages of high-deposition rate welding procedures could be realized.

Preliminary joining studies concentrated on use of the gas-shielded

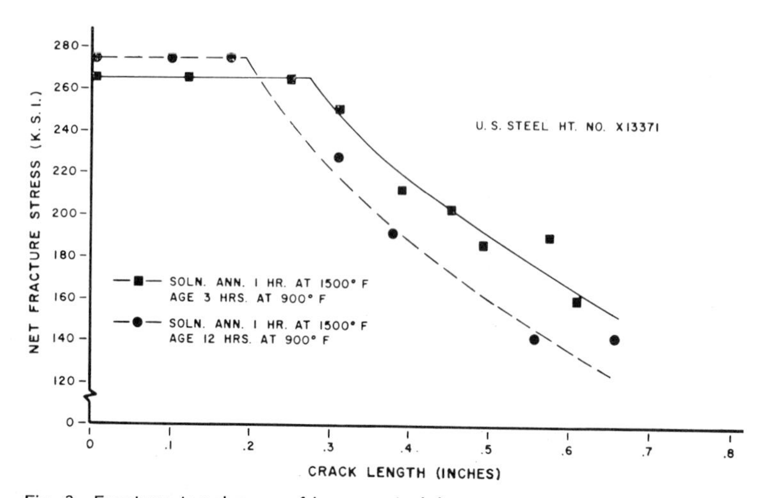

Fig. 3—Fracture toughness of base material

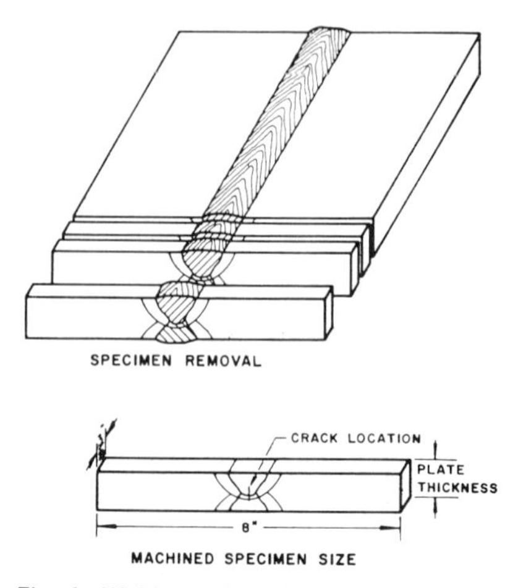

Fig. 4—Weld specimen configuration for fracture toughness test

processes (gas tungsten-arc and gas metal-arc), primarily because filler metals of suitable composition for use with these processes were commercially available. Both were found to be satisfactory, and a considerable amount of gas metal-arc welding was performed. However, use of this process for joining maraging steels is hampered by two factors when high-deposition rate techniques are used:

1. A heavy surface oxide is formed on each bead, making the use of trailing shields necessary.
2. Even when meticulous care is exercised, gas metal-arc deposits in this alloy are prone to unsoundness in the form of porosity and sidewall cold-shutting.

The shipbuilding industry has historically relied heavily on the submerged-arc process for joining of a variety of steels, and the authors' company has a tremendous backlog of experience in its use. It was felt that this process could be employed successfully for welding the 18% Ni alloy, but it was recognized that a significant amount of investigation would be required. Primarily, two problems demanded resolution.

1. Delineation of a suitable filler metal chemistry. Molten state flux/melt reactions are extremely complex and not too well understood, but it was known that readily volatilized elements such as titanium and aluminum, would be lost during arc transfer. It was believed, however, that compensation could be attained by use of special, high Ti filler metal which could accommodate high losses of this element.

2. Formulation of a sufficiently "neutral" welding composition or flux. Most materials commercially available for use in submerged-arc welding impart silicon and manganese to the deposits. The maraging alloy system requires that these elements be held to low limits if a proper balance of material properties is to be maintained.

Despite the above problems, the high deposition rates and over-all deposit soundness inherent in the submerged-arc process were sufficiently attractive to promote and sustain the program required to define satisfactory solutions.

Preliminary Submerged-Arc Studies

As a preliminary study, submerged-arc multilayer bead-on-plate welds were made, using a low Ti filler metal and a group of commercially available fluxes. Metallographic and hardness studies on these deposits permitted the following conclusions to be made:

1. The process was capable of producing sound, crack-free welds with most of the fluxes employed.
2. Silicon and manganese pickup was too high for *all* of the fluxes tried.
3. Losses of titanium during arc-transfer averaged from 40 to 70%.
4. The best titanium recovery resulted when fluxes were used which imparted the lowest Si and Mn contents.
5. Based on hardnesses obtained, most of the fluxes resulted in deposits which responded well to aging, with a cycle of 875° F for 6 hr appearing optimum.

The above test results were sufficient to convince the investigators that the process did in fact exhibit potentials. Accordingly, a quantity of filler metal containing 2.08% Ti was obtained, and manufacturers of submerged-arc welding composition were solicited to aid in the development of fluxes suitably low in silicon and manganese. While this development effort was getting under way, a group of experimental joints were welded utilizing the above filler metal and the more promising available fluxes in order to obtain an evaluation of joint mechanical properties. After considerable experimentation to optimize welding conditions, the 11 welds shown in Table 2 were executed. All were radiographically inspected and exhibited excellent soundness, exceeding the requirements to Standard 1 of Military Specification MIL-R-11468. In addition to determining uniaxial mechanical properties, the shallow, partial thickness crack, fracture toughness test was employed in studying toughness of the welds. The specimen configuration shown in Fig. 4 was employed, and the results of this testing are compiled in Figs. 5, 6 and 7. The first two of these show the net fracture stress curves for each individual weld, whereas Fig. 7 is intended to illustrate the remarkably slight spread of critical flaw length, for a group of welds of widely differing chemistries.

It is felt that a few of the more important observations made during the preliminary test welding should be discussed briefly at this point, inasmuch as they affected the planning of the remainder of the program.

Table 2—Submerged-Arc Weld Properties—Experimental Materials

					Weld chemistry, %				Uniaxial mechanical properties[b]				
Weld[a] no.	Base metal thickness (in.) and source	Flux type	No. passes	Polarity	Ti	Mn	Si	Co	Fty (0.2 offset)	Ftu	Elongation in 2 in., %	Reduction in area, %	Fracture[c]
D4924	1/2 Beth.	2A	2	Rev	0.61	0.05	0.27	7.4	257,800	260,000	5.0	30	HAZ
D4923	3/4 U.S.S.	2A	2	Rev	0.58	0.03	0.19	7.4	237,000	239,800	5.0	26	HAZ
D4920	1/2 Beth.	2A	2	Rev	0.61	0.11	0.29	7.7	244,000	245,500	4.5	30	Weld
D4918	3/4 U.S.S.	2A	4	Rev	0.64	0.12	0.25	7.6	237,000	241,000	4.0	17	Weld
D4925	3/4 U.S.S.	2A	2	Rev	0.64	0.08	0.20	8.0	239,000	244,000	4.5	25	HAZ
D1915	1/2 Beth.	2B	2	St	0.44	0.24	0.53	7.4	250,000	256,000	4.5	31	HAZ
D996	1/2 Beth.	2C	2	Rev	0.78	0.16	0.17	7.5	257,500	259,500	5.0	31	HAZ
D1914	3/4 U.S.S.	2B	4	St	0.46	0.27	0.64	7.5	244,000	250,000	3.6	13	Weld
D997	3/4 U.S.S.	2C	2	Rev	0.74	0.16	0.12	7.4	245,500	248,000	5.5	27	HAZ
D991	3/4 U.S.S.	2C	4	Rev	0.86	0.21	0.16	7.6	230,000	239,000	6.3	24	HAZ
D1913	3/4 U.S.S.	2B	2	St	0.55	0.18	0.38	7.5	243,000	247,000	4.5	19	HAZ

[a] All welds made in annealed plate, and directly aged @ 875° F/6 hr.
[b] All properties shown resulted from testing of standard 0.505 in. diam transverse tensile specimens; the values shown are averages of at least three tests. All test plates restrained in the flat position.
[c] Heat-affected zone (HAZ) failures occurred primarily in the light-etching, impaired-response zones.

Heat-Affected Zone Characteristics

The unique appearance of transverse weld macroetch specimens promoted an early interest in the metallurgical characteristics of the heat-affected regions adjacent to welds. When deposits were made in solution-annealed plate, it was noted that the joint area appeared to be composed of 4 differentially-etching areas. This is illustrated schematically for a 2 pass weld in Fig. 8, and the observed zones are described as follows:

1. Zone 1—composed of material extending outward from the fusion lines which has been exposed to temperatures ranging from near-molten at the fusion line to approximately 1300° F at its outer edge. This area is white etching, and exhibits hardnesses of 28 to 35 R_c, having been completely or partially re-solution annealed by conduction of welding heat.
2. Zone 2—a band of material which etches black, extending through both the base metal and a portion of the first-pass deposit. Hardnesses of 45 to 48 R_c are noted for this band, indicating that it has been exposed to temperatures in the aging range for sufficient time to produce an aged martensitic structure.
3. Zone 3—consists of base material at such a distance from the weld fusion lines that temperature reached in these areas was insufficient to cause permanent metallurgical changes. Hardnesses are the nominal 30 R_c of solution-annealed base material.
4. The weld deposits proper, which are grey-etching, and display the expected dendritic cast structure. Hardnesses here average 30 R_c.

The above observations tie in well with the material behavior one would expect from a study of the alloy system. From the weldability standpoint, the presence of a solution-annealed, and hence, relatively soft and ductile zone immediately adjacent to the weld deposits, aids in absorbing solidification stresses and eliminates the cracking propensity associated with such areas in weldments executed in quenched and tempered steels. If the weld joint areas were composed only of the metallurgical structure types mentioned above with clean demarcations between each, no problems would occur on aging, since all areas would respond fully to thermal treatment. Early in the program, however, testing of transverse joint tensile specimens indicated that fractures were occurring in unexpected locations.

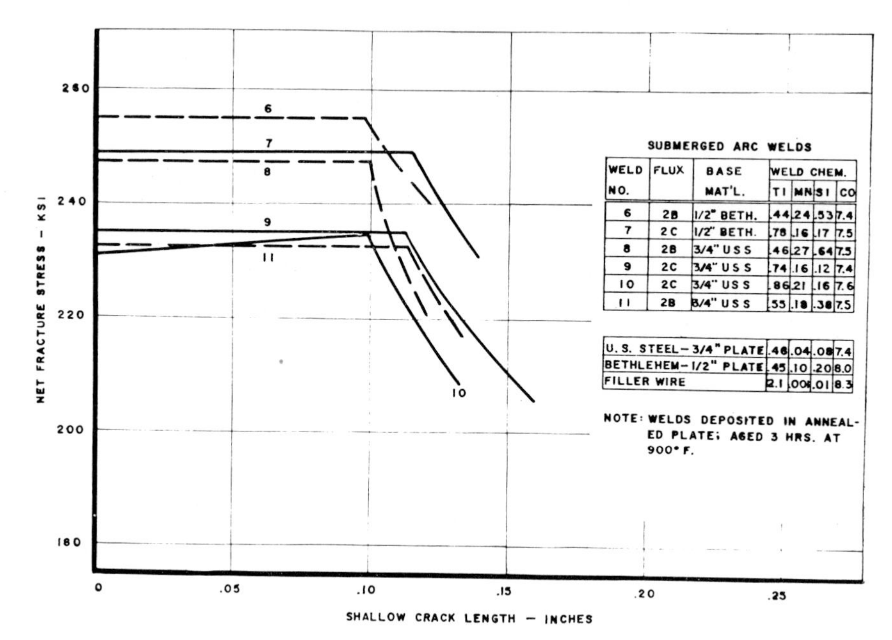

WELD NO.	FLUX	BASE MAT'L.	WELD CHEM. TI	MN	SI	CO
6	2B	1/2" BETH.	.44	.24	.53	7.4
7	2C	1/2" BETH.	.78	.16	.17	7.5
8	2B	3/4" USS	.46	.27	.64	7.5
9	2C	3/4" USS	.74	.16	.12	7.4
10	2C	3/4" USS	.86	.21	.16	7.6
11	2B	3/4" USS	.55	.18	.38	7.5

	TI	MN	SI	CO
U. S. STEEL—3/4" PLATE	.46	.04	.08	7.4
BETHLEHEM—1/2" PLATE	.45	.10	.20	8.0
FILLER WIRE	2.1	.004	.01	8.3

Fig. 5—Fracture toughness of preliminary submerged-arc welds

As soon as welds whose strength approached that of the base material were made, it was observed that "neck-down," and ultimately tensile failure, occurred predominantly along a plane best described as lying along a narrow band at the junction of Zones 1 and 2. Compared to tensile properties of non-heat-affected base material, it was noted that the selective failure described above resulted in a net loss in mechanical properties of from 5 to 10%.

In order to gain an understanding of the observed material degradation, a test bar of 18% Ni steel was suitably instrumented to monitor thermal gradients within the bar while a heat source was applied to one end to simulate the effect of welding. After heating, the bar was severed longitudinally, and etching revealed the presence of heat-affected zones identical to those found adjacent to welds. Metallurgical studies were made and related to the temperatures which were known to have prevailed in each zone. Slices from the bar were subjected to several direct aging cycles. It was found that in a narrow band between Zones 1 and 2, full hardening response was not realized, even for long aging cycles. This impairment resulted in a potential hardness loss of as much as 3 to 4 R_c. It was also found that re-solution annealing of the specimen, followed

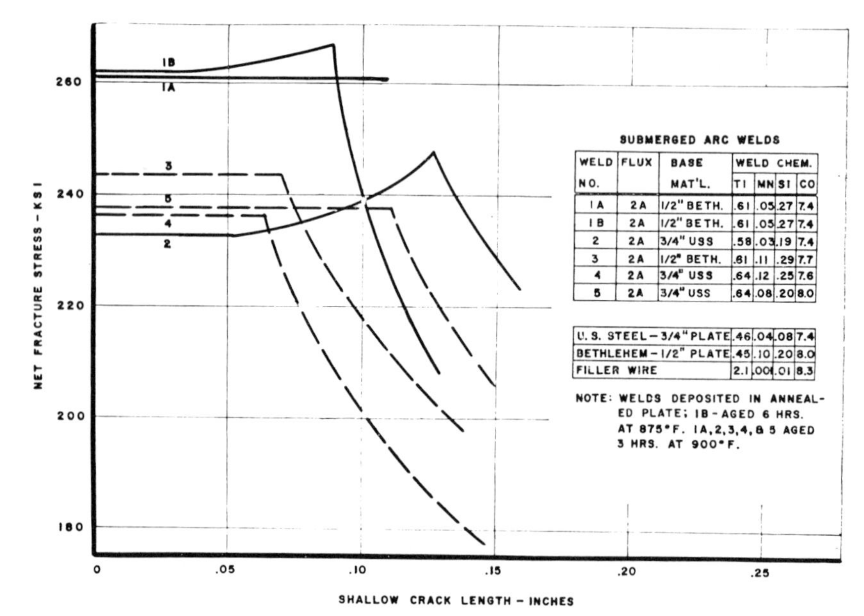

WELD NO.	FLUX	BASE MAT'L.	WELD CHEM. TI	MN	SI	CO
1A	2A	1/2" BETH.	.61	.05	.27	7.4
1B	2A	1/2" BETH.	.61	.05	.27	7.4
2	2A	3/4" USS	.58	.03	.19	7.4
3	2A	1/2" BETH.	.61	.11	.29	7.7
4	2A	3/4" USS	.64	.12	.25	7.6
5	2A	3/4" USS	.64	.08	.20	8.0

	TI	MN	SI	CO
U. S. STEEL—3/4" PLATE	.46	.04	.08	7.4
BETHLEHEM—1/2" PLATE	.45	.10	.20	8.0
FILLER WIRE	2.1	.001	.01	8.3

Fig. 6—Fracture toughness of preliminary submerged-arc welds

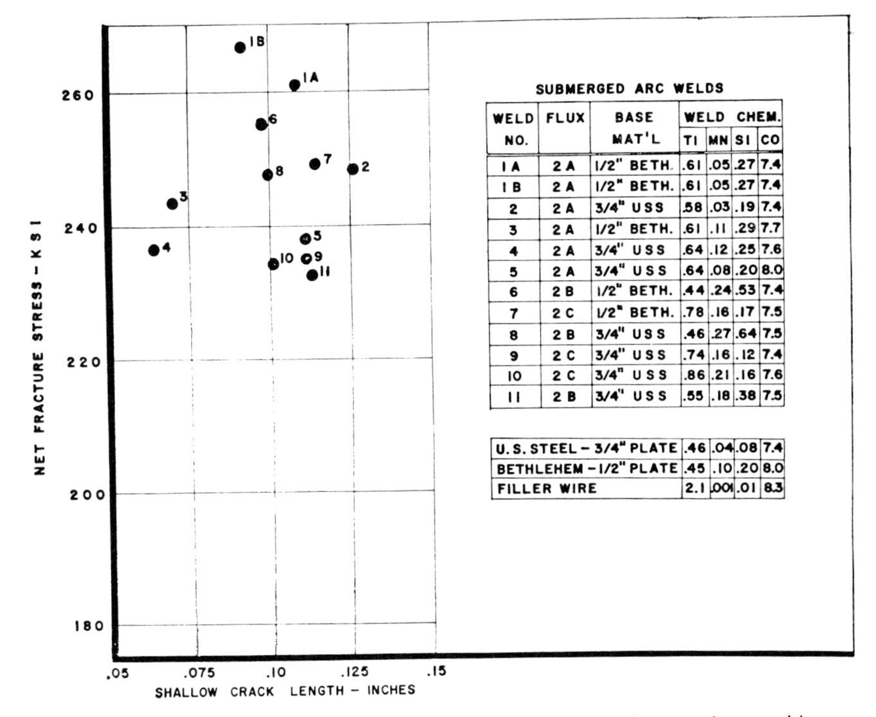

SUBMERGED ARC WELDS

WELD NO.	FLUX	BASE MAT'L	WELD CHEM. TI	MN	SI	CO
1 A	2 A	1/2" BETH.	.61	.05	.27	7.4
1 B	2 A	1/2" BETH.	.61	.05	.27	7.4
2	2 A	3/4" USS	.58	.03	.19	7.4
3	2 A	1/2" BETH.	.61	.11	.29	7.7
4	2 A	3/4" USS	.64	.12	.25	7.6
5	2 A	3/4" USS	.64	.08	.20	8.0
6	2 B	1/2" BETH.	.44	.24	.53	7.4
7	2 C	1/2" BETH.	.78	.16	.17	7.5
8	2 B	3/4" USS	.46	.27	.64	7.5
9	2 C	3/4" USS	.74	.16	.12	7.4
10	2 C	3/4" USS	.86	.21	.16	7.6
11	2 B	3/4" USS	.55	.18	.38	7.5

	TI	MN	SI	CO
U. S. STEEL – 3/4" PLATE	.46	.04	.08	7.4
BETHLEHEM – 1/2" PLATE	.45	.10	.20	8.0
FILLER WIRE	2.1	.001	.01	8.3

Fig. 7—Comparison of critical crack lengths for preliminary submerged-arc welds

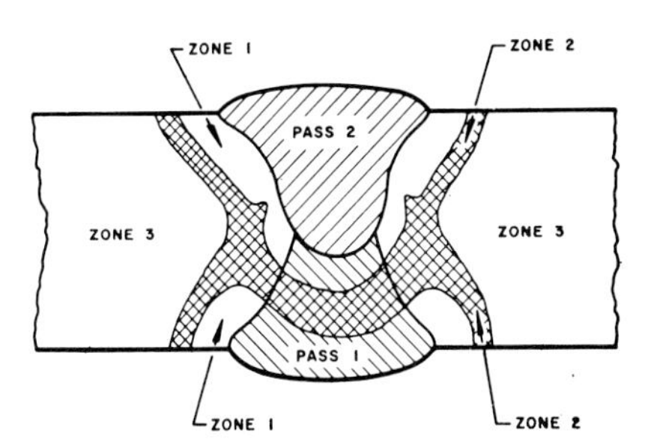

Fig. 8—Schematic of weld macro specimen

by aging, resulted in the reattainment of full hardening response.

Metallurgical definition of the impaired response band, which by now was being referred to as an "eyebrow" in deference to its appearance in macroetch specimens, was impossible by light microscopy. An excellent paper dealing with the condition was published,[8] which explained the failure to respond as attributable to a localized reversion to a metastable austenitic condition in zones exposed for rather brief periods to temperatures in the 1100 to 1300° F range. Comparative testing of welds made by 4 pass and 2 pass procedures indicated generally superior properties for the latter; it appears therefore that a greater number of intersecting impaired response zones is more damaging than a minimum number of larger size zones.

Joint Design

A number of joint designs compatible with fabrication requirements and material thicknesses were investigated during the program. For the 2 pass technique selected, it was found that conventional double J-groove preparations with sufficient land thickness to support the heat input imposed in depositing the first pass were satisfactory. This joint design was employed for all thicknesses through $1^1/_4$ in., with actual joint dimensions varying as a function of material thickness.

Hydrogen Embrittlement

Although various investigations have shown the 18% Ni maraging alloy to be relatively insensitive to hydrogen in the wrought form, a definite propensity to hydrogen-induced cracking has been noted for cast weld structures. Several instances of transverse weld cracking which occurred during the program were thoroughly investigated, and it was found that when interstitial hydrogen contents in excess of approximately 5 ppm are present in the deposits, cracking can be expected. Close control must be exercised to preclude hydrogen from the melt. This requirement has resulted in exceedingly tight limits on filler metal hydrogen content, scrupulous maintenance of joint cleanliness, and in the case of submerged-arc welding, baking of the welding composition to very low moisture contents. In this connection, it should be noted that for 2 pass welds no cracking has ever been encountered except in cases which were proved by vacuum fusion analytical methods to be due to excessive hydrogen contents in the defective welds.

Optimum Submerged-Arc Procedures

By the time the preliminary welding studies had been completed, several factors germane to the continuation of the program had been defined.

1. A special welding composition, low in Si and Mn contents, had been developed by the Linde Co. Two

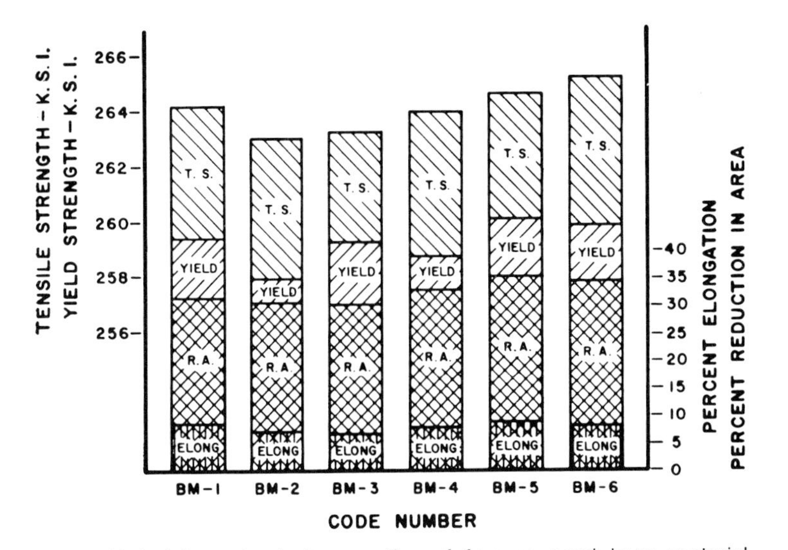

Fig. 9—Uniaxial mechanical properties of furnace aged base material panels. Special Notes: (1) Base metal is U.S.S. heat X14359 with 18 Ni, 7.4 Co, 4.8 Mo and 0.39 Ti. (2) Base metal properties were obtained from panels which had been aged in the 260 in. diam test cylinder. (3) Test cylinder was subjected to following thermal treatment: (a) charged into a cold furnace; (b) heated to 800° F within 3 hr; (c) heated from 800 to 900° F ± 10° F within 1 hr, 20 min; (d) heated at 900° F ± 10° F for 2 hr; (e) air cooled to room temperature

pass test welds utilizing this material exhibited reproducibly excellent properties, and a decision was made to employ this composition for all further work. (First given a laboratory designation, this flux is now marketed as "Unionmelt 105.")

2. Studies of the effect of welding parameters on the degree of Ti loss during arc transfer had shown that such loss could be reproducibly held to 50% when the optimum volt-ampere-polarity relationship was maintained.

3. Design studies for the proposed hardware indicated that welds therein would be satisfactory if yield strengths of 230 ksi were reproducibly attainable. Since deposit Ti contents of 0.50 to 0.60% would ensure meeting this strength requirement, a filler metal Ti level of approximately 1.2% was defined, and a quantity of this special analysis material was obtained.

Table 3—Deposit Chemical Compositions—Optimum Submerged-Arc Welds

	Element, wt %							
Code no.	Ni	Co	Mo	Ti	Mn	Si	Al	Cu
D-99B03	17.90	8.05	4.75	0.54	0.13	0.10	0.07	0.05
D-99BO4	17.80	8.15	4.80	0.56	0.12	0.10	0.08	0.04
D-99BO5	17.96	7.96	4.75	0.52	0.13	0.09	0.07	0.04
D-99BO6	18.00	8.00	4.75	0.53	0.12	0.09	0.07	0.04
D-99BO7	18.13	7.83	4.60	0.53	0.13	0.10	0.08	0.05
D-99BO8	18.00	8.10	4.73	0.54	0.12	0.09	0.08	0.05
U.S.S. Heat X14359	17.73	7.40	4.80	0.39	0.03	0.09	0.07	0.11

Reproducibility Tests

It is impossible to go into a detailed description of the large number of test welds executed using the optimum flux/filler metal combination without exceeding the space limitations of this paper. Excellent reproducibility has been obtained, however, by use of 2 pass procedures for joining material thicknesses ranging from $^1/_2$ to $1^1/_4$ in. One of the more interesting studies involved an investigation of the capability of oil-fired production furnaces for aging of 18% Ni maraging welded assemblies. In this test, 6 base material panels and 6 panels containing a longitudinal weld were prepared using material from Heat X14359. All were attached to a full-scale carbon steel mockup of a motor case assembly, and each panel was instrumented to record thermal history during aging. The mockup was then exposed to an optimum aging cycle in a large production furnace, using a low-sulfur fossil fuel. Excellent temperature control was maintained, and following cool-down, the panels were removed for evaluation studies. A metallurgical analysis of base material microstructure and surface condition before and after furnace exposure indicated normal material reaction to the aging cycle. Surface scale formation was considerably less than is noted for similar electric furnace cycles, due to the slightly reducing atmosphere present as a result of the oil-fired combustion products. Uniaxial mechanical properties of the aged base material panels are shown in Fig. 9, while the results of shallow crack fracture toughness testing is presented graphically in Fig. 10. Normal material response to the aging cycle is noted.

The deposit composition in the welded panels was determined, and is shown in Table 3, indicating the excellent reproducibility obtainable with the submerged-arc process under controlled conditions. Uniaxial mechanical properties of the 6 welds are shown in Fig. 11. Shallow-crack fracture toughness test results, as shown in Fig. 12, caused considerably concern, inasmuch as a critical flaw length of approximately 0.07 in. was indicated, whereas all previous tests had shown that 0.10 in. was the minimum which could be expected at this strength level. In prior work, mechanically-induced crack-starter notches had been employed, followed by fatigue extension with the specimens in the as-welded condition, after which they were aged and tensile tested. Since in the current tests the specimens were removed from previously aged panels, mechanically induced notching procedures were found to be unsatisfactory, and electrical discharge machining practices were employed for this purpose. In an effort to gain a better understanding of the effect of specimen preparation sequence on test results, a weld was executed from which shallow-crack specimens were prepared using 5 different sequences. Data thus generated is shown in Fig. 13, demonstrating that when cracks are induced in weld deposits in the aged condition, lower apparent critical flaw lengths will be obtained than when cracking is performed prior to aging. It is important to note, therefore, that in

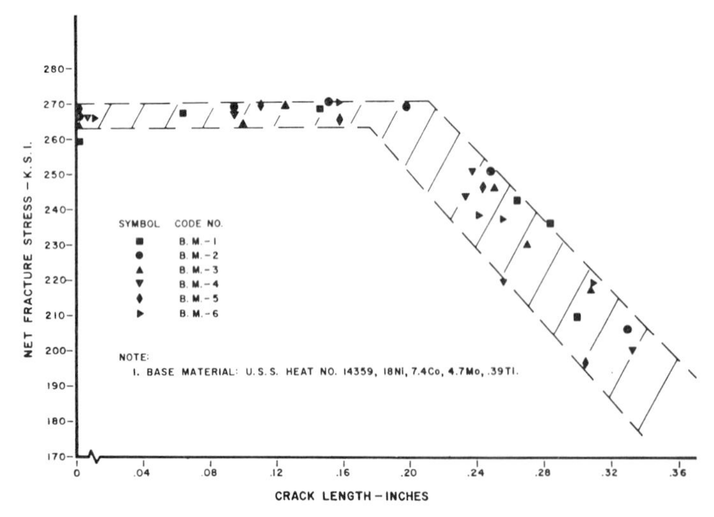

Fig. 10—Fracture toughness of furnace aged base material panels

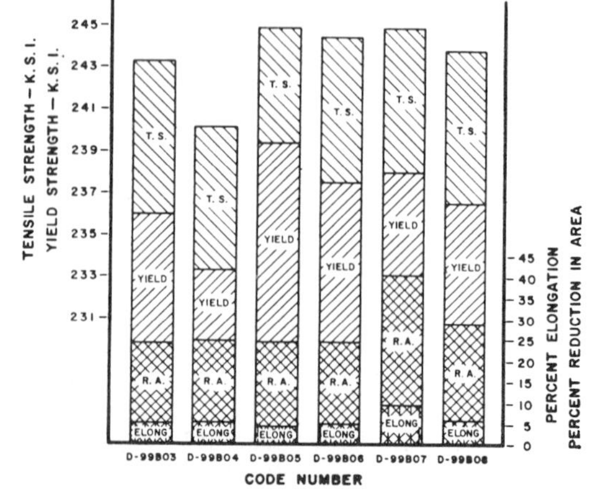

Fig. 11—Uniaxial mechanical properties of furnace aged welded panels. See special notes for Fig. 9

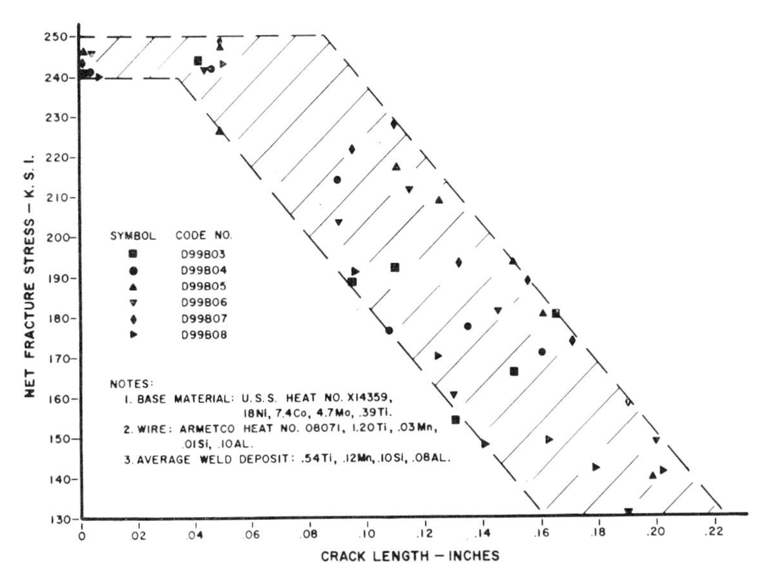

Fig. 12—Fracture toughness of furnace aged welded panels

any comparative assessment of the toughness of maraging weldments determined by the shallow-crack test, the specimen preparation sequence must be known in each instance.

Procedure Qualification Test Welds

Prior to the start of actual component fabrication, qualification tests had to be performed for the welding procedure to be used, and production welding operators had to be trained and qualified. A considerable amount of 2 pass submerged-arc welding was executed during operator training. Tests conducted on these welds added to the statistical evidence that this process was ideally suited for welding of 18% Ni maraging steel. Four welds executed to provide procedure qualification for $^3/_4$ in. plate were performed by separate, trained operators to provide simultaneous personnel qualification. Chemical analysis of the deposits showed them to be identical to those shown in Table 3. Average mechanical properties for the 4 welds were: 0.2% offset yield strength—235 ksi; ultimate tensile strength—246 ksi; elongation in 2 in. gage—5%; reduction in area—24.5%.

All aging was for 3 hr at 900° F without intermediate annealing, and all fractures occurred in the eyebrow area of the heat-affected zones. Some "splitting" or tensile delamination of the base material in the fracture areas was noted, but this did not adversely affect properties. The results of fracture toughness testing for these welds is given in Fig. 14, which re-confirms the critical flaw length of 0.10 in. for the cast weld structures. Figure 15 shows a slightly magnified view of surface bead appearance, while Fig. 16 indicates the typical transverse and longitudinal macroetch appearance with the specimen in the as-welded condition. Welds in this material provide a fertile area for metallographers and research metallurgists, and decisions as to just what structure a given weld microphotograph depicts must be carefully made, and based on knowledge of the metallographic techniques employed. This is particularly true of heat-affected zone studies, where additionally, an intimate knowledge of the specimen thermal history is mandatory. Figure 17 illustrates a typical microphotograph of submerged-arc deposit microstructure. The surface being viewed is parallel to the weld axis; i.e., it is normal to the dendritic fusion line-to-center solidification pattern. The cored structure normally encountered in cast weld deposits is noted, with a small amount of reverted austenite present.

Repair Welding

While the soundness of all 2 pass submerged-arc welds in the 18% Ni alloy made to date has been uniformly excellent, repair welding procedures have been studied. Suffice to say that the gas metal-arc process, using small diameter filler metal with Ti content in the 0.40 to 0.50% range will be employed for making minor repairs. Defects requiring major excavation of previous weld deposit will be repaired by the submerged arc process.

Local Aging of Weld Zones

During fabrication, certain welds will be made in aged base material. Aging of the weld zones, thus created, will be necessary without recourse to a furnace cycle for accomplishing this. Tests have been conducted employing several means for locally heating welds, and it has been found that attainment of a uniform, controllable cycle of 900°

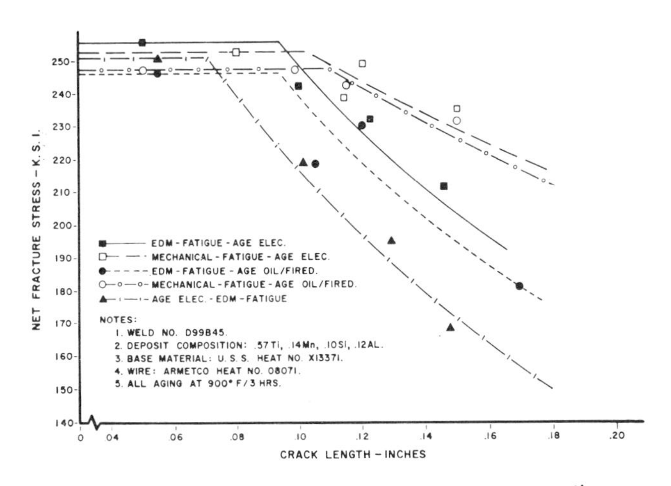

Fig. 13—Effect of fracture toughness specimen preparation sequence

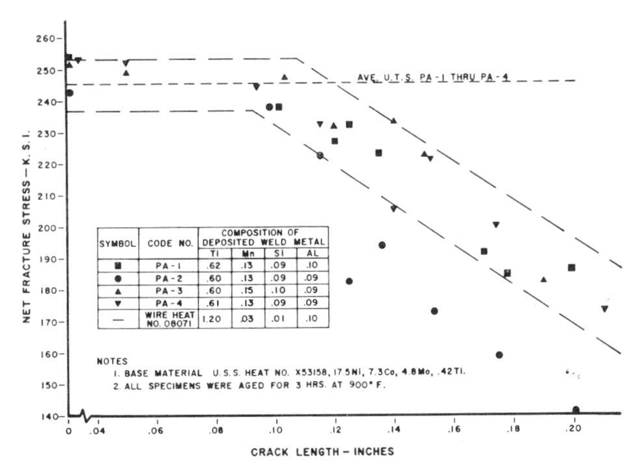

SYMBOL	CODE NO.	COMPOSITION OF DEPOSITED WELD METAL			
		Ti	Mn	Si	AL
■	PA-1	.62	.13	.09	.10
●	PA-2	.60	.13	.09	.09
▲	PA-3	.60	.15	.10	.09
▼	PA-4	.61	.13	.09	.09
—	WIRE HEAT NO. 08071	1.20	.03	.01	.10

Fig. 14—Fracture toughness of procedure qualification welds

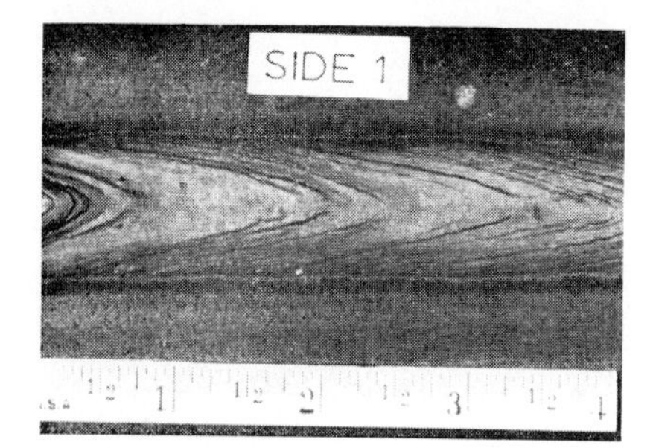

Fig. 15—Weld appearance of procedure qualification welds

F for 3 hr can be realized without significant "overreach" of the heated area into previously aged material. Two pass submerged-arc test welds so treated have been found to exhibit the same mechanical properties that are obtained when the welds are furnace aged. That such a concept is feasible is of particular interest to designers of very large maraging structures since welded subassemblies can be furnace aged, joined in the aged condition and these final welds can then be locally aged.

Heavy 18% Ni Plate

It was mentioned at the beginning of the paper that material thicknesses up to approximately 4 in. were involved in the proposed construction. It was recognized that plate of this thickness would probably generate problems not forseeable by an extrapolation of data generated in studying thinner material of the same type. Accordingly, an extensive investigation of heavy plate properties and weldability was undertaken.

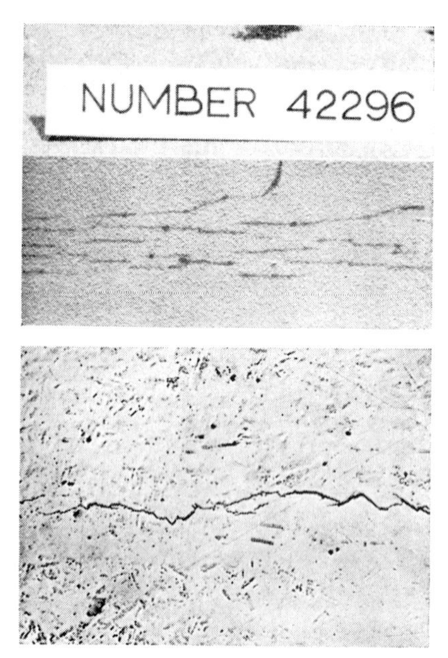

Fig. 18—Thermal shock de-laminations in $1^1/_4$ in. plate resulting from plasma cutting. A (top)—delaminations as detected by liquid penetrant; B (bottom)—microstructure of transverse section, mixed acids for etchant. ×100

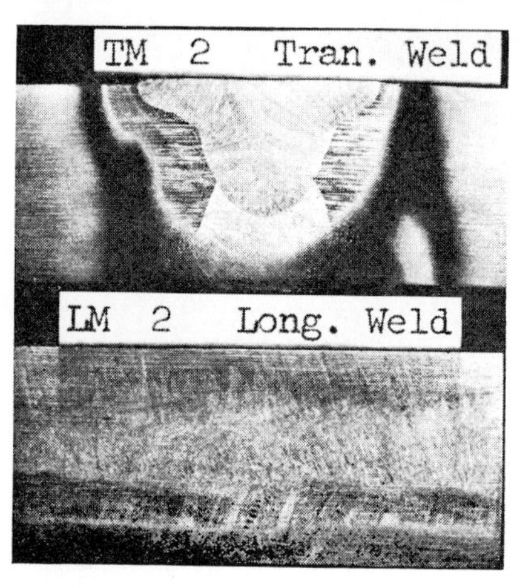

Fig. 16—Macroetch of $^3/_4$ in. 2 pass weld

Base Metal Limitations

The application of concentrated heat sources to thick sections of solution annealed 18% Ni maraging steel has been found to pose several problems. Heat generated by the plasma cutting torch, the welding arc or by other means, may cause this material to delaminate or even crack severely in the heat-affected areas. Figure 18 shows delaminations in $1^1/_4$ in. thick material which has been plasma cut. As can be seen, the delaminations are easily detected by liquid penetrant procedures, and also, they may be located by magnetic particle or ultrasonic methods.

The size, frequency, and nature of these defects vary widely, but they appear to be dependent upon the thickness of the material and to some extent on the degree of "banding" prevailing. A section of 4 in. thick material which has been plasma cut is shown in Fig. 19. Whereas the $1^1/_4$ in. thick material delaminated, the thicker material cracked severely. The cracks, which are primarily normal to the cut surface, appear to initiate in the dark etching heat-affected zone. Removal of $^1/_2$ in. of material from the plasma cut surface reveals a mosaic crack pattern as shown in Fig. 19.

Examination of the material at high magnification definitely shows that the cracks originate near the dark etching heat-affected zone and generally propagate normal to the cut surface. Examination of the microstructure along a typical crack length reveals a variance of structure. Such a change, progressing away from the plasma cut surface, is to be expected as the thermal cycle encountered during cutting diminishes in severity as the distance from the plasma cut surface increases. Starting from the plasma cut surface, 4 distinct regions are noted in Fig. 20. The first contains material which has been austenitized during cutting. In the second region, material which has been partially austenitized is noted. This zone does not become completely martensitic upon cooling, as some of the reverted austenite becomes metastable. The third region contains material which has been partially aged, and the crack is at its maximum width here.

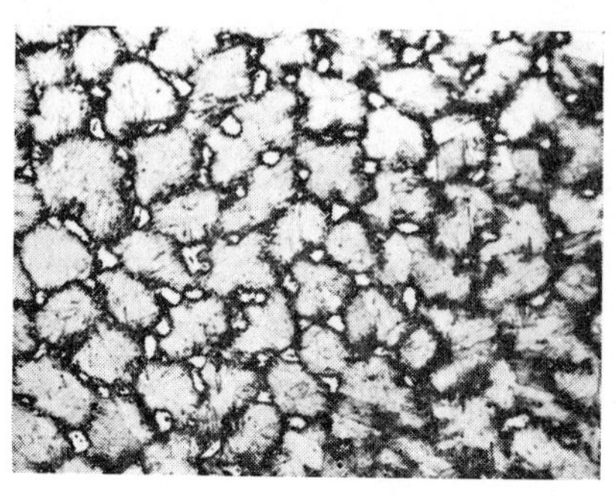

Fig. 17—Microstructure of a 2 pass submerged-arc weld. Aged for 3 hr at 900° F; Fry's etchant. × 500 (reduced 50% on reproduction)

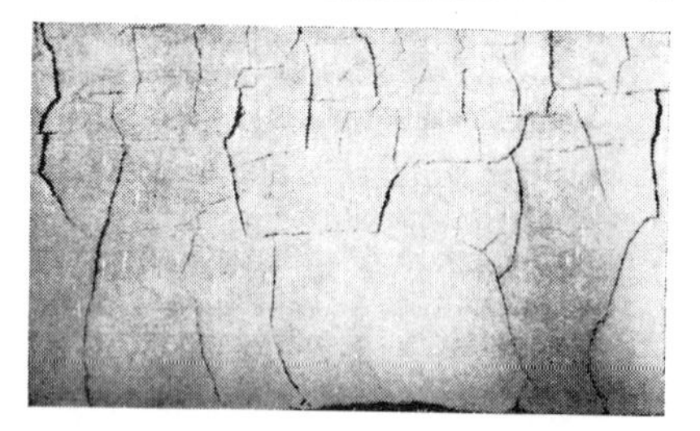

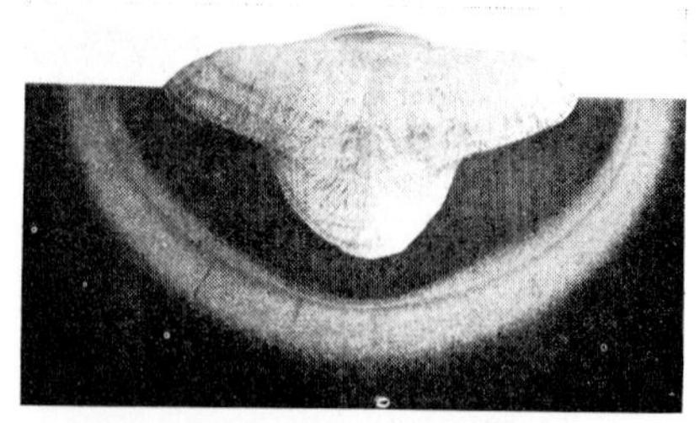

Fig. 19—Base material cracking resulting from concentrated heat sources. A (top)—the top edge of the 4 in. plate has been plasma cut; B (center)—plasma cut surface of 4 in. plate after removal of $^1/_2$ in. material; C (bottom)—submerged-arc weld deposit on 4 in. plate. Note heat-affected zone cracking

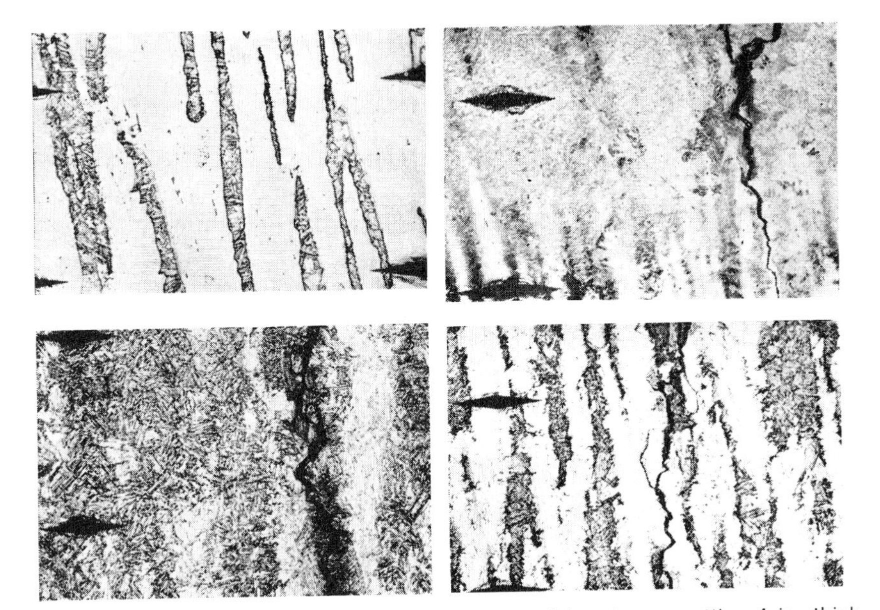

Fig. 20—Microstructure of heat-affected zone created by plasma cutting 4 in. thick material. A (upper left)—austenitized during cutting; B (upper right)—partially austenitized during cutting; C (lower left)—partially aged during cutting; D (lower right)—unaffected base material. Kroll's etchant. × 100 (reduced 50% in reproduction)

The fourth region is unaffected base material. Examination of Fig. 20 shows the pronounced banding of the base material, but does not delineate any pattern of preferential cracking.

Discovery of the base material cracking initiated an investigation to determine if cracking also occurred when plasma cutting other heats of material, and if the cracking was unique only to the plasma process. The results showed that all maraging steel was subject to varying degrees of delamination and cracking. Deposition of a submerged-arc weld bead on the same heat of 4 in. thick material resulted in heat-affected zone cracking similar to that encountered in plasma cutting, with cracks radiating from the dark etching heat-affected zone. This test definitely proved that some heats of maraging steel are crack sensitive and that the cracking is not unique to the plasma process.

The primary cause of this cracking has not been determined to date, but volumetric changes during cutting, inclusions, banding, microsegregation, and hydrogen are contributing factors. The term used to describe this occurrence is thermal shock.

The discovery of this base material problem served to further complicate the potentially difficult task of welding thick plate.

Experimental Heavy Plate Welding

Early work on welding 2 in. thick material concentrated on development of multipass submerged-arc techniques. Multipass welding was desirable because once a procedure was established any thickness of material could be welded. Also, because multipass techniques deposit a smaller weld nugget and leave a smaller heat-affected zone, it was felt that the properties of multipass welded joints would be superior to 2 pass joints. However, obtaining a satisfactory weld in the 2 in. thick material proved to be a difficult task. Attempts to deposit weld beads in this material led to weld cracking of three distinct types. These were:

1. Centerline restraint cracking.
2. Centerline shrinkage cracking.
3. Underbead cracking.

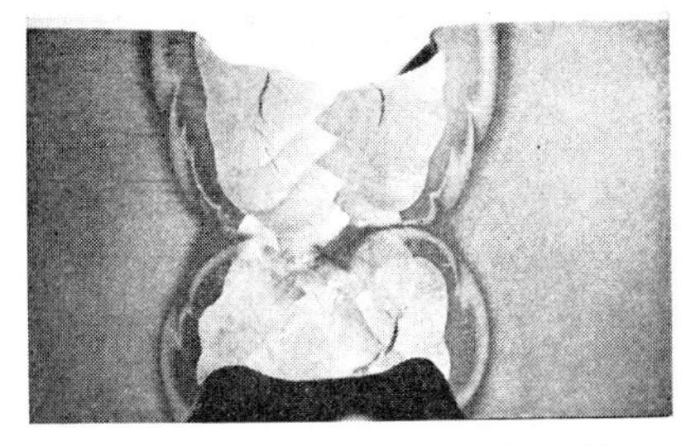

Fig. 21—Centerline weld metal cracking in multipass welds in 2 in. plate

An example of centerline cracking is shown in Fig. 21. Various methods, such as altering the basic joint design, addition of preheat, alteration of deposit chemistry, welding and holding above the M_s temperature, and oscillation of the welding head were all employed to combat the problem. No method or combination of methods successfully eliminated all three types of cracking.

At the same time that attempts were being made to multipass weld the 2 in. thick material, additional test results obtained from welding $^3/_4$ in. thick material proved some previous assumptions false; sufficient data had been gathered to prove conclusively that the properties of 2 pass weldments were superior to those obtained from multipass deposits. In light of this finding, a 2 pass submerged-arc weld was successfully executed in $1^1/_4$ in. thick material. The weld deposit was completely sound

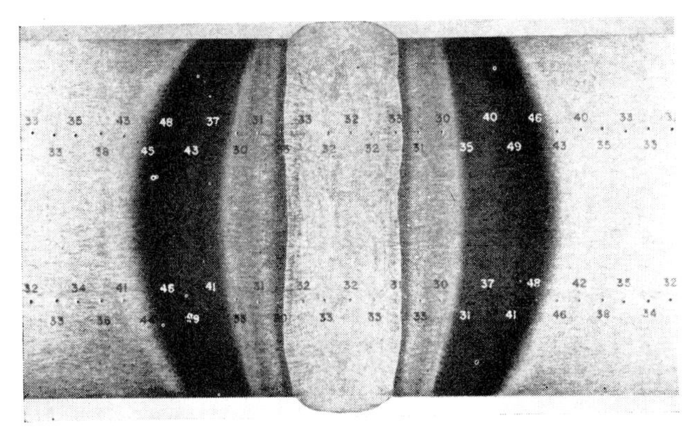

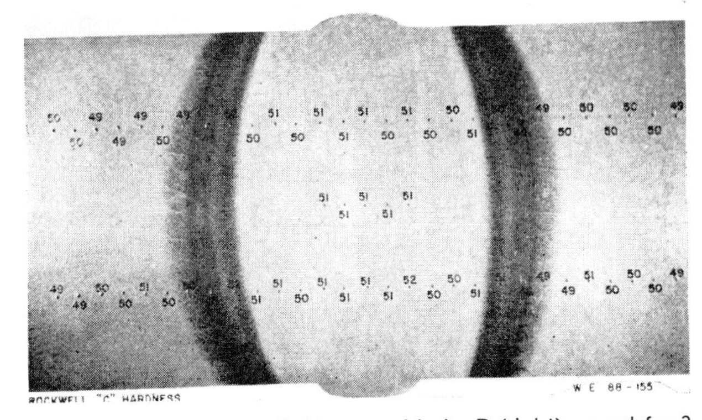

Fig. 22—Hardness survey conducted on a vertical submerged-arc weld in 3 $^3/_4$ in. plate. A (left)—as-welded; B (right)—aged for 3 hr at 900° F

and metallurgically identical to those made in $^{3}/_{4}$ in. thick material. The results were not surprising. They had been anticipated from experience gained from welding thinner material. However, a new problem had arisen in the form of base material delamination in the heat-affected zone. Again, the findings were not surprising as plasma cutting a piece of the $1^{1}/_{4}$ in. thick material had shown that the material was prone to delaminate.

Thus, an impasse was reached. The only proved successful method of obtaining a sound submerged-arc weld deposit was to fill the joint in 2 passes, but program requirements called for material up to approximately 4 in. thick to be welded. While 2 pass welding of this thickness is theoretically feasible, the method was not deemed practicable or even desirable in view of the heat-affected zone delamination problem; therefore, in order to circumvent the numerous problems, a new approach had to be taken.

Careful analysis of the situation narrowed the decision to two possible choices. The first involved using a conventional low heat-input, multipass, inert gas process which preliminary tests had already shown to be feasible. The second choice was a radical method that had never before been tried on this material involving welding in the vertical position. Economic considerations favored a welding process which could deposit metal at the fastest rate; ideally, this would be a continuous single pass process. Since the process is being applied to an increasing number of materials, it was decided that single-pass vertical submerged-arc welding would be the most desirable method for welding thick sections of maraging steel. Favoring this decision were the many inherent advantages of the process some of which are as follows:

1. The slow vertical welding speed allows heat to be conducted throughout the material far in advance of the slowly rising molten weld puddle, thus gently preheating the workpieces.
2. Any material thickness can be welded in a single pass.
3. The resultant weldment properties are comparable to those obtained by any other welding process.
4. The total time required to make a heavy plate weldment is much less than by any other procedure.

Vertical Submerged-Arc Welding

Once the welding process was chosen, a development program was undertaken to adapt standard submerged-arc equipment for vertical welding, obtain a welding composition, and establish welding conditions. With the help of the Arcos Corp., a trial weld was made in the vertical position which indicated that the concept was applicable to 18% Ni steel.

The following procedure was established for welds less than 18 in. long:

1. The workpieces are placed with the joint axis in the vertical position, separated by a 1 in. root gap, aligned, and fixed.
2. Starting and finishing blocks are tack welded to the assembly.
3. Water-cooled copper shoes are clamped across the open sides of the joint to contain the molten puddle.
4. The welding composition is added and the arc is started.

Once a welding composition was developed and welding conditions established, a number of laboratory and production welds were executed in material up to $3^{3}/_{4}$ in. thick and 9 in. high. All of the vertical submerged-arc welds executed to date have been of excellent quality. Weld metal and heat-affected zone cracking have been nonexistent. The weld metal cracking problem encountered in conventional welding of thick plate was eliminated because of the biaxial stress distribution and freedom of lateral movement in vertical welding. Lack of cracking in the heat-affected zone is attributed to the gentle preheating of the workpieces by the slowly rising molten weld puddle. It is significant to note that the same thick material which had cracked so severely on plasma cutting was successfully welded by the vertical submerged-arc process.

In addition, the weld deposits have been free of gas pockets and macro-inclusions. Defect-free welds typify vertical welding because the comparatively slow solidification rates facilitate the rising of slag particles and gas bubbles. Naturally, some micro-inclusions are found dispersed throughout the weld. An area of "worst" inclusions is shown in Fig. 25.

Hardness Survey

A hardness survey of a typical weld, in the as-welded and aged conditions, is shown in Fig. 22.

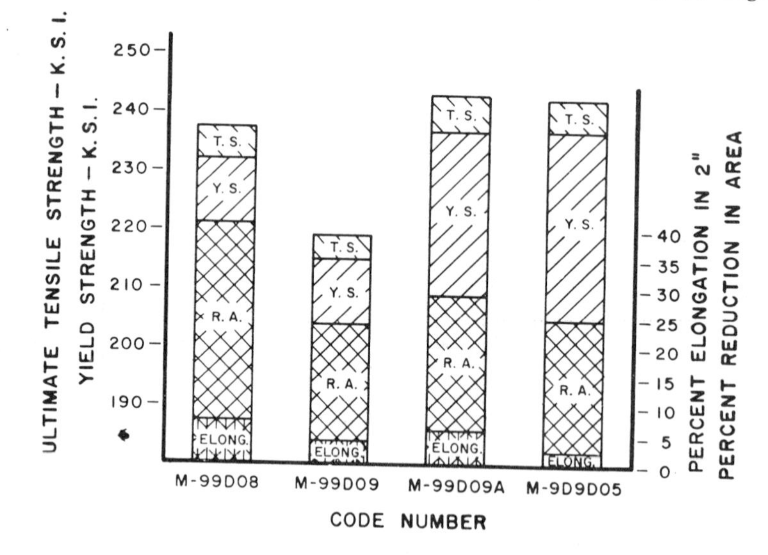

CODE NO.	COMPOSITION OF DEPOSITED WELD METAL				HEAT TREATMENT	FRACTURE LOCATION
	Ti	Mn	Si	AL		
M-99D08	.72	.07	.08	.12	I HR. AT 1550° F 3 HR.S AT 900° F	B. M.
M-99D09	.63	.06	.07	.07	3 HR.S AT 900° F	EYEBROW
M-99D09A	.75	.04	.05	.08	I HR. AT 1650° F 3 HR.S AT 900° F	EYEBROW
M-9D9D05	.43	.04	.38	.12	3 HR.S AT 900° F	EYEBROW
WIRE CHEMISTRY	1.20	.03	.01	.15	—	—

Fig. 23—Uniaxial mechanical properties and chemistry of vertical submerged-arc welds

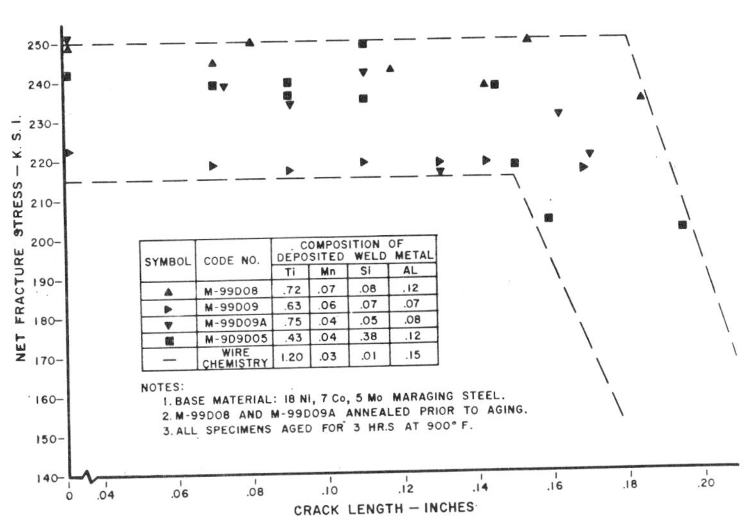

Fig. 24—Fracture toughness of vertical submerged-arc welds

Of interest, is the "dark eyebrow" region which does not completely respond to aging (R_c 45-48). This region has been subjected to the austenitic reversion temperature range (1100–1300° F), during welding, and some martensite has been reverted to metastable austenite; consequently, after aging, the hardness and the strength of this region is lower than either the weld or base material. Therefore, transverse tensile specimens fail in this region. The joint efficiency, however, is excellent.

Weld Properties

Chemistry. Weld deposit chemistries obtained from 2 flux types are shown in Fig. 23. The prime difference between the deposit chemistry of the 2 flux types are the Ti and the Si levels. The flux that was used to obtain weld M-9D-9D05 is a commerical flux, while the flux used to obtain the 3 other welds was specially developed by the authors' company for use with this alloy.

Mechanical Properties

The mechanical properties of test specimens obtained from representative weldments were excellent. The average results for testing elongated, 0.505 in. diam transverse round bar tensile specimens are graphically shown in Fig. 23. The data reveals that the weld strength is uniformly high, as no failures occurred within the weld. All test specimens failed either in the base material or the dark-etching heat-affected zone (eyebrow); therefore, a true test of the weld metal strength was not obtained. The one weldment, Code, M-99D09, which exhibited exceptionally low strength, failed in the eyebrow region, but a subsequent solution anneal after welding, (Code M-99D09A) restored the base material to full strength.

The ductility of the weldments is comparable to that obtained by use of conventional submerged-arc procedures in thinner materials.

Fracture Toughness

The fracture toughness properties of representative welds are graphically shown in Fig. 24. Noteworthy is the fact that the critical flaw lengths of these welds average approximately 0.165 in. long. This flaw size equals or exceeds that obtained from welds executed by the conventional submerged-arc process. The specimen used to obtain these results was the shallow, partial thickness crack configuration previously described as being employed for thinner materials. In all tests, the crack length was parallel to the weld axis.

Solution annealing after welding did not appear to affect the critical flaw size to any extent. Figure 24 is deceiving in that it does not denote the fracture location. Comparing welds M-99D09 and M-99D09A, the latter appears to possess superior properties, but specimens from weld M-99D09 failed in the heat-affected zone, and not in the weld metal fatigue crack, until a flaw length of 0.142 in. was reached. At this crack length and longer, the net fracture stress of both welds are approximately the same.

Also significant is the fact that the chemical composition of the weld metal did not seem to greatly affect the critical flaw length. Comparing welds M-99D09A and M-9D9D05, the latter weld deposit contains approximately × 7 the amount of Si and one-half the amount of Ti, yet the critical flaw size is approximately the same; therefore, it can be said that variations in silicon and titanium, within certain ranges, do not significantly affect the weld toughness.

Weld Microstructure

The microstructure of a typical weld is primarily martensite containing small pools of reverted austenite. It is believed that the austenite present is predominately

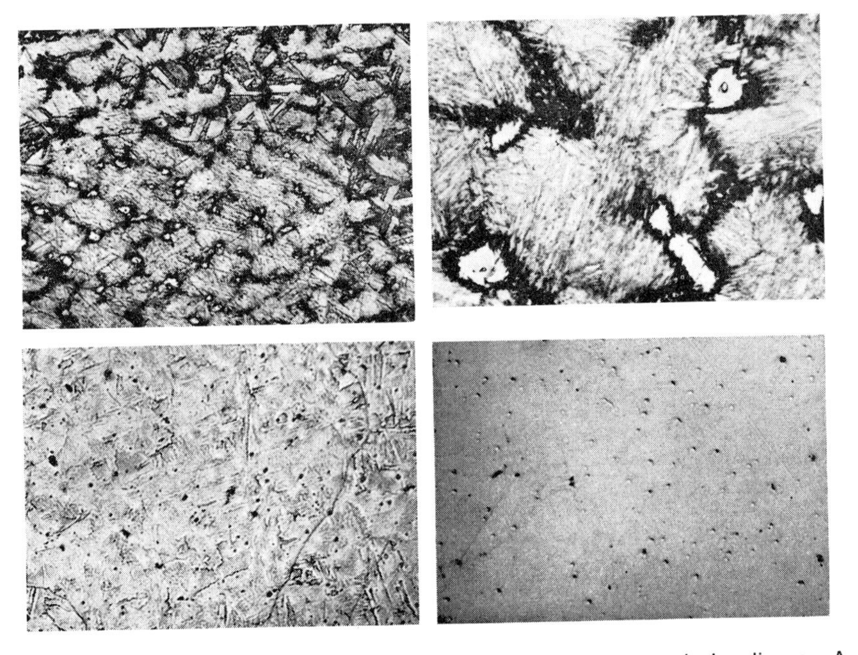

Fig. 25—Vertical submerged-arc weld microstructure, grain size and cleanliness. A (upper left)—aged 3 hr at 900° F; Fry's etchant; × 180. B (upper right)— aged 3 hr at 900° F; Kroll's etchant; × 500. C (lower left)—typical grain structure; mixed acids for etchant; × 100. D (lower right)—area of "worst" inclusions; no etchant. × 100. (Reduced 50% on reproduction)

reverted because examination of the microstructure prior to aging shows that little or no austenite is present. The dark etching areas in Fig. 25 represent coring of the weld metal and are not to be mistaken for grain boundaries. The grains themselves are extremely large and elongated as shown in Fig. 25. The grain size cannot be readily determined by conventional techniques.

Summary of Vertical Submerged-Arc Welding

The success of vertical submerged-arc welding was beyond what had been anticipated. Use of established techniques and procedure assures complete reproducibility of weld quality, soundness and appearance.

The results of uniaxial tensile and fracture toughness tests indicate that the properties of weldments, executed by the vertical submerged-arc process, equal or surpass those obtained by any other welding process.

Summary

The investigations reported briefly in this paper encompass but a small portion of the more important results which have been generated during a $2^1/_2$ yr period of intensive study of 18% Ni maraging steel. The initial selection of this material as a prime candidate for applications where very high strength, excellent toughness, ease of fabrication, good weldability and simplicity of heat treatment are needed, has been substantiated by our studies.

Selection of the submerged-arc process for joining of this alloy was based primarily on three factors:

1. The desire to capitalize on the base material's acceptance of high-heat inputs during welding in order to benefit from the economic advantages of a high-deposition rate process.
2. A knowledge of the characteristic soundness of properly executed welds made by this process.
3. The degree of confidence in the process which accrues from 25 yr experience in its use on a variety of materials.

The following conclusions can be made in summarizing the submerged-arc investigations performed.

1. Two pass procedures have been employed to join 18% Ni plate in the thickness range of $^1/_2$ to $1^1/_4$ in.

2. Vertical submerged-arc techniques are used for thicknesses above $1^1/_4$ in. Successful welds have been made in material $3^3/_4$ in. thick, and there is no upper thickness limitation on the use of this technique.

3. All of the welds display excellent soundness, which is highly desirable in a material expected to perform at high stress levels.

4. Using optimum welding parameters, a loss of approximately 50% of the filler metal Ti content is to be expected. Since deposit strength level is roughly proportional to Ti content, arc transfer losses must be compensated by the use of special analysis filler metal.

5. Submerged-arc deposits are prone to hydrogen-induced cracking, and strict control must be exercised to preclude entry of this element into the weld melt. Threshold limits appear to be approximately 5 ppm.

6. Welds in all plate thicknesses respond satisfactorily to the optimum base material aging cycle of 900° F for 3 hr.

7. Transverse weld tensile specimens fail routinely in the impaired response region of the base material heat-affected zones, due to the presence there of reverted, metastable austenite. When weldments can be solution annealed prior to aging, significantly higher joint strengths are realized due to the restoration of response potential in these areas.

8. Critical flaw length for weldments is roughly one-third to one-half that of the base material at equivalent stress levels. However, such flaw lengths are readily detectable by ultrasonic techniques.

9. Largely as a result of the impetus supplied by this program, a welding composition suitable for use with the submerged-arc process on 18% Ni maraging steel is commercially available.

10. When required, welds can be executed in aged material, using the same parameters employed for welding material in the solution-annealed condition. Localized heating of the weld and heat-affected zone to 900° F for 3 hr will result in raising these zones to the same strength level realized in furnace aging of weldments.

Acknowledgments

Much of the data on which this paper is based was generated by work performed under Thiokol Chemical Corp. subcontract. The authors are grateful for permission to use this material. They also wish to extend their thanks for the assistance given by U. S. Steel Corp., Linde Div., Union Carbide Corp., the Arcos Corp., and The International Nickel Co. in the form of materials and technical advice. Finally, the authors which to express their indebtedness to the many individuals in their own company without whose valuable assistance during the course of the research program, this paper would have been impossible.

References

1. Decker, R. F., Eash, J. T., and Goldman, A. J., "18% Nickel Maraging Steel," Trans. Am. Soc. Met., 55, No. 1, March 1962, pp. 58–76.
2. Decker, R. F., *et al.*, "The Maraging Steels," Materials in Design Eng., May 1962.
3. Yates, D. H., and Hamaker, J. C., "New Ultra High Strength Steels—The Maraging Grades," Metal Prog., Sept. 1962, pp. 97–100.
4. Smith, H. R., Anderson, R. E., and Bingham, J. T., "A User Evaluates Maraging Steels," Metal Prog., Nov. 1962, p. 103.
5. Irwin, G. R., "Fracture Testing of High Strength Materials Under Conditions Appropriate for Stress Analysis," Naval Research Laboratory Report No. 5486, July 27, 1960.
6. Pendleberry, S., and Yen, C. S., "Fracture Strength of High Strength Steels Containing Shallow Cracks," Douglas Aircraft Co. Engnr. Paper, Sept. 1, 1961.
7. Witherell, C. E., and Fragetta, W. A., "Weldability of 18% Ni Steel," Welding Journal, Research Suppl., 41 (11) 481-s to 487-s.
8. Floreen, S., and Decker, R. F., "Heat Treatment of 18% Ni Maraging Steel," Trans. Am. Soc. Met., Vol. 55, September 1962, pp. 518–530.

How to Weld Thick Plates of 18% Ni Maraging Steel

By PHILIP P. CRIMMINS *and* WALTER S. TENNER*

Engineers at Aerojet-General Corp. have defined welding procedures and filler metal compositions for joining thick plates of 18% Ni maraging steel. The work paved the way for the successful fabrication and hydrotesting of a 260 in. diameter rocket motor case.

MATERIALS AND WELDMENTS with high strength to weight ratios and sufficient fracture toughness to withstand service stresses in the presence of small defects are required for large diameter rocket motors now under development. To achieve this combination of properties in the 260 in. diameter solid rocket motor recently fabricated and hydrotested, we specified Grade 200, 18% Ni maraging steel plate and forgings with a yield strength of 200,000 to 235,000 psi.

Selection was based on data developed in previous Aerojet and Air Force† programs which indicated that if yield strength exceeded 240,000 psi, the fracture toughness of 18% Ni maraging steel plate and weldments was inadequate for the large chambers. However, while Grade 200 has adequate fracture toughness, it lacked a background of data on the tensile and fracture toughness properties of weldments and of the effects of varying filler wire composition and joint designs on these properties.

Before fabricating the 260 in. rocket motor (photos, p. 58), we had to establish an acceptable welding procedure and filler wire composition for joining plate 0.6 in. thick. The recommendations, judged from a thorough test program to determine aging response, tensile properties and fracture toughness of numerous weldments, are given in Table I.

The composition of the base metal (7 to 8% Co, 4 to 4.5% Mo, 0.05 to 0.25% Ti) was specified to produce a 0.2% offset yield strength of 200,000 to 235,000 psi. Both the parent metal and filler wire compositional limits were selected to provide maximum fracture toughness while still meeting a minimum 0.2% offset yield strength requirement of 200,000 psi. The lower alloy contents minimize segregation and the tendency toward banding which have caused delaminations in stronger compositions during forming, welding and cutting operations. The range of each alloying element was controlled to the minimum that could be maintained by steel producers. This reduces heat to heat variations in strength and fracture toughness.

Twelve Filler Metals Screened

Selection of the inert gas, tungsten arc process is the outcome of previous studies which revealed that it would produce sound weldments with superior fracture toughness in comparison with inert gas metal arc and submerged arc welding. In the evaluation to establish an optimum procedure and filler metal composition, 12 filler wires were screened, and this number was narrowed to the three (Table II) most likely to produce welds with yield strengths (0.2% offset) of 200,000 to 215,000 psi after aging. This range exceeds the minimum yield strength established as a

---

*Mr. Crimmins is supervisor, structural metals and metals joining section, and Mr. Tenner is manager, chamber materials and fabrication research and development department, Aerojet-General Corp., Sacramento, Calif.

†Contracts AF 33(657)-8740 and AF 04(695)-350.

Fig. 1 — Huge rocket motor chamber, 260 in. in diameter, was fabricated by Sun Shipbuilding and Dry Dock Co. for the Aerojet - General Corp.

Fig. 2 — Segments of head subassembly were joined in this fixture.

Fig. 3 — Inspectors radiograph welds in the head.

Fig. 4 — Girth welding fixture.

Fig. 5 — Crane lowers section of aging furnace over the motor chamber.

Table I — Recommended Procedures for Welding 0.6 In. Thick Grade 200 18% Ni Maraging Steel

Filler Metal
Composition: 0.03 C (max), 17.5 to 18.5 Ni, 7.5 to 8.0 Co, 3.6 to 3.8 Mo, 0.26 to 0.30 Ti, 0.10 Al, 0.10 Mn, 0.01 P, S, Si (max, each)
Melting practice: vacuum arc remelt or vacuum induction
Welding Process: inert gas, tungsten arc
Joint Design: single U or double V
Maximum Root Gap: 0.070 in.

Table II — Filler Wire Compositions

Designation	Composition, %									
	C	Ni	Co	Mo	Ti	Al	Mn	P	S	Si
A	0.010	18.4	7.5	4.4	0.29	0.09	0.02	0.008	0.008	0.04
B	0.026	18.4	8.0	4.2	0.35	0.06	0.02	0.005	0.007	0.033
C	0.003	18.7	4.2	3.5	0.45	0.30	0.02	0.004	0.004	0.020

goal (195,000 psi), yet it is not high enough to affect fracture toughness adversely.

As part of the preliminary tests, we used all 12 filler metals to make welds in single U and double V joints according to the conditions given in Fig. 6 and Tables III and IV. Table V lists properties of weld metal deposited by the three filler wires selected for further evaluation. As is evident, weldments in the single U design contained slightly higher titanium contents than those in the double V. This suggests that, as expected, base metal dilution is less in the single U joint. Both joint configurations are acceptable for fabricating large diameter chambers. However, the single U joint was used in producing most of the weldments for the 260 in. diameter chamber.

How Weld Metal Responds to Aging

Three of the selected filler wires were used to establish relationships between weld metal tensile properties and aging time and temperature and between properties of the weld deposit and the parent metal plate. Significantly, the weld metals responded to aging in the same way as the parent metal. Also, similar properties of the weld and base metals, obtained by aging the specimens in the as welded condition, attest that special heat treating procedures, such as intermediate solution annealing, are not required in processing welded components.

This accomplishment was considered a major milestone since it meant that the chamber could be assembled and aged as a single unit, as shown in Fig. 5, rather than by aging the individual components and then locally heat treating the weldments. The former approach is desirable because the components must be sized locally during girth welding to minimize mismatch. Such an

Table III — Welding Conditions for Single U Joints

Pass*	Amp	Volts	Wire Feed, In. Per Min
1	100	9	10
2	150	9	10
3	195	9	20
4	200	10	20
5	200	10	20
6-12	225	10	30

*Electrode, 1/8 in. thoriated tungsten, 1/16 in. diameter filler wire; travel speed, 8 in. per min; torch gas, argon and helium at 10 and 40 cu ft per hr, respectively; backup gas, argon at 15 cu ft per hr.

Table IV — Welding Conditions for Double V Joints

Pass*	Amp	Volts	Wire Feed, In. Per Min
1	325	10	8
2	250	10	20
3	225	11	25
4	225	11	25
5	230	11	25
6	230	11	25
7	250	10	10
8-12	240	11	25

*Electrode, 1/8 in. thoriated tungsten, 1/16 in. diameter filler wire; travel speed, 8 in. per min for passes 1 and 2, 20 in. per min for all others; torch gas, argon and helium at 8 1/2 and 35 cu ft per hr, respectively; backup gas, argon at 15 cu ft per hr.

Table V — Composition and Properties of Weld Metal

Filler Wire Designation*	Material	Joint Design	Composition, %				Strength, Psi		Elongation in 1 In., %	Reduction in Area, %
			Ni	Co	Mo	Ti	Tensile	Yield (0.2% Offset)		
A	Wire		18.4	7.5	4.4	0.29				
	Weld	Double V	18.2	8.7	3.8	0.25	224,300	215,000	13	53
	Weld	Single U	18.8	8.7	4.0	0.28	226,000	215,000	15	56
B	Wire		18.4	8.0	4.2	0.35				
	Weld	Double V	18.5	8.8	3.5	0.24	232,500	218,500	11	47
	Weld	Single U	18.0	8.5	3.8	0.26	226,000	219,000	16	56
C	Wire		18.2	4.3	3.5	0.45				
	Weld	Double V	18.5	4.4	3.2	0.37	224,100	206,800	14	56

*Base metal: 0.6 in. thick maraging steel containing 19 Ni, 9 Co, 3.5 Mo, 0.22 Ti. Properties listed were obtained by aging as welded specimens for 8 hr at 900 F.

Source: *Metal Progress*, May 1965

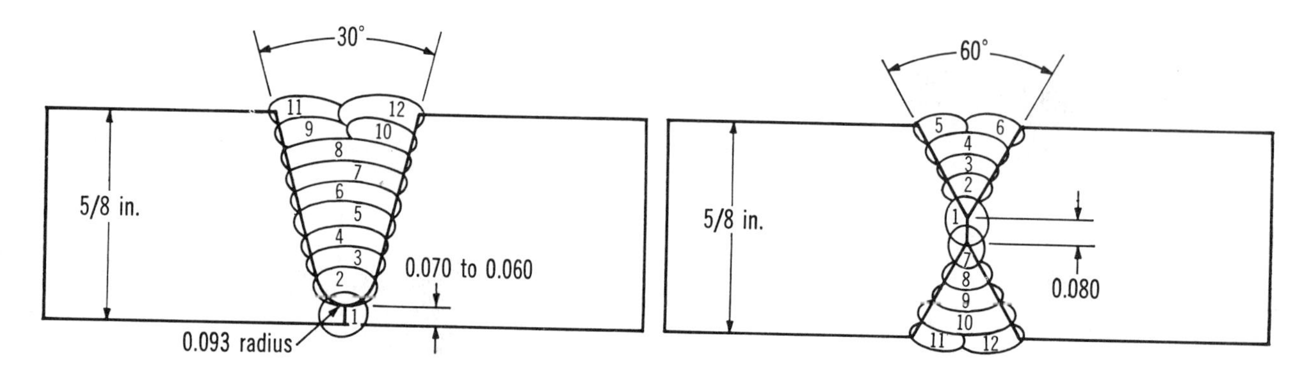

Fig. 6—Drawings of single U and double V joints illustrate welding sequence used in evaluation of the properties of welds in 18% Ni maraging steel and in fabrication of the 260 in. diameter motor chamber. Tables III and IV give the welding conditions.

operation would require expanding fixtures of significantly higher capacity when sizing components which had already been aged to a minimum yield strength of 200,000 psi. Also, the magnitude of increased residual stresses which would be anticipated when locally aging the weldment represented an unknown area as did their effect on subcritical flaw growth and performance of the chamber.

The general aging response of weld metal produced by filler metal B in Table II is typical of that experienced by the other two compositions. We related longitudinal tensile properties of welds, aging temperature (850, 900 and 950 F) and aging time (2 to 16 hr). When the deposit was aged at 850 F, its yield strength increased from 205,000 psi after a 4 hr aging time to a maximum of 221,000 psi after aging for 16 hr. At 900 F, yield strength went from 208,000 to 216,000 psi between 2 and 4 hr aging time, then remained relatively constant with aging times up to 16 hr. Overaging is encountered at 950 F when times exceed 4 hr.

Ductility of inert gas, tungsten arc weldments (filler metal B) is similar to that of the base metal at the same strength level. Elongations of 14 to 23% and reductions in area of 52 to 56% were observed; the higher strength levels accompanied the lower ductilities. Similar relationships between aging time and temperature were also exhibited by the tensile properties of weldments representing the other two filler wire compositions (A and C in Table II). Both produced yield strength levels in weld deposits which were slightly lower than provided by filler wire B.

In transverse weld tension tests, failure occurred in the weld metal and heat affected zone at stresses of 212,000 to 225,000 psi. Failures at these positions were expected since the filler wire compositions selected produced weld metal with yield strengths between 200,000 and 215,000 psi; the yield strengths of the base plate employed in the welding studies ranged from 215,000 to 242,000 psi after aging.

Fracture Toughness

Slow notch bend and precracked Charpy impact fracture toughness tests were em-

Table VI — Fracture Toughness and Critical Flaw Size of Welds

Defect	Flaw Depth to Length Ratio	Weld Metal From Filler Wire A*: Fracture Toughness (K_{Ic}), Psi $\sqrt{In.}$	Filler Wire A*: Critical Flaw Size, In. — Length	Filler Wire A*: Critical Flaw Size, In. — Depth	Weld Metal From Filler Wire B*: Fracture Toughness (K_{Ic}), Psi $\sqrt{In.}$	Filler Wire B*: Critical Flaw Size, In. — Length	Filler Wire B*: Critical Flaw Size, In. — Depth
Semicircular surface crack	0.5	102,000	0.26	0.132	90,000	0.22	0.110
Long shallow surface crack	0.1	102,000	0.53	0.053	90,000	0.42	0.042
Subsurface porosity	0.5	102,000	0.33†	—	90,000	0.26†	—

*Compositions are given in Table II. †Diameter.

How the Tests Were Made

Weldments evaluated in the testing program were made by joining two 12 by 14 by 0.6 in. thick test plates having machined edges. Degreased and then ground to remove scale, the plates were inert gas, tungsten arc welded immediately after the joint edges had been wiped with acetone to remove oil and other foreign materials.

Equipment and Procedures

As depicted in the drawing (right), the welding fixture included a rigid, water cooled, copper backup bar to which the plates were clamped. The bar was grooved to provide a means for introducing inert gas which was used to shield the root of the weld.

Precautions were taken to obtain fast weld cooling rates and to minimize the size of weld heat affected zones. Preheat and postheat were not used. Also, interpass temperature was not permitted to exceed 200 F. These are standard precautions taken in welding 18% Ni maraging steel plate to minimize austenite reversion in the heat affected and fusion zones.

Inert gas, tungsten arc welds were made with a Linde HWM-2 automatic welding head, equipped with a Linde HW-13 welding torch and a Linde EG103 governor controlled, cold wire feeder, and a 400 amp dc welding power source. Weld joint configurations and welding schedules are given in Fig. 6 and Tables III and IV. Welding and shielding conditions were established to give sound welds in accordance with applicable x-ray, ultrasonic and magnetic particle inspection specifications.

Tensile, slow notch bend and precracked Charpy impact tests were used to evaluate the aging response, tensile properties and fracture toughness of the weldments. From each welded plate we obtained tensile specimens (longitudinal and transverse), slow notch bend bars and precracked Charpy impact bars. The gage length of longitudinal tensile specimens consisted almost entirely of weld metal; that of the transverse bars extended across the welded joint and included the weld fusion zone, both weld heat affected zones and unaffected base metal. Fracture toughness specimens were machined transverse to the weld with the notch oriented in the weld metal or heat affected zones. To locate the notches, we etched the specimen and scribed it to mark the center line of the notch. The notches then were machined.

Machined from as welded plate, the specimens were age hardened in accordance with the selected aging cycle. Before the tests, surface scale on the tensile specimens was removed with abrasive paper and the fracture toughness specimens were subjected to cyclic stresses to form a fatigue crack at the root of the notch.

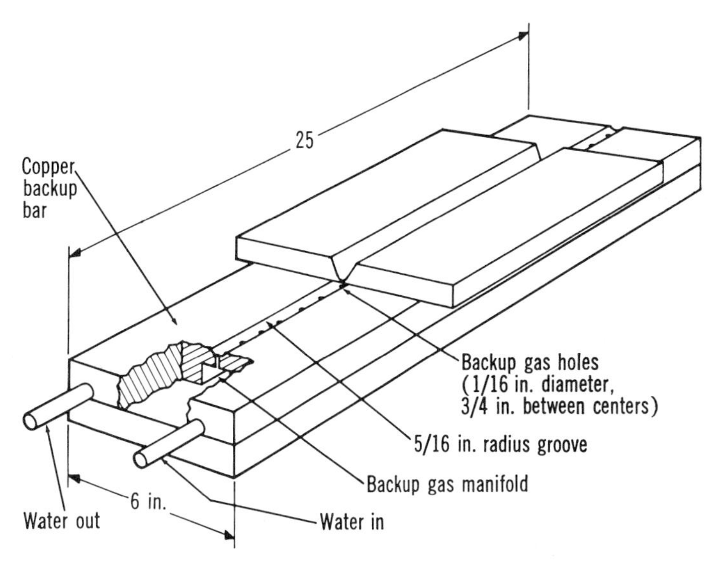

Evaluating the Data

Transverse tensile specimens were used primarily to determine the location of tensile failures. Tensile strength, yield strength, elongation and reduction in area were also recorded for the transverse tensile specimens. This data must be interpreted cautiously since transverse weld tensile specimens vary in thermal history and composition along the gage length. Longitudinal weld tensile specimens supplied data on the yield strength and ductility of the weld metals. The combined information from transverse and longitudinal tensile tests is necessary to evaluate the tensile properties of the weldments adequately.

We tested five slow notch bend specimens for each variable to determine qualitative plane strain fracture toughness, G_{Ic} and K_{Ic}, and the effects of aging cycle, filler wire composition and welding procedures on this property. Six precracked Charpy impact specimens provided quantitative data on the effects of each welding variable on the plane stress fracture toughness, G_c and K_c, of weldments. Duplicate impact specimens at each of three test temperatures (−40, room temperature and +200 F) established relationships between fracture toughness and test temperature. Specimens were tested in a subsize impact testing machine which provides energy readings with an accuracy of ±0.010 ft-lb in the low energy range of 1 to 5 ft-lb.

ployed in these studies to obtain qualitative plane strain (G_{Ic}) and plane stress (W/A) fracture toughness data. The plane stress and plane strain fracture toughness of weld metals varied for the different filler wire compositions and aging cycles investigated. These differences in fracture toughness may be related to the composition and tensile properties of the weld metals; however, reliable correlation cannot be established without additional data. In general, the fracture toughness of the weld metals and heat affected zones obtained with the selected filler wires came close to or exceeded those of the base metal in the transverse orientation after each of several aging cycles.

Weld metal deposited with filler wire A, which is considered representative of the filler wire composition recommended for the 260 in. rocket motor, exhibited slow notch bend fracture toughness (G_{Ic}) of 340 to 360 in.-lb per sq in. after being aged at 850, 900 and 950 F for 8 hr to a yield strength (0.2% offset) of 207,000 to 215,000 psi. These correspond to fracture toughness (K_{Ic}) values of 101,000 to 104,000 psi $\sqrt{in.}$ and critical surface flaw lengths listed in Table VI – assuming the indicated flaw shape and an operating stress of about 210,000 psi.

Weld metal deposited with filler wire B, had plane strain fracture toughness (G_{Ic}) values of about 260 in.-lb per sq in. at a yield strength level of 207,000 to 219,000 psi. This corresponds to a K_{Ic} of approximately 90,000 psi $\sqrt{in.}$ and the critical surface flaw lengths which are also shown in Table VI. Again, it is assumed that the operating stress will be about 210,000 psi.

Critical Defects Can Be Detected

A flaw shape corresponding to a surface flaw depth to length ratio of 0.1 is the most severe condition from a performance consideration because it represents a long, shallow defect. A ratio of 0.5 is the most critical in terms of nondestructive inspection capability since it represents a flaw of the shortest surface crack length which is normally encountered. As indicated by the data in Table VI, critical surface flaw lengths characteristic of filler wire compositions A and B and both defect configurations can be detected by magnetic particle, x-ray, dye penetrant and ultrasonic inspection. Also, the porosity size indicated in the table can be detected by x-ray inspection and is larger than the maximum size (0.060 in.) permitted by current x-ray criteria employed in the solid rocket motor program. Further, the critical surface flaw lengths of the weld deposits are approximately 0.06 to 0.1 in. smaller than those obtained for the base metal and heat affected zone. Judging from other studies, we consider the data in Table VI to be qualitative but conservative.

Establishing the Root Gap

Studies were also conducted to determine the effects of variable root gap on welding procedures and the tensile and fracture toughness properties of the weldments. Initially, plates machined from 0.6 in. thick 18% Ni maraging plate (0.02 C, 17.7 Ni, 7.9 Co, 4.2 Mo, 0.16 Ti) with a single U weld groove were welded to determine the maximum gap that could be tolerated before burn-through occurred. The joint, a 45° single U design with a 0.093 radius and a 0.050 to 0.060 in. thick root, widened gradually along its length from 0 to 0.187 in. Using weld schedules similar to those established as satisfactory for 0.6 in. thick plate, we found that successful welds could be made with root gaps of 0 to 0.070 in. However, when the gap widened to 0.100 in., severe burn-through became a problem. Consequently, with the procedures for welding the 260 in. diameter chamber, the maximum root gap that can be tolerated without encountering burn-through is approximately 0.070 in.

Three plates having a root gap of 0.070 in. extending the full length of each 12 in. joint were subsequently welded. Filler wire B was used to make the welds. All test specimens were tested in the as welded and aged (900 F, 8 hr) condition. The results of the tensile, precracked Charpy impact and slow notch bend fracture toughness tests revealed that the tensile strength and fracture toughness of the weldments were similar to or exceeded those obtained previously in evaluating double V welds made with the same filler wire. Also, the fracture toughness of the weld heat affected zone was similar to that of the base metal in both the longitudinal and transverse orientations.

Development of flux covered electrodes for welding 18% Ni–Co–Mo maraging steel

by C. Roberts

The development and testing of a flux covered electrode suitable for welding 18% Ni–Co–Mo maraging steel is described. A special flux has been developed to ensure low silicon pick-up in the weld bead, due to the reduction of silicates in the flux by titanium, and adequate titanium recovery. Tensile and impact tests have been carried out on butt welds produced in $\frac{3}{16}$ in. thick sheet and on all-weld-metal test pieces produced by butt welding $\frac{3}{4}$ in. thick material. Ageing for 3 h at 480°C. produced tensile strengths of 96·65 tonf/in² with impact values of about 10 ft-lb. From side fillet welds shear strengths approximately 0·4 of the tensile strength were obtained. Extensive radiography shows that sound welds can be produced over the range of thicknesses investigated. Adjustments in the core wire composition to obtain an electrode more suitable for use with the 90 tonf/in² grade of maraging steel are suggested.

THE 18% Ni–Co–Mo maraging steels have been shown to be readily weldable by conventional gas shielded processes.[1] Whilst this is generally the case some difficulties have been experienced in multipass welding by these techniques, particularly with MIG welding where lack of fusion, lack of penetration and inter-run porosity have been encountered. Because of this, multipass welding by the MIG process would not be recommended and far superior welds have been produced using TIG welding methods.

Quite extensive programmes of work have been carried out on submerged-arc[2,3] and electron beam[4–6] welding of these materials. Whilst sound welds have been produced their toughness has been found to be inferior to that of welds produced by the gas shielded processes. To date little attention has been given to the use of flux covered electrodes for welding maraging steel.

With the gas shielded processes, almost complete transfer of the alloying elements occurs in the arc; therefore the wire can be chosen to match the plate except in regard to titanium the minimum level of which is about 0·5% in the wire so as to ensure thorough deoxidation of the weld pool and eliminate porosity and cracking.[7] With submerged-arc and flux covered electrodes greater losses are likely to occur both in the arc and to the flux, particularly of titanium, due to its high volatility and strong reducing action, whilst pick-up of both desirable and undesirable elements from the flux can occur. The problem of producing a satisfactory flux covered electrode is therefore to control the recovery of titanium in the weld deposit and the pick-up from the flux of elements (*e.g.* silicon) likely to reduce the weld metal toughness. Witherall *et al*[7] recommend a titanium level of 2–2·5% in the core wire for flux covered electrodes to obtain complete deoxidation of the weld pool and sufficient recovery for hardening.

Starting with a core wire containing about 2·5% titanium an investigation was carried out into the development of a suitable flux to produce a weld deposit suitable for use with the 90 tonf/in² grade of maraging steel.

The development and testing of the experimental electrodes produced is reported below.

Manuscript received 22nd June, 1967.

The author is at the Military Engineering Experimental Establishment, Christchurch, Hants.

1178

Experimental

Materials. Welding trials were carried out on air melted 90 tonf/in² grade material of various thickness between $\frac{3}{16}$ and $\frac{3}{4}$ in. whilst two casts of induction melted material drawn down to 8 and 10 gauge were used as core wires. The analyses of these, a typical MIG wire and the parent material are given in Table I.

Flux. The following series of specially prepared coatings, highly basic in nature, have been developed and tested:

(a) Standard lime fluorspar with additions of ferro-titanium, ferro-molybdenum and nickel powder to compensate any anticipated losses
(b) Lime, crysalite, rutile with ferro-titanium, ferro-molybdenum and nickel powder
(c) Traditional lime titania with ferro-titanium, ferro-molybdenum and nickel powder
(d) Lime titania with reduced titania content and additions of ferro-titanium, ferro-molybdenum and nickel powder
(e) As (d) but with no ferro-titanium, bonded using dextrine
(f) As (e) but bonded with dilute silicate

Table I

Analyses of materials

Element %	*Parent Plate*	*10 gauge Wire*	*8 gauge Wire*	*$\frac{3}{64}$ in. MIG Wire*
C	0·028	0·008	0·017	0·017
Mn	0·03	0·05	0·05	0·03
Si	0·16	0·02	0·05	0·04
S	0·01	0·008	0·004	0·01
P	0·007	0·004	0·003	0·003
Ni	17·8	18·6	18·88	17·72
Ti	0·24	2·68	2·41	0·51
Co	7·98	8·9	8·68	6·00
Mo	3·30	5·02	5·04	3·47
Cr	—	—	—	—
Al	0·055	0·18	0·22	0·07
Bo	0·003	0·0002	0·0003	0·0002
Pb	0·002			
Zr	0·010	trace	trace	

Table II

Weld bead analyses

Deposit	*Composition, %*						
	C	*Mn*	*Si*	*Ni*	*Ti*	*Co*	*Mo*
MIG Argon	0·01	0·020	0·010	17·8	0·38	7·20	3·45
MIG Argon 1% oxygen	0·01	0·020	0·010	18·1	0·43	7·30	3·45
Electrode a			0·60	19·8	0·62		4·9
,, d			0·57	19·82	0·37		5·0
,, e			0·25		0·25		
,, f			0·30		0·25		
,, g	0·031	0·03	0·09	19·30	0·29	9·00	4·55
,, h	0·028	0·04	0·15	19·10	0·28	8·85	4·65
,, P.G.	0·03	0·04	0·13	18·9	0·24	9·20	4·80

(g) Lime titania but with more basic ingredients and bonded using normal silicate but no ferro-titanium

(h) As (g) but with a higher content of fluoride-bearing minerals

Production G: As (g) but with no ferro-molybdenum and no nickel powder.

Welding. Conditions for welding were determined by depositing a bead on a plate. Both a.c. and d.c. power supplies were used, but generally d.c. (electrode +ve) gave the better control.

Butt welds were produced in $\frac{3}{16}$ in. plates prepared with a single 60°V, a $\frac{1}{16}$ in. root face and no root gap. In each case a single run and backing run were used, the backing run being put down after grinding out the root of the original run to give a balanced weld. Welding was carried out without any preheat using a current of between 90 and 120 amp, and sealing runs were made after allowing the joint to cool down to room temperature.

Welding behaviour. Bead-on-plate tests indicated the (a) flux to have superior running to (b) whilst (c) did not show satisfactory running either. This was attributed to a large amount of titanium in the wire being oxidised during welding resulting in a slag very rich in titania. With a view to rectifying this trouble coating (d) was produced with which reasonable running was obtained. Further testing was therefore carried out using (a) and (d). Both these gave rather sticky pools which were

Table III

Mechanical properties of parent plate and butt welds

	Hardness Hv Unaged	*Hardness Hv Aged*	*Tensile Strength, tonf/in²*	*Charpy Impact (10 × 5 mm), ft–lb*	*Allison Bend Parameter 10^4 lb/in²*
Parent Plate	320	480	85·1	16	N.F.C.
MIG Weld	320	490	86·5	11	N.F.C.
a	370	580	87·3	—	—
d	360	560	84·0	—	—
e	360	540	86·4	—	—
f	360	500	63·0	—	—
g	360	530	84·8	6	0·57 min.
h	360	530	87·5	7	0·75 min.

Table IV

Fatigue of maraging steel

	Cycles to Failure, 8–48 tonf/in²					*Average*
Parent Plate	19,053	14,840				16,947
MIG Weld	2,853	2,483	3,099	1,727	1,682	2,369
a	4,180	1,560(S)	7,450(S)			4,400
d	1,460	2,140	4,220(S)			2,600
e	4,260	4,550	7,160			5,320
f	3,460	2,280(S)	2,870(S)			2,870
g	6,050	7,400	5,450	7,520	7,810	6,770
	6,070	7,120				
h	7,370	6,900	4,770	5,600	4,190	5,550
	4,450					

All failures occurred at toe except those marked (S) where weld failures occurred due to slag entrapment.

difficult to handle and resulted in undercutting. With practice however a reasonable weld could be laid down which slagged readily but radiography revealed extensive slag entrapment, and analyses indicated the weld beads to have high silicon and titanium contents (Table II).

Coatings (e) and (f) were made with the object of reducing the titanium recovery by omitting the ferro-titanium and reducing the silicon in the weld bead by reducing the available silica in the coating. Whilst deposits made with these electrodes were more fluid than the previous ones tested some slag entrapment was still obtained and (f) showed a tendency to cracking. Analyses revealed titanium contents of the correct order but silicon was still somewhat high. Further drawbacks with these coatings were that they were not sufficiently robust and were not impermeable enough. The latter resulted in a peculiar oxidation effect in the core wire at the tip of the electrode heated by the arc and was probably the cause of cracking observed in deposits made using coating (f). These were superseded by (g), another lime titania electrode but more basic than (d) and bonded with normal silicate, and (h) a modification of (d) with higher fluoride content. Electrodes produced with both these coatings ran well, slagged easily and produced sound welds. Analyses revealed higher cobalt, molybdenum and nickel than in the parent material and that nickel was higher in the bead than in the original wire. The latter was probably due to pick-up from the flux. Similarly, some pick-up of molybdenum probably occurred and omission of these elements from the flux would probably result in a bead nearer to the plate analysis. Of (g) and (h) the former had a lower silicon level; therefore a modified form of this termed Production

Table V
Tensile properties of butt welds in maraging steel

Welding Technique	*Condition*	*Ageing Treatment*	*0·2% Proof Stress, tonf/in²*	*Max Stress, tonf/in²*	*Elong., %*	*Gauge Length in.*	*Failure*
MIG Argon Shield	Undressed	88 h at 460°C.	85·7	92·0			W
			83·3	91·5			T
			86·7	93·2			T
MIG Argon 1% Oxygen Shield	Undressed	3 h at 480°C.	88·9	91·5	9·5	2	P/P
	,,	,, ,,	85·7	90·3	9·0	2	P/P
	,,	Unaged	59·0	66·7	10·0	2	P/P
	Dressed	Unaged	65·1	71·8	4·5	2	W
	,,	3 h at 480°C.	90·9	95·5	4·0	2	W
	,,	,, ,,	90·4	92·2	1·0	2	W
Flux Covered Electrodes	Undressed	3 h at 480°C.	89·1	92·6	8·0	2	P/P
	,,	,, ,,	81·1	92·6	3·0	2	T
	,,	Unaged	62·0	69·0	10·5	2	P/P
	Dressed	Unaged	59·0	64·3	2·0	2	W
	,,	3 h at 480°C.	91·6	97·8	2·5	2	W
	,,	,, ,,		82·3	1·0	2	W
All-weld-metal specimens produced using flux covered electrodes	Standard specimens to BS.639	Unaged	39·0	40·1	2·0	2	W
		3 h at 480°C.	95·5	96·65	1·5	2	W

T = toe failure; W = weld metal failure; P/P = parent plate failure

(G) was prepared and tested in detail even though it was realised that the cobalt content of the resulting deposit would be unaffected and could only be appreciably reduced by adjustment of the core wire.

Tee fillet welds have been produced with the minimum of undercutting and good penetration with both (G) and (h) fluxes.

Since there is such a great difference between titanium levels in the wire and deposit the possibility of high local titanium contents may exist which could result in hard spots and brittle areas. Microprobe analysis was carried out to determine the degree of segregation in weld deposit (h). Ultra slow scans for titanium were made along 350–400 micron lines across the interfaces with the weld and both plates. The plate material showed a fairly uniform titanium distribution (0·1–0·25%) while variations in the weld bead were between 0·2–0·6% but some particles were noted of the order of 10 microns diameter which contained greater than 1·0% titanium. The general silicon level was below the detection limit of about 0·1% and no silicon-rich particles were observed. For comparison, a scan was carried out on a MIG weld and a general titanium variation of 0·1–0·2% and 0·2–0·6% was observed in the parent plate and weld bead respectively, whilst levels of up to 1% were noted in particles of the order of 5 microns diameter. In both cases no significant variation in titanium content was observed along the length of the weld bead.

The results of mechanical and fatigue testing of these electrodes are presented along with those of parent plate and MIG welds in Tables III and IV. From Table III it will be noted that, whilst the tensile strength is about the same in all cases, the hardness of aged welds is much higher for the flux covered electrodes and the toughness is correspondingly reduced. The better fatigue results obtained with flux covered electrodes is probably due to the better weld profile and reduced amount of undercutting. Improved fatigue lives were often obtained even when slag had been trapped in the weld.

Table VI
Results of tensile tests on fillet welds

End Fillets		*Side Fillets*	
MIG Welds tonf/in²	*Flux Covered Electrode tonf/in²*	*MIG Welds tonf/in²*	*Flux Covered Electrode tonf/in²*
51·0	30·0	31·1	35·4
65·7	43·8	32·2	36·4
73·0	51·0	36·8	36·5
76·0	56·4	42·9 (unaged)	38·8

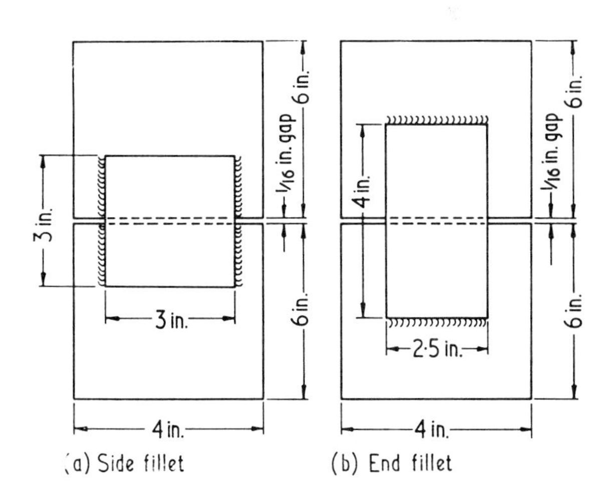

1—Specimen geometry of end and side fillet welds

Evaluation of the welding behaviour and properties of electrode Production G

Butt welds. Butt welds were produced in $\frac{3}{16}$ in. plate, as previously described. From these were produced tensile specimens with the weld at the centre of the gauge length and perpendicular to the direction of stressing.

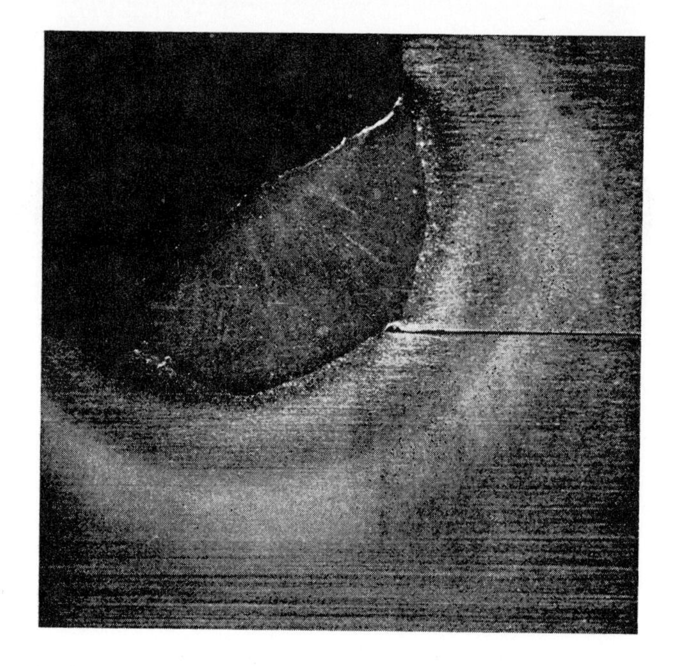

2—Section of weld produced in cold cracking test. Etched 10% nital

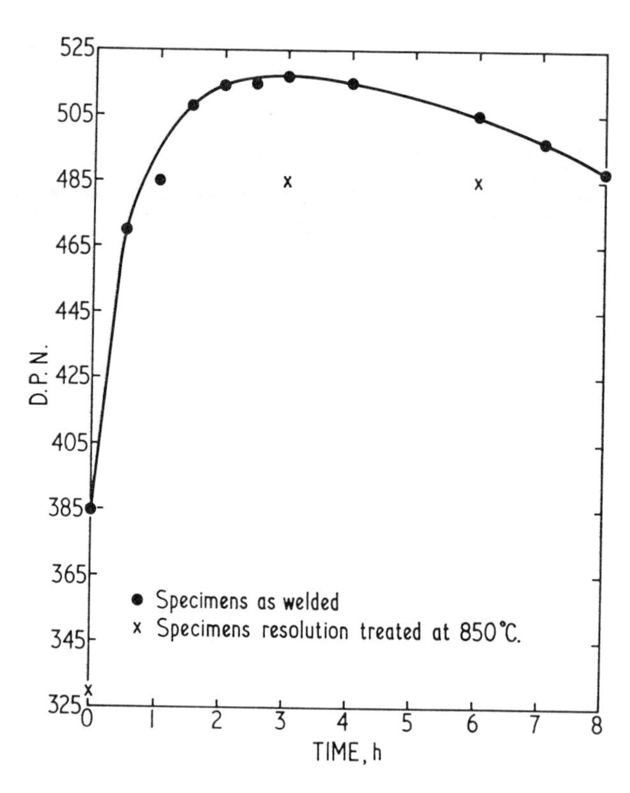

3—Ageing curves for all-weld-metal test specimens

Some of these specimens were tested as-welded, unaged and aged whilst others were tested after any weld bead reinforcement had been removed by dressing the weld flush with the parent material or by completely removing any undercutting. The results obtained in these tests are given in Table V along with those obtained with MIG welds.

Fillet welds. The results of tensile tests on side and end fillet welds of geometry shown in Fig. 1 are given in Table VI.

To establish the behaviour of this electrode under conditions of stress, hot and cold cracking tests were carried out in accordance with BS. 709 (Parts 8 and 9) except that 10 s.w.g. electrodes were employed instead of 6 s.w.g. as recommended.

Hot cracking test. After the welds had been allowed to cool the slag was removed and the surface examined for visible cracks. The weld was then broken open and the fracture examined for evidence of oxidation or temper colouring. Neither of these were observed, which would indicate the absence of hot cracking.

Cold cracking test. The test was carried out according to the specification and sectioned after about 120 h. The sections were polished initially on silicon carbide papers then finished using diamond paste down to 0·1 micron grade. After etching with 10% nital, examination was carried out using the optical microscope. No cracking was observed in either the HAZ or the weld bead on any of the sections examined. A typical section is shown in Fig. 2.

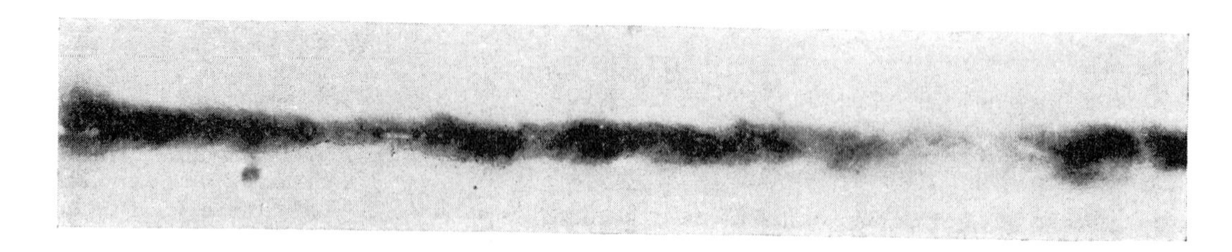

4a—Radiograph of butt weld produced using early electrode, showing slag entrapment

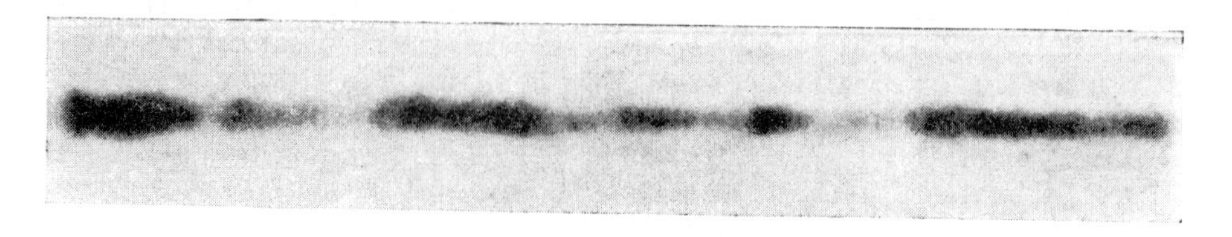

4b—Butt weld produced using electrode (G)

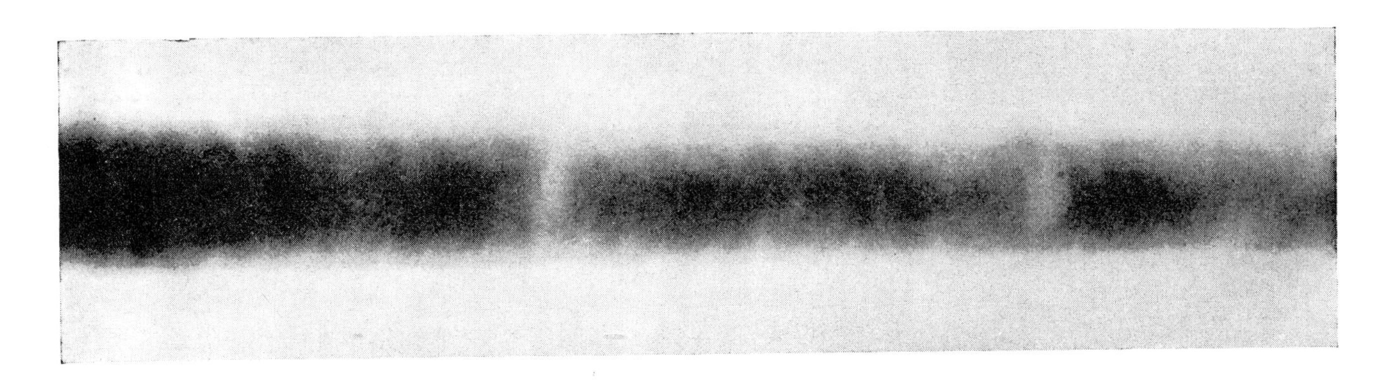

5—Butt weld produced in ¾ in. thick material from which all-weld-metal tensile pieces were taken

6—All-weld-metal ring

All-weld-metal tests

All weld test specimens were produced in accordance with Appendix D of BS. 639. Details of the mechanical properties obtained in the aged and unaged conditions are given in Table V, whilst the ageing behaviour is shown in Fig. 3.

Radiography. Many of the welds produced have been radiographed for soundness. Figure 4*a* shows a radiograph of a butt weld produced on $\frac{3}{16}$ in. thick plate which shows some slag entrapment. This can be compared with a similar weld produced using electrode G shown in Fig. 4*b*. A butt weld in ¾ in. thick plate used for producing all-weld-metal test specimens is shown in Fig. 5. Figure 6 shows an all-weld-metal ring ¼ in. wide produced from a pad of weld metal built up on a piece of maraging steel plate. This was produced to examine the electrical and magnetic properties of the weld metal.

Discussion

Production of a flux covered electrode suitable for use with maraging steels necessitates the use of a highly basic flux in order to minimise the silicon pick-up in the weld bead. Whilst highly basic fluxes have been developed in a granulated form, for submerged-arc welding of these materials[2,3] some complications are introduced by the properties the coating is required to exhibit. The main requirements are that it should have good adhesion, be robust, and in the present case should be impermeable. Normally the choice of binder lies between dextrine and silicates. With the former, pick-up of carbon in the weld is likely, which will tend to reduce the toughness by the formation of titanium carbides at the grain boundaries,[8,9] whilst silicon pick-up from the latter due to the reduction of silicates by titanium also reduces the weld metal toughness. Neither coatings appear entirely satisfactory due to the reduction in toughness produced and the choice was based on the superior properties of the silicate coating. Fluxes were therefore developed with the aim of keeping the silicon pick-up in the weld pool as low as possible. It has been recommended[10] that the Si + Mn content should not exceed 0·25%. Initially, flux development was aimed at achieving the correct titanium and silicon contents with an electrode that could be readily handled. At this stage more complete analysis revealed that the additions of nickel and ferromolybdenum to the flux were not necessary and they were omitted from the production G electrode. Comparing the analysis of production G deposit, taken from an all-welded-metal test piece, with the analysis of the 8 gauge wire it can be seen that complete transfer of nickel and manganese was obtained. Some increase in the cobalt was observed which cannot be explained, there was some silicon pick-up as expected and carbon increased probably due to the reduction of carbonate present in the flux. Losses to the flux are shown by only titanium and molybdenum with 10% and 95% transfer. Using these figures and tensile data it should be possible to adjust the core wire composition to give a better match with the parent material.

The presence of cobalt in this alloy results in acceleration of the precipitation reaction by increasing the super saturation of molybdenum.[11,12] It would appear that adjustment of only the molybdenum content of the core wire is necessary and a value of about 3·5–4·0% would appear to be the correct level.

The microsegregation observed by microprobe analysis indicates that whilst local variations exist on a microscale in these welds there is no evidence to suggest these variations were due to coring. Similar microsegregation has been observed over areas of a few microns in submerged-arc welds.[13]

Using the bead analysis the likely properties of the weld can be assessed. A guide to yield strengths can be

Table VII
Calculated yield strength, hardness and impact values for various materials

Material	*0·2% P.S. from A K.S.1.*	*0·2% P.S. from B K.S.1.*	*Average 0·2% P.S. (A + B)/2*	*Hardness As Welded, Hv*	*Hardness Aged 3 h. at 480°C., Hv*	*Charpy Impact ft–lb*
Parent Plate	200·6	204·1	202·4	300–360	460	30
MIG Weld Argon Shield	208·6	212·7	210·6	330	475	16–24
MIG Weld Argon 1% Oxygen Shield	212·5	216·9	214·7	330	475	16–24
Electrode g	249·1	245·7	247·4	340	522	12
Electrode h	249·7	245·9	247·8	340	524	14
Electrode P.G.	254·1	248·6	251·3	384	517	12–8
Vascomax Air melted	—	—	250 Nominal	280	520	18
Vascomax Vacuum melted	—	—	250 Nominal	280	520	22
Casting Vacuum melted	—	—	250 Nominal	280	520	20

A. 0·2% P.S. (K.S.1.) = 15·1 + 9·1 (% Co) + 28·3 (% Mo) + 80·1 (% Ti)
B. 0·2% P.S. (K.S.1.) = 38·1 + 8·8 (% Co) + 22·6 (% Mo) + 87·7 (% Ti)

obtained by a multiple regression analysis of the variation of 0·2% proof stress with chemical analysis. Two equations have been suggested:[14,15]

$$0{\cdot}2\%\ \text{P.S. (tonf/in.}^2) = 15{\cdot}1 + 9{\cdot}1\ (\%\ \text{Co})\ 28{\cdot}3\ (\%\ \text{Mo}) + 80{\cdot}1\ (\%\ \text{Ti})$$

$$0{\cdot}2\%\ \text{P.S. (tonf/in.}^2) = 38{\cdot}1 + 8{\cdot}8\ (\%\ \text{Co}) + 22{\cdot}6\ (\%\ \text{Mo}) + 87{\cdot}7\ (\%\ \text{Ti})$$

Calculations made using these equations together with hardness and impact data are presented in Table VII. From these it will be observed that the theoretical 0·2% P.S. of the flux covered electrodes is not realised and that the impact values are low even for material of the theoretical strength. Possible reasons for this low toughness are the structure of the weld and the possibility of grain boundary segregation. The weld is essentially a cast structure in which large columnar grains exist, many with their major axis perpendicular to the surface. Higher heat inputs result in coarser weld pool structures and reduction in the toughness.[16] Fracture has been observed to occur along either interdendritic boundaries or grain boundaries and is said to be associated with an unidentified phase.[17]

Electron fractography has revealed that a direct correlation exists between toughness and the size and distribution of particles on a fracture face.[18] Clean fractures are generally associated with high toughness. Figures 7 and 8 show carbon extraction replicas taken from fracture surfaces of the parent plate and a weld produced using electrode (G). A much higher particle density is observed on the latter fracture. These particles are not yet identified. Similar particles have been observed on the fracture surfaces of welds produced by other processes and have been identified as titanium nitrides, carbides, or sulphides.[19–23] Where these particles have been observed to form networks at the prior austenite grain boundaries their presence has been most deleterious.

It has also been suggested[23] that an excessive amount of titanium molybdenum rich inclusions may explain why the maraging steels do not reach their potential strength and this could be a contributory factor in the present case.

Using the present electrodes the welds overmatch the 90 tonf/in^2 grade parent material with the result that good tensile strengths are obtained but associated with these are low impact values. The impact values obtained from parent material and a casting of the same strength level are given in Table VII. The value quoted for the casting would suggest that, since the weld is essentially a small casting, impact values of the same order of magnitude should be obtainable. However, the casting quoted was firstly vacuum melted and secondly had been re-solution treated at 1150°C. and rapidly cooled to prevent the precipitation of titanium carbides. Both

7—Charpy V-notched specimen of parent plate. Carbon extraction replica

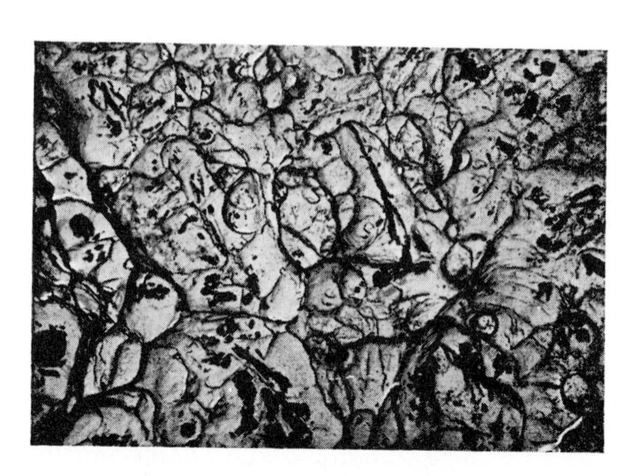

8—Charpy V-notched specimen fracture in weld produced using flux covered electrodes. Carbon extraction replica

these treatments would be expected to improve the toughness considerably. Such a treatment is not feasible with welded joints but by ensuring rapid cooling of the weld through the temperature range 760–980°C. precipitation of titanium carbide should be prevented and a tougher weld obtained.

Conclusions

A flux suitable for the production of covered electrodes for welding maraging steel has been developed. Using this flux with a core wire of the correct composition it should be possible to obtain an electrode with which a deposit of the desired strength level can be obtained. The electrode produced in the present work had a tensile strength higher than that required for welding the 90 tonf/in² grade of maraging steel. It is therefore proposed that molybdenum in the core wire should be reduced from 5% to 3·5%. This would result in a deposit of about the correct strength, which should have greater toughness than has been obtained with the present core wire. The improved weld bead profile obtained with these electrodes resulted in improved fatigue life, compared with similar MIG welds, which could be very important in view of the generally poor fatigue behaviour of maraging steels.

Acknowledgments

The author acknowledges the assistance of Metrode Products Ltd. in the development of fluxes, and I.N.C.O. and Bristol Aerojet for assistance with analysis and helpful discussions.

References

1. D. A. Corrigan: *Welding J.*, 1964, vol. 43, p. 292*s*.
2. E. M. Wilson and A. I. Wildman: *Brit. Welding J.*, 1966, vol 13, p. 67.
3. F. D. Duffey and W. Sutar: *Welding J.*, 1965, vol. 44, p. 251*s*.
4. T. Boniszewski and D. M. Kenyon: *Brit. Welding J.*, 1966, vol. 13, p. 415.
5. C. M. Adams and R. E. Travis: *Welding J.*, 1964, vol. 43, p. 193*s*.
6. M. L. Kohn and C. D. Schaper: *Metal Progress*, 1964, vol. 85, p. 93.
7. C. E. Witherall, D. A. Corrigan and W. A. Peterson: *Metal Progress*, 1963, vol. 84, p. 81.
8. "Influences of impurities and related effects on strength and toughness of high strength steels", paper at ASM Conf. Feb. 1964.
9. W. J. Jackson: Final report M.O.A. agreement No. PD/23/039, Brit. Steel Castings Res. Assn., Dec. 1966.
10. Lab. rep. D.4836 No. 1, International Nickel Co., 1965.
11. D. T. Peters and C. R. Cupp: *Trans. A.I.M.E.*, 1966, vol. 236, p. 1421.
12. G. P. Miller and W. I. Mitchell: *J. Iron Steel Inst.*, 1965, vol. 203, p. 899.
13. B. G. Reisdorf, A. J. Birkle and P. H. Salmon-Cox: Tech. rep. A.F.M.E., T.R.65–364, Nov. 1965, Air Force Systems Command, Wright Patterson Air Force Base, Ohio.
14. S. Floreen and R. F. Decker: *Trans. Quart. A.S.M.*, 1962, vol. 55, p. 518.
15. Rep. Contract A.F.33 (657)—8740, Aerojet Corp., Sacramento, Calif., April 1964.
16. J. B. Tobias and G. K. Bhat: Mellon Inst. W.R.L. Contract Non R 4595, Feb. 1965.
17. Z. P. Saperstein and B. V. Whiteson: Rep. No. S.M. 45959. indep. R & D programme, Douglas Aircraft Co., April 1964.
18. A. J. Birkle, D. S. Dabkowski, J. P. Paulina, L. F. Porter: *Trans. A.S.M.*, 1965, vol. 58, p. 285.
19. T. Boniszewski and E. Ledieu: *J. Iron Steel Inst.*, 1966, vol. 204, p. 360.
20. T. Boniszewski and D. M. Kenyon: Rep. C139H/1/65, Brit. Welding Res. Assn., 1965.
21. B. G. Reisdorf, A. J. Birkle and P. H. Salmon-Cox: Proj. No. 89 025–022 1/1/65—31/3/65, US Steel Corp. App. Res. Labs., 1965.
22. Contract N.A. 57–214 WOO–PR–63–172, Douglas Missile and Space Divn., Feb. 1964.
23. "Vascomax 250 and 300—18% nickel ultra high strength maraging steels", Vanadium-Alloys Steel Co., Latrobe, Pa., USA.

Fracture toughness and microstructure of the heat-affected zone of a welded maraging steel

The fracture toughness and microstructure of the heat-affected zone of welded 300 grade maraging steel have been studied. A wide range of welding processes has been covered using a thermal simulation technique. This approach has been justified by correlating real and simulated heat-affected zone properties and structures. Fracture toughness has been found to be a function of peak temperature, but independent of heating and cooling rates. Increases in fracture toughness were found and related to peak temperature, specific values being 75 $MNm^{-3/2}$ (67·8 ksi$\sqrt{in}$) at 650°C and 66–77·5 $MNm^{-3/2}$ (59·6–70·1 ksi$\sqrt{in}$) over 1 200°–1 400°C, as compared with the parent material value of 60 $MNm^{-3/2}$ (54·2 ksi$\sqrt{in}$). These increases in fracture toughness were accompanied by decreases in tensile strength in all cases. The metallographic, fractographic and fracture toughness results have been correlated to explain the observed changes in properties. It is shown that reverted austenite gives increased toughness at 650°C while acicular structures account for toughness increases over the 1 200°–1 400°C range. Constitutional liquation of inclusions was detected at the higher temperatures but did not produce grain-boundary embrittlement in this material. MT/16

 Manuscript received 5 July 1973. *Dr Jordan is with the Production Engineering Department, University of Aston, Gosta Green, Birmingham: Dr Coleman, formerly with the Metallurgy Department of the University of Aston, is now a Research Officer at the Marchwood Engineering Laboratories of the CEGB.*

M. C. Coleman
M. F. Jordan

The 300 grade maraging steel is one of the 18%Ni maraging steels described by Decker[1] and based on an alloy developed by Bieber.[2] Its physical metallurgy is reviewed in full by Floreen.[3] The alloy contains very little carbon and gains its strength from the precipitation hardening of its soft iron–nickel martensitic microstructure. As a consequence it possesses a combination of strength and toughness superior to other high-strength steels[4] while using a relatively simple heat treatment.[5] Welding of maraging steel can be carried out without preheat by processes ranging from electron beam[6] to submerged arc.[7,8] By using the material in the annealed condition the completed weld, heat-affected zone (HAZ), and parent metal can be aged in a single post-weld heat treatment. With such a simple procedure few problems have arisen in producing welded joints. However, difficulties have been reported in obtaining satisfactory properties, particularly fracture toughness in the heat-affected zone.[9–14]

Variation in properties in weld HAZs are generally attributable to changes in the microstructure brought about by the welding thermal cycle, and in the case of maraging steel two phenomena have been noticed, i.e. constitutional liquation and reverted austenite. Constitutional liquation is an effect proposed by Pepe and Savage,[15,16] who maintain that under the rapid heating conditions and high peak temperatures of the HAZ, intermetallic compounds such as titanium sulphide and titanium carbonitride liquate and penetrate grain boundaries. In this way they produce brittle grain-boundary films in the resulting microstructure. Reverted austenite is a decomposition product which forms on heating iron–nickel martensite.[17,18] Owing to its high nickel content, it does not retransform to martensite on cooling to room temperature, and results in incomplete precipitation hardening on aging. Sufficient segregation has been found in maraging steel plate[19] and weld metal[20] to produce reverted austenite on heating to the normal aging temperature of 485°C, but it forms most readily at 650°C.[18,20] Inevitably the regions of the weld HAZ experiencing peak temperatures up to 650°C will contain such reverted austenite and its effects on properties. The work of some investigators suggests that these two effects are not significant since they report that normal aging restores HAZ properties.[21,22] However, they either failed to carry out confirmatory property and structural examinations or tested the HAZ but 'made every effort to avoid the reverted austenite regions'. Recently Weiss *et al.*[23] made the correlation of HAZ properties and structures the prime objective of an investigation involving maraging steel, but unfortunately misinterpretation of the hardness results with respect to the associated structural phenomena undermined the value of the investigation.

The present investigation involves a detailed and systematic examination of the fracture toughness and microstructural relationships of the weld HAZ of an 18%Ni maraging steel. Most of the work was done on synthetic HAZ material produced by simulation techniques[24] since real weld HAZs are very narrow and non-uniform. The scope was limited to the conditions of single pass full penetration welding but the range of heat inputs employed covered the majority of fusion welding processes.

Experimental procedure

The material was a vacuum induction melted 300 grade maraging steel, supplied as nominally 6 mm thick plate in the solution-annealed condition (1 h at 820°C), and having the composition detailed in Table 1. Its plane strain fracture toughness and yield strength after aging for 3 h at 485°C were found to be 60 $MNm^{-3/2}$ (54·2 ksi$\sqrt{in}$) and 1975 MNm^{-2} (129 t/in^2) respectively, aging increasing the hardness from 320 to 590 HV. The as-received microstructure was Widmanstätten in the solution-annealed condition,

Table 1 Composition of the 18%Ni maraging steel parent material

Element	Ni	Co	Mo	Ti	C	Mn	Si	S	P
Composition, wt-%	18·5	8·1	5·0	0·8	0·010	0·04	0·15	0·006	0·005

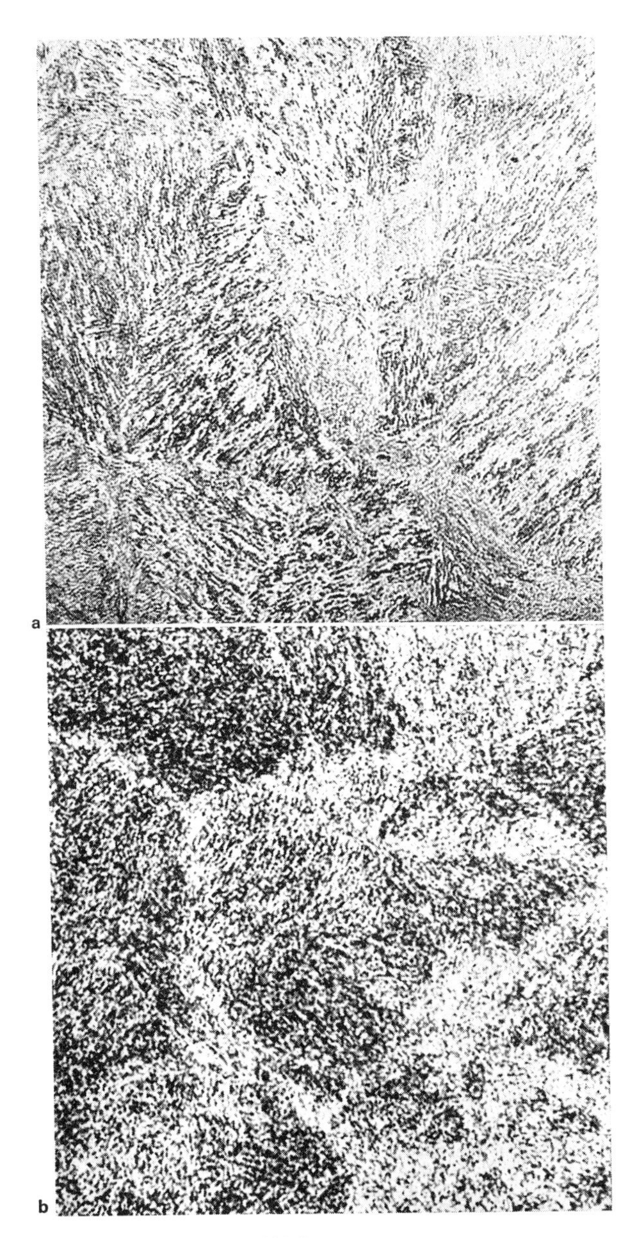

a as received; *b* aged 3 h at 485°C

1 Martensitic microstructures of the parent material; etchant 10% nital ×500

Fig.1*a*, and this darkened and became diffuse on aging, Fig.1*b*. The inclusions present were identified using electron probe microanalysis and electron diffraction as Ti_2S, in angular and rod shapes, and Ti(CN), as cuboids, both types commonly found in maraging steels.[25]

Experimental measurement of the thermal cycles in the weld HAZ is a lengthy and difficult task, so the thermal cycles used in the simulation studies were calculated from Rosenthal's two-dimensional heat flow equation.[26] This equation is known to give an acceptable estimate of practical values when applied to steels.[27,28] The practical range of process parameters for full penetration welding of sheet and plate was obtained for electron-beam, microplasma, tungsten inert gas, metal inert gas and submerged-arc welding. These values were then incorporated in a computer program which generated the thermal cycle appropriate to any preselected peak temperature.

Preliminary experiments which are reported elsewhere[29] indicated that a standard three-point bend specimen[30] based on a thickness of 6·35 mm was suitable both for thermal simulation and obtaining valid K_{IC} fracture toughness values. Consequently the procedure followed was to thermally cycle nominally 12·5×7×87·5 mm blanks in a thermomechanical simulator, varying the gripped length to give the appropriate cooling rate. Machining was then carried out before aging for 3 h at 485°C and the specimens were surface ground to the final dimensions. Fracture toughness testing was carried out in accordance with the recommended practice[31] and no.11 Hounsfield testpieces were used for tensile measurements. In all cases a minimum of three tests was carried out for each HAZ thermal cycle simulated.

To investigate the validity of the thermal simulation approach a number of real weld HAZs were examined. The submerged-arc and metal inert gas arc welding processes were selected because they could be used to produce full penetration welds in 6 mm thick plate in a single pass. This not only facilitated fracture toughness testing but ensured that two-dimensional heat flow conditions were generated, and that the HAZs were sufficiently wide to permit property measurements corresponding to preselected peak temperatures.

Experimental results

SIMULATED WELD HAZ PROPERTIES

Initial attention was confined to HAZ peak temperatures of 650° and 1200°C, the former to study reverted austenite, and the latter as typical of the temperature at which constitutional liquation of the Ti_2S inclusions might occur.[13] The fracture toughness results for these peak temperatures are plotted in Fig.2 against the different welding parameters, represented by heating rates. This shows that in these cases toughness is independent of the thermal cycling rates, but dependent on the peak temperature of the thermal cycle. A similar trend was found with the tensile results which were very consistent, 1865–1930 MNm^{-2} (121–125 t/in²) for 1200°C and, apart from one result of 1590 MNm^{-2} (103 t/in²) for the slowest heating rate, between 1715 and 1900 MNm^{-2} (111 and 123 t/in²) for 650°C. Further thermal cycles were then made to a range of other HAZ peak temperatures using slow, intermediate,

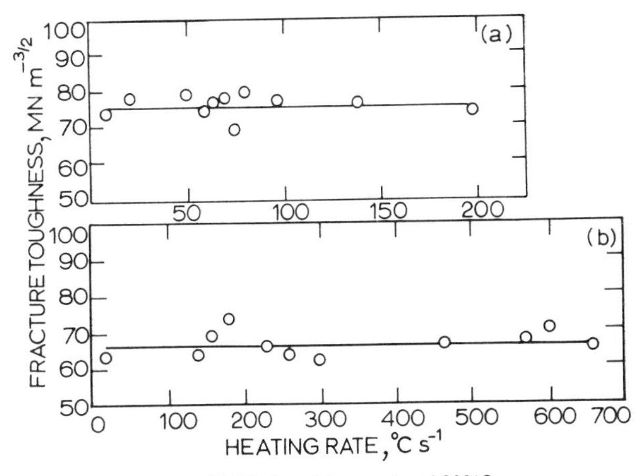

a peak temperature 650°C; *b* peak temperature 1200°C

2 Relationship between heating rate and fracture toughness for simulated HAZ thermal cycles to peak temperatures indicated

and fast heating and cooling rates. A similar consistency of properties, independent of the thermal cycling rates but dependent on peak temperature, was found in these tests. All the results are given in Table 2 where, apart from 650° and 1200°C, each result is the average of nine measurements, three values at each of three cycling rates. Details of the hardness values, before and after aging, are given in Fig.3 where it is evident that complete hardening was not achieved in every case.

REAL WELD HAZ PROPERTIES

The real welds were produced using conventional submerged-arc and MIG welding equipment with thermocouples embedded in the plate to monitor the HAZ thermal cycles. Macrosections were taken from the welds and specimen locations were selected in the HAZs on the basis of thermal cycle peak temperature and structural considerations. The submerged-arc weld macrosection is shown in Fig.4, and the selected testing positions are indicated. Position 1 covers one structure and corresponds to the 650°C isotherm. Positions 2 and 3 incorporate a range of structures with the corresponding peak temperatures estimated from the thermocouple readings to range from 1000° to 1300°C and 1200° to 1460°C plus, respectively. The MIG weld had the same general macrostructural appearance but the HAZ width was smaller. Consequently only two positions were selected for testing corresponding to peak temperatures of 650°C and a range from 1000° to 1460°C plus. The fracture toughness and tensile properties measured in the real HAZs are given in Table 3.

The results of a microhardness traverse across the submerged-arc weld HAZ, at the plate mid-thickness, are given in Fig.5. An identical distribution was found in the MIG weld HAZ, but the distance from the fusion boundary to the as-welded peak hardness position was about 7 mm as opposed to 9 mm in the submerged-arc weld HAZ. In each case this position was coincident with both the dark etching band and the 650°C isotherm.

METALLOGRAPHY

Detailed metallographic and fracture surface studies were made on the simulated HAZ specimens after aging and the results at each peak temperature are summarized.

- 600°C unchanged aged martensite
- 650°C white etching phase present in aged martensite giving mottled matrix and some grain-boundary films, Fig.6*a*; ductile dimples 1–10 μm in length on fracture surface, Fig.6*b*
- 800°C aged martensite comparable to initial condition
- 1000°C aged martensite slightly more acicular than initial condition
- 1200°C acicular martensite containing domains of differing orientation within prior austenite grains; grain size increased from 15 to 30μm with decreasing heating rate; undulating fracture paths, Fig.7, deflected by acicular matrix; ribbon-like white etching phase sometimes associated with fracture path; fracture surface contained dimples ~10 μm in length and flatter areas of fine dimples ~1 μm long, Fig.8; these fine dimple areas appear to be present when fracture path follows the white etching phase
- 1325°C acicular martensite similar to 1200°C; grain-size variation from 27 to 56 μm; larger amounts of white etching phase, appearing to increase with decreasing heating rate, and associated with

Table 2 The fracture toughness and tensile properties of simulated HAZ structures produced by thermal cycling to different peak temperatures

Thermal cycle peak temperature, °C	Fracture toughness, MNm$^{-3/2}$	Tensile strength, MNm^{-2}
600	63·0	1880
650	75·0	1775
800	63·0	1960
1000	58·7	1960
1200	66·0	1900
1325	67·6	1865
1400	77·5	1820

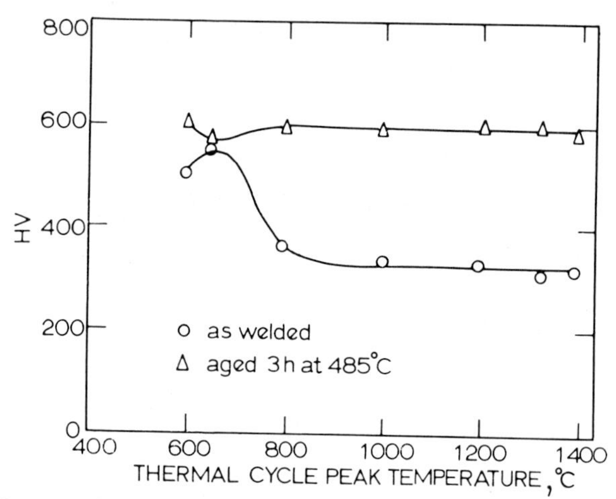

3 Variation of hardness with thermal cycle peak temperature in the simulated HAZs

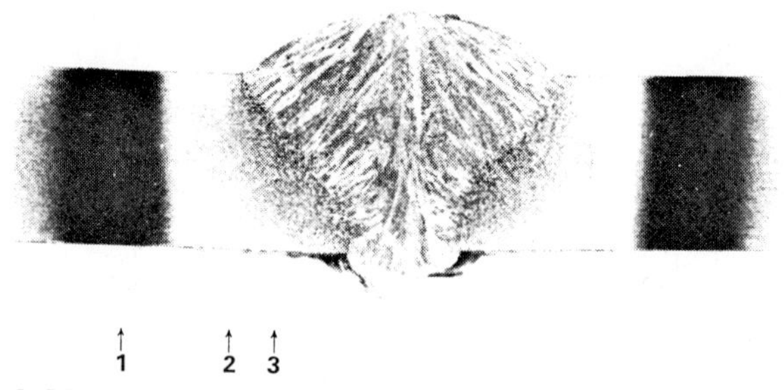

4 Macro cross-section of submerged-arc weld and HAZ; etchant 25% nitric acid ×2·5

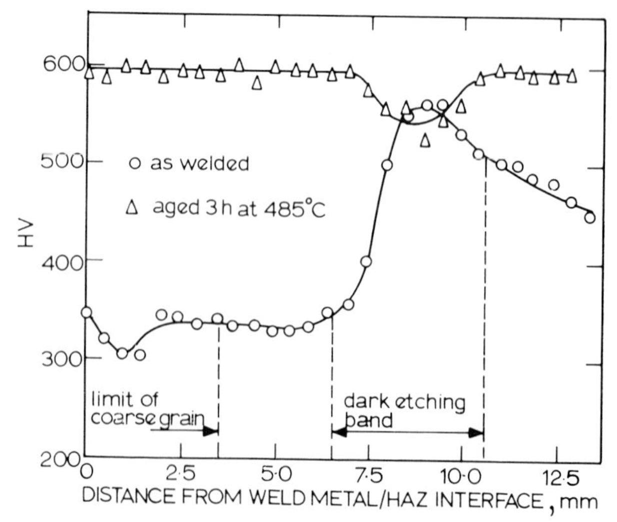

5 Variation in hardness across the maraging steel submerged-arc weld HAZ

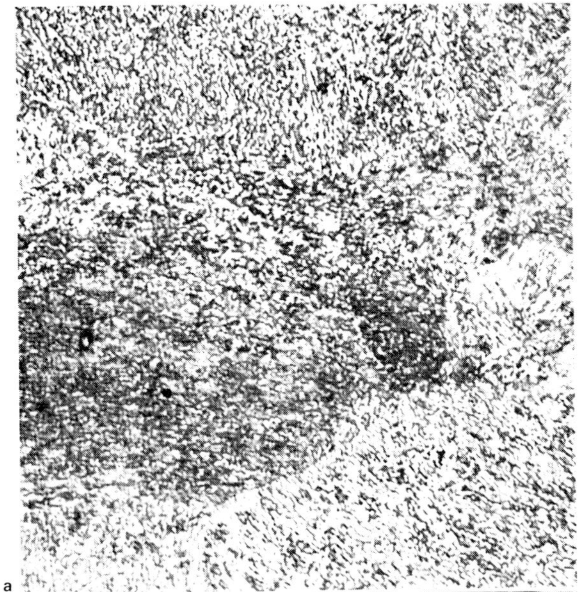

a

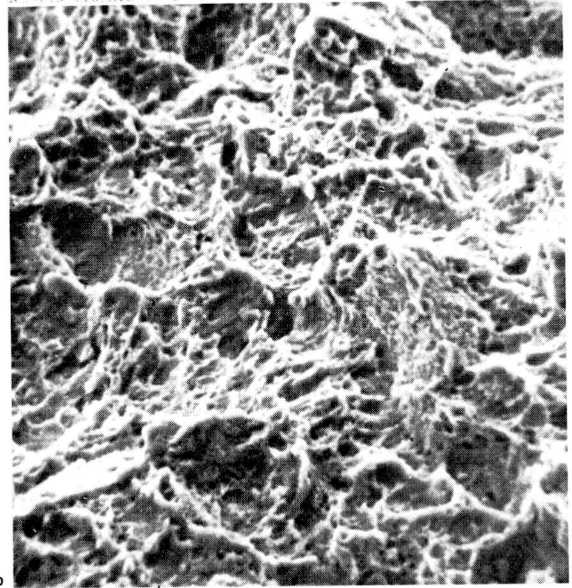

b

a mottled microstructure, etchant 10% nital ×500; *b* fracture surface ×1500

6 Microstructure and fracture appearance of the simulated weld HAZ; thermal cycle peak temperature 650°C

7 Fracture path through the simulated weld HAZ microstructure; thermal cycle peak temperature 1 200°C; etchant 10% nital ×500

liquated grey Ti_2S, Fig.9, but not with Ti(CN) inclusions; fracture surface containing dimples 5–10 μm and increased amounts of fine dimpled areas

1 400°C strongly acicular martensite with large domains and sharply undulating fracture paths, Fig.10; grain size increased with heating rate over range 35–66 μm; white etching ribbons from liquated Ti_2S were less apparent but liquation of Ti(CN) observed; fracture path not deviated by grain boundaries containing liquated material, Fig.11; fracture surface contained dimples 5–10 μm, less fine dimpled areas but severe undulations, Fig.12.

The submerged-arc and MIG weld HAZs each revealed three types of microstructure. In the macroscopically dark etching bands of the HAZs, coincident with the 650°C isotherm, microscopic examination revealed the mottled structure of white etching phase in a martensitic matrix. Immediately adjacent, on either side, the structures were similar to the unaffected parent plate, and comparable to the simulated structures at 600° and 800°C. Similarly, fracture toughness tests, limited to the 650°C structures, revealed fracture surfaces consisting of ductile dimples about 1 μm in size. As the fusion boundaries of the welds were approached the HAZ structures became acicular with well-defined prior austenite grain boundaries. It was difficult to relate grain size exactly to peak temperature, but from about 1 000°C to the fusion boundary, 1 460°C, it increased from 20 to 70 μm average. The white etching ribbon-like phase was observed generally throughout these structures indicating constitutional liquation of Ti_2S. However, liquated Ti(CN) was only observed close to the fusion boundary. No evidence of grain-boundary embrittlement or hot cracking was detected in any of the HAZs. Examination of the fracture surfaces traversing the high-temperature areas of the HAZs revealed similar features to the simulated HAZs. A range of ductile dimple sizes from 5–10 μm was found, in general with areas of fine dimples ~1 μm.

Table 3 The fracture toughness and tensile properties at various thermal cycle peak temperature positions in the submerged-arc and MIG weld HAZs

Welding process	HAZ peak temperature, range, °C	Fracture toughness, $MNm^{-3/2}$	Tensile strength, MNm^{-2}
Submerged arc	650	74·6	1 775
	1 000–1 300	64·5	1 930
	1 200–1 460+	66·3	—
MIG	650	74·0	1 790
	1 000–1 460+	65·5	—

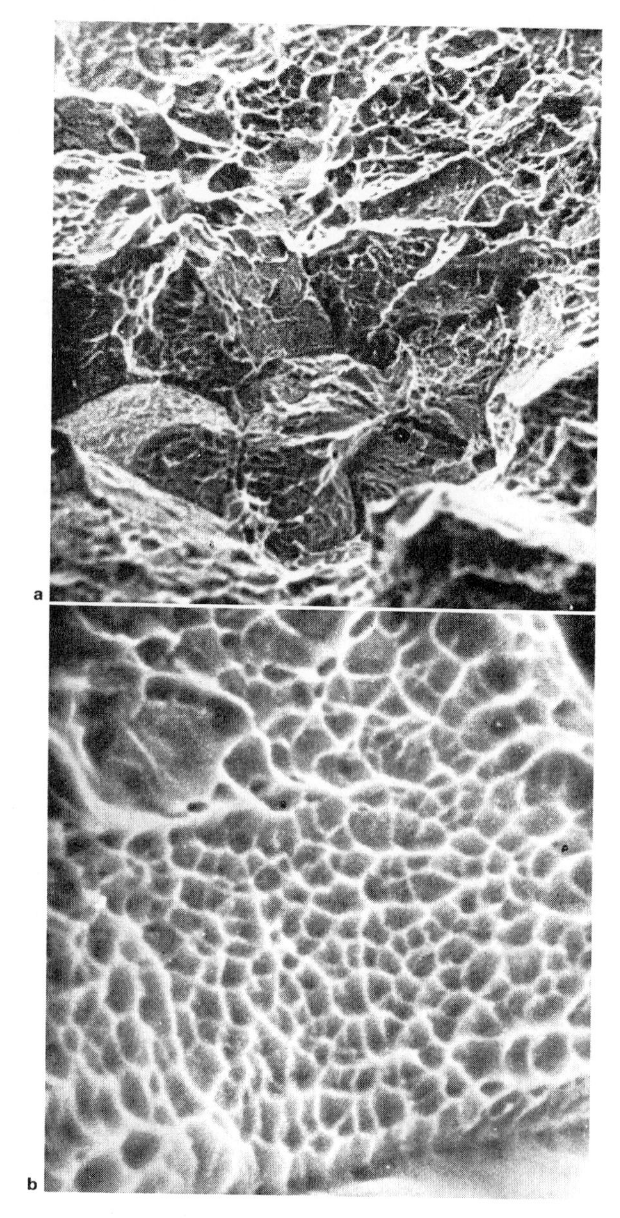

a ductile dimples and flatter 'intergranular' fracture ×700; *b* ductile dimples in the 'intergranular' fracture shown in *a* ×7000

8 Fracture surfaces in the simulated weld HAZ; thermal cycle peak temperature 1 200°C

Table 4 The volume fraction of reverted austenite and corresponding fracture toughness for a number of simulated and real HAZ microstructures (*real HAZ values)

Heating rate, °C s^{-1}	Fracture toughness, MNm$^{-3/2}$	Reverted austenite, vol.-%
7	78·2	53·7
7	85·0	42·2
22	81·0	17·5
30	73·3	18·2*
50	84·0	28·2
55	74·1	22·5*
64	71·5	29·8
64	72·0	11·8
70	79·2	14·9
76	66·2	17·4
80	75·5	17·3
98	81·2	32·5
200	71·5	7·9

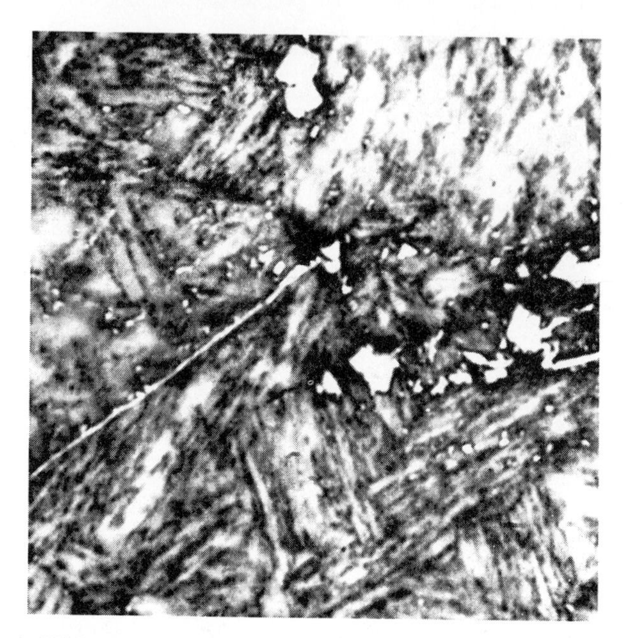

9 White etching ribbon-like phase formed by constitutional liquation of inclusions in the simulated weld HAZ; thermal cycle peak temperature 1 325°C; etchant 10% nital ×1000

The white etching phase observed in the mottled microstructure at the 650°C peak temperature was positively identified as reverted austenite by X-ray diffraction. Using this technique quantitative data were then obtained for simulated and real HAZ specimens produced with a range of heating rates and having differing fracture toughness values, Table 4. The white etching ribbons present in the high-temperature regions >1 000°C were not amenable to analysis by X-ray diffraction. However, electron probe microanalysis detected a local concentration of 0·97% titanium and this evidence, in conjunction with published data,[19] etching characteristics, morphology and ductile

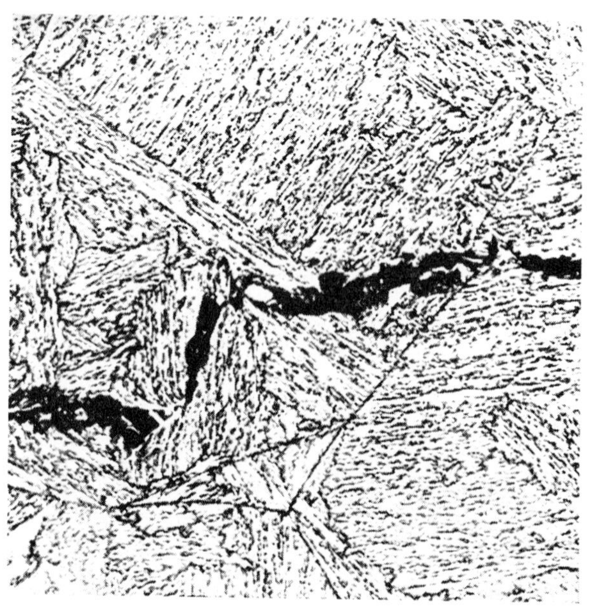

10 Fracture path through the simulated HAZ microstructure; thermal cycle peak temperature 1 400°C; etchant 10% nital ×500

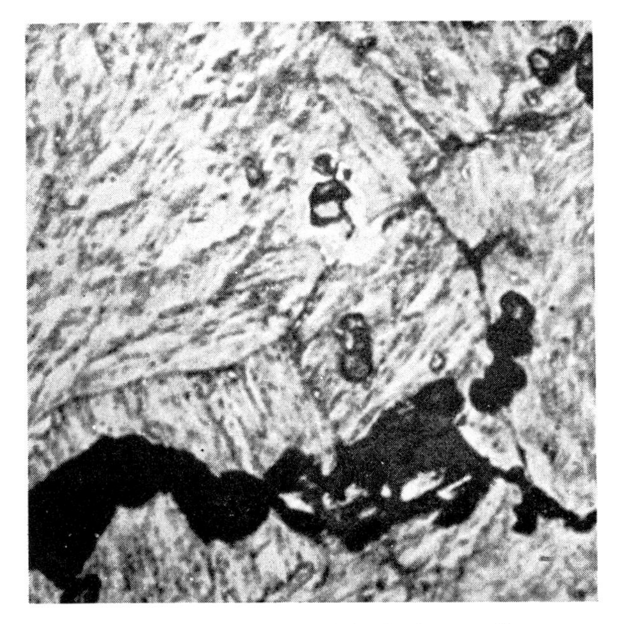

11 Constitutionally liquated inclusions adjacent to the fracture path in the simulated weld HAZ; thermal cycle peak temperature 1400°C; etchant 10% nital ×1000

failure mode, indicated that the ribbons were also reverted austenite. In this case they were associated with the liquation and dissolution of titanium-rich inclusions.

Discussion

VALIDITY OF THERMAL SIMULATION

The criterion for judging the validity of simulation must be the comparison of the fracture toughness and microstructure of real and simulated HAZs. Comparison of the results, Tables 2 and 3, shows that at 650°C very good

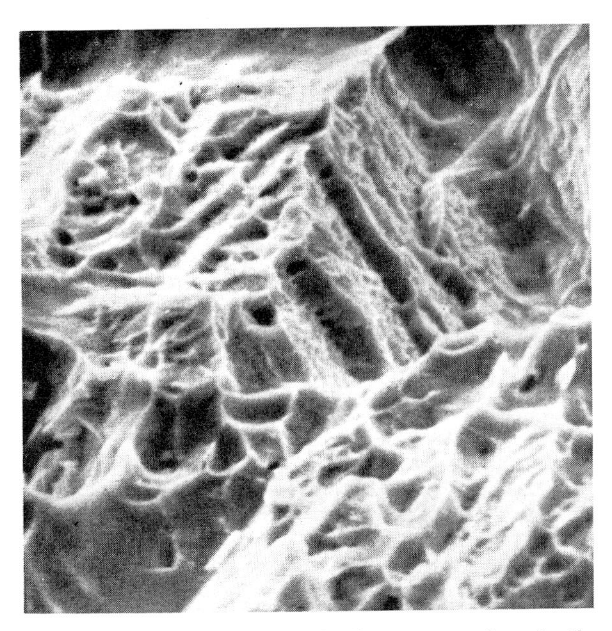

12 Acicular appearance of the fracture surface in the simulated weld HAZ; thermal cycle peak temperature 1400°C ×3000

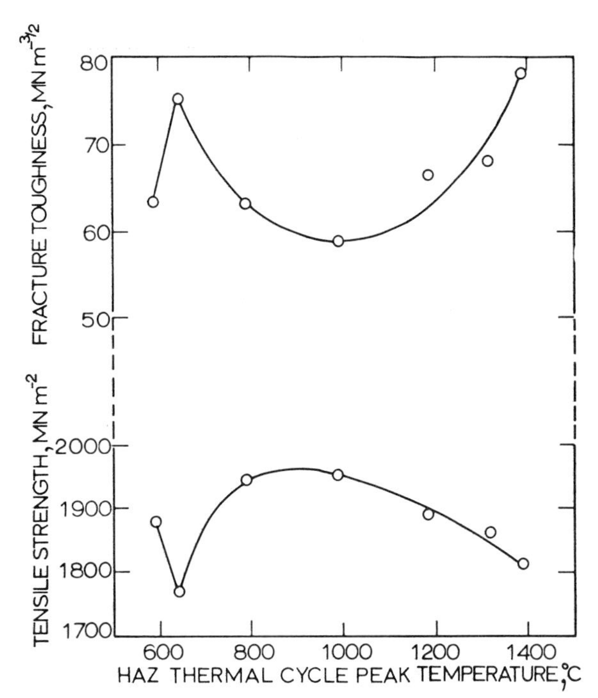

13 Variation in fracture toughness and tensile strength with HAZ thermal cycle peak temperature

agreement was obtained; 75·0 MNm$^{-3/2}$ by simulation with 74·6 MNm$^{-3/2}$ and 74·0 MNm$^{-3/2}$ respectively for submerged-arc and MIG welds. In the high-temperature regions such a direct comparison was not possible. However, 'averaged' results for a range of temperatures for simulated and real HAZs gave good agreement as can be seen in Table 5. The metallographic examination confirmed that these property results were supported by good agreement of the microstructures. The volume fractions of reverted austenite at 650°C for real HAZs were 18·2% and 22·5% which fell into the range for simulated specimens, *see* Table 4. Above 1000°C well-defined prior austenite grain boundaries, acicular matrices, undulating fractures and liquation of Ti_2S and Ti(CN) were evident in both cases. Furthermore, as Figs.3 and 5 show, there was also good agreement between the hardness values obtained. In view of these results it is considered that in this material the thermal simulation approach is a valid experimental technique for studying the HAZ microstructure and properties.

FRACTURE TOUGHNESS AND MICROSTRUCTURE

It has been shown clearly that there is no loss of fracture toughness in the weld HAZ in this material but the creation of zones of increased toughness, Fig.13. Thus values of

Table 5 Comparative fracture toughness values from the high-temperature regions of simulated and real HAZs

HAZ type	Temperature range, °C	Fracture toughness, MNm$^{-3/2}$
Simulated	1000–1325	64·1
Submerged arc	1000–1300	64·5
Simulated	1200–1400	70·4
Submerged arc	1200–1460+	66·3
Simulated	1000–1400	67·4
MIG	1000–1460+	65·6

75·0 $MNm^{-3/2}$ and 77·5 $MNm^{-3/2}$ were obtained at peak temperatures of 650° and 1400°C compared to 60 $MNm^{-3/2}$ in the parent material with smaller but still significant increases at 1200° and 1350°C. In all cases there were also the corresponding decreases in tensile strength expected from the basic inverse relationship between toughness and strength. Since the results were independent of heating and cooling rates the fracture toughness/peak temperature relationship of Fig.13 will be characteristic of HAZs in this material irrespective of the welding process used. The only difference will be the width of the zone over which these variations occur. The microstructure of the HAZ correlated very well with the observed fracture toughness values. Thus at 600°, 800° and 1000°C where the fracture toughness was equivalent to the parent material, the microstructure had been unaltered by the thermal cycle. The increases of fracture toughness at 650° and 1200°–1400°C were associated respectively with reverted austenite and the development of coarse acicular microstructures.

Formation of reverted austenite is due to segregation of the elements nickel and titanium, which form the major strengthening precipitates. Consequently this produces a decrease in strengthening on aging and therefore a corresponding increase in fracture toughness. The amount of reverted austenite decreased with increasing heating rate, *see* Table 4, but since the toughness was constant, the major effect must have been produced at low levels of reversion. It is interesting that in weld metal reverted austenite decreases toughness.[11,20] This difference in behaviour is attributable to a difference in morphology of the reverted austenite. In weld metal a distinct two-phase microstructure is produced and fracture takes place through the austenite pools. By contrast the HAZ distribution gives a mottled microstructure with no preferential fracture path.

In the acicular structures observed in the high-temperature regions from 1200° to 1400°C, the grain size increased with increasing temperature and decreasing heating rate and all contained evidence of constitutional liquation. However, none of the microstructural or fracture features could be used to describe quantitatively the progressive increase in toughness with peak temperature. Qualitatively the effect appears to be due primarily to the acicular domains of the matrix deflecting the fracture path, Fig.10. Other features such as grain boundaries and liquation films deflected the fracture but these varied in size and extent respectively with heating rate at any one temperature, while toughness and acicular morphology of the microstructure were constant. These increased toughness values in the high-temperature region of the HAZ agree with the findings of others[9,11] but no evidence was found of the reported embrittlement due to liquated inclusions.[14] When liquated grain-boundary films were observed, they yielded ductile areas with dimpled fracture surfaces, Fig.8. It would appear that in the present material the liquated films were sufficiently low in carbon and sulphur and high in titanium to give reverted austenite on aging. This is consistent with the observation that vacuum induction melted steels are sufficiently low in carbon and sulphur to eliminate grain-boundary embrittlement.[13,14]

Conclusions

1. The fracture toughness and microstructure produced by thermal simulation in this alloy correlate with those produced in real weld HAZs.

2. The HAZ fracture toughness increased significantly in certain regions, with mean values of 75 $MNm^{-3/2}$ (67·8 ksi$\sqrt{in}$) and 66–77·5 $MNm^{-3/2}$ (59·6–70·1 ksi$\sqrt{in}$) being measured as compared with the parent material value of 60 $MNm^{-3/2}$ (54·2 ksi$\sqrt{in}$). These increases were a function of peak temperature only and correspond to 650° and 1200°–1400°C respectively.

3. The increases in HAZ fracture toughness are related to the formation of reverted austenite at 650°C and acicular structures at 1200°–1400°C.

4. Thermal cycles to peak temperatures of 1200°–1400°C produce constitutional liquation in the HAZ, but low carbon, sulphur and consequent inclusion content ensure grain-boundary embrittlement is not a problem in vacuum induction melted material.

Acknowledgments

The authors wish to thank the University of Aston for providing laboratory facilities and the Science Research Council for their financial support. They also gratefully acknowledge the United Kingdom Atomic Energy Authority for assistance with the thermal simulation experiments and Bristol Aerojet Ltd for supplying material and advice. Finally they would thank their colleagues at the University of Aston for cooperation throughout the project.

References

1. R. F. DECKER *et al.*: *Trans. ASM*, 1962, **55**, 58.
2. C. G. BIEBER: *Met. Prog.*, 1960, **78**, 99.
3. S. FLOREEN: *Met. Rev.*, 1968, **13**, 115.
4. N. A. MATTHEWS *et al.*: *Met. Eng. Q., ASM*, 1963, **3**, 1.
5. S. FLOREEN and R. F. DECKER: *Trans. ASM*, 1962, **55**, 518.
6. T. BONISZEWSKI and D. M. KENYON: *Brit. Weld. J.*, 1966, **13**, 415.
7. F. D. DUFFEY and W. SUTAR: *Welding J.*, 1965, **44**, 251s.
8. E. M. WILSON and A. I. WILDMAN: *Brit. Weld. J.*, 1966, **13**, 67.
9. H. E. ROMINE: NASA, CR-302, Sept. 1965.
10. P. H. SALMON-COX *et al.*: *Trans. ASM*, 1967, **60**, 125.
11. W. A. PETERSEN: *Welding J.*, 1964, **43**, 428s.
12. A. I. WILDMAN: Bristol Aerojet Ltd, TR no.468, April 1968.
13. P. F. LANGSTONE: Bristol Aerojet Ltd, TR no.467, April 1968.
14. L. GROOME: Bristol Aerojet Ltd, TR no.480, August 1968.
15. J. J. PEPE and W. F. SAVAGE: *Welding J.*, 1967, **46**, 411s.
16. J. J. PEPE and W. F. SAVAGE: *ibid.*, 1970, **49**, 545s.
17. N. P. ALLEN and C. C. EARLEY: *JISI*, 1950, **166**, 281.
18. A. J. SEDRIKS and J. V. GRAIG: *ibid.*, 1965, **203**, 268.
19. P. H. SALMON-COX *et al.*: *Trans. Met. Soc. AIME*, 1967, **239**, 1809.
20. N. KENYON: *Welding J.*, 1968, **47**, 193s.
21. T. BONISZEWSKI *et al.*: 'High-strength steel', 100; 1962, London, The Iron and Steel Institute.
22. Z. P. SAPERSTEIN *et al.*: *AIME Met. Soc. Conf.*, 1966, **31**, 163.
23. B. Z. WEISS *et al.*: *Welding J.*, 1972, **52**, 449s.
24. D. J. WIDGERY: *Met. Constr.*, 1969, **1**, 328.
25. T. and E. BONISZEWSKI: *JISI*, 1966, **204**, 360.
26. D. ROSENTHAL: *Welding J.*, 1941, **20**, 220s.
27. D. ROSENTHAL: *Trans. ASME*, 1946, **68**, 849.
28. P. S. MYERS *et al.*: Welding Research Council Bulletin, no. 123, 1968.
29. M. C. COLEMAN: Ph.D. thesis, October 1971, University of Aston in Birmingham.
30. Bisra open report, MF/EB/312/67.
31. W. F. BROWN and J. E. SRAWLEY: ASTM STP 410, 1966.

SECTION IX: Machining

SURFACE INTEGRITY IN MACHINING 18% NICKEL MARAGING STEEL

John A. Bailey, Professor, and S. Jeelani, Graduate Student
Department of Mechanical and Aerospace Engineering
North Carolina State University

ABSTRACT

The effect of cutting speed and tool wear land length on the surface integrity of annealed 18% Nickel Maraging Steel machined under dry and lubricated orthogonal conditions is determined. The surface region of the machined workpiece is examined using optical microscopy, scanning electron microscopy, X-Ray microprobe analysis, microhardness measurements and profilometry. In addition tool forces are measured and tool temperatures calculated.

The results show that during machining a surface region is produced which is different from the bulk of the material. When cutting at low speeds with sharp cutting tools it is found that damage is produced which is restricted to a variety of geometrical defects associated with the surface. When cutting at high speeds or with tools having large artificially controlled tool wear lands it is found that surface damage is accompanied by a work hardened surface region. The results are interpreted in terms of the type of chip produced during machining, the temperatures generated and the interaction between the tool nose region and workpiece.

It is suggested that observations based on scanning electron microscopy are more indicative of the true surface condition than surface roughness measurements.

INTRODUCTION

Recent improvements in alloy development and heat treatment techniques have led to the production of materials with increased inherent strength, and strength to weight ratio. In turn, this has led to a reduction in the section thickness of many components used in weight sensitive situations with a corresponding increase in the surface area to volume ratio. In addition, advances in engineering design in recent years have necessitated the use of both new and existing materials under increasingly severe service conditions with respect to stress, temperature, and environment. These trends dictate that increased attention must be paid to the possibilities of service failures produced by creep, fatigue and stress corrosion cracking. Such failures invariably start at the surface of a component and depend very sensitively on the condition of the surface [1-3]. Clearly, it is important that design engineers have complete information on the surface characteristics of a component in addition to those mechanical and physical properties generally recognized as essential. In particular, the design engineer must ascertain whether or not a particular manufacturing process alters the surface of a component and makes it different from the bulk material. In addition, the effect of surface damage on surface sensitive mechanical properties must be understood so that remedial manufacturing procedures can be introduced.

Source: *SME Technical Paper IQ74 – 185*

In spite of the rapid advances being made in the development of new and improved production techniques the metal machining process still features either directly or indirectly in the manufacture of most of the items used in our present technological society. Clearly, knowledge of the nature and quality of machined surfaces produced under various conditions is of considerable industrial importance.

The quality of a machined surface of a component is usually assessed in terms of parameters such surface roughness, waviness, lay and dimensional tollerances. For medium and low strength materials available evidence suggests for a given surface quality that knowledge of the bulk mechanical properties can be used as a satisfactory guide to the performance of components in service. However, for high strength materials these properties serve as a poor guide to performance in service, particularly with respect to resistance to fatigue, creep and stress corrosion cracking. This has been attributed to subsurface damage produced in the component by the machining process. Thus, for high strength materials the aforementioned parameters appear insufficient to describe the true quality of a machined surface.

SURFACE INTEGRITY

Surface integrity is a relatively new term [4] used to describe the nature or condition of a surface. It is interpreted as including those elements which describe the actual structure of both the surface and subsurface. If the surface of a component possesses integrity then it is unimpaired by the action or impact of the machining process used in its fabrication. Conversely, if integrity is lacking then surface and subsurface damage has occurred as a result of the machining process.

In the past studies of the nature of machined surfaces have involved investigations of the effects of parameters such as cutting speed, work material, tool material, tool geometry, and lubricant on the surface finish and dimensional accuracy of components [5-7]. The investigations have involved those elements of surface integrity pertaining to the actual topography or geometry of the machined surface.

In recent years increasing attention has been directed towards studies of the structure of both the surface and subsurface. It has been found on many occasions that machining can produce changes in the metallurgical structure of the subsurface to the point where it becomes markedly different from that of the bulk material. In addition, it has been shown that these changes can lead sometimes to drastic reductions in the resistance of materials to fatigue, creep, and stress corrosion cracking [2-4, 8-10]. However, most of this recent work appears to have been conducted for the purpose of identifying the types of damage produced in a wide variety of high strength materials by various conventional and nonconventional machining processes, and establishing the optimum cutting conditions for minimizing surface damage. Only a small proportion has been conducted in an effort to attack the fundamental problem of determining in some systematic manner the effects of some of the aforementioned cutting parameters on the total surface integrity and mechanical properties of a given material.

Excellent reviews of previous work relating to surface integrity in metal machining, and the techniques used for surface and subsurface evaluation have been given recently in the published literature [11, 12].

The purpose of the present investigation is to determine in a comprehensive manner the effects of cutting speed and tool wear land length on the surface integrity of annealed 18% Nickel Maraging steel machined under both dry and lubricated orthogonal conditions. Examination of the machined surface is conducted using profilometry, optical microscopy and scanning electron microscopy. Examination of the subsurface structure is conducted using optical microscopy, X-Ray Diffraction Analysis, X-Ray Microprobe Analysis and Microhardness Measurements. In addition, quantitative determination of subsurface plastic strain is made from the distorsion of an orthogonal array of lines embossed on the side surface of the workpiece.

EXPERIMENTAL WORK

In the experimental investigation workpieces of 18% Nickel Maraging steel were used. The work material was received in the form of a solid bar which was machined into disc shaped test pieces 3.0 in. in diameter and 0.1 in. in width with a central hub on one side of the disc and a shallow recess on the other. A diagram illustrating the test piece geometry is given in Figure 1.

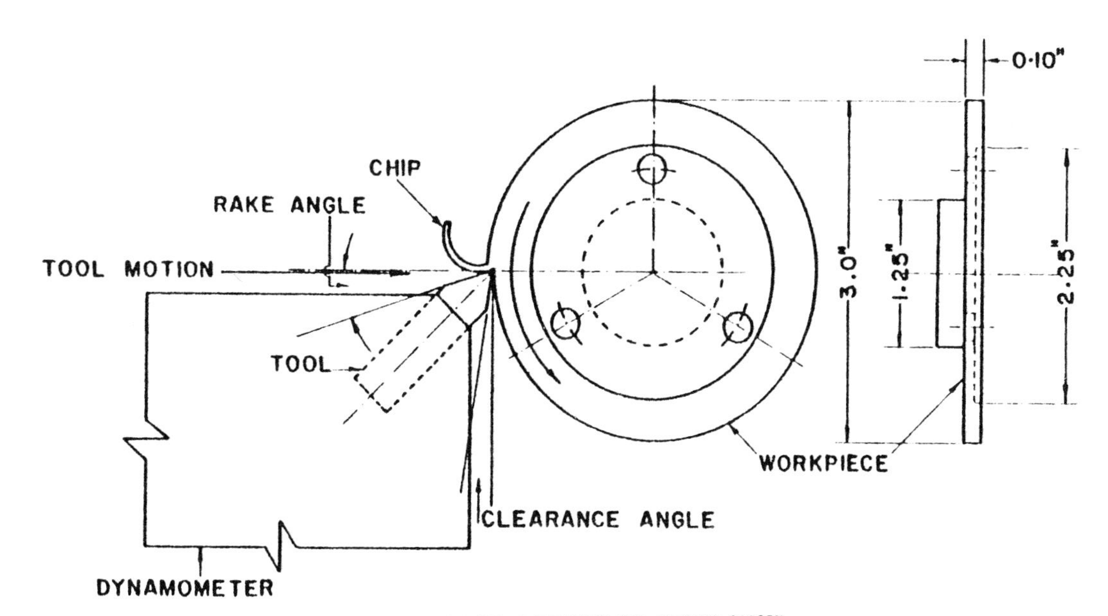

FIGURE 1. WORKPIECE GEOMETRY AND CUTTING ACTION

After preparation the test pieces were heated to 1550^{0}F in an atmosphere of nitrogen then air cooled to room temperature. The hardness of each specimen was determined after heat treatment and any with a hardness

Source: *SME Technical Paper IQ74 – 185*

different from 28Rc was rejected. The side surface of each disc was polished metallographically in the usual way. Final polishing was carried out with 0.03 micron alumina powder and water. Finally, a grid consisting of an array of orthogonal lines measuring approximately 0.080 in. by 0.080 in. was embossed on the polished surface of the specimen close to the periphery using a modified microtome. The line density was 1000 lines per linear inch. This technique has been described in detail elsewhere.[13,14].

In the cutting tests the discs were bolted to a specially made mandrel then machined on a lathe under both dry and lubricated orthogonal conditions. The direction of tool motion was perpendicular to the axis of rotation of the workpiece. The cutting action is illustrated in Figure 1. Tests were conducted at cutting speeds in the range from 10 to 400ft./min. using tungsten carbide grade C3 cutting tools with artificially controlled wear land lengths of 0.005 in., 0.010 in., and 0.020 in.; the feed or undeformed chip thickness being 0.005 in. per revolution in each test. A summary of the test conditions is given in Table 1. The artificial wear lands were ground on the tools with a nominal negative clearance angle of 0.5 degrees to ensure contact between the land and freshly machined workpiece surface. In addition, tests were conducted with a sharp cutting tool. A constant tool nose radius of 0.00025 in. and tool rake and clearance angles of 10 and 5 degrees respectively, were used throughout the work. Once steady state cutting had been established the cutting and thrust components of the resultant tool force were measured using a two component cutting force dynamometer [15]. The cutting action was then stopped suddenly leaving the chip attached to the workpiece. The chip thickness and cutting ratio were determined from the weight of a known length of chip sample.

TABLE I

MATERIAL - 18% NICKEL MARAGING STEEL			
SUMMARY OF TEST CONDITIONS			
Cutting Speed (ft/min)	Rake Angle (degrees)	Clearance Angle (degrees)	Tool Land
10, 20, 40	10	5	0.000, 0.005
80, 120, 160			0.010, 0.020
200, 240, 280			
320			
Heat Treatment	Lubricant	Tool Material (Carbide)	Feed in/rev
Annealed, (28 Rc)	None	C3	0.005
	Soluble		
	Oil		

Small pieces measuring approximately 1.0 in. in length were cut from the machined discs, cleaned with an aqueous solution of trisodium phosphate, then rinsed in hot water and methyl alcohol. The machined surfaces were then examined on an optical microscope, and a JSM-2 scanning electron microscope over a wide range of magnifications. Surface roughness measurements were made on these samples in a direction perpendicular to the direction of relative work tool motion using a profilometer. The chip "roots" and sections perpendicular to the machined surface and parallel to the direction of relative work tool motion were polished metallographically in the usual manner. Care was taken during preparation to obtain maximum resolution at the machined surface by preplating the specimens with copper. Microhardness measurements and X-Ray microprobe analyses were made across the surface of the specimens. The specimens were then etched in a ten per cent solution of ferric chloride in water and examined on an optical microscope.

The plastic strain in the surface region was determined from the displacements produced by machining of points on the rectangular grids which were originally embossed on the side surface of the workpieces. The displacements were determined from comparisons of grid points before and after deformation.

RESULTS AND DISCUSSION

(a) Tool Forces and Tool Temperatures

The cutting and thrust components of the resultant tool force were measured over the range of cutting conditions used in the investigation. Results, selected from these tests are presented in Figure 2. It was

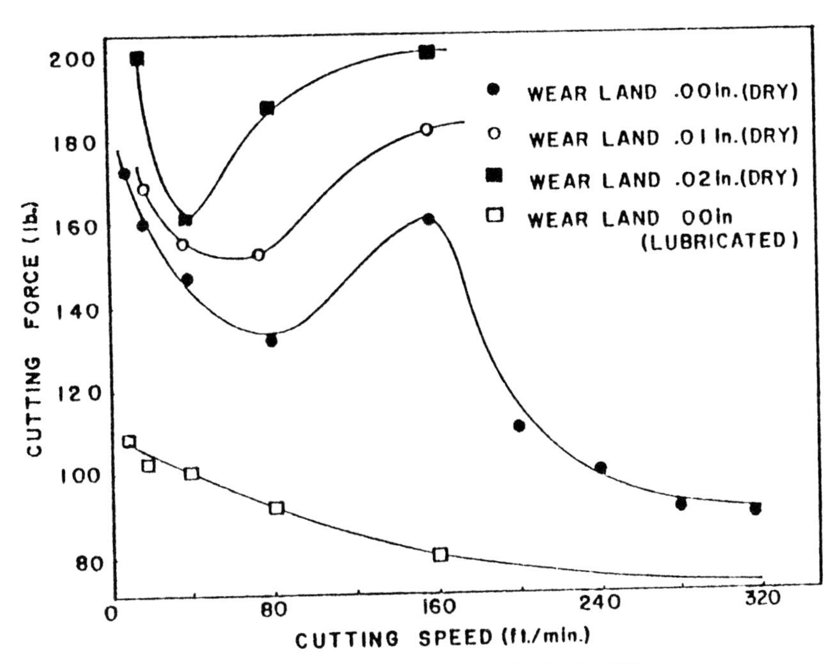

FIGURE 2. EFFECT OF CUTTING SPEED ON CUTTING FORCE.

Source: *SME Technical Paper IQ74 – 185*

found when cutting under dry, unlubricated conditions that the cutting and thrust components of the resultant tool force first decreased with an increase in cutting speed to a minimum then increased with a further increase in cutting speed to a maximum. Finally, the tool forces decreased continuously with an increase in cutting speed at high cutting speeds. At cutting speeds greater than 160ft/min. reliable cutting force data could be generated only when cutting with a sharp cutting tool. When cutting with tools having wear lands of finite dimensions cutting was unstable with erratic fluctuations in tool forces which caused in most instances rapid tool degradation. For lubricated cutting the tool forces decreased continuously with an increase in cutting speed tending to become constant at high cutting speeds. It was found also that an increase in tool wear land length produced an increase in tool forces for both dry and lubricated cutting.

The variation of tool forces with changes in cutting speed during machining is related to the type of chip produced and frictional conditions at the tool rake face. It was found from metallographic examination of suddenly stopped chip samples produced from tests conducted under dry, unlubricated conditions at cutting speeds up to and including 20ft/min that the chip formation process was partially discontinuous [16]. At other cutting speeds the chip formation process was continuous. In addition, at cutting speeds up to 120 ft/min chips were produced with the presence of a built-up edge on the tool rake face. It was found also that the built-up edge first increased in size with an increase in cutting speed to a maximum, then decreased with a further increase in cutting speed.

The built-up edge present at low cutting speeds forms as a result of work hardening of the chip at the tool rake face and the subsequent buildup of work hardened material. Growth is possible up to the point where the resolved shear stress parallel to the upper surface of the built-up edge causes the position of the shear zone to move into the built-up edge to retard growth. The partially discontinuous nature of the chip formation process at these cutting speeds arises because conditions of strain, strain rate and temperature in the primary deformation zone are such that the ductility of the chip material is exceeded, which results in fracture.

Calculation showed that an increase in cutting speed produced an increase in the frictional force and coefficient of friction at the tool rake face. It is believed that this is, in part, caused by an increase in the area of contact between the chip and tool. The effect is to allow growth of the built-up edge to a greater extent before the resolved shear stress parallel to the upper side of built-up edge can cause the shear zone to move into the built-up edge. Thus, the cutting force will decrease with an increase in cutting speed as the built-up edge grows because of a decrease in the effective tool rake angle. The tool forces reach a minimum when the built-up edge is maximum in size. This occurs at a cutting speed of approximately 80 ft/min.

It is believed that when the cutting speed is increased beyond 80 ft/min the temperatures generated produce softening and the gradual elimination of the built-up edge which produces an increase in the tool forces. The forces reach a maximum when the built-up edge is reduced to a shear zone. At high cutting speeds the tool forces decrease continuously with an increase in cutting speed because of a further increase in temperature at the tool rake face, which in turn, produces a reduction in the shear flow stress of the chip material in that region. Calculation of the temperature rise produced in machining from the tool forces for a sharp cutting tool showed that the shear zone and mean rake face temperatures generally increased with an increase in cutting speed. This is shown in Figure 3. Results very similar to the above have been reported by Zorev [17] and Williams et al. [18] for machining other materials.

It was found that an increase in tool wear land length did not produce any readily detectable change in the chip formation process. However, it would be anticipated that an increase in tool wear land length would produce an increase in temperature in the freshly machined workpiece surface and in the primary and secondary deformation zones.

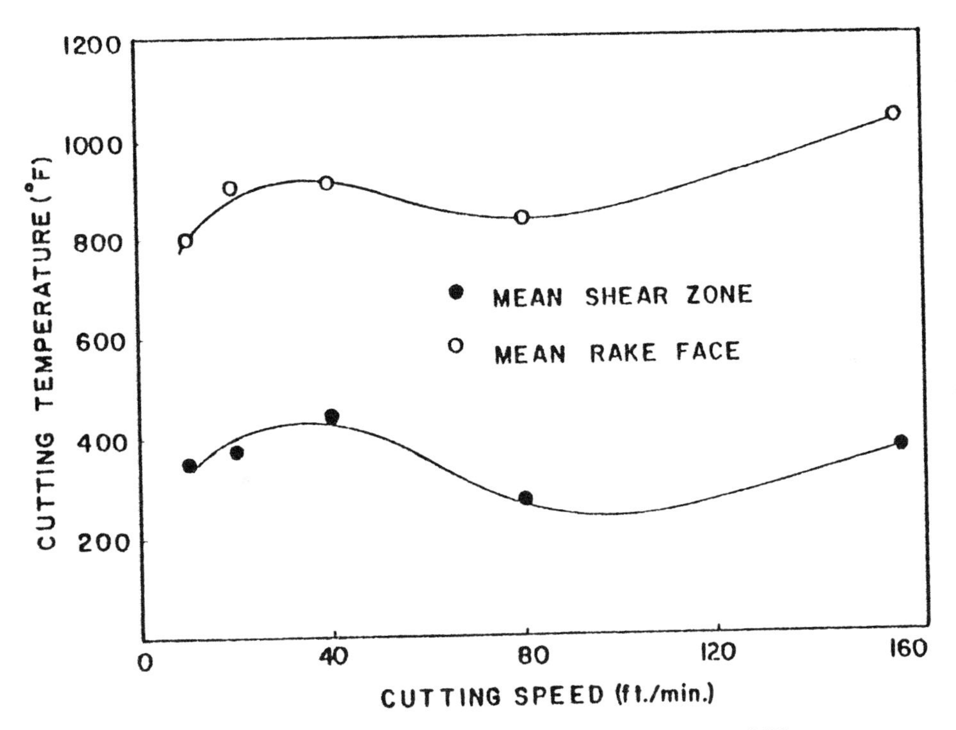

FIGURE 3. EFFECT OF CUTTING SPEED ON TEMPERATURE RISE.

It was found from an analysis of the cutting geometry that the cutting ratio first increased with an increase in cutting speed to a maximum then decreased with a further increase in cutting speed to a minimum; finally the cutting ratio increased with an increase in cutting speed tending to become constant at high cutting speeds. These results, which are shown in Figure 4 for machining with a sharp cutting tool clearly are consistant with the variation of tool forces with an increase in cutting speed. The cutting ratio did not vary with variations in the tool wear land length. This is interpreted as indicating that the cutting geometry is dependent on cutting speed but independent of tool wear land length.

Additional cutting tests were performed on the work material with the presence of a lubricant (Keystone - Keycut water soluble oil). The results are shown in Figure (2). It can be seen for tests conducted under lubricated conditions that the cutting component of the resultant tool force decreases continuously with an increase in cutting speed. Metallographic examination of suddenly stopped chip samples showed that for lubricated tests continuous chips were produced with an absence of a built-up edge on the tool rake face under all cutting conditions. Hence, the minimum observed in tool forces in the range of cutting speeds from 10 to 160 ft/min for tests conducted under dry conditions is absent. Calculation showed that the frictional force parallel to the tool rake face and coefficient of friction were lower for the lubricated tests.

In addition, it was found that the cutting ratio increased continuously with an increase in cutting speed tending to become constant at high speeds. This behavior is consistent with the variation of tool forces with cutting speed described above.

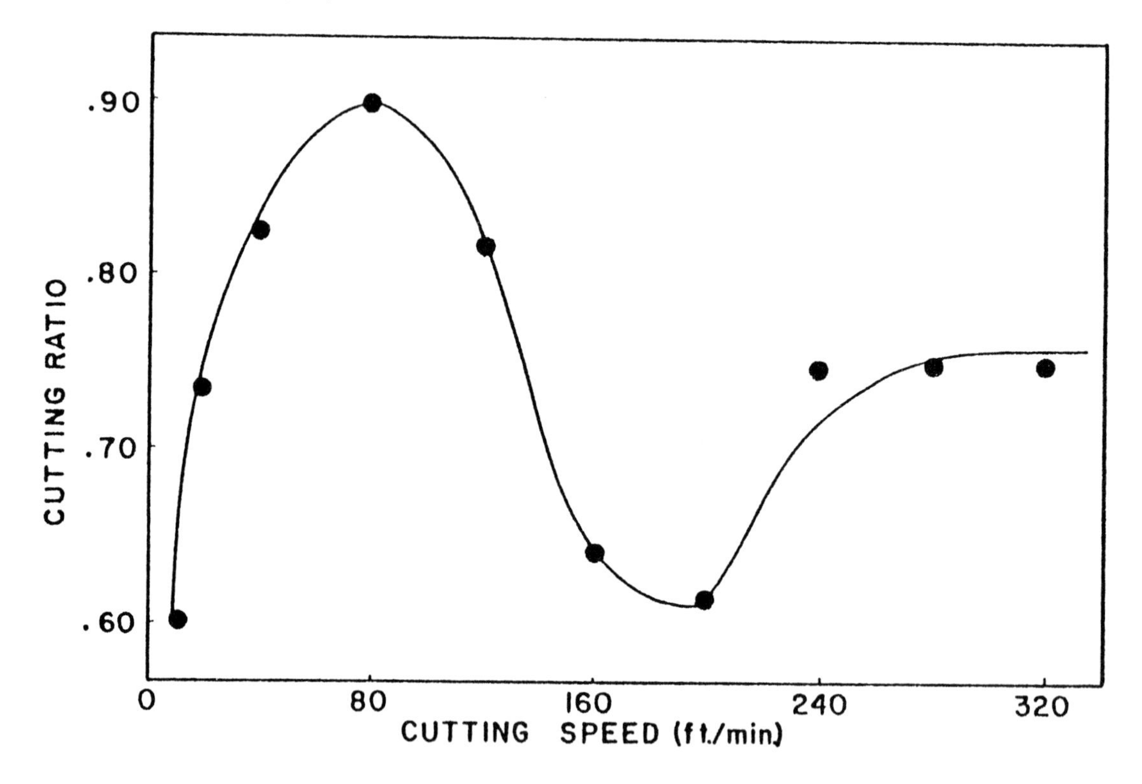

FIGURE 4. EFFECT OF CUTTING SPEED ON CUTTING RATIO

(b) Surface Examination

The results of the investigation showed that considerable variation in the appearance of the surface generated by machining can be produced by changes in cutting speed and tool wear land length. Unfortunately in some instances surface appearance varied within the same specimen. With this condition it becomes difficult to characterize the surfaces explicitly. However, an attempt will be made to interpret the results qualitatively and establish some general trends of behavior.

The machined specimens were examined on an optical microscope at magnifications up to 20 diameters to determine the variations in surface macrostructure with changes in cutting speed and tool wearland length. Figure (5) shows a selection of the generated surfaces produced when

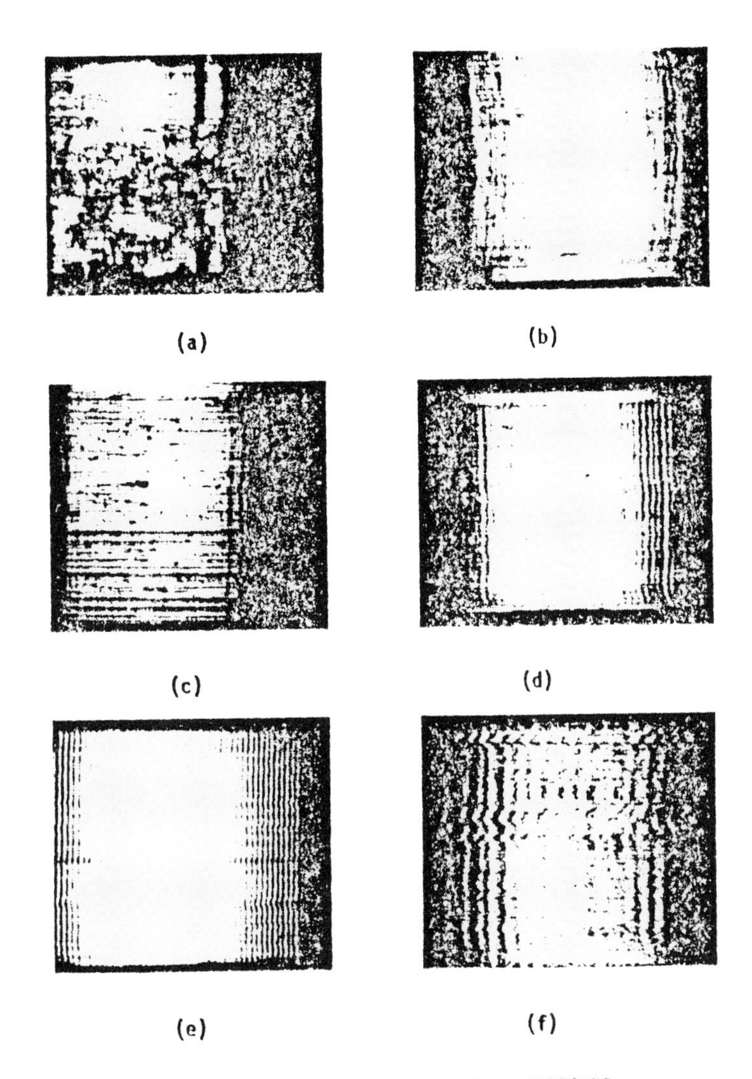

FIGURE 5 OPTICAL MACROGRAPHS OF MACHINED SURFACES
(Unlubricated Magnification 20 x)

LEGEND (a) 20ft./min., sharp tool (b) 20ft./min., wearland 0.020 in.
(c) 80ft./min., sharp tool (d) 80ft./min., wearland 0.020 in.
(e) 160ft./min., sharp tool (f) 160ft./min., wearland 0.020 in.

machining under dry, unlubricated conditions. It was observed when machining at low cutting speeds with a sharp tool that the surface consisted of chatter marks perpendicular to the direction of relative work-tool motion, well defined long straight grooves parallel to the direction of relative work-tool motion, intense local surface damage in the form of cavities, and cracks. It was found that an increase in cutting speed and tool wear land length produced an increase in the severity of the chatter marks.

Source: *SME Technical Paper IQ74 – 185*

It was found also that the intense surface damage, which appeared particularly severe in tests conducted as cutting speeds in the range from 10 to 80 ft/min. decreased in terms of intensity and total area affected. In addition, the long straight grooves parallel to the direction of relative work-tool motion become discontinuous and ill defined.

The machined specimens were examined next on the scanning electron microscope over a wide range of magnifications to determine the variations in the fine structure of the surface with changes in cutting speed and tool wear land length. The scanning electron microscope enables surfaces to be examined in much greater detail than is possible with optical microscopy because of the large depth of field and high resolution obtainable. The scanning electron microscope revealed surface characteristics not evident under the optical microscope.

Figures (6) to (9) show a selection of the generated surfaces. It was observed when machining at low cutting speeds with a sharp tool that the surface consisted of regions where cutting had occurred separated by regions where fracture had occurred. This is evident also in Figure (5). In the fractured areas large segments of the workpiece had been removed out of the surface leaving cavities. In the cavities there was evidence of plastic deformation, microcracks and voids. In the regions where cutting had occurred long straight grooves parallel to the direction of relative work-tool motion could be seen. These regions were populated with very small fractured areas where segments of the workpiece had been removed from the surface. This is illustrated in Figure (6). It was found that an increase in cutting speed led to the progressive reduction in the extent of cavity formation and the generation of a severely deformed surface region. The long straight grooves parallel to the direction of relative work-tool motion seen at low cutting speed became discontinuous and ill defined. In addition, an increase in cutting speed led to the generation of microchips and an appreciable amount of metal debris. Figure 7 shows a typical example of the surface generated at a cutting speed of 160 ft/min. with a sharp cutting tool.

Some changes in the fine structure of the surface were produced by changes in the length of the artificially controlled wear land. It was found at low cutting speeds that an increase in the length of the tool wear land produced an increase in the number of points at which surface damage produced by fracture occurred. However, the size of the fractured areas were smaller than those produced when cutting with sharp tools at the same cutting speed. In addition, the long straight grooves parallel to the direction of work-tool motion became discontinuous and ill defined. The surface showed evidence of severe plastic flow. Figure (8) shows a typical tool having an artificially controlled wear land of 0.020 in.

At high cutting speeds an increase in tool wear land length produced an appreciable change in the fine structure of the surface. At large tool wear lands the surface was irregular showing areas where cutting had occurred and areas where severe plastic deformation and ductile fracture had occurred. The cut areas were always traversed by ill defined grooves parallel to the direction of relative work-tool motion. The fractured

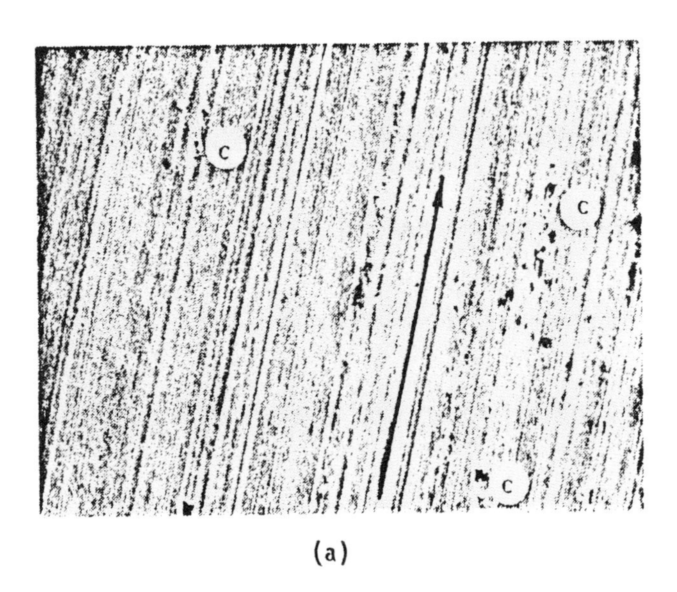

(a)

Magnification 250x

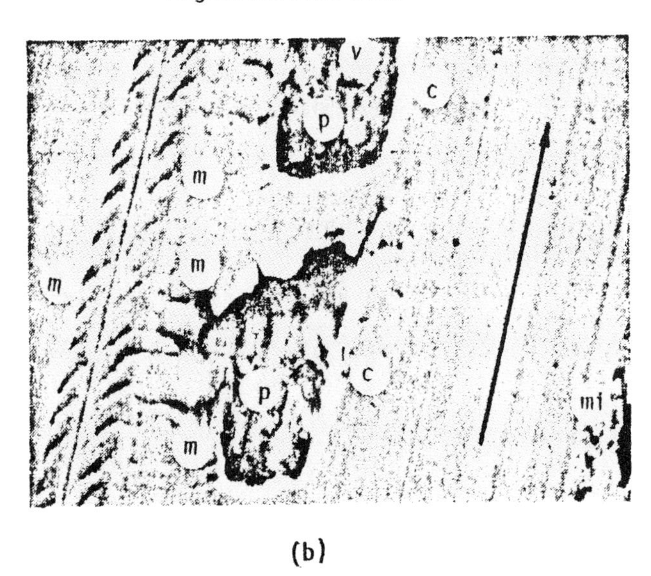

(b)

Magnification 1000x

FIGURE 6 SCANNING ELECTRON MICROGRAPH OF MACHINED SURFACE (Unlubricated, 20ft./min., Sharp tool, Direction of Tool Motion Given by Arrow).

LEGEND c - cavities, m - microcracks, mi - microchips
p - plastic deformation, v - voids.

areas always appeared dimpled. Figure (9) shows a typical example of the surface generated at a cutting speed of 160 ft/min with a tool having an artificially controlled tool wear land of 0.020 in. At both low and high cutting speed an increase in tool wear land length produced an increase in the amount of metal debris pressure welded to the freshly machined workpiece surface.

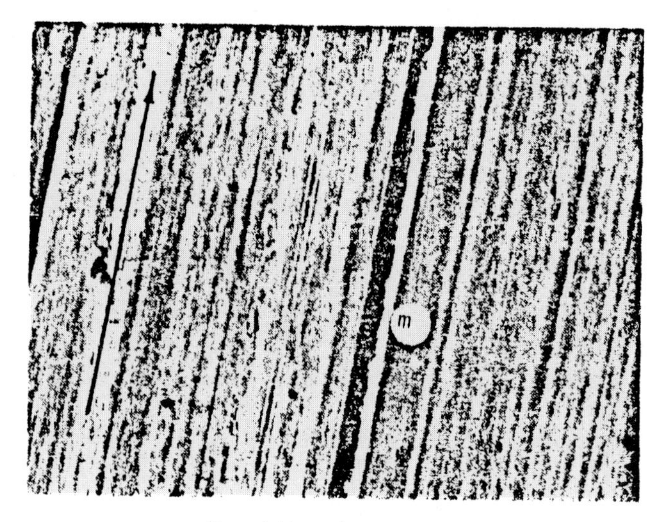

Magnification 550x

FIGURE 7. SCANNING ELECTRON MICROGRAPH OF MACHINED SURFACE. (Unlubricated, 160ft./min., Sharp Tool, Direction of Tool Motion Given by Arrow).

LEGEND m - microchip groove.

The nature of the surface generated can be related to the type of chip produced, interaction between the tool cutting edge and freshly machined workpiece surface and stability of the cutting process. It is believed that the chatter produced in machining arises because of the high tool loads and vibration produced in the cross feed mechanism of the lathe. It was found that the chatter could be reduced at any given cutting speed by reducing the tool forces through a reduction in the depth of cut of width of the workpiece.

It was mentioned in the preceding section that at low cutting speeds the chip formation process was partially discontinuous because the existing conditions of strain, strain rate and temperature in the primary deformation zone were such that the ductility of the chip material was exceeded. It is believed here that a crack is initiated at the tool cutting edge which propagates ahead and below the nose of the tool then intercepts the free surface. The result is that intermittent cutting and fracture occur with the production of a partially segmented chip and the removal of large fragments of the workpiece from below the general level of the generated surface forming cavities. Figure (5a). During crack propagation additional plastic deformation at the cavity surface can occur with the formation of microcracks and voids. This mechanism of cavity formation has been discussed previously in relation to the machining of AISI 4340 steel [19].

The regions where surface damage and fracture has occurred are connected by regions where cutting has occurred. In these latter regions the surface consists of long, straight, well defined grooves parallel to the cutting direction. It is believed that these grooves arise from irregularities or asperities along the tool cutting edge which are produced by the grinding used in tool preparation. The irregularities or asperities

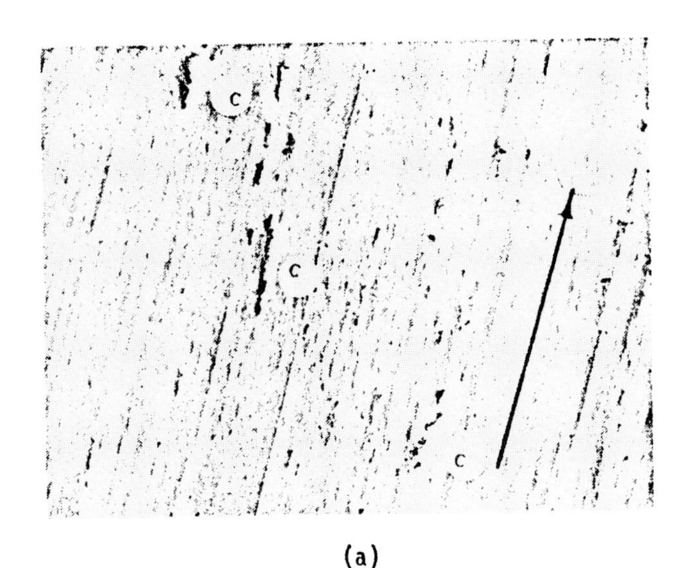

(a)

Magnification 300x

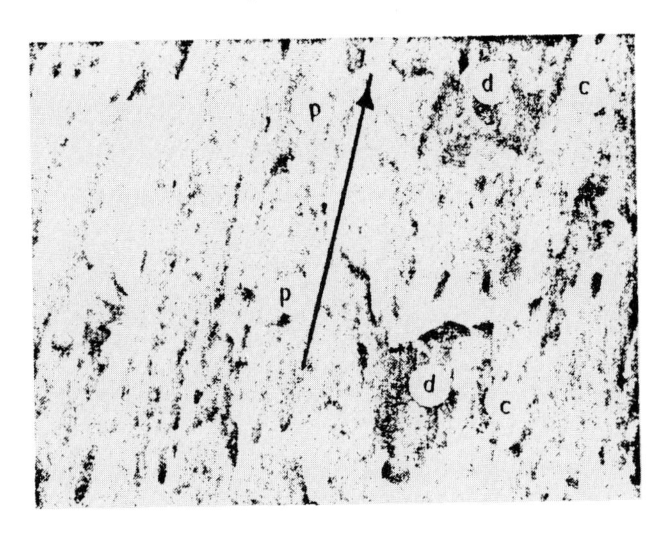

(b)

Magnification 1000x

FIGURE 8. SCANNING ELECTRON MICROGRAPH OF MACHINED SURFACE (Unlubricated, 20ft./min., Wearland 0.020 in., Direction of Tool Motion Given by Arrow).

LEGEND. c - cracks, p - plastic deformation
d - ductile fracture

at the tool cutting edge penetrate into the workpiece surface because of the high normal stresses transmitted across the tool-work piece interface. Relative motion between the tool and workpiece during cutting produces a considerable ploughing effect. At low cutting speeds the temperatures are such that adhesion or pressure welding between the tool and workpiece is limited; conditions of dry sliding friction exist and the grooves tend to remain sharp and well defined.

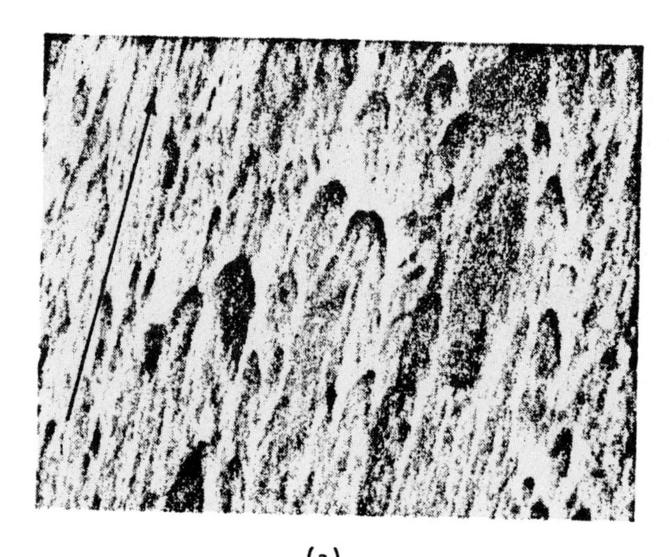

(a)

Magnification 5000x

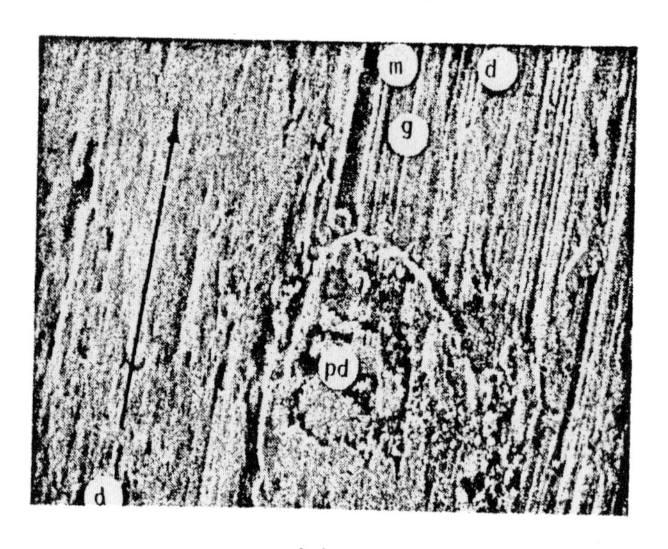

(b)

Magnification 550x

FIGURE 9. SCANNING ELECTRON MICROGRAPH OF MACHINED SURFACE (Unlubricated, 160ft./min., Wearland 0.020 in. Direction of Tool Motion Given by Arrow)

LEGEND m - microchip groove, d - dimpled region
g - grooves, pd - plastically deformed debris

An increase in cutting speed produces an increase in the temperatures generated in the various deformation zones (Figure (3)) and at the tool workpiece interface. There is a transition in the chip formation process from partially discontinuous to continuous and the generation of conditions of sticking friction at the tool-workpiece interface. The result is that there is a decrease in the amount of surface fracture and the gradual disappearance of the sharp grooves observed at low cutting speeds. At very high cutting speeds the surfaces become burnished showing evidence of severe plastic flow. The microchip grooves generated at high cutting speeds arise because of interaction between asperities at the tool flank and the freshly machined workpiece surface. A mechanism for microchip formation has been discussed previously in relation to the machining of AISI 4340 steel [20].

The changes in the appearance of the machined surfaces with an increase in tool wear land length can again be related to the temperatures generated. At low cutting speeds an increase in tool wear land length produces an increase in temperature at the clearance face and in the primary and secondary deformation zones. It is believed that this produces an increase in the ductility of the chip material and a reduction in the extent of gross fracture in the workpiece surface. In addition, there is a gradual transition from sliding to sticking conditions at the clearance face with an increase in subsurface plastic flow. The result is that the well defined grooves observed when cutting with a sharp tool gradually vanish and the surface appears burnished. At high cutting speeds an increase in tool wear land length again produces an increase in temperature at the clearance face and an increase in the extent of subsurface plastic flow with the production of a burnished surface. Thus, it would be expected that the fine scale surface damage should decrease with an increase in tool wear land length. However, data presented in Figure (5) and (9) show that the surface deteriorates as the wear land increases. It is believed that this is caused by the generation of metal debris. It is suggested here that debris can arise from several sources namely, fragments removed from beneath the general level of the workpiece surface by fracture, fragments of the built-up edge lost because of the excessive vibration induced at large tool wear lands, and removal of material as surface wear particles by ductile fracture within the bulk of the workpiece under sticking frictional conditions. Clearly, during machining the larger of these particles will be plastically deformed between the tool land and workpiece surface, to which they become pressure welded (Fig. (9b)).

Additional cutting tests were performed on the work material but with the presence of a lubricant. The specimens were examined first on an optical microscope at magnifications up to 20 diameters. It was found that the surfaces were extremely smooth with an apparent absence of surface damage and the chatter marks found in the tests conducted under dry conditions. The specimens were examined next on the scanning electron microscope over a wide range of magnifications. Figures (10) to (13) show a selection of the generated surfaces produced when machining under lubricated cutting conditions.

It was found that the general changes in the fine structure of the surface produced by changes in cutting speed and tool wear land length for the tests conducted under lubricated conditions were similar in many respects to those produced under dry unlubricated conditions. The major difference appeared to be that the amount of plastically deformed debris was reduced considerably for the tests conducted under lubricated conditions. The similarities in the surfaces generated under lubricated condition can be seen clearly by comparing Figures (6) to (9) with Figures (10) to (13).

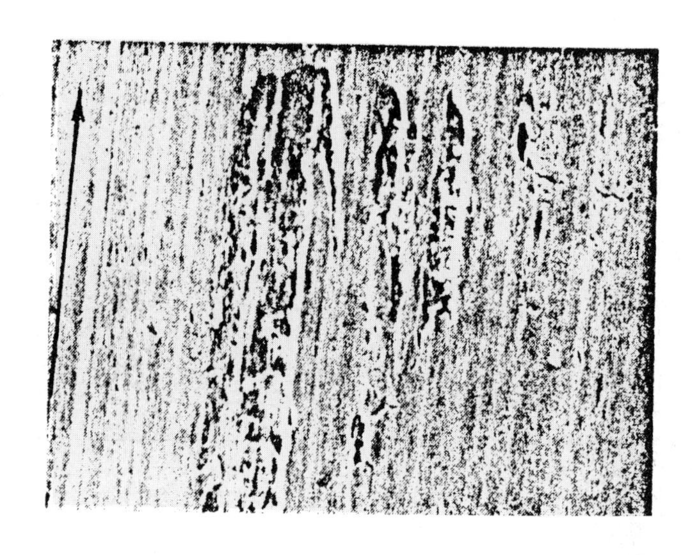

Magnification 300x

FIGURE 10. SCANNING ELECTRON MICROGRAPH OF MACHINED SURFACE (Lubricated, 20ft./min., Sharp Tool, Direction of Tool Motion Given by Arrow).

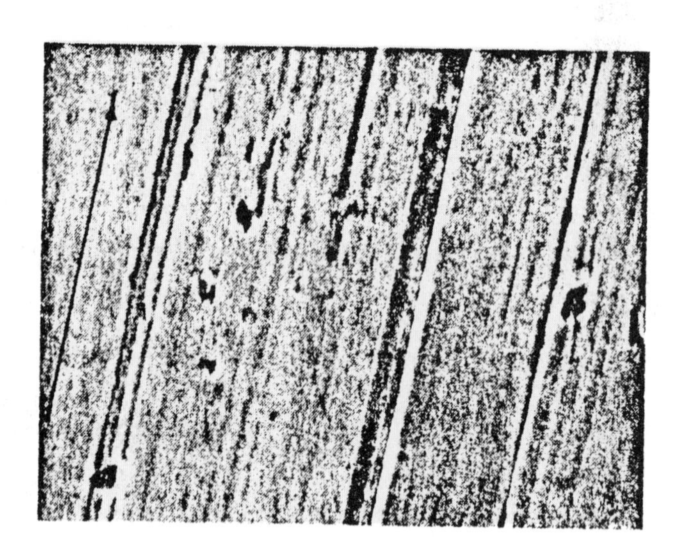

Magnification 1000x

FIGURE 11. SCANNING ELECTRON MICROGRAPH OF MACHINED SURFACE (Lubricated, 160ft./min., Sharp Tool, Direction of Tool Motion Given by Arrow).

It is believed that the reduction in the severity of tool chatter and elimination of much of the plastically deformed metal debris can be attributed, in part, to the elimination of the built-up edge, reduction in tool forces, reduction in the coefficient of friction and frictional force parallel to the tool rake face and increase in cutting ratio brought about by the application of the lubricant.

Magnification 300x

FIGURE 12. SCANNING ELECTRON MICROGRAPH OF MACHINED SURFACE (Lubricated, 20ft./min., Wearland 0.020 in. Direction of Tool Motion Given by Arrow).

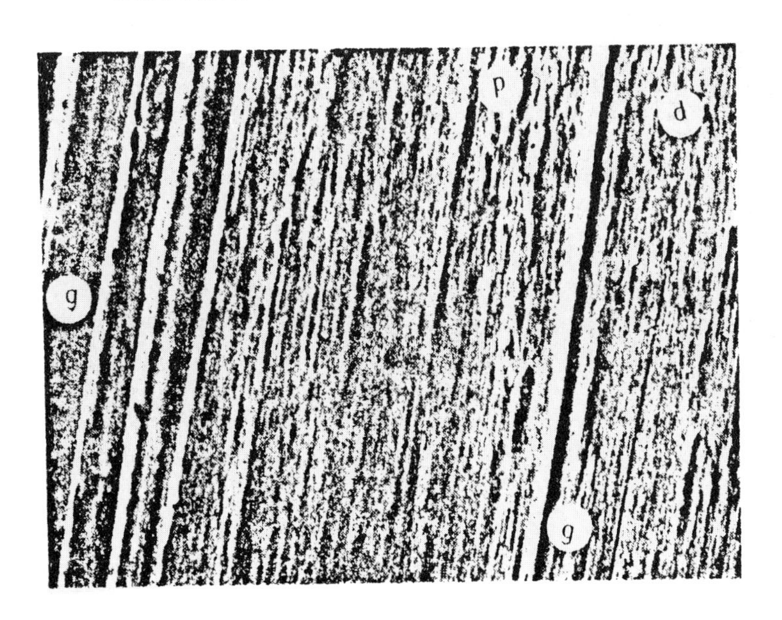

Magnification 300x

FIGURE 13. SCANNING ELECTRON MICROGRAPH OF MACHINED SURFACE (Lubricated, 160ft./min., Wearland 0.020 in. Direction of Tool Motion Given by Arrow).

(c) Subsurface Examination

Sections perpendicular to the machined surface and parallel to the direction of relative work-tool motion were prepared for metallographic examination in the manner described in a previous section. The specimens were examined in the etched and unetched condition and the structure analyzed. Observations were made for specimens machined under dry and lubricated conditions. Figures (14), (15) and (16) show a selection of the microstructures generated and microhardness variation through the surface region.

It was found when machining under dry, unlubricated conditions that a very thin disturbed or plastically deformed layer was produced at the machined surface. The hardness at the machined surface was high and decreased rapidly with increasing depth beneath the surface until the bulk hardness of the work material was attained. Some typical results are shown in Figure (16). It was found also that the thickness of the disturbed layer increased with an increase in cutting speed and tool wear length. This is shown in Figure 17. The presence of phase changes within the surface region was not detected by optical microscopy: a result which may be anticipated. When machining under lubricated conditions subsurface plastic flow occurred only at widely spaced isolated areas at the highest cutting speeds or with tools having the largest wear land lengths used in the investigation.

The hardness alterations and plastic deformation produced in the region of the surface of the workpiece occur largely as a consequence of the interaction between the tool nose region and freshly machined workpiece surface and the temperatures generated during machining. When cutting at low cutting speeds with a sharp tool the temperature rise produced in the freshly machined workpiece surface is small so that condition of sliding friction generally exist with little adhesion and subsurface plastic flow. An increase in cutting speed produces an increase in temperature in the various deformation zones which, in turn, produces an increase in temperature in the region of the surface of the workpiece and the generation conditions of sticking friction. Thus, the extent of subsurface plastic flow increases with an increase in cutting speed. The increase in hardness in the surface region is believed to arise because of plastic deformation and work hardening.

The effect of tool wear land length on the changes observed can again be interpreted in terms of the interaction between the tool nose region and the freshly machined workpiece surface. An increase in tool wear land length produces an increase in adhesion at the flank-workpiece interface. Clearly, the changes in the surface region are consistent with data generated using scanning electron microscopy.

When the work material was machined under lubricated conditions it was found with the exception of tests conducted at the highest cutting speeds or with tools having large wear lands, that the disturbed or plastically deformed layer in the surface region of the workpiece was eliminated. In addition, there was an absence of hardness changes in the region of the surface of the workpiece. These results may be anticipated because an application of a lubricant to the machining process will produce conditions of sliding friction both at the tool rake and clearance faces. This will eliminate the built up edge and prevent flow in the surface region of the workpiece. Again this data is consistent with that generated using scanning electron microscopy.

Attempts were made to investigate quantitatively the distribution of plastic strain in the surface region of the workpiece from an analysis of the distorsion produced in a fine grid embossed on the side surface of the workpiece. This technique has been discussed in detail elsewhere [13]. Figure (18a) shows the grid prior to machining and Figure (18b) the grid after machining at a cutting speed of 160 ft/min with a tool

Magnification 550x

FIGURE 14. SURFACE REGION (Unlubricated, 20ft./min. Sharp Tool)

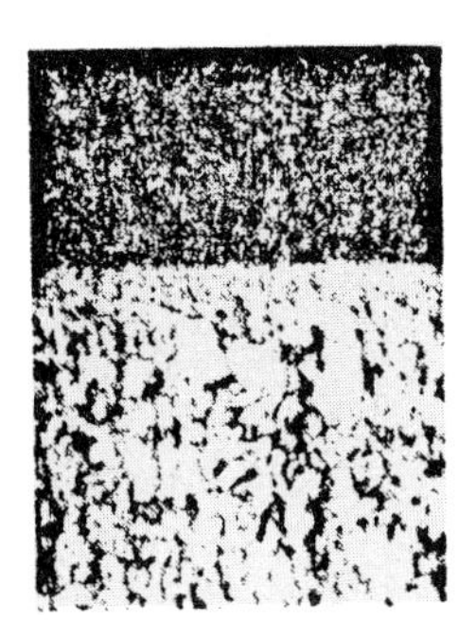

Magnification 550x

FIGURE 15. SURFACE REGION (Unlubricated, 160ft./min. Sharp Tool)

having an artificially controlled wear land of 0.020 in. It can be seen that the subsurface plastic deformation is restricted to a very narrow region having a width of approximately 0.0015 in. The grid spacing is "large" compared to the width of the zone to the point where quantitative determinations of the plastic strain in the surface region cannot be made accurately. However, a test was conducted at a cutting speed of 80 ft/min with a tool having zero clearance angle. The distorted grid produced when machining under this condition is shown in Figure (19). It can be seen clearly that appreciable subsurface plastic strain is produced.

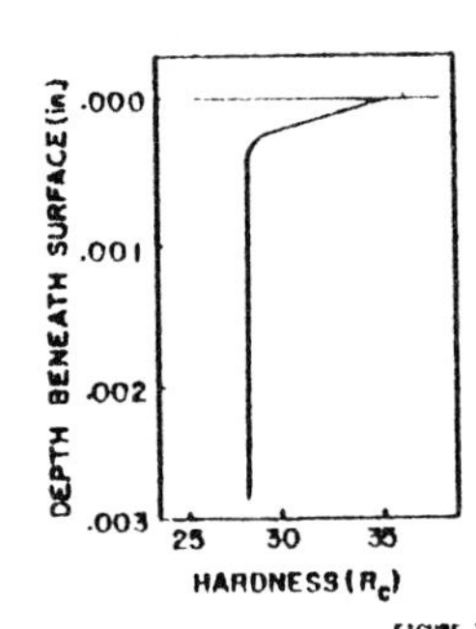

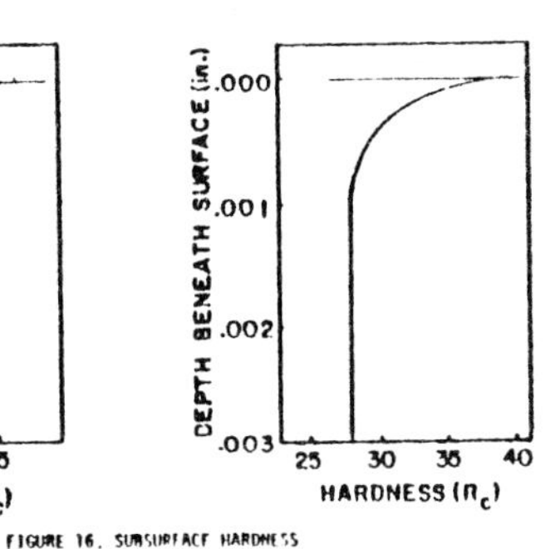

FIGURE 16. SUBSURFACE HARDNESS

The plastic strain distribution in the surface region was calculated from the distorsion produced in the grid and the assumption that deformation is plane strain. The result is presented in Figure (20). It can be seen that the shear strain at the surface is low. It increases with increasing depth beneath the surface to a maximum then decreases

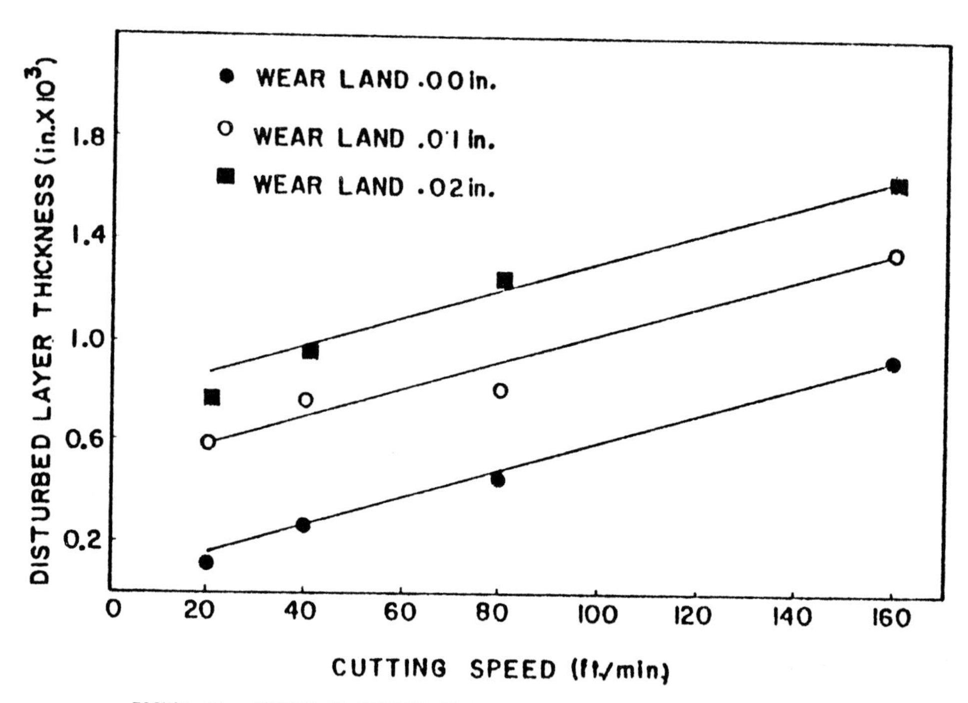

FIGURE 17. EFFECT OF CUTTING SPEED ON SUBSURFACE PLASTIC FLOW

continuously with a further increase in depth. The normal strain parallel to the direction of relative work-tool motion is tensile at the surface and decreases very rapidly with increasing depth beneath the surface. The normal strain perpendicular to the direction of relative work-tool motion is compressive at the surface. It increases with increasing depth beneath the surface to a maximum then decreases with a further increase in depth to a minimum. Finally, the strain increases with an increase in depth beneath the surface becoming negligible at large depths. The depth to which the plastic deformation extends is approximately 0.005 in. The sum of the normal strains is not zero and indicates that some lateral spread of material has occurred violating the assumption of plane strain deformation.

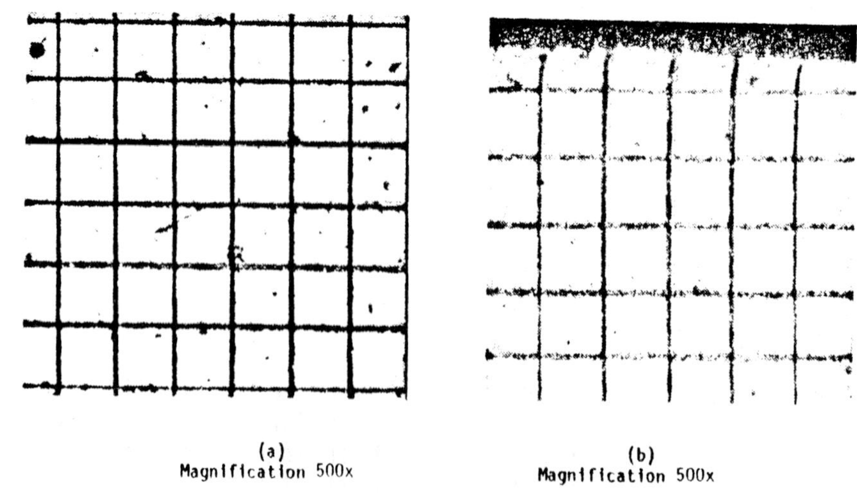

(a) Magnification 500x

(b) Magnification 500x

FIGURE 18. SUBSURFACE DEFORMATION.

FIGURE 19. SUBSURFACE PLASTIC FLOW

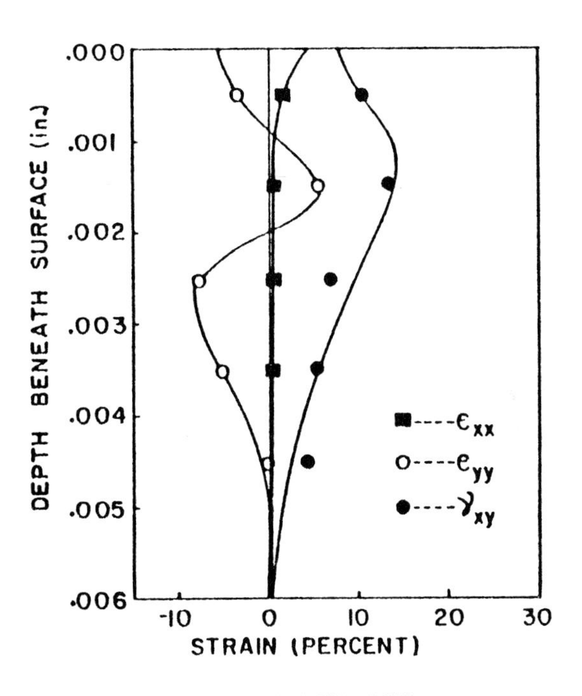

FIGURE 20. SUBSURFACE PLASTIC STRAIN

LEGEND ϵ_{xx} - Normal Strain Parallel to Direction of Tool Motion
ϵ_{yy} - Normal Strain Perpendicular to Direction of Tool Motion
γ_{xy} - Shear Strain

(d) Surface Roughness

Surface roughness measurements were made on the specimens in a direction normal to that of relative work-tool motion. It was found that the surface roughness decreased with an increase in cutting speed and decrease in tool wear land length. This is illustrated in Figure 21. Reliable surface roughness measurements could not be obtained in a direction

Source: *SME Technical Paper IQ74 – 185*

parallel to that of relative work-tool motion because of the general curvature of the surface of the workpiece. It is pointed out here that the ideal surface roughness is zero because cutting is conducted under orthogonal conditions. Therefore, surface roughness measurements reflect only instabilities within the cutting process.

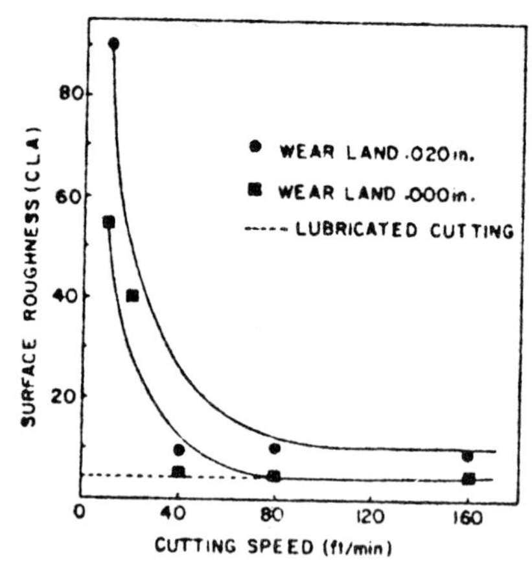

FIGURE 21. EFFECT OF CUTTING SPEED ON SURFACE ROUGHNESS

(e) X-Ray Microprobe and Auger Analysis

X-Ray Microprobe and Auger analyses were conducted at points on the surface and to a depth of approximately 0.005 in. beneath the surface. It was found that there was an absence of compositional changes in the surface region. The composition of material at the surface appeared identical to that of the bulk material. However, X-Ray Microprobe analysis did prove useful for identifying fragments of metal debris on the workpiece surface.

CONCLUSIONS

The following conclusions are drawn from the results of the investigation.

(1) A surface region is produced which may be quite different from the bulk material.

(2) When machining at low cutting speeds with sharp tools damage, which is restricted to a variety of geometrical defects associated with the surface, arises because of the presence of a built-up edge on the tool rake face and relative motion between the tool nose region and workpiece. The tool nose region is interpreted as including the tool cutting edge and tool land.

(3) When machining at high cutting speeds or with tools having large artificially controlled wear lands the surface damage is accompanied by a plastically deformed or disturbed surface region which arises because of seizure between the tool and workpiece in the tool nose region.

(4) There is an absence of phase changes of a metallurgical nature and compositional variations in the surface region.

ACKNOWLEDGEMENT

The authors are very grateful to the National Science Foundation f the aid received to support the work through Grant GH 33761.

REFERENCES

1. P.G. Fluck, The Influence of Surface Roughness on the Fatigue Life and Scatter of Test Results of Two Steels. Proc. Am. Soc. Test Mat., Vol. 51, p. 584, 1951.

2. W. P. Koster, L.J. Fritz and J.B. Kohls, Surface Integrity in Machining of 4340 Steel and Ti-6Al-4V. SME Technical Paper No. 1Q71-237, 1971.

3. W. P. Koster and L. J. Fritz, Surface Integrity in Conventional Machining, ASME Paper No. 70-GT-100, 1970.

4. M. Field and J.F. Kahles, The Surface Integrity of Machined and Ground High Strength Steels. DMIC Report 210, p.54, 1964.

5. A. J. Chisholm, The Characteristics of Machined Surfaces. Machinery (London), Vol. 74, p. 729, 1950.

6. C. T. Ansell and J. Taylor, The Surface Finishing Properties of Carbide and Ceramic Cutting Tools. Advances in Machine Tool Design and Research, Oxford, Pergamon Press Ltd., 1962.

7. A. F. Allen and R. C. Brewer, The Influence of Machine Tool Variability and Tool Flank Wear on Surface Texture. Advances in Machine Tool Design and Research, Oxford, Pergamon Press Ltd., 1965.

8. M. Field, W. P. Koster and J. F. Kahles, Effect of Machining Practice on the Surface Integrity of Modern Alloys. International Conference on Manufacturing Technology, ASTME, p. 1319, 1967.

9 J. F. Kahles and M. Field, Surface Integrity - A New Requirement for Surfaces Generated by Material Removal Methods. Proc. Inst. Mech. Eng., Vol. 182, p. 31, 1968.

10. M. Field, J. F. Kahles and W. P. Koster, The Surface Effects Produced in Nonconventional Metal Removal - Comparison with Conventional Machining Techniques. ASM Metals Engineering Quarterly, Vol. 6, No. 3, p. 32, 1966.

11. M. Field and J. F. Kahles, Review of Surface Integrity of Machined Components. Annals of CIRP, Vol. 20, No. 2, p. 153, 1971.

12. M. Field, J. F. Kahles and J. T. Gammett. A Review of Measuring Methods for Surface Integrity. Annals of CIRP, Vol. 21, p. 219, 1972.

13. J. A. Bailey and S. Jeelani, Determination of Subsurface Plastic Strain in Machining Using an Embossed Grid, paper submitted to ASME Journal of Engineering for Industry.

14. R. A. Douglas, C. Akkoc and C. E. Pugh, Strain Field Investigations with Plane Diffraction gratings. Proc SESA, Vol. 22, No. 2, p. 233, 1964.

15. G. Boothroyd, A Metal Cutting Dynamometer Engineer, London. Vol. 213, p. 351, 1962.

16. M. C. Shaw, The Assessment of Machinability. Proc. Conference on Machinability ISI, Inst. Metals, Inst. Mech. Eng., Inst. Prod. Eng., p. 1, 1965.

17. N. N. Zorev, Metal Cutting Mechanics (English Translation) Pergamon Press, 1966.

18. J. A. Williams, E. F. Smart and D. R. Milner, The Metallurgy of Machining, Part 2. The Cutting of Single Phase, Two Phase and Some Free Machining Alloys. Metallurgia, Vol. 81, p. 51, 1970.

19. J. A. Bailey, Surface Damage in the Machining of AISI 4340 Steel Wear, Vol. 27, p. 161, 1974.

20. J. A. Bailey and S. E. Becker, On Microchip Formation in the Machining of a High Strength Steel. To be published Transactions, ASME, J. Materials Engineering and Technology.

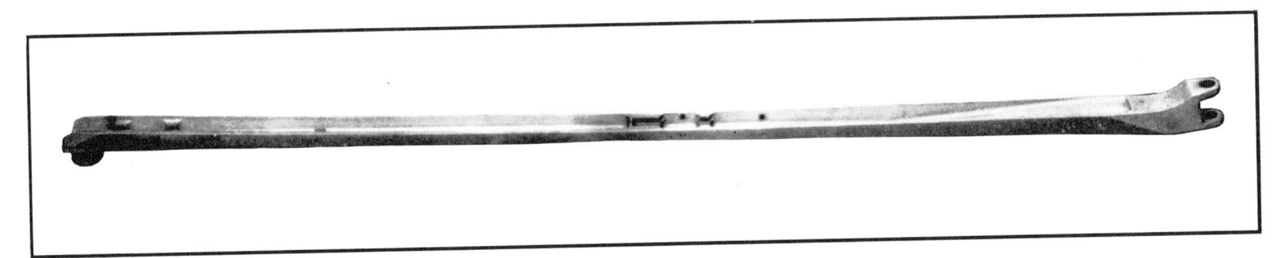

Preventing Embrittlement in Cadmium Plated Maraging Steel

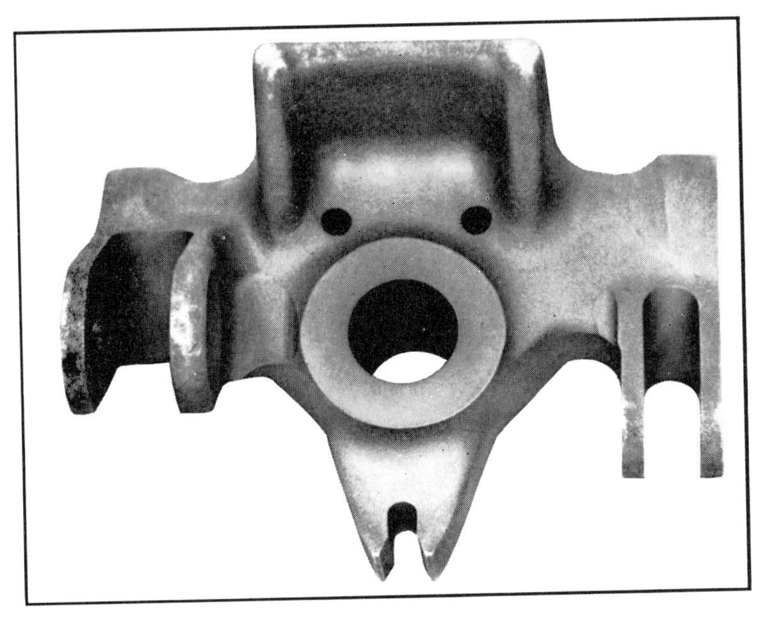

Made of 18% Ni maraging steel, shanks and fittings of arresting hooks are machined, heat treated and plated with cadmium to prevent corrosion. Baking for 24 hr at 375 F follows plating.

By EDWARD A. LAUCHNER and ROBERT E. HERFERT

"Plate components of 18% Ni maraging steels in low hydrogen baths, and follow by baking", say the authors.

Two sentences sum up the results of our research into the effects of cadmium plating on maraging steels:

- Although maraging steel reportedly resists hydrogen embrittlement, cadmium plating in a standard QQ-P-416 bath containing brighteners can produce embrittlement.
- Embrittlement is eliminated by plating in a special low hydrogen bath that does not contain

Mr. Lauchner is senior metallurgist, materials engineering, and Mr. Herfert is engineering specialist, Metallics Research Branch, Northrop Norair, Div. of Northrop Corp., Hawthorne, Calif.

Source: *Metal Progress*, Sept 1966

brighteners, followed by baking for minimum times.

This work was inspired by plans to manufacture aircraft parts of 18% Ni maraging steel. These shanks and fittings of arresting hooks, shown on the opposite page, were to be machined to finish dimensions, then maraged to desired tensile strength, 280,000 psi. Since maraging steel is not corrosion resistant, the parts had to be cadmium plated, then baked for 24 hr at 375 F.

According to previous research, maraging steel resists hydrogen embrittlement. Reportedly more resistant to absorption of hydrogen than is low alloy steel of comparable strength, maraging steel also releases hydrogen more readily when baked, restoring ductility.

For these reasons, we planned to use electrodeposited cadmium coatings. But it was not known whether a standard cadmium plating solution (QQ-P-416) would suffice. (A special low hydrogen plating bath might be needed instead.) Therefore, we decided to determine the effects, on maraging steels, of conventional sodium cyanide baths containing brighteners and of low hydrogen cyanide plating baths.

Details of Testing

Round transverse notch bars (0.357 in. reduced diameter; 0.252 in. diameter at a notch root of 0.002 in. radius) were machined from a bar of 18% Ni maraging steel. After these specimens were aged to tensile strengths of approximately 280,000 psi, they were cleaned by blasting with 150 grit aluminum oxide. One group was plated in a standard cadmium plating bath containing brighteners. The other set of specimens was plated in a low hydrogen cyanide bath (sodium cyanide and cadmium oxide dissolved in de-ionized water) that did not contain brighteners. After being plated (to 0.0003 to 0.0005 in. thicknesses), bars were given a chromate treatment. Various specimens were baked for 1, 3, 8 or 24 hr at 375 F; others were left unbaked as controls. All bars were then statically loaded on a machine at 260,000 psi until failure or until 200 hr had passed.

Results revealed that all bars plated in the standard bath failed, whether baked or not, in less than 130 hr. Conversely, of the specimens plated in the special bath, only one unbaked bar failed after 163 hr. Furthermore, no baked bars failed. Figure 1 shows actual times that all bars endured testing at 260,000 psi.

Fig. 1—Specimens plated in the bath containing brighteners failed under a static load at 260,000 psi before the 200 hr limit was reached. Conversely, bars plated in the low hydrogen bath resisted breakage at 260,000 psi, particularly when baked. Only one of these specimens, unbaked, failed this test.

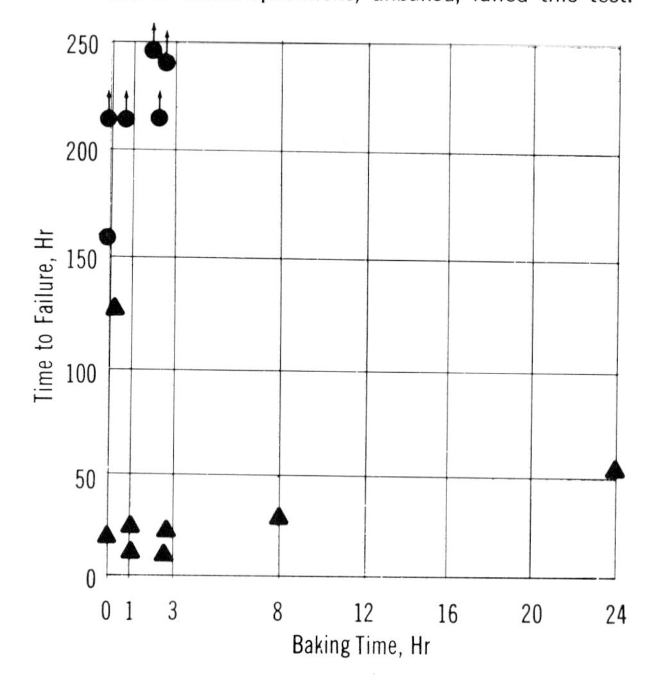

SECTION XI:
Powder Metallurgy

Type 350 maraging steel processed by powder metallurgy

S. Isserow

Prealloyed powder of Type 350 maraging steel was prepared by the rotating electrode process. Rods and tubes were extruded from this powder and also from cast-and-wrought bar stock from two sources. Ageing for 6 h at 510°C brought the various extrusions to full strength without a re-solution heat-treatment. Quantitatively the mechanical behaviour was similar for rods and tubes from either powder or cast-and-wrought stock; strength levels in tension tests were about the same, ductility tended to be lower for powder-derived stock, and Charpy impact values showed an advantage for powder-derived stock at −40°C but not at room temperature. Qualitatively, the powder-derived material showed a different fracture mode in both tension and impact tests, especially the former. The wooden fracture of powder-derived material is related to the greater banding of this material.

Serial No. 454. Manuscript received 13 April 1977; in final form 4 July 1977. S. Isserow, BA, MS, PhD, is with the Army Materials and Mechanics Research Center, Watertown, Massachusetts, USA. This subject has been discussed by the author in Report No. AMMRC TR 76–29 which includes a greater number of figures.

Processing of prealloyed powders offers several advantages in highly alloyed materials such as the maraging steels. Alloy powders often provide the benefit of improved homogeneity relative to cast material. In addition, preparation of powder by rapid cooling of molten globules imparts finer distributions of microstructural features in the powder which can have a marked effect on mechanical properties. Examples of this improvement are shown in titanium-,[1] magnesium-,[2] aluminium-,[3] and nickel-base superalloys.[4] Application of this approach to the elimination of segregation in maraging steel is the subject of a pending patent application by AMMRC personnel.[5]

Different methods are available for large-scale preparation of prealloyed powders. For rapid quenching, two methods have been applied to a variety of alloys. One method, atomization, depends on one of the inert gases dispersing a stream of molten metal into globules which freeze as they dissipate heat to the gas. The other method, the rotating electrode process (REP) invented by A. R. Kaufmann,[6] utilizes the formation of molten metal at the tip of a rapidly rotating electrode when an arc is maintained between this and another electrode, the latter generally being a nonconsumable tungsten rod. The molten globule is whirled off by centrifugal force and freezes in flight. The choice of powder preparation method for a specific situation is determined by factors such as powder size and morphology required, reactivity of the molten alloy with crucible or nozzle materials, ease of electrode preparation, and economic considerations.

This paper uses the 'short' ton (2000 lb).

APPLICATION TO MARAGING STEELS

The maraging steel family offers a remarkable combination of high strength and toughness, even at strengths beyond 2100 MPa (300 ksi [lbf × 10^3 in^{-2}]). In applications such as aircraft structures, considerable benefit accrues from even a modest increase in either strength or toughness. Previous work has indicated that the powder approach offers a means of enhancing one of these properties without compromising the other. In preliminary work on Types 300 and 400 maraging steels,[7,8] Abrahamson investigated rods extruded from REP powder and demonstrated that the best combination of strength and toughness is achieved by direct ageing of the extruded rod without intermediate heat-treatment or homogenization.

Maraging steel powder can also be prepared by gas atomization. Snape and Veltry[9] extruded Type 350 powder at higher temperatures than used by Abrahamson. The extruded rods were homogenized and aged before testing. The strengths were somewhat lower (YS 2206 MPa (320 ksi); UTS 2289 MPa (332 ksi)) than for cast-and-wrought (c/w) material, but very high impact strengths of 37–42 J (23·5–27 ft lbf) were observed. These values were attributed to benign heterogeneity in the extruded rod, as shown also by the woody fracture.

Van Swam, Pelloux, and Grant investigated different methods of powder preparation and consolidation for Type 300 maraging steel,[10] and found that powder-derived material was superior to conventional material in tensile properties but fracture toughness and fatigue life were not improved. Commercial stock was improved in both tensile properties and toughness by grain refinement through heavy hot rolling.

Work with elemental or partially prealloyed powders has yielded inferior mechanical properties compared with fully prealloyed powders (see for example Ref. 9), and is therefore not cited here.

EXPERIMENTAL PROGRAMME

The programme reported here concentrated on Type 350 maraging steel (nominal composition 17·5Ni–12·5Co–3·8Mo–1·7Ti–0·15Al; in the materials used Ni and Mo are higher, Ti and Al are lower—see Table 1). Rods and tubes were extruded under conditions guided by Abrahamson's experience with Types 300 and 400 maraging steels.[7,8] Both rods and tubes were prepared from powder and also from c/w stock from two sources. Two types of powder were used in each case, REP and gas-atomized. The gas-atomized powders gave unsatisfactory extrusions, which were not processed further. Both tube and rod stock were subjected to several ageing treatments before mechanical testing, and were tensile tested. The rod stock was also used for Charpy impact tests.

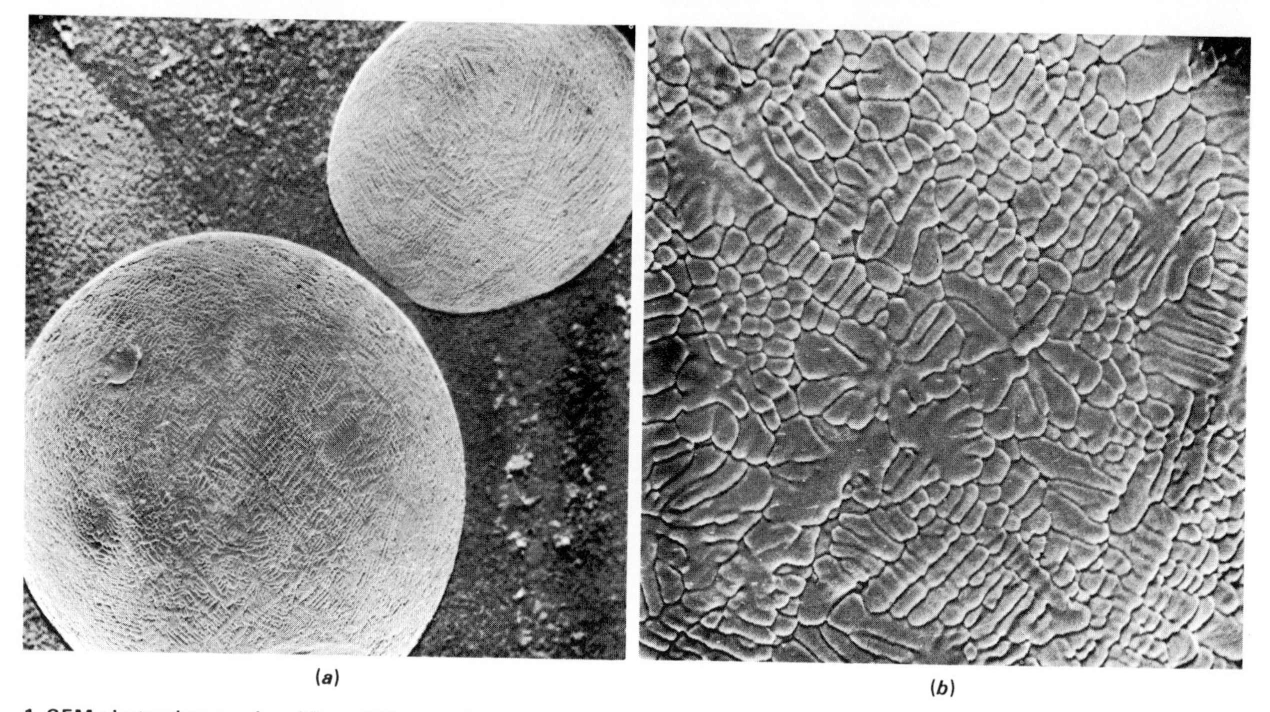

1 SEM photomicrographs of Type 350 maraging steel powder by rotating electrode process (REP). Magnifications ×225 and ×1125.

Preparation of materials

Powder

Standard 254 mm (10 in) long electrodes were machined from commercial 76 mm (3 in) dia. bar. Each rod was provided with a 63·5 mm ($2\frac{1}{2}$ in) dia. hub, 50·8 mm (2 in) long. The remaining length was machined to a diameter near 68·6 mm (2·7 in) to maximize the REP powder yield while remaining within the 7·25 kg (16 lb) limit specified for each electrode. Twenty-one electrodes gave 118 kg (261 lb) of powder. The sieve analysis obtained for the powder is shown in Table II.

Powder was used directly for the extrusion billets except that the fines (−325 mesh) were first screened out. Chemical analyses of the powder and the c/w stock at AMMRC gave the results listed in Table I, along with the analyses reported by the suppliers. The surface of the powder is seen in the scanning electron microscope (SEM) photomicrographs at different magnifications in Figs. 1(*a*) and (*b*). Microhardness measurements with a 100 g load averaged MH_{100} 398 corresponding to 39 HRC or less, indicative of the solution-treated structure.

One tube and one rod were extruded from each of two different lots of gas-atomized powder. Unfortunately, both lots proved to be contaminated, resulting in defective extrusions containing excessive numbers of inclusions and some cracks. For this reason the processing of these powders is mentioned only incidentally in the remainder of the paper.

Extrusions

A set of six tubes and a set of five rods were extruded on a 1270 t (1400 ton) press under essentially the same conditions (see Table III). All billets (except for Rod 5) consisted of mild steel cans in which the maraging steel stock was contained.

Table I Chemical analysis of Type 350 maraging steel (wt.-%)

	114 mm ($4\frac{1}{2}$ in) RCS bar		76 mm (3 in) dia. bar (electrode stock)		REP powder
Element	Source	AMMRC	Source	AMMRC	AMMRC
Ni	18·65	18·44	18·39	18·54	18·43
Co	12·03	12·44	12·03	12·18	12·28
Mo	4·70	4·57	4·78	4·68	4·64
Ti	1·38	1·47	1·35	1·68	1·56
Al	0·10	0·14	0·13	0·08	0·16
C	0·008	0·008	0·011	—	0·016

Table II Sieve analysis of REP powder

Sieve (US series)	35	45	60	80	120	170	230	325	PAN
Microns	500	354	250	177	125	88	63	44	<44
Retained on screen, %	0	1·71	11·79	28·44	22·54	21·17	7·99	5·27	1·09

The cans were evacuated, outgassed at 427°C, and sealed. The powder charges were weighed and showed a packing density of ~68% of theoretical for REP powder vs. ~63% for gas-atomized. The powder billets were not subjected to any separate prior compacting step. The upsetting of the preheated billet within the extrusion tools was considered to provide the prior compaction. All billets except for Rod 1 were preheated to 900°C and the extrusions were rapidly water-quenched.

Tubes with nominal dimensions of 46·48 mm (1·83 in) o.d. x 5·08 mm (0·20 in) wall thickness were sought in the series of six extrusions. Two billets were prepared from each of the three types of Type 350 maraging steel stock: cast and wrought; gas-atomized powder; and REP powder. The second tube (B) had different c/w stock in the front and rear halves (Table III). The extrusion forces, especially at the upset or breakthrough of the billets, approached the capacity of the press. For example, the final billet, containing REP powder, needed ~12 MN (1300 tonf) for the upset. These high forces compounded problems arising from the small diameter of the mandrel and the difficulty of maintaining lubrication between billet and mandrel. Each of the four powder-containing billets thus manifested a deficiency associated with the mandrel. Each type of powder gave one billet in which the load was sufficient to cause failure of the mandrel under tension, so that the inward deformation of the billets was not controlled by the mandrel; the resulting tubes (C and F) were unsatisfactory. Each type of powder did give one satisfactory tube (D and E) which, however, was bound on its mandrel so that the tube did not clear the die but had to be cut in the press before it could be placed in a quench tank. The two billets of solid stock (c/w) extruded with no difficulty and were quickly water-quenched.

Rods with diameters near 25 mm (1 in) were obtained in the series of five extrusions. Four rods represented different starting materials extruded under identical conditions, two powders and two c/w bars. (One of these bars was extruded bare since it was large enough for extrusion in the same press liner without any can). An additional REP powder billet was prepared for extrusion at lower temperature. After this billet stalled in an attempt to extrude at 760°C, it was turned down and extruded about a week later at 843°C. The other four extrusions, all at 900°C, were uneventful. The high reduction ratio (13:1) and the high extrusion constants (K_u and K_r) necessitated ram stresses very close to the limit of 1500 MPa (110 tonf in^{-2}).

Post-extrusion processing

Surface clean-up before heat-treatment was necessary to remove lubricants, canning, or oxidation product. For rods, machining is the simplest means, especially for test specimens. For tubes, chemical means were sought to overcome any difficulty in machining the inside and to conserve the rather thin wall (~5 mm (0·2 in)). To reduce the generation of fumes and also the possibility of hydrogen pick-up by the core metal, a substantial portion of the outer can was machined off before removal of the balance by acid. Initially 254 mm (10 in) lengths of tubing were pickled in a 1:1 aqueous solution of concentrated nitric acid. For a subsequent section, E3, the 1:1 solution used concentrated hydrochloric acid, which left a thicker wall, indicating that the nitric acid had attacked the maraging steel at a slower rate than the mild steel jacket.

The tubes posed another problem in preparation of test specimens: namely, the matching of the curvature to a grip or fixture. It was decided to flatten tube sectors for flat serrated grips. This procedure was first checked on a strip cut from Tube B prepared from c/w stock. Strips ~19 mm (¾ in) wide were cut from cleaned 229 mm (9 in) lengths of the tube. At

Table III Extrusion of Type 350 maraging steel billets

Material ident.	Material	Temp., °C	FORCES Upset MN (tonf)	K_u,* MPa (tonf in^{-2})	Running MN (tonf)	K_r,* MPa (tonf in^{-2})
Tubes	*Tool diameter: liner 118·24 mm (4·655 in); mandrel 35·6 mm (1·4 in); die 50·8 mm (2·0 in); R=9·6*					
A	Cast/wrought 114 mm (4·5 in) RCS bar	904	10·85(1220)	471(34·2)	9·52(1070)	420(30·5)
B	Front: c/w 76 mm (3 in) electrode stock, upset to 102 mm (4 in) Rear: as A	904	10·23(1150)	452(32·8)	9·25(1040)	408(29·6)
C	Gas-atomized powder	904	10·76(1210)	475(34·5)	9·34(1050)	412(29·9)
D	Gas-atomized powder	904	11·12(1250)	491(35·6)	9·56(1075)	422(30·6)
E	REP powder	904	11·34(1275)	500(36·3)	10·23(1150)	452(32·8)
F	REP powder	904	11·56(1300)	510(37·0)	10·45(1175)	462(33·5)
Rods	*Tool diameter: liner 92·84 mm (3·625 in); die 25·4 mm (1 in);* R = *13·1*					
1	REP powder	760	10·00(1125)†			
		843	9·87(1110)	576(41·8)	8·94(1005)	522(37·9)
2	C/w 76 mm (3 in) electrode stock	904	8·98(1010)	525(38·1)	8·45(950)	493(35·8)
3	REP powder	904	8·72(980)	508(36·9)	8·45(950)	493(35·8)
4	Gas-atomized powder	904	8·54(960)	499(36·2)	8·10(910)	473(34·3)
5	C/w 114 mm (4·5 in) RCS bar‡	904	9·34(1050)	546(39·6)	8·45(950)	493(35·8)

* $P=F/A=K \ln R$
† Stalled (1610 MPa [117 tonf in^{-2}] stress on 88·9 mm [3·5 in] dia. ram vs. usual 1500 MPa [110 tonf in^{-2}] limit)
‡ Extruded bare, omitting mild steel canning and evacuation

this stage the tubing was still relatively soft (~32–35 HRC). The strips were flattened at room temperature in a forging press with subsequent straightening on an anvil. Substantial machining close to final dimensions is best performed at this stage, before the ageing treatment for rods as well as for tubes. Similarly, the Charpy specimens were milled from the rods, aged, and then notched.

The following heat-treatments, designated A–D, were used for final ageing:

A. Aged 6 h at 510°C
B. Aged 3 h at 510°C
C. Aged 6 h at 482°C
D. Solution-treated 1 h at 816°C; aged 6 h at 510°C.

Heat-treatment D added solution-treatment before final ageing mainly because this STA treatment (solution-treat and age) is the standard sequence for conventional material. All heat-treatments were in air and the samples were air-cooled.

Since the results for the tubes showed that Treatments B and C produced underaged material, these treatments were included only for Rod 1, on the chance that the lower extrusion temperature might modify its ageing response.

Mechanical tests

The rods were used to prepare standard round tension specimens 12·827 mm (0·505 in) gauge diameter and standard Charpy V-notch specimens. The flattened tube sectors were used to obtain standard flat tension specimens, but the width of the gauge section was reduced from 12·7–7·6 mm (0·5–0·3 in).

The results of the tests are summarized in Table IV for the tubes and Table V for the rods. Elongation data are not included for the tubes since almost all the specimens gave very low values as a result of failure outside the gauge marks. This behaviour is conceivably related to the flattening of the tube sectors. Yet an effect due to straining in the flattening operation seems to be ruled out by the similarity of properties for strips aged directly (Heat-treatment A) and strips solution-treated at 816°C before ageing (Heat-treatment D).

Metallography

The powder specimens were metallographically prepared using the following etchant: 1 g $CuCl_2$, 150 ml water, 50 ml concentrated HCl.

Sections from the tubes and rods were examined, both as-extruded and heat-treated. All sections were cut longitudinally, i.e. parallel to the extrusion direction, to permit observation of fibres or banding. The etchant was 5 g $FeCl_3$, 2 ml concentrated HCl, ~300 ml ethyl alcohol.

As-extruded tubes are seen in Fig. 2. Banding is seen at ×60 in both materials; it is most extreme in the powder-derived material (Fig. 2(*b*)).

Table IV Mechanical properties of strips from Type 350 maraging steel tubes

Tube identification*	Material	0·2 YS, MPa(ksi)	UTS, MPa(ksi)	RA, %	HRC	H content, ppm
B2A	C/w 76 mm (3 in) bar (electrode stock)	2384(345·7)	2395(347·3)	45·2	58·2	
B2B		2248(326·1)	2293(332·5)	43·8	56·9	
B2C		2240(324·9)	2286(331·5)	41·0	57·5	
B2D		2404(348·6)	2428(352·2)	36·5	58·1	
B4A	C/w 114 mm (4·5 in) RCS bar	2378(344·9)	2450(355·3)	35·7	59·3	0·1
B4B		2375(344·4)	2402(348·3)	45·3	56·8	0·1
B4C		2417(350·6)	2430(352·4)	39·0	58·0	
B4D		2377(344·7)	2413(350·0)	30·9	58·5	
E2A2	REP powder extruded	2437(353·5)	2438(353·6)	25·2	58·8	0·2
E3A1		2273(329·6)	2423(351·4)	17·6	58·2	1·0
E3A2		2387(346·2)	2427(352·0)	13·8	58·3	0·4
E2B		2270(329·2)	2373(344·1)	20·9	57·7	0·2
E2C		2269(329·1)	2297(333·2)	25·8	56·8	
E2D		2366(343·2)	2410(349·6)	25·6	57·7	

* First letter refers to tube, second letter is code for heat-treatment (see text).

Table V Mechanical properties of Type 350 maraging steel rods

Rod identification.*	Material source	0·2 YS MPa(ksi)	UTS MPa(ksi)	Elong., %	RA, %	Charpy V-notch impact	
						Room temp., J (ft lbf)	−40°C, J (ft lbf)
2A	C/w 76 mm (3 in) dia. bar (electrode stock)	2502(363·0)	2544(369·0)	8·5	36·0	11·0(8·1)	6·8(5·0)
2D		2517(365·0)	2554(370·5)	6·0	20·5	8·7(6·4)	8·4(6·2)
5A	C/w 114 mm (4·5 in) RCS bar	2403(348·5)	2475(359·0)	7·5	32·5	11·0(8·1)	7·3(5·4)
5D		2413(350·0)	2451(355·5)	8·0	27·5	11·4(8·4)	11·0(8·1)
3A	REP powder extruded at 899°C	2372(344·0)	2482(360·0)	8·0	25·0	11·4(8·4)	11·0(8·1)
3D		2437(353·5)	2468(358·0)	5·0	12·0	11·7(8·6)	8·4(6·2)
1A	REP powder extruded at 843°C	2413(350·0)	2436(353·3)	4·5	14·0	11·7(8·6)	11·4(8·4)
1B		2365(343·0)	2413(345·0)	7·5	25·4	9·5(7·0)	11·0(8·1)
1C		2317(336·0)	2413(345·0)	8·5	19·5	11·0(8·1)	7·1(5·2)
1D		2475(359·0)	2496(362·0)	4·5	17·5	11·7(8·6)	10·6(7·8)

* Number refers to rod, letter is code for heat-treatment (see text).

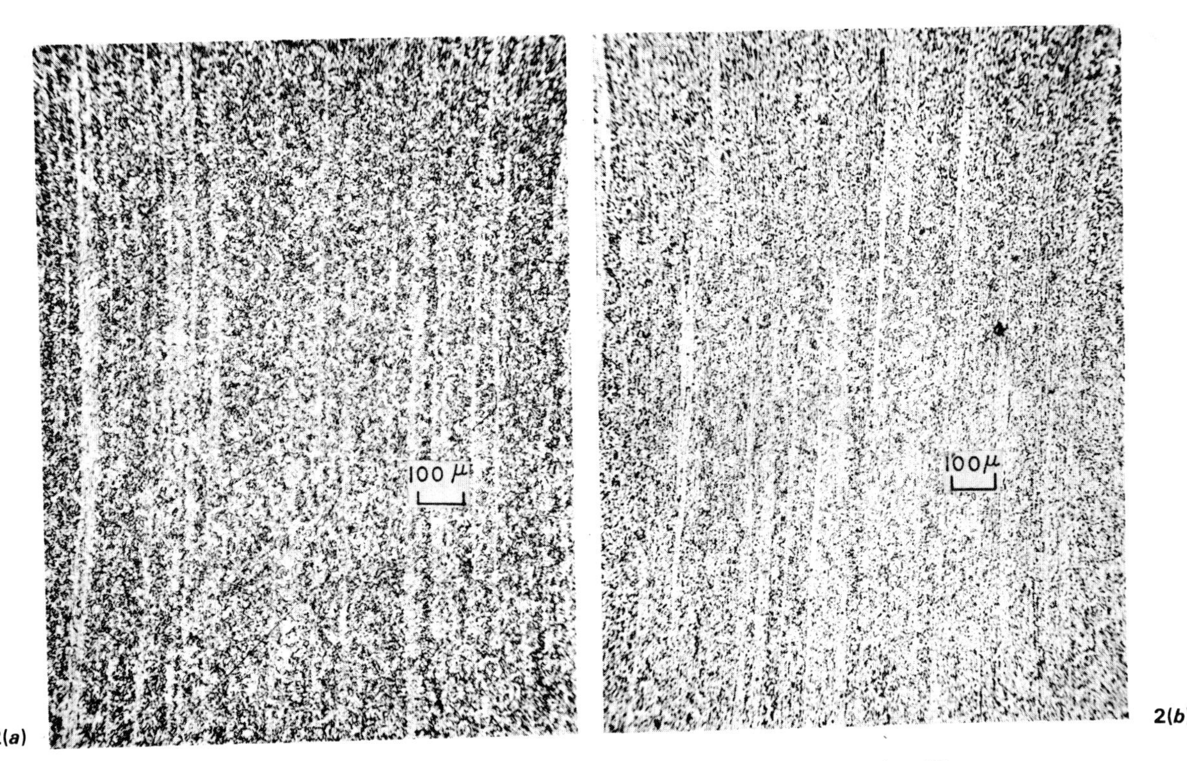

2 As-extruded Type 350 maraging tubes: (*a*) 114 mm (4½ in) RCS bar stock; (*b*) powder stock. × 60.

Banding in the heat-treated tubes follows a similar pattern. In tubes aged directly at 510°C (Fig. 3), banding is absent from the electrode stock but marked in the powder-derived tube. Solution-treatment and then ageing resulted in some banding in the electrode stock but the banding was much more marked in powder material.

The structures of the as-extruded rods were similar to that of the tube shown in Fig. 2. In the aged rods, banding is not visible in the electrode stock at × 600 (Fig. 4(*a*)), but is marked in the powder stock (Fig. 4(*b*)). As in the tubes (Fig. 2(*a*)), the heavier c/w stock (Tube B rear or B4 and Rod 5, not shown) had banding intermediate to that of the electrode stock and the powder.

DISCUSSION OF RESULTS

The test data in Tables IV and V permit assessment of the factors affecting the mechanical behaviour of the tubes and rods. These assessments consider strength, ductility, and impact resistance. Hardness values are not cited, but in practically every case they support conclusions based on strength. Yield and tensile strengths are discussed interchangeably; their values are quite close, as the maraging steels undergo very little work-hardening.

Heat-treatment

The specimens were subjected to the four heat-treatments described earlier. Ageing for 6 h at 510°C (Heat-treatment A) is necessary to achieve full strength. This need was established for tubes from the electrode stock B and from powders E2 and E3, but not from the 114 mm (4½ in) bar stock B4. Rod 1, the only rod subjected to the full set of heat-treatments, confirmed this. Reduction of time to 3 h (Heat-treatment B) or of temperature to 482°C (Heat-treatment C) left the material underaged, preventing realization of full strength (except for the B4 stock). This underageing did not affect impact resistance nor, except possibly for Rod 1, did it enhance ductility.

Solution-treatment (Heat-treatment D) before the optimum ageing did not affect the strength of the tubes or rods made from either c/w or powder stock. The only exception is Rod 1 (lower-temperature extrusion), whose strength was greater by 62 MPa (9 ksi) after Heat-treatment D. The elongation and reduction-of-area data indicated that ductility may be lowered by the insertion of solution-treatment between extrusion and ageing. The Charpy impact data are all so close that they do not show any effect of this solution-treatment.

It is significant that conventional c/w material can also be directly aged after the 899°C extrusion, dispensing with the usual 816°C solution-treatment. More detailed examination of the effects of fabrication variables on such material may be worth pursuing.

Material source

The comparison of c/w bars and powder as starting materials was the principal point of interest in both the tubes and the rods. No evidence can be presented of greater strength as a result of the use of powder. In fact, the highest strength values of nearly 2550 MPa (370 ksi) were observed in Rod 2, which was prepared from 76 mm (3 in) dia. electrode stock.

The powder-derived materials were lower in ductility (RA and also elongation, the latter value available only for the rods). The low RA in some of the tubes may relate to another factor, hydrogen content, which was checked by analysis. The results in the last column of Table IV are consistent with attributing reduced ductility to hydrogen content,[11] with a threshold content of ~0·4 ppm needed to degrade ductility. The hydrogen is not necessarily inherent to powder-derived material but may relate to the pickling of the tubes in concentrated acid. Note the lower hydrogen level for Stock E3 pickled in hydrochloric acid.

Greater significance is assigned to differences (or their absence) in impact resistance values for the various rods and also in the appearance of the fracture surfaces in both rods and tubes. The room-temperature Charpy values are in-

Source: *Powder Metallurgy*, 1977, No. 3

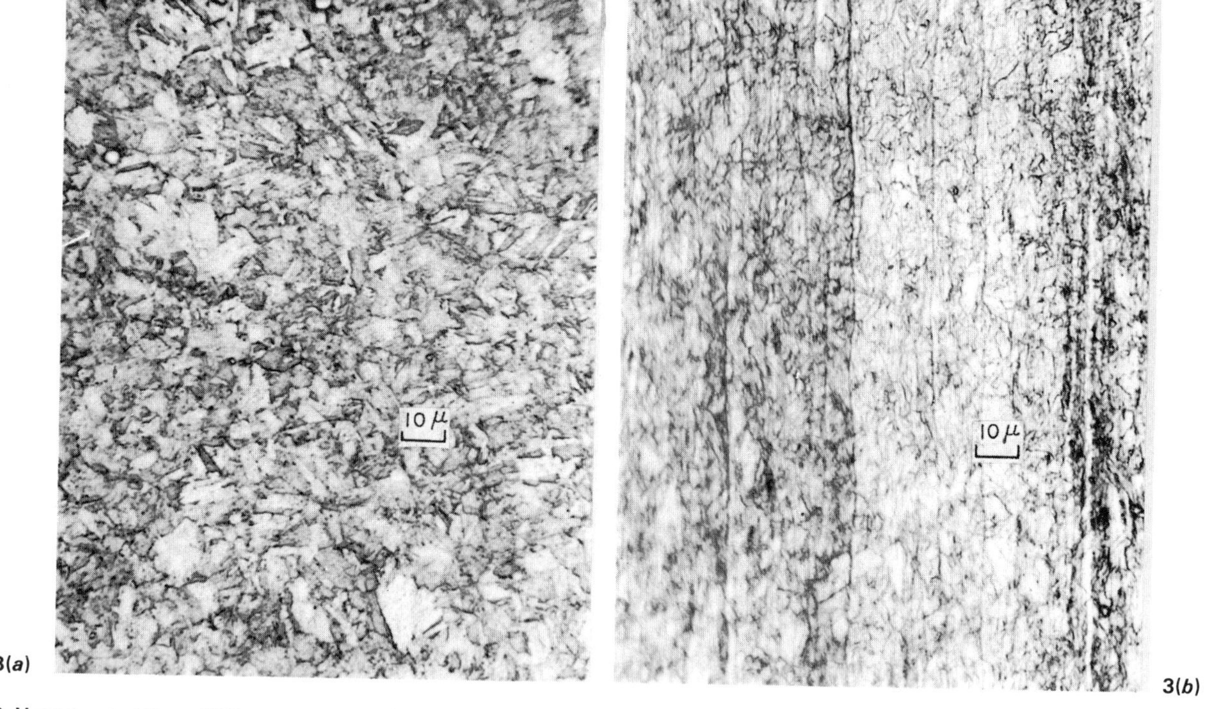

3 Heat-treated Type 350 maraging tubes, aged 6 h at 510°C: (*a*) electrode stock (B2A); (*b*) powder stock (E3A1). ×600.

distinguishable—a remarkable numerical agreement considering the marked differences not only in the material source but also in microstructure and the appearance of the fracture surfaces. The subsequent Charpy data at −40°C differentiated somewhat between the materials. The results for powder-derived Rods 1 and 3 showed very little change from the room-temperature values; the results for c/w Rods 2 and 5 decreased by up to ~4 J (3 ft lbf), so that they were now lower than those for the powder-derived rods.

The impact specimens at both test temperatures showed smooth fracture surfaces in the c/w rods compared with rougher, perhaps wooden, surfaces in the powder rods. The

4 Aged Type 350 maraging rods, aged 6 h at 510°C: (*a*) electrode stock (Rod 2); (*b*) powder stock (Rod 3). ×600.

(*a*)

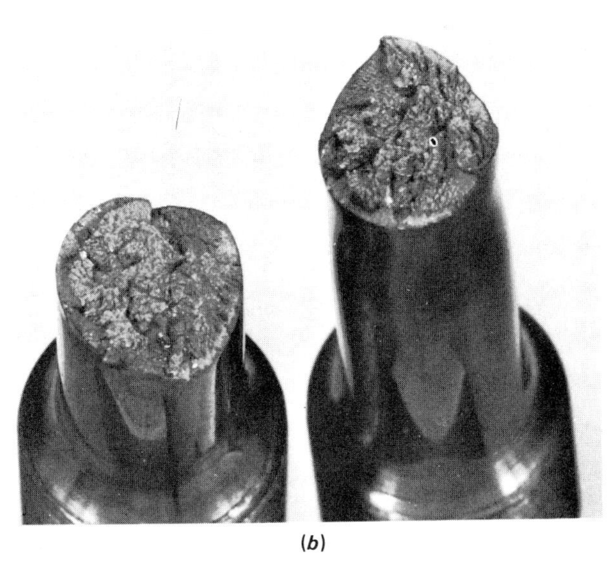

(*b*)

5 Fracture surfaces of tension specimens: (*a*) solution-treated and aged c/w electrode stock 2D; (*b*) solution-treated and aged powder stock 3D.

difference in fracture surfaces is much more marked in the tension specimens (Fig. 5). The c/w rods show a smooth cup/cone fracture in contrast to the wooden fracture of the powder rods. Differences resulting from the material source and reflected in the tensile fracture appearance might show up more strongly in other mechanical properties.

The outstanding strength of Rod 2 deserves separate discussion. The tubing B2 prepared from the same electrode stock did not stand out. Microstructurally the rod and the tube were similar (Figs. 3(*a*) and 4(*a*)); they differed from all the other extrusions in the absence of banding. Conceivably the heavier extrusion reduction of the rod (13·1 vs. 9·6) may have effected greater grain refinement. A possibility exists that the rods were extruded at a lower true billet temperature, as suggested by the extrusion constants (Table III). Differences between rods and tubes from the same stock are discussed in the next section. In no case were they anywhere near as large as those between Rod 2 and Tube B2; note the large, aligned carbides in Fig. 4(*a*) (the white etching phase). Whatever the explanation, the outstanding strength of Rod 2 confirms that a powder source is not necessary to produce the very fine grain sized, strong Type 350 maraging steel. Such material can apparently be obtained by heavy working, which the 76 mm (3 in) bar stock must have received. The benefits of heavy working of commercial Type 300 maraging stock have been cited by Van Swam *et al.*[10] More generally, these maraging steels already have such a fine distribution of microstructural features that little room remains for further contribution from powder processing.

Rods vs. tubes

The data in Tables IV and V permit direct comparisons to be made between tubes and rods from the same starting material: Tube B2 vs. Rod 2; Tube B4 vs. Rod 5; Tubes E2 and E3 vs. Rod 3. The difference of ~140 MPa (20 ksi) between the first pair (i.e., the outstanding strength of Rod 2) has been discussed above. For the other pairings, the rod is almost always stronger but only by ~35 MPa (5 ksi). This difference may be due to the higher reduction ratio (13:1) of the rods, possibly accompanied by a lower extrusion temperature as suggested by the unexpectedly higher extrusion constants. The achievement of a yield strength of 2380 MPa (345 ksi) and ultimate strength of 2415 MPa (350 ksi) in all the tubes (when adequately aged) supports the expectation that extrusion can be scaled up for larger components.

Extrusion of powder at different temperatures

Rods 1 and 3 provide the only comparison of the effect of different extrusion temperatures on the same starting material. Since the strength data overlap no conclusion can be reached. Inferences regarding the effect of the lower temperature of Rod 1 would have to be qualified by its more complicated schedule of being heated twice, the first time for an unsuccessful attempt to extrude at 760°C.

SUMMARY AND CONCLUSIONS

Rods and tubes of Type 350 maraging steel were extruded from rotating electrode process powder and also from cast-and-wrought bar stock from two sources.

The various extrusions of rods and tubes from powder or bar stock were directly aged to full strength without an intermediate heat-treatment between extrusion and ageing.

Ageing for 6 h at 510°C was necessary for achievement of full strength. Reduction of the time or lowering of the temperature gave underaged material, lower in strength and unimproved in ductility or impact resistance.

Strength levels were approximately the same for all the rods and tubes with the exception of one stronger rod prepared from bar stock. Ductility values (elongation and reduction of area) tended to be slightly lower for the powder-derived rods. Greater lowering of RA was found in the powder-derived tubes but this lowering may be due to another factor such as hydrogen pick-up during acid pickling.

Charpy values at −40°C showed a slight but distinct advantage for the powder-derived rods. This advantage is related to the different fracture modes, seen most noticeably in tension specimens (Fig. 5). These differences are related to the greater banding of powder-derived material.

The rod prepared from one type of bar stock (used for electrodes for the REP powder) showed exceptional strength, nearly 2550 MPa (370 ksi) in the tension tests. Such strength is presumably the result of heavier working of the starting material, also indicated by the absence of banding in the extrusions (both rod and tube). The tube prepared from this stock resembled tubes prepared from the other materials in showing strengths near 2415 MPa (350 ksi).

REFERENCES

1. S. ABKOWITZ: 'High-strength wrought weldable titanium alloy mill product manufacture,' US Patent 3,343,998, September 26, 1967.
2. S. ISSEROW and F. J. RIZZITANO: *Int. J. Powder Metall.*, 1974, **10**, 217.
3. L. A. JACOBSON, C. M. PIERCE and M. M. COOK: 'Microstructures of powder and conventionally processed 7075 aluminum alloy,' Air Force Materials Laboratory, AFML-TR-71-240, December 1971.
4. G. FRIEDMAN and E. KOSINSKI: *Met. Eng. Q.*, 1971, **11**, 48.
5. F. J. RIZZITANO and E. P. ABRAHAMSON II: 'An 18%Ni–Co–Mo maraging steel having improved toughness and its method of manufacture' (patent pending).
6. A. R. KAUFMANN: 'Method and apparatus for making powder,' US Patent 3,099, July 30, 1963.
7. E. P. ABRAHAMSON II: 'Processing and properties of 18Ni maraging steel by powder metallurgy', Army Materials and Mechanics Research Center, AMMRC TR 73-4, February 1973, AD 758 439.
8. E. P. ABRAHAMSON II: 'Processing and properties of 13Ni (Type 400) maraging steel by powder metallurgy'. Army Materials and Mechanics Research Center, AMMRC TR 74-14, June 1974.
9. E. SNAPE and F. J. VELTRY: *Int. J. Powder Metall.*, 1972, **8**, 193.
10. L. F. VAN SWAM, R. M. PELLOUX, and N. J. GRANT: *Powder Met.*, 1974, **17**, 33.
11. D. P. DAUTOVICH and S. FLOREEN: *Metall. Trans.*, 1973, **4**, 2627.

Ductility in Hot Isostatically Pressed 250-Grade Maraging Steel

R. M. GERMAN AND J. E. SMUGERESKY

Prealloyed 250-grade maraging steel powder produced by the rotating electrode process was fully consolidated by hot isostatic pressing (HIP) at 1100 and 1200°C. The strength following aging (3 h at 480°C) equalled that of wrought material; however, ductility was negligible. This lack of ductility in the powder metallurgy product was traced to titanium segregation which occurred at the powder surface during powder production. The formation of a titanium intermetallic at the prior particle boundaries during aging caused failure at low plastic strains. Altered aging treatments successfully broke up the embrittling film and resulted in a significant ductility recovery for the HIP material. Analysis of the fracture process indicates that further ductility gains are possible by reducing the titanium content, refining the particle size, and optimizing the thermal cycles.

THE unique combinations of toughness and strength (*e.g.*, 95 MPa $\sqrt{m}$ and 1585 MPa) available with the maraging steels have made them desirable for structural applications. However, problems such as high material cost, casting segregation, and lack of a refined grain size have hindered wide-spread use of these alloys. Attempts to solve these problems have explored various powder metallurgy processes.[1-15] The powder metallurgy approach is especially attractive for fabricating structural components having large diameters and thin walls. Prior investigations into processing maraging steels by powder metallurgy have included a variety of approaches. Both prealloyed and mixed elemental powders consolidated through extrusion, hot isostatic pressing (HIP), or conventional press and sinter cycles have been investigated. Unfortunately, efforts to develop the powder metallurgy for these alloys have shown that the powder compacts must undergo considerable plastic deformation in order to achieve significant ductilities.

BACKGROUND

The initial approaches to processing the maraging steels by powder metallurgy attempted to consolidate elemental powder blends by the conventional press and sinter approach.[1-7] In the aged condition these materials invariably had low ductility, even though high strength was obtained. When such powder products were subjected to extensive plastic deformation, both the strength and ductility improved.[7] Hot extrusion of prealloyed powder is another alternative that has succeeded in producing a material with excellent strength and ductility.[9-11,13-15] However, attempts to consolidate prealloyed powder by HIP have resulted in poor ductility, unless the compact is subsequently plastically deformed.[9,13,14] Such evidence suggests that disruption of the prior particle surfaces is necessary to achieve ductile interparticle bonding. This conclusion is substantiated by the apparent fracture that occurs along prior particle boundaries in undeformed HIP material.[13,14]

In the fabrication of large structural components of a complex shape, HIP processing of a prealloyed powder offers several advantages over wrought material, including lower cost, homogeneity, and uniform properties. In the current program, prealloyed 18 pct Ni (250-grade) maraging steel has been densified by the HIP process. Various aging treatments were investigated in the hope of developing a ductile product which did not require post-consolidation plastic deformation. As a result, an alternative aging treatment has been developed which provides a ductile maraging steel by HIP. The success of this approach, which is described below, makes it feasible to use HIP to fabricate these unique alloys into complex parts having nearly net shape.

EXPERIMENTAL APPROACH

Bar stock of 250-grade maraging steel obtained from Latrobe Steel Company was converted into spherical powder by the rotating electrode process using an inert cover gas.* The powder was screened to a −297 + 74 μm (−50 + 200 mesh) size fraction, giving a mean diameter on a weight basis of 190 μm. The

*Nuclear Metals Inc., Concord, Massachusetts.

Table I. 250-Grade Maraging Steel Powder Chemistry

Element	Wt Pct
Ni	18.35
Co	8.22
Mo	4.85
Ti	0.43
Al	0.06
Si	0.03
Mn	0.01
C	0.006
S	0.003
P	0.003
O	0.006
N	0.004
	Bal Fe

R. M. GERMAN, formerly with Sandia Laboratories, Livermoe, California, is now Director of Research and Development, Mott Metallurgical Corporation, Farmington, CT 06032. J. E. SMUGERESKY is Member Technical Staff, Sandia Laboratories, Materials Development Division 8312, Livermore, CA 94550.

Manuscript submitted April 11, 1977.

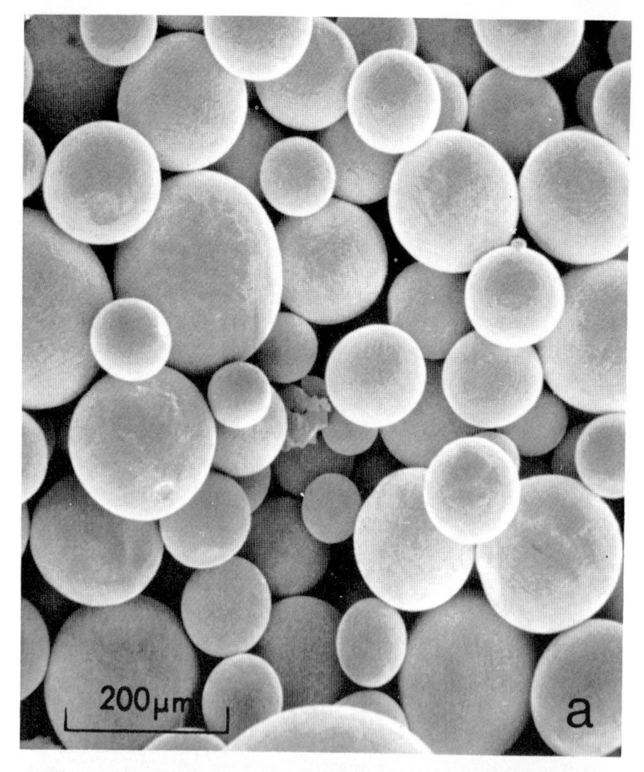

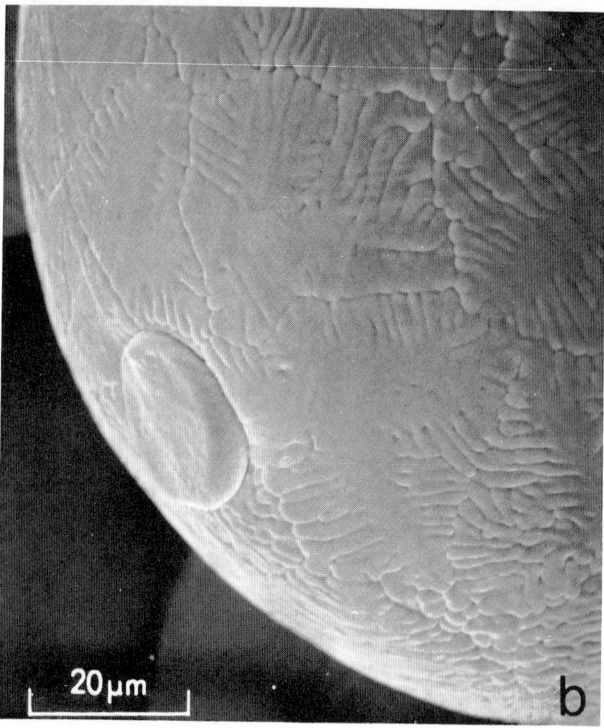

Fig. 1—Scanning electron micrographs of the 250-grade maraging steel powder showing the spherical shape (*a*) and surface dendritic pattern (*b*).

resulting powder had a bulk density of 4.39 g/cm^3. Chemistry for the screened powder is listed in Table I. Figure 1(*a*) shows a scanning electron micrograph of the powder, while Fig. 1(*b*) shows a high-magnification micrograph of the dendritic structure on the powder surface. Metallographic cross sections showed the powder to be pore-free.

The powder was vibrationally loaded into outgassed 304 stainless steel cans approximately 2 cm in diam by 10 cm long. The loaded cans were evacuated for four hours prior to being sealed by electron beam welding. Hot isostatic pressing was performed at 1100 and 1200°C with 210 MPa inert gas (Ar) pressure. Both the pressure and temperature were maintained for three hours. The resulting compacts were fully densified by the HIP treatment. Following consolidation the specimens were machined into tensile bars and solutionized. Solutionizing was achieved in a conventional two-stage treatment in argon: 925°C for one hour, followed by 815°C for one hour, each stage was followed by air cooling.[16] Various aging treatments, also in argon, were performed on both the HIP specimens and wrought materials to determine the cause of ductility losses in HIP materials. In all cases, wrought specimens from the same alloy heat were given equivalent aging treatments to provide a basis for comparing mechanical properties. Several standard analytical instruments were used in the analysis of the microstructural changes induced by the altered aging cycles. These analytical techniques included microprobe analysis, scanning electron microscopy, X-ray diffraction, Auger analysis, optical metallography, and transmission electron microscopy.

RESULTS

Although two different HIP temperatures were used, no difference could be detected in either the microstructure or the mechanical properties of the HIP specimens; hence no distinction is made between HIP temperatures in the results.

The conventional aging treatment for the 250-grade maraging steel is three hours at 480°C.[17,18] Table II compares the average mechanical properties of the wrought and HIP specimens following this treatment. The HIP material reached full strength but had negligible ductility. A similar lack of ductility for HIP 300-grade maraging steel has been reported by Van Swam, Pelloux, and Grant.[14] Figure 2 compares the fracture surfaces for the wrought and HIP materials following the conventional aging process. In the powder product, failure is along the prior particle boundaries. Close examination of the HIP fracture surface shows each of the planar surfaces to be composed of small, finely spaced dimples as shown in Fig. 3. Associated with each dimple is a second-phase particle (0.5 to 1.0 µm in size). Metallographic examination of the powder product shows a second phase associated with the prior particle boundaries and large titanium-rich inclusions, as shown in Fig. 4. The second phase at the boundaries produced failure at low strains, resulting in limited tensile ductilities.

Thin foil examination of the aged HIP material re-

Table II. Properties Following Conventional Aging (3 H at 480°C, Air Cool), Average of Three Specimens

Sample	0.2 Pct Yield Strength (MPa)	Strength (MPa)	Notched Tensile Breaking Strength (MPa)	Elongation (Pct)	Reduction in Area (Pct)
Wrought	1680	1750	2590	14	62
HIP	1700	1730	1680	<1	0.2

vealed evidence of austenite, sigma phase (FeTi), Ni_3Mo, and free titanium. The presence of austenite and free titanium was indicated by diffraction analysis, while FeTi and Ni_3Mo were revealed by dark field imaging. FeTi was present as 50- to 100-nm precipitates, while Ni_3Mo was present as 4 to 15-nm precipitates. Thus the 0.5 to 1.0 μm particles observed at the bottom of the dimples on the fracture surfaces do not appear to be these phases. X-ray examination of fracture surface extraction replicas shows the presence of Ni, Ti, and Al, suggesting Ni_3 (Ti, Al) as the prior

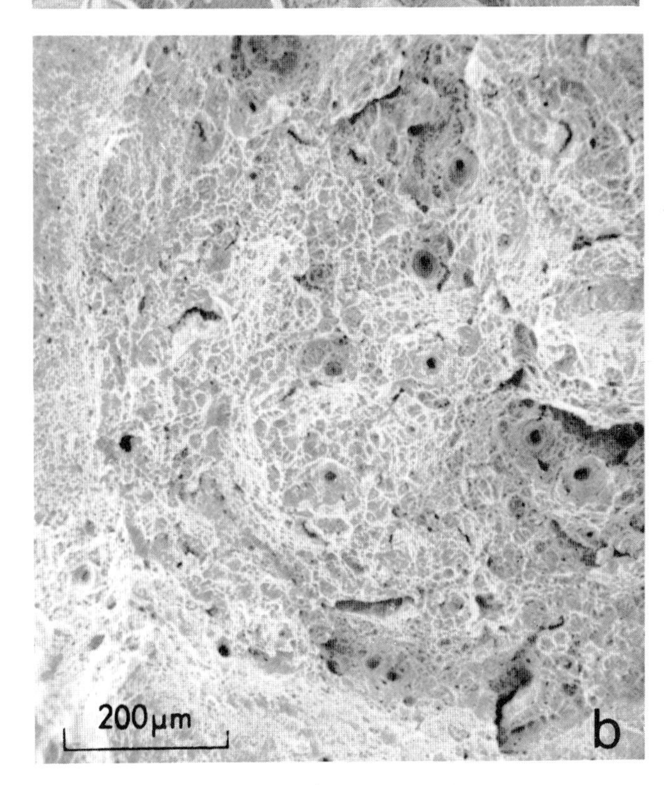

Fig. 2—Fracture surfaces of HIP (*a*) and wrought (*b*) samples following conventional aging at 480°C for three hours.

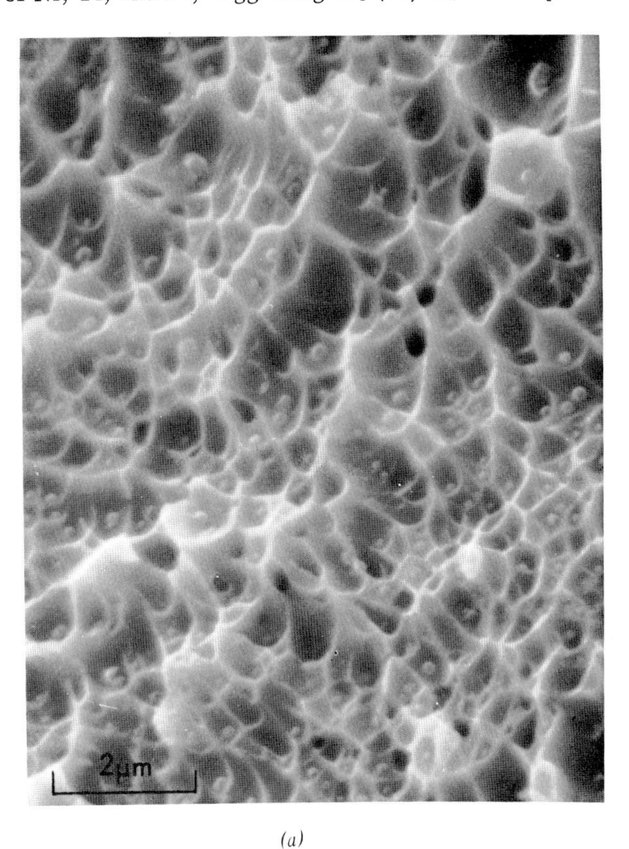

(*a*)

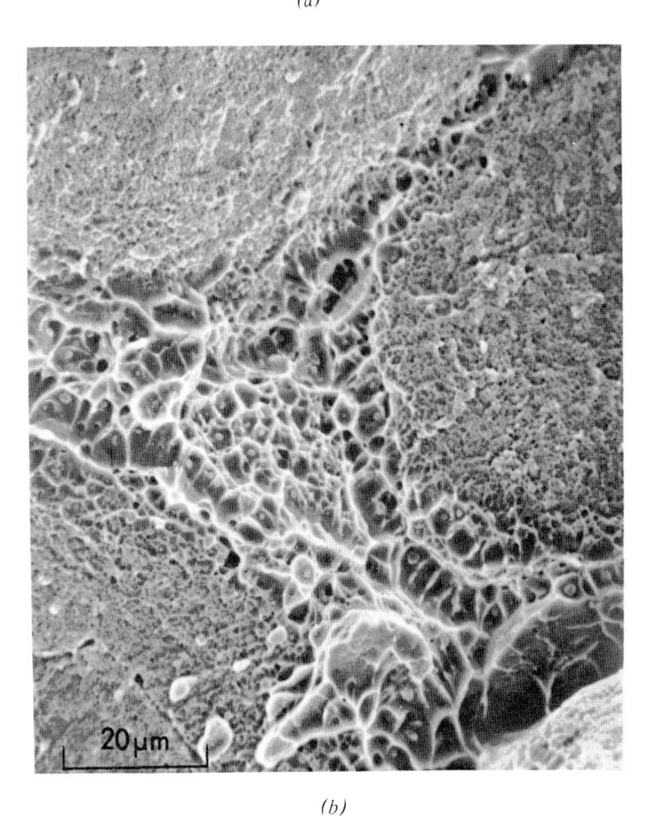

(*b*)

Fig. 3—(*a*) The dimples at a powder particle junction on the fracture surface of the Hip sample shown in Fig. 2(*a*), and (*b*) the finer dimples on a faceted region. Note the association of second-phase precipitate with each dimple.

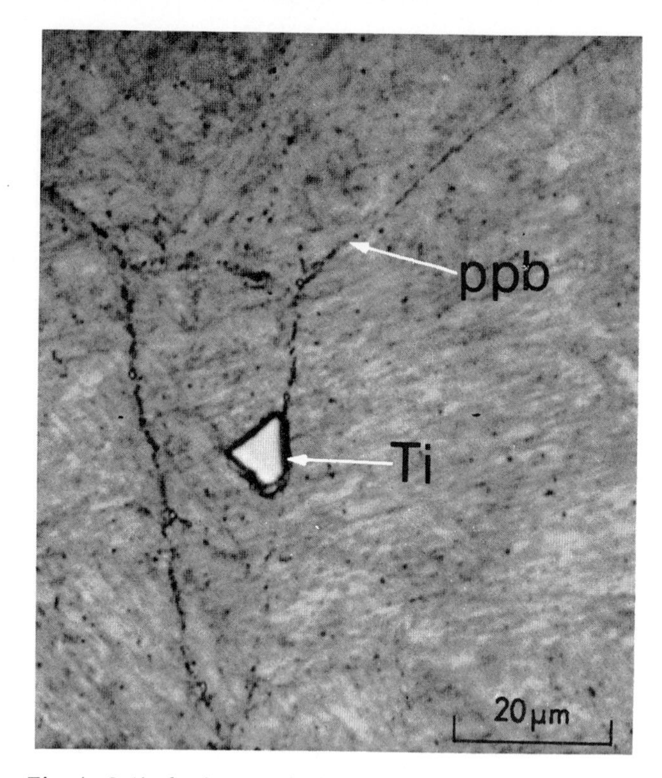

Fig. 4—Optical micrograph showing a typical Ti-rich inclusion and prior powder boundary (ppb) decoration following conventional aging of the HIP material.

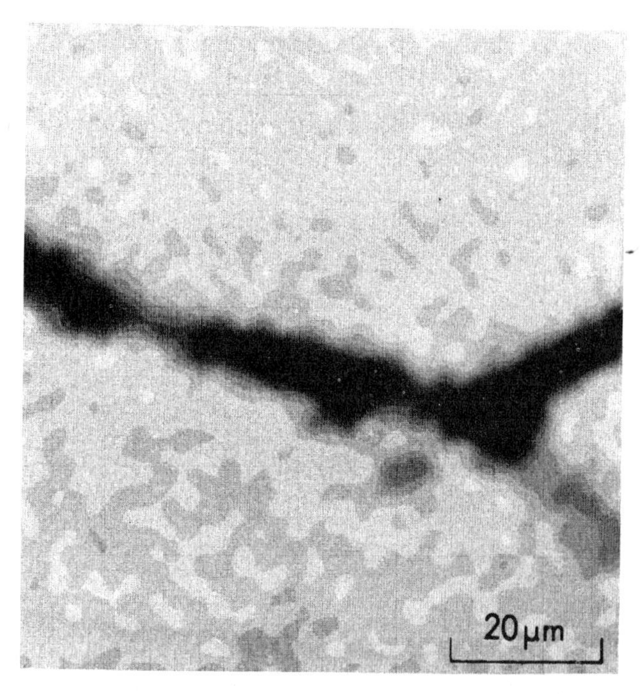

Fig. 5—Computer-enhanced microprobe scan for Ti over an 80 × 80 μm area at the junction of three particles following hot isostatic pressing. Darker areas correspond to higher Ti concentrations.

particle boundary phase, but no diffraction data supports this.

Previous investigations of maraging steels have documented the formation of an embrittling Ti(C, N) phase on prior austenite boundaries with certain thermal cycles.[19-25] It is reasonable to expect the prior particle boundaries of the HIP material to behave like prior austenite boundaries in wrought material. Although embrittlement is commonly attributed to Ti(C, N) formation, it is not the only observed cause of low ductilities in these alloys. Typically, Ti segregation to the prior austenite boundaries is associated with either compound formation or austenite retention at these boundaries.[24,26-30] Thermal embrittlement due to extensive Ti(C, N) formation is detrimental to the ductility in the solutionized condition,[24,25] but the mechanical properties for the HIP and wrought materials in the solutionized condition were equivalent, as shown in Table III. The comparable wrought and HIP ductilities in the solutionized condition differ significantly from the results of Van Swam, Pelloux, and Grant.[14] Thus it appears that the ductility loss for the present HIP maraging steels is not due to Ti(C, N). Instead there is evidence that an intermetallic phase formed at the prior particle boundaries during aging results in the ductility loss.

Microprobe analysis of the aged HIP material confirmed Ti segregation to the prior particle boundaries. Furthermore, no N, C or O segregation was detected. Figure 5 shows a computer-enhanced microprobe scan for Ti over an 80 × 80 μm area at a three-particle junction. Since the microprobe beam size is of the order of 2 μm, no discrete precipitates were resolvable. Cox, Reisdorf, and Pellissier[27] identified similar Ti-enriched regions in the interdendritic areas of cast 250-grade maraging steel. The appearance of a Ti-rich second phase also agrees with the findings of Chilton and Barton.[28] Thus it appears that the prior particle boundaries are enriched with titanium and act as favorable heterogeneous nucleation sites for Ti-rich precipitates in a manner similar to that of prior austenite boundaries. The formation of these precipitates results in a high density of second-phase particles which causes failure at low strains.

Because it was recognized that the prior particle boundaries act like prior austenite boundaries, studies were launched to alter the second-phase morphology through modified aging treatments. The phase located at the prior particle boundary is due to the localized supersaturation in titanium. On aging, this titanium supersaturation leads to the formation of a fine, nearly continuous second-phase film at the boundaries. The localized supersaturation can be lowered if the

Table III. Strength and Ductility for Various Aging Conditions

Treatment	Material	0.2 Pct Yield Strength (MPa)	Reduction in Area (pct)
Solutionized (S)	Wrought	795	85
	HIP	780	80
S + 3h at 480°C	Wrought	1680	62
	HIP	1700	0.2
S + 3h at 511°C	Wrought	1655	53
	HIP	1765	1
S + 500°C/3h at 480°C	Wrought	1640	62
	HIP	1515	40
S + 525°C/3h at 480°C	Wrought	1630	67
	HIP	1655	47
S + 540°C/3h at 480°C	Wrought	1540	69
	HIP	1600	40
S + 560°C/3h at 480°C	Wrought	1630	59
	HIP	1655	27

aging treatment is begun at temperatures above 480°C (the conventional aging temperature) to reduce the tendency for the Ti-based second phase to form. However, undesirable austenite reversion will occur with prolonged aging above 480°C.[16,30-32] An alternative was found following the suggestion of Peters[32,33] that a brief preage at a temperature over 480°C will reduce the supersaturation, thereby reducing the tendency for heterogeneous nucleation. Thus several aging treatments were performed involving preaging at temperatures in excess of 480°C. After a 10-min specimen equilibration at the preage temperature, the furnace temperature was decreased to 480°C. Approximate cooling rate to 480°C was 1.5°C/min. After the specimen had reached 480°C, it was kept at this temperature for three hours and then air-cooled to room temperature. The resulting mechanical properties for both the wrought and HIP specimens are compared in Table III. The ductility, as represented by the tensile reduction in area, is shown as a function of preage temperature in Fig. 6. The wrought ductility increases slightly with preaging temperature and maximizes at approximately 540°C. However, a dramatic increase in the ductility of HIP material with preage temperature was observed before a maximum was seen at approximately 525°C. The fracture surface for a 525°C preage HIP specimen, shown in Fig. 7, reflects a change in fracture mode compared to Fig. 2(*a*). A mixed-mode failure, incorporating both interparticle and transparticle fracture, is now evident. The incomplete transition to transgranular failure (see Fig. 2(*b*)), coupled to a tensile reduction in area less than that for wrought, indicates that further ductility could possibly be reclaimed from the HIP material. Unfortunately, because of insufficient HIP material, variable hold times and cooling rates could not be investigated. Potentially, either of these factors could further increase the HIP product ductility. It is significant to note that increased ductility was achieved without plastic deformation of the powder compacts, as has been found necessary in several previous studies.

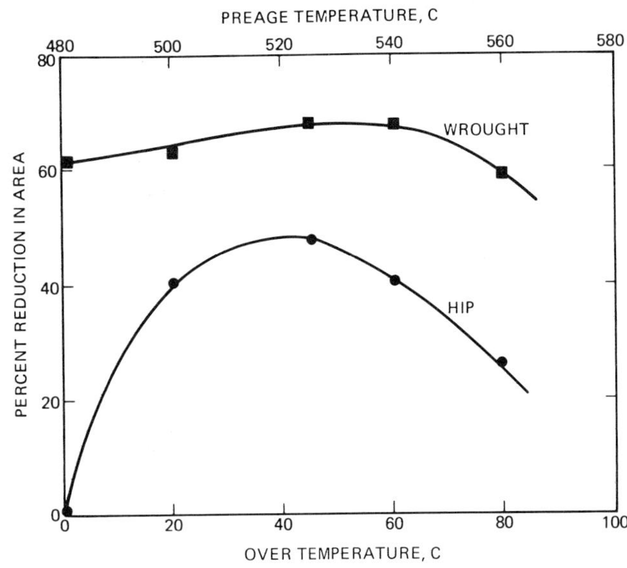

Fig. 6—Ductility variations (as measured by the percent reduction in area) *vs* preage temperature for both HIP and wrought material. All samples solutionized, heated to preage temperature, held 10 min, cooled at 1.5°C/min to 480°C, held 3 h at 480°C with subsequent air cool.

DISCUSSION

Prior to this investigation, powder metallurgy processing of maraging steels suffered from one very significant property detriment; lack of ductility unless mechanical deformation was incorporated either after or during consolidation. Although HIP consolidation of prealloyed rotating electrode powders should produce properties comparable to those of a wrought product, premature fracture occurs along prior particle boundaries.[13,14] The lack of ductility has been attributed to thin oxide layers on the powder particles,[13,14] but no conclusive experimental evidence has been reported to support this concept. Achievement of ductilities equivalent to those of a wrought product had been obtained,[11,13-15] but only after extrusion of the powder. Consequently the advantage of powder processing to a nearly net shape is not realized when some type of deformation processing is also required.

There are two important differences between this study and that of Van Swam, Pelloux, and Grant;[14] 1) this HIP material had significant ductility after solutionizing (80 pct RA), and 2) this alloy had a lower Ti content. The former is a crucial point suggesting that Ti(C, N) thermal embrittlement did not occur in this HIP material.[24,25] The large >1 μm blocky titanium carbonitrides and sulfides were visible on the fracture surface, but their size precludes a dominant role.[34] The lower titanium content is also significant because microprobe examination shows segregation of titanium to prior particle boundaries (Fig. 5). We believe such Ti segregation during powder production is the major cause of poor ductility in the HIP product rather than oxide layers. In contrast to the suggestion of Van Swam, Pelloux, and Grant,[14] microprobe analysis of the present HIP material showed no O, C or N segregation to the prior particle boundaries.

Previous correlations between fracture behavior and solute segregation in the maraging steels have identified Ti as the key element.[21-27] Figure 8 shows a plot of the Ti and Ni Auger intensities on the fracture surfaces *vs* the measured reduction in area. A generalized inverse relation between the Auger intensities and ductility illustrates the detrimental role of Ti. Johnson and Stein[25] have shown a similar plot for toughness in a 250-grade maraging steel. In the wrought material, the Ni intensity variations on the fracture surface have no significance because the Ni is not associated with Ti. However, in the HIP material, the Ni and Ti segregations both correlate with the ductility loss. Extraction replicas of the fracture surfaces successfully removed several of the particles (similar to those shown in Fig. 3). These particles were found to be of two types; the predominant particles were Ti-rich, with Ni, Fe, and Mo present. The other type of particle was Al-rich, with some Ni and Ti (no analysis of C, N, or O was possible). For the lower ductilities, Fe and Cr were also detected in extraction replicas. In the wrought state (67 pct reduction in area), Mo- and Ti-enriched precipitates of 0.2 μm were found. Energy-dispersive X-ray analysis using a scanning electron microscope further confirms these compositions. Optical metallography revealed a change in both the size and extent of second phase at

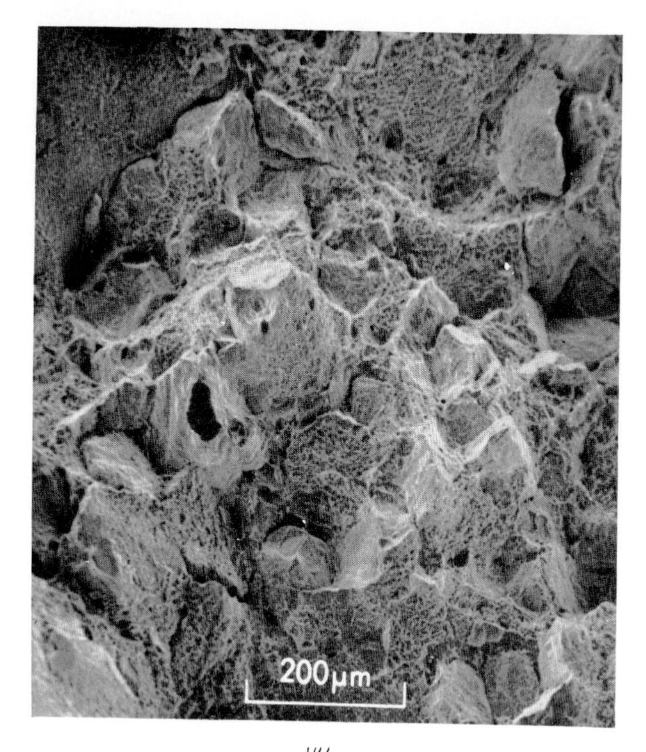

(a)

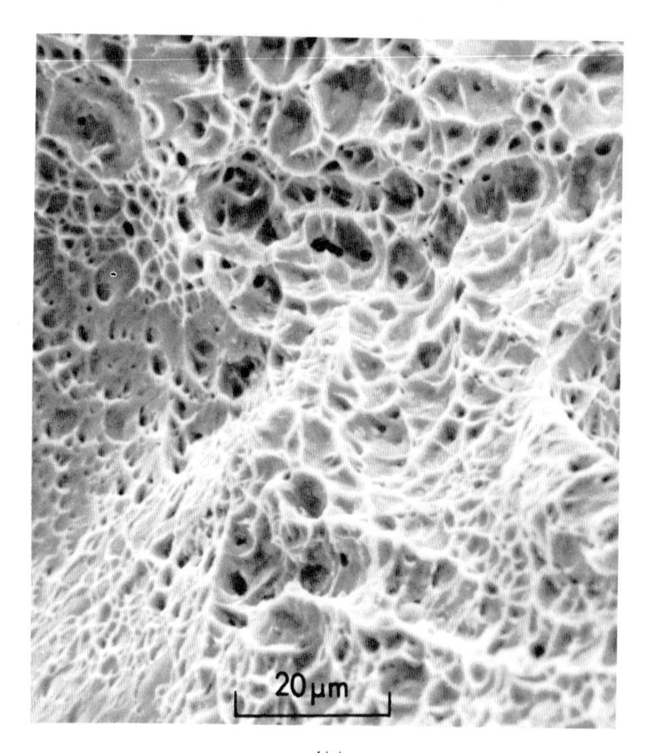

(b)

Fig. 7—Mixed-mode fracture for HIP material given the 525°C preage treatment; (*a*) low-magnification micrograph corresponding to Fig. 2(*a*), (*b*) high-magnification micrograph of the junction of two powder particles corresponding to Fig. 3(*a*).

the prior particle boundaries with aging treatment. This variation is demonstrated in Fig. 9. Conventional aging at 480°C with no preage produces a nearly continuous film on the boundaries. Inclusion of a preage in the hardening process produces a different second-phase morphology at the boundaries. This result

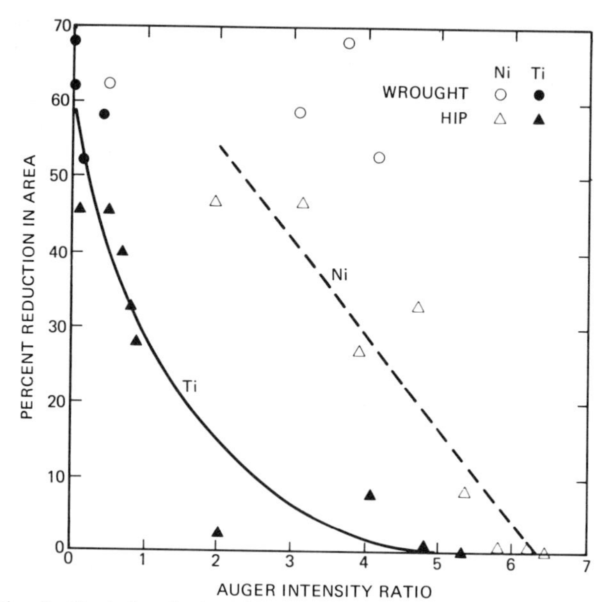

Fig. 8—Variation in tensile reduction in area with nickel and titanium Auger intensities (in ratio to the iron intensity) on the fracture surfaces of both HIP and wrought samples.

coupled to the Auger results 35 suggests low ductility in the HIP material is associated with fracture along prior particle boundaries enriched with Ti. The preage treatments act to change the nature of the prior particle boundary second phase such that the fracture path no longer follows the boundaries. Such treatments are at too low a temperature to appreciably change the distribution of alloying elements in the compact. In fact, it appears that geometrical rather than chemical changes are responsible for the ductility increases with the preage treatments.

In general, the low ductility conditions were characterized by Ti-rich fracture surfaces following along prior particle boundaries. Using the preage led to improved ductilities and mixed-mode fracture surfaces. Three possible explanations for this behavior were considered: 1) reversion to austenite and corresponding solutionizing of the titanium; 2) thermal embrittlement via Ti(C, N); and 3) reaction of titanium with other alloying elements to form particles that act as void initiation sites.

Austenite reversion results in reduced yield strength, but this explanation does not seem likely. Although changes in strength are observed in both wrought and HIP material, these reductions are not systematically accompanied by strong changes in ductility.

Thermal embrittlement has also been ruled out, because after the powder product had been solutionized, the ductility was no different from that of the solutionized wrought product.

Consequently, segregated titanium present at prior particle boundaries (as a result of powder production) reacting with other alloying elements can explain the observed ductility recovery. After HIP and solutionizing, the boundary regions remain enriched in titanium, giving a localized supersaturation. This supersaturation results in a larger number of nucleation sites during aging at 480°C. The resultant precipitate distribution consists of a high concentration of very small sizes at the prior particle boundaries con-

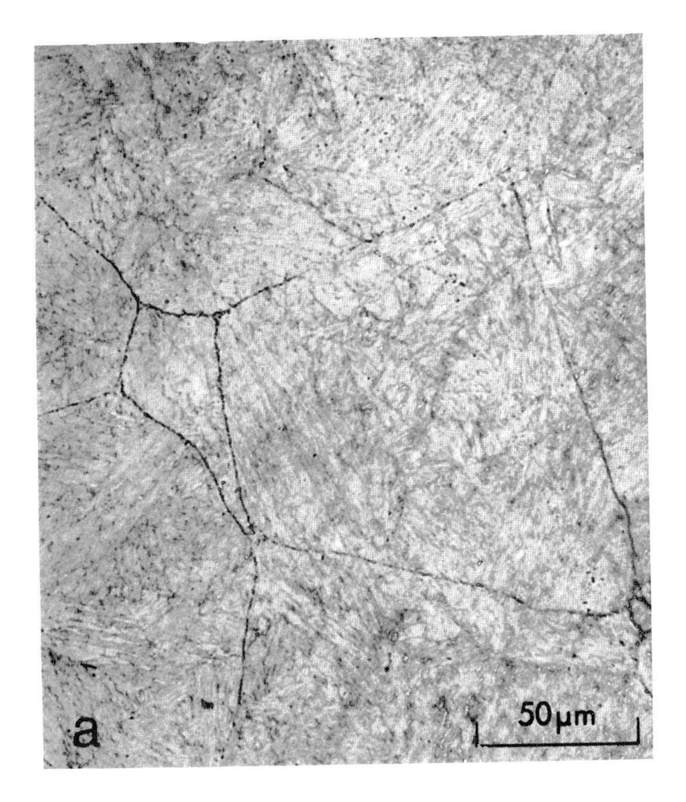

stituting a near continuous second-phase film. Since each precipitate acts as a void initiation site there is a preference for fracture to begin in this region. In addition the matrix also contains a large number of smaller strengthening precipitates. With a large concentration of void initiation sites at the prior particle boundaries, the interface fails at low strains. Thus, the fracture is characterized by a large number of fine dimples localized on the prior powder particle boundaries.

With the preage, the number of nucleation sites is

Fig. 9—Optical micrographs of the HIP material showing the effect of aging treatment on the prior particle boundary precipitates; (*a*) conventional age 480°C/3 h, 0 pct RA, (*b*) preage 525°C followed by 480°C/3 hr, 47 pct RA, and (*c*) preage 560°C followed by 480°C/3 h, 27 pct RA.

reduced and the subsequent aging at 480°C results in fewer but larger precipitates. At the boundary a smaller fraction of the interfacial area is occupied by the second phase. This provides greater interparticle bonding across the boundary and therefore less tendency for interparticle failure. As the same time the strength of the matrix is reduced, also due to fewer and larger strengthening precipitates. It is quite possible that the altered aging cycles are also changing the fracture mechanism, leading to increased ductility.

Confirmation of these concepts was not possible because of a lack of sufficient quantities of material. In order to test this hypothesis, additional experiments are in progress. These include three new features: 1) reduction in powder particle size; 2) atomization to minimize segregation; and 3) a reduction in the Ti content. With less segregation and a lower Ti content, less titanium will be available to form void-initiating precipitates. By reducing the powder particle size, we will be increasing the particle boundary area. Both features should result in reduced density of titanium-enriched precipitates at prior particle boundaries and lead to reduced interparticle failure and increased ductility in nonworked powder metallurgy processed maraging steel.

SUMMARY

The unique combination of toughness and strength which is possible in maraging steels has been extended to include material processed by powder metallurgical techniques. The ductility loss observed in underformed powder metallurgy products has been reversed by preaging at temperatures above the standard aging temperature. This ductility loss in the HIP material was caused by Ti segregation at particle surfaces during powder production. The local Ti saturation resulted in intermetallic formation at the prior particle boundaries during conventional aging. Altering the aging treatments changed the intermetallic morphology and resulted in improved ductilities. It is speculated that a finer powder size (more rapidly solidified) and lower Ti content would further enhance the properties of HIP maraging steels.

ACKNOWLEDGMENT

This work was supported by the United States Energy Research and Development Administration, Contract Number AT-(29-1)-789.

REFERENCES

1. J. J. Fischer: *Inter. J. Powder Met.*, 1966, vol. 2, no. 4, pp. 37-42.
2. Toshihiro Kinoshita, Yoichi Tokunaga, and Hiromasa Kobayashi: *J. Japan Soc. Powder and Powder Met.*, 1966, vol. 13, pp. 228-35.
3. Toshihiro Kinoshita, Yoichi Tokunaga, and Hiromasa Kobayashi: *J. Japan Soc. Powder and Powder Met.*, 1966, vol. 13, pp. 236-42.
4. Toshihiro Kinoshita, Yoichi Tokunaga, and Isao Taniguchi: *J. Japan Soc. Powder and Powder Met.*, 1967, vol. 14, pp. 213-20.
5. V. A. Tracey and R. S. K. Raman: *Powder Met.*, 1969, vol. 12, pp. 131-56.
6. Joel S. Hirschhorn and David A. Westphal: *Modern Developments in Powder Metallurgy*, vol. 5, H. H. Hausner, ed., Plenum Press, NY, 1971, pp. 481-90.
7. V. N. Antsiferov and Yu. M. Kolbenev: *Sov. Powder Met. Metal Ceram.*, 1972, vol. 11, pp. 287-89.
8. R. Widmer: *Powder Metallurgy for High-Performance Applications*, J. J. Burke and V. Weiss, eds., Syracuse University Press, Syracuse, NY, 1972, pp. 69-84.
9. R. M. Pelloux: *Powder Metallurgy for High-Performance Applications*, J. J. Burke and V. Weiss, eds., Syracuse University Press, Syracuse, NY, 1972, pp. 351-63.
10. E. Snape and F. J. Veltry: *Inter. J. Pwder Met.*, 1972, vol. 8, pp. 193-302; or *Powder Met.*, 1972, vol. 15, pp. 332-45.
11. Ernest P. Abrahamson, II: "Processing and Properties of 18 Ni Maraging Steel by Powder Metallurgy," Army Materials and Mechanics Research Center, Watertown, MA, Report No. AMMRC TR 73-4, February 1973.
12. J. K. Mukherjee: *Inter. J. Powder Met.*, 1973, vol. 9, pp. 95-100.
13. N. J. Grant, R. M. Pelloux, M. C. Flemings, and A. S. Argon: *Structure and Property Control Through Rapid Quenching of Liquid Metals*, Massachusetts Institute of Technology, Cambridge, MA, Report No. AD-775 225, June 1973.
14. L. F. Van Swam, R. M. Pelloux, and N. J. Grant: *Powder Met.*, 1974, vol. 17, pp. 33-45.
15. Ernest P. Abrahamson, II: *Processing and Properties of 13 Ni (400) Maraging Steel by Powder Metallurgy*, Army Materials and Mechanics Research Center, Watertown, MA, Report No. AMMRC TR 74-14, June 1974.
16. C. S. Carter: *Met. Trans.*, 1971, vol. 2, pp. 1621-26.
17. R. F. Decker, J. T. Eash, and A. J. Goldman: *Trans. ASM*, 1962, vol. 55, pp. 58-76.
18. S. Floreen: *Met. Rev.*, 1968, vol. 13, pp. 115-28.
19. A. J. Birkle, D. S. Dabkowski, J. P. Paulina, and L. F. Porter: *Trans. ASM*, 1965, vol. 58, pp. 285-301.
20. T. Boniszewski and E. Boniszewski: *J. Iron Steel Inst.*, 1966, vol. 204, pp. 360-65.
21. G. J. Spaeder, R. M. Brown, and W. J. Murphy: *Trans. ASM*, 1967, vol. 60, pp. 418-25.
22. L. Roesch and G. Henery: *Electron Microfractography*, ASTM Special Technical Publication 453, pp. 3-29, ASTM, Philadelphia, PA, 1969.
23. G. J. Spaeder: *Met. Trans.*, 1970, vol. 1, pp. 2011-14.
24. D. Kalish and H. J. Rack: *Met. Trans.*, 1971, vol. 2, pp. 2665-72.
25. W. C. Johnson and D. F. Stein: *Met. Trans.*, 1974, vol. 5, pp. 549-54.
26. B. G. Reisdorf: *Trans. ASM*, 1963, vol. 56, pp. 783-86.
27. P. H. Salmon Cox, B. G. Reisdorf, and G. E. Pellissier: *Trans. TMS-AIME*, 1967, vol. 239, pp. 1809-17.
28. J. M. Chilton and C. J. Barton: *Trans. ASM*, 1967, vol. 60, pp. 528-42.
29. W. R. Bandi, J. L. Lutz, and L. M. Melnick: *J. Iron Steel Inst.*, 1969, vol. 207, pp. 348-52.
30. T. Suzuki: *Trans. Iron Steel Inst. Japan*, 1974, vol. 14, pp. 67-81.
31. S. Floreen and R. F. Decker: *Trans. ASM*, 1962, vol. 55, pp. 518-30.
32. D. T. Peters and C. R. Cupp: *Trans. TMS-AIME*, 1966, vol. 236, pp. 1420-29.
33. D. T. Peters: *Trans. TMS-AIME*, 1967, vol. 239, pp. 1981-88.
34. E. Nes and G. Thomas: *Met. Trans. A*, 1976, vol. 7A, pp. 967-75.
35. H. J. Jack and P. H. Holloway: *Met. Trans. A*, 1977, vol. 8A, pp. 1313-15.

SECTION XII:
Physical Metallurgy

Structure and Strengthening Mechanisms in 300-Grade 18Ni–Co–Mo–Ti Maraging Steel

W. A. SPITZIG, J. M. CHILTON AND C. J. BARTON

ALTHOUGH a considerable amount of literature exists concerning the structure and strengthening mechanisms in 250-grade 18Ni maraging steels (Reference 1 includes an extensive list of references), only limited data are available on the fine structure of 300-grade 18Ni maraging steels (2, 3). The strengthening precipitates in 250-grade 18Ni – Co – Mo – Ti maraging steels have been identified, through selected-area electron-diffraction analyses (measuring the angles between planes and the *d*-spacings), as Ni_3Mo and a sigma phase designated FeTi (1). An advantage of selected-area electron-diffraction analysis is that when specimens in the aged condition are examined, overaging is not needed to produce precipitates large enough to be extracted. In addition, in the selected-area electron-diffraction method, the angles between planes can be measured, and consequently, a distinction can be made among compounds having similar *d*-spacings.

The authors are associated with the Applied Research Laboratory, U. S. Steel Corp., Monroeville, Pa. Technical Note received 16 May 1968.

In previous investigations of 300-grade 18Ni – Co – Mo – Ti maraging steels, the strengthening precipitates were identified by means of extraction replication techniques as Ni_3Mo and possibly Ni_3Ti (2, 3). However, to more conclusively identify the strengthening precipitates, a selected-area electron-diffraction analysis was conducted on metal foils of a 300-grade 18Ni – Co – Mo – Ti maraging steel, and both the angles between planes and the *d*-spacings of the precipitates were measured.

MATERIALS AND EXPERIMENTAL WORK

The material analyzed in this investigation was obtained from a high-purity 300-lb laboratory heat that was vacuum-melted and vacuum-cast into a 5 by 12 by 17-in. ingot. The ingot was soaked at 2200 F for 6 hr, reduced to a 1.6-in. thick plate between 2200 and 1800 F, and water-spray-quenched. The plate was subsequently sectioned into 4 pieces, heated to 2200 F for 4 hr, cross-rolled to $\frac{1}{2}$-in. thick plate (finishing between 1700 and 1800 F), and water-

Source: *Transactions of ASM*, Vol 61, 1968

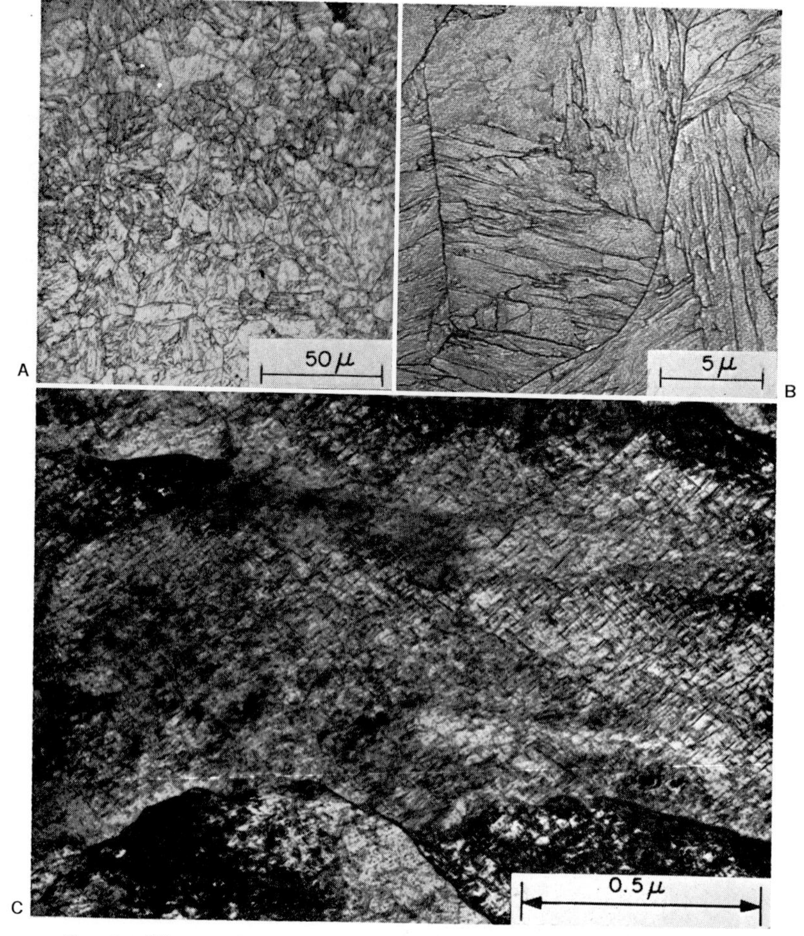

FIG. 1. Micrographs showing structure of aged 300-grade maraging steel.

TABLE 1. Composition and Mechanical Properties of the 300-Grade Steel Investigated

Composition, wt %										
C	Mn	P	S	Si	Ni	Mo	Co	Ti	Al	N
0.002	0.002	0.0006	0.007	0.052	18.1	4.98	8.84	0.77	0.033	0.001

Mechanical Properties*				
Yield strength, 0.2% offset, ksi	Tensile strength, ksi	Elongation in 1 in., %	Reduction of area, %	Charpy impact energy, ft-lb
272	278	12	60	24

* After the steel was aged for 3 hr at 900 F.

spray-quenched. Sections, $\frac{1}{2}$ by $\frac{1}{2}$ by 3 in. were machined from the plates, solution-annealed for 1 hr at 1500 F in a neutral protective atmosphere, and cooled in air. After cooling to room temperature, the sections were aged for 3 hr at 900 F and air-cooled. This is a typical aging treatment for the maraging steels. The chemical composition and mechanical properties are given in Table 1.

Metallographic samples were etched with a 2% nital solution. Formvar replicas were used for high-magnification analysis of the microstructures. Thin foils for examination by electron microscopy were prepared from the $\frac{1}{2}$ by $\frac{1}{2}$ by 3-in. sections by a technique similar to that described previously (1). The thin-foil specimens were examined at 100 kv.

The precipitate phases were identified with the aid of a computer program (1) that resulted in the calculation and listing of *d*-spacings and angles between planes for the tetragonal, orthorhombic, and hexagonal systems when the computer was supplied with the appropriate lattice parameters. High-resolution dark-field techniques were employed to distinguish between precipitates and, therefore, to aid in the analysis of the electron-diffraction patterns.

Diffraction patterns were indexed with an estimated accuracy of ±0.05 angstrom (A) for the *d*-

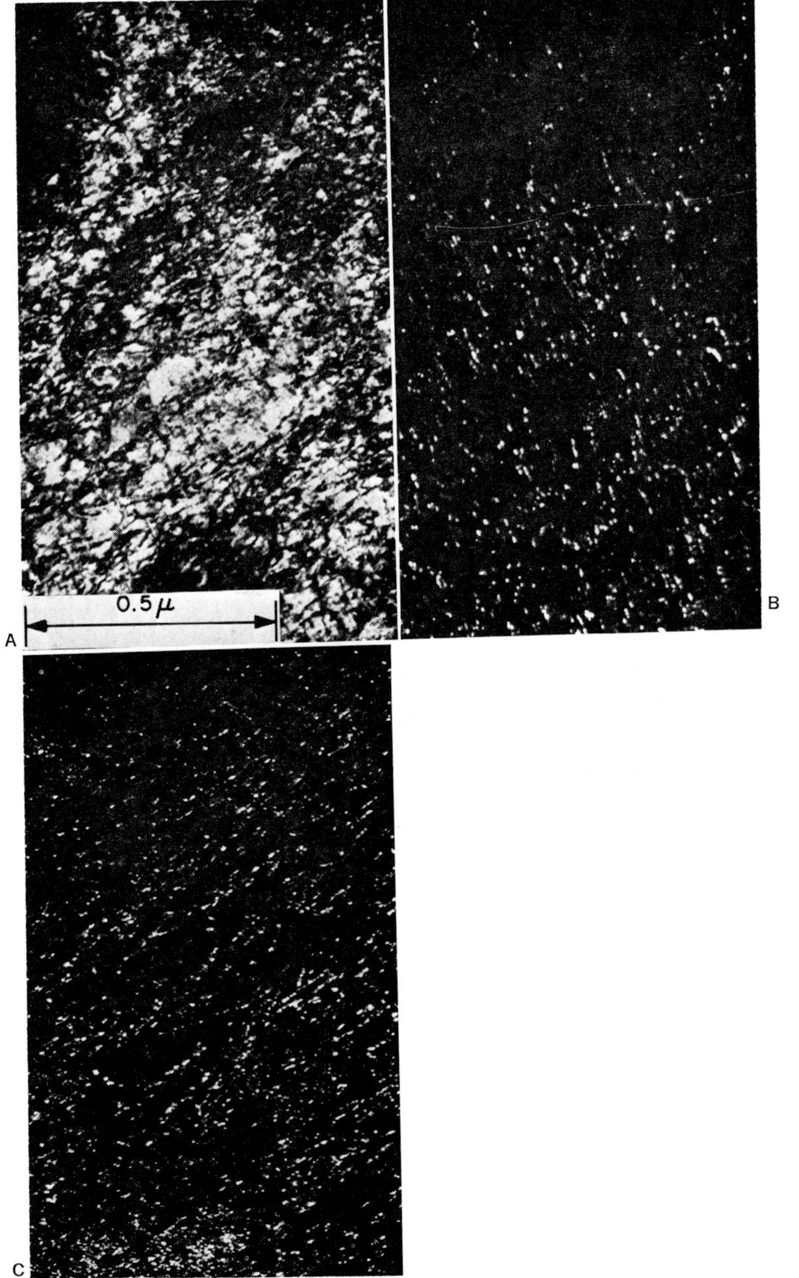

FIG. 2. Thin foil electron micrographs of aged 300-grade maraging steel. (A) Bright-field; (B) and (C) dark-field of region shown in (A) using the (211) Ni_3Mo (B) and the (022) sigma-phase; (C) reflections shown in Fig. 3.

spacings and ±2 degrees for the angles between planes. The camera constant was determined for each plate by means of the matrix diffraction pattern.

RESULTS AND DISCUSSION

The structure of the aged 300-grade maraging steel is shown in Fig. 1 at three different magnifications. Figure 1A is a light micrograph showing numerous prior austenite grain boundaries, along with the martensite lath substructure present within the individual grains. Figure 1B is a higher magnification electron micrograph of a replica of a polished and etched surface, showing more clearly the martensite lath substructure within a prior austenite grain. Figure 1C is a considerably higher magnification electron-transmission micrograph of a thin foil, showing the defect structure and precipitates within an individual martensite lath.

Thin foil examination of individual martensite laths having different matrix orientations showed that two different precipitates were present within the martensite laths. To identify the precipitates, the diffraction patterns from martensite laths with six different matrix orientations were analyzed in detail. These precipitates were identified as an orthorhombic

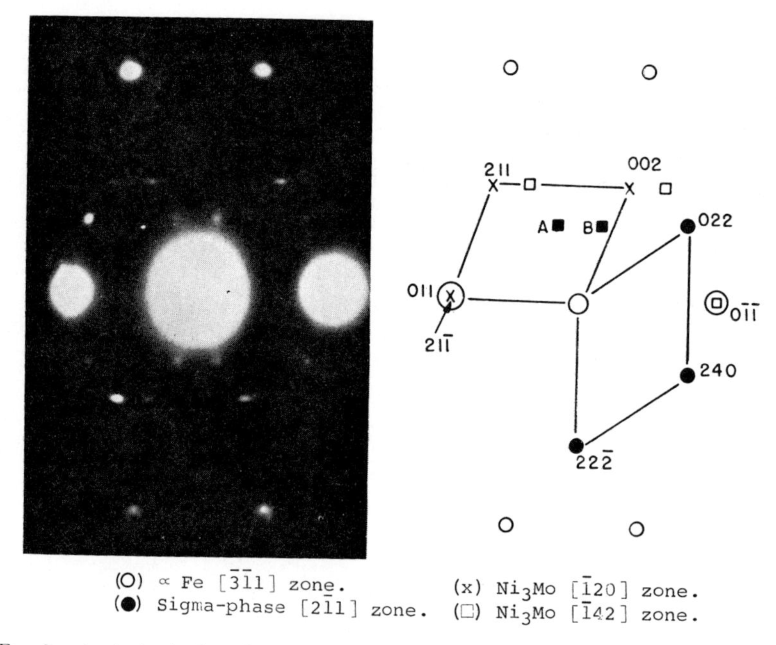

FIG. 3. Analysis of selected-area electron-diffraction pattern of region shown in Fig. 2.

Ni_3Mo phase (4) and a tetragonal phase, the latter phase having lattice parameters of 8.9 and 4.6 A and corresponding to a sigma-phase-type structure. Diffraction patterns from the tetragonal sigma phase were not consistent with Ni_3Ti, Fe_2Ti, cubic FeTi, or any of the other possibilities that have been suggested in previous investigations of 250-grade maraging steel (1).

An example of one of the matrix orientations analyzed in detail is shown in the series of bright- and dark-field images in Fig. 2. Figure 2A is a bright-field micrograph of a region in a foil with a fairly high density of precipitates. The diffraction pattern from this region is shown in Fig. 3 along with its analysis. The matrix zone is [$\bar{3}11$], the Ni_3Mo zone is [$\bar{1}20$], and the sigma-phase zone is [$2\bar{1}1$]. Two Ni_3Mo precipitate zones are in this region, and one spot from each Ni_3Mo zone is coincident with a {011} matrix reflection. The spots labeled A and B in the diffraction pattern could not be accounted for by precipitates or oxides, and dark-field images from these spots confirmed that they resulted from double diffraction (5); the (011) matrix spot is the origin of spot A and the ($0\bar{1}\bar{1}$) matrix spot is the origin of spot B. Sometimes reflections with high *d*-spacings were observed, which could only be indexed as Fe_3O_4. This oxide probably formed on the foil either during electropolishing or during examination in the electron microscope. Figure 2B is a dark-field image from the (211) spot of Ni_3Mo, and Fig. 2C is a dark-field image from the (022) spot of the sigma phase.

In accordance with the previous work (1), the tetragonal phase was designated as sigma FeTi. The designation of the sigma phase as FeTi appears probable because titanium enters into the strengthening reaction very strongly (6), and because the sigma-phase FeV was identified (1) when vanadium was substituted for titanium. Because the properties of the titanium and vanadium containing 250-grade maraging steels are similar (1), it appears reasonable that titanium would substitute for vanadium in the sigma phase. However, it is also possible that the sigma phase may be different from FeTi (for example, CoTi, NiTi) because the lattice parameters of all sigma phases are similar (7, 8). Therefore, it cannot be specified with certainty from the diffraction data which sigma phase is present. It should also be pointed out that sigma phases FeTi, CoTi, and NiTi have not been reported in equilibrium diagram studies, although they have been postulated as being likely to occur (9).

There is a very good possibility that substitution of other elements may occur (for example, cobalt or nickel for iron), so that the sigma phase in the maraging steels may be more complex (that is, a ternary or higher order phase) than the binary phases considered (7, 10). All that can be stated with certainty is that the second precipitate in the 300-grade maraging steel, as in the 250-grade maraging steel, can be consistently indexed as the sigma phase.

A comparison of Fig. 2C with a similar dark-field micrograph for 250-grade maraging steel (1) indicates that more sigma phase is present in the 300-grade maraging steel. This supports the argument that the sigma phase is a titanium compound because the 300-grade maraging steel contains more titanium. It appears probable that the higher strength of the 300-grade maraging steel results from the increased amount of sigma phase. It also appears that both the Ni_3Mo phase and the sigma phase were nucleated on dislocations in 300-grade maraging steel, since there are indications in both Fig. 2B and 2C that precipitates lie along lines. In 250-grade maraging steel, however, it appeared that only the Ni_3Mo was nucleated on dislocations and that the sigma phase was uniformly distributed throughout the matrix (1).

The analyses of diffraction patterns of the ferrite matrix and Ni_3Mo precipitate zones were in agreement with the orientation relationships observed between these two phases in 250-grade maraging steel (1). However, the orientation relationships between the sigma phase and ferrite matrix were not always in agreement with those observed in the 250-grade maraging steel (1). The reason for this is not apparent. It may be that the higher supersaturation of the sigma phase in the 300-grade maraging steel, because of the higher titanium content, results in a competition between Ni_3Mo and sigma phase for nucleation on dislocations. The apparent ability of the sigma phase in the 300-grade maraging steel to nucleate on dislocations may be the reason that the orientation relationships between the sigma phase and the matrix are not in general agreement with those observed for the 250-grade maraging steels.

SUMMARY

The results of this investigation show that in the 300-grade 18Ni – Co – Mo – Ti maraging steels, the strengthening precipitates are the orthorhombic Ni_3Mo phase and a tetragonal phase, the latter phase having lattice parameters of 8.9 and 4.6 A and corresponding to a sigma-phase-type structure. These strengthening precipitates are the same as those observed in 250-grade 18Ni – Co – Mo – Ti maraging steels. The sigma phase has been designated as FeTi but is probably more complex than a simple binary phase.

The higher strength in the 300-grade steel compared with the 250-grade steels appears to result from the greater amount of the sigma-phase precipitate in the 300-grade material.

REFERENCES

1. J. M. Chilton and C. J. Barton, Identification of Strengthening Precipitates in 18Ni(250) Aluminum, Vanadium, and Titanium Maraging Steel, ASM Trans Quart, 60 (1967) 528.
2. B. G. Reisdorf and A. J. Baker, The Kinetics and Mechanisms of the Strengthening of Maraging Steels, Contract AF33 (657)-1149, Technical Report No. AFML-TF-64-390 (January 1965).
3. B. R. Banerjee, J. M. Capenos and J. J. Hauser, Aging Kinetics in 300-Grade Maraging Steel, Advances in Electron Metallography, Vol. 6, ASTM STP No. 396, ASTM, Philadelphia, Pa. (1966) 115.
4. S. Saito and P. A. Beck, The Crystal Structure of $MoNi_3$, Trans AIME, 215 (1959) 938.
5. P. B. Hirsch, A. Howie, R. B. Nicholson, D. W. Pashley and M. J. Whelan, Electron Microscopy of Thin Crystals, Butterworths, London (1965) 148.
6. R. F. Decker, J. T. Eash and A. J. Goldman, 18% Nickel Maraging Steel, ASM Trans Quart, 55 (1962) 58.
7. P. Duwez, Intermediate Phases in Alloys of the Transition Elements, Theory of Alloy Phases, ASM, Metals Park, Ohio (1956) 243.
8. W. B. Pearson, A Handbook of Lattice Spacings and Structures of Metals and Alloys, Vol. 2, Pergamon Press, New York, N. Y. (1967) 93.
9. T. V. Philip and P. A. Beck CsCl-Type Ordered Structures in Binary Alloys of Transition Elements, Trans AIME, 209 (1957) 1269.
10. H. J. Goldschmidt, A Survey of the Occurrence of Sigma-Phases Amongst Alloy Systems, The British Iron and Steel Research Association, MG/CG/194/51 (1951).

A Study of Austenite Reversion During Aging of Maraging Steels

D. T. PETERS

ABSTRACT. Austenite reversion during aging of maraging steels has been measured over a range of aging times up to 1000 hr on a series of binary and ternary alloys containing a systematic variation of Ni, Co, Mo, and Ti. Increasing the Ni content from 11 to 30% and the Mo content from 1.1 to 7.4% markedly increased the rate of formation and the apparent equilibrium amounts of austenite. Addition of 8% Co to the Fe – 18 Ni base alloy somewhat retarded the reaction, but 20% Co considerably accelerated it. A Ti addition of 1.4% to the base alloy drastically reduced the tendency toward austenite formation. The effect of the precipitation hardening elements molybdenum and titanium on the reversion reaction appears to be the result of the change in nickel content of the matrix effected by the precipitation reaction. The austenitic "banded" regions sometimes found in maraging steels are suggested to arise from regions of high Mo content.

The effect on reversion of prior austenite grain size and of pretreatments involving an 1100 F heat treatment were investigated in an 18Ni – 250 maraging steel. Increasing austenitic grain size delayed the start of reversion slightly but had no effect at longer aging times. Austenite retained in the structure after re-austenitizing following an exposure at 1100 F markedly accelerated the reversion reaction on subsequent aging. A marked acceleration was also noted in specimens of similar history where the re-austenitizing treatment left no retained austenite, but was still not sufficiently long to completely eliminate the nickel-enriched regions.

Reverted austenite is shown to be the principal cause of magnetic hardening of maraging steels and the martensitic Fe – Ni and Fe – Ni – Co alloys.

Although the precipitation hardening reaction in the 18% Ni maraging steels has received a great deal of attention in recent years, the formation of austenite during aging of these steels has been little studied. This reaction is commonly referred to as austenite reversion. The presence of austenite is probably in part responsible for softening at long aging times (1) and may be significant to fatigue and stress-corrosion cracking (2, 3). Magnetic properties of the maraging steels, especially at elevated temperatures, are profoundly affected by austenite in the structure.

The purpose of this work was to evaluate the effect of several alloying elements on the rate of austenite reversion in maraging steel. The effect of prior thermal history was also investigated.

Allen and Earley (4) suggested that, on heating to within the two-phase region of the iron-rich side of the iron-nickel diagram, austenite is formed by the following diffusion-controlled decomposition reaction:

$$\alpha_2 \rightarrow \alpha' + \gamma'$$

where α_2 is the martensite, α' is a low-nickel bcc phase and γ' refers to a nickel-enriched fcc phase which has subsequently been called reverted austenite in the maraging steel literature. This is in contrast to the transformation which occurs when heating to above a temperature known as the A_s temperature. Austenite of the same composition of the martensite is formed under these conditions by an instantaneous reaction which evidently involves shear. (This latter reaction is not considered in this paper.)

The reverted nickel-enriched austenitic phase will transform wholly or in part to martensite on cooling to room temperature if its nickel content is below about 30%, corresponding to an M_s above room temperature. Aging temperatures in this study were chosen to be sufficiently low so that the reverted austenite which formed would contain sufficient nickel to be stable at room temperature. The temperature range studied (900 – 1100 F) is of interest in age hardening the maraging steels.

The author is associated with The International Nickel Co. Inc., Paul D. Merica Research Laboratory, Suffern, N. Y.
Manuscript received 26 October 1967.

EXPERIMENTAL PROCEDURE

ALLOY PREPARATION

Compositions of the alloys studied are given in Table 1. Thirty-pound heats were prepared by vacuum induction melting. The raw materials were electrolytic iron and cobalt, carbonyl nickel pellets, molybdenum pellets, and titanium sponge. No titanium or aluminum was used to deoxidize the melts because of the marked effect of these elements on subsequent austenite formation. Deoxidation was accomplished instead with carbon.

TABLE 1. Compositions of Alloys Studied

Alloy designation	Composition, wt %					
	Ni	Co	Mo	Ti	C	Fe
A	11.0	...	...	...	0.011	bal
B	17.9	...	...	...	0.005	bal
C	24.3	...	...	...	0.011	bal
D	30.0	...	...	...	0.013	bal
E	17.0	8.15	...	...	<0.03	bal
F	20.3	23.6	...	...	<0.03	bal
G	18.2	...	1.07	...	<0.03	bal
H	17.6	...	3.10	...	<0.03	bal
I	17.3	...	5.40	...	0.016	bal
J	18.2	...	7.40	...	<0.03	bal
K	17.0	...	...	1.39	0.026	bal
L	17.4	7.80	4.82	...	0.020	bal
18Ni – 250	18.6	7.65	4.97	0.45 + 0.10 Al	0.020	bal

The 4 by 4-in. ingots were soaked at 2300 F for 1 hr and forged to 2 by 2-in. bars. These were cooled to room temperature and reheated to 2100 F for hot rolling to 1-in. square bar. The finishing temperature was kept above 1800 F to assure a completely recrystallized structure free of texture. The commercial 18Ni – 250 maraging steel was obtained in the form of 1-in. diam hot rolled bar stock.

Specimens $\frac{3}{16}$ in. thick for aging studies were homogenized in argon at 1800 F for 4 hr, air cooled, and then water quenched from a 1-hr austenitizing treatment at 1500 F. The only exception was the high titanium alloy (alloy K) which was quenched from 2100 F. All aging was in salt baths controlled to ±3 F.

DETERMINATION OF AUSTENITE CONTENT

The volume per cent austenite was determined by the usual "direct comparison" x-ray diffraction technique (5), where the integrated intensity of an austenite line is compared to that of a martensite line. The measure of integrated intensity used throughout this work was the product of the peak height times the width at half peak height. Specimens were prepared by wet grinding through 320-grit paper. Several variations in surface preparation including polishing through Linde "B" and electropolishing were found to have no effect on measured austenite content.

The integrated intensities were compared assuming that the two phases had identical composition i.e., that of the alloy. This approximation is valid in an Fe – Ni or Fe – Ni – Co alloy because of the small difference in atomic scattering factors of these elements. In the case of the molybdenum- and titanium-containing alloys, the precipitation reaction has progressed to such a large extent before significant amounts of austenite form, that little molybdenum or titanium remain in solution in the martensite to contribute to the intensity of a martensite peak (1). The austenite peak intensities

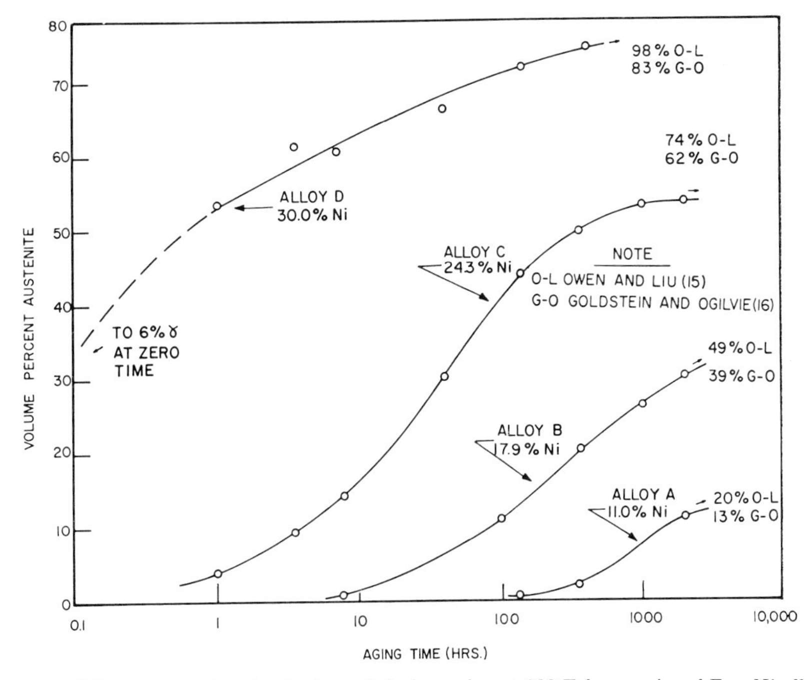

Fig. 1. Volume percent austenite formed during aging at 900 F for a series of Fe – Ni alloys.

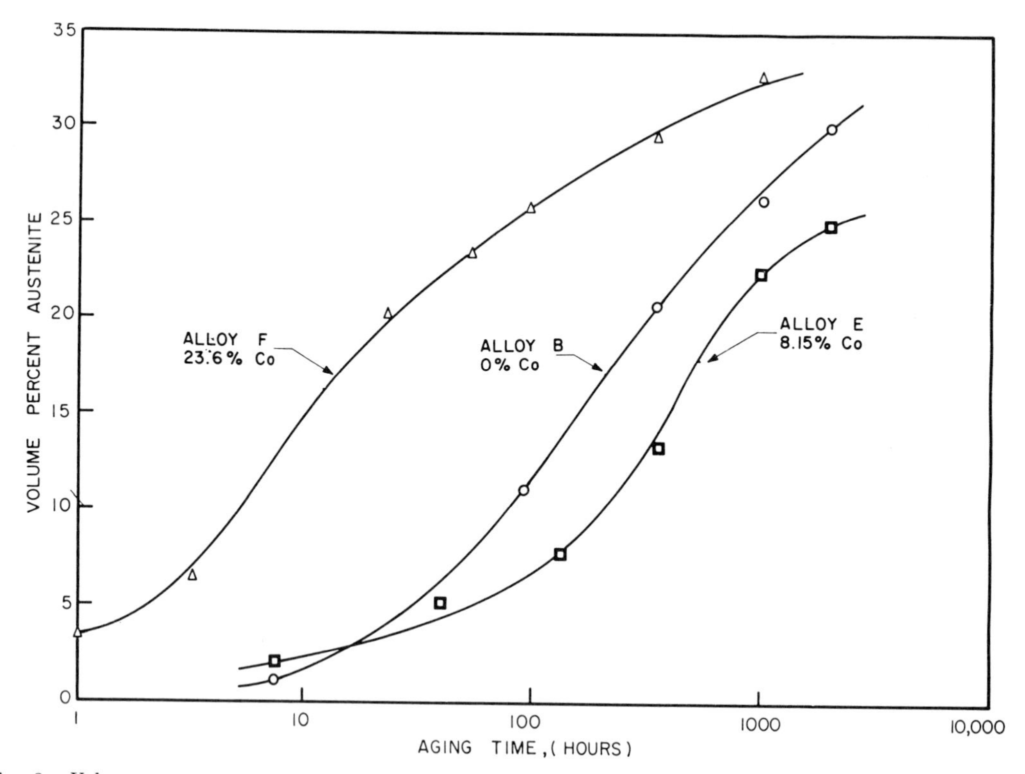

FIG. 2. Volume percent austenite formed during aging at 900 F for Fe – 18%Ni alloys with varying cobalt content.

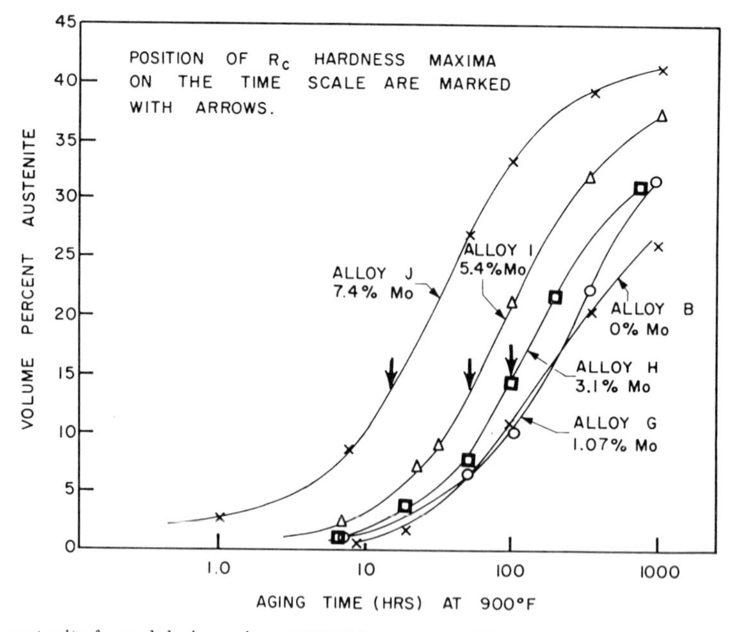

FIG. 3. Volume percent austenite formed during aging at 900 F for a series of Fe – 18 Ni alloys with varying molybdenum content.

would be slightly raised, however, if molybdenum or titanium were in solid solution in the austenite. A correction for this has not been made, but would amount to about 1.5% austenite in 40% in the alloy containing the largest amount of molybdenum, the element with the highest scattering factor (alloy J). The volume fraction of the precipitate is not known in the molybdenum- and titanium-containing alloys. Not taking this fraction of precipitate into account could result in a few per cent over-estimation of the volume fraction of austenite.

A more important error in the austenite determination could arise from preferred orientation in the samples. For the most part, the volume fraction austenite was calculated using the $(110)\alpha$-$(111)\gamma$ peak intensities obtained with vanadium filtered $CrK\alpha$ radiation, but checks were made using other combinations of peaks. Excellent agreement was

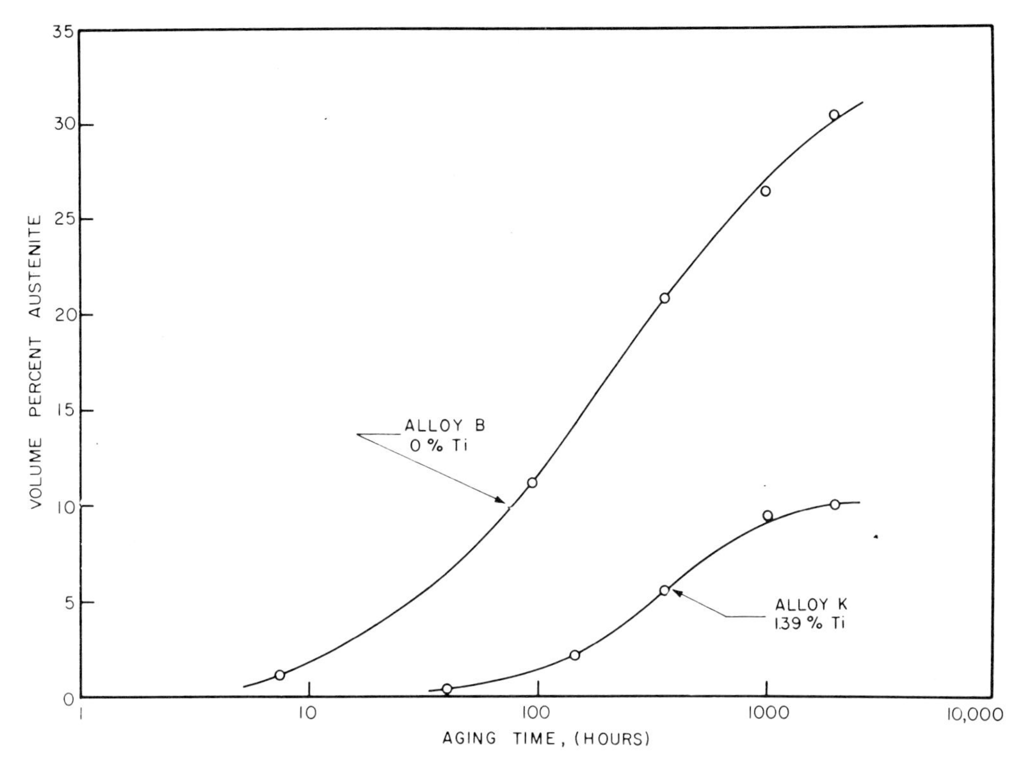

FIG. 4. Effect of titanium on austenite reversion in an Fe – 18 Ni base alloy aged at 900 F.

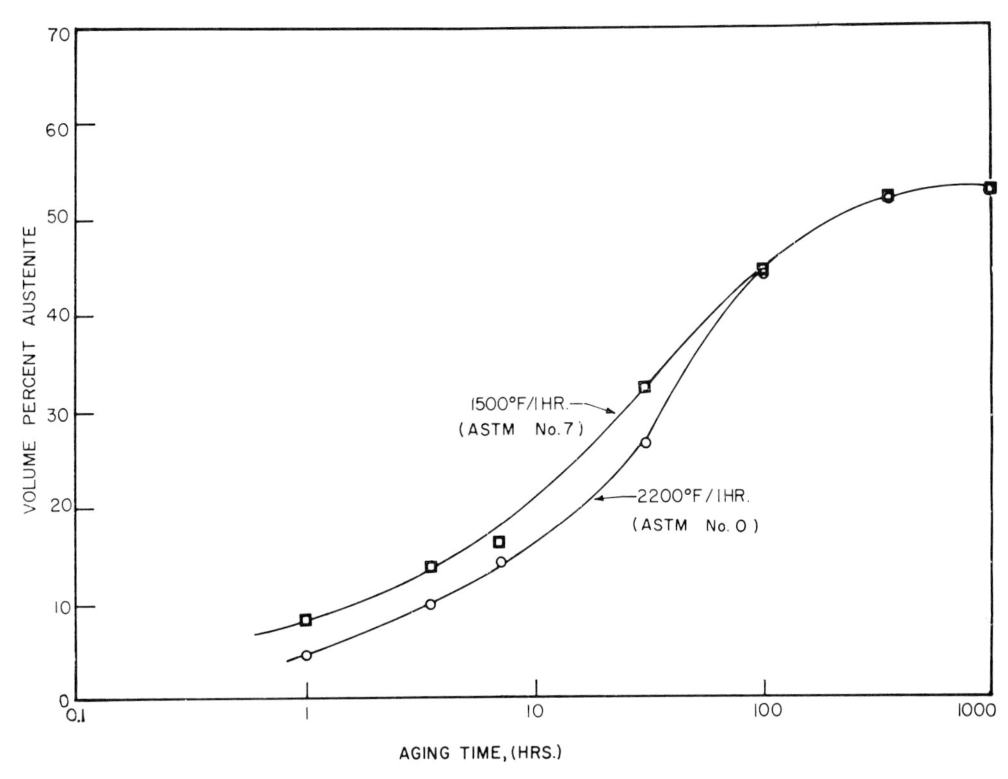

FIG. 5. Effect of annealing temperature on austenite reversion in an Fe – 25 Ni alloy aged at 900 F.

obtained using the (211)γ-(310)α-(220)γ set of peaks obtained with MoKα radiation filtered with zirconium. A representative comparison of results from the two sets of peaks is as follows: the Fe – 30-Ni alloy annealed 4 hr at 1800 F, air cooled, followed by 1 hr at 1500 F and air cooled showed 77% austenite with MoKα and 75.4% with CrKα radiation. After the specimen had been cooled to liquid ni-

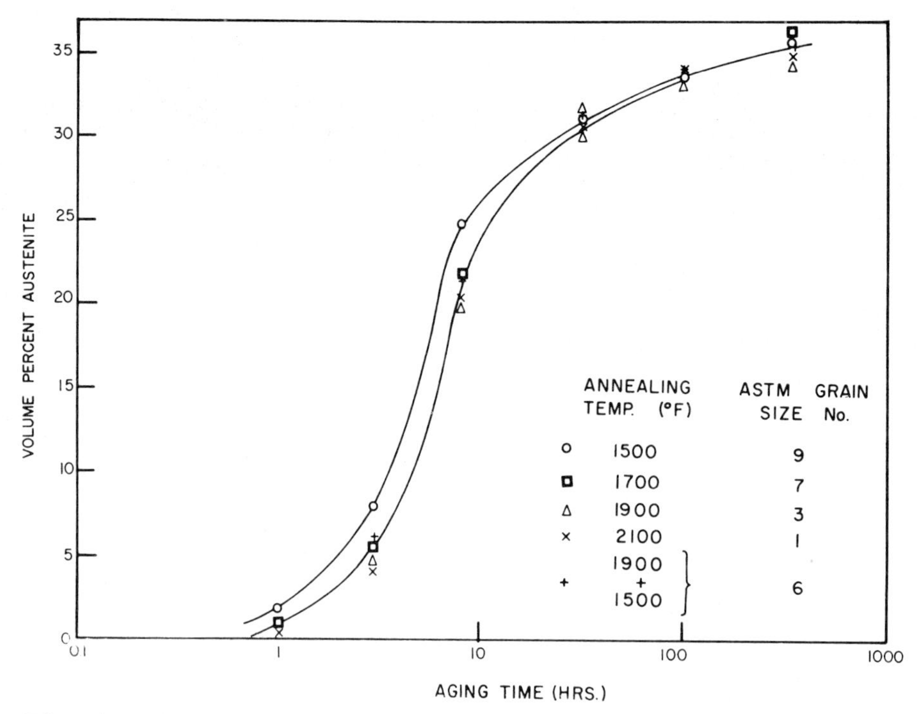

FIG. 6. Effect of annealing temperature on austenite reversion in an 18 Ni – 250 ksi maraging steel aged at 1000 F.

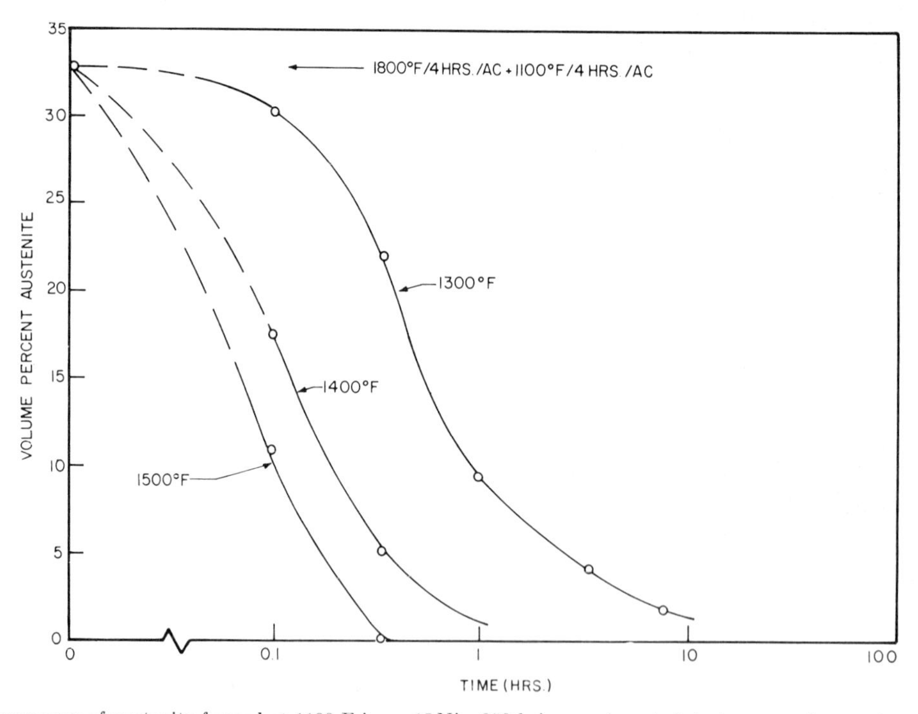

FIG. 7. Disappearance of austenite formed at 1100 F in an 18 Ni – 250 ksi maraging steel during annealing at three temperatures.

trogen temperature, both procedures indicated 6 ± 1% austenite.

MEASUREMENT OF COERCIVE FORCE

The coercive force was measured to assess the effect of reverted austenite on magnetic hardness of the maraging steels. In the method of measurement employed, the specimen, 2 in. long by 0.1 in. in diameter, is first magnetized by application of several hundred oersteds in one direction. After reducing the field strength to zero, small magnetizing fields are applied in the opposite direction. On removing the specimen from the search coil, a deflection is observed on the fluxmeter until that field strength corresponding to the coercive force is reached where the induction of the specimen is zero.

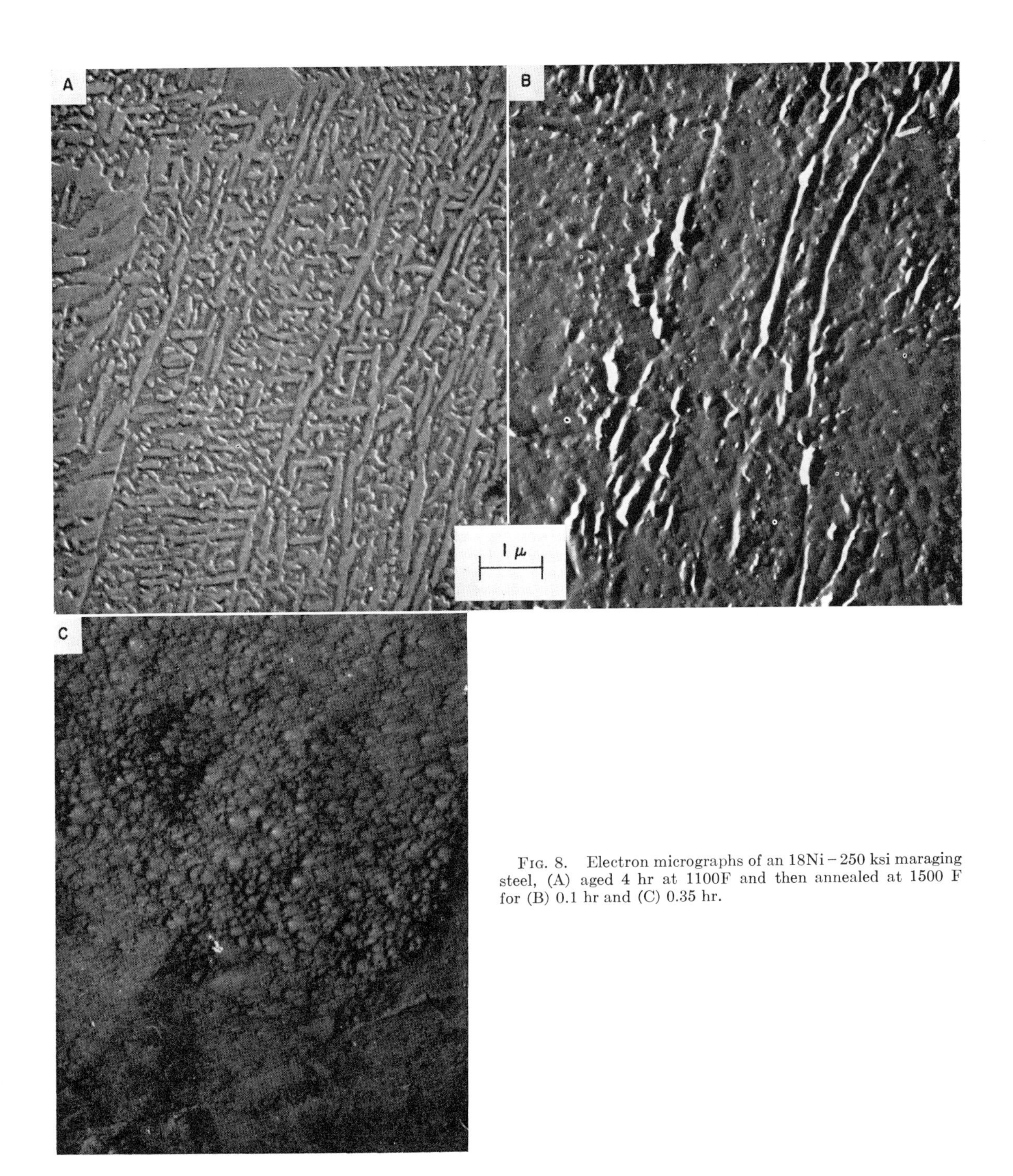

FIG. 8. Electron micrographs of an 18Ni – 250 ksi maraging steel, (A) aged 4 hr at 1100F and then annealed at 1500 F for (B) 0.1 hr and (C) 0.35 hr.

The search coil consisted of 600 turns of #32 Formvar or ceramic insulated copper magnet wire wound on a 6-mm OD vycor tube. Elevated-temperature measurements were made by placing a noninductively wound heating element between the search coil and magnetizing coil.

RESULTS

EFFECT OF COMPOSITION OF SIMPLE ALLOYS

The time-dependence of austenite reversion at 900 F in a number of simple alloys is shown in Fig. 1 through 4.

Figure 1 shows the drastic effect of nickel content on increasing the amount of austenite present at any time. The relationship between the austenite content at the longest aging time used here and that predicted by the presently accepted phase diagram is discussed below.

Figure 2 indicates that cobalt in the range used in presently commercial maraging steels (about 8%) slightly retards the austenite reaction, while at higher levels the reaction is accelerated considerably. The slight retardation at 8% Co was also found in alloys containing 5% Mo (1).

Figure 3 shows the effect of molybdenum in the

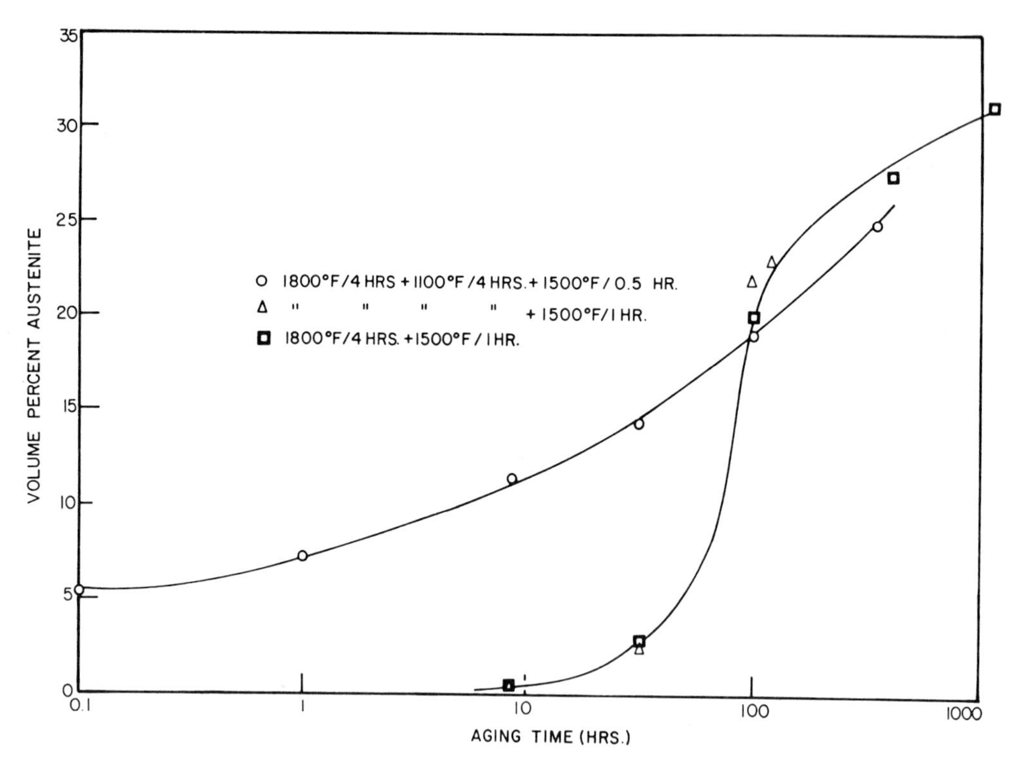

FIG. 9. Formation of austenite in an 18Ni – 250 ksi maraging steel at 900 F following a "reversion" heat treatment.

Fe – 18Ni base alloys both in accelerating the reaction at short times and in increasing the austenite present as equilibrium is approached. The addition of 1% Mo had no marked effect but an increase in reversion tendency at 3% Mo and above is evident. Peak hardness occurs in all the age hardenable Fe – 18Ni – Mo alloys with about 14% reverted austenite present.

A titanium addition of 1.39% greatly retards reversion in the same base composition (Fig. 4).

EFFECT OF GRAIN SIZE

Figure 5 shows the effect of a large difference in prior austenite grain size produced by single annealing treatments at 1500 (ASTM 7) and 2200 F (ASTM 0). The 25% Ni binary alloy was selected for this experiment because the austenite reaction is nearly completed within 1000 hr at 900 F. Increased grain size does retard the early reversion to austenite, but as expected, the curves come together as equilibrium is approached. This retardation is not unexpected since much of the austenite appears at prior austenite and martensite platelet boundaries.

Microsegregation of alloying elements to austenite grain boundaries in a maraging steel would be expected to alter the rate of austenite reversion. Both the degree of microsegregation and grain size will vary with austenitizing temperature. To determine a possible effect of microsegregation, specimens of the 18Ni – 250 steel were quenched from 1500, 1700, 1900, and 2100 F. The resulting volume per cent austenite-aging time curves obtained on aging at 1000 F are shown in Fig. 6. The results are similar to those of the binary alloy. Grain growth was evident from micrographs of specimens austenitized at 1700 F and above and these curves are identical within experimental error. Re-annealing the 1900 F specimen at 1500 F refined the austenitic grain size considerably (ASTM 3 to 6), but this had no apparent effect on the rate of reversion. Only the 1500 F single annealing treatment producing the finest grain size (ASTM 9) showed any tendency for accelerated austenite reversion.

EFFECT OF PRIOR EXPOSURE IN THE 1100–1200 F RANGE

A maraging steel can be exposed in the 1100 to 1200 F temperature range at a heat-affected zone of a weld, or purposely, as in certain reversion heat treatments which have been attempted to improve certain properties (2, 3, 6). The question arises whether a 1-hr austenitizing treatment following an 1100-1200 F exposure is sufficient to eliminate the nickel-enriched regions in the austenite resulting from the formation of reverted austenite.

Experiments were performed to answer this question. In one experiment, the retained austenite content at room temperature in the 18Ni – 250 steel was measured following treatments for various lengths of time at a particular austenitizing temperature on specimens previously given a treatment of 4 hr at 1100 F. This treatment produced about 33 vol % austenite. The disappearance of the reverted austenite with time at three austenitizing temperatures is shown in Fig. 7. Apparently less

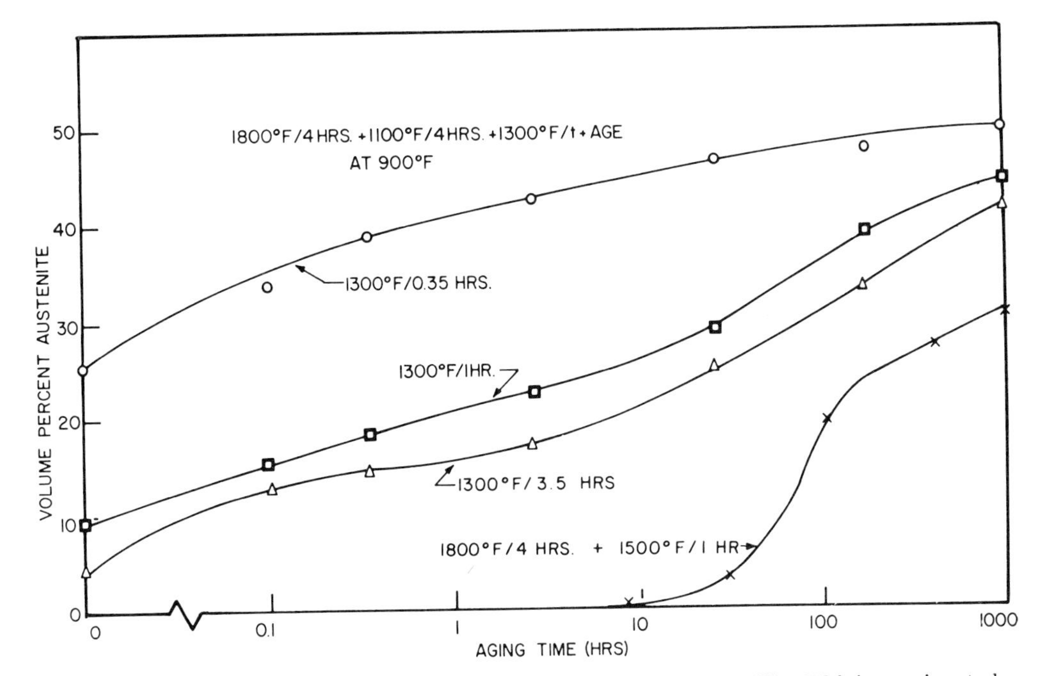

FIG. 10. Effect of residual austenite on austenite reversion at 900 F in an 18Ni – 250 ksi maraging steel.

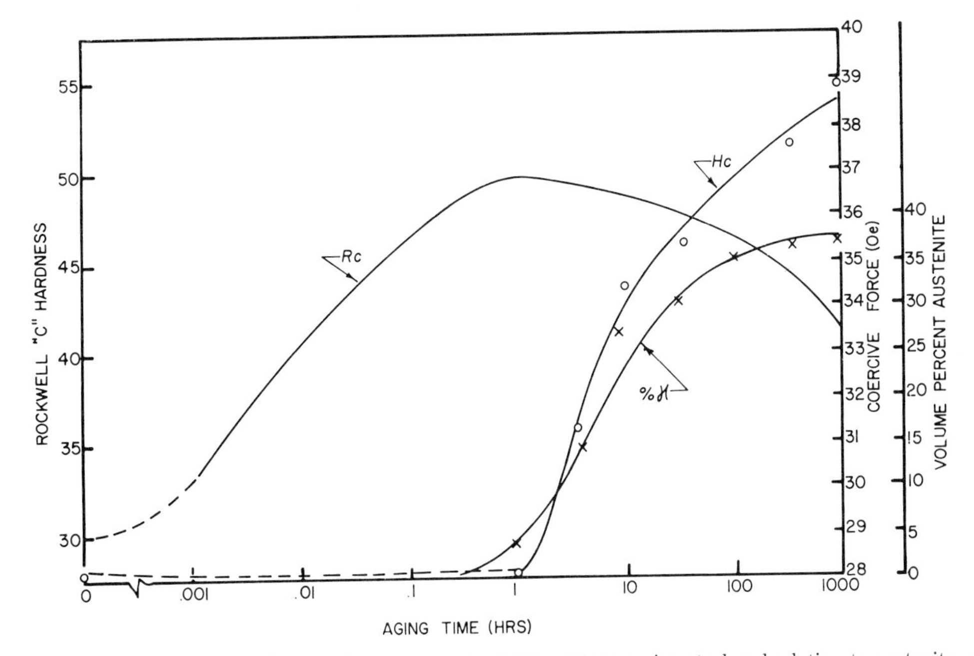

FIG. 11. Comparison of mechanical and magnetic hardening in 18Ni – 250 maraging steel and relation to austenite content. Aged at 1000 F.

than 1 hr at 1500 F is sufficient to eliminate the austenite formed at 1100 F. Replica electron micrographs from the specimen treated for 4 hr at 1100 F and those austenitized for 0.1 and 0.35 hr at 1500 F are shown in Fig. 8. Figure 8C has the appearance of material in the fully austenitized condition.

The existence of nickel-enriched regions remaining after an austenitizing treatment might be detected by their effect on the rate of austenite reversion during a subsequent aging treatment where such regions would be expected to accelerate reversion. That this is true is seen by the results plotted in Fig. 9. The 18Ni – 250 steel specimens were aged 4 hr at 1100 F and austenitized for either 0.5 or 1 hr at 1500 F. Although the specimen austenitized 0.5 hr at 1500 F showed no retained austenite, substantial reverted austenite was present after very short aging times in the subsequent 900 F aging treatment. Austenitizing for 1 hr at 1500 F appears to have completely erased the effect of the previous treatment at 1100 F for 4 hr. Subsequent work has shown that even 1 hr at 1500 F is not sufficient to erase the effect of a very long aging at a lower temperature (e.g., 1000 hr at 1000 F) where the differ-

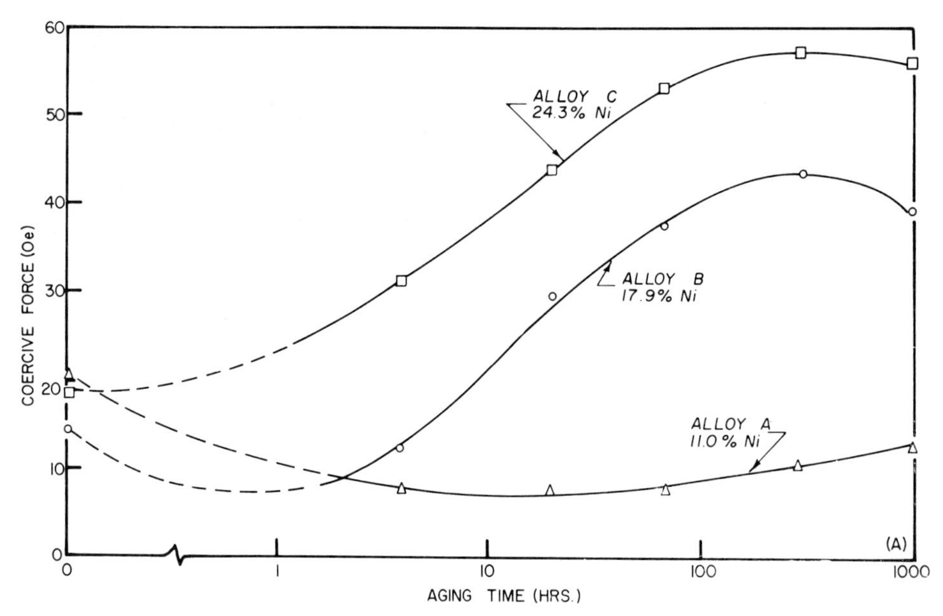

FIG. 12. Change of coercive force with aging time at 1000 F for three Fe – Ni binary alloys.

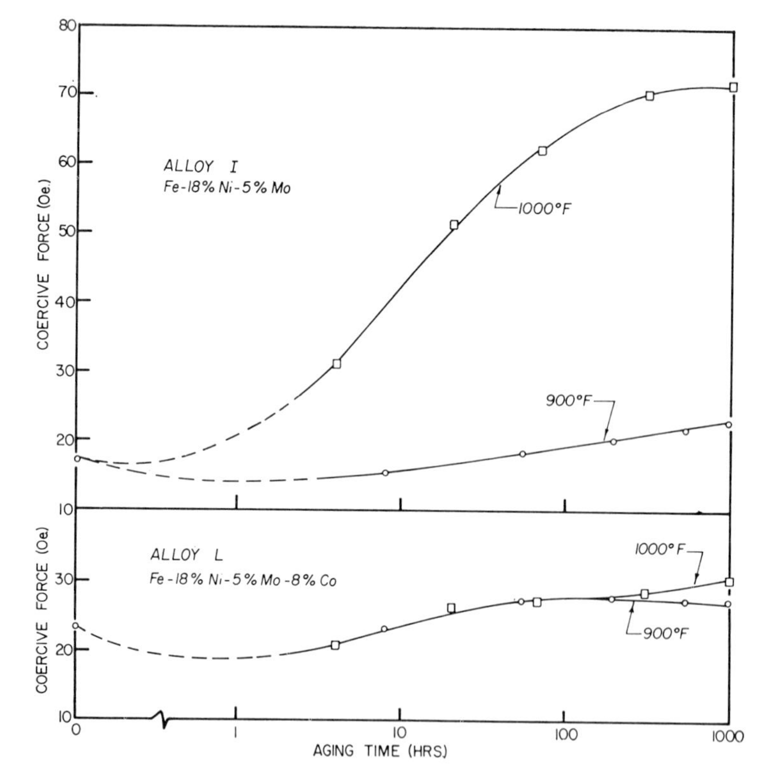

FIG. 13. Coercive force vs aging time for 18% Ni ternary and quaternary alloys.

ence in nickel content between the two phases is greater.

In Fig. 10 the presence of retained austenite from an incomplete austenitizing treatment is shown to have a pronounced effect on austenite reversion in a subsequent aging treatment. Again the 18Ni – 250 maraging steel specimens were aged 4 hr at 1100 F and then given incomplete austenitizing treatments of 0.35, 1, and 3.5 hr at 1300 F. These treatments left 26, 9.5 and 4% retained austenite, respectively. In each case, the amounts of austenite increased due to reversion with less than 1 hr aging at 900 F. In the absence of retained austenite prior to aging, reversion would not be encountered in up to 10 hr at 900 F. The amounts present after aging for 1000 hr appear to be roughly the amount formed normally in that time plus that initially retained. The shape of the curves suggests that the specimens with the greatest amounts of retained austenite would actually decrease in austenite content if aged for several thousand hours.

EFFECT OF AUSTENITE CONTENT ON COERCIVE FORCE

Figure 11 shows a comparison of mechanical and magnetic hardening in the 18Ni – 250 maraging steel and the volume fraction austenite formed at the aging temperature of 1000 F. Clearly the increase in coercive force with aging time is related to the increase in the austenite content. Detert reached the same conclusion from other considerations for a 15% Ni maraging steel (7). The contribution of the age hardening precipitate appears to be quite small at peak hardness and small compared to that of the austenite after considerable overaging. Aging the simple Fe – Ni binary alloys should lead to a similar increase in coercive force if austenite formed on aging controls this property. The coercive force aging time curves are shown in Fig. 12. Aging Alloys A, B, and C at 1100 F gave a large increase in coercive force, but aging at 900 F gave only a 1 or 2 oersted increase. A similar behavior can be seen in the isochronal data of Fedash (8) and Zel'dovich and Sadovskiy (9) in Fe – Ni and Fe – Mn alloys.

The change in coercive force with aging time at 900 and 1000 F in Alloys I and L shows the effect of the addition of Mo and Co to the Fe – 18Ni base alloy (Fig. 13). The rise in coercive force is more rapid in Alloy I compared to the 18Ni – 250 steel because, in the absence of titanium and cobalt, austenite reversion proceeds more rapidly (Fig. 2 and 4). Comparing the peak Hc values in Alloys B and I suggests that the molybdenum-rich precipitate makes a contribution of nearly 30 oersteds. Another explanation is suggested by the electron micrographs of Fig. 14. The austenite distribution in the precipitation hardenable alloys is much finer than in the binary alloys. Grain boundaries and martensite platelet boundaries appear to be the predominant locations for austenite in the binary alloy with some coarse globular particles in the platelets themselves. The alloys containing molybdenum-rich precipitates show a Widmanstätten distribution of austenite particles in addition to the boundary-nucleated austenite.

Dijkstra and Wert (10) have shown that for a particle to make a sizable contribution to the coercive force, its diameter should be on the order of the domain wall thickness. This is on the order of 1000 A (0.1 μ) in iron. Transmission electron microscopy of maraging steels has shown the dimensions of the precipitate particles to be at least an order of magnitude smaller than this even with considerable overaging (11 – 13). The austenite particles of Fig. 14 are near the optimum size.

A ferromagnetic particle is expected to make a much smaller contribution to the coercive force (10). This provides an explanation for the observation

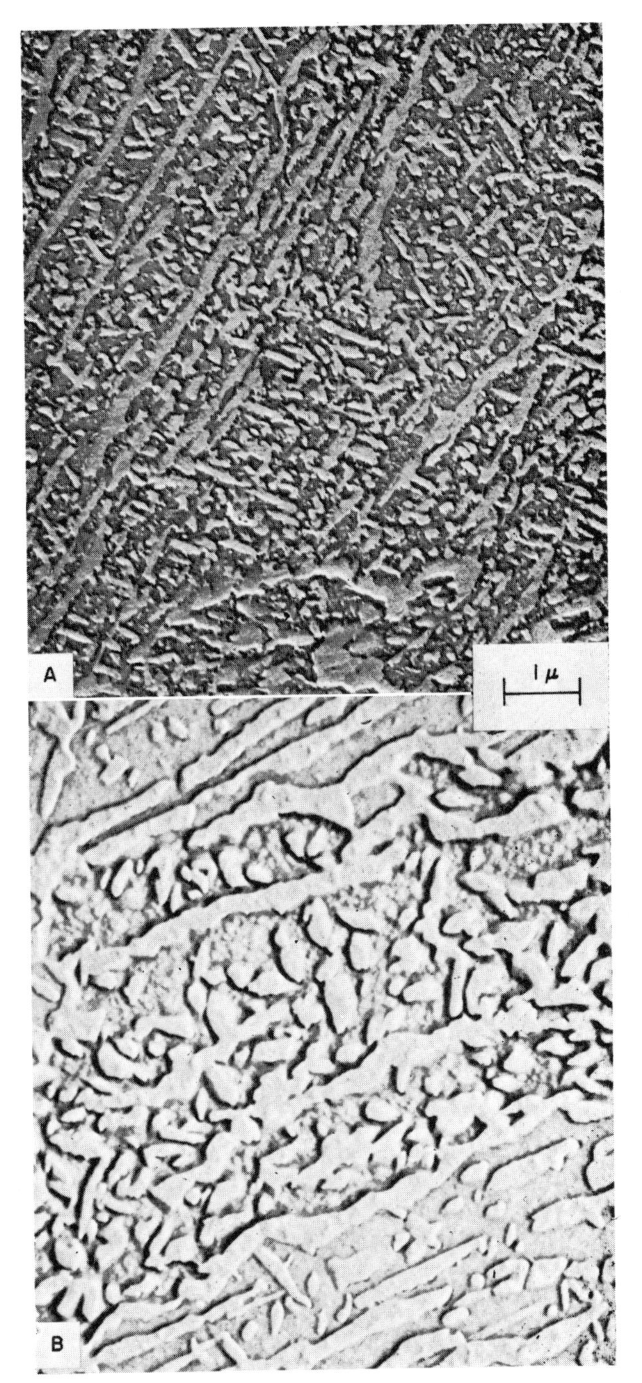

Fig. 14. Electron micrographs of negative parlodian replicas of: (A) 18Ni – 250 steel aged 1000 hr at 900 F, and (B) the 18% Ni binary alloy aged 350 hr at 1000 F.

that on aging at 900 F, the Fe – Ni binary alloys and the Fe – 18Ni – 5Mo alloys showed only a very small increase in Hc. Furthermore, the Fe – 18Ni – 5Mo – 8Co alloy showed no response to aging at 900 or 100 F (Fig. 13). The higher nickel austenite formed at 900 F would be expected to have a Curie temperature above room temperature (14). Bozorth's data indicate this is the case for the cobalt-containing alloy at both aging temperatures as cobalt raises the Curie tem-

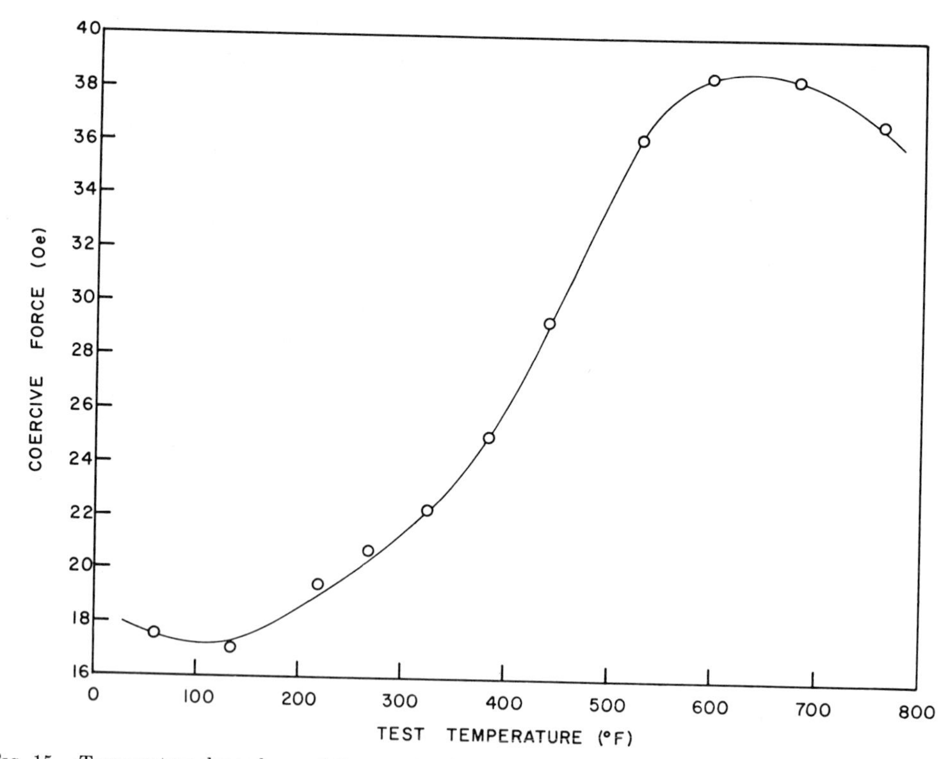

FIG. 15. Temperature dependence of the coercive force in the Fe – 18 Ni – 5%Mo alloy aged 540 hr at 900 F.

perature of iron. Evidently the austenite in the Fe – 18Ni – 5Mo alloy is ferromagnetic when formed at 900 F but is paramagnetic when formed at 1000 F. Figure 15 shows the sharp increase in Hc as the temperature is increased, indicating the Curie temperature of the austenite formed in this alloy at 900 F to be about 600 F.

On the other hand, the coercive force of the 18Ni – 250 steel increases on aging at both 900 and 1000 F. This alloy differs significantly from the quaternary Alloy L only in its 0.55% Al plus Ti content. The composition of the austenite formed in the steel after dissolution of the molybdenum- and titanium-containing precipitates is evidently such as to make it paramagnetic at room temperature. How this comes about is not clear since this small quantity of aluminum plus titanium would not be expected to depress the Curie temperature of the austenite by as much as 600 F. Evidently, there is considerable partitioning of the elements aluminum, titanium, and molybdenum during or after austenite formation, resulting in austenite which is paramagnetic at room temperature.

DISCUSSION

IMPLICATIONS TO THE Fe – Ni PHASE DIAGRAM

Although the decomposition reaction in the Fe – Ni alloys has not reached equilibrium at the maximum aging times indicated in Fig. 1, the large discrepancy between the final amounts of austenite which are apparently being approached especially in Alloys C and D and that predicted by the currently accepted phase diagram of Owen and Liu (15) is disturbing. Either there is a consistent error in the measurement of the volume fraction austenite tending to underestimate the amount present, or the positions of the phase boundaries are incorrect. Recently, Goldstein and Ogilvie re-determined the $\alpha/\alpha + \gamma$ and $\gamma/\alpha + \gamma$ phase boundaries in the Fe – Ni system by microprobe measurements of the interface compositions of diffusion couples and two-phase mixtures (16). The repositioning of the phase boundaries they propose would account for most of the discrepancy in the x-ray data. Possibly, extremely long time at temperature would bring the curves up to the levels they propose. The equilibrium levels proposed by both diagrams are indicated on Fig. 1.

A higher nickel content in the austenite than predicted by the Owen-Liu diagram is required also to account for the Curie temperature of about 600 F inferred from Fig. 15 for the austenite in the Fe – 18Ni – 5Mo alloy aged at 900 F. Since precipitates are not seen in the austenite in the aged structure (13), about 5% Mo must be in solution in the austenite after dissolution of the precipitate particles. Electrical resistivity measurements of this and similar alloys are also best interpreted on this assumption.* Therefore, it would be expected that the Curie temperature of this austenite would be lower than the value of about 400 F indicated from the nickel content of the austenite predicted by the Owen-Liu diagram. The Goldstein-Ogilvie diagram predicts about 35% Ni, corresponding to a

* D. T. Peters, unpublished work.

Curie temperature of 600 F. To account for the effect of molybdenum in solution in the austenite, a still higher nickel content in the binary alloy austenite is required.

EFFECT OF COBALT

The effect of cobalt in lowering the amount of austenite formed in the Fe – 18Ni base alloy at low concentrations and raising it at high concentrations is reminiscent of its effect on the M_s temperature of the Fe – 22.5Ni base alloy as reported by Yeo (17). Cobalt at first raises M_s and then lowers it at higher concentrations. This suggests that the M_s and A_s lines have the same general shape as the equilibrium $\alpha/\alpha + \gamma$ and $\gamma/\alpha + \gamma$ phase boundaries in the Fe – Ni – Co diagram just as in the Fe-Ni diagram. These must be concave downwards and might be an exaggeration of the shape of the equilibrium lines on the Fe – Co system.

EFFECT OF THE PRECIPITATION HARDENERS Mo AND Ti

In alloys where a hardener such as molybdenum or titanium is present, the additional complicating factor of a change in the matrix composition as a result of the much faster precipitation reaction must be considered in rationalizing the tendency for austenite reversion. In the case of a titanium hardened alloy (Fig. 4), the precipitate has been identified as Ni_3Ti (13). The 1.4 wt % Ti alloy (1.64 at. %) would then absorb about 5.0 wt % Ni since the titanium solubility is very low. The matrix would then behave as if it were an 11 or 12% Ni alloy. This results in a drastic reduction in the amount of austenite present as can be seen in Fig. 1.

Evidently the molybdenum-containing alloys are the converse of this for molybdenum increases the tendency for austenite formation. This suggests that much of the precipitate present during austenite formation is an Fe – Mo compound, thus enriching the matrix in nickel, rather than the Ni – Mo compound thought to be present early in the aging process. Recent identification of the precipitate present in overaged specimens as Fe_2Mo by transmission electron microscopy and microprobe analysis of extraction replicas supports this explanation (13, 18). The acceleration of reversion caused by molybdenum relates to the problem of "banding" sometimes observed in inadequately worked and homogenized maraging steels. These regions, which are observed to be austenitic, become pronounced during aging and have been found to contain a high molybdenum content.

The 1.1% Mo alloy showing no increase in rate of reversion over the base alloy also does not age harden (1). This alloy is probably near the limit of solid solubility at 900 F.

SUMMARY AND CONCLUSIONS

1. In Fe – Ni binary alloys, increased nickel content in the range 11 to 30% greatly accelerates the rate of austenite reversion. The volume fraction austenite obtained in these binary alloys aged for a sufficiently long time time to approach equilibrium support other recent work which proposes to shift the $\alpha/\alpha{+}\gamma$ and $\gamma/\alpha{+}\gamma$ phase boundaries of the Fe-rich portion of the Fe – Ni equilibrium diagram in the direction of higher nickel.
2. In Fe – 18Ni – Co ternaries, 8% Co slightly retards the austenite reaction, but at 20% Co the reaction is considerably enhanced. The retarding action at 8% Co is also evident in alloys containing 5% Mo.
3. In an Fe – 18Ni alloy, increasing the molybdenum in the range 1.1 to 7.4% markedly accelerates austenite reversion, while 1.4% Ti has the opposite effect. These compositional effects are thought to arise from the change in nickel content of the matrix resulting from the precipitation hardening reaction. Titanium precipitated as Ni_3Ti lowers this content while molybdenum precipitated in the later stages of aging as an Fe – Mo compound raises it. The austenitic, high-molybdenum, "banded" regions in an inhomogeneous maraging steel are explained on this basis.
4. Increased prior austenitic grain size slightly retards the beginning of reversion in both binary Fe – Ni alloys and 18Ni – 250 maraging steel.
5. The rate of austenite reversion during aging is greatly accelerated by the presence of either retained austenite or nickel-enriched regions in the martensite resulting from an incomplete austenitizing treatment following a prior treatment which formed a significant quantity of austenite.
6. Magnetic hardening of maraging steels, Fe – Ni and Fe – Ni – Co alloys is shown to be caused by paramagnetic austenite formed in the structure during aging and possibly only indirectly related to the age hardening precipitate. A ferromagnetic austenite, such as that formed on aging the iron – nickel binary alloys at 900 F, will not contribute significantly to the coercive force measured at room temperature.

REFERENCES

1. D. T. Peters and C. R. Cupp, The Kinetics of Aging Reactions in 18 Pct Ni Maraging Steels, Trans AIME, 236 (1966) 1420.
2. R. A. Covert and E. P. Sadowski, Effect of Heat Treatment on Stress Corrosion Cracking Behavior of Cast 17 Percent Nickel Maraging Steel, Presented at 22nd Annual NACE Conference, Miami Beach, Fla., April 1966; to be submitted to Corrosion.
3. E. P. Sadowski, Effect of Heat Treatment on Properties and Microstructure in Thin and Heavy Sections of 17 Percent Nickel Maraging Steel, Presented at AFS Congress, Pittsburgh, Pa., May 1966; submitted to Modern Castings.

4. N. P. Allen and C. C. Earley, The Transformations $\alpha \rightarrow \gamma$ and $\gamma \rightarrow \alpha$ in Iron-Rich Binary Iron-Nickel Alloys, J Iron Steel Inst, 166 (1950) 281.
5. B. D. Cullity, Elements of X-Ray Diffraction, Addison-Wesley Publishing Co., Inc. Reading, Mass. (1956) 391.
6. North American Aviation, Maraging Casting Quality Improved by Revised Thermal Processing Treatment, Western Machinery and Steel World (March 1966) 26–27.
7. K. Detert, Investigation of Transformation and Precipitation in 15% Ni Maraging Steel, ASM Trans Quart, 59 (1966) 262.
8. G. M. Fedash, Study of Coercivity of Cold Worked and Annealed Iron Alloys, Fiz Metal Metalloved, 4 (1957) 257.
9. V. I. Zel'dovich and V. D. Sadovskiy, Effect of Heat Treatment on the Magnetic Properties of Certain Alloys of the Systems Fe – Mn and Fe – Ni, Fiz Metal Metalloved, 20 (1965) 406.
10. L. J. Dijkstra and C. Wert, Effect of Inclusions on Coercive Force of Iron, Phys Rev, 79 (1950) 979.
11. B. G. Reisdorf, Identification of Precipitates in 18, 20 and 25% Nickel Maraging Steels, ASM Trans Quart, 56 (1963) 783.
12. A. J. Baker and P. R. Swann, The Hardening Mechanism in Maraging Steels, ASM Trans Quart, 57 (1964) 1008.
13. G. P. Miller and W. I. Mitchell, Structure and Hardening Mechanisms of 18% Ni – Co – Mo Maraging Steels, J Iron Steel Inst, 203 (1965) 899.
14. R. N. Bozorth, Ferromagnetism, D. Van Nostrand Co., Inc , Princeton, N. J. (1951) 111.
15. E. A. Owen and Y. H. Liu, Further X-Ray Study of the Equilibrium Diagram of the Iron-Nickel System, J Iron Steel Inst, 163 (1949) 132.
16. J. I. Goldstein and R. E. Ogilvie, Fe – Ni Phase Diagram, Goddard Space Flight Center Report (NASA) X-640-65-117 (March 1965).
17. R. B. G. Yeo, The Effect of Some Alloying Elements on the Transformation of Fe-22.5 Pct Ni Alloys, Trans AIME, 227 (1963) 884.
18. M. J. Fleetwood, G. M. Higginson, and G. P. Miller, The Identification of Precipitates in Maraging Steels by Electron Microscopy and Electron Probe X-Ray Microanalysis, Brit J Appl Phys, 16 (1965) 645.

The Kinetics of Aging Reactions in 18 Pct Ni Maraging Steels

D. T. Peters and C. R. Cupp

Aging of commercial 18 pct Ni maraging steel and of related binary, ternary, and quaternary alloys has been studied by electrical resistivity, X-ray diffraction, and hardness measurements. The aging process is in three stages:

1) recovery of martensite;
2) two precipitation reactions;
3) formation of nickel-rich austenite.

The main objective of this work was to discover the nature of the Co-Mo interaction. Addition of cobalt to the Fe-18 pct Ni-5 pct Mo alloy results in acceleration of the precipitation reaction by increasing the supersaturation of molybdenum. The cobalt-containing alloys show no incubation period for the precipitation reaction, but the alloy without cobalt shows an initial slow reaction. These results are interpreted in terms of precipitation on dislocations. At low aging temperatures, the presence of cobalt leads to the appearance of a second type of precipitate which is a particularly effective hardener. Cobalt was found to retard the formation of austenite. Overaging is a combination of classical overaging and the formation of austenite in the structure. The activation energy for the precipitation reaction determined from resistivity measurements is 42.8 kcal per mole and the growth of the precipitate is found to obey a $t^{1/2}$ *law at high temperatures and a* $t^{1/3}$ *law at low temperatures.*

D. T. PETERS, Junior Member AIME, and C. R. CUPP, Member AIME, are Research Metallurgist and Supervisor of Metal Physics Section, respectively, Paul D. Merica Research Laboratory, The International Nickel Co., Sterling Forest, Suffern, N.Y.

Manuscript submitted December 29, 1965. IMD

THE mechanism of strengthening of the 18 pct Ni maraging steels has been studied extensively since 1962.[1-7] The strengthening is due to precipitation hardening of the martensitic matrix. The main precipitate is an intermetallic compound involving molybdenum, but there is disagreement about its identity.

Source: *Transactions of the Metallurgical Society of AIME*, Vol 236, Oct 1966

Another precipitate, Ni_3Ti, is generally agreed to be present also. Cobalt has not been found to any significant extent in the extracted precipitate examined by microprobe analysis.[4-7]

The central objective of this work was to explain the interaction of cobalt and molybdenum in the strengthening of the alloy. Floreen and Speich[8] have shown the unique interaction between cobalt and molybdenum and the absence of such an interaction in other Fe-18 pct Ni-8 pct Co-X quaternaries. They showed that a significant but nearly constant solid-solution hardening contribution was present in the cobalt-containing alloys. It has been suggested that localized rearrangements to form ordered regions without the aid of long-range diffusion may be a contributing factor in the strengthening.[1] This could account for the rapid hardening that has been observed at short aging times. Cobalt might be responsible for the early hardening due to its tendency to promote ordering within the matrix. Mihalisin has obtained evidence for ordering in an Fe-22.7 pct Ni-19.3 pct Co alloy.[9]

The phase transformations during aging have been followed largely by electrical resistivity in the present work. Because of their chemical similarity, substituting cobalt for iron leads to only a small change in resistivity. Therefore any cobalt precipitation will go unnoticed, and the resistivity is only an indicator of the amount of molybdenum in solid solution. Also formation of long-range order of the FeCo type does not change the resistivity appreciably.[10] Thus this ordering cannot be proven or disproven by this technique, But the considerable effect of cobalt on the precipitation of molybdenum from solid solution can be and has been studied here. The effect of cobalt and molybdenum on the kinetics of formation of austenite and how its formation is related to the precipitation reaction are questions for which an answer was sought in this investigation.

EXPERIMENTAL PROCEDURE

The alloys that were studied are shown in Table I. Alloys A, B, and C were studied most intensely. Alloys B and C are simplifications of the commercial alloy, prepared in order to study the basic hardening reaction.

Several other compositions over a range of molybdenum contents and an Fe-18 pct Ni binary alloy were also examined. The laboratory heats were made by vacuum induction-melting electrolytic iron, carbonyl nickel pellets, electrolytic cobalt, and molybdenum chips. No titanium or aluminum was added so that precipitation of the $Ni_3(Ti,Al)$ phase would not confuse the study of the main reaction. Deoxidation was accomplished with carbon. The material was forged and hot-rolled to $\frac{1}{4}$-in.-thick strip, and then cold-rolled to $\frac{1}{8}$-in.-thick strip. Pieces, $\frac{1}{8}$ in. square in section and 12 in. long, were cut from the strip in the rolling direction. The corners were rounded on a belt sander and the strips cold-rolled, swaged, and drawn to $\frac{1}{32}$-in. wire with appropriate intermediate anneals. The final draws totaled over 50 pct reduction in area. Hardness specimens $\frac{1}{4}$ by $\frac{1}{2}$ in. were cut from the $\frac{1}{8}$-in. strip. Some of these were cut in half for hardness specimens to be used at aging times of less than 0.1 hr. They were annealed in argon at 1500°F for 1 hr, unless otherwise noted, and water-quenched contrary to usual practice, for reasons to be discussed later. Aging was performed in sodium nitrate-nitrite salt baths at selected temperatures in the range from 675° to 1100°F. The temperature was controlled to within ±2°F of the chosen aging temperature.

Resistivity vs aging time curves at given temperatures were obtained on single wires aged intermittently to a total time of about 1000 hr. Aging times were measured from the time the specimen entered the salt until it was quenched into water. Resistivity measurements were made by the four-probe potentiometric technique with a specimen gage length of 6 in. In the early stages of this work, three specimens were run concurrently to determine the reproducibility of the results. Reproducibility was excellent, and the number of specimens per aging response curve was ultimately reduced to one.

Separate specimens were used for each point on the hardness curves. The specimens were wet-ground through 320 paper following the aging treatment. The average of 10 Rockwell "*C*" readings was used for each point on the curve. The scatter in the readings amounted to ±1 R_C after short time aging, and less after longer times.

The volume percent of austenite that formed by reversion of the martensite during aging was determined from the relative intensities of the $\{110\}$ α and $\{111\}$ γ X-ray diffraction peaks. The scattering factors of the two phases are assumed to be identical. The integrated intensity of the peaks was taken to be the product of the peak height and width at half maximum. The volume fraction of the precipitate is unknown so the values of volume percent austenite reported here are slightly high. The specimens were irradiated with vanadium-filtered CrKα radiation. The largest uncertainty in this determination was expected to be due to preferred orientation. Fortunately, preferred orientation was not found in the laboratory heats, but the cold-rolled commercial plate had such a pro-

Table I. Composition of Alloys Studied

Alloy Designation	Wt Pct					
	Ni	Mo	Co	C	Ti	Al
A) Commercial 18 Ni-250 maraging steel	18.39	4.82	7.83	0.02	0.35	0.07
B) Quaternary (nominal 18 pct Ni-8 pct Co-5 pct Mo)	17.40	5.67	7.80	0.03	–	–
C) Ternary (nominal 18 pct Ni-5 pct Mo)	17.25	5.40	–	0.016	–	–
D) Binary (nominal 18 pct Ni)	18.30	–	–	0.015	0.096	0.05
E) Other Fe-18 pct Ni-Mo ternaries	23.9	0.95	–	<0.03	–	–
F) Other Fe-18 pct Ni-Mo ternaries	17.6	3.1	–	<0.03	–	–
G) Other Fe-18 pct Ni-Mo ternaries	18.2	7.4	–	<0.03	–	–
H) Other Fe-18 Ni-Mo-Co quaternaries	18.4	1.0	1.3	<0.03	–	–
I) Other Fe-18 Ni-Mo-Co quaternaries	17.4	3.0	4.7	<0.03	–	–
J) Other Fe-18 Ni-Mo-Co quaternaries	16.9	7.75	12.2	<0.03	–	–
K) Ternary (nominal 18 pct Ni-8 pct Co)	17.0	–	8.15	0.017	–	–

nounced texture that austenite determinations would have been meaningless and were not taken.

The A_s and M_s temperatures were determined dilatometrically on $\frac{1}{8}$-in.-diam specimens, Table II. A furnace at 1600°F was raised about the mounted specimen and the length change recorded. The specimen was then held as austenite for 15 min before the furnace was removed and the specimen allowed to air-cool while the length change was again recorded.

EXPERIMENTAL RESULTS

The isothermal resistivity and hardness curves are shown in Figs. 1 and 2 for the commercial maraging steel, Alloy A. The resistivity curves have three characteristic features:

1) a small decrease at short aging times;
2) a large decrease coinciding with the increase in strength;
3) a final increase and leveling off at long aging times.

Stages 1 and 3 are also apparent in the curve obtained from the Fe-18 pct Ni binary, Fig. 3. The decrease clearly seen in the 800°F curve must be due to partial recovery of the martensitic defect structure. The contribution of this effect to the total aging curve is always small (about 1 microhm-cm). The increase at long aging times, as shown in Fig. 3 on the curve from the specimen aged at 1000°F, is seen to be due to the diffusion-controlled formation of austenite by comparing the resistivity data to the volume percent austenite measurements of Speich[11] on a similar Fe-20 pct Ni alloy. Speich's measurements suffice to show the austenite-resistivity relationship, although the alloy and aging temperature used are slightly different. The shift of the austenite curve to longer times is consistent with the lower aging temperature (932°F) used by Speich.

The large decrease in resistivity in Fig. 1 is characteristic of precipitation of solute from solid solution. When the data from the curves of Fig. 1 are plotted on a linear time scale, no incubation period is seen, Fig. 1(*b*). No evidence of a point of inflection is found for either Alloys A or B, and the rate of decrease of resistivity is always much faster than could be accounted for by recovery of the martensitic structure alone. The ternary Alloy C shows a somewhat slower reaction at very short aging times.

From Figs. 1 and 2 it can be seen that the increase in strength is accompanied by a decrease in resistivity, even at very short aging times. Overaging is due at least in part to the formation of austenite as characterized by the increase in resistivity at long

Table II. A_s and M_s Temperatures of Four Alloys

Alloy and Nominal Composition	A_s, °F	M_s, °F
A) Commercial 18 Ni-250 maraging steel	1203	267
B) Fe-18 pct Ni-8 pct Co-5 pct Mo quaternary	1162	329
C) Fe-18 pct Ni-5 pct Mo ternary	1048	218
D) Fe-18 pct Ni binary	1122	543

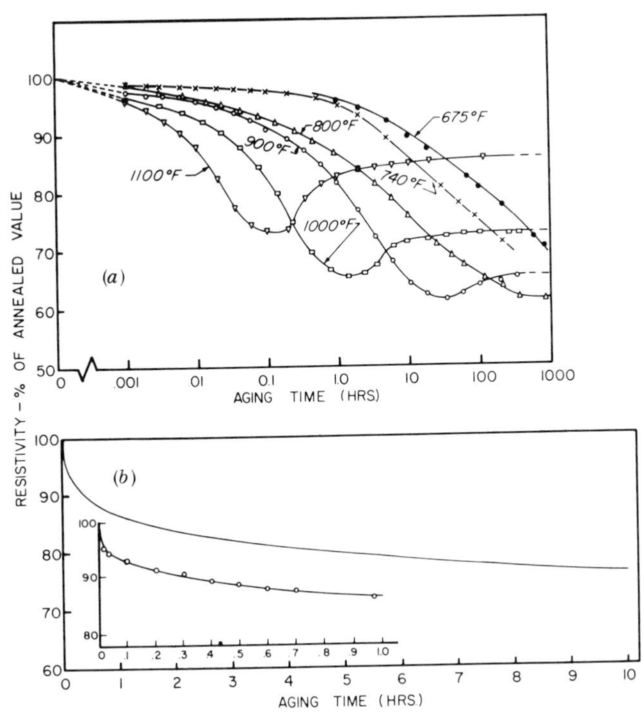

Fig. 1—(*a*) Effect of aging time and temperature on the resistivity of commercial maraging steel, Alloy A. (*b*) Linear time plot of the resistivity decrease at 800°F from the data of Fig. 1(*a*).

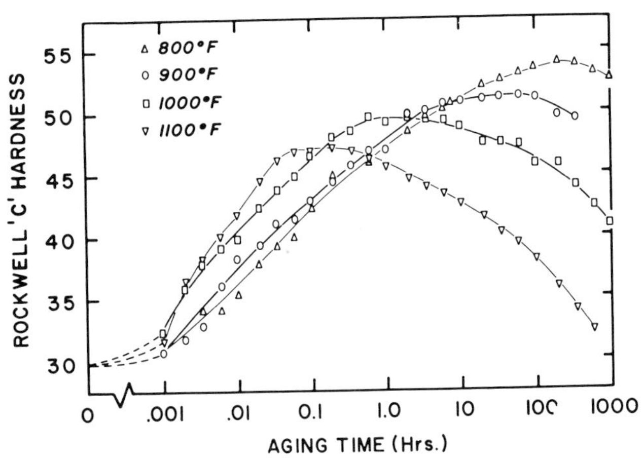

Fig. 2—Hardening response as a function of time for commercial maraging steel, Alloy A, at four aging temperatures.

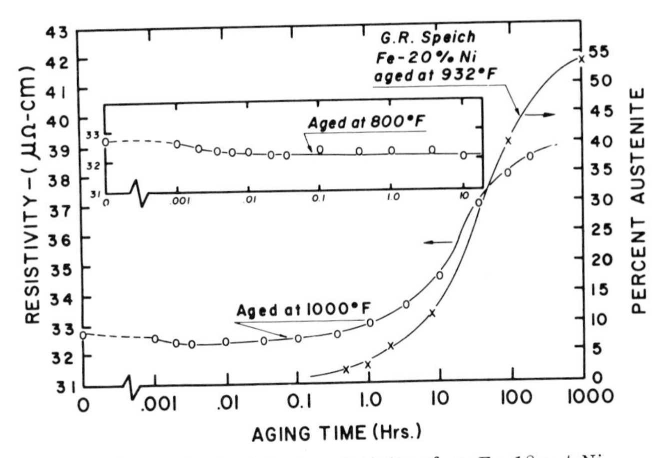

Fig. 3—Change in electrical resistivity of an Fe-18 pct Ni alloy as a function of aging time.

Source: *Transactions of the Metallurgical Society of AIME*, Vol 236, Oct 1966

aging times. But at lower aging temperatures, the resistivity decrease proceeds further before the upturn, suggesting that austenite formation is retarded at a faster rate by lowering the temperature than is the precipitation reaction.

Typical examples of the resistivity* and hardness curves for Alloys B and C aged at 850°F are shown in Fig. 4. The resistivity minimum is not as low in the ternary Alloy C indicating more rapid austenite formation in this alloy. Comparison of the curves of austenite content in these two alloys as a function of aging time at a given aging temperature shows this to be true, Figs. 5 and 6. The most noticeable effect of the presence of the 8 pct Co is the acceleration of the age-hardening reaction. This is most apparent in the resistivity curves of Fig. 4(*a*), but the time to reach peak hardness is also shorter in the cobalt-containing alloy, Fig. 4(*b*). The large effect of cobalt on the strength of the alloy is also evident in this figure.

*For the 18 pct Ni-Mo ternaries, a straight-line relationship was found between the annealed resistivities and molybdenum content. The line had a slope of 5.8 microhm-cm per weight percent molybdenum up to about 5.5 pct Mo. Thus there is a one-to-one equivalence between the concentration of molybdenum dissolved in the matrix and the resistivity. Cobalt has a very much smaller effect on the resistivity of the martensite, with 8 pct Co raising it only 3 microhm-cm at the 5 pct Mo point. Therefore, the resistivity curve is not much affected by whether or not cobalt is precipitated with the molybdenum, and is only important in its effect on the resistivity of the austenite. Nickel too has very much less effect than molybdenum on the resistivity of the martensite. The resistivity curve for Fe-Ni alloy in the range from 12 to 18 pct Ni is practically flat.[12]

The combined effect of increased rate of age hardening and decreased rate of austenite formation in the presence of cobalt leads to less austenite at peak hardness. This is indicated by the arrows of Figs. 5 and 6 and in Table III.

It was first thought that the small hardness peak preceding the main hardening reaction, Fig. 4(*b*), might

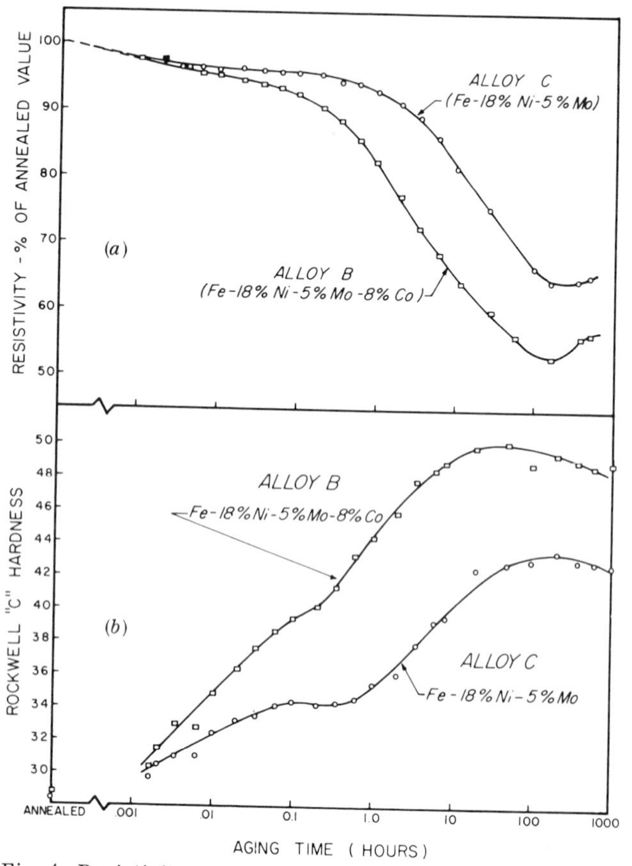

Fig. 4—Resistivity (*a*) and hardness (*b*) as a function of aging time at 850°F for the quaternary Alloy B and ternary Alloy C.

Table III. Volume Percent Austenite at Peak Hardness

Aging Temp, °F	Alloy: Fe-18 pct Ni-5 pct Mo Alloy C	Alloy: Fe-18 pct Ni-8 pct Co-5 pct Mo Alloy B
1050	19	4.5
1000	9.5*	–
950	14	3
900	9*	–
850	12	~2
800	~1	~1
740	0	0

*These somewhat lower values were obtained from specimens that had been given an intermediate 1100°F aging treatment before resolutioning at 1500°F and then aging at the indicated temperature.

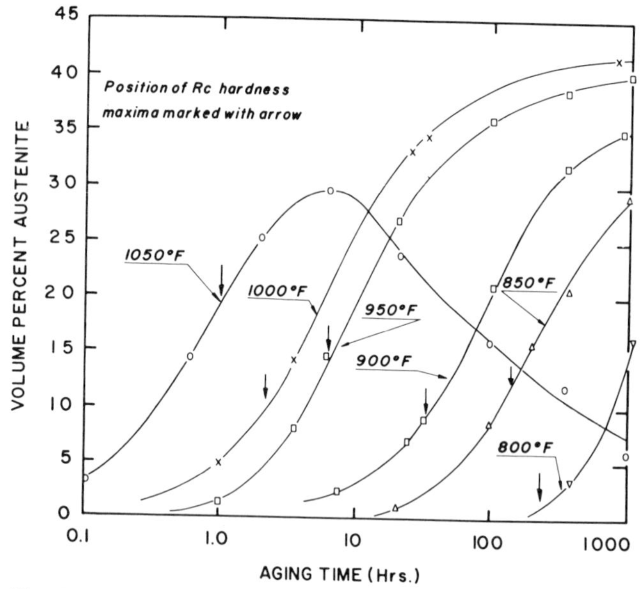

Fig. 5—Volume percent austenite as a function of aging time at several temperatures for the ternary Alloy C.

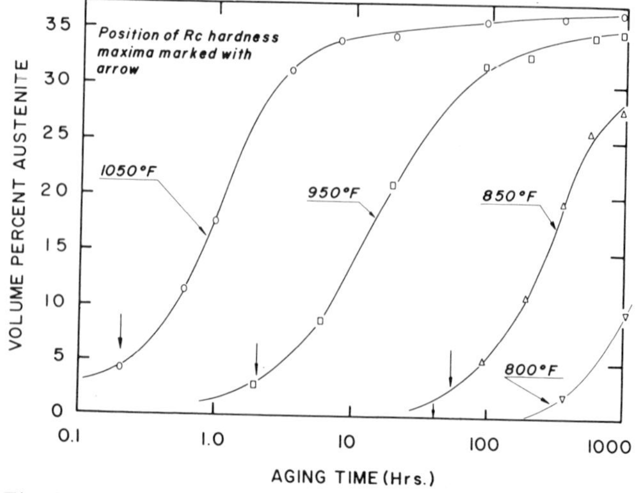

Fig. 6—Volume percent austenite as a function of aging time at several temperatures for the quaternary Alloy B.

be significant in terms of the molybdenum precipitation sequence, but it was subsequently found in Fe-18 pct Ni and Fe-18 pct Ni-8 pct Co alloys as shown in Fig. 7. This is evidently due to the precipitation of a carbide.[13] A small amount of hardening during the period of austenite formation is also evident in these simple alloys. Mihalisin[9] has reported a very pronounced hardening of this type in an Fe-20 pct Ni-20 pct Co alloy.

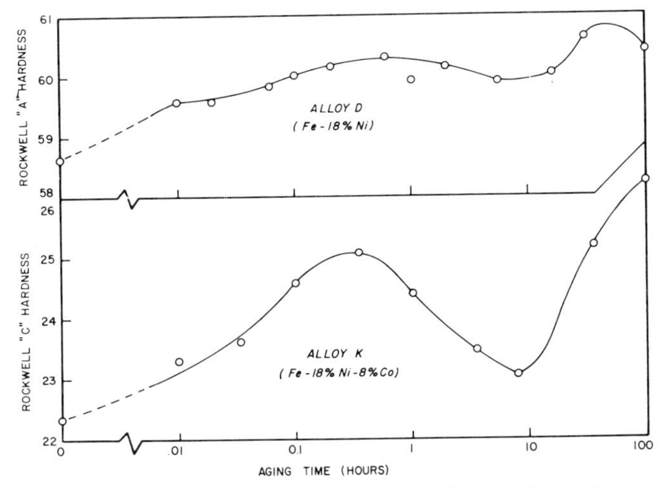

Fig. 7—Hardening response as a function of aging time at 800°F of the binary Alloy D and ternary Alloy K.

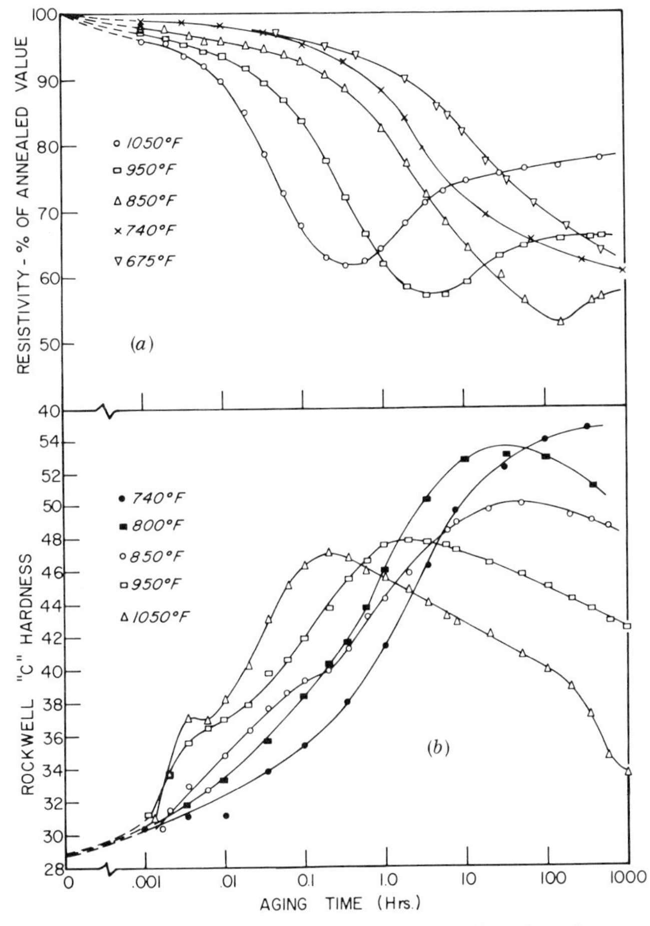

Fig. 8—Resistivity (*a*) and hardness (*b*) as a function of aging time at several aging temperatures for the quaternary Alloy B.

Figs. 5 and 6 show that increased temperatures appear to lead to an increase in the "equilibrium" amount of austenite present, as expected, although for all but the highest aging temperatures the curves do not level off in times that are practical for experimental studies.

The irregular temperature dependency to be seen in Fig. 5 indicates the sensitivity of austenite formation to specimen history. The specimens used to obtain the curves at 1000° and 900°F had previously been used to obtain an aging curve at 1100°F. Because of the specimen shortage, they were reannealed at 1500°F before aging at 1000° and 900°F. Table III shows the effect of the 1100°F pretreatment on the austenite content in the ternary alloy at peak hardness. A complete explanation of this sensitivity awaits further experimental work. The behavior of the 1050°F curve of the ternary alloy in Fig. 5 will be discussed later.

Resistivity and hardness curves for several aging temperatures for the quaternary Alloy B are shown in Fig. 8. The peak hardnesses for Alloys B and C are shown in Fig. 9. The presence of 8 pct Co increases the hardness by about 5.5 points R_C for aging temperatures above about 850°F. Below this temperature there is a change in behavior in the quaternary alloy B. The slope of the lines of Fig. 9 cannot be explained by the small amount of austenite present at the low aging temperatures. The slope of the resistivity curves decreases at longer aging times. X-rays show little or no austenite in this region.

Arrhenius plots of the time to reach ρ/ρ_0 = 0.8 or a given R_C hardness are shown in Figs. 10 and 11. The activation energy from the resistivity data at higher aging temperatures is 42.8 kcal per mole for Alloys A, B, and C. The quaternary Alloy B has a low-temperature segment of about the same slope. The points which deviate from the high-temperature line correspond to the aging temperatures which showed more pronounced hardening in Figs. 8 and 9. The low aging temperature points for Alloy A also deviate from the high aging temperature line in the direction of shorter times. The ternary Alloy C shows no deviation over the range of aging temperatures studied. When the quaternary Alloy B is given a short preage of 0.05 hr at 1000°F and then aged at

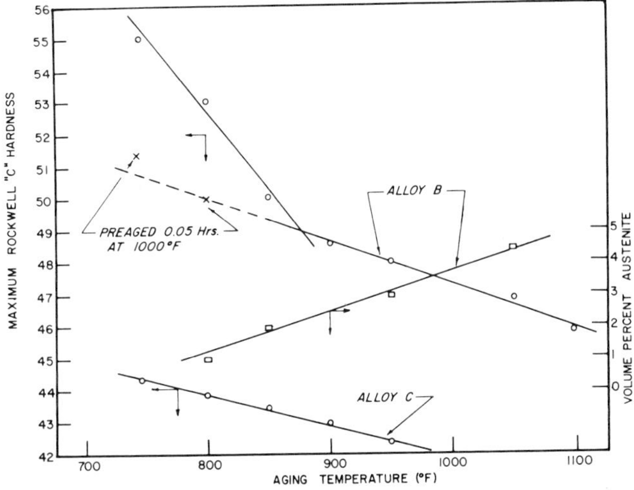

Fig. 9—Effect of aging temperature on peak hardness of the quaternary Alloy B and ternary Alloy C.

lower temperatures, the line has the same slope and the lowest temperature point (at 740°F) is not below the line as before. The Arrhenius plot from the hardness data from Alloy B, Fig. 11, shows the clearly defined low-temperature segment. Alloy A showed deviation to low times for the 800°F aging curve and Alloy C showed slightly low points for the 675°F aging curve. The activation energy obtained from hardness data was consistently about 10 kcal per mole lower than that from resistivity and increased with increasing hardness in all alloys.

The growth exponent, n, has been determined for Alloys A, B, and C by plotting the resistivity data on a log-log coordinate chart, Fig. 12, according to the equation

$$\bar{C} = C_0 - Kt^n$$

where $\bar{C}$ is the average and C_0 is the initial molybdenum concentration of the matrix. The resistivity is taken to be proportional to the molybdenum concentration. The exponent n varies from 0.5 at high aging temperatures to 0.33 at the lowest temperature.

The effect of from 1 to 7.5 pct Mo on the Fe-18 pct Ni matrix was determined with and without cobalt. The cobalt-containing series maintained a nominal weight percent ratio of cobalt to molybdenum of 1.6 as in the commercial alloy. Hardness, resistivity curves, and austenite content from the Fe-18 pct Ni-Mo series are presented in Figs. 13, 14, and 15, respectively.

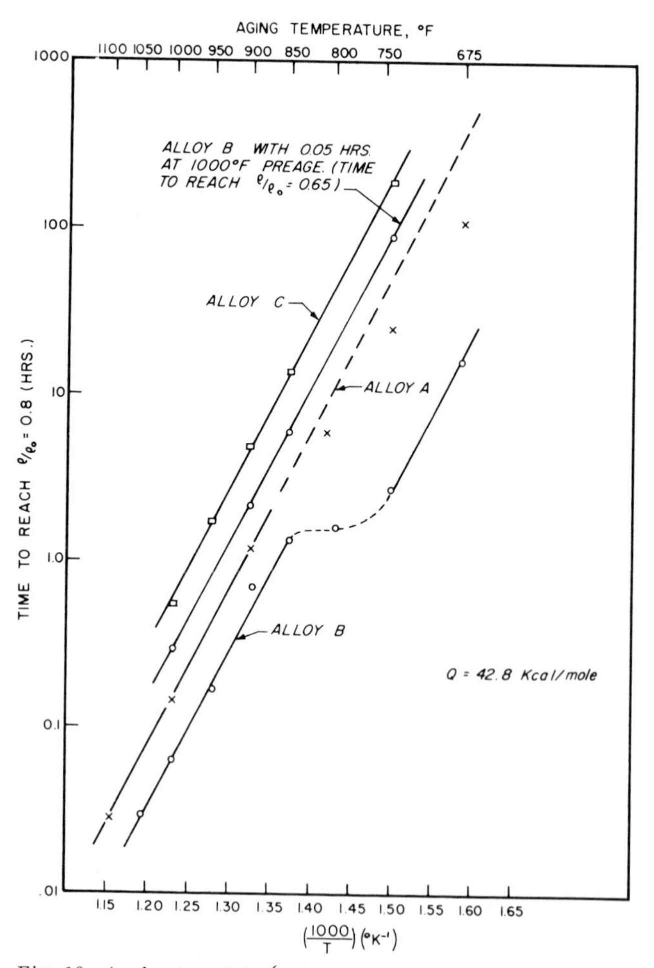

Fig. 10—Arrhenius plot constructed from resistivity data of Alloys A, B, and C.

Increasing the molybdenum increases the hardness and accelerates both the age hardening and austenite formation reactions. It also increases the equilibrium austenite content. Precipitation of molybdenum from the 18 pct Ni solid solution is barely evident at the 1 pct level from resistivity changes although the unintentionally high nickel content of this alloy led to excessive austenite formation as seen by the large rise in resistivity.

Identical trends were seen in the cobalt-containing

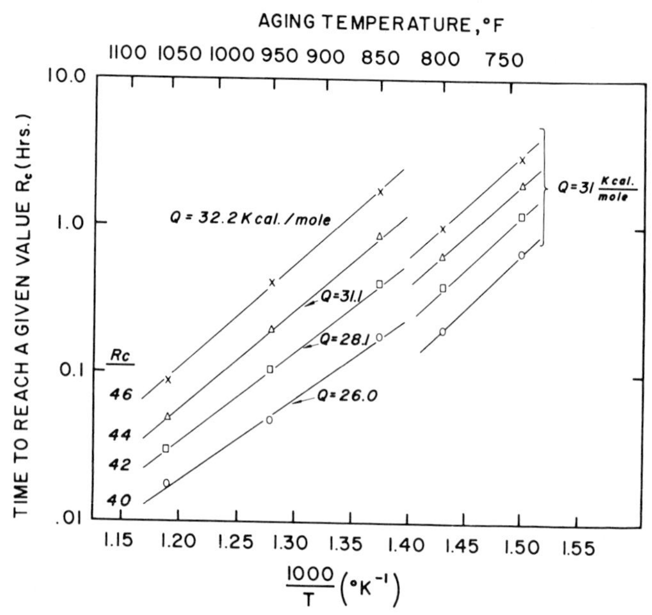

Fig. 11—Arrhenius plot from hardness data of the quaternary Alloy B.

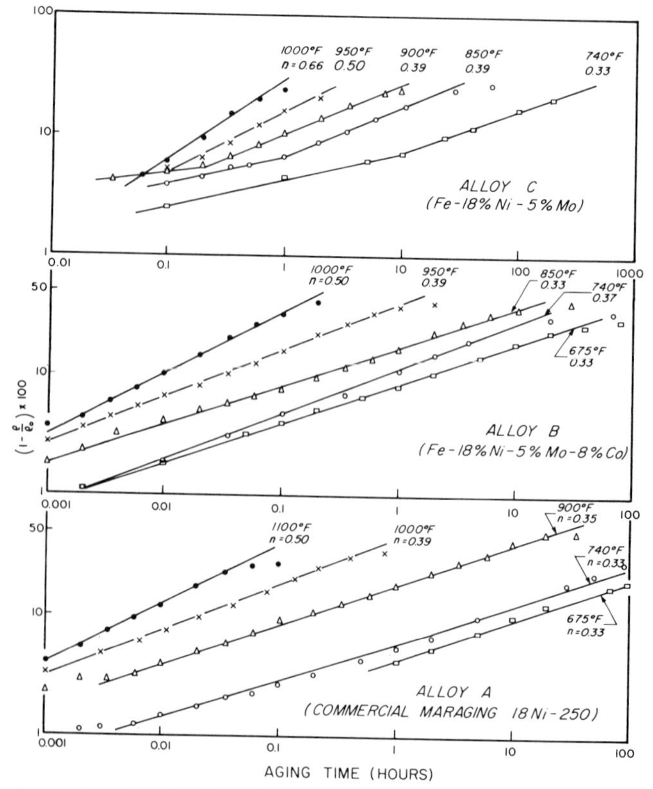

Fig. 12—Determination of precipitate growth exponent, n, from resistivity data.

series except that the hardness increased much faster with increasing molybdenum content. Also the resistivity minima were lower, suggesting more molybdenum precipitation before the austenite formation became appreciable.

Thus, at a given aging temperature, both cobalt and molybdenum gave added strength and moved the resistivity curves and hardness peaks to shorter times.

The possibility that the rate of aging could be affected by the solute or vacancy distribution in the austenite prior to the martensitic transformation was investigated by furnace-cooling specimens of Alloy B from 2100° to 1900°, 1700°, 1550°, and 1300°F and holding (longer times at the lower temperatures) to equilibrate prior to water quenching. This treatment was done in a vertical-tube furnace in a dry-hydrogen atmosphere followed by a direct water quench. The as-quenched resistivity values are shown in Fig. 16. Evidently there is some clustering or precipitation of solute in the austenite at 1300°F. The rise at 2100°F is probably due to the dissolution of a carbide or some other phase. Accelerated aging is discernible in specimens annealed at 1300°F and possibly 1500°F, as shown in Fig. 17, but the effect is small and just barely outside the limit of experimental error.

DISCUSSION

The apparent lack of a free-energy barrier as evidenced by the lack of an incubation period in the cobalt-containing alloys leads one to conclude that precipitation is largely on dislocations. Cahn[14] has shown that the free-energy barrier to nucleation can be reduced by accommodation of the precipitate-matrix lattice misfit in the dislocation-strain field. Under conditions of high supersaturation the free-energy barrier is eliminated. The present work has shown that the addition of cobalt apparently acts in the same way as increasing the molybdenum, *i.e.*, by raising the molybdenum supersaturation at a given aging temperature. Apparently at the supersaturation effective in the 5 pct Mo-8 pct Co composition, no nucleation barrier is

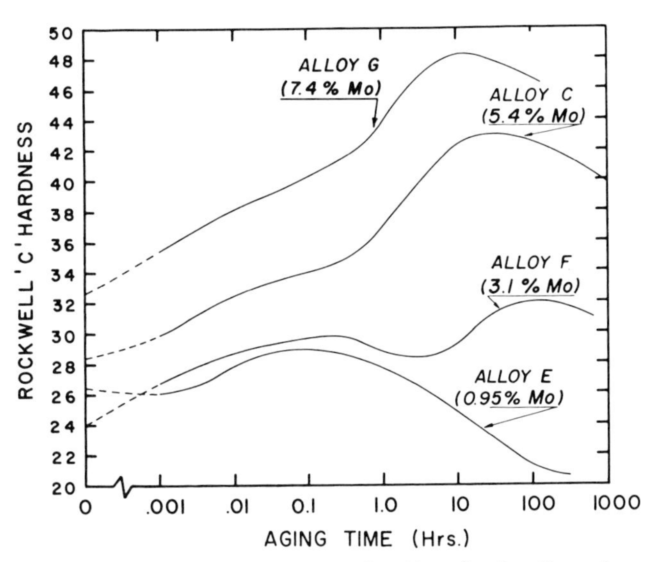

Fig. 13—Hardening response as a function of aging time at 900°F for a series of Fe-18 pct Ni-Mo alloys.

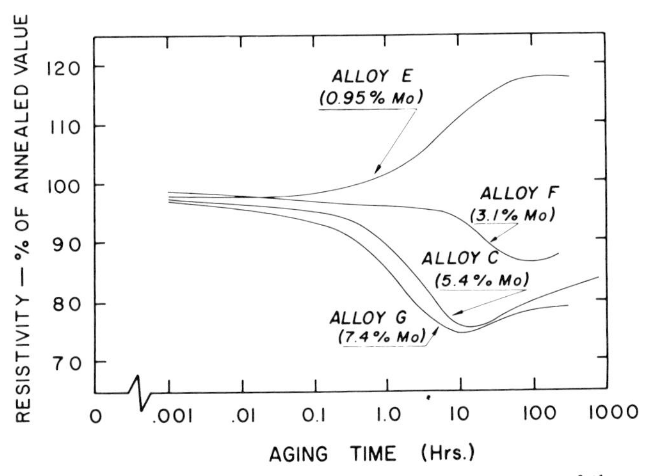

Fig. 14—Effect of aging time and molybdenum content of the resistivity of a series of Fe-18 pct Ni-Mo alloys aged at 900°F.

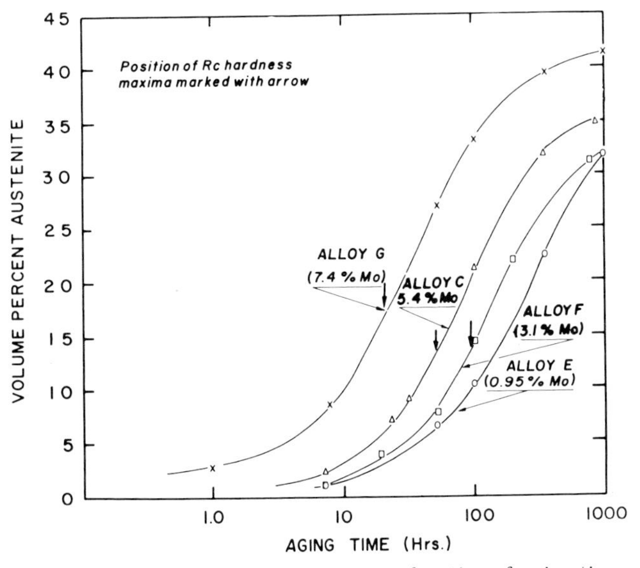

Fig. 15—Volume percent austenite as a function of aging time at 900°F for a series of Fe-18 pct Ni-Mo alloys.

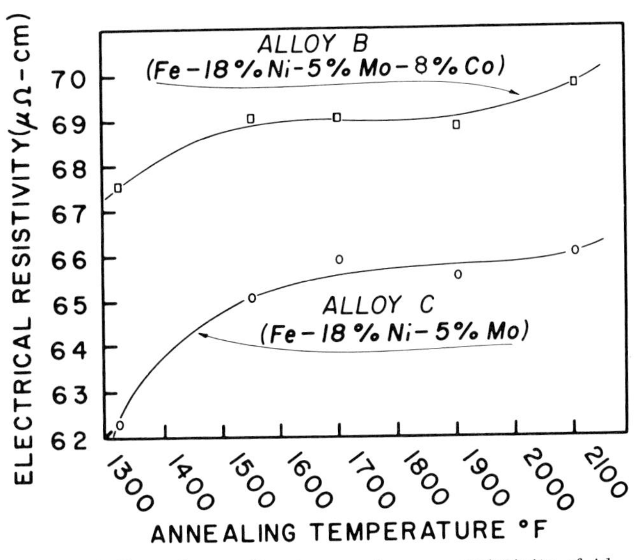

Fig. 16—Effect of annealing temperature on resistivity of Alloys B and C.

present. Without the cobalt, the alloy goes through an early, slow stage before assuming the normal growth law for the alloy, whereas the cobalt-containing alloys obey it from the shortest aging times of about 4 sec, Fig. 12. This behavior is generally consistent with the Cahn model.

Thus the effect of cobalt on the strength of the alloy can be explained in part by the lower solubility of molybdenum in the presence of cobalt, thus giving a larger volume fraction of precipitate. From the data on the effect of molybdenum on peak hardness, Fig. 13, the 5.5 points R_C effected by 8 pct Co corresponds to about an additional 2 pct Mo precipitated from solution. But the change in the slope of the peak hardness-aging time curve of Fig. 9 suggests that cobalt has another effect on the alloy. At the high supersaturation resulting from lower aging temperatures, a marked additional hardening component is found. This component is absent if the supersaturation is reduced by a short preaging treatment at 1000°F. This is shown by the two points marked by crosses in Fig. 9. Evidently another precipitate with a higher free-energy barrier to nucleation is present at these aging temperatures. A marked change in precipitate reversion behavior with aging temperature has been observed in this alloy, and will be described in a subsequent paper. The unexpectedly rapid precipitation observed in Alloy B at the low aging temperatures, Figs. 10 and 11, suggests shorter average diffusion distances and thus a finer precipitate distribution.

We can only speculate on the nature of this finely distributed low-temperature precipitate. Evidently it is nucleated in the matrix away from a dislocation. At lower aging temperatures, the slower growth rate of the dislocation-nucleated phase maintains the high supersaturation necessary for nucleation of the finer precipitate. It may well be a bcc molybdenum-rich cluster, coherent with the matrix, such as Hornbogen observed in a binary Fe-Mo alloy.[15] This cluster might correspond to the spherical particles seen in the transmission electron micrographs of Reisderf and Baker.[4] These authors described the particles as titanium-rich compound but were unable to verify this by electron diffraction. The fine precipitate is evidently unstable relative to the dislocation-nucleated phase. The decrease in slope of the low-temperature resistivity curves, Fig. 8(*a*), at longer aging times strongly suggests the dissolution of this precipitate.

Hereafter, the low-temperature precipitate will for convenience be referred to as the "matrix precipitate".

The activation energy for the precipitation reaction is much lower than the expected 57.7 kcal per mole based on the diffusion of molybdenum in α iron.[16] This could arise by several means:

1) the number of precipitate particles is a function of aging temperature;[17]

2) there is a high supersaturation of vacancies;

3) there is a large variation of $\partial^2F/\partial C^2$ with temperature;[18]

4) pipe diffusion is important.

In this case, the number of particles per unit volume is thought to be considerably increased below the transition temperature already discussed. The results of the preaging experiment, Fig. 10, seem to indicate that, when nucleation is primarily on dislocations, the particle density is independent of temperature, *i.e.*, at the higher aging temperature. The preage was intended to establish the precipitate distribution characteristic of 1000°F and lower the supersaturation sufficiently to discourage the nucleation of the matrix precipitate at the lower aging temperatures. The disappearance of the low-temperature anomaly suggests that this was accomplished. The resulting line on the Arrhenius plot, Fig. 10, has the same slope as without the preaging treatment. Thus, possibility No. 1 above is eliminated on this basis.

Quenching the alloy from various temperatures in the austenite range had very little effect on the rate of precipitation, Fig. 17, suggesting that differences in the vacancy concentration in the austenite were not maintained in the martensite. The vacancy-migration energy in the bcc lattice is so low as to ensure this. Therefore possibility No. 2 above can be eliminated.

It is difficult to test the third possibility experimentally. Little is known of the details of the phase diagram, but it seems unlikely that this composition is near an inflection in the free energy-composition curve as found to be the case in the Al-Zn system by Herman, Cohen, and Fine.[18]

It is likely that pipe diffusion is largely responsible for the low activation energy and corresponding high atom mobility at relatively low temperatures. This makes maraging possible, because aging must necessarily be carried out at temperatures below A_s. The activation energy determined from hardness curves is particularly low at low hardness, Fig. 11, values probably because an appreciable part of the early hardening comes from a carbide reaction.

Because of the absence of an incubation period, it is felt that nucleation occurs at the very beginning of aging. Therefore, the shape of the reaction curve reflects the growth of any one particle. At high aging temperatures, the growth exponent is 0.5, which corresponds to the diffusion-controlled growth of platelets by thickening.[19] As the aging temperature is lowered, the exponent decreases to 0.33 at the lowest aging temperatures used. An exponent lower than $\frac{1}{2}$ can mean either that some of the particles are dissolving, or that there is stress-assisted attraction of solute in the strain field of the dislocations of a partially coherent platelike precipitate. The latter concept can be thought of as a one-dimensional form of

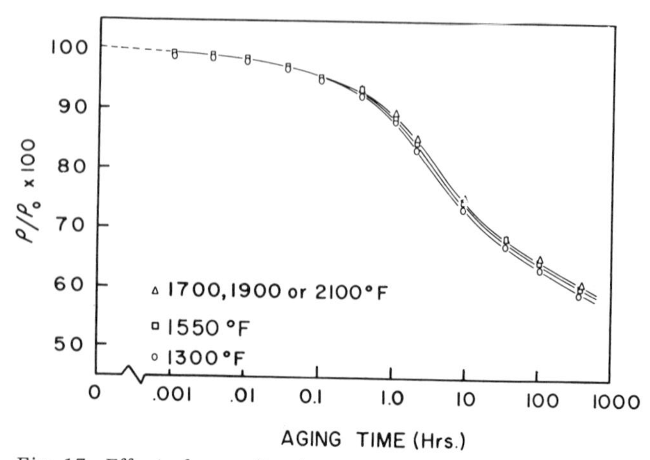

Fig. 17—Effect of annealing temperature on rate of precipitation at 740°F in the quaternary Alloy B.

the Cottrell-Bilby $t^{2/3}$ law[20] and has been used to explain the observed time law for the first stage of tempering,[21] zone formation in Al-2 at. pct Cu,[22] and the precipitation of iron from copper solid solutions.[23] Since n equals exactly $\frac{1}{3}$, it might be postulated that such a mechanism would come into play at lower temperatures where concentration gradients would not be established so rapidly and the stress gradients would remain unrelaxed over larger distances. But it would be dangerous to suggest that this is the only possibility here, because it has been shown that another precipitate comes into play at lower aging temperatures. This matrix precipitate may well be unstable relative to the precipitate at dislocations and would dissolve during the aging process. This would reduce the exponent below $\frac{1}{2}$.

A conclusion concerning the softening process at long aging times can be at least tentatively drawn by relating austenite formation data to hardening. From Fig. 6 the onset of overaging seems to correspond to formation of significant amounts of austenite. But, in some instances, Fig. 5, hardening is still taking place coincidentally with the formation of up to about 18 vol pct austenite. Also the drop in hardness at 1000 hr does not seem to be closely related to the austenite content. Evidently classical overaging (*i.e.*, excessive precipitate particle growth) is responsible for a large part of the softening. Rapid softening is observed at low aging temperatures probably due to rapid overaging of the matrix precipitate.

The unusual behavior of the 1050°F curve of Fig. 5 can probably be explained in the following way. The A_S temperature is initially just above 1050°F. Dilatometry shows this temperature to be 1048° ± 5°F. Dilatometry indicated that the A_S temperature rises slightly in the early part of aging up to 1 hr and then decreases rapidly. As aging proceeds high nickel austenitic regions are formed but eventually the remainder of the martensitic structure shears over to austenite as the A_S temperature goes below the aging temperature. The high and low nickel and molybdenum regions then require considerable time to smooth out. On cooling to room temperature, the highest nickel regions are not converted to martensite. Thus the smooth decrease in austenite content is observed tending toward zero as the concentration gradients in the austenite are eliminated.

The results show that the effect of cobalt is the sum total of several phenomena all working in the same direction to increase strength. As pointed out in the beginning, the present approach could not hope to show any ordering or precipitation reactions involving cobalt, so this possibility cannot be completely dismissed and may be another additive effect. Recent neutron-diffraction data obtained by Mihalisin on an Fe-20 pct Ni-25 pct Co alloy provide evidence for ordering around the Fe_2NiCo composition.[9] Whether or not ordering is present and if it contributes significantly to the strength at the 7 to 8 pct Co level is not known. The most obvious reason for increased hardness due to cobalt is the few points Rockwell "C" due to solid-solution hardening and seen in the quenched alloys. More subtly cobalt evidently reduces the metastable solubility for molybdenum and thus has the same effect as raising the molybdenum content as shown in Fig. 13. This leads to a finer particle dispersion and the possibility of nucleation at sites other than at dislocations. This may in part be due to a small reduction in the lattice parameter of the solid solution caused by the substitution of 7 or 8 pct Co for iron which increases the misfit for a molybdenum atom. The strain around a partially coherent precipitate is also increased.

Since premature austenite formation appears to detract from the strength level that can be reached, the relative position in time at a given aging temperature of these two reactions is very important in determining strength. Here cobalt seems to help by two means. First, because of the increased supersaturation it effects, the whole precipitation reaction is speeded up. Precipitation can proceed further with cobalt than without before it is interfered with by austenite formation. Second, the austenite reaction itself is retarded when cobalt is present. The combined effect is to have only 3 pct austenite present at peak hardness in the cobalt-containing alloy and 15 pct in the ternary alloy.

SUMMARY

1) Electrical resistivity is a sensitive monitor of the recovery, precipitation, and austenite formation reactions that occur during the aging of nickel martensites in general and maraging steels in particular.

2) Resistivity measurements started at very short aging times (about 4 sec) show that there is no incubation period for the precipitation in the cobalt-containing alloys. The Fe-18 pct Ni-5 pct Mo alloy shows a somewhat slower reaction in the beginning. These facts are consistent with a model of nucleation of precipitate particles on dislocations.

3) Cobalt accelerates the precipitation reaction in a way similar to increasing the molybdenum content, suggesting that cobalt increases the supersaturation of molybdenum. This leads to the appearance of a second type of precipitate at low aging temperatures.

4) Overaging of maraging steels is due to a combination of classical overaging and softening due to formation of austenite.

5) The activation energy for the precipitation reaction is 42.8 kcal per mole from resistivity measurements and is best explained as due to pipe diffusion.

6) The growth exponent varies from 0.5 at high aging temperatures to 0.33 at low temperatures indicating growth by thickening of platelets and dissolutioning of the low-temperature precipitate.

7) The alloys contain up to about 45 pct austenite after aging for 1000 hr at 900°F. Cobalt substituted for iron tends to retard austenite formation and reduce the equilibrium amount present. This probably accounts for part of the effect of cobalt in increasing peak hardness. Increasing molybdenum contents accelerate austenite formation and increase the amount present as equilibrium is approached.

REFERENCES

1. S. Floreen and R. F. Decker: *Am. Soc. Metals, Trans. Quart.*, 1962, vol. 55, p. 518.
2. A. J. Baker and P. R. Swann: Electron Microscope Society of America, 21st Annual Meeting, Denver, Colo., August 28, 1963.
3. B. G. Reisdorf: *Am. Soc. Metals, Trans. Quart.*, 1963, vol. 56, p. 783.
4. B. G. Reisdorf and A. J. Baker: Applied Research Laboratory, U.S.

Some Observations on the Strength and Toughness of Maraging Steels

S. FLOREEN AND G. R. SPEICH

ABSTRACT. The strength and toughness of a series of quaternary Fe−18Ni−8 Co-base alloys with additions of aluminum, beryllium, columbium, manganese, molybdenum, silicon, or titanium have been determined. These data are compared with earlier results obtained with ternary Fe − 18Ni-base alloys without cobalt. The combination of molybdenum and cobalt produced much higher strengths than predicted from the ternary data. The added strengthening appears to be due to the formation of a finer dispersion of precipitates when cobalt is present. With all of the other hardening elements studied the addition of 8% Co produced only a minor strength increase. The alloys containing cobalt and molybdenum showed superior fracture toughness. Additions of 2% Mo to brittle quaternary alloys markedly improved the toughness. The beneficial effect of molybdenum is attributed to the prevention of grain-boundary segregation and/or precipitation in a manner analogous to the influence of molybdenum on temper embrittlement in low-alloy steels. *ASM-SLA Classification: Q27, Q23p, N7, 2-60; 55.*

THE DEVELOPMENT of the 18% Ni maraging steels has shown that excellent combinations of strength and toughness can be achieved by age-hardening an iron-nickel martensite (1). At yield strengths of 250 to 300 ksi these steels nominally contain 18% Ni, 5% Mo, 7 to 9% Co, 0.4 to 0.7% Ti and very low carbon (0.03% max). Normally they are hardened by aging 3 hr at 900 F.

Recent transmission electron microscopy studies have indicated that very fine precipitates of Ni_3Mo and possibly Ni_3Ti form during age-hardening (2–5). It seems reasonable to assume that the resultant strength after aging is primarily due to these precipitates. In fact, Baker and Swann (5) have shown that the yield strength can be accounted for by the Orowan relationship:

$$\sigma = \sigma_o + \frac{G\mathbf{b}}{\lambda} \qquad \text{Eq 1}$$

σ is the yield strength after aging, σ_0 the yield strength of the precipitate-free matrix, G the shear modulus of the matrix, $\mathbf{b}$ the Burgers vector, and λ the interparticle spacing. The values of λ measured by Baker and Swann were such that yield strengths of 300 ksi would be predicted by Eq 1.

However, two very important points still must be settled before the strengthening of these alloys can be satisfactorily explained. First, the hardening produced when cobalt, molybdenum, and titanium are co-present is much greater than the sum of the strengthening increments due to cobalt, molybdenum and titanium used singly (1, 6). There is, in other words, some interaction between these alloying

S. Floreen is associated with the Research Laboratory of International Nickel Co., Inc., Suffern, N. Y.; G. R. Speich is associated with the Fundamental Research Laboratory, U. S. Steel Corp., Monroeville, Pa. Manuscript received May 15, 1964.

elements that results in a much higher strength.

Of greater importance is the toughness. A study of an Fe – 18 Ni binary alloy showed that a low-carbon 18% Ni martensite had excellent ductility (7). A further study of a number of age-hardenable iron – 18 Ni-base ternary alloys showed, however, that regardless of the hardening element the alloys became brittle at a tensile strength of approximately 220 ksi (6). The nickel-cobalt-molybdenum steels, though, maintain excellent toughness up to tensile strengths on the order of 300 ksi. These results make it clear that the hardening by the cobalt-molybdenum-titanium combination produces significantly better toughness at high strengths. Until now no explanation for the superior toughness of the maraging steels has been made.

Since both molybdenum and titanium apparently form precipitates during aging, one might assume that the special properties of the maraging steels result from an interaction of cobalt with one or both of the precipitation hardening reactions. If this were true, then one would also expect to find a rather strong effect of cobalt when other age-hardening elements besides molybdenum or titanium were used. To study this question a series of Fe – 18 Ni – 8 Co-alloys containing various hardening elements have been examined. The results were compared to the hardening of Fe – 18 Ni-base alloys in order to determine the effects of 8% Co on the strength characteristics.

Experimental Procedure

Materials. Thirty-pound melts were prepared by induction melting under an argon blanket. Additions of 0.1 wt % Ti, Al, Si, and Mn were made for refining purposes. The melting procedure was the same used in a previous ternary alloy study (6). The ingots were homogenized and forged at 2300 F and cut into halves. One-half of each ingot was hot rolled at 1600 F to $\frac{3}{4}$-in. round bar stock. The remaining half was one-direction hot rolled, starting at 1800 F, to $\frac{5}{8}$-in. thick plate stock. After hot rolling both the plate and the bar stock were annealed 1 hr at 1500 F and air cooled.

Specimen Preparation and Testing. Smooth and notched tensile specimens were machined from the bar stock. The smooth specimens had a $\frac{1}{4}$-in. diam and a 1-in. gage length. The notched tensile specimens had a 0.300-in. major diam, 0.212-in. root diam., notch radius of approximately 0.0005 in. and a notch acuity factor of 10.

Charpy V-notch impact specimens were machined from the plate stock. All specimens were taken in the transverse direction.

The finished specimens were aged at 800 F for either 3 or 24 hr. Aging studies have shown that the maraging steels (8) and all the Fe – 18 Ni-base ternary alloys (6) obey essentially identical aging kinetics. Therefore reasonably valid evaluations of the strength properties can be made by comparing the results from different alloys after any given aging treatment. The 800 F aging temperature was selected in order to minimize austenite formation. All specimens were tested at room temperature.

Microscopy. A number of the alloys were examined by light and also transmission electron microscopy. The fracture surface of broken Charpy specimens were examined by carbon replica fractography. The fractured Charpy specimens were also nickel plated and examined metallographically to determine whether the fracture was transgranular or intergranular.

Results and Discussion

Strength Properties

Before considering the cobalt-containing alloys it will be useful to demonstrate that there is no strong hardening interaction between titanium and molybdenum. Table 1 gives the tensile properties of an 18 Ni – 5 Mo – 0.4 Ti alloy and an 18 Ni – 5 Mo – 0.7 Ti alloy. These alloys

Source: *Transactions of ASM*, Vol 57, 1964

TABLE 1. Composition and Mechanical Properties of Fe – 18 Ni – 5 Mo – Ti Alloys

Alloy	Composition, wt %*							
	Ni	Mo	Ti	C	Al	Si	Mn	Fe
1	17.5	5.02	0.35	0.016	0.07	0.04	0.07	Bal
2	17.7	5.00	0.74	0.029	0.07	0.10	0.09	Bal

Alloy	Aging time, hr	Tensile strength Yield, ksi	Tensile strength Ultimate, ksi	Elongation, %	Reduction area, %	Charpy V-notch impact energy, ft-lb	Calculated yield strength, ksi*
1	3	175	184	16	61	30.0	180
	24	192	201	14	64	22.7	210
2	3	211	221	14	55	15.0	200
	24	227	237	13	53	10.5	235

* Calculated on the basis of the hardening produced by Ti or Mo in Fe – 18 Ni-base ternary alloys, and assuming strengths are additive.

are essentially the 250 and 300 ksi varieties of the maraging steel, but without cobalt. The yield and ultimate strengths of these alloys are approximately 80 ksi below the strengths that would be obtained in the cobalt-containing maraging steel compositions using these heat treatments.

Also shown in Table 1 are the estimated strengths based on adding the hardening observed in Fe – 18 Ni – 5 Mo and Fe – 18 Ni – 0.4 Ti or Fe – 18 Ni – 0.7 Ti ternary alloys. These estimated strengths compare quite well with the measured strengths of the two alloys. Thus it is

TABLE 2. Compositions and Mechanical Properties of Quaternary Alloys

Alloy	Composition, wt %							
	Ni	Co	Al	Ti	Si	Mn	C	Bal Fe, other
				Aluminum Series				
3	18.1	8.2	0.25	0.10	0.09	0.08	0.019	
4	18.1	8.2	0.73	0.15	0.14	0.08	0.012	
5	17.9	8.3	1.29	0.12	0.15	0.11	0.013	
6	18.0	8.3	2.57	0.13	0.16	0.11	0.010	
				Beryllium Series				
7	17.2	8.3	0.03	0.05	0.05	0.07	0.026	0.15 Be
8	17.9	8.4	0.08	0.10	0.21	0.14	0.021	0.34 Be
				Columbium Series				
9	17.8	8.1	0.07	0.08	0.08	0.09	0.020	2.37 Cb
10	18.0	7.9	0.11	0.08	0.09	0.08	0.016	3.10 Cb
				Manganese Series				
11	18.8	8.0	0.07	0.08	0.07	1.85	0.015	
12	18.0	7.9	0.05	0.06	0.05	1.90	0.016	
13	17.1	7.4	0.07	0.09	0.04	2.45	0.009	
				Molybdenum Series				
14	17.6	7.7	0.03	0.05	0.05	0.08	0.020	0.98 Mo
15	17.9	8.0	0.05	0.07	0.08	0.13	0.018	2.92 Mo
16	18.0	7.9	0.03	0.05	0.05	0.08	0.020	4.95 Mo
				Silicon Series				
17	18.2	8.3	0.08	0.09	2.05	0.10	0.015	
18	17.4	8.3	0.08	0.09	2.91	0.17	0.013	
				Titanium Series				
19	18.2	8.2	0.05	0.41	0.08	0.06	0.014	
20	18.2	8.3	0.04	0.87	0.08	0.07	0.013	
21	17.3	7.7	0.08	1.43	0.08	0.10	0.014	
22	17.8	8.0	0.06	2.50	0.11	0.09	0.013	

TABLE 2 (cont.). Compositions and Mechanical Properties of Quarternary Alloys

Alloy	Aging time, hr	Tensile strength, ksi 0.2% Offset yield	Tensile strength, ksi Ultimate	Elongation, %	Reduction area, %	Notch tensile strength, ksi	NTS/UTS	Charpy V-notch impact energy, ft-lb
				Aluminum Series				
3	3	159	161	18	72	260	1.61	37.5
	24	169	169	15	69	270	1.60	23.0
4	3	191	197	15	62	275	1.40	5.2
	24	204	208	13	61	273	1.31	5.3
5	3	226	235	4	16	95	0.40	2.5
	24	227	236	(a)	(a)	111	0.47	3.0
6	3	244	258	1	4	79	0.30	1.5
	24	258	267	(a)	(a)	82	0.31	1.5
				Beryllium Series				
7	3	213	221	15	61	324	1.47	12.0
	24	224	228	11	59	334	1.46	9.3
8	3	239	242	10	35	250	1.03	4.8
	24	248	255	9	31	205	0.80	4.8
				Columbium Series				
9	3	192	198	18	52	303	1.53	14.5
	24	207	211	11	46	303	1.44	14.8
10	3	199	208	10	33	280	1.30	10.2
	24	216	221	8	29	281	1.22	10.0
				Manganese Series				
11	3	203	207	11	54	147	0.71	2.5
	24	186	188	11	57	265	1.41	5.5
12	3	208	210	10	48	219	1.04	5.0
	24	186	186	13	57	281	1.51	9.5
13	3	175(b)	175	0	0	68	0.39	1.0
	24	209	215	12	50	144	0.67	2.0
				Molybdenum Series				
14	3	152	154	19	73	252	1.64	43.5
	24	157	157	20	71	256	1.63	44.7
15	3	171	176	16	66	290	1.64	32.2
	24	201	205	15	65	323	1.58	26.3
16	3	203	214	15	61	333	1.56	22.7
	24	251	263	11	54	371	1.42	17.0
				Silicon Series				
17	3	212	215	13	56	270	1.25	4.2
	24	226	230	10	51	199	0.87	4.0
18	3	247	254	(a)	(a)	91	0.36	1.0
	24	261	268	(a)	(a)	117	0.44	1.0
				Titanium Series				
19	3	169	175	16	64	278	1.59	16.0
	24	176	180	13	64	290	1.61	10.0
20	3	195	205	14	57	282	1.38	5.2
	24	210	218	13	54	207	0.95	6.8
21	3	236	251	3	1	87	0.35	1.0
	24	254	266	(a)	(a)	101	0.38	1.0
22	3	290(b)	290	(a)	(a)	118	0.40	1.0
	24	217	217	(a)	(a)	121	0.56	1.0

(a) Broke into several pieces.
(b) Broke before 0.2% offset.

clear that there is no significant titanium-molybdenum interaction, and cobalt must be necessary to produce the extra hardening observed in the maraging steels.

Table 2 summarizes the compositions and mechanical properties of the Fe – 18 Ni – 8 Co-base quaternary alloys. In Fig. 1 the yield strengths of each alloy have been plotted against the strengths of the corresponding cobalt-free ternary alloy, based on the results of the earlier ternary alloy study (6). The dashed line with a slope of 45° is a curve for σ quaternary $= \sigma$ ternary.

The results for both the 3 and the 24-hr aging treatments are shown for each alloying element. In general, a smooth curve can be drawn through all the points for each

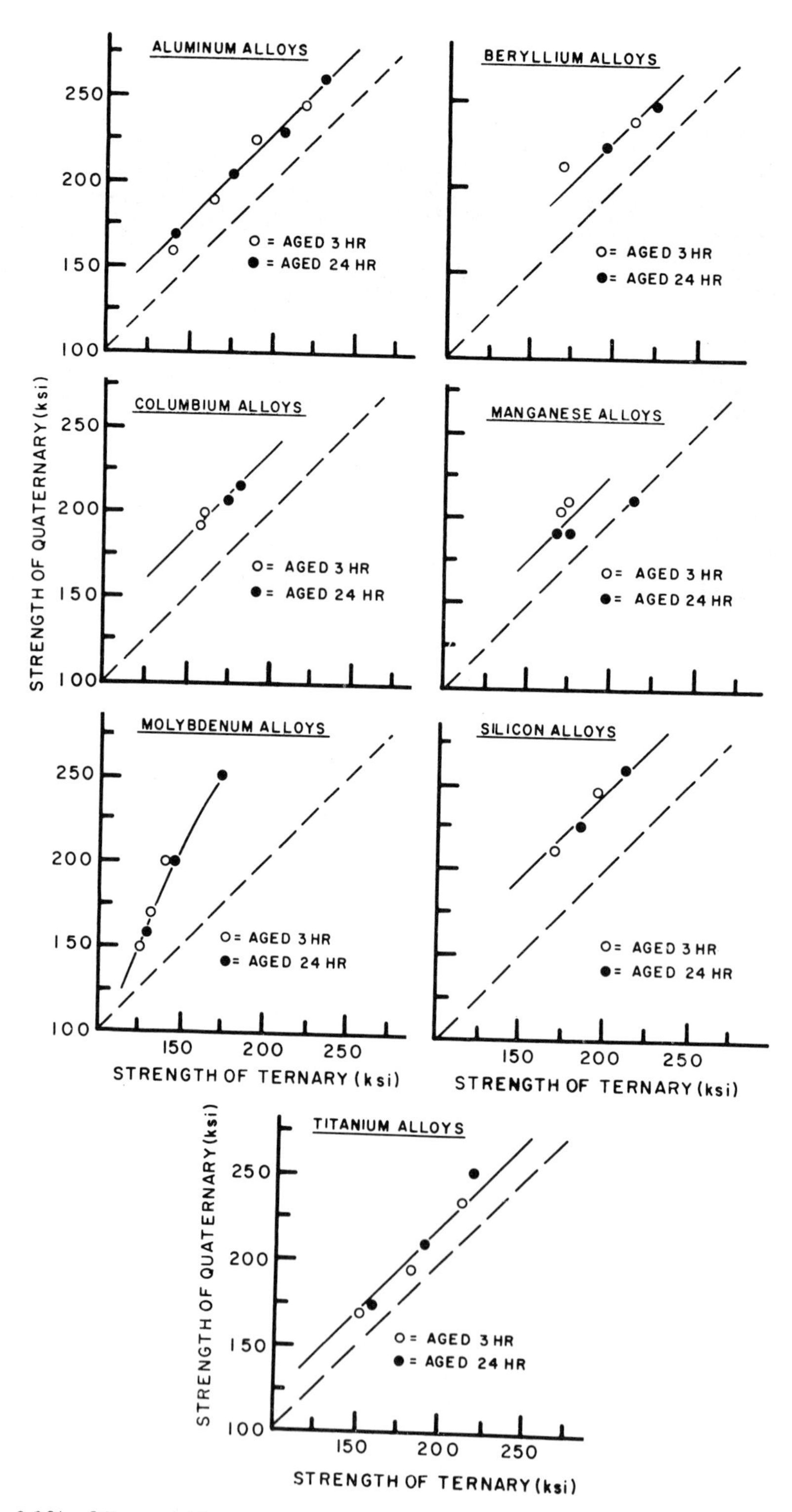

FIG. 1. 0.2% Offset yield strength of Fe-18Ni-8 Co-base quaternary alloys vs yield strengths of Fe-18 Ni-base ternary alloys. Specimens annealed 1 hr at 1500 F, air cooled, and aged either 3 or 24 hr at 800 F.

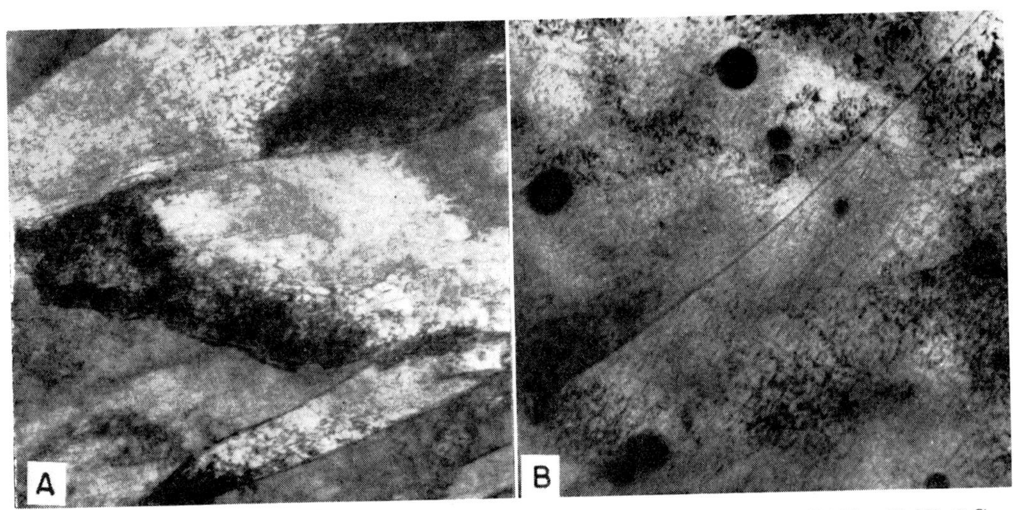

FIG. 2. Comparison of transmission micrographs of (A) Fe–20 Ni–5 Mo and (B) Fe–18 Ni–8 Co–4.95 Mo. Aged 24 hr at 800 F. ×40,000.

element studied, as would be expected from the similar aging kinetics of the alloys. An exception is one of the manganese alloy results, in which the strength of the quaternary was lower than expected. This erratic result may have been due to reversion to austenite during aging of this alloy.

For six of the seven elements studied the quaternary alloy results lie on a curve parallel to and slightly above the σ ternary = σ quaternary dashed curve. Thus in these alloys the addition of 8% Co to the ternary composition produced a constant strength increment that was independent of the original strength of the ternary. The magnitude of this increment ranged from approximately 20 to 45 ksi for the hardening elements studied. Of special interest in this group are the titanium-containing alloys. Adding 8% Co to Fe–18 Ni–Ti alloys raised the strength by only approximately 20 ksi. Thus, there was no significant cobalt-titanium interaction, and the major interaction in the maraging steels must be only between cobalt and molybdenum.

The ternary alloy study showed that adding 18% Co to an Fe–18 Ni-base alloy raised the strength about 25 ksi. This strength increment compares reasonably well with the 20 to 45 ksi increment observed in the quaternary alloys. It is likely that this strengthening results from solid-solution hardening, and can be considered equivalent to increasing the term σ_0 in Eq. 1. Clearly there was no strong cobalt hardening, such as found in the maraging steels, in these alloys.

In contrast to these results was the marked effect of cobalt in the molybdenum-containing alloys. At lower molybdenum levels the effect of cobalt was relatively weak. At the 1% Mo level, for example, cobalt contributed about only 25 ksi to the strength. At 5% molybdenum, however, the alloy with cobalt was 75 ksi stronger than the cobalt-free ternary alloy.

Several quaternary alloys were examined by transmission electron microscopy and compared with ternary alloys from an earlier study (3). Figures 2A and 2B compare an Fe–20 Ni–5 Mo alloy with an Fe–18 Ni–8 Co–5 Mo alloy. Figures 3A and 3B compare the structure of an Fe–20 Ni–1 Ti alloy (3) with an Fe–18 Ni–8 Co–1 Ti alloy. The precipitates appear to be Ni_3Mo and Ni_3Ti in the alloys with molybdenum and titanium, respectively, both with and without cobalt (3, 5).

Careful examination of Fig. 2A and 2B

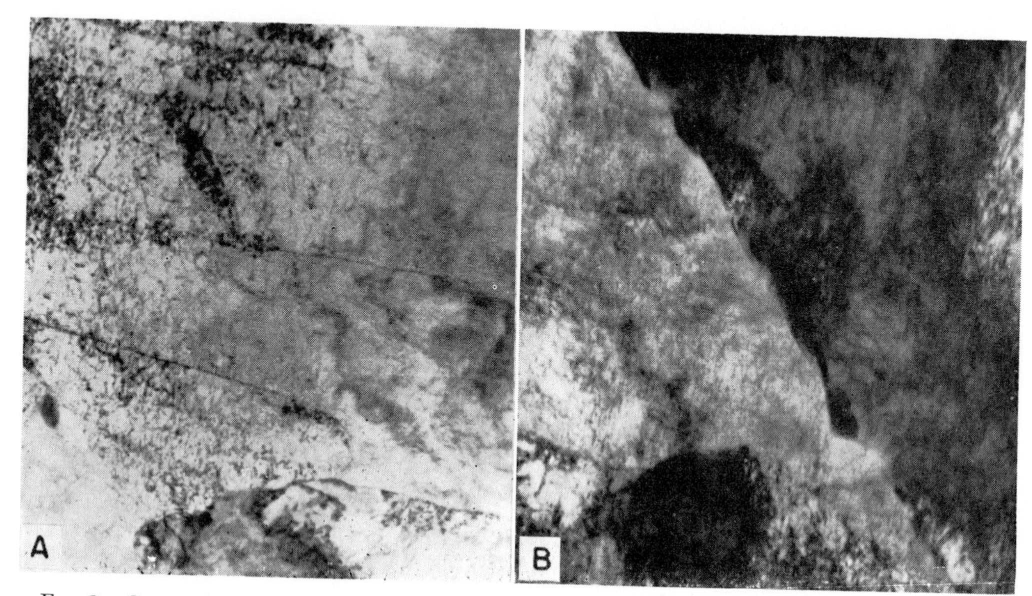

FIG. 3. Comparison of transmission micrographs of (A) Fe−20 Ni−1 Ti and (B) Fe−18 Ni−8 Co−0.87 Ti. Aged 24 hr at 800 F. ×40,000.

shows that precipitation is clearly evident in the alloy containing cobalt and molybdenum (Fig. 2B) but not extensive in the alloy with only molybdenum (Fig. 2A). In the case of the alloys with titanium, this difference also appears to be true, but to a lesser degree (Fig. 3A and 3B). Attention is also called to the former austenite grain boundaries in Fig. 2B and 3B. In the alloy containing cobalt and molybdenum (Fig. 2B) the former austenite grain boundaries appear free of precipitate, while in the alloy with cobalt and titanium (Fig. 3B) precipitation has occurred there. This grain-boundary precipitate will be discussed further in the section on toughness.

It thus appears that the addition of cobalt produces a finer dispersion of precipitates in the molybdenum-containing alloys, but may not significantly alter the dispersion in the other alloys. The results in Fig. 1 also indicate that this effect of cobalt apparently is more pronounced at higher molybdenum contents.

The simplest explanation of this effect is that cobalt lowers the solubility of Ni_3Mo in the iron-nickel martensite. This would tend to produce a finer dispersion of Ni_3Mo, and the resultant strength increase could than be due to a smaller value of the interparticle spacing, λ, in the Orowan relationship (Eq. 1).

It is not yet certain, however, that deformation in the maraging steels proceeds by dislocations bowing between particles, as required if the Orowan relationship is to be applied. An alternative deformation mechanism, proposed by Ansell and Lenel (9), is that the dislocations shear through the particles. Here again a smaller value of λ would also increase the yield strength. Further strengthening might also result if cobalt were present in the particles and increased the shear strength of the particles. This latter effect, however, is probably not of major importance in the present case because the results in Fig. 1 show that cobalt had a progressively greater strengthening effect at increasingly higher molybdenum levels. If the primary effect of cobalt were to strengthen the precipitates, then one would expect to see much more hardening due to cobalt at lower levels of molybdenum.

Solid-solution hardening by cobalt, based on the results in Fig. 1, probably con-

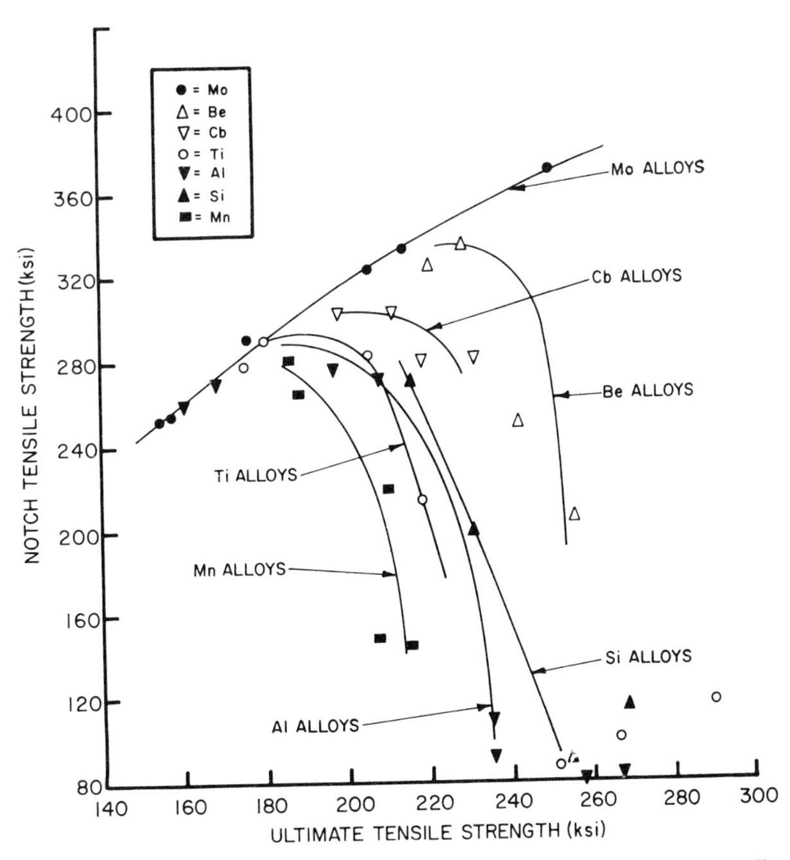

Fig. 4. Notch tensile strength vs ultimate tensile strength of quaternary alloys.

tributes a strength increase on the order of 25 ksi in the cobalt-molybdenum alloys. The remaining increase in strength, as discussed above, is most simply explained by a finer interparticle spacing of the Ni_3Mo precipitates. Until the question of whether the dislocations bow between the precipitates or shear through them is answered, there is little value to speculate further about the hardening mechanism.

Toughness Properties

Figure 4 shows the changes in notch tensile strength with ultimate strength for the various quaternary alloys. With the exception of the alloys hardened by molybdenum the notched strengths decreased at higher ultimate strengths.

Figure 5 shows the Charpy V-notch impact energies vs the yield strengths of the alloys. The much better toughness of the molybdenum-containing alloys is clearly evident.

Metallographic examination was made of several of the brittle alloys (Ti, Al, or Si additions) and the tougher molybdenum alloys to see if there was any obvious reason for the differences in toughness. One obvious difference noticed after etching for 20 sec in 2% nital was the presence of a more rapid grain-boundary attack in the brittle alloys (Fig. 6A, 6B, 6C) than in the molybdenum alloy (Fig. 6D). This more rapid attack might be due to either a precipitate or segregation at the grain boundaries. The transmission micrographs discussed earlier (Fig. 2B and 3B) indicated that grain-boundary precipitates might be

Source: *Transactions of ASM*, Vol 57, 1964

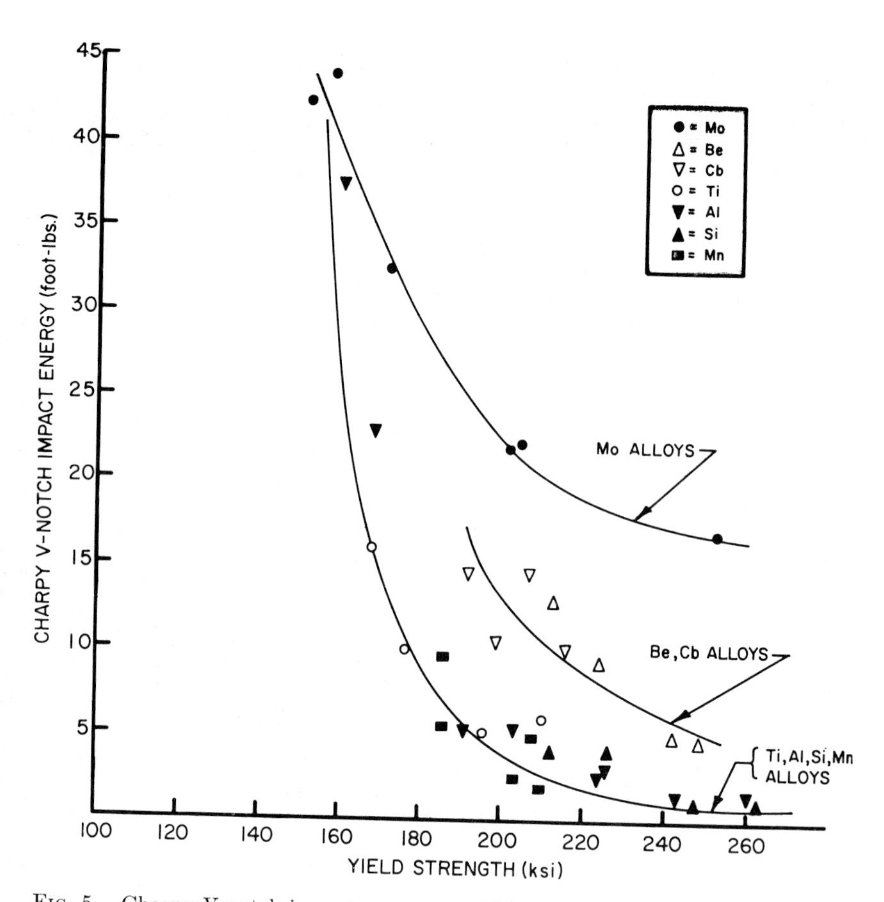

FIG. 5. Charpy V-notch impact energy vs yield strength of quaternary alloys.

TABLE 3. Composition and Mechanical Properties of Alloys Containing 2% Mo and Corresponding Molybdenum-Free Alloys

Alloy	Composition, wt %								
	Ni	Co	Mo	C	Ti	Al	Si	Mn	Fe
5	17.9	8.3	0	0.013	0.12	1.29	0.15	0.11	Bal
23	17.6	7.8	1.98	0.017	0.11	1.03	0.12	0.12	Bal
20	18.2	8.2	0	0.013	0.87	0.04	0.08	0.07	Bal
24	17.5	8.0	1.95	0.010	0.88	0.08	0.07	0.08	Bal

Alloy	Aging time, hr	Tensile strength, ksi 0.2% Offset yield	Ultimate	Elongation, %	Reduction area, %	Notch tensile strength, ksi	Charpy V-notch impact energy, ft-lb
5	3	225	235	4	16	95	2.5
	24	226	236	(a)	(a)	111	3.0
23	3	216	225	15	57	343	17.7
	24	232	241	15	56	355	18.7
20	3	195	205	14	57	282	5.2
	24	210	218	13	54	207	6.3
24	3	220	230	14	59	349	16.5
	24	235	247	14	57	364	18.7

(a) Broke into several pieces.

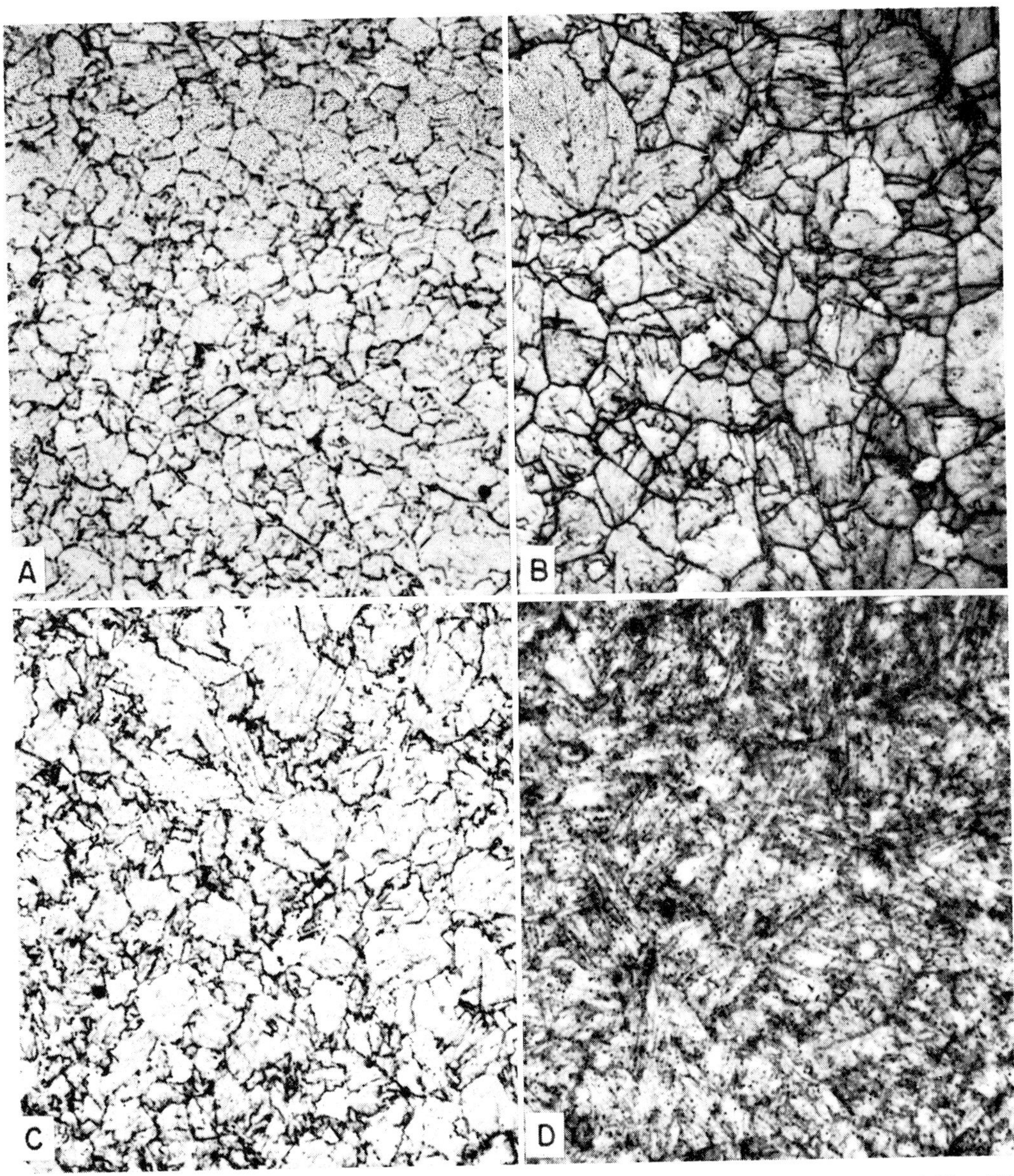

FIG. 6. Microstructures of several quaternary alloys, all specimens aged 24 hr at 800 F. (A) 1.29% Al alloy; (B) 0.87% Ti alloy; (C) 2.05% Si alloy; (D) 4.95% Mo alloy. Etched in 2% Nital, ×1000.

formed in titanium-hardened alloys, but not in the molybdenum-containing alloys. The correlation of brittleness with etching response is similar to the effects observed by Keh and Porr (10) in a study of temper embrittlement.

Figure 7 shows carbon replica fractographs of the fractured surfaces of the Charpy bars of a molybdenum and of a titanium-hardened alloy. A rather striking difference can be noted between the molybdenum-containing alloy showing a dimpled-type fracture surface characteristic of ductile fracture, and the brittle titanium-

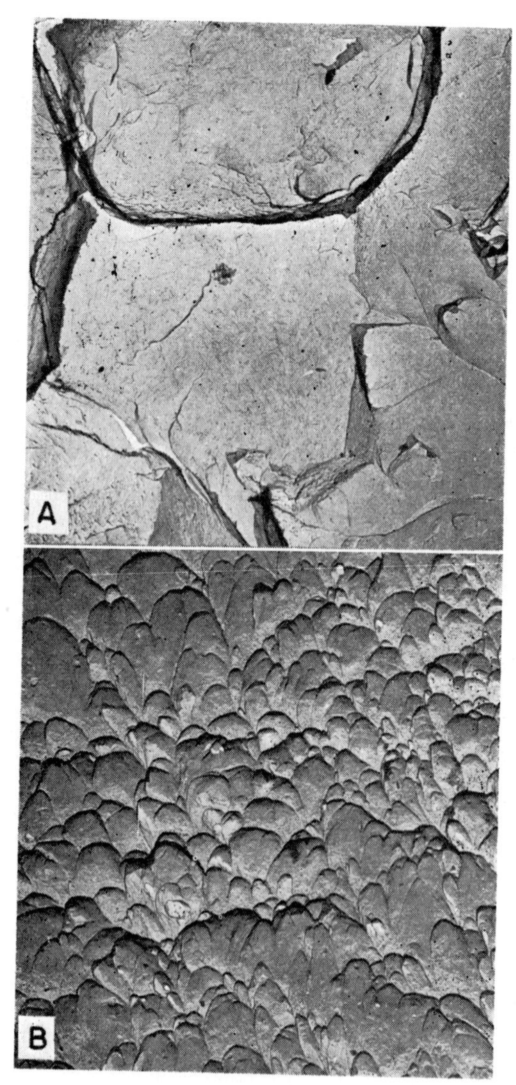

FIG. 7. Carbon replicas of fractured surface of Charpy V-notch impact bars; specimens aged 24 hr at 800 F. (A) Fe−18 Ni−8 Co−0.87Ti; (B) Fe−18 Ni−8 Co−4.95 Mo. ×3600.

containing alloy showing a flat fracture surface typical of a brittle intergranular fracture (11, 12). This latter appearance was typical of all the brittle alloys examined.

The fractured surface of a number of Charpy bars were also nickel plated and a polished surface examined normal to the fractured surface. Figure 8 shows the results for the Fe−18 Ni−8 Co−1 Ti and the Fe−18 Ni−8 Co−5 Mo alloy. It is quite evident that the fracture is intergranular in the brittle alloy (Fig. 8A) but of the ductile fibrous nature in the tough alloy (Fig. 8B).

To further study the influence of composition on toughness, several alloys were made in which 2% Mo was added to brittle (cobalt plus titanium, and cobalt plus aluminum) alloys. The properties of these alloys are compared with the corresponding molybdenum-free alloys in Table 3. The results show that molybdenum did not change the strengths noticeably. However, adding 2% Mo considerably improved the toughness.

Examination of these molybdenum-containing alloys revealed much less grain-boundary attack after etching and a ductile fibrous-type fracture. Electron transmission microscopy also indicated that the former austenite grain boundaries were free of precipitate.

All of these results indicate that it is not the hardening mechanism *per se*, but the relative presence or absence of precipitation or segregation at the prior austenite grain boundaries that is of major importance to toughness. If precipitation or segregation to the former austenite boundaries can be minimized, then the toughness is improved.

This beneficial affect of molybdenum in improving toughness by preventing the formation of a grain-boundary segregation or precipitation is similar to the effects of molybdenum in the temper embrittlement of steel. Steven and Balajiva (13) have shown that the temper embrittleness caused by the addition of a number of different trace elements is greatly reduced by the addition of 0.5% Mo. A similar beneficial effect appears to occur in maraging steels. Whether temper embrittleness is a result of the formation of discrete precipitates at the former austenite grain boundaries or simply a segregation to these boundaries is not clear (14). Also, the exact mechanism by which molybdenum prevents embrittlement has not been clarified. It has been

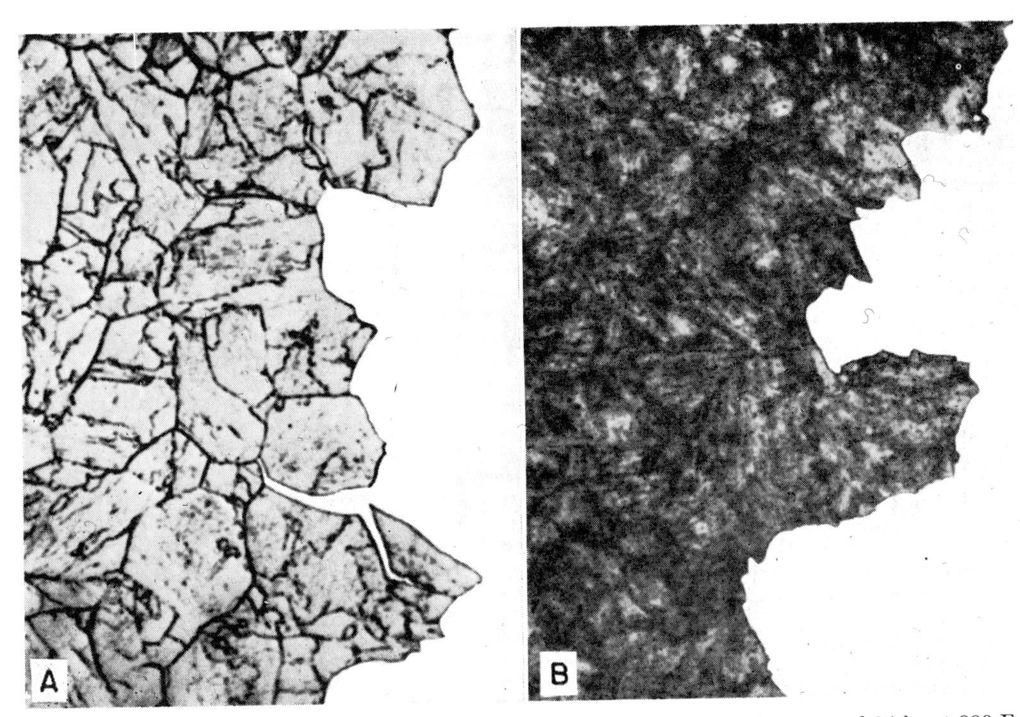

Fig. 8. Fractured surfaces of Charpy V-notch impact specimens; specimens aged 24 hr at 800 F. (A) Fe - 18 Ni - 8 Co - 0.87 Ti; (B) Fe - 18 Ni - 8 Co - 4.95 Mo. Etched in 2% Nital, ×1000.

well established, however, that continuous grain-boundary films or precipitates cause embrittlement and that when these are removed, toughness is markedly improved.

The 18% Ni maraging steel development (1) showed that at high molybdenum levels the toughness became poor. When molybdenum is used with cobalt, high strengths can be achieved with only 5% Mo. Thus, indirectly, the cobalt-molybdenum interaction improves the toughness because less molybdenum is required to achieve high strengths. More directly, the lack of grain-boundary embrittlement must be of primary importance. The lack of grain boundary embrittlement will not in itself guarantee good toughness, because there are other well-known ways to cause embrittlement. Other things being equal, however, the ability to achieve strengthening without grain-boundary embrittlement appears to be responsible for the superior toughness of the maraging steels.

Conclusions

1. The combination of cobalt and molybdenum in an Fe - 18 Ni martensite produces a much higher strength than would be expected from ternary alloy data. This added strengthening appears to result from the formation of a finer dispersion of precipitates when cobalt is present. In alloys containing aluminum, beryllium, columbium, manganese, silicon or titanium as hardeners, the addition of 8% Co only slightly increases the strength.

2. The quaternary Fe - 18 Ni - 8 Co - 5 Mo alloy has distinctly superior notched tensile strengths and Charpy impact energies at strengths above 200 ksi than other Fe - 18 Ni - 8 Co-base quaternary alloys. Segregation and/or precipitation at former austenite grain boundaries was observed in the brittle quaternary alloys, but not in the molybdenum-containing alloys. Fracture was intergranular through the former austenite grain boundaries in the brittle

Source: *Transactions of ASM*, Vol 57, 1964

alloys, and transgranular in the molybdenum alloys.

3. The addition of 2% Mo to Fe – 18 Ni – 8 Co – 1 Al or Fe – 18 Ni – 8 Co – 1 Ti alloys did not markedly change the yield strengths but gave significant improvement in the impact energies. The alloys containing 2% Mo showed much less grainrbounday attack on etching, and ductile fractures. Thus, it is not the cobalt-molybdenum hardening mechanism *per se,* but the presence of molybdenum that provides superior toughness in the maraging steels. This effect of molybdenum in preventing grain-boundary embrittlement is analogous to the effect of molybdenum in minimizing temper embrittlement of low-alloy steels.

Acknowledgment

The authors thank L. S. Torressen, R. C. Glenn, and R. Poliak for their aid in the experimental work. Discussions with W. C. Leslie, R. F. Decker, and A. S. Keh were also helpful.

REFERENCES

1. R. F. Decker, J. T. Eash and A. J. Goldman, 18% Ni Maraging Steels, ASM, Trans Quart, 55 (1962), 58. (U. S. Patent 3093519)
2. W. C. Leslie, A. S. Keh and G. R. Speich, Contributions to the Metallurgy of Steels, AISI (March 1963).
3. G. R. Speich, Age Hardening of Fe-20 pct Ni Martensites, Trans AIME 227, (1963), 1426.
4. B. G. Reisdorf, Identification of Precipitates in 18, 20 and 25% Ni Maraging Steels, ASM Trans Quart, 56 (1963), 783.
5. A. J. Baker and P. R. Swann, submitted to Nature (1963).
6. S. Floreen, Hardening Behavior of Alloys Based on Iron-18% Nickel, ASM Trans Quart, 57 (1964), 38.
7. S. Floreen, Deformation Characteristics of an Iron-18 Nickel Binary Alloy, Trans AIME, to be published.
8. S. Floreen and R. F. Decker, Heat Treatment of 18% Ni Maraging Steel, ASM Trans Quart, 55 (1962), 518.
9. G. S. Ansell and F. V. Lenel, Criteria for Yielding of Dispersion Strengthened Alloys, Acta Met, 8 (1960), 612.
10. A. S. Keh and W. C. Porr, Effect of Cold Work on Temper Brittleness, Trans ASM, 52 (1960), 81.
11. J. R. Low, *Fracture,* Averbach, Felbach, Hahn and Thomas, Editors, John Wiley, New York (1959), 68.
12. C. Crussard, J. Plateau, R. Tamhankar, G. Henry and D. Lajeunesse, *Fracture,* Averbach, Felbach, Hahn and Thomas, Editors, John Wiley, New York (1959), 524.
13. W. Steven and K. Balajiva, The Influence of Minor Elements on the Isothermal Embrittlement of Steels, J Iron Steel Inst, 193 (1959), 141.
14. B. C. Woodfine, Temper Brittleness; A Critical Review of the Literature, J Iron Steel Inst, 173 (1953), 229.

A Study of Precipitation in Stainless and Maraging Steels Using the Mössbauer Effect

HARRIS MARCUS, LYLE H. SCHWARTZ AND MORRIS E. FINE

ABSTRACT. Mössbauer spectroscopy was used to study the precipitation reactions in 17-7 PH stainless steel and in an 18Ni – 8Co – 5Mo maraging steel. In the 17-7 PH stainless steel the changes in internal field were the same after 399, 510, and 593 C aging treatments indicating that the final matrix compositions were the same within the precision of the measurements. No paramagnetic Fe-bearing precipitates were formed. Cr did not diffuse to or away from the precipitate. A quantitative analysis of austenite content vs aging treatment is given. During aging at 480 and 535 C, the 18Ni – 8Co – 5Mo maraging steel had at least a two-step precipitation reaction. The first step did not involve paramagnetic precipitate containing Fe. The second step was the formation of Fe_2Mo. Co remained in solution in the martensite.

A NEW AND REVEALING technique of elucidating precipitation reactions in Fe bearing alloys has been made available with the development of Mössbauer spectroscopy. The effect was discovered first in Ir^{191} by Rudolf Mössbauer (1) in 1958 and later in Fe^{57} (2, 3). The Mössbauer effect can be used to study the local environment about Fe^{57} and changes in this local environment when Fe is either the host material or an alloying element. Reviews of the theory and applications of the Mössbauer effect are given, for example, in books by Frauenfelder (4) and Wertheim (5).

The purpose of this study was to use this technique to determine the precipitation reactions and changes in austenite content which occur in 17-7 PH stainless steel and in an 18Ni– 8Co – 5Mo maraging steel during heat treatment. 17-7 PH stainless steel was selected for investigation because it represents a system in which the austenite content changes drastically during heat treatment (6, 7). When in the quenched state, it is essentially all austenite, it retains austenite when transformed to martensite and reverts in part to austenite when aged at temperatures above 525 C. Information about the role of Fe in the precipitation reaction in 17-7 PH stainless steel was also of interest.

18Ni – 8Co – 5Mo maraging steels are hardened by aging in the martensitic condition. Previous studies (8–10) in overaged alloys using the electron microprobe and diffraction from extraction replicas have shown that the precipitate is either Ni_3Mo or Fe_2Mo. The role of cobalt in the hardening process is not clear. Changes in the local environments of the Fe^{57} atoms during aging change the Mössbauer spectra and provide partial answers to these questions.

This research was supported by ARPA through the Northwestern Materials Research Center and NASA by a traineeship to Harris Marcus.

The authors are Graduate Student, Assistant Professor and Professor, respectively, Materials Science Dept. and Materials Research Center, Northwestern University, Evanston, Ill. Manuscript received April 22, 1966.

Theory

A general description of the theory of the Mössbauer effect can be found in the references (4, 5). A short qualitative treatment of the theory will be given here using Fe^{57} as the isotope of interest.

Figure 1 shows the nuclear decay scheme for Co^{57}. In the final step Fe^{57} in the first excited state (nuclear spin $\frac{3}{2}$) decays to Fe^{57} in the ground state (spin $\frac{1}{2}$) with emission of 14.39 Kev γ-rays. When radiation of this energy interacts with another Fe^{57} nucleus in the ground state, resonant absorption may occur; that is, the absorbing nucleus may be excited to the $\frac{3}{2}$ state with absorption of a 14.39 Kev photon. When the emitting nucleus and γ-rays are considered as the total system, conservation of energy and momentum require a reduction in the energy of the emitted γ-ray from 14.39 Kev due to recoil of the nucleus. Thus resonant absorption would be impossible since such a γ-ray does not have sufficient energy to raise the absorbing nucleus from the $\frac{1}{2}$ to the $\frac{3}{2}$ state and in addition supply its recoil energy.

Mössbauer found that when the nuclei are rigidly bound in a crystal there is a finite probability (called the recoilless fraction) that all of the momentum change will be transferred to the center of mass of the lattice with the result that the recoil energy loss of the γ-ray is completely negligible and that resonant absorption is possible. Thus the Mössbauer effect is in fact the recoilless resonant emission and absorption of γ-rays by nuclei. The recoilless fraction is a function of temperature, the Debye temperature of the nucleus in the lattice in question, the mass of the isotope, and the energy of the γ-ray involved. By a fortunate combination of these parameters, the recoilless fraction for Fe^{57} in Pt, Pd or Cu is roughly 0.75 at room temperature. Thus studies of the Mössbauer effect in Fe can be made at room temperature without the experimental complications of cryogenics required for other Mössbauer nuclides.

The usual Mössbauer experiment is done in transmission. For the source of γ-rays, Co^{57} is diffused into a nonmagnetic foil such as Cu, Pd or Pt. Since in these materials the energy of the $\frac{3}{2}$ and $\frac{1}{2}$ states of Fe^{57} is unsplit by crystal fields, a monoenergetic γ-ray source is produced. A thin specimen containing Fe^{57} is then placed between the source and a detector for 14 Kev γ-rays. Since Fe^{57} has a natural abundance of 2.25%, enrichment is usually not necessary.

The energy dependence of the transmission through the sample, and thus absorption, is measured by moving the sample relative to the source, which results in a Doppler energy shift of the γ-ray energy given by:

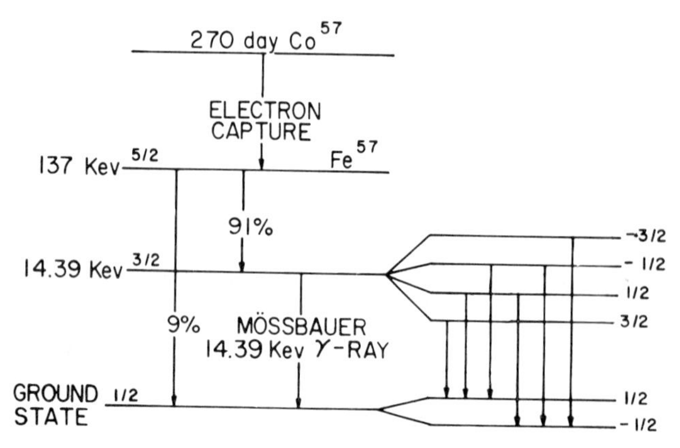

Fig. 1. Decay scheme of Co^{57} and hyperfine splitting of Fe^{57}.

$$E(v) = E_o(1 + v/c)$$

where E_o = 14.39 Kev for Fe^{57}; v = velocity of sample in mm/sec; c = velocity of light = 3×10^{11} mm/sec.

By convention the velocity is considered positive when the motion of the sample is toward the source. Since the energy change of the γ-ray is directly proportional to the velocity, the usual method of conducting an experiment is to measure transmitted intensity vs relative velocity. As it is also conventional to report data in this form, some confusion arises in terminology. In this paper minima in transmission will be referred to as *peaks* (as they would appear if per cent *absorption* vs velocity were reported).

Figure 2 shows transmitted intensity vs velocity in a 17-7 PH stainless steel sample in the austenitic condition, with Co^{57} in Cu as the source. The austenite is paramagnetic resulting in a single absorption line. With the exception of Fig. 3, velocity in all figures in this paper is relative to the center of the pattern of pure Fe taken as zero. The deviation of the transmission minimum from zero velocity is called the isomer shift provided there is no second order Doppler effect (temperature shift, etc.). This shift is a function of the chemical environment of the Fe^{57} atoms and will vary for different chemical and crystallographic conditions. Using the isomer shift it is possible in many cases to distinguish between phases in the same sample.

The shape of the transmission curve is Lorentzian, given by

$$I(v) = I_o - I_m\left[1 + 4\left(\frac{v - v_o}{\Gamma}\right)^2\right]^{-1}$$

I_o is the background 14.39 Kev γ-ray count. (The count far away from absorption peaks.) I_m is the difference between I_o and the count at the minimum of the curve, v_0 is the velocity at this minimum and Γ is the breadth of the curve at half minimum. The theoretical value for Γ in pure Fe is 0.194 mm/sec.

The value of Γ = 0.47 mm/sec observed in the stainless steel represents an energy spread of (Γ_c/E_o = 2.25×10^{-8} ev. It is this remarkable sharpness of the absorption peaks that enables investigation of small changes in the nuclear energy levels associated with changes in local chemical environment.

Figure 3 is a spectrum for a pure Fe sample enriched to 50% Fe^{57}. Velocity is relative to Fe^{57} in Cu. This ferromagnetic material shows a characteristic six-line pattern due to the nuclear Zeeman effect. The degeneracy of the $\frac{3}{2}$ and $\frac{1}{2}$ energy levels in Fe^{57} is lifted by a magnetic field as shown in Fig. 1. The width of this splitting is a measure of the internal field (hyperfine field) at the iron nucleus which has been found to be −330 Koe in pure Fe. As indicated by the negative sign, a reduction of the splitting occurs when an external magnetic field is applied. When alloying elements other than Co and small amounts of Ni (11, 12) are added to Fe, the internal field and the resulting splitting is reduced.

Changes in the value of the internal field can also be used to describe differences in the local environments of the Fe^{57} atoms. In general, for a complex alloy one observes the superposition of ferromagnetic six-line patterns (made complex by alloying elements) on any single line paramagnetic patterns which may be present. Decomposition of the experimental pattern into its various parts will give information about the structure of the alloy. Any subsequent changes in the patterns during heat treatment of an alloy will then give information about changes in the structure.

The 0.47 mm/sec value of Γ for austenite as opposed to the ideal 0.19 for pure Fe arises for four reasons. The first is an instrumentation problem due to the presence of vibration that smears the velocity spectrum. A second source is finite specimen and source thickness (15, 16). The experimental half widths obtained for the best conditions range from 0.21 to 0.24 mm/sec for Fe. The third contribution to

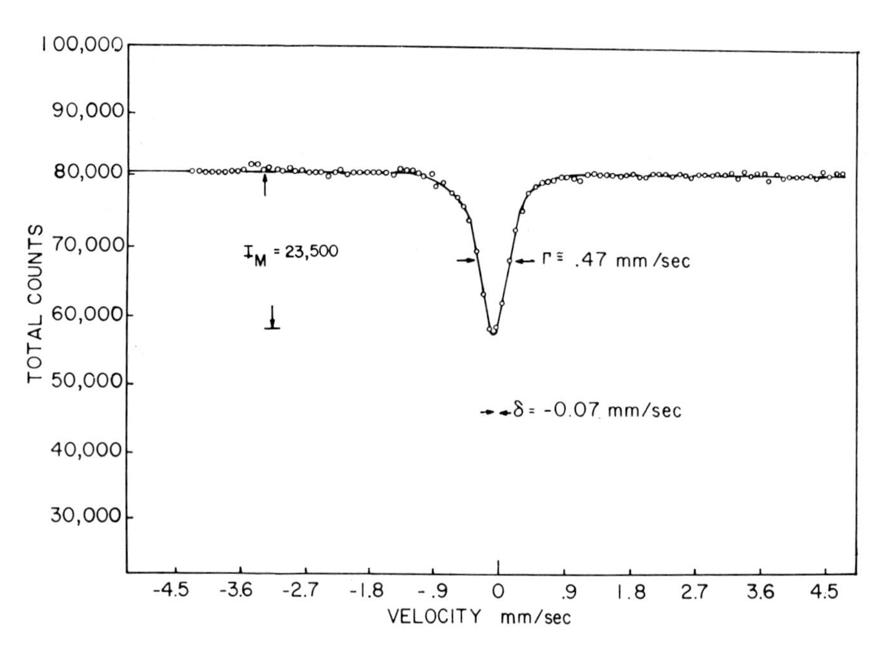

FIG. 2. Mössbauer pattern of paramagnetic 17-7 PH stainless steel. Solution treated 0.5 hr 1065 C, air cooled, Condition A.

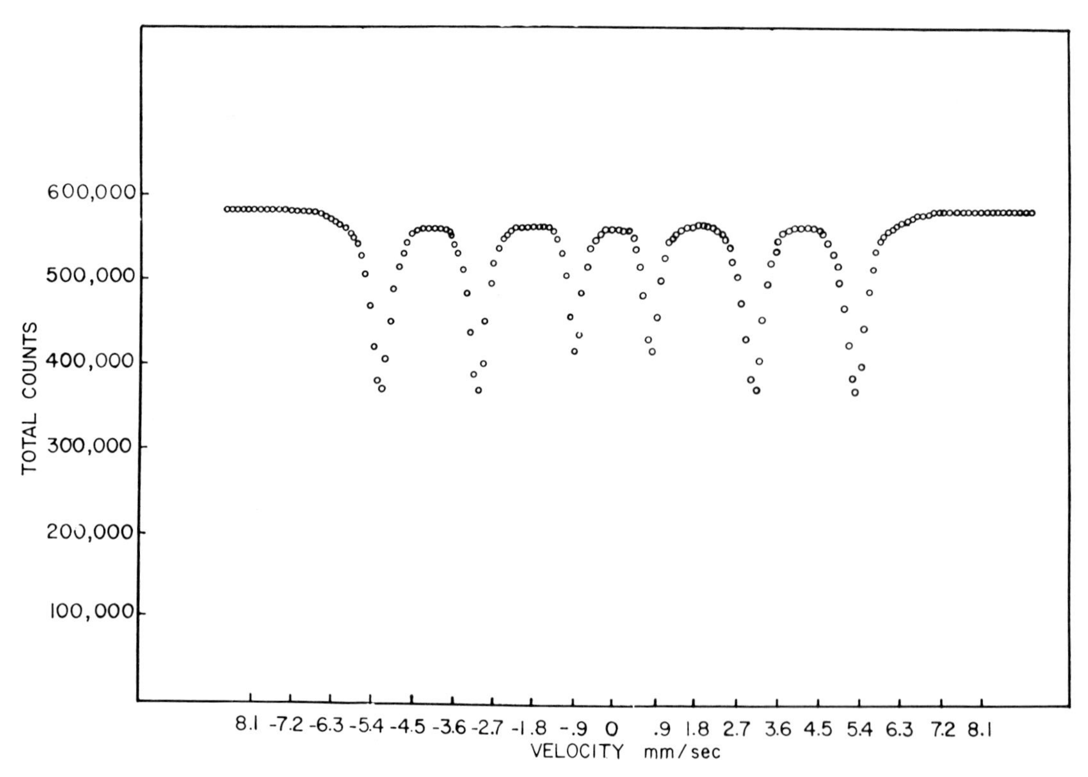

FIG. 3. Mössbauer pattern from an enriched pure Fe standard.

the broadening in the austenite comes from the fact that all iron atoms in the austenite do not have the same local environment, which leads to variation of the isomer shift. The main contribution to broadening comes from the removal of local symmetry about the Fe^{57} atoms due to the presence of alloying elements which induces an electric field gradient giving rise to quadrupole splitting which broadens the peaks. These variations are the same for several austenitic stainless steels. Measurements made on 301, 303, and 316 stainless steels show isomer shifts of -0.08 ± 0.03 mm/sec and a value of $\Gamma = 0.46 \pm 0.03$ mm/sec for 0.001-in. samples. For this reason it is reasonable to expect that any small change of composition of reverted austenite from the original austenite composition would not affect either the isomer shift or the half breadth of the peak in the work reported here.

It is important to remember that only the Fe^{57} is seen in a Mössbauer experiment. Any analysis of the contribution of the various phases present must taken account of the percentage of Fe in each phase before direct comparison of intensities is possible.

Experimental Procedure

There are two different methods in general use for obtaining Mössbauer patterns. The first involves a constant acceleration drive unit that scans the total velocity scale each cycle and stores the count for each velocity increment in one channel of a multichannel analyzer. This is continued until sufficient counts are obtained for good statistics. Description of this type of electromechanical Mössbauer spectrometer is given in the literature (13, 14). Part of the results with the 17-7 PH stainless steel samples were obtained with this type of spectrometer at Northern Illinois University.

The second type of spectrometer is a constant velocity instrument (14). The transmitted intensity is determined for a fixed time period for each velocity. Two scalers are operated and counts for both positive and negative values for the velocity setting are determined simultaneously. The velocity is then changed by a fixed increment and the procedure repeated. In this manner the complete spectrum of transmitted intensity vs velocity is determined. The patterns for the maraging steel and several for the 17-7 PH stainless steel were obtained with an instrument of this type manufactured by Nuclear Science and Engineering Corp. In all measurements reported here, sample and source were at room temperature.

The experimental data were corrected for background from high energy γ-rays by placing an 0.095-in. Al sheet in front of the counter and also for the exponential time decay of the source intensity in the constant velocity measurements. The data was analyzed using a CDC 3400 computer. Theoretical curves were synthesized and compared with experimental curves. Some of the experimental curves were decomposed making use of a program written by Davidon (17).

A problem that arises from the use of thin foils of ferromagnetic materials as Mössbauer specimens is that the magnetic domains prefer to be in the plane of the foil. This results in a partial polarization of the incoming γ-rays and relative intensities of the six absorption peaks in the Fe pattern different from those from random domain orientation. For this reason the total pattern must always be integrated in each case to get the true fraction of Fe present.

The 17-7 PH stainless steel samples were supplied by Armco Steel Co. as cold drawn and centerless-ground $\frac{1}{4}$-in. rods. (0.065% C, 0.55% Mn, 0.016% P, 0.014% S, 0.47% Si, 17.04% Cr, 7.22 % Ni, 1.17% Al, balance Fe.) The 18Ni – 8Co – 5Mo maraging steel was supplied by The United States Steel Corp. as 0.050-in. sheet (0.017% C, 0.04% Mn, 0.003% P, 0.008% S, 0.05% Si, 0.13% Cr, 17.21% Ni, 7.97% Co, 0.18% Al, 4.87% Mo, 0.52% Ti, 0.005% Zn, 0.01% Cu). All compositions are given in weight per cent. Both samples were then cold rolled with intermediate vacuum anneals to 0.001-in. thickness. The 17-7 PH stainless steel specimens were annealed at 1065 C ± 10 C for $\frac{1}{2}$ hr and then air cooled to room temperature. They were then "conditioned" at 760 C ± 10 C for $1\frac{1}{2}$ hr. All treatments were carried out in dynamic vacuum. The samples were then aged at 400 or 510 C for $33\frac{1}{3}$ hr or 595 C for 16 hr in dynamic vacuum. Mössbauer patterns were run at each of the described conditions on both the electromechanical Mössbauer spectrometer and the constant velocity spectrometer with excellent reproducibility. The maraging steel specimens were annealed 1 hr at 815 C and aged at 480 C for 3 and 41 hr and at 540 C for 34 hr. The 17-7 PH stainless steel treatments were taken at points of interest previously determined by magnetic and hardness measurements (6). The maraging steel aging times were similar to those used by Reisdorf (8), Baker and Swann (9), and Floreen and Speich (10).

Results and Discussion

17-7 PH Stainless Steel

Figure 2 is the spectrum for 17-7 PH stainless steel after a $\frac{1}{2}$ hr austenizing treatment at 1065 C. This has an isomer shift relative to pure Fe of −0.07 mm/sec. The breadth at half depth, Γ, is 0.47 mm/sec. The sample was then given a $1\frac{1}{2}$ hr conditioning treatment at 760 C to precipitate the carbides and raise the M_s temperature above room temperature. Figure 4A shows a pattern for the sample in this condition, condition T. The points of interest here are the presence of a complicated six-peak pattern associated with the magnetic martensite phase and the single peak in the center of the pattern associated with the retained austenite. The isomer shift associated with the austenite remains at −0.07 mm/sec. The hyperfine splitting of the six-peak pattern is less than that of the pure iron spectrum. The splitting of the pure iron spectrum measured between the outside peaks is 10.64 mm (18). This corresponds to an internal field of −330 Koe. The outside peaks in the 17-7 PH stainless steel spectrum do not have well-defined Lorentzian shapes but represent the superposition of many peaks. The splitting for the approximate minimum of the outside peaks corresponds to a field of −260 Koe.

Figure 4B is for a sample heat-treated $33\frac{1}{3}$ hr at 400 C. Again the complex six-peak pattern and the single peak associated with the retained austenite are seen. There is no change in the amount of retained austenite. The internal field associated with the approximate minimum of the outside peaks has increased to −290 Koe. There is no sign of an additional peak that could be associated with a paramagnetic iron-bearing precipitate. Figure 4C is for a sample aged at 510 C for $33\frac{1}{3}$ hr. The austenite peak has now increased in intensity compared with the sample in condition T. The general shape and the field associated with the martensite phase is the same as for the 400 C treatment. Figure 4D is for a sample aged 16 hr at 595 C. Now the pattern is dominated by the austenite peak; a large amount of reversion to austenite has taken place. The location of the minima in the outside peaks is the same as after aging at 400 and 510 C.

The absorption data for the seven-peak patterns were pointwise integrated and the center three peaks were synthesized by computer methods. The results of the integration make it possible to determine the fraction of the total iron present in the austenite phase. There was 6 ± 1.0% austenite after condition T and in the sample aged at 400 C, 8 ± 1.0% retained and reverted austenite after aging $33\frac{1}{3}$ hr at 510 C and 17 ± 1.0% retained and reverted austenite after aging 16 hr at 595 C. Computer-generated spectra have demonstrated that the presence of less than $\frac{1}{2}$% austenite in a martensite matrix can be determined when sufficient counts are taken to reduce the random error.

The results of the 17-7 PH experiments indicate that determination of austenite content of thin foil specimens by Mössbauer spectroscopy is feasible. There is a dis-

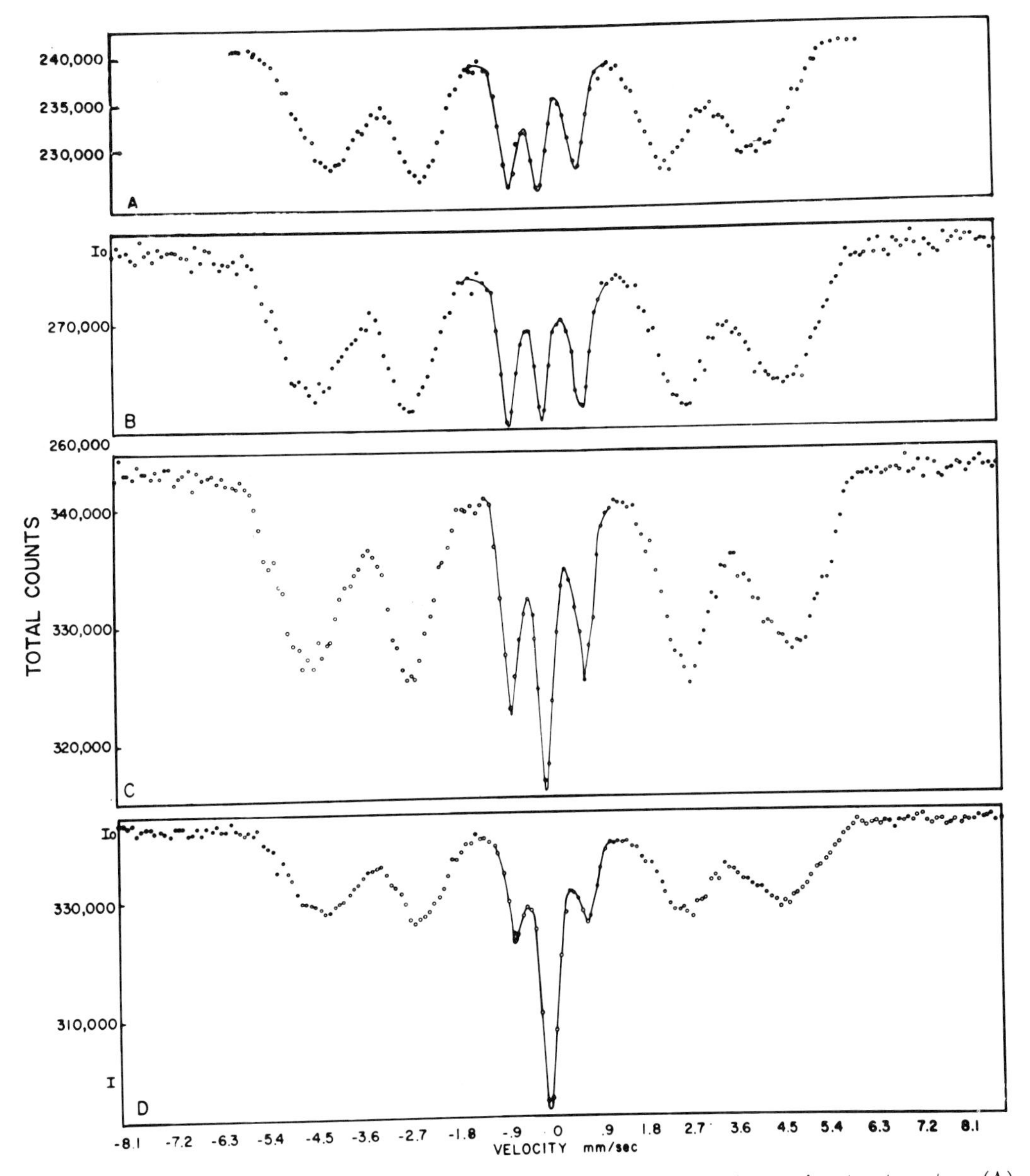

Fig. 4. Mössbauer pattern from 17-7 PH stainless steel after various aging treatments. (A) Annealed 1.5 hr at 760 C, Condition T; (B) aged 33.3 hr at 400 C; (C) aged 33.3 hr at 510 C; (D) aged 16 hr at 593 C.

tinct advantage over the standard x-ray technique since the Mössbauer absorption is unaffected by preferred orientation, which usually has limited the accuracy of the x-ray technique.

The change in the internal magnetic field at the nucleus of the martensite phase during precipitation is consistent with the precipitate previously proposed (6). Johnson, Ridout and Cranshaw (12) determined the change in the internal field for various alloying elements added to iron. Their results, and the results of Yamamoto (19) for Cr in iron, give an internal field for an

Fe — 17% Cr alloy of −290 Koe. Johnson et al, and Stearns (20) have shown that Al reduces the internal field. Ni has only a small effect on the field (18). These results for the binary alloys are consistent with the observed changes in the internal field in 17-7 PH stainless steel. The field associated with condition T is less than would be expected in an Fe — 17% Cr matrix. Precipitation of the proposed $(Ni, Cr, Fe)_{1+y}Al$ phase, with the virtual depletion of Al from the matrix, would then raise the field closer to that expected in Fe — 17% Cr.

The inability to locate a peak associated with iron in the precipitate is due to the small amount of iron and the weak ferromagnetism in the precipitate. This would have a six-peak pattern of low intensity that would be lost in the martensite pattern.

The study of Yamamoto also indicates that Cr does not enter into the reaction. A higher final internal field in the martensite would be expected if the Cr were concentrated into the precipitate. Yamamoto aged an Fe — Cr alloy for 150 hr at 500 C. This resulted in an increase in the hyperfine field and the appearance of a paramagnetic peak from a Cr-rich precipitate. These results indicate that a Mössbauer study would be useful in determining σ phase formation.

The lack of any distinct characteristic pattern showing the presence of δ ferrite can be explained by the very small intensities one would expect from only a small amount of a ferromagnetic phase. The total pattern for an austenitized sample does have fine structure associated with the presence of a ferromagnetic phase, but is too diffuse for quantitative analysis. The outside peaks for this pattern, which are the most intense, occur outside the velocity range covered in Fig. 2.

Maraging Steel

The Mössbauer patterns for the 18Ni − 8Co − 5Mo maraging steel are shown in Fig. 5. Figure 5A is for annealing 1 hr at 820 C and air cooling. A characteristic six-peak pattern for the martensite is observed. The internal field corresponding to the minimum in the outside peaks is −340 Koe. The peaks are more distinct with less overlap than in the 17-7 PH stainless steel patterns. Note the fine structure on the inside of the outside peaks. Figure 5B is for a sample aged at 480 C for 3 hr. The internal field has increased to −350 Koe. There is no evidence for a precipitate peak. The fine structure of the outside peaks has been greatly reduced.

For a sample aged 41 hr at 480 C, Fig. 5C, there is a definite indication of an iron-containing precipitate in the center of the pattern. Its isomer shift is -0.20 ± 0.05 mm/sec. About 2% of the total Fe is present in the precipitate phase. The internal field associated with the martensite phase is just slightly larger than that for the 3 hr aging at 480 C. The outside peaks have become sharper and their fine structure has disappeared.

The data for a sample aged 34 hr at 535 C are shown in Fig. 5D. The outside peaks have similar shapes and locations to those in Fig. 5C but an increase to about 4% is noted in the fraction of total Fe in the precipitate phase.

Preliminary results of experiments in progress at this laboratory indicate that Mo causes only a minor change in the internal field for Fe with no Mo near neighbors but reduces it 12% when there is one Mo near neighbor. The reduction in field corresponding to the extra peaks associated with the outside peaks for the maraging steel as quenched is also 12%. The precipitation process removes most of the Mo from the matrix as shown by the elimination of the extra peaks.

The lack of a paramagnetic peak in the initial aging seems to confirm the idea that the early precipitate is Ni_3Mo; however, this technique would not easily detect a slightly ferromagnetic precipitate.

The depletion of the alloying elements Mo, Al and Ti, which combine with Ni to form precipitates, increases the internal

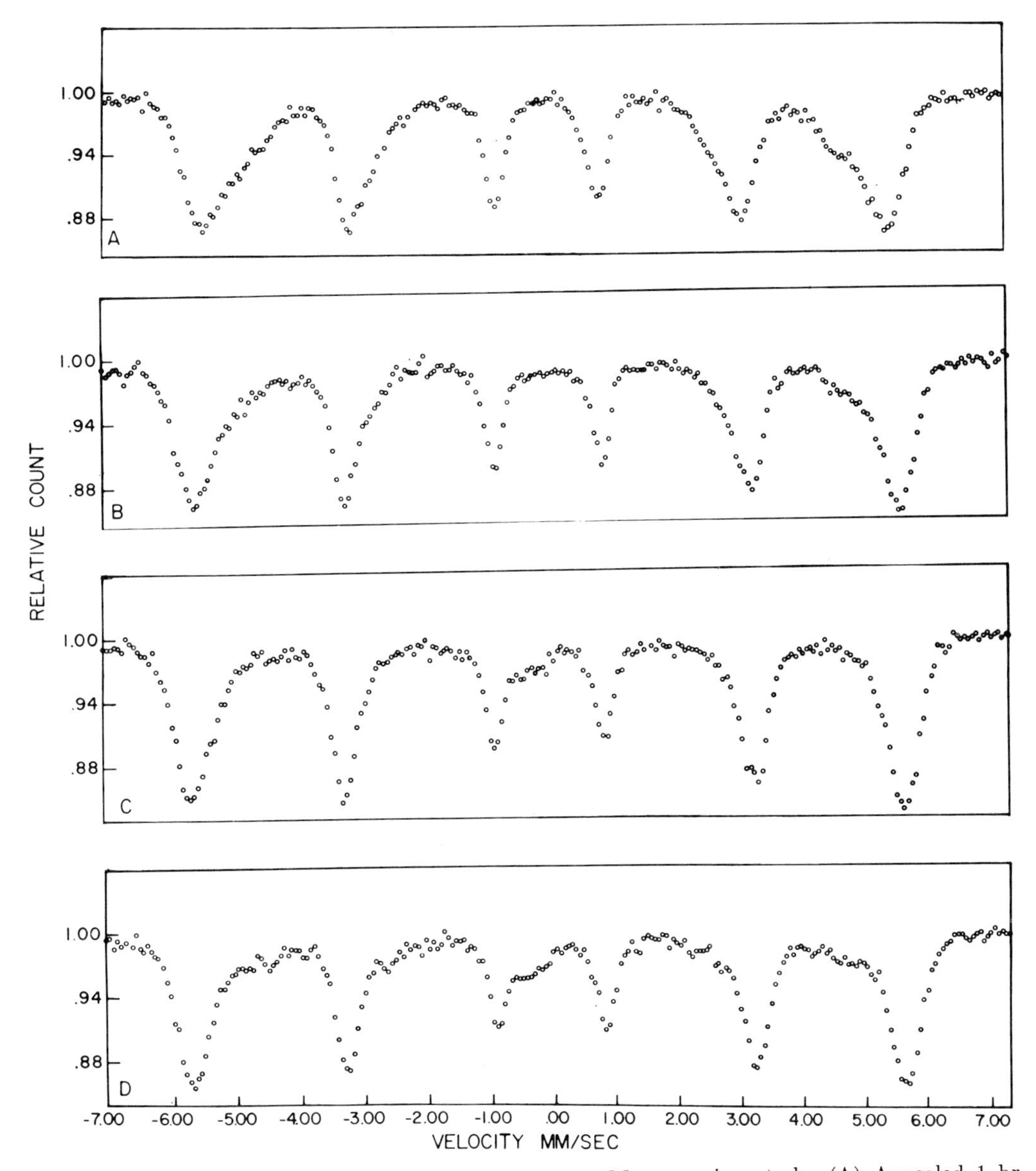

FIG. 5. Mössbauer patterns from the 18Ni – 8Co – 5Mo maraging steel. (A) Annealed 1 hr at 820 C; (B) aged 3 hr at 480 C; (C) aged 41 hr at 480 C; (D) aged 34 hr at 535 C. The I_0's were in the range of 250,000 to 300,000 counts per point.

field close to that expected for an Fe — 8% Co alloy, –350 Koe (11, 12). The Co does not seem to enter into the precipitation reaction since the field after precipitation is larger in magnitude than in pure Fe. The absence of an appreciable field change during longer aging indicates that the depletion of alloying elements from the martensite matrix is about complete after 3 hr at 480 C. During overaging the more stable Fe_2Mo (21) phase replaces the initial precipitate as evidenced by the appearance of absorption due to a paramagnetic phase. The isomer shift of –0.20 mm/sec is identical to that found for Fe_2Mo samples produced in this laboratory.

In the solution treated sample, Fig. 5A, there is no indication of a peak due to paramagnetic austenite. In fact, there is little evidence for paramagnetic austenite after aging 34 hr at 535 C where extensive reversion to austenite has been reported (22). X-ray diffraction study of our sample did show the presence of a substantial quantity of austenite, but the measurement was not quantitative due to preferred crystal orientation in the sample. The austenite here evidentally contains sufficient Ni and Co to make it ferromagnetic at room temperature. Peters and Cupp (23) concluded from magnetic mass balance that the austenite formed by reversion at 538 C is indeed ferromagnetic. The internal field is small compared to that in the martensite so the pattern expected from the austenite is a diffuse six-line pattern, with only a small amount of splitting compared to the martensite, centered about −0.08 mm/sec, the isomer shift noted in several stainless steels. In Fig. 5D a small general depression of the pattern near 0 velocity is noted.

Conclusions

This investigation demonstrates the value of using Mössbauer spectroscopy to study the structures of steels. There are three main results. First, a quantitative measurement of retained and reverted austenite when it is paramagnetic can be made, unaffected by texture. Second, a paramagnetic Fe-bearing precipitate can be identified during a precipitation process. Third, by careful analysis of the changes in the magnetic internal field it is possible to infer which alloying elements are entering into the precipitation reactions.

The results for the 17-7 PH stainless steel support the previous suggestion (6) that the precipitate is ferromagnetic $(Ni, Cr, Fe)_{1+y}Al$, containing the same fraction of Cr as the alloy composition. Reversion to austenite takes place at as low a temperature as 510 C. The isomer shift for the reverted austenite is the same as for the retained austenite. The precipitate process is the same at 400, 510 and 595 C, as indicated by the similar shapes and splitting of the patterns.

The precipitation reaction in the 18Ni − 8Co − 5Mo maraging steel proceeds in at least two steps. The first does not give a paramagnetic Fe^{57} absorption peak even though the matrix is depleted of solute. The overaged precipitate is Fe_2Mo. Co does not seem to enter into the precipitation reaction.

Acknowledgments

The authors would like to express their thanks to Dr. C. Kimball and R. Hannon at Northern Illinois University for their aid in doing the electromechanical Mössbauer experiments and to E. Hall for his aid in preparing the computer programs. Helpful discussions with Drs. C. Cupp and D. Peters of the International Nickel Co. Research Laboratory concerning austenite content of maraging steels are also gratefully acknowledged.

REFERENCES

1. R. L. Mössbauer, Kernresonanzfluoreszenz von Gammastrahlung in Ir^{191}, Z. Physik, 151 (1958) 124.
2. R. V. Pound and G. A. Rebka, Jr., Resonant Absorption of the 14.4 KeV Gamma Ray from 0.10-Microsecond Fe^{57}, Phys Rev Letters, 3 (1959) 554.
3. J. P. Schiffer and W. Marshall, Recoiless Resonance Absorption of Gamma Rays in Fe^{57}, Phys Rev Letters, 3 (1959) 556.
4. H. Frauenfelder, The Mössbauer Effect, W. A. Benjamin, Inc., New York, 1963.
5. G. K. Wertheim, Mössbauer Effect: Principles and Applications, Academic Paperbacks, New York, 1965.
6. H. L. Marcus, J. N. Peistrup, and M. E. Fine, Precipitation in 17-PH Stainless Steel, ASM Trans Quart, 58 (1965) 176.
7. H. C. Burnett, R. H. Duff, and H. C. Vacher, Identification of Metallurgical Reactions and Their Effect on the Mechanical Properties of 17-7 PH Stainless Steel, J of Res of Nat Bureau of Standards, 66 C (1962) 113.

8. B. G. Reisdorf, Identification of Precipitates in 18, 20, and 25% Nickel Maraging Steels, ASM Trans Quart, 56 (1963) 783.
9. A. J. Baker and P. R. Swann, The Hardening Mechanism in Maraging Steels, ASM Trans Quart, 57 (1964) 1008.
10. S. Floreen ond G. R. Speich, Some Observations on the Strength and Toughness of Maraging Steels, ASM Trans Quart, 57 (1964) 714.
11. G. K. Wertheim, V. Jaccarino, J. H. Wernick, and D. N. E. Buchanan, Range of the Exchange Interaction in Iron Alloys, Phys Rev Letter, 12, #1 (1964) 24.
12. C. E. Johnson, M. S. Ridout and T. E. Cranshaw, The Mössbauer Effect in Iron Alloys, Proc Phys Soc, 81 (1963) 1079.
13. F. J. Lynch and J. B. Baumgardner, Argonne National Laboratory Report ANL-6391 (1961) 10.
14. Mössbauer Effect Methodology, Volume 1, Edited by I. J. Gruverman, Plenum Press, New York (1965) 47.
15. S. Margulies and J. R. Ehrman, Transmission and Line Broadening of Resonance Radiation Incident on a Resonance Absorber, Nuclear Instr and Methods, 12 (1961) 131.
16. D. A. Shirley, M. Kaplan and P. Axel, Recoil-Free Resonant Absorption in Au^{197}, Phys Rev, 123, #3 (1961) 816.
17. W. C. Davidon, Variable Metric Method for Minimization, ANL-5990, AEC Research and Development Report (1959).
18. J. G. Dash, R. D. Taylor, D. E. Nagle, P. P. Craig, and W. M. Visscher, Polarization of Co^{57} in Fe Metal, Phys Rev, 122 (1961) 1116.
19. H. Yamamoto, A Study of the Nature of Aging of Fe-Cr Alloys by Means of the Mössbauer Effect, Japanese J of Appl Phys, 3, #12 (1964) 745.
20. M. B. Stearns, Variation of the Internal Fields and Isomer Shifts at the Fe Sites in the FeAl Series, J of Appl Phys, 35, #3 (1964) 1095.
21. G. P. Miller, and W. I. Mitchell, Structure and Hardening Mechanisms of 18% Ni - Co - Mo Maraging Steels, JISI, 203, Sept 1965, 899.
22. D. T. Peters and C. R. Cupp, The Kinetics of Aging Reactions in 18% Ni Maraging Steels, accepted for publication Trans AIME.
23. D. T. Peters and C. R. Cupp, private communication.

SECTION XIII: Appendix

The 18 per cent nickel maraging steels

Engineering properties

Introduction

The development of the nickel maraging steels began in the Inco research laboratories in the late 1950s and was based on the concept of using substitutional elements to produce age-hardening in a low-carbon iron-nickel martensitic matrix. Hence the term 'maraging' was given to them to signify this strengthening mechanism.

The work led to the discovery that balanced additions of cobalt and molybdenum to iron-nickel martensite gave a combined age-hardening effect appreciably greater than the additive effects of these elements used separately. Furthermore, the iron-nickel-cobalt-molybdenum matrix was found to be amenable to supplemental age-hardening by small additions of titanium and aluminium. Thus the 18 per cent nickel-cobalt-molybdenum family of maraging steels was developed.

There are basically four wrought commercial maraging steels of the 18 per cent nickel family and one cast grade currently available from special steel manufacturers. The nominal compositions and 0·2 per cent proof stress values are presented in Table 1. The reader should note that the numbers ascribed to the various grades in this publication correspond to the nominal proof stresses given in SI units, whereas the identical grades in some countries have designations with lower numbers corresponding to the units traditionally used for proof stress values, e.g. in the U.S.A. the maraging steel numbers refer to the nominal 0·2 per cent proof stress values in kilopounds/inch^2.

These steels have been designed to develop high proof stress with optimum toughness for the various strength levels. In contrast to conventional ultra-high-strength alloy steels in which carbon is an essential constituent and the formation of hard carbon-martensite is necessary for the development of high strength, nickel maraging steels have a very low carbon content and their high strengths are derived by age-hardening of relatively soft low-carbon martensite. In consequence, the toughness of the maraging steels is distinctly superior to that of conventional steels at the same strength levels, as shown for example by the comparison of notched tensile strengths of the several steels illustrated in Figure 1.

Because the physical metallurgy and properties of maraging steels are unique they have many commercial advantages which are summarized in Table 2 (*page 4*). Since the production of the first commercial heat in December 1960 the range of applications has steadily grown and is ever widening. A sample of typical applications is given in Table 3 (*page 4*).

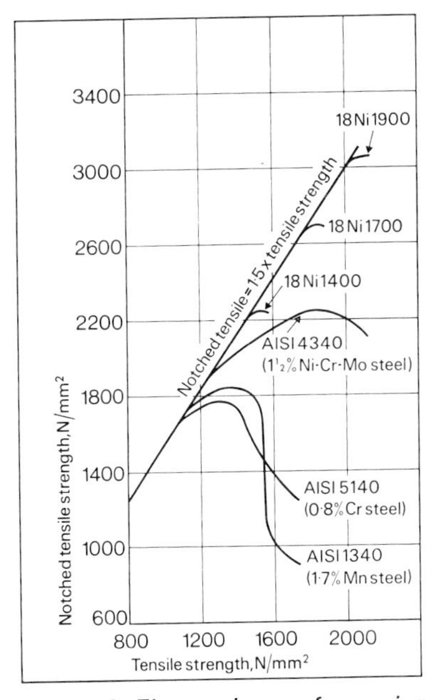

Figure 1. *The toughness of maraging steels is demonstrated by their adherence to the relationship Notched tensile strength/Tensile strength ≙ 1·5, up to higher strength levels than for conventional steels.*

Commercial and national specifications

The composition ranges developed by Inco to provide several combinations of properties are detailed in Table 4 (*page 4*). These have formed the basis for commercial production throughout the world with minor variations sometimes being adopted in the manufacture of proprietary designated grades and in some authoritative specifications. Tables 5 and 6 (*pages 5–9*) summarize the requirements of several commercial specifications used in various countries as national and international standards.

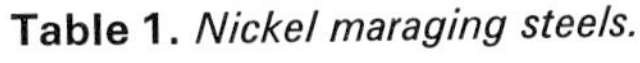

Table 1. *Nickel maraging steels.*

Type	Nominal 0·2% proof stress					Nominal composition, Weight %				
	N/mm²	10³lbf/in²	tonf/in²	kgf/mm²	hbar	Ni	Co	Mo	Ti	Al
18Ni1400	1400	200	90	140	140	18	8·5	3	0·2	0·1
18Ni1700	1700	250	110	175	170	18	8	5	0·4	0·1
18Ni1900	1900	280*	125	195	190	18	9	5	0·6	0·1
18Ni2400	2400	350	155	245	240	17·5	12·5	3·75	1·8	0·15
17Ni1600(cast)	1600	230	105	165	160	17	10	4·6	0·3	0·05

* *This steel is generally designated the 300 ksi grade in the U.S.A., the 0·2 per cent proof stress normally ranging from 260,000 to 300,000 lbf/in².*

Melting practice

Nickel maraging steels are generally produced by vacuum melting, or by a double melting and refining procedure involving both air and vacuum melting, while double vacuum melting is often employed. Whatever the process the objective is (i) to hold composition within the prescribed limits with close control over impurities, (ii) minimize segregation, (iii) obtain a low gas content and a high standard of cleanness. The degree to which these objectives are reached will influence the toughness and to some extent the strength of finished mill products.

Small amounts of impurities can decrease the toughness significantly. In particular sulphur should be kept as low as possible and silicon and manganese must not exceed a combined level of 0·2 per cent. The elements P, Pb, Bi, O_2, N_2 and H_2 are all maintained at low levels in good melting practice.

Ingot sizes and shapes and pouring practice should be selected to ensure sound ingots with minimum alloy segregation.

Source: *INCO Databook*, 1976

Table 2. *Advantages of nickel maraging steels.*

Excellent Mechanical Properties	Good Processing and Fabrication Characteristics	Simple Heat Treatment
1. High strength and high strength-to-weight ratio. 2. High notched strength. 3. Maintains high strength up to at least 350°C. 4. High impact toughness and plane strain fracture toughness.	1. Wrought grades are amenable to hot and cold deformation by most techniques. Work-hardening rates are low. 2. Excellent weldability, either in the annealed or aged conditions. Pre-heat not required. 3. Good machinability. 4. Good castability.	1. No quenching required. Softened and solution treated by air cooling from 820–900°C. 2. Hardened and strengthened by ageing at 450–500°C. 3. No decarburization effects. 4. Dimensional changes during age hardening are very small – possible to finish machine before hardening. 5. Can be surface hardened by nitriding.

Table 3. *Typical applications.*

Aerospace	Tooling and Machinery	Structural Engineering and Ordnance
Aircraft forgings (e.g. undercarriage parts, wing fittings). Solid-propellant missile cases. Jet-engine starter impellers. Aircraft arrestor hooks. Torque transmission shafts. Aircraft ejector release units.	Punches and die bolsters for cold forging. Extrusion press rams and mandrels. Aluminium die-casting and extrusion dies. Cold reducing mandrels in tube production. Zinc-base alloy die-casting dies. Machine components: gears index plates lead screws	Lightweight portable military bridges. Ordnance components. Fasteners.

Table 4. *Composition ranges – weight per cent – of the 18 per cent Ni-Co-Mo maraging steels.*(1)

	Wrought				Cast
Grade	18Ni1400	18Ni1700	18Ni1900	18Ni2400	17Ni1600
Nominal 0·2% proof stress:					
N/mm² (MPa)	1400	1700	1900	2400	1600
tonf/in²	90	110	125	155	105
10³lbf/in²	200	250	280(2)	350	230
kgf/mm²	140	175	195	245	165
hbar	140	170	190	240	160
Ni	17–19	17–19	18–19	17–18	16–17·5
Co	8·0–9·0	7·0–8·5	8·0–9·5	12–13	9·5–11·0
Mo	3·0–3·5	4·6–5·1	4·6–5·2	3·5–4·0	4·4–4·8
Ti	0·15–0·25	0·3–0·5	0·5–0·8	1·6–2·0	0·15–0·45
Al	0·05–0·15	0·05–0·15	0·05–0·15	0·1–0·2	0·02–0·10
C max.	0·03	0·03	0·03	0·01	0·03
Si max.	0·12	0·12	0·12	0·10	0·10
Mn max.	0·12	0·12	0·12	0·10	0·10
Si+Mn max.	0·20	0·20	0·20	0·20	0·20
S max.	0·010	0·010	0·010	0·005	0·010
P max.	0·010	0·010	0·010	0·005	0·010
Ca added	0·05	0·05	0·05	none	none
B added	0·003	0·003	0·003	none	none
Zr added	0·02	0·02	0·02	none	none
Fe	Balance	Balance	Balance	Balance	Balance

(1) *The composition ranges given are those originally developed by Inco which broadly cover current commercial practice. Slight changes in these ranges have been made in some national and international specifications.*

(2) *See footnote of Table 1.*

Castings

The 17Ni1600 grade developed for castings has slightly different composition compared with the wrought grades (see Tables 1 and 4) in order to minimize retention of austenite that might otherwise occur in more highly alloyed regions of the structure due to micro-segregation which tends to persist in the absence of hot working. This steel has good fluidity and pouring temperatures much higher than 1580°C are generally not desirable if segregation is to be minimized.

Mechanical properties of wrought maraging steels

The usual heat treatment applied to the 18Ni1400, 1700 and 1900 grades of maraging steels comprises solution annealing at 820°C followed by ageing 3 hours at 480°C. Similar annealing is applied to the 18Ni2400 steel, but ageing is effected by heating for 3 hours at 510°C or longer times at 480°C. The normally expected mechanical properties after these treatments are given in Table 7 (*page 10*) while Tables 8–11 (*pages 10–11*) present typical mechanical test data for material of various section sizes obtained from production casts.

Elastic and plastic strain characteristics

The uniaxial tensile deformation behaviour of maraging steels is shown by the stress-strain and true stress-true strain curves of Figs. 2 and 3 (*page 11*), respectively. Plastic yielding occurs at stresses of the order of 95 per cent of the ultimate tensile stress. The strain hardening moduli of maraging steels subjected to plastic strain are:

Steel Type	*Strain Hardening Modulus*
18Ni1400	724 N/mm²
18Ni1700	758 N/mm²
18Ni1900	793 N/mm²
18Ni2400	827 N/mm²

Fracture toughness

The ability of metallic alloys to resist rapid propagation of a crack or unstable fracture originating at an imperfection (i.e. brittle fracture), particularly in materials of high strength, is of great importance in determining their utility for engineering purposes. As with other high-strength alloys, various methods have been employed to evaluate the fracture characteristics of maraging steels with the results described in the sections on pages 10 and 12.

Table 5. *18 per cent nickel maraging steels. Chemical compositions quoted in national and international specifications.*

Country	Specifying body	Specification	Method of manufacture	Form of product	Composition, per cent Ni	Co	Mo	Ti	C max.	Other
United Kingdom	Ministry of Technology. Aerospace Material Specification.	DTD 5212 (Jan. 1969)	Double vacuum melted (induction +vacuum arc remelt)	Billets Bars Forgings	17·0–19·0	7·0–8·5	4·6–5·2	0·30–0·60	0·015	a, c d
		DTD 5232 (Aug. 1969)	Single vacuum melted (air melt +vacuum arc remelt)							
International	Association Internationale des Constructeurs de Material Aerospatiale	AICMA–FE–PA95 (provisional recommenda-tion Dec. 1965)	Vacuum melted or vacuum remelted	Bars Plates Forgings	17·0–19·0	7·5–8·5	4·6–5·2	0·30–0·50	0·03	a, b
Germany	Normenstelle Luftfahrt. (Aeronautical standards)	I.6359 (Nov. 1973)	Consumable electrode remelt	Sheet Plate Bars Forgings	17·0–19·0	7·0–8·5	4·6–5·2	0·30–0·60	0·03	a, c
		I.6354 (Nov. 1973)	Consumable electrode remelt	Bars Forgings	17·0–19·0	8·0–9·5	4·6–5·2	0·60–0·90	0·03	a, c
		I.6351 (Draft specification Aug. 1973)	Melted and cast under argon or vacuum* (Shaw or precision casting methods)	Precision castings	16·0–18·0	9·5–11·0	4·5–5·0	0·15–0·45	0·03	a, c
U.S.A.	A.S.T.M.	A538–72a: Grade A	Electric furnace air melt. Vacuum arc remelt. Air – or vacuum – induction melt	Plates	17·0–19·0	7·0–8·5	4·0–4·5	0·10–0·25	0·03	a, e
		Grade B			17·0–19·0	7·0–8·5	4·6–5·1	0·30–0·50	0·03	a, e
		Grade C			18·0–19·0	8·0–9·5	4·6–5·2	0·55 0·80	0·03	a, e
		A579–70: Grade 71	Electric arc air melt. Air – or vacuum – induction melt. Consumable electrode remelt or combination of these	Forgings	17·0–19·0	8·0–9·0	3·0–3·5	0·15–0·25	0·03	a, f
		Grade 72			17·0–19·0	7·5–8·5	4·6–5·2	0·30–0·50	0·03	a, f
		Grade 73			18·0–19·0	8·5–9·5	4·6–5·2	0·50–0·80	0·03	a, f
	S.A.E. Aerospace Material Specification	AMS 6512 (May 1970)	Vacuum arc remelt or consumable electrode remelt in air using vacuum induction melted electrodes	Bars Forgings Tubes Rings	17·0–19·0	7·0–8·5	4·6–5·2	0·30–0·50	0·03	a, e, g
		AMS 6514 (May 1970)			18·0–19·0	8·5–9·5	4·6–5·2	0·50–0·80	0·03	a, e, g
		AMS 6520 (May 1969)		Sheet Strip Plate	17·0–19·0	7·0–8·5	4·6–5·2	0·30–0·50	0·03	a, e, g
		AMS 6521 (May 1969)			18·0–19·0	8·5–9·5	4·6–5·2	0·50–0·80	0·03	a, e, g

** Casting in air may be adopted if agreed between supplier and purchaser.*
a. Si 0·10 max, Mn 0·10 max, S 0·010 max, P 0·010 max, Al 0·05–0·15.
b. B 0·003 max, Zr 0·02 max.
c. Ca, B and Zr may be added in amounts of 0·05 per cent max, 0·003 per cent max, and 0.02 per cent max, respectively.
d. Cr 0·25 max (DTD 5212) and 0·20 max (DTD 5232).
e. The following specified additions shall be made to the melt: B 0·003 per cent, Zr 0·02 per cent and Ca 0·05 per cent.
f. the following specified additions shall be made to the melt: B 0·003 per cent, Zr 0·02 per cent and Ca 0·06 per cent.
g. Cr 0·5 max, Cu 0·50 max.

Source: *INCO Databook*, 1976

Table 6. *18 per cent nickel maraging steels. Mechanical properties quoted in the specifications listed in Table 5. (Minimum values except as stated otherwise)*

Specification	Form of product	Heat treatment condition	Hardness	0·2% proof stress				Tensile strength				Elongation %		Reduction of area %		Impact energy value				Radius for 180° bend test
				Longit.		Transv.		Longit.		Transv.		Longit.	Transv.	Longit.	Transv.	Test-piece	J	ft lbf	kgf/cm²	
				N/mm² (MPa)	† h bar	N/mm² (MPa)	† h bar	N/mm² (MPa)	† h bar	N/mm² (MPa)	† h bar	Lo = 5·65√So								
DTD5212	Billets	S.A. 810°–830°C	≤321HB ≤335HV	—	—	—	—	—	—	—	—	—	—	—	—	—	—	—	—	—
	Bars Forgings	S.A. 810°–830°C+ ≥3h 475°–485°C	520–620HV	1700	170	1700	170	1800–2000	180–200	1800–2000	180–200	8	5	40	25	Izod		†		—
																Longit. Transv.	24 11	18 8	— —	
																KCU	J/cm²			
																Longit. Transv.	39 20	— —	4(a) 2(a)	
DTD5232	Billets	S.A. 810°–830°C	≤321HB ≤335HV	—	—	—	—	—	—	—	—	—	—	—	—	—	—	—	—	—
	Bars Forgings	S.A. 810°–830°C+ ≥3h 475°–485°C	520–620HV	1700	170	1700	170	1800–2000	180–200	1800–2000	180–200	6	4	35	20	Izod		†		—
																Longit. Transv.	16 8	12 6	— —	
																KCU	J/cm²			
																Longit. Transv.	29 20	— —	3(a) 2(a)	
					† kgf/mm²				† kgf/mm²								J/cm²			
AICMA-FE-PA95	Bar test-coupon, 16 mm dia.	S.A. 810°–830°C+ 3h 475°–485°C	—	1620	165	—	—	1720–1910	175–195	—	—	8	—	45	—	KCU Longit.	39	—	† 4	—
	Bars	S.A. 810°–830°C (as-supplied)	331HB max.	—	—	—	—	—	—	—	—	—	—	—	—	—	—	—	—	—
	Plates		350HB max.																	
	Forgings		—																	
	Bar ≤100 mm dia. Forgings	S.A. 810°–830°C+ 3h 475°–485°C	—	1620	165	—	—	1720–1910	175–195	—	—	6	4	40	25	KCU Longit.	20	—	† 4	—
	Plate ≤10 mm thickness	S.A. 810°–830°C+ 3h 475°–485°C	—	1620	165	—	—	1720–1910	175–195	—	—	5	—	—	—	—	—	—	—	≤3× thickness for plate ≤8 mm thick

				† N/mm² (MPa)		† N/mm² (MPa)		† N/mm² (MPa)		† N/mm² (MPa)										
Normenstelle Luftfahrt 1.6359: Werkstoff Nr. I.6359.9	Sheet 0·5–6·0mm thickness Plate >6 ≤10mm thickness (cold- or hot-rolled, polished or descaled)	S.A. 810°–830°C, air-cooled	≤350HV	—	—	—	—	—	—	—	—	—	—	—	—	—	—	—	—	3 × thickness for sheet or plate ≤8 mm thick (transv. test)
Werkstoff Nr. I.6359.4	Sheet and plate:												Lo= 50 mm (f)							
	2–3 mm thick	S.A. and aged 3h 470°–490°C air-cooled	(480 HV)	—	—	1620	—	—	—	1720	—	—	3	—	—	—	—	—	—	—
	>3 ≤6 mm												4	—	—	—	—	—	—	—
	>6 ≤10 mm												5	—	—	—	—	—	—	—
Werkstoff Nr. I.6359.9	Bar 2–100 mm dia. Forgings ≤70 mm thickness	S.A. 810°–830°C, air-cooled	≤353HB	—	—	—	—	—	—	—	—	—	—	—	—	—	—	—	—	—
Werkstoff Nr. I.6359.4												Lo=5·65√So					† J			
	Bar 2–100 mm dia. Forgings ≤70 mm thickness	S.A. and aged 3h 470°–490°C, air-cooled	(480 HV)	1620	—	1620	—	1720	—	1720	—	6	4	40	25	DVM	30 Longit. 20 Transv.	— —	— —	—
Normenstelle Luftfahrt I.6354: Werkstoff Nr. I.6354.9				† N/mm² (MPa)		† N/mm² (MPa)		† N/mm² (MPa)		† N/mm² (MPa)		Lo=5·65√So								
	Bar 2–100 mm dia. Forgings ≤70 mm thickness	S.A. 810°–830°C, air-cooled	≤353HB	—	—	—	—	—	—	—	—	—	—	—	—	—	—	—	—	—
Werkstoff Nr. I.6354.4																	† J			
	Bar 2–100 mm dia. Forgings ≤70 mm thickness	S.A. and aged 3–5h 480°–500°C, air-cooled	(542 HV)	1910	—	1910	—	1960	—	1960	—	4·5	3·5	30	20	DVM	20 Longit. 12 Transv.	— —	— —	—

Table continued

Source: *INCO Databook*, 1976

Table 6. *continued*

Specification	Form of product	Heat treatment condition	Hardness	0·2% proof stress Longit.		Transv.		Tensile strength Longit.		Transv.		Elongation % Longit.	Transv.	Reduction of area % Longit.	Transv.	Impact energy value Test-piece	J	ft lbf	kgf/cm²	Radius for 180° bend test
				Test-piece obtained from cast test coupon																
				N/mm²† (MPa)				N/mm²† (MPa)				Lo=5·65√So								
Normenstelle Luftfahrt I.6531: Werkstoff Nr. I.6531.9	Un-machined castings ≤50 mm section thickness	Homogenized 1150°±15°C, air-cooled	(≤32 HRC)	(600)				(900)				(8)								
																Charpy V-notch				
																	J			
Werkstoff Nr. I.6531.4	Finished cast parts ≤25 mm	Homogenized and aged 3h 475°–495°C, air-cooled	(52HRC)	1450				1600				4		10			(16)			
	>25 ≤50 mm section thickness			1450				1600				3		8			(14)			
ASTM A538–72a:				† N/mm² (MPa)	† 10³ lbf/in²	† N/mm² (MPa)	† 10³ lbf/in²	† N/mm² (MPa)	† 10³ lbf/in²	† N/mm² (MPa)	† 10³ lbf/in²	Lo=2 in. or 50 mm								
Grade A	Plate	S.A. 815°–982°C(b) + 3h 468°–500°C	—	1380–1620	200–235	1380–1620	200–235	1450	210	1450	210	8	8	40(c) 35(d)	40(c) 35(d)	—	—	—	—	—
Grade B	Plate		—	1580–1790	230–260	1580–1790	230–260	1650	240	1650	240	6	6	35(c) 30(d)	35(c) 30(d)	—	—	—	—	—
Grade C	Plate		—	1900–2100	275–305	1900–2100	275–305	1930	280	1930	280	6	6	30(c) 25(d)	30(c) 25(d)	—	—	—	—	—
												Lo=4·5√So				Charpy V-notch				
ASTM A579–70(e):																	† J	† ftlbf		
Grade 71	Forgings	S.A. and aged	—	1380	200	—	—	1450	210	—	—	12	—	55	—		48	35	—	—
Grade 72	Forgings		—	1725	250	—	—	1760	255	—	—	10	—	45	—		27	20	—	—
Grade 73	Forgings		—	1800	275	—	—	1930	280	—	—	9	—	40	—		20	15	—	—

SAE AMS 6512	Bars Forgings Tubes Rings	S.A. 816°–927°C	<34HRC <321HB†	—	—	—	—	—	—	—	—	—	—	—	—	—	—	—	—	—
	Cross-section thickness:				†		†		†		†					Charpy V-notch	J	† ftlbf		
	<2·5 in. (64 mm)	S.A. 816°–927°C+ 3–6h 477°–488°C	>52HRC† >560HV	1860	270	—	—	1930	280	—	—	5	—	30	—	Longit.	14	10	—	—
	2·5–4·0 in. (64–102 mm)			1860	270	1860	270	1930	280	1930	280	5	4	30	25	Longit. Transv.	11 8	8 6	— —	—
	4·0–10·0 in. (102–254 mm)			1860	270	1860	270	1900	275	1900	275	4	2	25	20	Longit. Transv.	8 5	6 4	— —	—
	>10·0 in. (254 mm)			to be agreed between purchaser and vendor.																
SAE AMS 6520	Sheet Strip Plate	S.A. 802°–830°C	<34HRC† <321HB	—	† —	—	† —	—	† —	—	† —	—	—	—	—	—	—	—	—	—
	≤9 in. wide (229 mm)	S.A. 802°–830°C+ 3–5h 477°–488°C	>48HRC† >500HV	1690	245	—	—	1760	255	—	—	(g)	—	—	—	Fracture toughness values are negotiated between purchaser and vendor. The ASTM sharp-notch test for high strength sheet materials is suggested				—
	>9 in. wide (229 mm)			—	—	1690	245	—	—	1760	255	—	(g)	—	—					—
SAE AMS 6521	Sheet Strip Plate	S.A. 802°–830°C	<34HRC† <321HB	—	—	—	—	—	—	—	—	—	—	—	—	—	—	—	—	—
	≤9 in. wide (229 mm)	S.A. 802°–830°C+ 3–5h 477°–488°C	>50HRC† >530HV	1860	270	—	—	1930	280	—	—	(g)	—	—	—	Fracture toughness values are negotiated as stated above				—
	<9 in. wide (229 mm)			—	—	1860	270	—	—	1930	280	—	(g)	—	—					—

† *Specified value. Other values have been obtained by conversion.*

S.A. Solution annealed.

( ) *Figures in parentheses are approximate values to be expected.*

$\overline{(a)}$ *The KCU test is not a mandatory requirement of the specification, but impact values which may be expected are quoted in the specification for guidance.*

(*b*) *A double solution anneal is permissible. It is recommended that the lowest temperature is used, within the range specified, which will effect recrystallization at the mid-thickness position.*

(*c*) *Round test-piece. For plate thickness >0·75 in (19 mm) a standard 0·5 in (12·7 mm) diameter test-piece shall be used. For plate thickness ≤0·75 in (19 mm) sub-size round test-pieces or rectangular test-pieces may be used.*

(*e*) *Vacuum melting is normally required to achieve the listed properties. The indicated 0·2 per cent proof stress values can usually be achieved at a depth = ¼ thickness in section sizes up to 12 in (305 mm) in the direction of maximum hot-working. Because of variations in forging configuration and processing it does not follow that the ductility and impact strengths listed can always be obtained at these depths.*

(*f*) *Lo = 5·65√So for plate.*

(*g*) *The minimum elongation value specified varies according to the thickness of the product and the test-piece gauge length as shown in the table below:*

Product thickness: inch† mm	<0·030 <0·8	0·030–0·045 0·8 –1·1	>0·045–0·065 >1·1 –1·6	>0·065–0·090 >1·6 –2·3	>0·090–0·125 >2·3 –3·2	>0·125–0·250 >3·2 –6·4	>0·250–0·375 >6·4 –9·5	>0·375 >9·5
	Minimum elongation, per cent							
Test-piece gauge length: 2 in. (51 mm) or 4D	—	—	—	2·5	3·0	4·0	5·0	6·0
1 in. (25 mm)	—	—	2·0	5·0	6·0	8·0	—	—
0·5 in. (13 mm)	1·0	2·0	—	—	—	—	—	—

Source: *INCO Databook*, 1976

Table 7. *Mechanical properties of wrought 18 per cent nickel maraging steels.*

Property	18Ni1400	18Ni1700	18Ni1900	18Ni2400
	Solution-annealed 1h 820°C			
0·2% proof stress, N/mm²	800	800	790	830
Tensile strength, N/mm²	1000	1010	1010	1150
Elongation, Lo=4·5$\sqrt{So}$., %	17	19	17	18
Reduction of Area, %	79	72	76	70
Hardness, HRC	27	29	32	35
	Solution-annealed 1h 820°C, aged 3h 480°C			Solution-annealed 1h 820°C, aged 12h 480°C
	Normal ranges of properties			Typical properties
0·2% proof stress, N/mm²	1310–1550	1650–1830	1790–2070	2390
Tensile strength, N/mm²	1340–1590	1690–1860	1830–2100	2460
Elongation, Lo=4·5$\sqrt{So}$., %	6–12	6–10	5–10	8
Reduction of Area, %	35–67	35–60	30–50	36
Modulus of Elasticity (E), GN/m²	181	186	190	191–199
Modulus of Rigidity (G), (torsional shear), GN/m²	—	71·4	—	74·5
Hardness, HRC	44–48	48–50	51–55	56–59
Charpy V-notch impact value, J (daJ/cm²)	35–68 (4·4–8·5)	24–45 (3·0–5·6)	16–26 (2·0–3·3)	11 (1·4)
Poissons ratio	0·264	0·30	0·30	0·26
Notched tensile strength (Kt=10) N/mm²	2390	2350–2650	2700–3000	1430
(Kt=3·5) N/mm²	—	—	—	2700

Table 8. *Typical mechanical properties of double vacuum melted 18Ni1400 maraging steel.*
(Solution-annealed and aged 3h 480°C, except as stated otherwise).

Section size	Direction of test	0·2% proof stress N/mm²	Tensile strength N/mm²	Elong. Lo=5·65$\sqrt{So}$ %	R of A %	Charpy V-notch impact value J	Charpy V-notch impact value daJ/cm²
150×150mm	Longit.	1450	1480	12	65	81	10·1
150×150mm*	Longit.	1510	1540	12	64	91	11·4
100×100mm	Longit.	1420	1470	12	68	73	9·1
100×100mm*	Longit.	1450	1510	13	67	92	11·5
75×75mm	Longit.	1420	1450	13	70	115	14·4
150×150mm	Transv.	1390	1470	10	55	38	4·8
150×150mm*	Transv.	1450	1510	10	55	35	4·4

** Aged 10h 460/465°C (Data by courtesy of Firth Brown Limited, Sheffield, U.K.)*

Table 9. *Typical mechanical properties of double vacuum melted 18Ni1700 maraging steel.*
(Solution-annealed and aged 3h 480°C).

Section size	Direction of test	0·2% proof stress N/mm²	Tensile strength N/mm²	Elong. Lo=5·65$\sqrt{So}$ %	R of A %	Izod impact value J
300×300mm	Transv.	1730	1860	7	35	20
125mm dia.	Transv.	1750	1860	7	35	20
100mm dia.	Transv.	1780	1860	6	30	16
75mm dia.	Transv.	1800	1860	5	30	14
100mm dia.	Longit.	1820	1980	9	50	27
75mm dia.	Longit.	1840	1960	9	50	41
50mm dia.	Longit.	1730	1860	10	55	41
25mm dia.	Longit.	1730	2000	12	55	47
13mm dia.	Longit.	1800	1960	12	55	47

(Data by courtesy of Firth Brown Limited, Sheffield, U.K.)

Impact transition temperature

The temperature range of transition from ductile to brittle fracture in notched impact tests conducted at various temperatures, and the corresponding change in impact energy absorption values, provides a comparative assessment of the fracture behaviour of various materials. In the case of maraging steels the fall in energy absorption values with decreasing temperature is small and gradual with useful toughness being retained at low temperatures, as shown in Figure 4. This lack of an abrupt transition in impact energy absorption with fall of temperature is a significant measure of the relatively high resistance of these steels to unstable fracture propagation.

Notched tensile properties

18 per cent Ni maraging steels exhibit high ratios of notched tensile strength to tensile strength, these being unobtainable in conventional steels at the same high strength levels. Figure 5 presents notched tensile data for bar and sheet. In general, longitudinal tests on bar material of the 18Ni1400, 1700 and 1900 grades give NTS/UTS ratios of about 1·5. Deviations below that value shown for the 18Ni1700 and 18Ni1900 steels are, in general, from the transverse direction of large forgings. The values for sheet follow a ratio of about 1·0, although lower ratios may be obtained depending on processing variables. Notched tensile strengths of sheet at temperatures as low as −196°C are only 10–20 per cent lower than the room-temperature values. This retention of high notched tensile strength at low

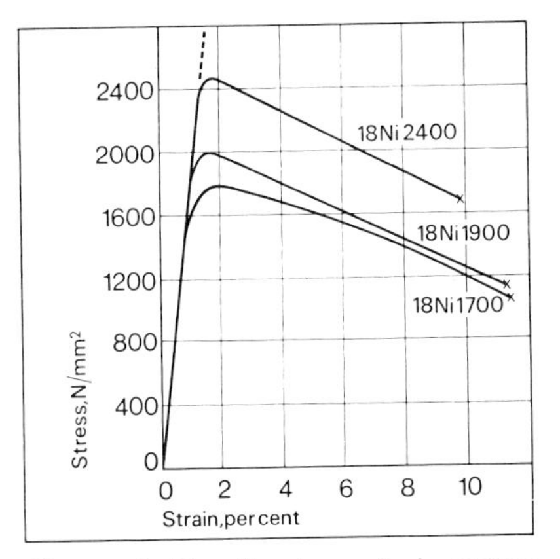

Figure 2. *Tensile stress-strain curves of maraged steels.*

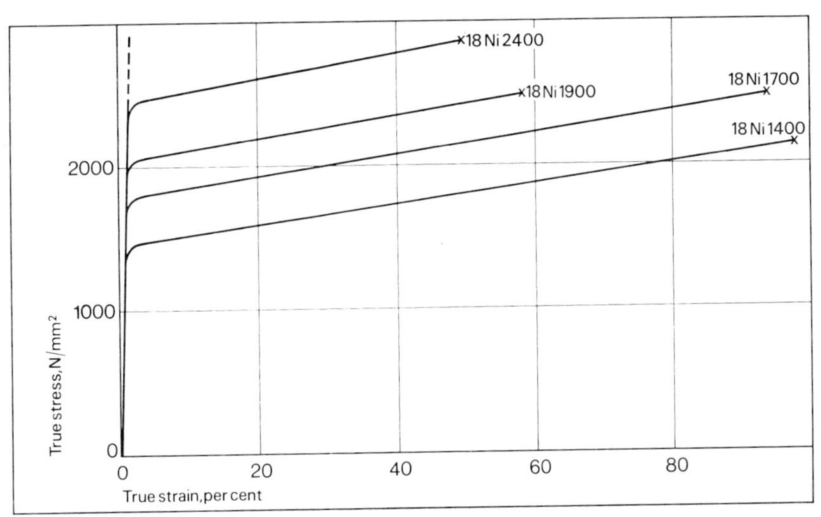

Figure 3. *True stress-true strain curves of maraged steels.*

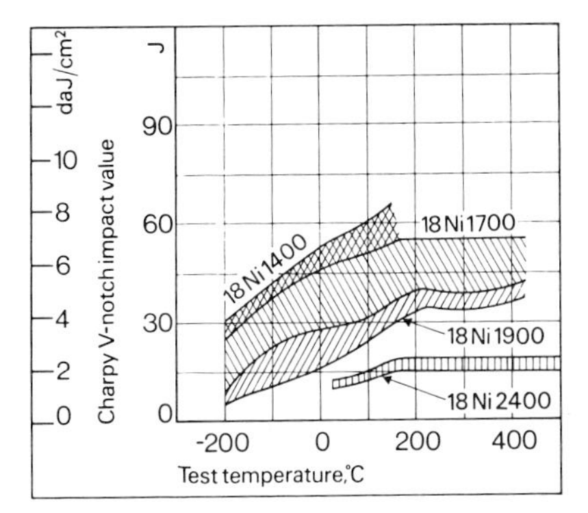

Figure 4. *Charpy V-notch impact value of 18% nickel maraging steels as a function of test temperature.*

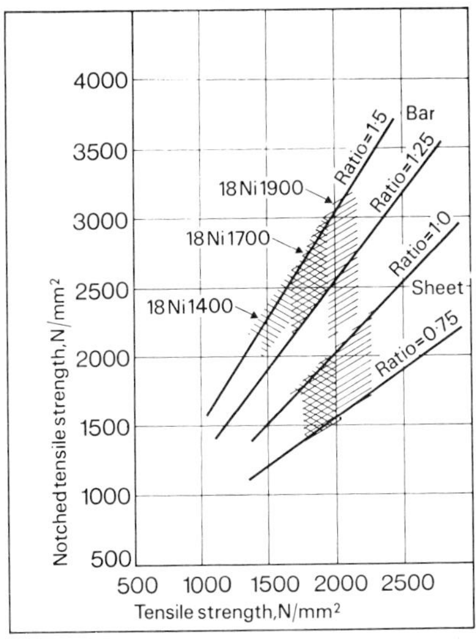

Figure 5. *Notched tensile/tensile strength ratios of maraging steels in bar and sheet form.*

Table 10. *Typical mechanical properties of double vacuum melted 18Ni1900 maraging steel. (Solution-annealed and aged 3h 480°C, except as stated otherwise).*

Section size	Direction of test	0·2% proof stress N/mm²	Tensile strength N/mm²	Elong. Lo= $5{\cdot}65\sqrt{So}$ %	R of A %	Izod impact value J
200mm dia.	Transv.	1930	2030	4	25	9·5
125mm dia.	Transv.	1930	2020	4	20	11
115mm dia.	Transv.	2080	2140	6	25	16
280×100mm	Transv.	1910	1970	5	25	—
250×57mm*	Transv.	1970	2000	8	35	—
190×38mm	Transv.	2020	2120	4	25	13·5
200×19mm	Transv.	2020	2140	6	35	16
70×70mm	Transv.	1970	2020	4	25	—
83mm dia. ×44mm thick	Tangential	1970	2020	9	50	16
280×100mm	Longit.	1950	2010	10	45	—
60×60mm	Longit.	2000	2040	8	55	—
19mm dia.	Longit.	2010	2170	8	50	27

** Aged 3h 500°C. (Data by courtesy of Firth Brown Limited, Sheffield, U.K.)*

Table 11. *Typical mechanical properties of double vacuum melted 18Ni2400 maraging steel. (Solution-annealed 1h 820°C, air cooled, and aged 6h 500°C)*

Section size	Direction of test	0·2% proof stress N/mm²	Tensile strength N/mm²	Elong. Lo= $5{\cdot}65\sqrt{So}$ %	R of A %	Charpy V-notch impact value J	daJ/cm²
250×250mm	Longit.	2390	2470	6	31	5	0·6
250mm dia.	Longit.	2390	2490	5	24	—	—
105×105mm	Longit.	2380	2440	7	34	8	1·0
55×55mm	Longit.	2410	2490	8	54	14	1·8
20mm dia.	Longit.	2390	2490	8	51	11	1·4
Tube	Longit.	2390	2470	6	40	7	0·9
150mm dia.	Transv.	2290	2390	3	17	8	1·0
105×105mm	Transv.	2410	2470	4	19	5	0·6
55×55mm	Transv.	2390	2470	6	34	11	1·4
100×25mm	Transv.	2380	2460	4	34	5	0·6

(Data by courtesy of Firth Brown Limited, Sheffield, U.K.)

Source: *INCO Databook*, 1976

temperatures is unique at the high strength levels of the maraging steels.

The 18Ni2400 steel exhibits somewhat lower ratios of NTS/UTS than the lower-strength grades as shown in Figure 6.

Fracture mechanics

Metallic components with limited toughness can support tensile loads up to the yield stress if they are sufficiently flawless. Cracks or harmful discontinuities of sufficient size must pre-exist, or develop, for unstable fracture to occur at a stress below the tensile yield (0·2 per cent proof) stress. Fracture mechanics provides a means of determining the quantitative relationship between crack size, the elastic stress field surrounding the crack and the fracture properties of the material.

The controlling parameter in representing the combined effect of crack dimensions and stress field at the leading edge of a crack is K, the stress field intensity. In crack propagation, as the stress increases, the crack will first grow to a critical size determined by the toughness of the material before rapid crack propagation or unstable fracture is initiated. The stress intensity factor at the onset of rapid crack propagation is taken as a measure of fracture toughness and is designated K_c, the 'plane stress fracture toughness'.

Under conditions where the material is relatively ductile, K_c is the controlling factor in determining toughness and a shear type of fracture is obtained. For a given material, K_c varies with thickness and temperature. K_c tends toward a minimum value as the thickness increases and temperature decreases. This minimum value is called the 'plane strain fracture toughness', K_{Ic}, and is associated with a brittle mode of rapid crack propagation.

The value of K_{Ic} is of great practical importance in evaluating the toughness of high-strength materials. When the stress intensity factor reaches the value of K_{Ic}, slow crack growth is followed by rapid crack growth and brittle fracture. Of further practical importance is the use of K_{Ic} to calculate the critical dimensions of a crack, or flaw, below which the steel may be used without danger of the crack initiating catastrophic failure. Thus it is useful in determining an allowable design stress to prevent propagation of the minimum flaw detectible by inspection.

Representative K_{Ic} values of commercial heats of 18Ni1700, 1900 and 2400 maraging steels in the solution-treated and aged condition are presented in Table 12. Similar data for cold-rolled and aged sheet of 18Ni2400 are given in Table 13.

Typical K_{Ic} values for weldments in 18 per cent Ni maraging steels are given in a later section.

Table 12. *Representative plane strain fracture toughness values (K_{1c}) of double vacuum melted commercial heats of 18Ni1700, 18Ni1900 and 18Ni2400 maraging steels.*
(3-point bend tests on 127mm × 25mm × 12.7mm specimens).

Material	Product section size	Direction of test	Tensile strength	K_{1c}
			N/mm²	$MNm^{-3/2}$
18Ni1700	127×127mm	Longit.	1820	101·5
	127×127mm	Transv.	1800	93·1
	200×200mm	Transv.	1845	89·9
	250×250mm	Transv.	1860	89·8
18Ni1900	130×19mm	Longit.	2020	68·7
	130×19mm	Transv.	2020	66·5
18Ni2400	105×105mm	Longit.	2440	33

(Data by courtesy of Firth Brown Limited, Sheffield, U.K.)

Table 13. *Plane strain fracture toughness values (K_{1c}) of 18Ni2400 maraging steel sheet, cold rolled and aged 3h 480°C.*
(Transverse tensile tests)

Sheet thickness mm	Cold rolling reduction in thickness %	0·2% proof stress N/mm²	Tensile strength N/mm²	Elong. %	K_{1c} $MNm^{-3/2}$
3·7	25	2540	2570	2·5	64
3·0	45	2510–2570	2530–2610	3	58
2·0	60	2590	2600	2	59
1·24	75	2700	2740–2790	2	54

Fatigue properties

Conflicting data from fatigue tests on maraging steels have been reported. This is largely attributable to variations in the procedures used for the production of the test-pieces. The conventional preparation generally applied to high-strength alloy steels, of machining and polishing test-pieces from fully heat-treated material, has sometimes been used, while in other instances the finished test-pieces have been prepared from solution-annealed maraging steel with ageing applied as the final operation. Thus wide ranges of fatigue properties have been obtained. Examples of the effects of test-piece preparation are shown by the fatigue curves in Figures 7–11. These curves also show the effects of notch stress concentration factor and of testing method, viz. rotating bend, tension-tension, and tension-zero stress on given heats of maraging steel. Figures 12–16 summarize comprehensive data on the effects of these factors based on tests of many heats.

The highest fatigue strengths are obtained in shot-peened material of high purity and, where fatigue is an important criterion, the use of vacuum-refined material and application of shot peening to the finished component is recommended.

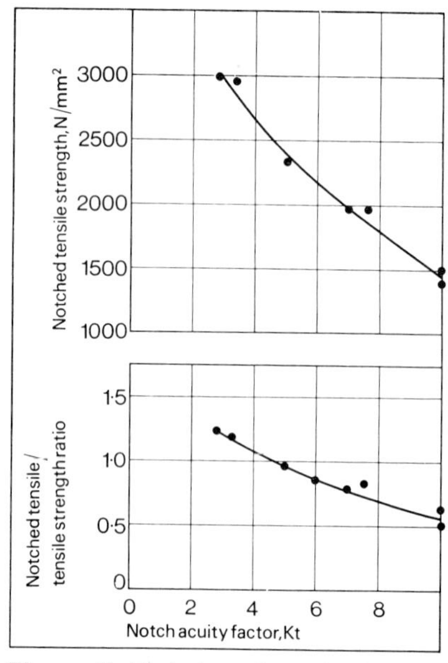

Figure 6. *Variation of notched tensile strength of 18Ni2400 maraging steel and notched tensile/tensile strength ratio, with notch acuity factor, K_t.*

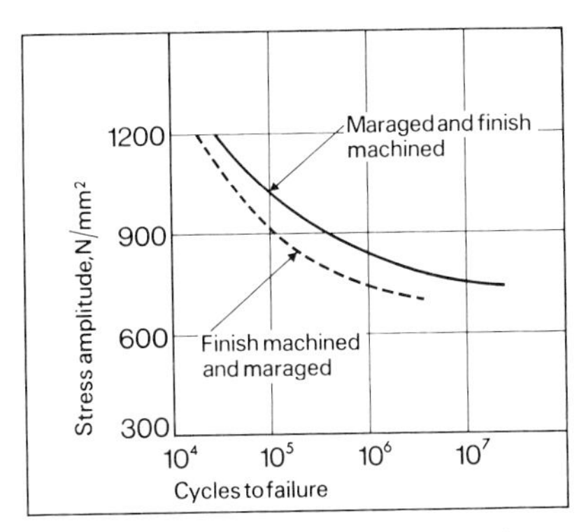

Figure 7. *Smooth-bar rotating-beam fatigue curves of 18Ni400 maraging steel.*

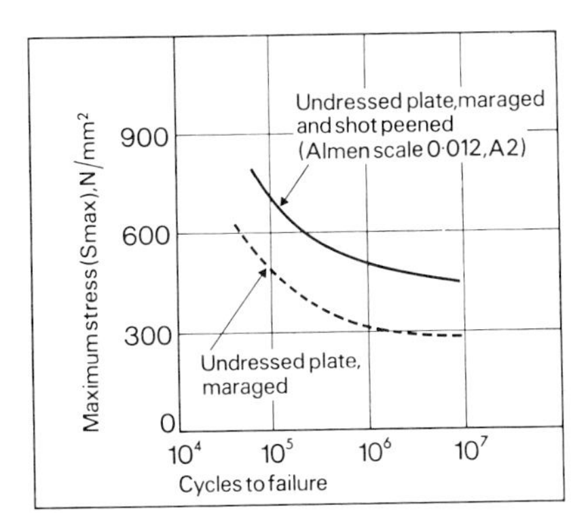

Figure 8. *Tension-zero stress* $(R = \frac{S\ min.}{S\ max.} = 0)$ *fatigue curves of 18Ni400 maraging steel plate, 5mm thick.*

*(After Brine F. E., Webber D. and Baron H. G., British Welding Jnl., 1968, **15**, (11), 541–546).*

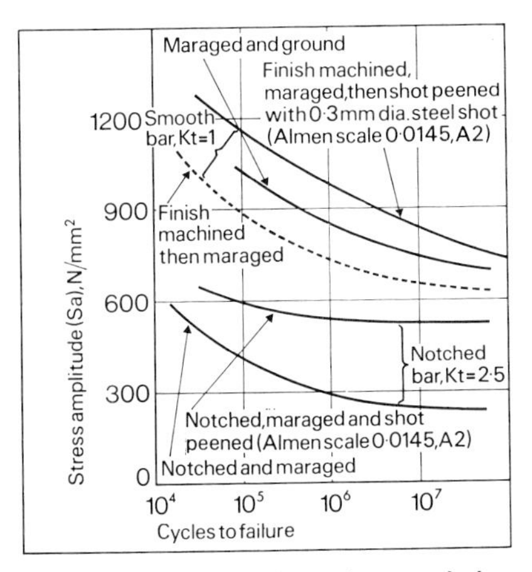

Figure 9. *Notched- and smooth-bar rotating-beam fatigue curves of 18Ni700 maraging steel. (K_t=Notch stress concentration factor).*

(Data supplied by W. M. Imrie, Dowty Rotol Limited, Gloucester, U.K.).

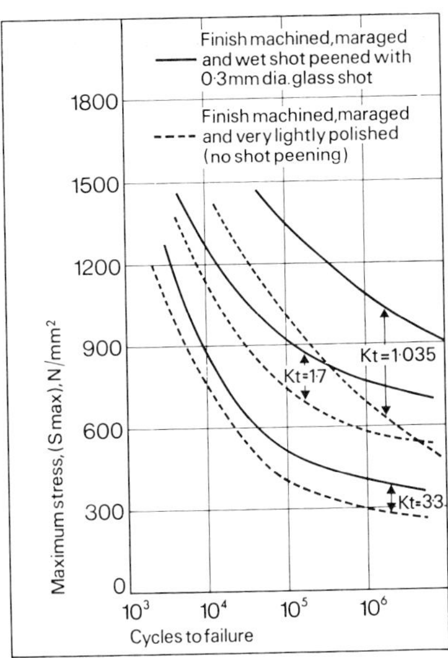

Figure 10. *Tension – tension* $(R = \frac{S\ min.}{S\ max.} = 0{\cdot}1)$ *notched-bar fatigue curves of 18Ni1700 maraging steel. (K_t=Notch stress concentration factor).*

(After Souffrant M. P. 'Perspectives d'utilisation du maraging en grosses pieces forgees', Conference on Les Aciers Speciaux au Service de l'Aviation, June 1967, Aéroport du Bourget).

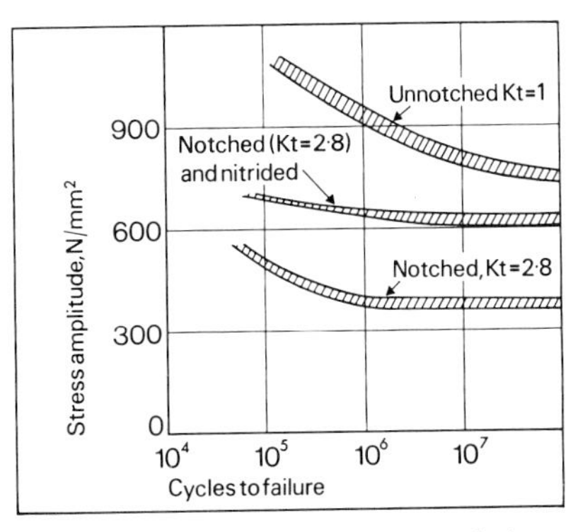

Figure 11. *Rotating-beam fatigue curves of 18Ni2400 maraging steel. Nitriding the surface of maraging steels improves the resistance to fatigue.*

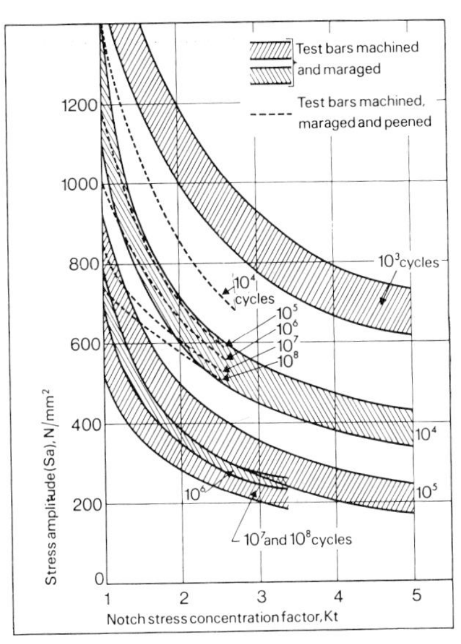

Figure 12. *Effect of notch stress concentration factor on the fatigue strengths of 18Ni1700 and 18Ni1900 maraging steels for various cycles to failure.*

$$R = \frac{Smean - Sa}{Smean + Sa} = -1$$

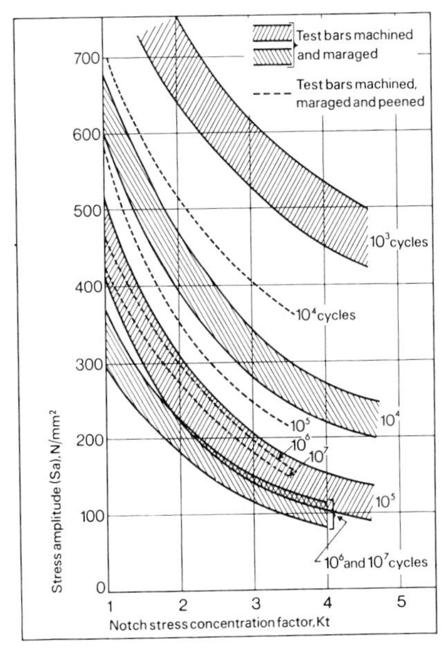

Figure 13. *Effect of notch stress concentration factor on the fatigue strengths of 18Ni1700 and 18Ni1900 maraging steels for various cycles to failure.*

$$R = \frac{Smean - Sa}{Smean + Sa} = 0 \text{ to } 0{\cdot}1$$

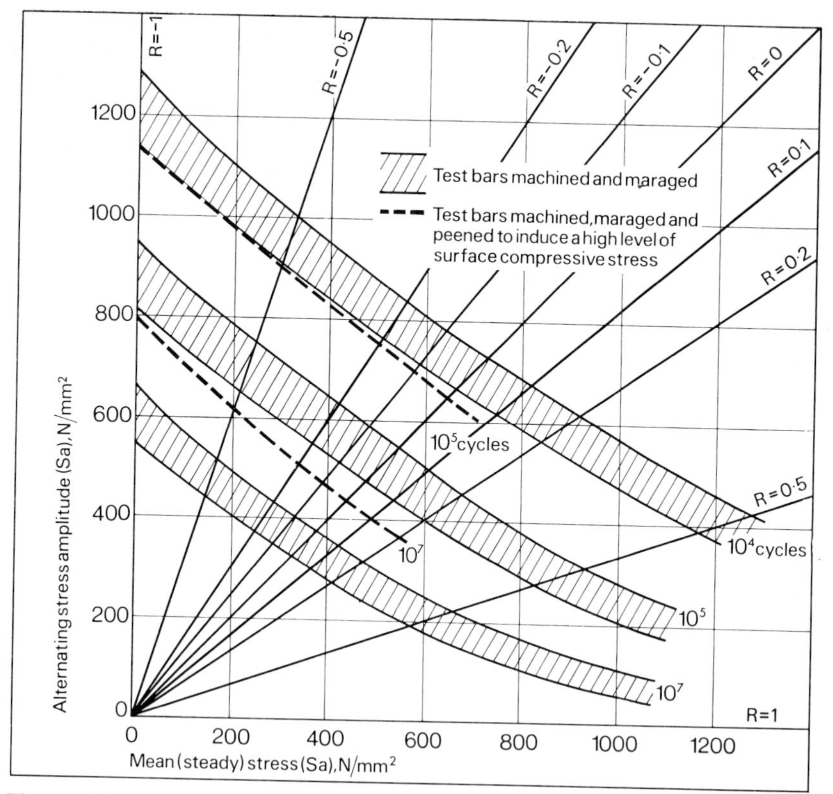

Figure 14. *Fatigue strengths of 18Ni1700 and 18Ni1900 maraging steels for various cycles to failure in smooth-bar tests, $K_t=1{\cdot}0$, under alternating stress + steady stress conditions,* $R=\dfrac{Sm-Sa}{Sm+Sa}$

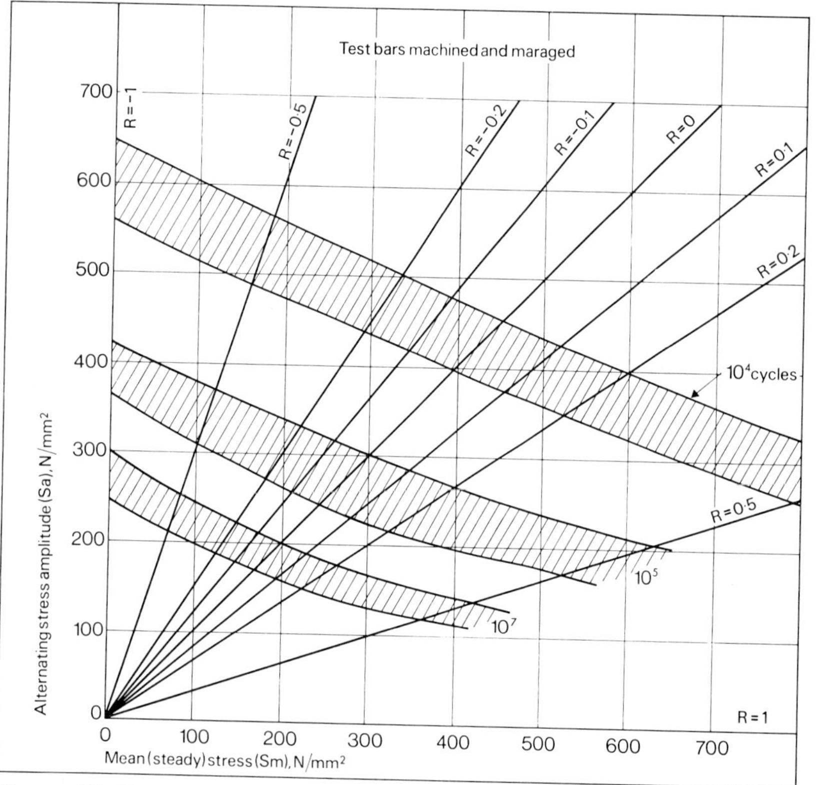

Figure 15. *Fatigue strengths of 18Ni1700 and 1900 maraging steels for various cycles to failure in notched-bar tests, notch stress concentration factor $K_t=2{\cdot}4$–$2{\cdot}5$, under alternating stress + steady stress conditions,* $R=\dfrac{Sm-Sa}{Sm+Sa}$

Properties at elevated temperatures

The effects of temperature on the mechanical properties of the various grades of maraging steel are presented in Figures 17–22 and in Tables 14 and 15.

The dynamic modulus of elasticity values for the 18Ni1400 grade tend to be on the low side of the band shown in Figure 18 and those for the higher strength grades at the high side of that band. The vertical line at 20°C in Figure 18 shows the variation in modulus of elasticity according to the direction of testing in 4 mm thick plate which has been cold rolled 60 per cent followed by ageing. The modulus increases from about 180 to 213 GN/m² as the angle to the rolling direction is varied between 0° and 90°. This variation indicates that some preferred crystal orientation exists in the plate. In most applications it is not expected to be of engineering significance.

The effects of long-time exposure at elevated temperatures on tensile and impact properties are shown in Tables 16 and 17 (*page 16*), while stress-rupture and creep properties are presented in Figure 23 (*page 17*) as a Larson-Miller plot. The latter is based on limited data currently available and should be considered as tentative.

Mechanical properties of cast maraging steel (17Ni1600)

The cast grade of nickel maraging steel is normally solution annealed by homogenizing for 4 hours at 1150°C, followed by air cooling. Maraging of the solution-annealed material is effected by heating for 3 hours at 480°C. Typical mechanical properties after these treatments are presented in Table 18 (*page 16*). That table also shows that use of a double solution-annealing treatment at 1150°C and 820°C, with an intermediate reheating at 595°C, provides a significant improvement in toughness without impairment of tensile properties. Solution annealing at 820°C without prior homogenizing generally results in appreciable loss of ductility.

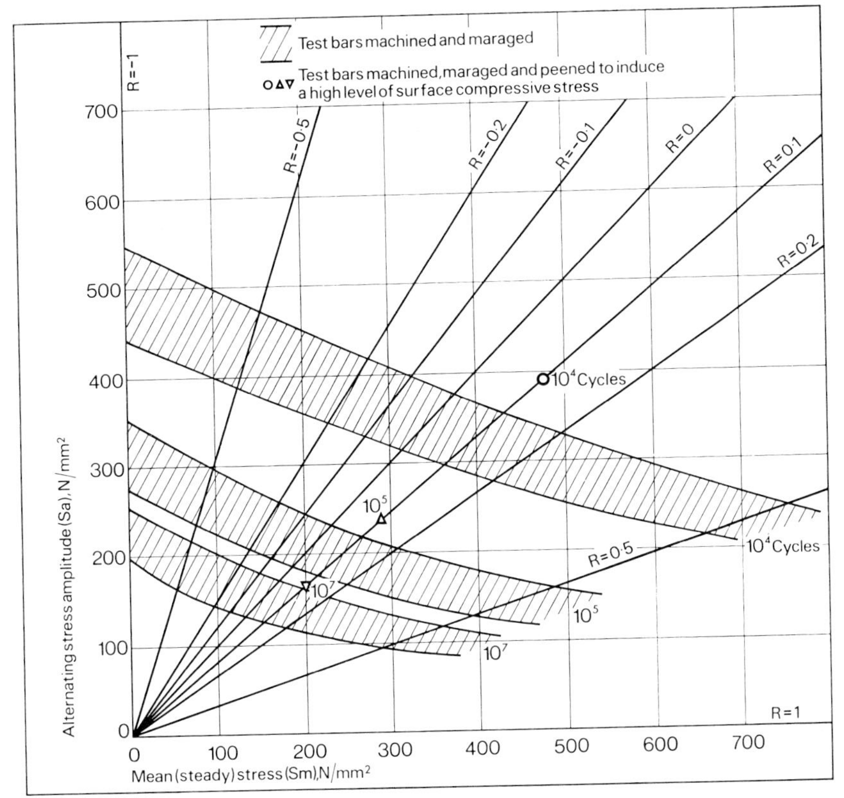

Figure 16. *Fatigue strengths of 18Ni1700 and 1900 maraging steels for various cycles to failure in notched-bar tests, notch stress concentration factor K_t=3·2–3·3, under alternating stress + steady stress conditions,* $R = \frac{Sm - Sa}{Sm + Sa}$

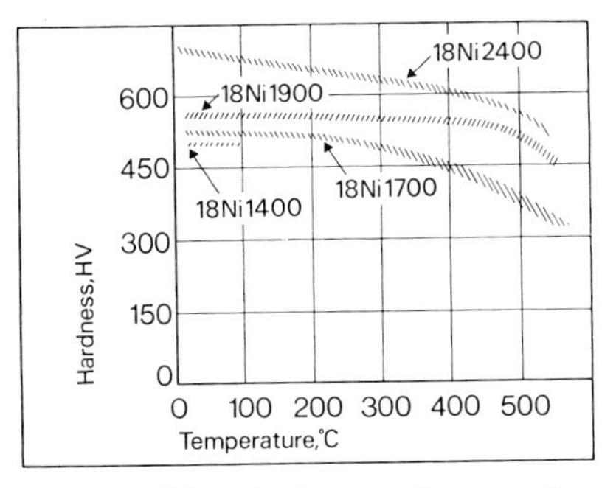

Figure 17. *Hardness of maraging steels as a function of temperature.*

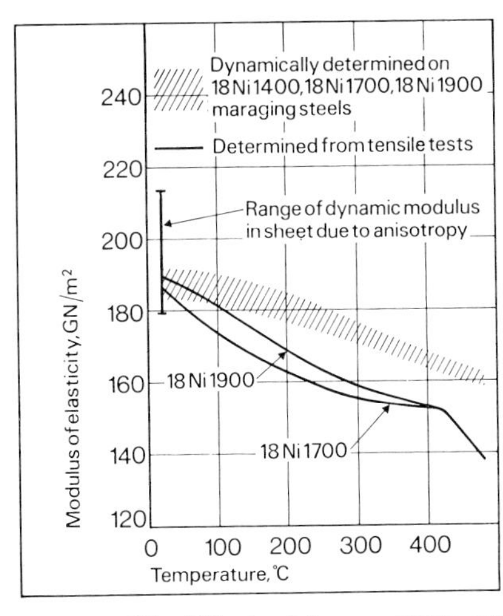

Figure 18. *Effect of temperature on the modulus of elasticity of maraging steels, determined dynamically and by tensile measurements (the variation in the room-temperature dynamic modulus of cold-rolled and aged sheet as a function of angle to the rolling direction is indicated by the range of values shown).*

Table 14. *Effect of temperature on 0.2 per cent proof stress and tensile strength of 18 per cent nickel maraging steels.*

Test temperature °C	Approximate ratio of strength at test temperature to strength at 20°C	Change in 0·2% proof stress and tensile strength between 20°C and test temperature	
		18Ni1700 and 18Ni1900 grades N/mm²	18Ni2400 grade N/mm²
−100	1·11	+170/250	—
− 40	1·03	+50/80	—
20	1·00	0	0
100	0·95	− 80/140	−140
200	0·90	−160/220	−240
300	0·87	−170/280	−350
400	0·82	−280/420	−440
480	0·73	−420/550	−570

Table 15. *Effect of temperature on compressive yield stress and shear strength of 18 per cent nickel maraging steels.*

Test temperature °C	18Ni1700		18Ni1900			
	Compressive yield stress		Compressive yield stress		Ultimate shear stress	
	N/mm²	Per cent of value at 25°C	N/mm²	Per cent of value at 20°C	N/mm²	Per cent of value at 20°C
20	—	—	1980	100	1130	100
25	1840	100	—	—	—	—
150	1650	90	—	—	—	—
175	—	—	1790	90	—	—
315	1540	84	—	—	—	—
345	—	—	1670	84	940	83
430	1450	79	1590	80	860	76
540	1170	76	—	—	—	—

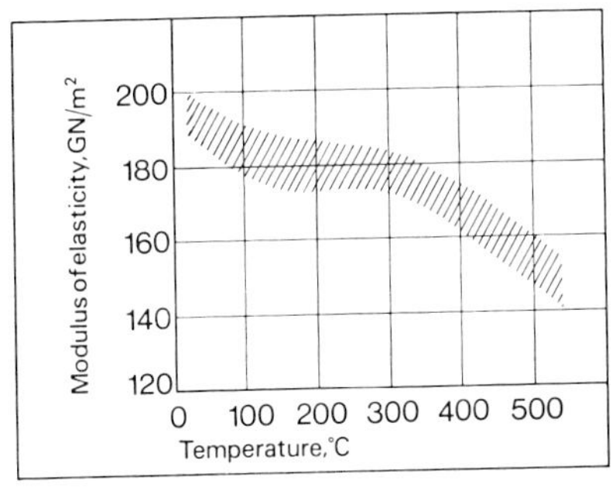

Figure 19. *Effect of temperature on the modulus of elasticity of 18Ni1700 maraging steel in compression.*

Source: *INCO Databook*, 1976

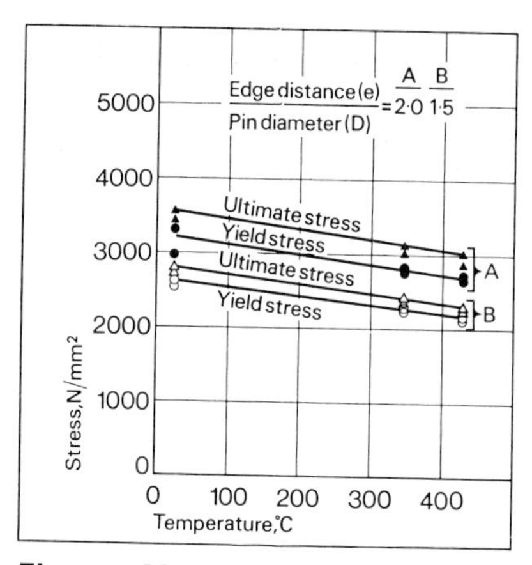

Figure 20. *The bearing strength (ASTM.E238 test) of 18Ni1900 maraging steel as a function of temperature.*

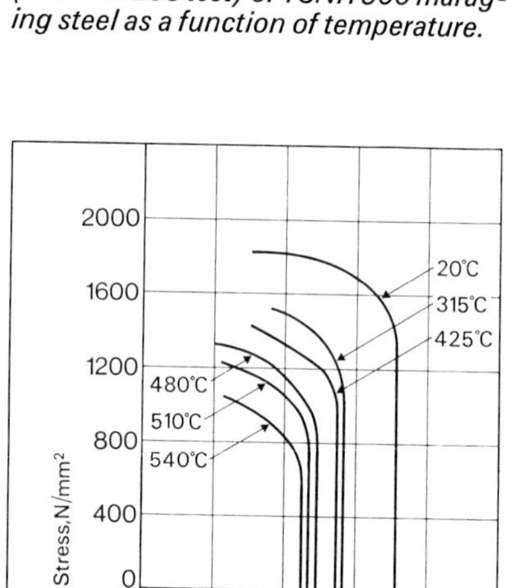

Figure 21. *The secant modulus of 18Ni1700 maraging steel as a function of applied stress at various temperatures. (Data derived from static tests).*

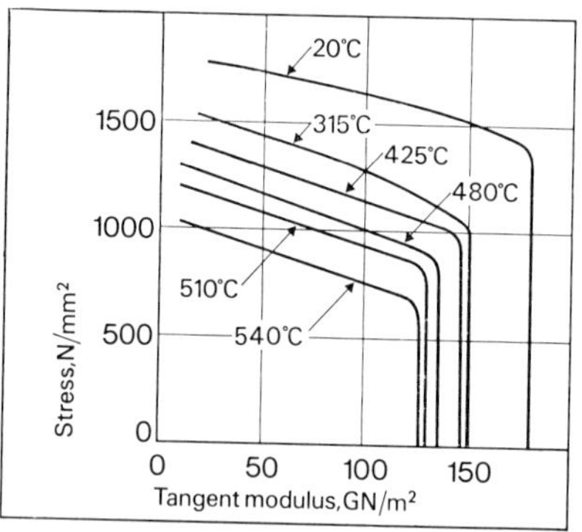

Figure 22. *The tangent modulus vs. stress for 18Ni1700 maraging steel at various temperatures. (Data derived from static tests).*

Table 16. *Effect of long-time exposure at elevated temperatures on room-temperature 0.2 per cent proof stress of 18 per cent nickel maraging steels.*

Condition	0·2% proof stress			
	18Ni1400	18Ni1700		18Ni1900
	N/mm²	Steel 1 N/mm²	Steel 2 N/mm²	N/mm²
As-maraged	1450	1730	1860	2000
Maraged+200h 150°C	—	1710	—	—
Maraged+200h 260°C	—	1730	—	—
Maraged+200h 370°C	—	1730	—	—
Maraged+200h 480°C	—	1140	—	—
Maraged+200h 540°C	—	830	—	—
Maraged+200h 650°C	—	280	—	—
Maraged+1000h 315°C	1540	—	1950	2060
Maraged+1000h 370°C	1590	—	2000	2180

Table 17. *Effect of long-time exposure at elevated temperatures on room-temperature impact strength of 18 per cent nickel maraging steels.*

Condition	Charpy V-notch impact value					
	18Ni1400		18Ni1700		18Ni1900	
	J	daJ/cm²	J	daJ/cm²	J	daJ/cm²
As-maraged	56	7·0	24	3·0	18	2·3
Maraged+1000h 315°C	43	5·4	22	2·8	18	2·3
Maraged+1000h 370°C	52	6·5	16	2·0	18	2·3

Table 18. *Typical mechanical properties of cast 17Ni1600 maraging steel.*

Heat treatment: a – Homogenized 4h 1150°C, air cooled.
b – 4h 1150°C+4h 595°C+1h 820°C, air cooled.
c – 4h 1150°C+3h 480°C.
d – 4h 1150°C+4h 595°C+1h 820°C+3h 480°C.

Property	Solution annealed		Maraged			
	a	b	c		d	
0·2% proof stress, N/mm²	740	750	1650			
Tensile strength, N/mm²	960	990	1730			
Elongation, Lo=5·65$\sqrt{So}$. %	12	13	7			
Reduction of area, %	58	62	35			
Notched/smooth tensile strength ratio	—	—	1·25			
Modulus of elasticity(E). GN/m²	—	—	188			
Modulus of rigidity (G) (torsional), GN/m²	—	—	72			
Poissons ratio	—	—	0·30			
Hardness: HV	295	320	530			
HRC	29	32	49			
Plane strain fracture toughness (K_{1c}), $MNm^{-3}/_2$	—	—	82·5			
Charpy V-notch impact value at:			J	daJ/cm²	J	daJ/cm²
−196°C	—	—	3	0·38	8	1·0
−100°C	—	—	9	1·1	18	2·3
−40°C	—	—	16	2·0	20	2·5
20°C	—	—	18	2·3	22	2·8

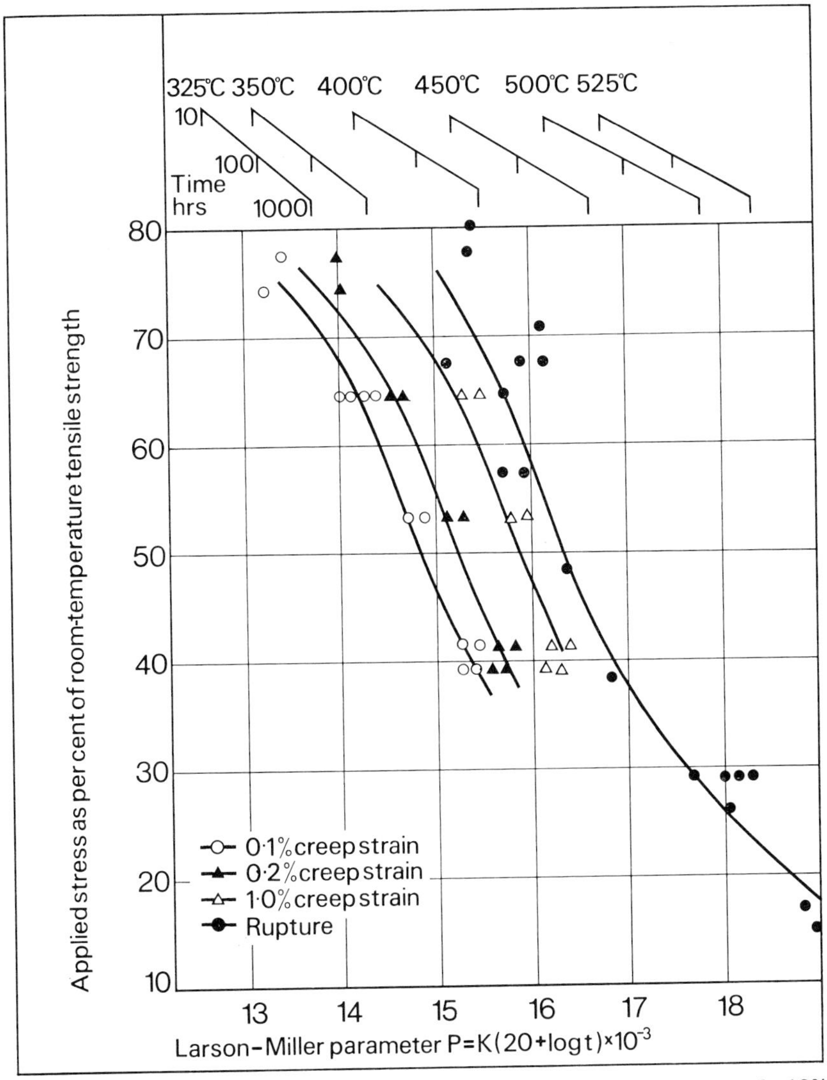

Figure 23. *Larson-Miller plot of creep- and rupture-strength of 18% nickel maraging steels. The plotting of applied stress as per cent of the room-temperature tensile strength normalizes the data for all grades. The time-temperature conversion scales along the top of the diagram enable the Larson-Miller parameter to be converted into time and temperature values.*
K = Absolute temperature (Kelvin)   t = Time in hours

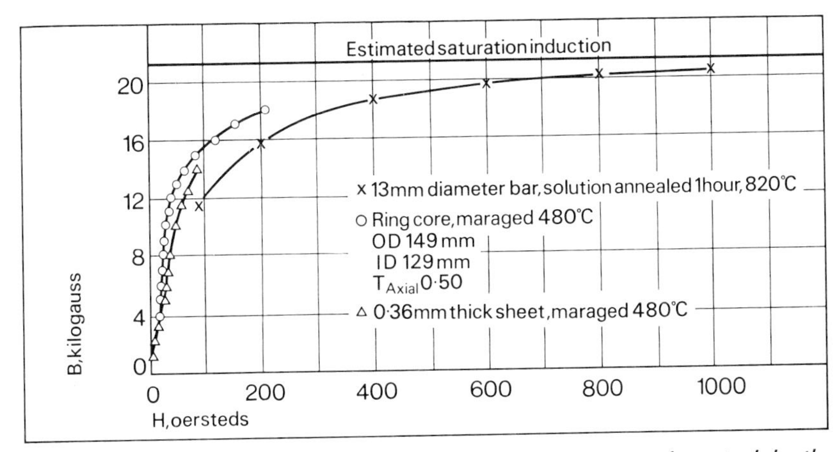

Figure 24. *D.C. magnetization curves of 18Ni1700 maraging steel in the solution-annealed and the maraged conditions. Normal coercive force values (Hc) of 18Ni1700 are: Solution-annealed 22–34 oersteds, 1750–2700 A/m. Maraged 21–54 oersteds, 1670–4300 A/m.*
Approximate remanence (Br) = 5·5 kilogauss or 0·55 teslas.

Physical properties

These are presented in Table 19 (*page 18*) and Figure 24.

Processing and forming

Hot working

The maraging steels are readily hot worked by conventional rolling and forging operations. A preliminary soak at 1260°C may be used for homogenization, except for the 18Ni2400 grade for which soaking in the range 1200–1230°C has been found to be more satisfactory. If the latter steel is inadvertently soaked at 1260°C or higher, it is necessary to hold for a time at the lower suggested temperature to eliminate any possible damage that may have been caused by the high-temperature exposure. For all grades adequate working should be applied to break up the as-cast structure and minimize directionality. For optimum properties a minimum reheating temperature of 1100°C should be used prior to final hot working. Hot working can continue to 870°C for 18Ni2400 and to 820°C for the lower-strength grades. A fine grain size is obtained by application of adequate reductions at the lower temperatures during final hot working and this enhances toughness. The transverse ductility and toughness of thick sections receiving only limited amounts of hot work may be reduced compared with thinner material, this effect being partly attributable to an embrittling reaction which can occur in the prior austenite grain boundaries if the steels are cooled slowly through the critical temperature range 980–760°C. The mechanical properties of material embrittled by this mechanism can be restored by reheating to 1200°C and cooling rapidly to room temperature.

Examples of the effects of hot-work finishing temperature and of variation in cooling rate after hot working are shown in Figures 25–30 (*pages 18–19*). For the 13mm plate represented in these diagrams, finishing temperature and mode of cooling had no marked effect on the tensile properties after solution annealing and ageing (Figures 26 and 27). On the other hand, fracture toughness (K_{Ic}) measurements on fatigue-precracked edge-notched bend specimens showed greater effects of finishing temperature and cooling rate (Figures 28–30), illustrating that relatively rapid cooling is desirable from any finishing temperature and that the need for rapid cooling is greater for the higher finishing temperatures. If rapid cooling cannot be achieved (e.g. if air-cooling must be used), a low finishing temperature with significant reduction at that temperature is desirable.

Source: *INCO Databook*, 1976

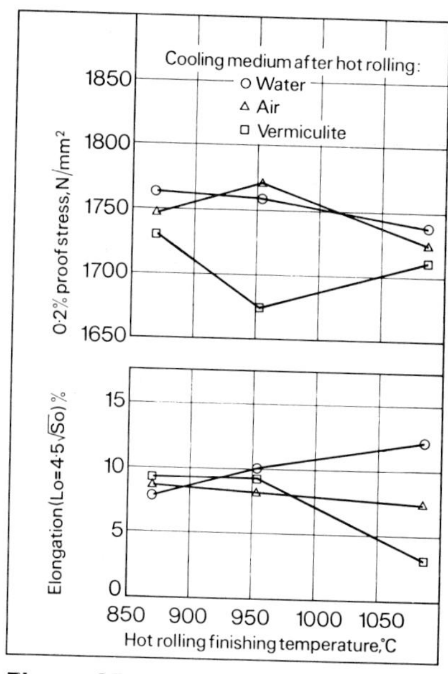

Figure 25. *Effect of hot-rolling finishing temperature and subsequent mode of cooling on the tensile properties of 18Ni1700 maraging steel plate (13mm thick, 35% reduction in final pass at finishing temperature).*
Condition: hot-rolled and aged.

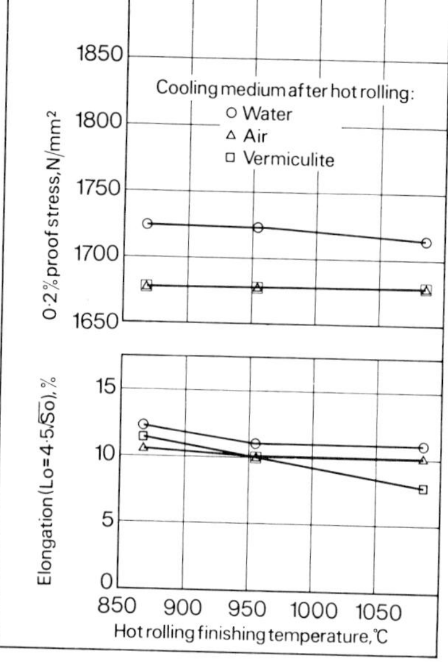

Figure 26. *Effect of hot-rolling finishing temperature and subsequent mode of cooling on the tensile properties of 18Ni1700 maraging steel plate (13mm thick, 35% reduction in final pass at finishing temperature).*
Condition: annealed at 815°C and aged.

Table 19. *Physical properties of nickel-cobalt-molybdenum maraging steels.*

Property	Wrought grades: 18Ni1400, 18Ni1700, 18Ni1900		Wrought grades: 18Ni2400		Cast grade: 17Ni1600	
Density g/cm³	8·0		8·1		8·0	
Specific heat	kJ/kgK	cal/g°C	kJ/kgK	cal/g °C	kJ/kgK	cal/g °C
	0·46	0·11	0·46	0·11	0·46	0·11
Thermal conductivity at:	W/mK	cal/cm s °C	W/mK	cal/cm s °C	W/mK	cal/cm s °C
20°C	21	0·050	—	—	29	0·070
100°C	23	0·054	—	—	32	0·077
200°C	26	0·061	—	—	37	0·088
300°C	27	0·064	—	—	—	—
400°C	28	0·066	—	—	41	0·099
480°C	28	0·067	—	—	—	—
Mean coefficient of thermal expansion: 10^{-6}/K						
20–100°C	9·9		—		9·6	
20–200°C	10·2		—		10·0	
20–300°C	10·6		—		10·5	
20–400°C	11·0		—		10·8	
20–480°C	11·3		11·4		11·0	
Linear contraction on ageing, % approx.	18Ni1400: 0·04; 18Ni1700: 0·06; 18Ni1900: 0·08		0·09		0·03	
Electrical resistivity,* μ Ω cm:	18Ni1700					
Solution annealed 820°C	60–70		—		—	
Maraged 3h 480°C	35–50		—		—	

* *The electrical resistivity increases within the ranges given primarily with the titanium content of the steel.*

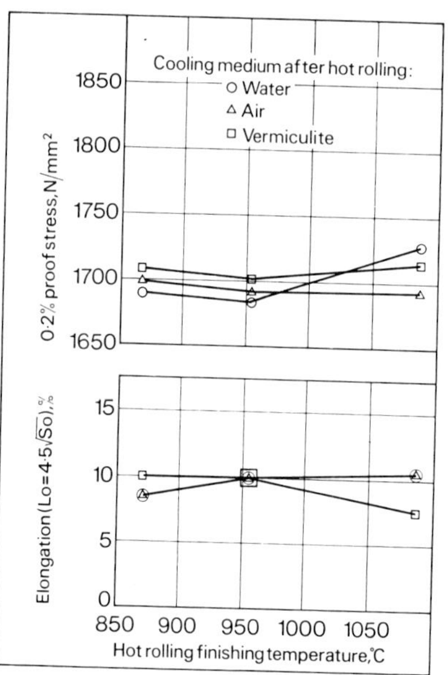

Figure 27. *Effect of hot-rolling finishing temperature and subsequent mode of cooling on the tensile properties of 18Ni1700 maraging steel plate (13mm thick, 35% reduction in final pass at finishing temperature).*
Condition: annealed at 870°C and aged.

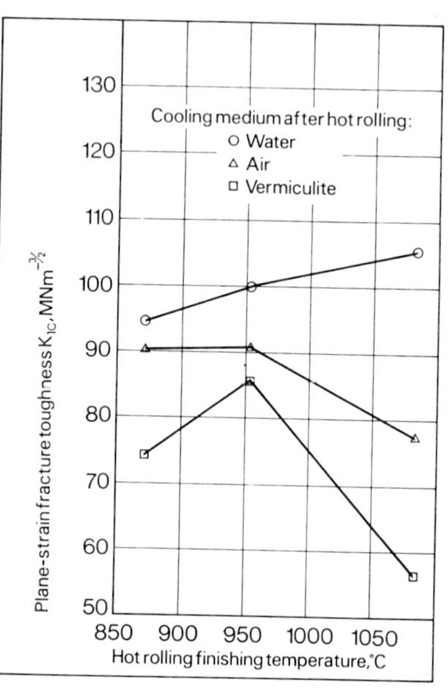

Figure 28. *Effect of hot-rolling finishing temperature and subsequent mode of cooling on plane-strain fracture toughness (K_1c) of 18Ni1700 maraging steel plate (13mm thick, 35% reduction in final pass at finishing temperature).*
Condition: hot-rolled and aged.

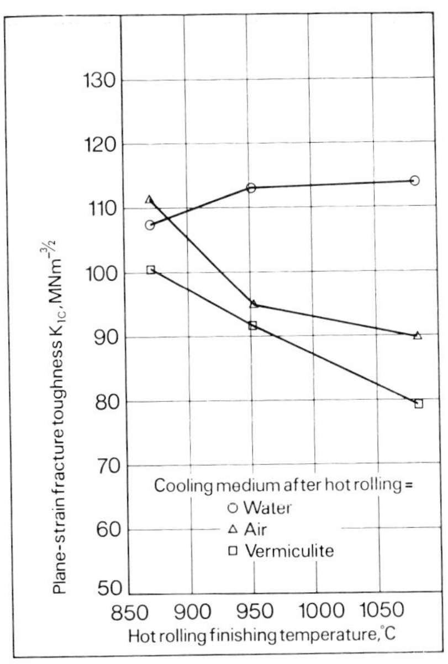

Figure 29. *Effect of hot-rolling finishing temperature and subsequent mode of cooling on plane-strain fracture toughness (K_1c) of 18Ni1700 maraging steel plate (13mm thick, 35% reduction in final pass at finishing temperature).*
Condition: annealed at 815°C and aged.

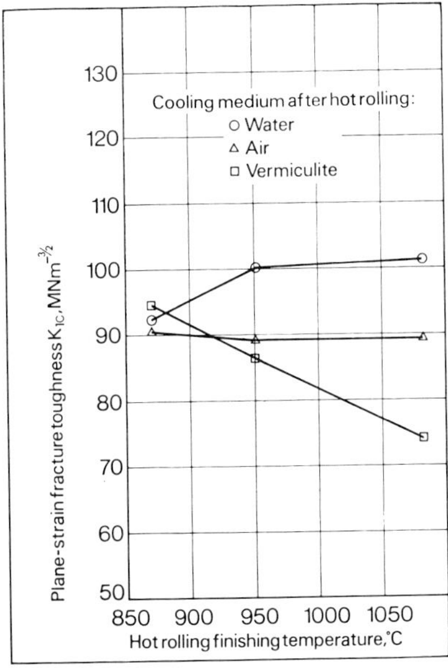

Figure 30. *Effect of hot-rolling finishing temperature and subsequent mode of cooling on plane-strain fracture toughness (K_1c) of 18Ni1700 maraging steel plate (13mm thick, 35% reduction in final pass at finishing temperature).*
Condition: annealed at 870°C and aged.

Cold working

Hot-rolled or annealed maraging steels are easily cold worked. They work-harden very slowly and can be reduced substantially (up to 85 per cent reduction) before intermediate annealing is required (Figure 31). The low work-hardening rate facilitates production of sheet, strip and wire. For forming and drawing applications, cold-rolled solution-annealed material is preferred because of its relative freedom from surface irregularities. However, hot-rolled solution-annealed material with good surface quality has performed satisfactorily. Tube spinning, shear forming, deep drawing, hydroforming, heading of fasteners, bending and shearing can all be accomplished by cold processes.

Cold work prior to ageing can be used to increase strength after maraging. Figures 32–34 show the change in the 0·2 per cent proof stress as a function of per cent cold reduction in 18Ni1700, 18Ni1900 and 18Ni2400. In all cases the proof stress for a given ageing treatment increases at an approximately constant rate up to about 50 per cent reduction, but thereafter strength increases at a higher rate for the 18Ni2400 steel, whereas for the lower-strength steels there is no further gain in strength with increase of cold work above 50 per cent and in some instances the strength may fall below the peak value corresponding to about 50 per cent reduction. Ductility also declines in maraged material above about 40 per cent reduction and when maximum toughness is required in the end product it is desirable to limit cold reduction to 40 per cent.

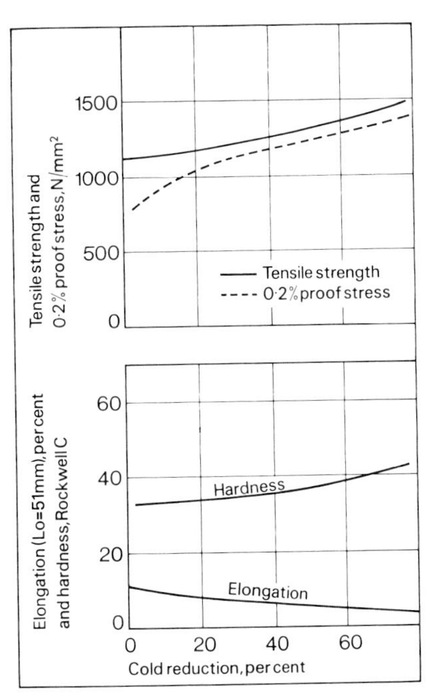

Figure 31. *Effect of cold work on the tensile properties and hardness of solution-annealed maraging steel.*

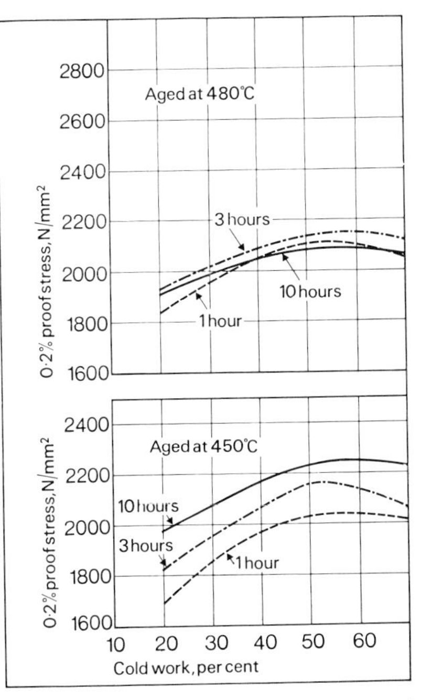

Figure 32. *The 0·2% proof stress of 18Ni1700 maraging steel is improved by cold working prior to ageing at 450°C and 480°C.*

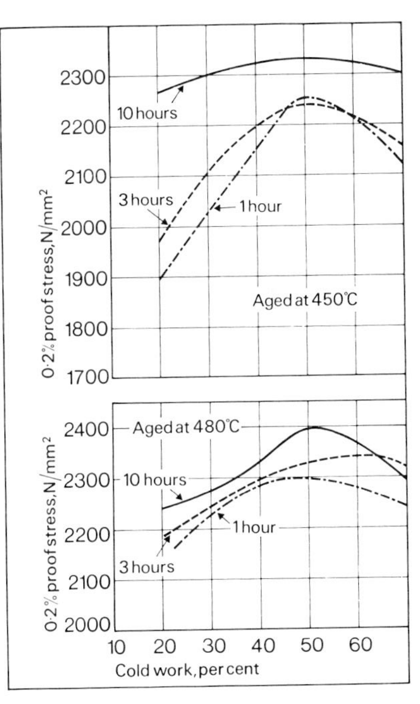

Figure 33. *Effect of cold work prior to ageing at 450°C and 480°C on the 0·2% proof stress of 18Ni1900 maraging steel.*

Source: *INCO Databook*, 1976

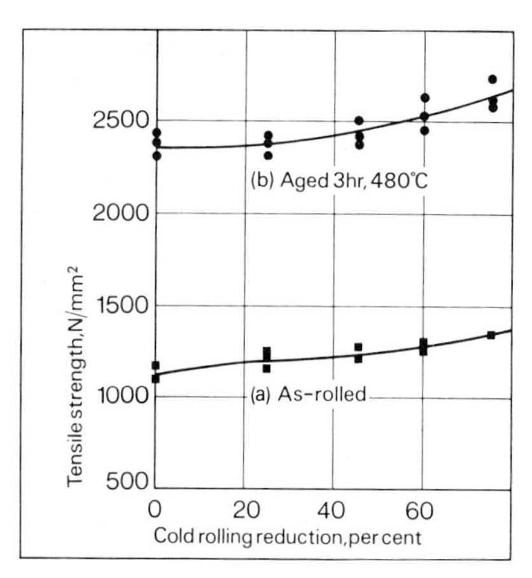

Figure 34. *Tensile strength of 18Ni2400 maraging steel sheet (a) annealed and cold rolled, (b) annealed, cold rolled and aged.*

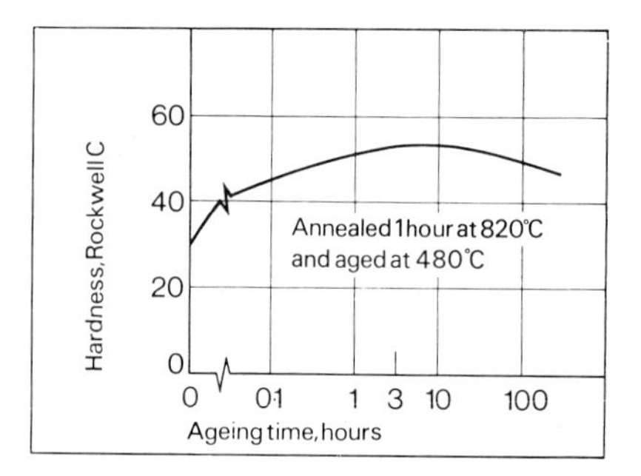

Figure 35. *Typical age-hardening curve of maraging steels. The hardening response is initially very rapid, while at the optimum ageing temperature the effect of overageing is slight even after 200 hours.*

Heat treatment

The simple heat treatment cycle of the maraging steels is one of the major advantages of these materials. The alloys are normally solution-annealed at 820°C for a minimum time of 15–30 minutes for 1·3 mm thick sections and for one hour per 25 mm for heavier sections, followed by air cooling to room temperature. Air cooling is an adequate quenching rate to induce complete transformation to martensite throughout the heaviest sections because at temperatures above the martensite formation range the austenite has high stability and does not transform to other decomposition products such as ferrite, peralite or bainite, even at relatively slow cooling rates.

Solution-annealing at 820°C is generally sufficient to give complete recrystallization of hot-worked structures and to ensure the formation of a fully-austenitic structure from which martensite can form on cooling, whereas heating below that temperature may not achieve these requirements for the attainment of optimum strength and toughness. However, with some heats and product forms of the 18Ni1400, 1700 and 1900 grades heating above 820°C may be necessary to promote recrystallization in order to obtain optimum transverse properties. On the other hand, increasing the annealing temperature above 820°C for the 18Ni2400 grade causes a gradual but definite drop in all properties.

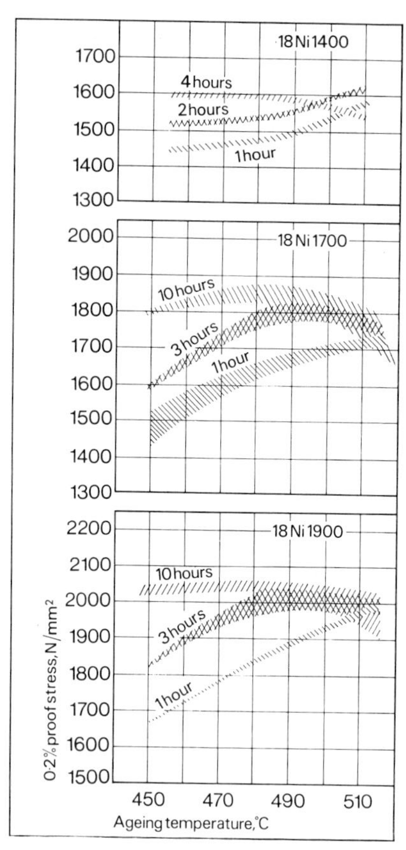

Figure 36. *Effect of ageing temperature and time on the 0·2% proof stress of maraging steels.*

The high strength properties of the 18Ni1400, 1700, 1900 and cast 17Ni1600 maraging steels are developed within relatively short time when the annealed steels are aged at 480°C, the standard ageing time being 3 hours, while the effect of overageing is slight even after 200 hours (Figure 35). The 18Ni2400 steel age-hardens at somewhat slower rate and an ageing temperature between 480°C and 540°C for times up to 12 hours at 480°C, or shorter times at the higher temperatures, is suggested to achieve the nominal 0·2 per cent proof stress of 2400 N/mm². Ageing this higher-strength steel at 510°C for 3 hours usually gives the optimum strength and toughness.

The effects of variations in maraging temperature and time on mechanical properties are presented in Figures 36–39.

Special furnace atmospheres to prevent decarburization during heat treatment are not required because of the low carbon content of the maraging steels. Normal precautions to prevent carburization, sulphurization or excessive oxidation are required. Fuel low in sulphur is preferred for fuel-fired furnaces. Fuel oil containing not more than 0·75 per cent (w/w) sulphur is satisfactory and fuel gas should contain no more than 2·3 grams of total sulphur per cubic metre (1 grain/cubic foot). All carbonaceous- and sulphur-containing impurities should be removed from the surfaces of material before annealing. Direct flame impingement on work material should be avoided. To produce a surface free of oxide, the material is heated and cooled to room

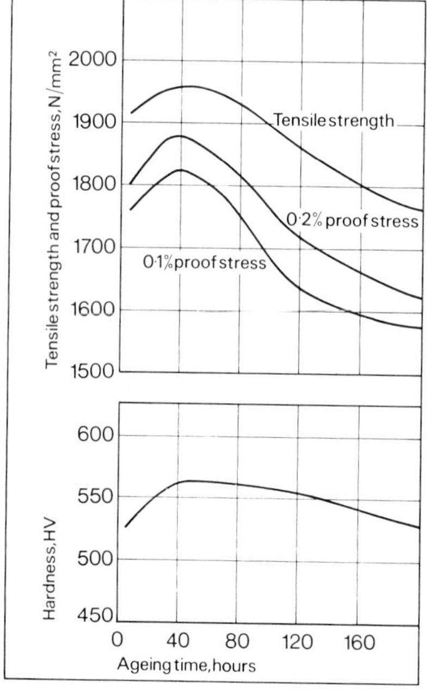

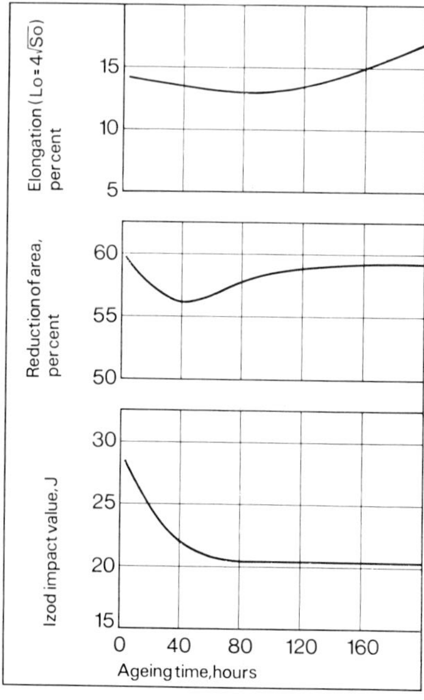

Figure 37. *Effect of ageing time at 480°C on the mechanical properties of 18Ni1700 maraging steel.*
(Data by courtesy of Firth Brown Limited, Sheffield, U.K.).

temperature in an atmosphere of either pure dry hydrogen with a dew point of −43°C or dissociated (completely dissociated) ammonia with a dew point of −45° to −50°C.

A variant of the normal maraging heat treatment may be applied to finished products of the 18Ni2400 steel to provide additional hardening. This involves cooling the material slowly from 820°C or higher, or interrupting the cooling cycle in the temperature range 510–650°C and holding for several hours prior to cooling to room temperature. Ageing of the austenitic matrix occurs, 'Ausaging', which may be due to precipitation of the eta phase (Ni_3Ti). Subsequent maraging then increases the hardness to 61–62 Rc compared with about 58 Rc for normally annealed and maraged material. The hardening response induced by interrupted cooling from 820°C is shown in Figure 40.

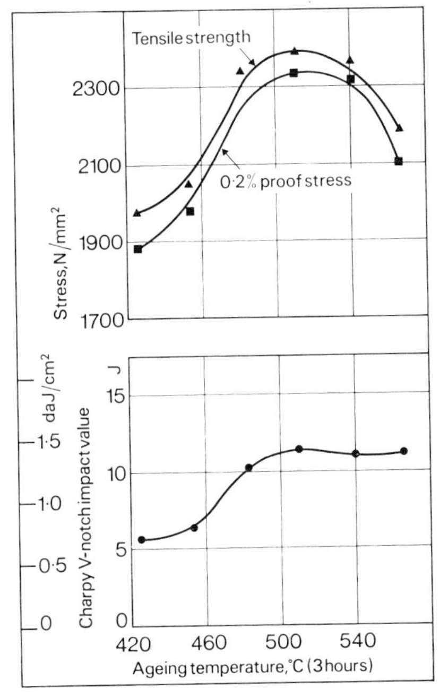

Figure 38. *Variation of tensile and impact properties of 18Ni2400 maraging steel with ageing temperature (steel annealed 1h 820°C before ageing).*

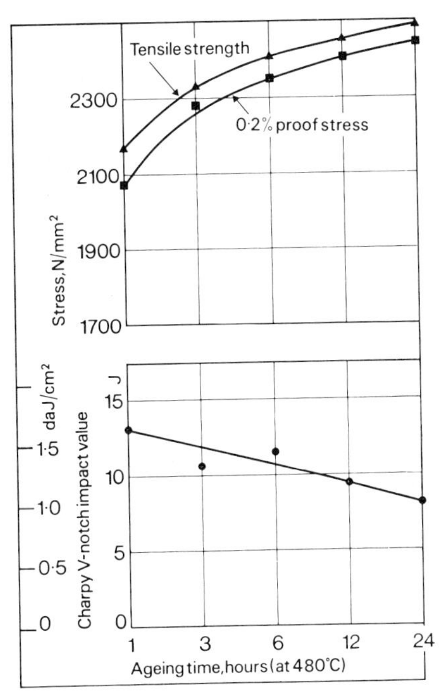

Figure 39. *Variation of tensile and impact properties of 18Ni2400 maraging steel with ageing time (steel annealed 1h 820°C before ageing).*

The ausaging reaction can also occur during cooling after hot working, but generally it should be avoided at that stage by ensuring that there are no delays in cooling through the temperature range 650–510°C, since it may impair subsequent processing (e.g. cold working and machining) and might cause unsuspected and nonuniform properties in hot-rolled or heavy sections. An anneal of one hour at 820°C is usually sufficient to eliminate this effect.

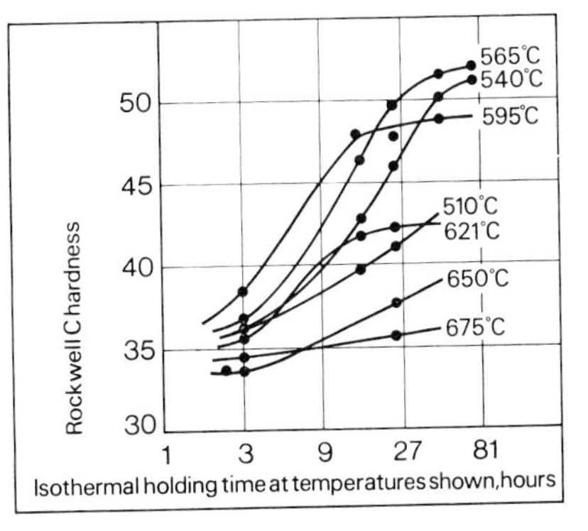

Figure 40. *Hardening induced in 18Ni2400 maraging steel by interrupted cooling (Ausaging) from 820°C. Subsequent ageing (Maraging) gives further hardening.*

A further heat treatment variant may be applied to 18Ni2400 steel in the processing of ultra-high-strength wire to produce the best combination of strength and toughness at a given strength level. Instead of applying an intermediate anneal at the normal temperature of 820°C, between wire drawing steps, the material is annealed for 1 hour at 620°C. This results in partial reversion of the low-carbon martensite to austenite with retention of the latter phase on cooling to room temperature. Subsequent working creates instability of the austenite and, after final maraging, maximum strength is developed combined with improved ductility. These effects on properties are shown by the data given in Table 20 which compares material given intermediate annealing at both 820°C and 620°C.

Dimensional stability

The nature of the maraging hardening mechanism is such that close dimensional control can be maintained in components that are finish-machined in the soft, annealed, condition and subsequently hardened. Absence of retained austenite also ensures that the alloys will be free of further dimensional change in service. During maraging a very small uniform contraction occurs; the percentage changes for the various wrought grades of steel are shown in Figure 41. As illustrated by the curves for the 18Ni2400 steel, overageing by heating at higher temperatures or longer times than those normally used for maraging may result in higher-than-normal shrinkage and should be avoided in finished products made to close tolerances.

Table 20. *Effects of intermediate annealing on the mechanical properties of 18Ni2400 wire.*

Drawing procedure:			
Pre-drawing annealing treatment	1h 815°C	1h 815°C	1h 815°C
Initial diameter, mm	6·4	6·4	5·8
First reduction, %	87	87	85
Size, mm	2·3	2·3	2·3
Intermediate anneal or reversion heat treatment	**1h 815°C**	**1h 627°C**	**1h 627°C**
Final reduction, %	87	87	87
Final size, mm	0·8	0·8	0·8
Tensile properties of as-drawn wire:			
0·2% proof stress, N/mm²	1260	1990	2030
Tensile strength, N/mm²	1540	2530	2540
Elongation, %	0·86	1·2	1·7
Tensile properties of drawn and aged wire:			
0·2% proof stress, N/mm²	—	2530	2600
Tensile strength, N/mm²	2710	2960	2960
Elongation, %	—	1·2	0·9

Nitriding

Maraging steels can be simultaneously nitrided and aged at 430–480°C to provide a shallow but hard case to improve wear resistance and/or fatigue properties. At the lower nitriding temperatures, 430–450°C, a surface hardness of about 860 HV can be obtained with a total case depth of about 0·15 mm after 48 hours treatment. A typical hardness traverse through a nitrided case is shown in Figure 42. Nitriding above 450°C produces some stable austenite in the nitrided zone and decreases the hardness. However, at 480°C a hardness of about 800 HV can still be achieved and the total case depth may be increased to about 0·25 mm after 70–90 hours treatment.

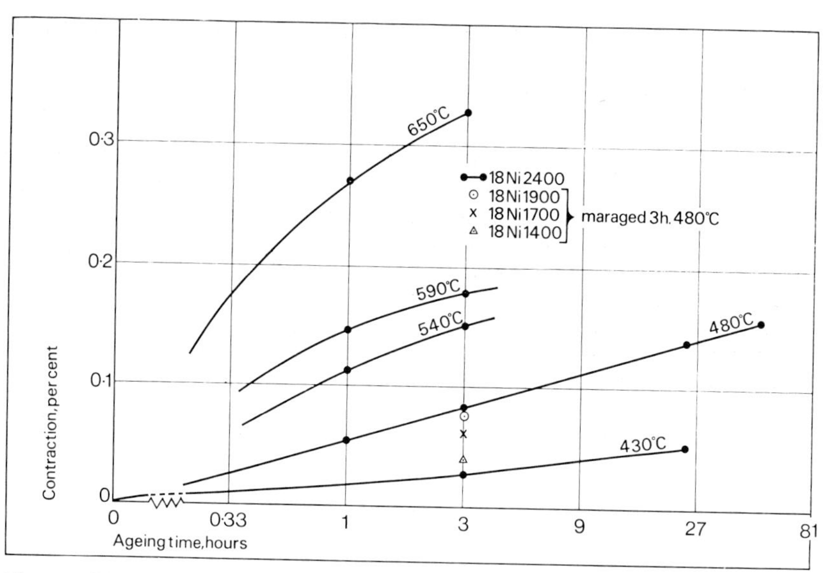

Figure 41. *Dimensional changes, due to ageing or ageing and reversion, of 18Ni2400 maraging steel aged at various temperatures, and of the lower-strength wrought grades given the standard maraging treatment at 480°C. The steels were initially annealed one hour at 820°C.*

Welding and joining

Nickel maraging steels are readily weldable without preheat in either the solution-annealed or fully-aged conditions. Gas-shielded arc processes are preferred; considerable experience and confidence has been generated with the TIG process. A machined groove preparation is desirable and the joint design may be the same as is commonly used for carbon steels. An argon gas shield is recommended for TIG and MIG welding, while pure helium is recommended for MIG short-arc welding. The maximum interpass temperature should be 120°C. Thorough interpass cleaning by power wire brushing is recommended. No postheat is required, but a post-weld ageing treatment such as 3 hours at 480°C is used to strengthen the weld and the heat-affected zone.

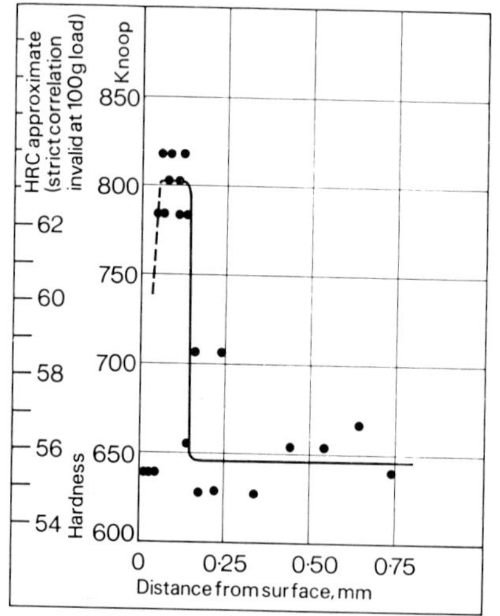

Figure 42. *Hardness of nitrided case on 18Ni2400 maraging steel billet, 95mm diameter. Steel annealed one hour 820°C then simultaneously aged and nitrided 48 hours at 450°C in 25–30% dissociated ammonia (Knoop hardness determined with 100 gram load).*

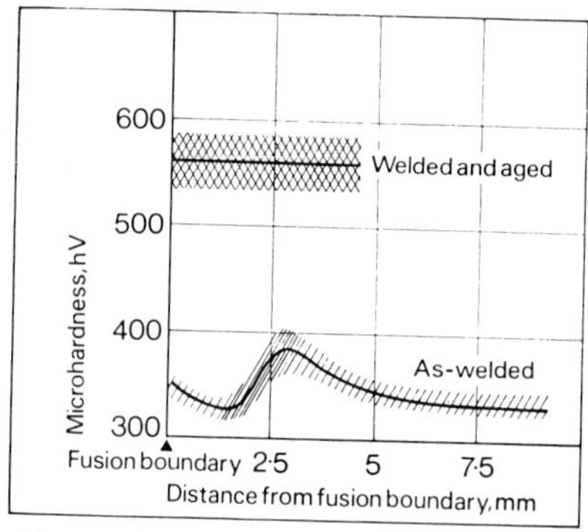

Figure 43. *Typical heat-affected-zone hardness distribution in maraging steels welded in the solution-annealed condition. As-welded and after subsequent ageing.*

Heat-affected zone

The behaviour of the heat-affected zone of maraging steels contributes to their good welding characteristics. There is no cracking problem as in conventional high-strength low-alloy steels, and very little distortion occurs whether welding maraging steel in the solution-annealed or aged condition. The hardness distribution in the HAZ of solution-annealed material is shown in Figure 43. After welding, very little softening occurs in the parent metal and the joint has virtually the same strength as the annealed parent material. Following post-weld ageing negligible variation in hardness is observed in the heat-affected zone.

The HAZ hardness distribution of maraging steel welded after ageing is shown in Figure 44. Softening may be observed in the grain-coarsened area near the fusion line and in regions which experience peak temperatures of approximately 650°C where partial austenite reversion and stabilization occurs. Tests of welded joints show that the fracture occurs in the weld metal, and that the characteristic notch toughness of the base metal is retained in the heat-affected zone. After re-ageing the weld area virtually uniform hardness is obtained in the weld zone.

Gas-shielded processes

An essentially matching composition filler wire is used to weld the 18Ni1700 grade and has also been used for welding 18Ni1400 and 18Ni1900. Vacuum-melting and vacuum-annealing of the wire product is used in production of the filler material to provide the low impurity and hydrogen levels (H_2 is typically <5 ppm) required for maximum weld properties. Filler wire

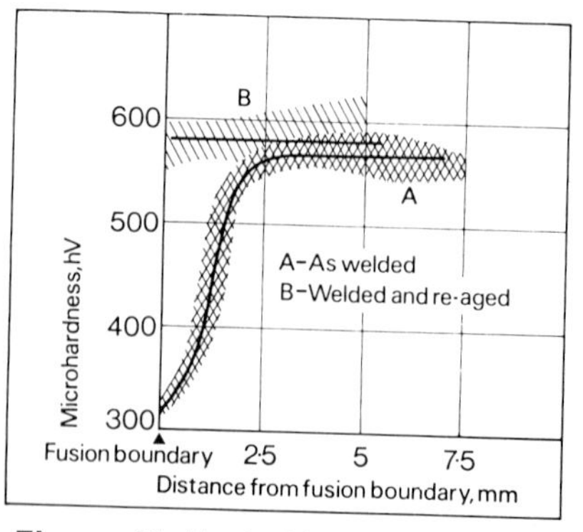

Figure 44. *Typical heat-affected-zone hardness distribution in maraging steels welded in the aged condition. As-welded and after subsequent re-ageing.*

compositions closely matching those of the parent metals have also been developed for welding the 18Ni1400, 18Ni1900 and 18Ni2400 maraging steels. However, most work has been done with the 18Ni1400 and 18Ni1700 grades. Typical wire compositions are shown in Table 21.

Heat inputs are maintained as low as possible to achieve optimum strength and toughness in the weld deposits and rather low travel speeds are employed to avoid contamination of the weld metal by atmospheric gases. Typical welding parameters are shown in Table 22.

Representative mechanical properties of TIG, MIG and short-circuiting arc weldments are presented in Table 23. The fracture toughness measurements of welded maraging steels are lower than those of the parent material but are better than those of conventional quenched and tempered steels at the same proof stress levels. Joint efficiencies of 95–100 per cent are usual in welds of maraging steels having nominal proof stress values of up to 1700 N/mm², but welds made with high heat inputs can have efficiencies down to 85 per cent and occasionally may be lower. Tensile failures in such welds often occur in the heat-affected zone as the result of formation of relatively large amounts of stable austenite. As the strength of the parent material is raised above 1700 N/mm² proof stress, joint efficiencies decrease. Small beads and low input help to keep strength reductions to a minimum. Solution-annealing a weld prior to ageing can increase strength significantly, but this may often be impractical.

The toughness of TIG weld-deposits is generally superior to that of MIG weld-deposits made either by spray or short-circuiting techniques.

Table 21. *Typical weld filler wire compositions.*

Grade	Composition. Weight per cent					
	Ni	Co	Mo	Ti	Al	Other
18Ni1400	18·2	7·7	3·5	0·24	0·10	0·03 max. C 0·01 max. S 0·01 max. P 0·10 max. Si 0·10 max. Mn <50ppm O <50ppm N <5ppm H Bal Fe (all grades)
18Ni1700	18·1	8·0	4·5	0·46	0·10	
18Ni1900	17·9	9·9	4·5	0·80	0·12	
18Ni2400	17·4	12·4	3·7	1·6	0·17	

Table 22. *Typical parameters for gas-shielded arc welding.*

Process	Volts	Amps	Travel speed mm/min.	Wire diameter mm	Wire feed rate m/min.	Shielding gas dm³/h
TIG	11–15	200–240	100–180	1·6	0·5–0·76	Argon 850
MIG	28–30	280–300	250	1·6	5	Argon 1400
Short Arc	24	140	Manual	0·8	—	He 1400

Flux-coated electrode welding

Electrodes for manual metal-arc welding the 18Ni1400 and 1700 maraging steels have been developed, but have not been

Table 23. *Representative mechanical properties of gas-shielded arc weldments. (Welds aged 3h 480°C, except as stated otherwise)*

Material welded	Thickness(a)	Filler wire type and source	Welding process	0·2% proof stress	Tensile strength	Elong. Lo=25mm	Red. of area	Charpy V-notch impact value		K1c
				N/mm²	N/mm²	%	%	J	daJ/cm²	MNm$^{-3}/_{2}$
18Ni1400	Plate	18Ni1400(C)	TIG	1370	1430	13	60	47	5·9	—
		18Ni1400(C)	MIG	1430	1480	6	34	28	3·5	—
		18Ni1700(L)	MIG	1380	1480	7	30	24	3·0	—
		18Ni1700(C)	TIG	—	1690(bd)	—	—	—	—	122
		18Ni1700(C)	MIG	—	1670(bd)	—	—	—	—	78
		18Ni1700(C)	Short Arc	—	1690(bd)	—	—	—	—	67
18Ni1700	Plate	18Ni1700(C)	MIG	1560	1670	4	7	14	1·8	—
		18Ni1700(C)	MIG	1610	1660	6	2·8	—	—	—
		18Ni1700(C)	MIG	—	1650(bd)	—	—	—	—	82
		18Ni1700(C)	TIG	—	1720(bd)	—	—	—	—	92
		18Ni1700(C)	Short Arc	—	1720(bd)	—	—	—	—	77
		18Ni1700(C)	TIG	1800	1850(d)	4	—	—	—	—
		18Ni1700(C)	TIG	1790(f)	1850(d)	3	—	—	—	—
		18Ni1700(L)	Short Arc	1570	1690	4	14	16	2·0	—
		18Ni1700(L)	MIG(e)	1660	1680	6	21	—	—	—
		18Ni1700(L)	MIG(e)	1520	1620	7	30	19	2·4	—
	Sheet	18Ni1700(L)	TIG	1690	1700	—	—	—	—	210 (Kc)
18Ni1900	Plate	18Ni1700(L)	MIG	1660	1680	6	21	—	—	—
		18Ni1900	TIG	1390	1680	8	40	—	—	65
		18Ni1900	MIG	1600	1690	3	13	—	—	59
	Sheet	18Ni1900	TIG	1800–1930	1830–1970	—	—	—	—	124–178 (Kc)
		18Ni1900	TIG	1650	1670	3 (on 51mm)	—	—	—	—
18Ni2400	Sheet	18Ni2400	TIG	1970	2030	1·5 (on 51mm)	—	—	—	36

(a) Plate=10–25 mm thick. Sheet=1·5–2·5 mm thick.
(b) Converted from Vickers Hardness value.
(d) Welds aged 4h 490°C.
(e) All weld metal properties.
(f) Weld locally aged.
(C) Commercially produced.
(L) Laboratory produced.

Source: *INCO Databook*, 1976

used commercially to any great extent. Core wire compositions are basically the same as those of the parent alloys, except that the titanium contents are increased to allow for high losses of this element across the arc. However, under normal fabrication conditions the titanium losses are variable and unpredictable.

In studies of MMA welding there has been no real difficulty in achieving welds of the required high strength, but they have been associated with a greater propensity to weld cracking and poorer toughness than has been observed in joints made by inert-gas welding.

Submerged-arc and electroslag welding

Both processes provide good joint strength, but weld ductility and toughness are low. In addition, multipass welds in submerged-arc welds have been subject to cracking. Neither of these processes is currently recommended for welding 18 per cent Ni maraging steels where maximum weld properties are required.

Other joining methods

Limited work on conventional spot- and seam-welding has shown promise, but more work is needed before these processes can be recommended for maraging steels. Ageing can be accomplished in the machine, although the relatively high-temperature short-time ageing cycles promote the formation of some austenite.

Excellent friction-welded joints have been achieved. This process is especially useful for joining shafting, tubing and bar products. Table 24 presents typical tensile properties of friction welds.

Flash welding has not been successful to date; low tensile strength and especially low tensile ductility have sometimes been encountered. Additional work on optimizing the welding parameters may overcome this difficulty.

Satisfactory electron-beam welds have been made by a number of fabricators in several thicknesses of maraging steels from thin sheet to moderately thick plate. However, published information to date suggests that it is difficult to make sound welds consistently, particularly in thick plate. Edges to be welded must be machined to close tolerances to ensure the proper fit-up that is an essential part of making good electron-beam welds. A variety of welding conditions that have been used is summarized in Table 25 together with the mechanical properties of the joints obtained. The ductilities, as measured by reduction of area, were generally about half those of the base-plate values.

Dissimilar metal butt joints have been made in 13 mm plate between 18Ni1700 maraging steel, type 304 (18Cr-10Ni) stainless steel, mild steel, HY-80 (2¾ per cent Ni-Cr-Mo) and SAE4340 (1¾ per cent Ni-Cr-Mo) steels. The results showed, Table 26, that sound joints can be produced without difficulty and, as may be expected, the strength of a particular joint is governed by the weaker material of the combination.

Table 24. *Tensile properties of friction welds in 25mm diameter bar of 18Ni1700 maraging steel. (Welds aged 3h 480°C)*

Specimen	0·2% proof stress N/mm²	Tensile strength N/mm²	Elong. Lo=51mm %	R of A %
Unwelded bar	1760	1810	12	58
Friction weld[a]	1730	1790	7	35
Friction weld[a]	1730	1780	7·5	36
Friction weld[a]	1740	1780	7·5	40
Friction weld[b]	1800	1820	7·3	38

(a) 13 mm diameter specimen.
(b) 19 mm diameter specimen.

Table 25. *Mechanical properties of electron-beam welds* (a) *in maraging steel plates.*

Grade and thickness of plate	Welding conditions				Heat treatment[b]	Mechanical properties				Charpy V-notch impact value		
	No. of passes	Voltage kv	Current mA	Travel speed cm/min.		0·2% proof stress N/mm²	Tensile strength N/mm²	Elong[c] %	R of A %	J	daJ/cm²	K_{Ic} MNm$^{-3/2}$
18Ni1700 2·5mm	1	Unwelded	—	—	SA	1760	1830	5·7	28	—	—	—
		30	65	100	SAW	1140	1140	2·5	21	—	—	—
		30	65	100	SAWA	1870	1890	4·1	13	—	—	—
		30	65	100	SWSA	1780	1820	3·2	13	—	—	—
18Ni1700 7·5mm	1	Unwelded	—	—	SA	1820	1840	15	—	—	—	—
		150	20	150	SAW	1670	1670	4	—	—	—	—
		150	20	150	SAWA	1810	1830	4	—	—	—	—
		150	20	150	SWSA	1880	1900	5	—	—	—	—
18Ni1700 13mm	1	Unwelded	—	—	SA	1710	1790	22	51	—	—	—
		150	17	43	SAW	1250	1300	8	30	—	—	—
		150	17	43	SAWA	1720	1800	14	25	—	—	—
18Ni1700 25mm	1	Unwelded	—	—	SA	1740	1790	12	50	—	—	96
		50	320	100	SAWA	1770	1790	4	28	—	—	75
		50	320	100	SWSA	1760	1770	6·5	41	—	—	{80, 94}
18Ni1400 25mm	1	Unwelded	—	—	SA	1430	1480	23	56	—	—	—
		50	400	100	SAW	1010	1040	7	31	19 (defects)	2·4	—
		50	400	100	SAWA	1350	1370	4	13	27	3·4	—
18Ni1400 25mm	2	150	13	25	SAW	1100	1160	10	32	—	—	—
		150	13	25	SAWA	1460	1500	7	14	—	—	—

(a) All samples failed in the weld.
(b) S=Solution-annealed, A=Aged, W=Welded.
(c) The tensile elongations must be treated with caution since the weld metal sometimes constitutes only a small portion of the gauge length.

Brazing has been successfully accomplished using palladium-containing and silver-copper-zinc alloys. Brazing temperatures in the range 800–870°C are considered to be the most suitable since temperatures in the austenite reversion range should be avoided unless the component is to be re-solution-treated prior to final ageing, while at temperatures above 870°C, times should be restricted to avoid grain coarsening and loss of ductility and toughness in the steel. For furnace brazing high vacuum is satisfactory, while dry hydrogen or argon atmospheres can also be used successfully providing a flux or pre-plated (e.g. with iron) surfaces are employed to achieve good flow and braze coverage. For torch brazing, use of a flux is mandatory, whether or not a pre-plate is used, although the latter does improve braze coverage and strength.

The degree of overlap in single-lap brazed shear specimens has a marked effect on shear strength, as shown in Figures 45 and 46, with strength decreasing rapidly as the overlap is increased. This is characteristic of brazements, because as the overlap increases, the greatest part of the load is carried by the ends of the joint while the central portion carries little load. In spite of the decrease in specific shear stress the tensile strength of the joint increases as the overlap increases (see Figures 45 and 46).

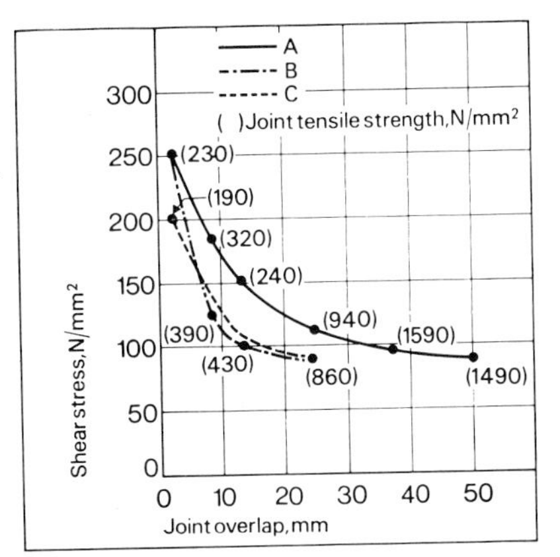

Figure 45. *Effect of degree of overlap on shear strength of single-lap furnace-brazed joints in maraging steels. Specimens aged 3 hours 480°C after brazing.*
A – 68% Ag, 27% Cu, 5% Pd brazing alloy; approximate liquidus temperature 810°C. Specimens brazed in argon.
B – 48% Ni, 31% Mn, 21% Pd brazing alloy; approximate liquidus temperature 1120°C. Specimens brazed in vacuum.
C – 54% Ag, 40% Cu, 5% Zn, 1% Ni brazing alloy; approximate liquidus temperature 855°C. Specimens brazed in argon.

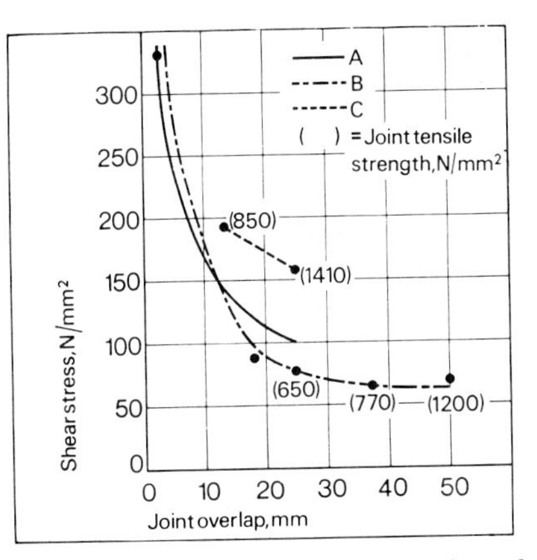

Figure 46. *Shear stress vs. overlap of torch-brazed single-lap joints in maraging steels. Specimens were pre-plated with iron and brazed with use of a flux, then aged 3 hours 480°C.*
A – 82% Cu, 18% Pd brazing alloy; approximate liquidus temperature 1090°C.
B – 53% Cu, 38% Zn, 9% Ag brazing alloy; approximate liquidus temperature 870°C.
C – 27% Cu, 68% Ag, 5% Pd brazing alloy; approximate liquidus temperature 810°C.

Table 26. *Mechanical properties (transverse)[(a)] of 13 mm thick weld joints between 18Ni1700 maraging steel and various dissimilar metals (post-weld heat treatment – 3h 480°C).*

Metal being joined	Filler material	Welding process	Condition	0·2% proof stress N/mm²	Tensile strength N/mm²	Elong.[(a)] Lo= $4{\cdot}5\sqrt{So}$ %	R of A %	Location of fracture[(b)]	Charpy V-notch impact value[(c)] J	daJ/cm²
Mild steel	INCO-WELD* 'A'	Manual Arc	As-welded	390	480	16	67	P	108	13·5
			Aged	320	470	16	66	P	133	16·6
HY-80 steel	INCO-WELD* 'A'	Manual Arc	As-welded	380	610	14	43	W	106	13·3
			Aged	400	610	10	29	W	98	12·3
AISI 4340 steel	INCO-WELD* 'A'	Manual Arc	As-welded	390	650	13	45	W	108	13·5
			Aged	420	690	12	34	W	94	11·8
AISI 304 stainless	INCO-WELD* 'A'	Manual Arc	As-welded	350	610	26	34	W	129	16·1
			Aged	370	630	36	79	P	132	16·5
Mild steel	18% Ni maraging steel[(d)]	MIG	As-welded	430	480	6	65	P	34	4·3
			Aged	340	460	5	65	P	22	2·8
HY-80 steel	18% Ni maraging steel[(d)]	MIG	As-welded	750	800	8	59	P	30	3·8
			Aged	740	800	7	58	P	14	1·8
AISI 4340 steel	18% Ni maraging steel[(d)]	MIG	As-welded	1010	1110	4	13	W	22	2·8
			Aged	1170	1240	4	13	P	11	1·4
AISI 304 stainless	18% Ni maraging steel[(d)]	MIG	As-welded	430	620	23	76	P	38	4·8
			Aged	420	620	32	75	P	19	2·4

Weld Conditions	*Manual Arc*	*MIG*
Current	125 A	280–290 A
Voltage	24 V	32 V
Travel speed	150mm/min.	250mm/min.
Wire feed	Manual	
Heat input	12kJ/cm	20–22kJ/cm
Filler size	4mm rod	1·6mm wire
Preheat	None	None

(a) 6·4 mm diameter tensile bar.
(b) Location of tensile fracture: W=Weld, P=Dissimilar metal being joined.
(c) Notched through weld, axis of notch normal to plane of plate.
(d) Filler composition: 17·9 Ni, 8·1 Co, 4·9 Mo, 0·5 Ti, 0·13 Al, balance essentially Fe.

*Trade Mark of Henry Wiggin & Company Limited.

Source: *INCO Databook*, 1976

Joint clearance, like overlap, affects joint strength: Figure 47 shows that the narrower the gap, the higher the strength. 'Contact clearance', i.e. between 0 and 35 μm, provides the best strength, while a gap of about 0·15 mm reduces the strength to about 70 per cent of the maximum level.

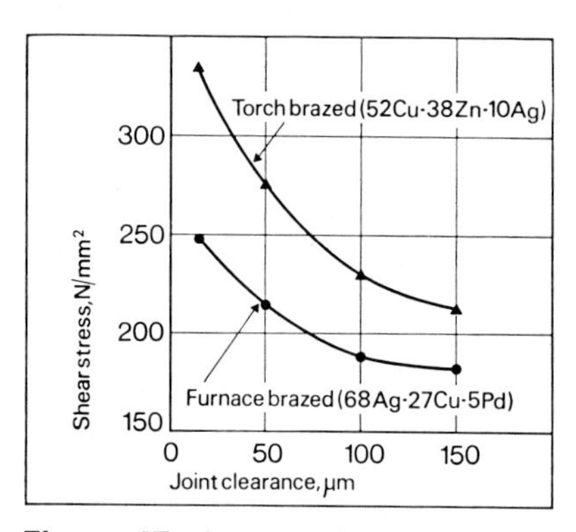

Figure 47. *Average shear stress as a function of joint clearance in brazed maraging steel joints at an overlap of 2·5mm.*

Machining and grinding

Maraging steels are machined most easily in the annealed condition. However, machining of maraged material is also possible and, in general, for either condition the same procedures should be used as are employed for conventional alloy steels having equivalent hardness levels. Rigid equipment and firm tool supports are essential. Tools should receive a copious stream of cutting fluid with every effort being made to get the lubricant to the tool cutting edge.

Grinding is essentially the same as grinding conventional constructional steels when using a heavy-duty water-soluble grinding fluid as is employed with stainless steels. It is essential to use the heavy-duty grinding fluid as wheel wear is substantially greater with an ordinary water-soluble oil.

Suggested machining procedures are given in Table 27.

Table 27. *Some suggested machining procedures for maraging steels.* [1]

TURNING. Single point and box tools.

Condition of Stock	Tool material	Depth of cut mm	Speed m/min.	Feed mm/rev.	Lubricants**
Annealed	HSS*, AISI types T15, M33, M41–47	0·64–3·8	21–27	0·13–0·25	2, 5, 7
	Carbide, ISO type K10	0·64–3·8	107–145	0·18–0·38	0, 4, 6
Maraged	HSS*, AISI types T15, M33, M41–47	0·64–3·8	14–18	0·13–0·25	3, 5, 7
	Carbide, ISO type K10	0·64–3·8	32–49	0·13–0·25	0, 4, 6

REAMING

Condition of stock	Tool material	Reamer diameter, mm 3	6	13	25	38	51	Speed, m/min.	Lubricants**
		Feed, mm/rev.							
Annealed	HSS*, AISI types T15, M33, M41–47	0·08	0·13	0·20	0·30	0·38	0·46	17	2, 5, 7
	Carbide, ISO type K20	0·08	0·15	0·20	0·30	0·38	0·46	49	1, 4, 6
Maraged	HSS*, AISI types T15, M33, M41–47	0·03	0·03	0·03	0·03	0·03	0·03	3	3
	Carbide, ISO type K20	0·03	0·03	0·03	0·03	0·03	0·03	15	2, 3

DRILLING

Condition of stock	Tool material	Nominal hole diameter, mm 3	6	13	19	25	38	51	Speed m/min.	Lubricants**
		Feed, mm/rev.								
Annealed	HSS*, AISI types T15, M33, M41–47	0·08	0·13	0·18	0·23	0·25	0·33	0·38	17	2, 5, 7
Maraged	HSS*, AISI types T15, M33, M41–47	0·05	0·08	0·10	0·10	0·10	0·10	0·10	6	3

TAPPING

Condition of stock	Tool material	Speed m/min.	Lubricants**
Annealed	HSS*, AISI types M10, M7, M1	6	3, 5, 7
Maraged	HSS, nitrided M10, M7, M1	1·5	3

1) Machining data are largely reproduced from the 'Machining Data Handbook' of the Machinability Data Center, Metcut Research Associates Inc., U.S.A., with their kind permission.

* *HSS - High Speed Steel*
** *Lubricants: 0 Dry.*
1 Light duty oils (general purpose).
2 Medium duty oils (sulphurized or chlorinated).
3 Heavy duty oils (sulphurized or chlorinated).
4 Soluble oils (light duty).
5 Soluble oils (heavy duty).
6 Synthetic (light duty, general purpose).
7 Synthetic (heavy duty).

SAWING (Annealed stock)					
Circular sawing. Tool material HSS* AISI types M2, M7	Thickness or bar diameter, mm				Lubricants**
	6–76	76–150	150–230	230–380	
Pitch, mm Cutting speed, m/min. Feed, mm/rev.	5·1–16·5 14 50–100	15·2–24·1 12 38–76	22·9–29·2 11 25–50	27·9–39·4 8 19–38	3, 5, 7
Power hacksawing HSS* blade	Material thickness, mm				Lubricants**
	<6	6–20	20–50	>50	
Teeth/dm Speed, strokes/min. Feed, mm/stroke	40 or more 85 0·13	40 85 0·13	24 85 0·13	16 85 0·13	2, 5, 7

PLANING								
Condition of stock	Tool material	Cutting speed m/min.	Feed, mm			Depth of cut, mm		Lubricants**
			Rough planing	Finish planing	Parting	Rough planing	Finish planing	
Annealed	Tungsten or molybdenum HSS*	12–15	0·4	5 max.	0·2 max.	5	0·25	0 for rough planing, 2 or 3 for finish planing and parting
Maraged	Tungsten or molybdenum HSS*	7·5	0·4	5 max.	0·1 max.	5	0·25	0 for rough planing, 2 or 3 for finish planing and parting

MILLING							
	Face milling		Slab milling	End milling – peripheral		End milling – slotting	
Tool material for:	HSS* AISI type	Carbide ISO type	HSS* AISI type	HSS* AISI type	Carbide ISO type	HSS* AISI type	Carbide ISO type
Annealed stock	T15, M33, M41–47	K20	T15, M33, M41–47	M2, M7	P20	M2, M7	
Maraged stock	T15, M33, M41–47	K20	T15, M33, M41–47	T15, T17, M33, M41–47	P20	—	K10
Depth of cut, mm	0·64–3·8	0·64–3·8	0·64–3·8	0·64–3·8	0·38–1·3	1·3–6·4	1·3–6·4
Speed m/min: Annealed stock	26–34	79–94	20–26	23–29	84–107	18–21	—
Maraged stock	9–12	20–26	—	8–9	23–30	—	12–15
Feed: mm/tooth: Annealed stock	0·08–0·13	0·13–0·15	0·10–0·13	0·03–0·05(a) 0·08–0·10(b)	0·04–0·05(a) 0·10–0·13(b)	0·013(c) 0·05–0·06(d)	—
Maraged stock	0·08–0·13	0·08–0·10	—	0·03(a)(b)	0·03–0·05(a) 0·08–0·10(b)	—	0·03–0·04(d)
Lubricants**: Annealed stock	2, 5, 7	0, 4, 6	2, 5, 7	2, 5, 7	0, 4, 6	2, 5, 7	0, 4, 6
Maraged stock	3	0, 2	3	3	3	3	3

(a) Cutter diameter 12·5 mm.
(b) Cutter diameter 25–51 mm.
(c) Width of slot 6·4 mm.
(d) Width of slot 25–51 mm.

Source: *INCO Databook*, 1976

Descaling and pickling

The oxide on hot-worked or thermally-treated materials may be removed by blasting or pickling by the procedures described below. Fused salt-bath pickling processes that operate at or above 320°C should not be used on maraging steels. The mechanical properties may be modified by treatments at or above that temperature.

Duplex pickling

Immerse work in Solution No. 1. The time required for the removal of oxide with a specific chemical pickling solution is dependent upon the nature and the amount of oxide. To avoid over-pickling the work should be frequently inspected during pickling.

Solution No. 1

	Parts by volume
Water	3
Hydrochloric acid (20° Baumé)	4
Temperature	70°C
Time	20 to 40 min. (additional time might be required for loosening heavy oxide)
Containers	earthenware crocks, glass, ceramic or acid-proof brick-lined vessels

The work coming from Solution No. 1 should be rinsed in cold water and immersed in Solution No. 2. The work from Solution No. 2 should be rinsed in cold water and then neutralized in a 1 to 2 per cent (by volume) ammonia solution.

Solution No. 2

	Parts by volume
Water	14
Nitric Acid (70%)	5
Hydrofluoric Acid (52%)	1
Temperature	25–30°C
Time	1½ to 2 min.
Containers	carbon or brick-lined tank

Single bath pickling procedure

Solution No. 3 is a rapid pickling solution that should be used cautiously to avoid over-pickling. This solution leaves a black smut on the surface of the pickled material. Following pickling the work should be rinsed in cold water and then neutralized in a 1 to 2 per cent (by volume) ammonia solution.

Solution No. 3

	Parts by volume
Water	20
Sulphuric Acid (66° Baumé, 93%) or (60° Baumé, 78%)	3 4
Temperature	65–75°C
Time	Approx. 15 min. (avoid over-pickling by inspecting the work for time of withdrawal)
Containers	earthenware crocks, glass, or ceramic vessels or rubber-lined tanks

Corrosion characteristics

In atmospheric exposure unprotected 18 per cent nickel maraging steels corrode in a uniform manner and become rust-covered. Pit depths tend to be more shallow than for conventional low-alloy high-strength steels, while corrosion rates of the maraging steels are about half those of low-alloy steels as shown by the data presented in Figures 48–50.

The corrosion rates for both maraging and low-alloy steels in sea water are similar initially, but from about six months onwards the former corrode more slowly (Figure 51).

In tap water and some neutral salt solutions the maraging steels are susceptible to pitting, but have lower average corrosion rates than low-alloy steels. Similarly the corrosion rates of maraging steels in acid solutions, although substantial, are lower than those of low-alloy steels. In general, protection of maraging steels from corrosive solutions is advisable.

Figure 52 shows that maraging steel has substantially greater resistance to oxidation in air at 540°C than a 5 per cent chromium tool steel.

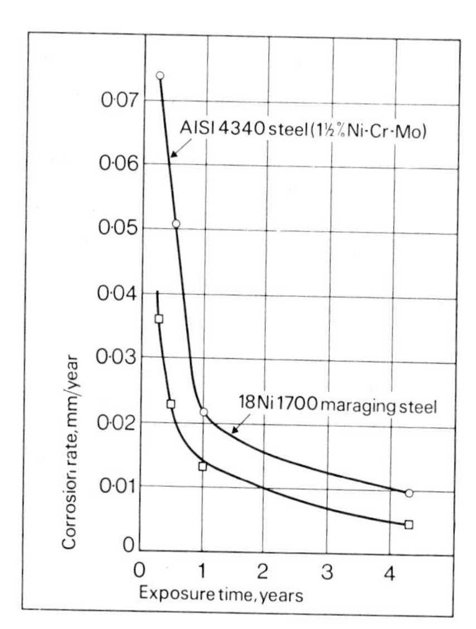

Figure 48. *Corrosion rates of maraging and low-alloy steels in an industrial atmosphere (at Bayonne, New Jersey, U.S.A.).*

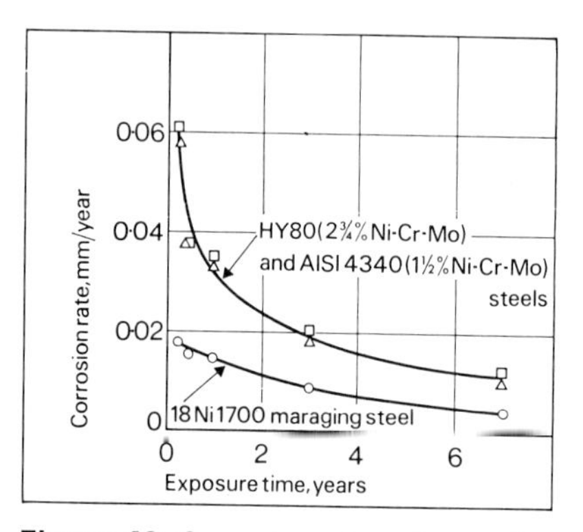

Figure 49. *Corrosion rates of maraging and low-alloy steels, 24 metres from the sea, at Kure Beach, North Carolina, U.S.A.*

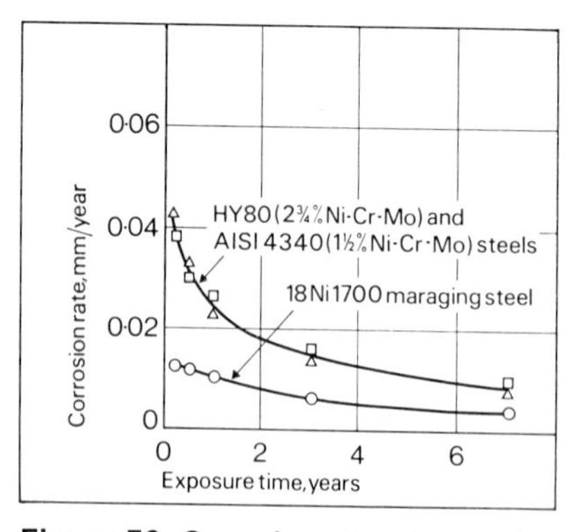

Figure 50. *Corrosion rates of maraging and low-alloy steels, 240 metres from the sea, at Kure Beach, North Carolina, U.S.A.*

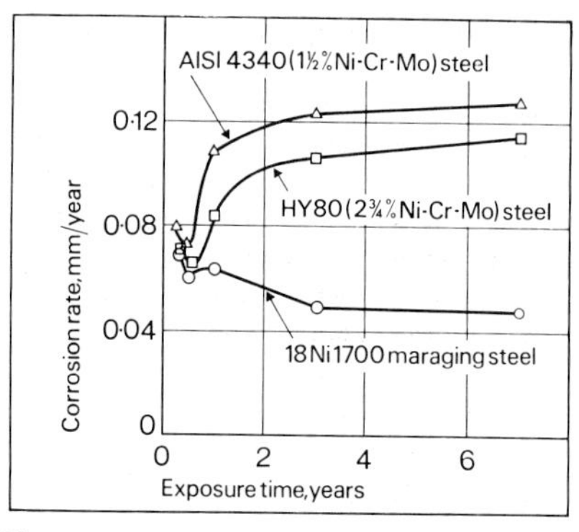

Figure 51. *Corrosion rates of maraging and low-alloy steels in sea water flowing at 0·61 metres per second.*

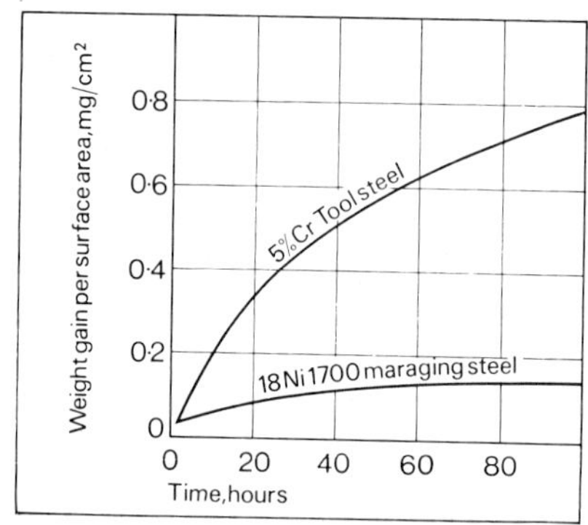

Figure 52. *The oxidation rate at 540°C of 18Ni1700 maraging steel compared with that of 5% chromium steel. Tests performed on 6·4mm cubes in refractory crucibles, exposed to still air for total times of 5, 25 and 100 hours.*

Stress-corrosion characteristics

In an earlier section the measurement of Plane Strain Fracture Toughness, K_{Ic}, as an engineering criterion of toughness, has been described. This fracture mechanics approach to engineering design can also be used to measure the resistance of a material to propagation of a pre-existing crack in the presence of a corrosive environment. Thus it is possible to determine the critical stress intensity factor, designated K_{Iscc}, below which a crack will not propagate under static loading conditions. Values of K_{Iscc} are invariably less than those for K_{Ic}.

Figure 53 presents K_{Iscc} data for unprotected 18 per cent nickel maraging steels in relation to the 0·2 per cent proof stress, with comparative data for some high-strength low-alloy steels and stainless steels. These data embrace exposure in aqueous environments with and without NaCl, but the variations in environments are not distinguished since there is no distinct difference between their effects on K_{Iscc}. Included in Figure 53 are lines showing critical crack depth, a_{cr}. The regions above these lines correspond to combinations of strength and K_{Iscc} for which a long crack of the specified depth will not propagate when stressed to the 0·2 per cent proof stress, while the regions below the lines correspond to strength/K_{Iscc} combinations for which the crack would propagate. The critical crack depth for cracks whose length greatly exceeds their depth is given by the equation:

$$a_{cr} = 0{\cdot}2\left[\frac{K_{Iscc}}{\sigma y}\right]^2,$$

where σy is the 0·2 per cent proof stress, and assuming that the applied stress $= \sigma y$.

In general, it is clear that maraging steels compare favourably with other high-strength steels, and offer relatively high K_{Iscc} values over a wide range of strengths. It is also clear that maraging steels can withstand greater crack depth without crack propagation under a given static stress.

There is some uncertainty regarding the roles of active path corrosion and hydrogen embrittlement in relation to stress-corrosion cracking of maraging steels. However, it is known that they are susceptible to hydrogen embrittlement, but to lesser extent than other high-strength steels. This can result in fracture under static load above certain values depending on the level of hydrogen absorbed during corrosion, pickling, plating, from lubricants, heat-treatment atmospheres or welding. The time to fracture decreases with increase of stress above the threshold value. Removal of absorbed hydrogen and recovery of properties can be achieved by baking treatments as shown in Figure 54. The maraging steels exhibit fast recovery characteristics and a bake of 24 hours at 150–300°C is usually sufficient to recover the full mechanical properties of the material.

Electroplating and surface coating for corrosion protection

Maraging steels can be electroplated with chromium, nickel or cadmium to provide suitable protection for use in severe environments or in highly critical parts where even slight corrosion would be unsatisfactory. Hydrogen which forms during cleaning and plating operations may be absorbed by the steels causing some embrittlement and baking for 24 hours at 200–320°C after plating is recommended to effect its removal.

Inorganic protective coatings of the various black oxide types used on conventional steels can be applied also to maraging steels. Both chromate and phosphate types have been successfully applied. Saturating the coating with oil enhances the protection afforded. Oxalate coatings of the types applied to stainless steels may also be of interest.

Organic coating systems which include polyurethanes and elastomeric neoprenes, such as are applied to conventional alloy steels, are being evaluated for nickel maraging steels.

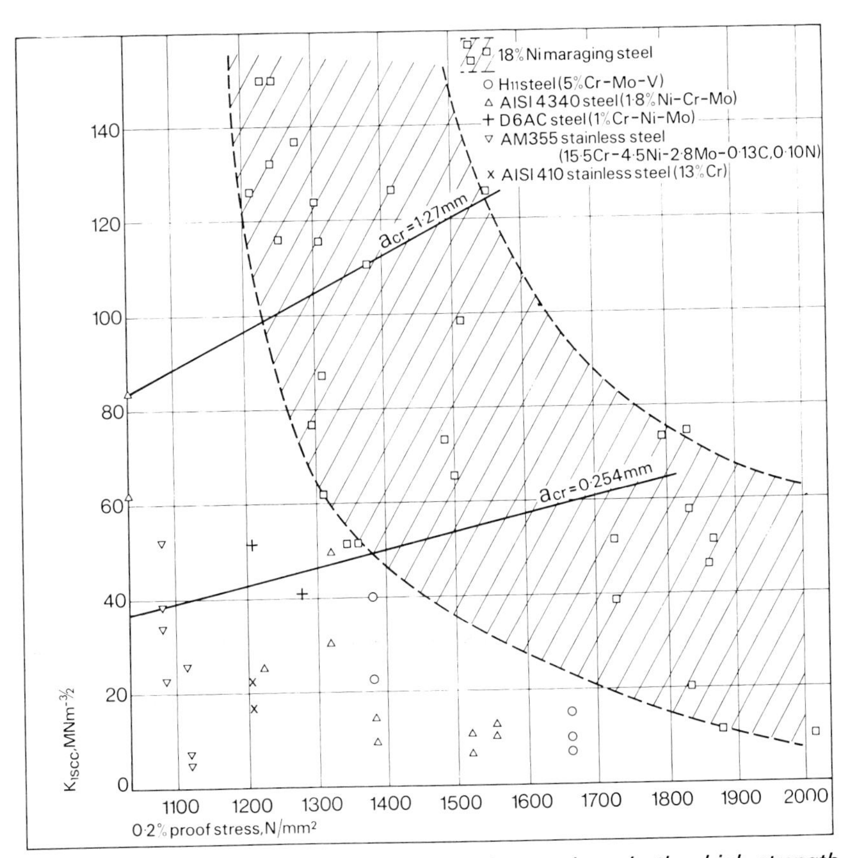

Figure 53. *K_{Iscc} values of 18% nickel maraging steels and other high-strength steels as a function of 0·2% proof stress. Data from various tests in aqueous environments with and without sodium chloride.*

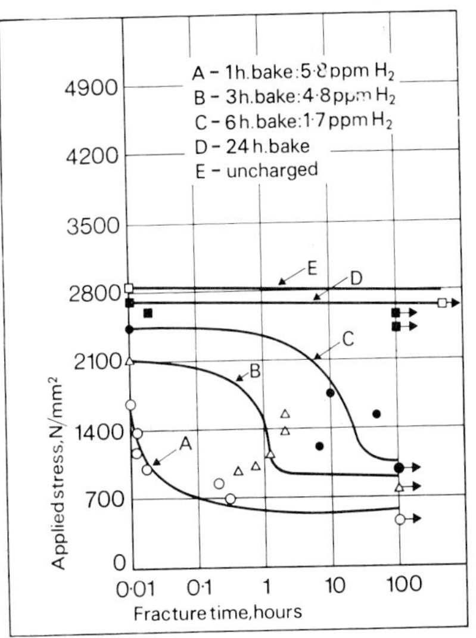

Figure 54. *Hydrogenated 18% nickel maraging steel regains static fatigue resistance rapidly when baked at 150°C. Tests were made on statically-loaded notched tensile specimens hydrogenated by electrolytic charging at 0·011 amp/cm² for 21 hours.*

Progress With 25% Nickel Steels for High-Strength Applications

Nickel steels containing 18 to 27% Ni have high strength and excellent cold working characteristics. Properties result from a combination of precipitation and transformation hardening. (A-general, Q-general; Fe-b, *Ni*, SGB-a)

WORK HAS BEEN UNDER WAY for some time at the Bayonne Research Laboratory of the International Nickel Co., Inc., on the evaluation of experimental low-carbon (less than 0.05%) nickel steels for high-strength applications. The new series is responsive to combined precipitation and transformation hardening when annealed, air cooled and aged. Nickel content may be varied from about 18 to 27%. Response to precipitation hardening derives from controlled additions of titanium and aluminum.

In this progress report, all information is based on laboratory heats. Alloy compositions, treatments and properties discussed are the results of a continuing program of evaluation.

20% Nickel Gives High Strength

A 20% nickel steel of the type under development will undergo transformation from austenite on air cooling from a solution annealing temperature of about 1500° F. and will have an annealed hardness of about Rockwell C-35. Aging at 850 to 1000° F. results in precipitation hardening. Composition, heat treatment and resultant properties are as follows:

Composition: 30 Ni, 1.7 Ti, 0.2 Al, 0.4 Cb.

Heat Treatment: Anneal 1 hr. at 1500° F., air cool; age 1 hr. at 950° F. air cool

Tensile properties (standard 0.252-in. diameter specimen):

Yield strength (0.2% offset)	270,000 psi.
Tensile strength	279,000 psi.
Elongation	12%
Reduction in area	57%
Notched tensile strength	323,000 psi.

Data on notched tensile strength were obtained on specimens with a parallel section diameter of 0.300 in. and a 60° circumferential notch at midlength. Root radius of the notch was 0.0006 in., depth about 0.045 in.

The 20% nickel steels may be cold worked and they are characterized by a particularly low work hardening rate. One steel with an annealed hardness of Rockwell C-36 had essentially the same hardness (Rockwell C-37) after being cold worked 70%. Stress-strain curves also reflect this low work hardening behavior.

Two Aging Treatments Needed

A steel containing 25% Ni has a somewhat lower M_S temperature than does one with 20% Ni; consequently it can be solution annealed to a lower hardness (Rockwell C-5 to 15). After annealing and prior to final aging, the 25% Ni steels must be heated to about 1300° F. for 4 hr. and air cooled to provide the desired hardening and transformation. At this temperature, the titanium and nickel held in solid solution are precipitated as a complex. The depletion of these two elements from the austenite matrix raises the M_S temperature of the steel so that it

---

Data cited in this article were furnished by C. G. Bieber, head, special alloys section, Research Laboratories, Development and Research Div., International Nickel Co., Bayonne, N.J.

Table I – Notched and Smooth Bar Tensile Properties of 25% Nickel Steels

Treatment	Smooth Bar				Notched Bar Tensile★	Notched to Smooth Tensile Ratio
	Yield	Tensile	Elongation	Reduction in Area		
Composition: 24.3% Ni, 1.6% Ti						
1500° F., 1 hr.†; –100° F., 16 hr.; 850° F., 4 hr.	234,000 psi.	253,000 psi.	14%	58%	—	—
1500° F., 1 hr.; cold reduced 50%; –100° F., 16 hr.; 850° F., 1 hr.	267,000	275,000	12	61	402,000 psi.	1.46
Composition: 27.6% Ni, 1.5% Ti, 0.15% Al, 0.43% Cb						
1500° F., 1 hr.‡; 1250° F., 4 hr.; –100° F., 16 hr.; 850° F., 1 hr.	249,000	281,000	13	61	273,000	0.97
1500° F., 1 hr.; cold reduced 50%; 850° F., 1 hr.	254,000	280,000	13	61	364,000	1.30

★All notched tensile data were obtained on specimens with a parallel section diameter of 0.300 in. and a 60° circumferential notch at midlength. Root radius of the notch was 0.0006 in., depth about 0.045 in. This gave a stress concentration factor, K, of 10.

†Annealed hardness: Rockwell C-25. ‡Annealed hardness: Rockwell C-15.

undergoes at least partial transformation when cooled to room temperature. Refrigeration at about –100° F. may be applied, if necessary, to promote further transformation.

This treatment is followed by a final age of 1 hr. at 850 to 950° F. Composition, treatment and representative properties achieved for a given steel of this type are cited below:

Composition: 24.8 Ni, 1.6 Ti, 0.18 Al, 0.4 Cb

Heat treatment: Anneal 1 hr. at 1500° F., air cool; intermediate age 4 hr. at 1300° F., air cool; age 1 hr. at 950° F., air cool.

Tensile properties (standard 0.252-in. diameter specimen):

Yield strength (0.2% offset)	252,000 psi.
Tensile strength	277,000 psi.
Elongation	12%
Reduction in area	44%
Notched tensile strength	298,000 psi.

In the annealed condition, the same steel exhibited these properties:

Treatment	Strength	
	0.2% Yield	Tensile
1 hr. at 1500° F., air cool	67,500 psi.	120,600 psi.
1 hr. at 1600° F., air cool	34,700	114,900

Elongation was 45%; reduction in area, 68%.

Good Cold Working Properties

Nickel steels containing 25% Ni are easily cold worked. These steels have very good tensile strength ratios, notched to smooth bar, when they have been cold reduced prior to final aging. Typical data are shown in Table I.

Section size appears to have little effect on properties developed after full heat treatment. Final properties are about the same whether the steel is furnace cooled or quenched in brine from the annealing temperature. The insensitivity of final tensile properties to such a spread in cooling rate from the annealing temperature indicates that the steel should respond satisfactorily to heat treatment over a wide range of section sizes.

Welding Studies in Progress

Filler metals and welding techniques are currently being evaluated. Fully heat treated steels are capable of withstanding the thermal shock associated with a welding arc. No preheat is necessary. A narrow region in the heat-affected zone of the parent metal experiences fairly complete solution-anneal softening during welding and post-weld treatment is required to restore this zone to its original strength level.

High-Hardness Alloys

Of more than passing interest are some preliminary hardness tests made on steels of about 25% Ni with a combined aluminum and titanium content of 6 to 8%. When such alloys are given a special heat treatment, they develop a hardness of Rockwell C-66 to 67. As with the alloys of lower titanium and aluminum content, these steels are readily hot workable in the solution annealed condition. Initial working, at least of heavy sections, should preferably be performed on material stripped hot from the mold and thermally equalized at the desired working temperatures.

Source: *Metal Progress*, Nov 1960

INDEX

X

Y